# Smart Innovation, Systems and Technologies

Volume 163

The Smart Innovation, Systems and Technologies book series encompasses the topics of knowledge, intelligence, innovation and sustainability. The aim of the series is to make available a platform for the publication of books on all aspects of single and multi-disciplinary research on these themes in order to make the latest results available in a readily-accessible form. Volumes on interdisciplinary research combining two or more of these areas is particularly sought.

The series covers systems and paradigms that employ knowledge and intelligence in a broad sense. Its scope is systems having embedded knowledge and intelligence, which may be applied to the solution of world problems in industry, the environment and the community. It also focusses on the knowledge-transfer methodologies and innovation strategies employed to make this happen effectively. The combination of intelligent systems tools and a broad range of applications introduces a need for a synergy of disciplines from science, technology, business and the humanities. The series will include conference proceedings, edited collections, monographs, handbooks, reference books, and other relevant types of book in areas of science and technology where smart systems and technologies can offer innovative solutions.

High quality content is an essential feature for all book proposals accepted for the series. It is expected that editors of all accepted volumes will ensure that contributions are subjected to an appropriate level of reviewing process and adhere to KES quality principles.

**** Indexing: The books of this series are submitted to ISI Proceedings, EI-Compendex, SCOPUS, Google Scholar and Springerlink ****

More information about this series at http://www.springer.com/series/8767

John Littlewood · Robert J. Howlett ·
Alfonso Capozzoli · Lakhmi C. Jain
Editors

# Sustainability in Energy and Buildings

## Proceedings of SEB 2019

Springer

*Editors*
John Littlewood
School of Art and Design
Cardiff Metropolitan University
Cardiff, UK

Alfonso Capozzoli
Politecnico di Torino
Turin, Italy

Robert J. Howlett
Bournemouth University
Poole, UK

KES International Research
Shoreham-by-sea, UK

Lakhmi C. Jain
University of Canberra
Canberra, Australia

University of Technology Sydney
Sydney, Australia

Liverpool Hope University
Liverpool, UK

KES International Research
Shoreham-by-sea, UK

ISSN 2190-3018 ISSN 2190-3026 (electronic)
Smart Innovation, Systems and Technologies
ISBN 978-981-32-9867-5 ISBN 978-981-32-9868-2 (eBook)
https://doi.org/10.1007/978-981-32-9868-2



This Springer imprint is published by the registered company Springer Nature Singapore Pte Ltd.
The registered company address is: 152 Beach Road, #21-01/04 Gateway East, Singapore 189721, Singapore

# International Programme Committee

Dr. Mohamed Abbas, UDES/CDER, Algeria
Dr. Kouzou Abdellah, University of Djelfa, Algeria
Prof. Abdel Ghani Aissaoui, University of Bechar, Algeria
Dr. Mahmood Alam, University of Brighton, UK
Dr. Nader Anani, University of Chichester, UK
Dr. Martin Anda, Murdoch University, Australia
Prof. Shady Attia, University of Liege, Belgium
Prof. Ahmad Taher Azar, Benha University, Egypt
Dr. Magda Baborska-Narożny, Wroclaw University of Technology, Poland
Dr. Gabriele Bernardini, Universita Politecnica delle Marche, Italy
Dr. Stephen Berry, University of South Australia, Australia
Prof. Frede Blaabjerg, Aalborg University, Denmark
Dr. Samuel Brunner, Empa, Switzerland
Prof. Alfonso Capozzoli, Politecnico di Torino, Italy
Prof. Francesco Causone, Politecnico di Milano, Italy
Dr. Boris Ceranic, University of Derby, UK
Prof. Mohammed Chadli, University of Picardie Jules Verne, France
Prof. Christopher Chao, The University of Hong Kong, Hong Kong
Dr. Fathia Chekired, UDES/CDER, Algeria
Dr. George Zhen Chen, University of Strathclyde, UK
Dr. Giacomo Chiesa, Politecnico di Torino, Italy
Dr. Alfonso Chinnici, The University of Adelaide, Australia
Dr. Marta Chinnici, ENEA, Italy
Prof. Francesco Calise, Universita degli Studi di Napoli Federico II, Italy
Prof. Dulce Coelho, Polytechnic Institute of Coimbra, ISEC, Portugal
Dr. Stefano Cascone, University of Catania, Italy
Prof. Pooya Davari, Aalborg University, Denmark
Prof. Mohamed Djemai, Universite de Valenciennes et du Hainaut Cambresis, France
Prof. Tomislav Dragicevic, Aalborg University, Denmark
Dr. Sonja Dragojlovic-Oliveira, University of West England, UK

# Preface

The 11th International Conference on Sustainability and Energy in Buildings 2019 (SEB-19) was a major international conference held in Budapest during 4–5 July 2019 organised by KES International in partnership with Cardiff Metropolitan University, Wales, UK.

SEB-19 invited contributions on a range of topics related to sustainable buildings and explored innovative themes regarding sustainable energy systems.

The aim of the conference was to bring together researchers and government and industry professionals to discuss the future of energy in buildings, neighbourhoods and cities from a theoretical, practical, implementation and simulation perspective. The conference formed an exciting chance to present, interact and learn about the latest research and practical developments on the subject.

The conference featured General Tracks chaired by experts in the field, and in addition, 13 Invited Sessions were proposed by prominent researchers.

SEB-19 featured two keynote speakers: Philippe Moseley, Senior Project Advisor from the Executive Agency for Small and Medium-sized Enterprises (EASME), at the European Commission, who gave a talk entitled *EU support to decarbonise the building stock* and Prof. Fernanda Rodrigues, University of Aveiro, Portugal, who gave a talk entitled *Climate Changes' Impact on the durability and energy performance of buildings*.

The conference attracted submissions from around the world. Submissions for the Full-Paper Track were subjected to a blind peer-review process. Only the best of these were selected for presentation at the conference and publication in the proceedings in a volume in the KES-Springer 'Smart Innovation, Systems and Technologies' series. Submissions for the Short-Paper Track were subjected to a 'lighter-touch' review and may be published elsewhere.

Thanks are due to the very many people who have given their time and goodwill freely to make SEB-19 a success. We would like to thank the members of the International Programme Committee who were essential in providing their reviews of the conference papers. We thank the high-profile keynote speakers for providing interesting talks to inform delegates and provoke discussion. Important contributors to the conference were made by the authors, presenters and delegates without whom the conference could not have taken place, so we offer them our thanks.

It is intended that this volume provides a useful and informative snapshot of recent research developments in the important and vibrant area of *Sustainability in Energy and Buildings*.

SEB-19 Conference Chairs

Poole, UK — Prof. Robert J. Howlett
Cardiff, Wales, UK — Dr. John Littlewood
Turin, Italy — Dr. Alfonso Capozzoli
Canberra, Australia — Prof. Lakhmi C. Jain

# Contents

# About the Editors

**John Littlewood** graduated in Building Surveying and holds a PhD in Building Performance Assessment. He is Head of the Sustainable and Resilient Built Environment group in Cardiff School of Art & Design at Cardiff Metropolitan University (UK). He Coordinates three Professional Doctorates in Art & Design, Engineering and Sustainable Built Environment; plus contributing to teaching in Architectural Design & Technology. John's research is industry focused, identifying and improving fire and thermal performance in existing and new dwellings, using innovative materials and construction, and also improving quality of life. He has authored and coauthored 120 peer-reviewed papers, and was also co-editor for the 'Smart Energy Control Systems for Sustainable Buildings' book published in June 2017.

**Dr. Robert J. Howlett** is the Executive Chair of KES International, a non-profit organization that facilitates knowledge transfer and the dissemination of research results in areas including Intelligent Systems, Sustainability, and Knowledge Transfer. He is a Visiting Professor at Bournemouth University in the UK. His technical expertise is in the use of intelligent systems to solve industrial problems. He has been successful in applying artificial intelligence, machine learning and related technologies to sustainability and renewable energy systems; condition monitoring, diagnostic tools and systems; and automotive electronics and engine management systems. His current research work is focussed on the use of smart microgrids to achieve reduced energy costs and lower carbon emissions in areas such as housing and protected horticulture.

**Alfonso Capozzoli** graduated in Mechanical Engineering and holds a Ph.D. in Engineering of Mechanical Systems. He works as an Associate Professor at the Department of Energy at Politecnico di Torino (Italy). He teaches HVAC systems and building physics at the Faculty of Engineering and Architecture. He is involved in various international research projects on building energy performance. His research is focused on energy saving strategies in air conditioning systems, data analytics for building energy management and super insulating materials in building components. His research activity is summarised in about 100 scientific papers published in international journals and conference proceedings.

**Dr. Lakhmi C. Jain** PhD, ME, BE(Hons), Fellow (Engineers Australia) is with the University of Technology Sydney, Australia, and Liverpool Hope University, UK.

Professor Jain founded the KES International for providing a professional community the opportunities for publications, knowledge exchange, cooperation and teaming. Involving around 5,000 researchers drawn from universities and companies world-wide, KES facilitates international cooperation and generate synergy in teaching and research. KES regularly provides networking opportunities for professional community through one of the largest conferences of its kind in the area of KES. His interests focus on the artificial intelligence paradigms and their applications in complex systems, security, e-education, e-healthcare, unmanned air vehicles and intelligent agents.

# Chapter 1
# The Utilisation of Smart Meter Technology to Increase Energy Awareness for Residential Buildings in Queensland, Australia

**Olusola Charles Akinsipe, Domagoj Leskarac, Sascha Stegen, Diego Moya and Parasad Kaparaju**

**Abstract** The paper aims to sensitise electricity subscribers on the significance of adopting smart meters in managing the energy consumption of residential buildings in Queensland, Australia. This paper examines the power consumption of residential buildings and the time-of-use energy tariffs across four climatic conditions. The analysis also involves applying statistical tools to understand the energy profiles of the study areas. The results show habitual and significant energy consumption of the study areas during the period under study. For instance, energy use during the spring and winter seasons peaked around 30 MWh as residential buildings consumed considerable electricity during the peak periods when the energy tariffs are high. The results also show that the time-of-use of energy consumption can impact the electricity bills as well as the electricity use of customers. Furthermore, there is a correlation between energy use and energy consumption time of the case study areas. Our results present the need to create awareness on the essence of adopting smart meters that will provide real-time information and energy tariffs at a different time of the day in order to optimise electricity consumption and expenses in Queensland. The intelligent machine alongside other technologies can broadcast electricity consumption and display real-time energy prices at frequent intervals thereby supporting energy consumers to make informed choices about deploying their electrical devices when the energy tariffs are affordable and economical.

O. C. Akinsipe · D. Leskarac · S. Stegen · P. Kaparaju (✉)
School of Environmental & Built Environment, Griffith University, Brisbane, QLD 4111, Australia
e-mail: p.kaparaju@griffith.edu.au

D. Moya
Department of Chemical Engineering, Imperial College London, London SW7 2AZ, UK

D. Moya · P. Kaparaju
Institute for Applied Sustainability Research (ISUR), Av. Granados E13-55 e Isla Marchena, No. 44, Quito 170503, Ecuador

D. Moya
Carrera de Ingeniería Mecánica, Facultad de Ingeniería Civil y Mecánica, Universidad Técnica de Ambato, Av. Los Chasquis y Rio Payamino, Ambato 1801314, Ecuador



J. Littlewood et al. (eds.), *Sustainability in Energy and Buildings*, Smart Innovation, Systems and Technologies 163, https://doi.org/10.1007/978-981-32-9868-2_1

## 1.1 Introduction

Energy consumption in the residential building sector has continued to increase alongside the transport and industry sectors [1]. The principal factors stimulating an increase in energy use include climatic conditions and wealth among others [2]. Energy consumption during peak periods if left unregulated can exceed both the generating and the distributing network capacity. Hence, developing and deploying renewable energy technologies and implementing energy efficiency measures will reduce emissions, address climate change, improve energy consumption characteristics and promote energy security [3–5]. These can be achieved by subsidising the costs of clean energy technologies and offering financial incentives to manufacturers of these technologies thereby making it available and affordable to energy consumers in the residential buildings. According to Britton and Stewart [3], the introduction of smart meters has made a significant improvement to residential buildings water consumption by mitigating water loss and enhancing water consumption management. An automated smart meter reading (AMR) system installed in the residential homes for Elster V100 meters connected with FIREFLY® data loggers supports the calibration of hourly water consumption data and broadcasts information to display technologies for monitoring purpose. The results showed a drastic reduction of water leakages as the consumers were able to detect water loss and make timely repairs of water supply systems. Further, this innovation has offered a wide range of solutions and benefits to both the utility companies and consumers by promoting quality service and water consumption management. Currently, Queensland adopts power meters to ascertaining usage patterns and evaluating electricity demand in residential houses [4]. These meters are exclusively designed to record data at intervals for the utility companies and electricity retailers but do not broadcast data to consumers. Contrary, smart meters support the growing and complex power demand [5] and also uphold bi-directional communication between consumers and energy providers via either the power line carrier (PLC) or the wireless communication system thereby facilitating end-users electricity intelligent consumption. In addition, energy data provided by smart meters is significant for power quality analysis, electrical load chart, and energy users' billing. In the coming days, intelligent meters will be able to determine total harmonic distortion (THD) which is pertinent for studying power quality [6].

Presently, in Queensland, electronic billing systems and paper are the principal modes of communicating electricity information and consumption. This approach is conservative and lacks dynamism and progressiveness. With this in mind, the aim of this paper is to sensitize consumers about adopting smart meters to optimise electricity consumption and minimise electricity expenses. Section 1.2 provides the method applied in this research, Sect. 1.3 presents electricity profiles of the case study areas and their statistical results. Section 1.4 discusses the functions of smart meters and Sect. 1.5 is the conclusion.

## 1.2 Methodology

The half-hourly energy consumption of three study areas which have been previously recorded by the electronic meters installed in the residential buildings at Sunnybank, Milton, and Indooroopilly in Brisbane, Queensland, is selected for this study. The suburbs are well established with a similar proportion of older homes in the 1980s and the new homes in 2010s. The year 2014 substation data provided by Energex Company are utilised in studying the suburbs electricity use. An extensive appraisal and the evaluation of electricity consumption in the studied suburbs are conducted over four seasons. Based on census statistics and land area, Sunnybank population with a weekly household income of $AUD 1,322 was 9697, Milton population and a weekly household income were 10,788 and $AUD 2,190, and Indoorpilly were 12,242 and $AUD 1,724 [7] (Fig. 1.1).

### *1.2.1 Time of Use Analyses*

To investigate energy consumption pattern of the consumers in the studied suburbs, three distinctive energy consumption regimes consisting of the off-peak periods 10 p.m.–6 a.m.; shoulder periods 6 a.m.–4 p.m. and the peak periods 4 p.m.–10 p.m., as stipulated by utility companies, are applied across the four seasons. Table 1.1 presents a summary of residential time-of-use (ToU) tariffs and price systems in Brisbane, Australia. The impact of energy tariffs on the electricity consumption of customers per day is analysed to determine the energy profiles of the studied areas

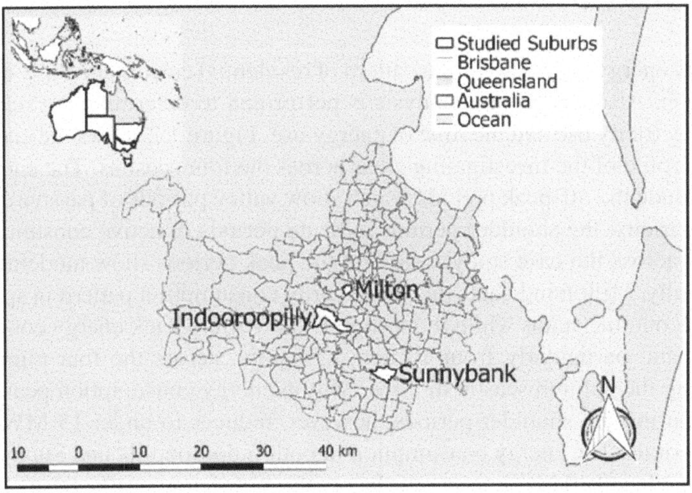


**Fig. 1.1** Chosen suburbs for the residential dwellings analysis in Brisbane, Queensland, Australia

**Table 1.1** Summary of residential ToU tariff and price systems [10]

| Residential ToU NTC8900 (Tariff 12) | Units (Australian currency) | Price excluding goods and services tax (GST) |
|---|---|---|
| Peak usage (weekdays 4 p.m.–8 p.m.) | Cents per kWh | 18.315 |
| Shoulder usage (weekdays 7 a.m.–4 p.m., 8 p.m.–10 p.m., weekends 7 a.m.–10 p.m.) | Cents per kWh | 10.624 |
| Off-peak usage cents per kWh (weekdays and weekends 10 p.m.–7 a.m.) | Cents per kWh | 6.943 |
| Supply charge per day | Cents per kWh | 50.20 |

[8]. The energy use is plotted against the time of the day which coincides with different energy tariffs in determining the energy profiles. Further, the impact of time on the electricity use across the four seasons is evaluated by analysis of variance (ANOVA), a Statistical Package for the Social Science (SPSS). Linear regression analysis is applied to the time of day of electricity use as the independent variables while energy consumption is the dependent variables [9]. The period of electricity consumption by the residents in the studied areas is crucial to determine the degree of the consumers' consciousness of the variability of energy prices at different tariff periods as well as the users' capabilities to regulate consumption and electricity expenses.

## 1.3 Results

Half-hourly energy curves show the pattern of residential customers' daily electricity consumption. Also, regression analysis is performed to determine the relationship between electricity use and the time of energy use. Figure 1.2 shows the energy consumption profile of the investigating areas across the four seasons. The energy tariff periods include the off-peak periods which show valley periods of passive consumption of electricity, the shoulder periods indicate periods of active consumption and occupancy across the case study areas and the peak periods show moderate energy use. Generally, Milton indicates the least energy consumption pattern in spite of the area socioeconomic status while Indooroopilly and Sunnybank energy consumption are significant, particularly from 10 a.m. to 4 p.m., across the four climatic seasons. During the autumn season in 2014, Milton energy consumption peaks around 20 MWh during the shoulder periods, however, reduces to under 15 MWh during the peak periods. The energy consumption in both Indooroopilly and Sunnybank are about 30 MWh and 40 MWh respectively, at different energy tariffs. Also, during the summer season, it can be seen that Milton electricity consumption increases from 7 MWh to around 20 MWh throughout the shoulder periods while Indooroopilly

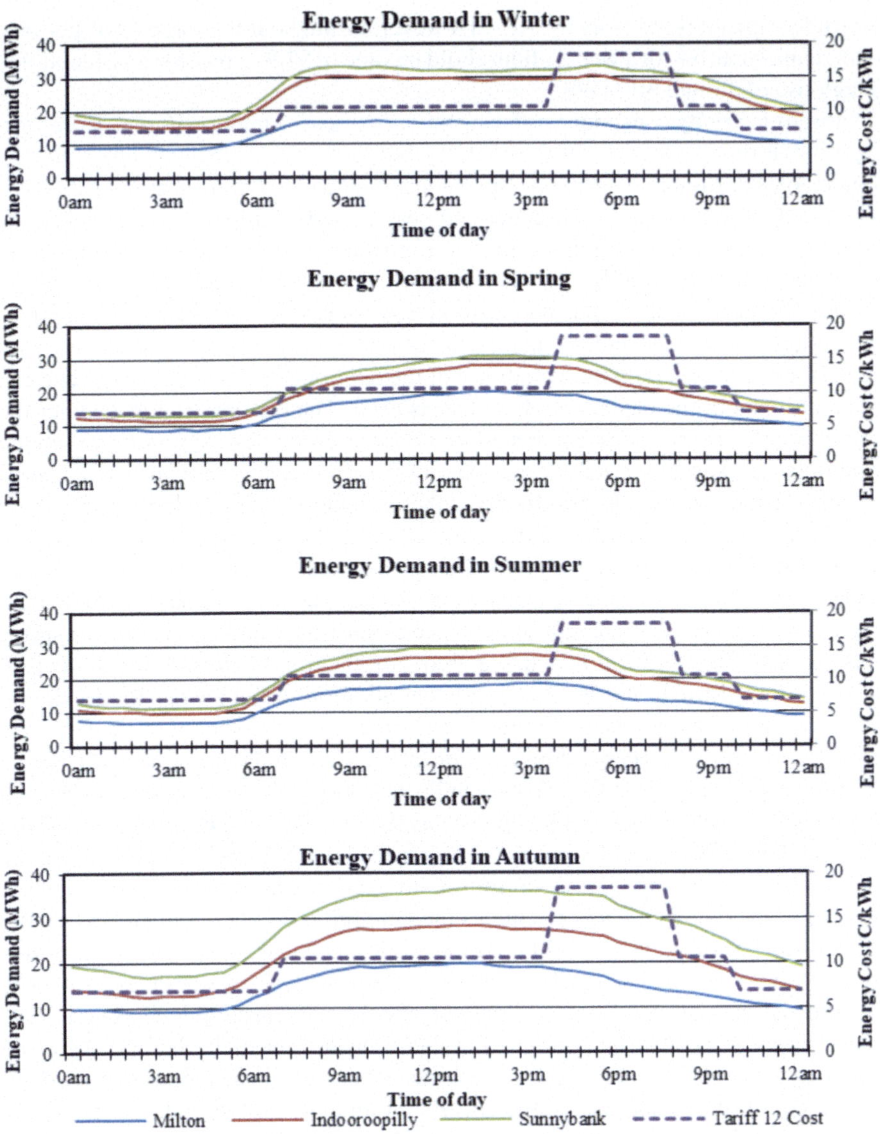


**Fig. 1.2** Energy Fig. 1.2. Energy consumption profile of the investigating areas across the four seasons

electricity consumption rises by 50%. However, throughout the three tariff periods, Sunnybank suburb with a weekly household income of $1,322 records an exceptional energy use of around 30 MWh.

Milton electricity consumption during the spring season under different energy tariff periods remains constant. On the contrary, Indooroopilly and Sunnybank record an appreciable consumption during the spring season and increase from around 15 MWh to about 30 MWh during the shoulder and peak periods. Further, Milton energy use during the winter shows similar patterns of consumption. However, Indooroopilly and Sunnybank have similar periods of significant consumption during the spring season. The periods of significant consumption suggest the increase in the deployment of electrical loads and occupancy of the residents. Further, statistical analyses are performed to ascertain the possibility of the relationship between energy use and different hours of the day when electricity is used. The big question is whether the amount of electricity use is influenced by the time of the day. Across four climatic conditions, the substation half-hourly data of the studied areas are analysed against the time of the day and the results show the correlation coefficients, $R^2$ with high probability values greater than 0.05. This indicates that the null hypothesis cannot be rejected. Statistically, the Null hypothesis is defined by Ho as the hypothesis that model observations decision basically from chance, and the *p*-value is defined as an index that measures the strength of evidence against the null hypothesis in a single analysis [11]. The implication is that a predictor, the time of the day when electricity is consumed might not have a significant addition to the energy model due to the variability of the predictor's value not correlating with changes in the response variable, which is the energy consumption of the case study areas. Nevertheless, a low probability value less than 0.05 is indicated during the winter season showing that the null hypothesis cannot be ignored as the results are significant at $p = 0.02$, 0.01 and 0.001. The possible drivers of increased electricity consumption during the winter season including thermal comfortability and heating systems. Nevertheless, the time of the day with a low *p*-value is meaningful to the energy model due to the relationship between the variability of time of the day and the response variable during the winter season. The results of a series of the analysis of variances (ANOVA) performed across the four seasons show that the time of consumption has no impact on energy use. A positive correlation exists between the time of the day and aggregate electricity consumption during the winter season because of the variation of the daily electricity consumed and the coefficients of variation of Milton, Indooroopilly, and Sunnybank indicating 10%, 13%, and 20% respectively. The general energy model equation; Eq. (1.1), is the representation of simple linear regression.

$$E = \beta o T + \varepsilon \quad (1.1)$$

where E (Energy) is the response variable, T (time of the day) is the predictor variable and $\beta o$ is the regression coefficient and $\varepsilon$ is an error to account for the discrepancy between the predicted data from Eq. (1.1) and the observed.

## 1.4 Discussion

This study aims to sensitise electricity consumers in Queensland, Australia about minimising electricity charges and billings tariffs through deploying smart meters. To achieve this objective, the pattern of the studied areas energy consumption is analysed to understand how residences respond to electricity use at a different time of the day. Further, statistical analyses are performed to determine the relationship between electricity demand at various hours of the day and different tariffing periods. The study explores different study areas to determine how consumers respond to electricity consumption at different tariffs' periods. Surprisingly, Milton shows constant electricity consumption style during the summer, spring and the winter seasons in spite of suburb economic potential whereas Sunnybank energy consumption accounts for 20% of the total energy use, possibly due to low active occupancy of the residents [12], demographic size and the nature of building apartments [13]. The energy pattern of this study appears to be similar to Chen et al. [14] findings which revealed a variation in the energy use throughout a day according to changes in the seasons. Also, the results show a decrease in energy consumption between 12 a.m. and 6 a.m. being the off-peak periods. Based on the scale of study in this paper, substation energy half-hourly data and time of the day of electricity use of the studied areas are analysed compare to Chen et al. [14] analysis based on energy consumption of some families using various appliances in seven cities during the winter season. In this study, the half-hourly charts show a combination of passive and active occupancy of customers in Milton, Indooroopilly, and Sunnybank. The active occupancy is used to indicate when customers are deploying their appliances while the passive occupancy is a period when people are either sleeping or not using their electrical loads [12]. The active occupancy fits in the shoulder and peak periods while the passive occupancy is appropriate for the off-peak periods. Moreover, it offers information such as the condition of occupancy and the frequency of appliances use, and the characteristics of appliances suggest a combination of low-capacity and high-capacity electrical loads. Similarly, at different tariff periods, the response of the residences is indicated on the demand profiles across the four seasons. The findings are successful as they can identify the need to optimise the electricity consumption and reduce expenses thereby assisting to understand the significance of introducing smart metering technologies in Queensland, Australia. Previous studies discussed how smart meters deployed technical and economic measures to minimise energy consumption [15]. Smart meters function with different communication networks ranging from home area network (HAN) to a wide area network (WAN) to broadcast electricity data. The WAN supports energy data transmission from the downstream via the technological communication technologies such as Global System for Mobile Communication (GSM) to energy provider [16] whereas the HAN backstops the connection among appliances and smart meters. The intelligent systems function with communication protocols such as Wi-Fi, and Zig-Bee, facilitating electricity data communication between neighbouring meters [17], as well as offering information to energy users. Based on the capability of smart meters, the power company in agreement with the consumer can connect or disconnect energy supply to maximise electricity use [6]. The economic models are the representations of the demand-side response measures to maximising energy consumption. Generally, the economic measures are represented by the time-dependent such as the time-of-use (ToU) among other measures [18]. According to Table 1.1, the utility company implemented the ToU pricing system to curtail peak hours use, particularly during shoulder and peak periods when the price of electricity generation increases exponentially [19]. The customers deserve awareness about smart meters capability to reducing peak hours energy use as well as minimizing electricity use and expenses. In Fig. 1.2, residences in the investigating areas consume considerable electricity during the shoulder periods as well as the peak periods. This result indicates the limitations of the contemporary electronic metering technologies to provide real-time and frequent information and electricity billing system to customers. Thus, electricity users are unaware of the extent of their power consumption and the costs incurred on electricity until after 90 days, approximately. A lack of real-time information can influence residential users to increase electricity consumption with attendant impacts on their electricity budgets. The question is why using automated technologies such as dishwashers and washing machines without smart meters to regulate and monitor energy use in a place like Queensland, Australia. The residential customers will be more willing to use their washing machines during off-peak hours when the electricity prices are economical and more affordable by deploying innovative technologies with the capability to facilitate and provide a real-time and bi-directional flow of information between the energy vendors and users regarding electricity use critical to energy management [20]. Further, implementing smart meters will support local and remote feedbacks by broadcasting and transmitting energy information [12]. On top of this, energy subscribers will have the opportunity to participate in the wholesale electricity market [15]. Unlike other developed nations, power meters are applied in the residential buildings to calibrate aggregate electricity consumption over an extended time in Queensland, Australia. The technologies depend on rotating discs for calibration but without functional memory compartments for data storage and management. Electricity consumption is computed by calibrating the position of the dial from the last measurement. Consequently, these technologies cannot record and broadcast real-time energy use because the technologies are not incorporated with power and voltage sensors that collect, transfer real-time energy data and support bi-directional real-time energy profiles collection and reporting. This role is performed by an authorized energy vendor who accesses the meter to take the electricity readings periodically [19]. While Energex offers customers the choices of energy prices at different hours of the day and different tariff schemes, there is still a lack of innovative technologies to broadcast the information to both the utility and customers. Consequently, optimising electricity use and minimising cost expended on electricity is unattainable [15]. As noted by Faruqui et al. [21], the deployment of In-Home Displays (IHDs) and AMI in New South Wales, Australia, facilitated energy savings by 8% and minimised energy consumption by around 30% during the summer and the winter seasons at critical tariffs and charges. Therefore, it is important that electricity consumers in

Queensland are cognisant of smart meters technologies operations to minimise their budgets on electricity use and power consumption.

## 1.5 Conclusion

This paper aims to sensitise electricity consumers in Queensland about deploying smart meters to minimise energy use and reduce energy charges and billings tariffs thereby minimizing electricity expenses. The research shows the limitations of power meter to record and broadcast real-time information. Based on energy plots across the four seasons, the results show that electricity is consumed more during the shoulder periods and the peak periods, and during the most expensive times of the billing cycle. Substituting electronic meters for smart meters is a more sustainable approach to stimulating residential buildings energy management because customers will have the opportunity to regulate their appliances and also monitor their consumption. The technology is adapted with information and communication technologies to promote energy prices at different hours of the day. This study suggests that implementing smart meters will promote electricity consumption and minimise electricity costs of the study areas. In future studies, more data regarding the electricity consumption of different households, the number of appliances per household, the present condition of the buildings and the frequency of usage of appliances may have to be explored.

**Acknowledgements** This paper acknowledges Energy Australia and Energex for providing the substation energy data and residential data used in this research. This research did not receive any specific grant from funding agencies in the public, commercial, or not-for-profit sectors.

## References

1. Yang, L., Yan, H., Lam, J.C.: Thermal comfort and building energy consumption implications: a review. Appl. Energy **115**(15), 164–173 (2014)
2. Zhao, H.-X., Magoulès, F.: A review on the prediction of building energy consumption. Renew. Sustain. Energy Rev. **16**(8), 3586–3592 (2012)
3. Britton, T.C., Stewart, R.A., O'Halloran, K.R.: Smart metering: enabler for rapid and effective post meter leakage identification and water loss management. J. Clean. Prod. **54**, 166–176 (2013)
4. Sharma, K., Saini, L.M.: Performance analysis of smart metering for smart grid: an overview. Renew. Sustain. Energy Rev. **49**, 720–735 (2015)
5. Kara, S., Bogdanski, G., Li, W.: Electricity metering and monitoring in manufacturing systems. In: Globalized Solutions for Sustainability in Manufacturing, pp. 1–10. Springer (2011)
6. Aziz, A.F.A., Khalid, S.N., Mustafa, M.W., Shareef, H., ALiyu, G.: Artificial intelligent meter development based on advanced metering infrastructure technology. Renew. Sustain. Energy Rev. **27**, 191–197 (2013)
7. ABS Statistics (n.d., 02/05/2019) Census QuickStats. (2016). https://quickstats.censusdata.abs.gov.au/census_services/getproduct/census/2016/quickstat

8. Jablonsky, J.: MS Excel based software support tools for decision problems with multiple criteria. Proced. Econ. Finance **12**, 251–258 (2014)
9. Fumo, N., Biswas, M.A.R.: Regression analysis for prediction of residential energy consumption. Renew. Sustain. Energy Rev. **47**, 332–343
10. Energex. (n.d., 23/03/2017) Residential tariffs and prices. https://www.energex.com.au/home/our-services/pricing-And-tariffs/residential-customers/residential-tariffs-and-prices
11. Silva-Ayçaguer, L.C., Suárez-Gil, P., Fernández-Somoano, A.: The null hypothesis significance test in health sciences research (1995–2006): statistical analysis and interpretation. BMC Med. Res. Methodol. **10**, 44–52 (2010)
12. McKenna, E., Richardson, I., Thomson, M.: Smart meter data: balancing consumer privacy concerns with legitimate applications. Energy Policy **41**, 807–814 (2012)
13. Kavousian, A., Rajagopal, R., Fischer, M.: Determinants of residential electricity consumption: using smart meter data to examine the effect of climate, building characteristics, appliance stock, and occupants' behavior. Energy **55**, 184–194
14. Chen, S., Li, N., Yoshino, H., Guan, J., Levine, M.D.: Statistical analyses on winter energy consumption characteristics of residential buildings in some cities of China. Energy Build. **43**(5), 1063–1070 (2011)
15. Siano, P.: Demand response and smart grids—a survey. Renew. Sustain. Energy Rev. **30**, 461–478 (2014)
16. Aswathi, M., Gandhiraj, R., Soman, K.: Application and analysis of smart meter data along with RTL SDR and GNU radio. Proced. Technol. **21**, 317–325 (2015)
17. McKerracher, C., Torriti, J.: Energy consumption feedback in perspective: integrating Australian data to meta-analyses on in-home displays. Energy Effi. **6**, 387–405 (2013)
18. Venkatesan, N., Solanki, J., Solanki, S.K.: Residential demand response model and impact on voltage profile and losses of an electric distribution network. Appl. Energy **96**, 84–91 (2012)
19. McLaughlin, S., McDaniel, P., Aiello, W.: Protecting consumer privacy from electric load monitoring. In: Proceedings of the 18th ACM Conference on Computer and Communications Security, pp. 87–98 (2011)
20. Carroll, J., Lyons, S., Denny, E.: Reducing household electricity demand through smart metering: the role of improved information about energy saving. Energy Econ. **45**(9), 234–243 (2014)
21. Faruqui, A., Sergici, S., Sharif, A.: The impact of informational feedback on energy consumption—a survey of the experimental evidence. Energy **35**(4), 1598–1608 (2010)

# Chapter 2
# Impact of Building Massing on Energy Efficient School Buildings

**Yasemin Afacan and Ali Ranjbar**

**Abstract** To produce energy-efficient buildings, optimization process for all design stages is necessary. Optimization starts with the massing of the building. This study investigates the impact of the five school massing typologies on energy efficiency: (i) spine/street; (ii) city/town; (iii) atrium; (iv) strawberry/cluster; and (v) courtyard. The chosen massing typologies respond to the question of what an optimum spatial organization of massing is to (i) maximize the use of renewable resources; (ii) utilize thermal inertia of buildings; and (iii) consider the relationship between inside and outside, both existing and future. For each massing type, Sefaira program was used, and simulations were run for annual energy use, annual energy cost and annual carbon dioxide ($CO_2$) emissions. The energy use indices (EUI) of the alternatives are around 86 kWh/m2/yr. In the spine massing, the EUI value is much higher than the other four buildings. The highest annual net $CO_2$ emissions are obtained in atrium type of building, which has more floors compared to other massing type. The courtyard type has the most efficient annual electricity cost per area. These findings showed that the goal of the building massing should be not only limited to achieve the low EUI. Thus, this study suggests that an energy-efficient massing should address the questions beyond well-known ASHRAE standards, and define a new holistic model that considers the ratio of surface area to volume more for reducing energy loads than a typical high-performance schools.

## 2.1 Introduction

It is an undeniable fact that the built environment is one of the main factors in global energy consumption [1]. So, achieving energy efficiency in current building stocks is crucial because sustainable design, planning, and construction decrease energy consumption by reducing environmental pollution, controlling energy waste patterns, as well as material waste [2]. U.S. Environmental Protection Agency Glossary of

---

Y. Afacan (✉) · A. Ranjbar
Department of Interior Architecture and Environmental Design, Bilkent University, 06800 Ankara, Turkey
e-mail: yasemine@bilkent.edu.tr



J. Littlewood et al. (eds.), *Sustainability in Energy and Buildings*, Smart Innovation, Systems and Technologies 163, https://doi.org/10.1007/978-981-32-9868-2_2

Climate Change Terms defined energy efficient buildings as systems that use less energy to perform as well or better than standard systems [3, 4]. Although rules and criteria for energy efficiency should be decided according to building type and end-use, Jones identified three key points to apply across all building types: (i) maximize the use of renewable resources from wind, sunlight, daylight taking into account the potential noise and pollutants; (ii) utilize thermal inertia of buildings and (iii) consider the relationship between inside and outside, both existing and future [5]. The wide range of assessment systems are also available all over the word [1], such as Leadership in Energy and Environmental Design (LEED) introduced by the US Green Building Council [6], and the Building Research Establishment Environmental Assessment Method (BREEAM) developed by Building Research Establishment (BRE) in the early 1990s as the first assessment method for office building in UK [7].

Both energy use and building first costs are correlated to the efficiency of a building massing, which is measured by the ratio of surface area (envelope) to volume known as the shape factor [8]. Building massing relates to shapes, proportions, layout, interior and exterior relationship and orientation of buildings [9]. The optimal massing reduces energy, eases solar shading strategies and increases effectiveness of natural ventilation [8, 9]. To produce energy efficient buildings, optimization process for all design stages is necessary [9, 10]. However, architects are still struggling to design energy-efficient buildings, especially in earlier stages of architectural design process, during which building massing is decided [11, 12]. School buildings have some unique features that distinguish them from other typologies [13]. In a school building, occupancy can be very high, reaching up to four times more occupants per square meter than in a typical office building [14]. Moreover, occupants spend much of their time inside classrooms. Because of this occupancy schedule, school buildings require special attention on building environmental quality managements [15]. Although there are numerous studies on the relationship among energy consumption, renewable energy production, and massing [16–20], there is a lack of studies on defining massing based on its relationship with sustainable interior planning strategies. The majority of the articles are focusing on the building geometry solely [21], rather than its close association with interior design parameters. Thus, this present study attempts to fill this gap by studying and comparing five school massing typologies based on sustainable interior functioning in terms of their impact on energy efficiency.

## 2.2 Energy-Efficient School Building Typologies

### *2.2.1 The School Typologies*

The typology is the fundamental aspect of the discipline of architecture. It means type, model, and primary form of a building [22]. The type plays a key role in defining preconceived entities for the conceptual design process [23, 24]. Graca et al. defined seven school typologies based on solely plan layout; row, double, two sets, two sets of L, U shape, L shape and two sets covered with patio area. They proposed multi-criteria optimization for achieving environmental comfort early in design process. Taylor [25] explored impact of school building typologies on visual, thermal, and energy performances by considering three basic typologies; linear, corridor, and concentrated model with different classroom proportions. They observed that compact typologies had better energy performance with heating, whereas linear typologies had better visual performance. So, as stated by Hawkes [26], the compactness of the shape would not be the optimal energy solution for schools.

Quan et al. [27] designed an experimental framework for computing energy performance of four building typologies; pavilion, slabH, slabV, and courtyard. Their results suggested that even with the same typology energy consumption could vary. Zhang et al. [28] studied the thermal performance of school buildings in China by performing energy simulations for eight different typologies; rectangle shape, L shape, C shape, H shape, H shape with atrium, courtyard, high-rise, and irregular shape. Their typologies were derived from geometry parameters based on the number of classrooms, number of stories, window-to-wall ratio and room depth. A key distinction between geometry and massing is the level of input detail, which means that building massing is the overall configuration of the building, whereas geometry is usually originated from proportions of a 2D drawing [21, 29]. So, building massing have not only a marginal effect on floor plan geometries of varying complexity, but also on energy consumption patterns for each interior space [29]. Thus, this study complements other studies in terms of addressing interior efficiency as well by focusing on massing typologies rather than on solely geometry.

### 2.2.2 Energy Efficient School Building

Energy efficient schools aims to open buildings to daylight and views to save lighting energy costs, improve indoor air quality and satisfy user needs and demands [30]. According to Kats, this interior and exterior relationship is not only aesthetically pleasing, but also increases user performance so that workers with views have performed 10–25% better on tests than those without views [31]. LEED also highlights the connection between indoor and outdoor environments [6]. Designers can maximize interior and exterior relationship through proper decisions on site, orientation, shading structures, interior divisions, glazing and circulation patterns.

Schools in the 1970s were designed with fewer windows, which created dark buildings with extreme opacity [25]. Following sustainability definition by United Nations Brundtland Commission in 1987, UN Commission Report in 1992 on sustainable development and Kyoto Conference by UN Framework on climate change, there were opportunities to discuss how to reduce carbon emissions and improve energy efficiency in buildings [3]. At European level, first action on energy regulation started with European Energy Efficiency Directive in 2002 [32] with a focus on optimized design of new buildings and reducing their impact on long-term energy consumption; later in 2012 the concept of Net Zero Energy Building was introduced [33]. This was followed by action in 2018 amending several points of the previous two actions [34]. Thus, school building sector had chance to apply energy efficiency issues after those dates. In Mediterranean region of Europe and in most areas of the world, many school buildings were constructed in the second half of the 20th century so that they fail to meet the most recent criteria in energy efficiency field [35]. There are lots of studies about energy efficiency in school building ranging from passive to active measures. Passive measures focus on building envelope to enhance sustainability through energy reduction in long term life of the building [36], whereas active measures are related to the use of cold and hot water systems, electrical equipment, lighting, heating, cooling, ventilation and air conditioning systems [20]. According to the review study by Dias Pereira et al. it is not possible to have well-specified energy measures for schools since there are uncertainties in energy consumption values depending on the country and location [15]. However, Taylor described the energy efficiency strategy for an ideal school building as a carefully designed physical location composed of natural, built, and cultural parts that work together to accommodate active learning across body, mind, and spirit [25]. Current architecture practices in educational buildings and learning theories are in line with this statement and indicate the importance of environmentally responsive design [37].

## 2.3 The Chosen Massing Typologies

To better understand the significance of interior space organization on building massing, this research differed from the other typology and energy-efficiency studies by choosing the five proposed typologies based on the above-mentioned Jones's three key points of energy efficiency [5]. The chosen massing typologies tries to respond to the question of what an optimum spatial organization of massing is to (i) maximize the use of renewable resources; (ii) utilize thermal inertia of buildings for lighting, heating, cooling, and ventilation; and (iii) consider the relationship between inside and outside, both existing and future. The five proposed typologies are based on architectural design guidelines for schools [38]. The typology classification of this guideline reference is significant because most schools around the world exhibit characteristics of one or more of these five massing typologies. Although there are many local classifications of school typologies in different countries as explained in Sect. 2.2.1, current study defined these five typology alternatives based on a global guide, which "provides an overview of best practices in contemporary school design from around the world, and is intended to serve as a point of departure for architects and owners in the holistic, artistic, and humanistic design of modern educational facilities" [38, p. 6]. (1) *Spine/street*: In spine massing, major school functions are placed along a central linear space that makes way finding, orientation and access easier. A strong axis could be achieved by placing entrance at one end of the axis. (2)

*City/town*: This massing type is defined as a loose type of massing. Classrooms could be located around the main functions, such as library, sports hall, multipurpose room, and cafeteria. This type of massing typology has more potential in terms of legibility and familiarity of facilities, which are essential parameters for considering inside outside relationship. (3) *Atrium*: This massing suits well for multi-storey schools. A full height atrium has potential to serve for all the three key points of energy efficiency; passive solar design, thermal inertia of the school and also for access outside views. (4) *Strawberry/cluster*: This massing type is similar to spine type. Like spine massing, there is a central core providing circulation. However, in this typology, classroom and other learning facilities offer more opportunities for passive energy efficiency measures, such as solar gain, shading, etc., and also close student/teacher relationship. (5) *Courtyard*: Courtyard massing offers flexibility in terms of organizing different functions around the courtyard. Although courtyard massing increases the amount of building envelope, there are many energy efficiency benefits of this type including access to natural light, views, ventilation, etc.

## 2.4 Methodology

To investigate the impact of the five school massing typologies on energy efficiency, in the study, the optimum school massing is assumed a square footage of 1500 $m^2$ with reference of the Advanced Energy Design Guide for K-12 schools [8]. So, the five baseline buildings with 30 m × 50 m footprint are studied. The climatic location of the study is determined as cold climate, which is Ankara, Turkey, according to ASHRAE 2013 [8]. The five alternatives are schematized in Fig. 2.1.

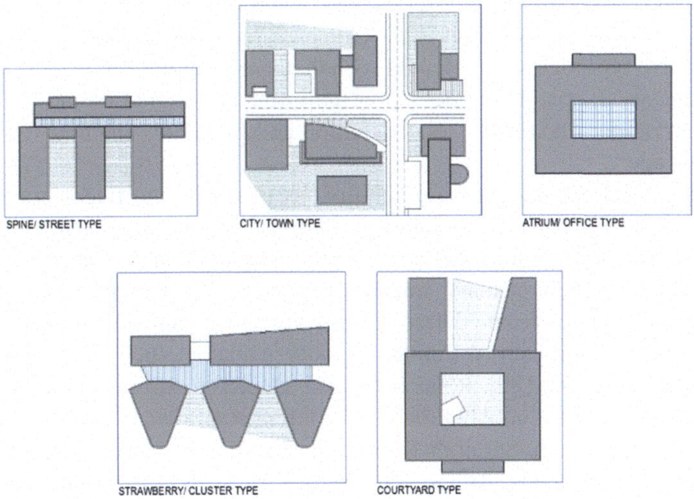


**Fig. 2.1** The five school massing typologies

**Table 2.1** Properties of construction of the five school massing typologies

| Building item | Sefaira ID | 203268, Ankara, TR |
|---|---|---|
| Location-orientation | deg | |
| *Exterior wall*—Assembly type | | Brick |
| | W/m$^2$K | 0.17 (U-factor/R-value) |
| N-E-S-W—*Facade glazing* | W/m$^2$K | 1.11 |
| *Floors*—Assembly type | – | Tiles |
| | W/m$^2$K | 0.32 (U-factor/R-value) |
| *Roof glazing* | W/m$^2$K | 2.4 (U-factor/R-value) |
| SHGC | – | 0.6 |
| *Roofs*—Assembly type | | Metal deck |
| | W/m$^2$K | 0.14 (U-factor/R-value) |
| *Infiltration*—Air changes | | |
| Facade area | m$^3$/m$^2$ h | 7.2 |

Rather than being climate or country based, in this study, the choice of alternatives is based on most contemporary school design from around the world including simplicity, solar potential, natural ventilation, envelope potential to energy usage and strong connections between outdoors and indoors [8]. The variation of shape and form allows to analyze variations in energy performance and $CO_2$ emissions resulted from geometries. The floor-to-floor height of the buildings is set at 4 m, with a floor-to-ceiling height of 2.7 m. The spine massing is one story high. The city, strawberry and courtyard are two stories high, where as the atrium is six stories high. The window-to-wall ratios are set at.% 30 on each façade. For optimal solar orientation, all the five typologies are oriented such that a rectangular footprint is elongated along an east-west axis. This orientation minimizes unwanted radiation from east and west surfacing, maximizes solar radiation and facilitates shading strategies on the south facade.

The study used Sefaira program simulations to model the five school massing typologies. Sefaira's Real-Time Analysis Plugins, which are dynamic simulation tools for energy assessment based on architecture, lighting, and mechanical systems, use EnergyPlus as their primary simulation engine [29]. EnergyPlus is validated thoroughly and it is distributed free of charge, which makes it one of the most accessible professional energy simulation tools available, such as Design Builder, eQUEST, IDA ICE, etc. [39]. It has been also adapted to stimulate energy balance and thermal comfort of school buildings in different countries, including Turkey [28, 29]. Design parameters in Sefaira's Real-Time Analysis provide constant feedbacks on envelope and material U values. Thus, Sefaira as a validated simulation program has been chosen among other simulation engines. Table 2.1 lists values, which were used in the simulation of the five school massing typologies.

The occupancy in the buildings is from 7 a.m. to 6 p.m. on weekdays. Weekends and summer holidays, from 1st June to 1st September, are assumed unoccupied with the primary heating, ventilation and cooling (HVAC) system off. For the HVAC system, fan coil units with central plant system are used. The HVAC system also uses ground-source heat pumps for higher efficiency. It is assumed that there is no hot water demand. Effect of shading by neighborhood trees, buildings and other structures is assumed negligible, whereas there is a self-shading in the massing types. Except the atrium massing type, which has five stories, all the four buildings have two stories.

## 2.5 Results and Discussion

For each typology, simulations were run for annual energy use, annual energy cost and annual $CO_2$ emissions. Table 2.2 lists annual energy use in terms of ventilation, cooling and heating. Annual energy costs and $CO_2$ emissions of the five buildings are presented in Table 2.3. The energy use indices (EUI) of the typologies are around 86 kWh/m$^2$/yr. Only, the spine has the value of 105.21 kWh/m$^2$/yr. These values, even the high one, are good values, because according to ASHRAE 2018 Advanced Energy Design Guide for K-12 School Buildings for achieving 50% energy savings [8], the targeted value for EUI is 104 kWh/m$^2$/yr for 4B cold climate zone. The reason for these low EUI values is the high performance envelope considering lighting and ventilation. However, it should be noted that in buildings, where daylighting is not as much as available as the others, the EUI is much higher. For example, due to shading of streets in the spine, the EUI value is higher than the other typologies.

According to Table 2.3, the highest annual net $CO_2$ emissions are obtained in atrium type, which has more floors compared to other types. Although this building type is the least efficient in terms of emissions, it is the most efficient massing in terms of annual energy cost per area. This result confirms the idea that compactness of the building shape is ideal for energy preservation [27]. The reason for this emission result is also the high-rise structure of the atrium massing, which reduces the opportunity to use passive solar energy in an integrated manner. However, the low-rise courtyard massing supports this finding by having the most efficient annual electricity cost per area, but lower $CO_2$ emissions. It highlights the importance of daylighting in massing decisions. A properly designed low-rise courtyard could use sunlight to offset artificial lighting loads and save energy. The most efficient annual cooling energy per unit area is again achieved in courtyard type.

Regarding the proportions of total energy use in city and atrium massing typologies, respectively, in the study, city massing type is the most efficient and the atrium type is the least efficient. However, it is seen that the largest load is the interior lighting at about 30%, and the second highest load is the equipment. It is typical for schools with an increased technology infrastructure. When comparing the five alternatives, the heating proportion in annual energy use in spine massing type has the highest

**Table 2.2** Annual energy use in terms of ventilation, cooling and heating of the typologies

| Massing type | Air handling (AHU) | Cooling | Heating | Energy use | | | |
|---|---|---|---|---|---|---|---|
| | AHU design airflow | Cooling equip. design cap. | Heating equip. design cap. | HVAC energy per unit area | Annual cooling energy per unit area | Annual heating energy per unit area | EUI |
| | L/s | kW | kW | kWh/m$^2$/yr | kWh/m$^2$/yr | kWh/m$^2$/yr | kWh/m$^2$/yr |
| 1. Spine | 5637 | 229.0 | 289.9 | 59.8 | 7.19 | 29.20 | 105.2 |
| 2. City | 6199 | 207.7 | 253.4 | 41.4 | 6.25 | 13.34 | 86.8 |
| 3. Atrium | 16,974 | 532.8 | 670.4 | 39.8 | 5.55 | 12.5 | 85.2 |
| 4. Strawberry | 7276 | 222.0 | 320.1 | 44.4 | 5.47 | 18.5 | 89.8 |
| 5. Courtyard | 8002 | 247.6 | 316.8 | 41.3 | 5.27 | 15.8 | 86.7 |

**Table 2.3** Annual energy costs and $CO_2$ emissions of the typologies

| Massing type | Annual energy cost | Annual elect. cost | Annual gas cost | Annual energy cost per area | Annual elect. cost per area | Annual gas cost per area | Annual net $CO_2$ emission |
|---|---|---|---|---|---|---|---|
| | $ | $ | $ | $\$/m^2$ | $\$/m^2$ | $\$/m^2$ | kg $CO_2e$/yr |
| 1. Spine | 51850.8 | 45901.6 | 5949.3 | 23.00 | 20.36 | 2.64 | 99,922 |
| 2. City | 51833.8 | 48846.4 | 2987.5 | 20.90 | 19.70 | 1.20 | 98,344 |
| 3. Atrium | 139836.9 | 132161.3 | 7675.6 | 20.60 | 19.47 | 1.13 | 265,112 |
| 4. Strawberry | 60430.9 | 55548.5 | 4882.4 | 20.76 | 19.09 | 1.68 | 115,386 |
| 5. Courtyard | 65330.5 | 60737.9 | 4592.6 | 20.81 | 18.98 | 1.43 | 124,383 |

load. The long north facade makes the heating energy more. The largest east façade the school has, the more energy consumption for cooling is required, such as atrium versus city massing. To better extrapolate the findings with the existing knowledge, the annual energy results support the literature [22], and the idea of compactness as the optimal solution for a school typology. However, the simulation results of total energy in terms interior lighting show that the compactness of the shape is not only the optimal solution, because linear typologies would have better daylighting and indoor air quality performance compared to compact typologies. This is also in line with the literature [26], and explains why energy consumption and $CO_2$ emissions are not directly correlated in atrium and courtyard typologies. Moreover, the climatic context of the study is Ankara, Turkey, a cold climate, where lower buildings (2–3 stories) have high annual energy consumption [27]. So, spine massing offered the worst energy performance regarding the climatic context, whereas the city massing has the best $CO_2$ emission performance because of the greater solar gains and less heat dissipation through envelope created by efficient indoor–outdoor relations.

## 2.6 Conclusion

As reported above, knowledge on massing is essential to guarantee energy efficient schools. How a school is shaped regarding the amenities in and out affect the energy and cost of the building. The goal of the building massing should be not only limited to achieve the low EUI. As supported by this study, the massing should be optimized to integrate school's programmatic elements along interior space global performance requirements.

Massing type should be also configured for minimal energy use without neglecting multiple usage and occupancy schedule. To achieve such efficiency, each design decision should have space planning, programming, specifications and installation of proper interior elements in addition to form, orientation, façade design, shading, heating/cooling/ventilation and lighting control. Thus, this study defines massing in schools with an emphasis on environmentally responsible school interior design, which is a comprehensive understanding of high-performance energy efficiency [5]. An energy-efficient massing should address the questions beyond well-known ASHRAE standards, and define a new holistic model that uses the optimum surface to volume ratio of the building more for reducing energy loads than a typical high-performance school. To extend the contribution of this study in design practice and generalize the results, a detailed analysis, where combinations of building shapes, window-to-wall ratio, room depth and orientation parameters, could be performed. In addition, both design and construction of schools should consider the subjective preference of users even during massing decisions. Proper school massing will not only improve energy savings, but also enhance subjective feeling of all users.

## References

1. Luther, M.B., Rajagopalan, P.: Defining and developing an energy retrofitting approach. J. Green Build. **9**(3), 151–162 (2014)
2. Douglas, J.: Building Adaptation. Butterworth-Heinemann, Edinburgh (2006)
3. Appleby, P.: Integrated Sustainable Design of Buildings. Earthscan, London (2011)
4. U.S. Environmental Protection Agency. Glossary of Climate Change Terms (2016). https://www.epa.gov/sites/production/files/signpost/cc.html
5. Jones, L.: Environmentally Responsible Design: Green and Sustainable Design for Interior Designers. Wiley, New Jersey (2008)
6. LEED. For existing buildings. http://www.usgbc.org/resources/leed-2009-existing-buildings-current-version (2009). Accessed 1 March 2015
7. Building Research Establishment (BRE). Building research establishment environmental assessment method (BREEAM). http://www.breeam.com (1982). Accessed 20 June 2014
8. ASHRAE. Advanced Energy Design Guide for K-12 School Buildings. U.S. Green Building Council (2018)
9. Parasonis, I., Keizikas, A., Kalibatiene, D.: The relationship between the shape of a building and its energy performance. Archit. Eng. Des. Manag. **8**(4), 246–256 (2012)
10. Dogan, T., Saratsis, E., Reinhart, C.: The optimization potential of floor-plan typologies in early design energy modelling. In: Proceedings of BS2015: 14th Conference of International Building Performance Simulation Association, pp. 1853–1860 (2015)
11. Shi, X., Tian, Z., Chen, W., Si, B., Jin, X.: A review on building energy efficient design optimization from the perspective of architects. Renew. Sustain. Energy Rev. **65**, 872–884 (2016)
12. Attia, S., Hamdy, M., O'Brien, W., Carlucci, S.: Assessing gaps and needs for integrating building performance optimization tools in net zero energy buildings design. Energy Build. **60**, 110–124 (2013)
13. Graca, V., Kowaltowskia, D., Petrecheb, J.: An evaluation method for school building design at the preliminary phase with optimisation of aspects of environmental comfort for the School system of the State Sao Paulo in Brazil. Build. Environ. **42**, 984–999 (2007)
14. Ching, F.D.K.: Architecture: Form, Space and Order. Wiley, New Jersey (2007)
15. Almeida, R., Freitas, P., Delgado, J.: School Buildings Rehabilitation: Indoor Environmental Quality and Enclosure Optimization. Briefs in Applied Sciences and Technology. Springer (2015)
16. Dudek, M.: The Architecture of Schools and the New Learning Environments. Architectural Press, Oxford and Woburn, MA (2000)
17. Granadeiro, V., Correia, J.R., Leal, V.M.S., Duarte, J.P.: Envelope-related energy demand: a design indicator of energy performance for residential buildings in early design stages. Energy Build. **61**, 215–223 (2013)
18. Gratia, E., De Herde, A.: Design of low energy of office buildings. Energy Build. **35**(5), 473–491 (2003)
19. Hemsath, T., Bandhosseini, K.A.: Sensitivity analysis evaluating basic building geometry's effort on energy use. Renew. Energy **76**, 526–553 (2015)
20. Ndiaye, D.: The impact of building massing on net-zero achievability for office buildings. Build. Simul. **11**, 435–448 (2018)
21. Jaeger, I., Reynders, G., Ma, Y., Saelens, D.: Impact of building geometry description within district energy simulations. Energy **158**, 1060–1069 (2018)
22. Montenegro, E., Potvin, A., Demers, C.: Impact of school building typologies on visual thermal and energy performances. In: Proceedings of the 28th International Conference on Passive and Low Energy Architecture, Lima, Perú, pp. 2012–2017 (2012)
23. Schneekloth, L., Franck, K.: Ordering Space: Types in Architecture and Design. Van Nostrand Reinhold, New York (1994)
24. Crowe, N.: Studies in typology. J. Archit. Edu. **38**(1), 10–13 (1984)

25. Taylor, A.: Linking Architecture and Education: Sustainable Design of Learning Environments. University of New Mexico Press (2009)
26. Hawkes, D.: The Enviromental Tradition: Studies in the Architecture of Environment, 1st edn. E & FN Spon, London (1996)
27. Quan, S.J., Economou, A., Grasl, T., Yang, P.P.-J.: Computing energy performance of building density, shape and typology in urban context. Energy Proced. **61**, 1602–1605 (2014)
28. Zhang, A., Bokel, R., van den Dobbelsteen, A., Sun, Y., Huang, Q., Zhang, Q.: The effect of geometry parameters on energy and thermal performance of school buildings in cold climates of China. Sustainability **9**, 1708–1727 (2017)
29. Dogan, T., Reinhart, C., Michalatos, P.: Automated multi-zone building energy model generation for schematic design and urban massing studies. In: Presented in IBPSA eSim Conference, Ottawa, Canada (2014)
30. Erhorn-Kluttig, H., Erhorn, H.: School of the future—towards zero emission with high performance indoor environment. Energy Proced. **48**, 1468–1473 (2014)
31. Kats, G.: Greening America's schools: costs and benefits. www.usgbc.org/ShowFile.aspx?DocumentID=2908 (2006). Accessed 30 Oct 2008
32. EU Directive 2002/91/EC of the European parliament and of the council of 16 December 2002 on the energy performance of buildings. Off. J. Eur. Union 65–71 (2002). https://doi.org/10.1039/ap9842100196
33. EU Directive 2012/27/EU of the European parliament and of the council of 25 October 2012 on energy efficiency. Off. J Eur. Union Dir. 1–56 (2012). https://doi.org/10.3000/19770677.L_2012.315.eng
34. EU Directive 2018/844 of the European parliament and of the council of 30 May 2018 amending directive 2010/31/EU on the energy performance of buildings and directive 2012/27/EU on energy efficiency
35. Krawczyk, D.A.: Theoretical and real effect of the school's thermal modernization—a case study. Energy Build. **81**, 30–37 (2014)
36. Balogun, A.A., Morakinyo, T.E., Adegun, O.B.: Effect of tree-shading on energy demand of two similar buildings. Energy Build. **81**, 305–315 (2014)
37. Dias Pereira, L., Raimondo, D., Corgnati, S.P., Gameiro Da Silva, M.: Energy consumption in schools—a review paper. Renew. Sustain. Energy Rev. **40**, 911–922 (2014)
38. Facility Planning and Architecture Section. Architectural Design Guidelines for Schools. http://www.infrastructure.alberta.ca/content/doctype486/production/architecturalguidelines.pdf (2012)
39. Garwood, T.L., Hughes, B.R., Oates, M.R., O'Connor, D., Hughes, R.: A review of energy simulation tools for the manufacturing sector. Renew. Sustain. Energy Rev. **81**, 895–911 (2018)

# Chapter 3
# Solar Home System with Peak-Shaving Function and Smart Control in Hot Water Supply

**Bin-Juine Huang, Po-Chien Hsu, Shen-Jie Sia, Min-Han Wu, Zi-Ming Dong, Jia-Wei Wang, Ming-Jia Lee, Jen-Fu Yeh, Min-Tso Wu, Ji-Ding Wu, Yan-An Pan, Ming-Shian Chen, Po-Hsien Wu, Kang Li and Kung-Yen Lee**

**Abstract** The hybrid solar PV system (HyPV) with dual energy storage and peak-shaving function was developed. The solar power is stored as heat using an electric water heater when the battery is full. The electric water heater was modified using a smart control and turned it into the main hot water supply system to eliminate a backup heater. Two HyPVs were built for field test. The long-term performance shows that the peak-shaving function is satisfactory and the smart hot water supply was also achieved.

## 3.1 Introduction

High penetration of solar PV system is a trend in renewable energy dissemination. However, grid instability will be serious if all the solar power is fed into grid. Solar PV power generation for self-consumption and with storage device is one way for high-penetration of solar PV application.

National Taiwan University has been devoted to the development of the hybrid solar PV system (HyPV) [1] as shown in Fig. 3.1. HyPV is a solar home system which operates in stand-alone PV mode or grid mode automatically using switching technique. No solar power is fed into grid. When solar power generation and battery storage is sufficient, it operates in PV mode and the load is powered completely by solar energy as a standalone PV system. When solar power generation and battery

B.-J. Huang (✉) · S.-J. Sia · M.-H. Wu · Z.-M. Dong · J.-W. Wang · M.-J. Lee · J.-F. Yeh · M.-T. Wu · J.-D. Wu · Y.-A. Pan · M.-S. Chen · P.-H. Wu · K. Li
Department of Mechanical Engineering, National Taiwan University, Taipei 106, Taiwan
e-mail: bjhuang@seed.net.tw

P.-C. Hsu
Department of Mechanical Engineering, National Yunlin University of Science and Technology, Yunlin 640, Taiwan

K.-Y. Lee
Department of Engineering Science and Ocean Engineering, National Taiwan University, Taipei 106, Taiwan

 
J. Littlewood et al. (eds.), *Sustainability in Energy and Buildings*, Smart Innovation, Systems and Technologies 163, https://doi.org/10.1007/978-981-32-9868-2_3

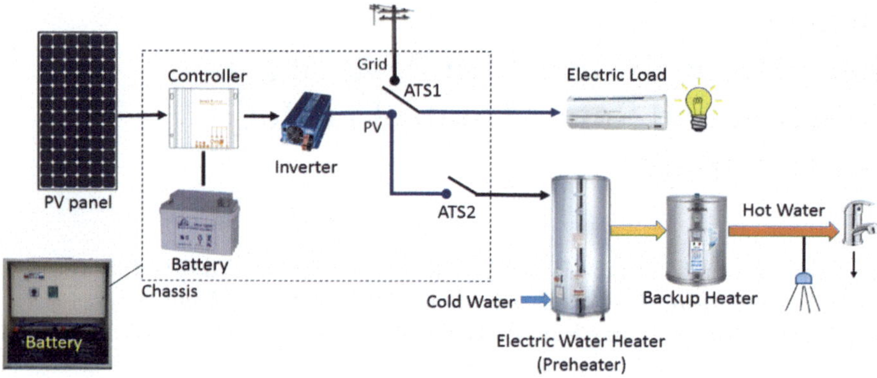


**Fig. 3.1** HyPV solar home system (with backup water heater)

storage is low, it switches to grid mode through ATS (Automatic Transfer Switch). The load is supplied completely by grid (grid mode) and the battery is charged again by solar PV. HyPV will switch back to PV mode when battery is charged to certain amount in grid mode. HyPV can also be utilized as a stand-alone solar system in off-grid area.

Since no solar PV power is fed into the grid, the HyPV system design match between the load power and the energy storage capacity needs to be optimized in order to reduce the PV generation loss and system cost. An electric water heater is added to store solar PV energy as heat (preheater) to replace part of battery storage (Fig. 3.1). HyPV is basically a stand-alone PV combined with grid power through switching technique. The dual energy storage in HyPV would reduce the system installation cost since the thermal storage (hot water) is much cheaper then battery storage. Not much studies on solar PV home system with dual storage was reported [2–6].

Since solar PV power generation is in phase with peak demand of grid, HyPV can be modified to provide peak-shaving function. That is, HyPV supplies the electric load only at peak time (13:00–15:00 and 19:00 after).

The electric water heater of HyPV is a water preheater to store excess solar PV energy when the battery is fully charged. A backup water heater is thus needed for stable hot water supply. In the present study, the water preheater is modified using a smart control. This will turn the preheater into main water heater for hot water supply all the time and eliminate the backup heater.

In the present paper, two HyPVs with peak-shaving function and smart hot water supply were built for long-term field test to show the performance.

## 3.2 Design and Installation of HyPVs

### *3.2.1 Design of HyPV*

Two HyPV systems were designed and built in the present study. The system design specification is shown in Table 3.1.

**Table 3.1** System design specification of two HyPVs

| HyPV Name | D3 | D8 |
|---|---|---|
| Installation location | Taipei | Taipei |
| Specific FIT PV energy generation with MPPT ($S_{pv}$), kWh/kWp per day | 2.47 | 2.47 |
| *PV Module* | | |
| Size ($P_{pv}$), kWp | 1.47 | 1.08 |
| Expected FIT PV energy generation with MPPT ($E_{pv}$), kWh/day | 3.63 | 2.67 |
| *Battery* | | |
| Type | LA | LA |
| Voltage, V | 48 | 24 |
| Storage capacity ($E_{bat0}$), kWh | 4.8 | 2.4 |
| Usable storage capacity ($E_{bat}$), kWh @ DOD 60% (LA) | 2.88 | 1.44 |
| Cost of battery, TWD | 20,000 | 10,000 |
| *Electric water heater* | | |
| Tank volume, L | 120 | 120 |
| Total power input, kW | 3.0 | 3.7 |
| Heat storage ($E_w$), kWh @ $\Delta T = 30$ °C | 4.2 | 4.2 |
| Cost of water heater, TWD | 8,400 | 8,400 |
| Total usable energy storage capacity ($E_{tot} = E_{bat} + E_w$), kWh | 7.08 | 5.64 |
| Inverter rated output, kW @ 220 V | 1.5 | 1.5 |
| *Load* | | |
| Type | Cooling, lighting | Cooling |
| Load power, kW | 0.2–1.4 | 0.2–1.2 |
| Load pattern | 24 h a day | 24 h a day |
| Ratio of usable battery storage to PV energy generation with MPPT $E_{bat}/E_{pv}$ | 0.79 | 0.54 |
| Ratio of total usable energy storage to PV energy generation with MPPT $E_{tot}/E_{pv}$ | 1.95 | 2.11 |
| Total cost of dual energy storage, TWD | 28,400 | 18,400 |
| Total cost of energy storage if using Li battery, TWD | 177,000 | 160,000 |
| 1 USD = 31 TWD | | |

**Design of HyPV D3**. HyPV D3 was installed in a house to collect real application data as a home appliance. Six solar panels with 1.47 kWp in total were installed. The installed lead-acid (LA) battery capacity is 4.8 kWh (48 V). The solar PV power is used to drive three air conditioners and LED lighting in the house. The city water is preheated by solar PV power using a 1.0 kW electric heater when the battery is fully charged. A 120 L water heater with an additional 2 kW electric heater is installed for boost heating. This provides additional 4.2 kWh thermal storage ($E_w$) at $\Delta T = 30$ °C which is larger than the usable battery capacity ($E_{bat}$= 2.88 kWh at 60% DOD).

**Design of HyPV D8**. HyPV D8 was installed in the laboratory to collect test data for long-term performance analysis. Four solar panels with 1.08 kWp total were installed. The installed lead-acid (LA) battery capacity is 2.4 kWh (24 V). PV power is used to drive an air conditioner in the laboratory. A 120 L water heater with 0.7 kW electric heater was installed for PV energy storage. An additional 3 kW electric heater was installed to provide boost heating. The thermal storage ($E_w$) is 4.2 kWh at water temperature difference $\Delta T = 30$ °C which is larger than the usable battery capacity ($E_{bat}$= 1.44 kWh). Excess solar PV energy is stored as heat by the 0.7 kW electric heater. A hot water load supply was simulated using a control valve to discharge the hot water according to a designated pattern.

**Design of Energy Storage**. It is known that the FIT (feed-in-tariff) solar PV system generates solar power and directly fed into grid at optimal energy conversion efficiency since the MPPT (maximum-power-point-tracking) was always used. As shown in Table 3.1, the ratio of the battery storage capacity to the PV energy generation, $E_{bat}/E_{pv}$, are 0.79 and 0.54 for D3 and D8, respectively, all less than 1.0. This may cause solar PV energy generation loss if only battery storage is used.

In the present HyPV design, the total usual energy storage capacity ($E_{tot}$) including battery and hot water heater are 7.08 kWh for D3 and 5.64 kWh for D8. With this storage combination, the ratio of total energy storage to expected solar PV energy generation ($E_{tot}/E_{\mathrm{pv}}$) is 1.95 for D3 and 2.11 for D8. This means that the total energy storage capacity ($E_{tot}$) including battery and water heater is about twice as much as the daily average solar PV energy generation. This will satisfy the energy storage demand in a clear day without PV energy generation loss. The hot water demand (55 °C) for a family can also be satisfied since the hot water consumption per person is 1.74–3.48 kWh for 50–100 L usage per day.

**Controller Design**. The operation of HyPV is automatically controlled by a MCU (microprocessor-controlled-unit) [1]. The MCU regulates the energy supply from PV or grid and controls the battery charge/discharge. A 1.5 kW-inverter (with conversion efficiency 0.93) was used in the two HyPVs to convert the DC power into AC to drive the load such as air conditioners, etc. The MCU has the measuring functions for voltages of battery and PV module, currents of PV and battery charge/discharge, and ac load power, etc. This will provide an important information for switching control between PV Mode and Grid Mode [1]. All the performance data are recorded by a notebook PC.

The power input of MCU can be switched to battery or grid power. For off-grid application, MCU is driven by battery. Hence, HyPV can be easily converted into a stand-alone solar PV system in off-grid applications. The MCU also provides a fault detection technique on battery performance. A warning signal will be displayed when the battery is decayed and needs maintenance.

The MCU provides a smart control function for hot water supply at temperature between 45 and 70 °C all day long. The highest water temperature for PV energy storage is 70 °C. The highest temperature for boost heating at sunset (powered by grid) is also set at 70 °C. The water supply temperature at the rest of the day (powered by grid) is around 45–50 °C. The temperature recovery rate is around 15 min at a high discharge rate of hot water, using full-speed heating by grid power. With this smart control, the house can obtain a steady hot water supply between 45 and 70 °C with only one electric heater.

HyPV can also be connected with the neighbor ones to build a pyramid solar micro-grid [7, 8]. The solar energy can be shared each other to reduce the cost.

### *3.2.2 Installation of HyPV*

**Installation of D3**. HyPV D3 was installed in a house (Fig. 3.2) in Taipei City to collect field application data. Six solar PV panels (1.47 kWp) were installed using vertical poles: four are installed nearby the street and the other two are mounted on the roof. The lead-acid (LA) battery capacity is 4.8 kWh/48 V. The MCU, inverter and battery are assembled in a chassis which is put in the garage. The solar energy provides about 15% of the total electric demand of the house, which is cheaper than the grid power price [6]. The PV power drives three air conditioners and LEDs in the house. A 120 L water heater with 3.0 kW total input was installed to supply hot water.

**Installation of D8**. HyPV D8 was installed in the laboratory (Fig. 3.3) to collect test data for long-term performance analysis. Four solar panels with 1.08 kWp in total were installed on the roof of the building using self-weight structure. The lead-acid (LA) battery capacity is 2.4 kWh (24 V). PV power is used to drive an air conditioner in the laboratory. A 120 L water heater with 3.7 kW total input was installed.

**Fig. 3.2** HyPV D3 installed in a house (Taipei)

**Fig. 3.3** HyPV D8 installed in laboratory (Taipei)

## 3.3 Field Test Results

### *3.3.1 Peak-Shaving Performance*

Peak-shaving in power grid is very important in improving grid efficiency and reduce electricity cost. In 2016, the peak power supply is 45% higher than the off-peak power in Taiwan. There are two grid power peaks in Taiwan: 13:00–15:00 and 19:00–21:00. The former peak occurs in summer season and the latter one in winter. Taiwan power company provides a demand response (DR) policy for their users during the daytime peak 13:00–15:00 in summer to cope with the shortage of peak power supply.

HyPV provides the capability to output the power from the battery storage during the peak hours. This increases the economic benefit of the grid. The control logics of HyPV for peak-shaving are as follows:

(1) Battery will discharge a small amount to the load before 13:00 only when the battery is fully charged.
(2) Solar energy discharge is activated during 13:00–15:00 and after 19:00.
(3) Solar energy discharge is terminated as long as the battery storage is low.

During the test of D8, the load (air conditioner) is turned on during 8:00–22:00. The power supply of the air conditioner is switched to grid if solar energy is not enough. Figure 3.4 shows the peak-shaving performance of D8 on 2018/5/27. The solar energy discharge begins when the battery is fully charged at 8:00. Afterwards, solar energy discharge continues since solar power generation is able to cover the load demand until 15:00. After 15:00, the solar energy discharge is terminated and the battery starts to store solar energy. Solar energy discharge is activated again only after 19:00 until the battery is fully discharged (at low voltage 23.6 V). It is shown that HyPV completely supplies the power demand during the day including daytime peak 13:00–15:00 and part power demand during night peak 19:00–19:30.

A test was performed on 2018/5/28 for HyPV stopping PV power output in the period 12:00–13:00. HyPV is switched to Grid Mode and the load is supplied by grid.

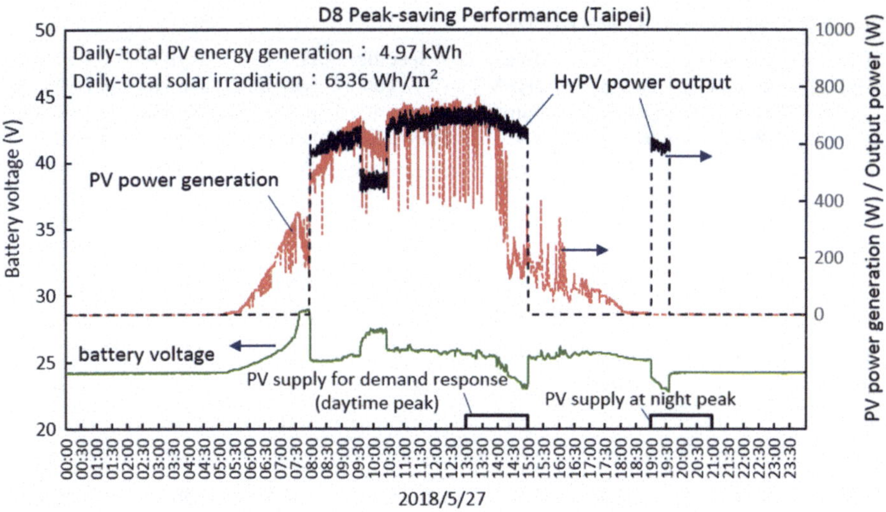


**Fig. 3.4** Peak-shaving performance of HyPV D8 (2018/5/27)

Figure 3.5 show that the battery reaches the maximum voltage (constant-voltage charge state) near 13:00. HyPV however supplies peak load during 13:00–14:30, because the battery is not fully charged. The battery is fully discharged at 14:30 and 19:30.

Table 3.2 shows that the average HyPV output time in peak hours is 121 min which covers the daytime peak. The average ratio of HyPV output in peak hours to

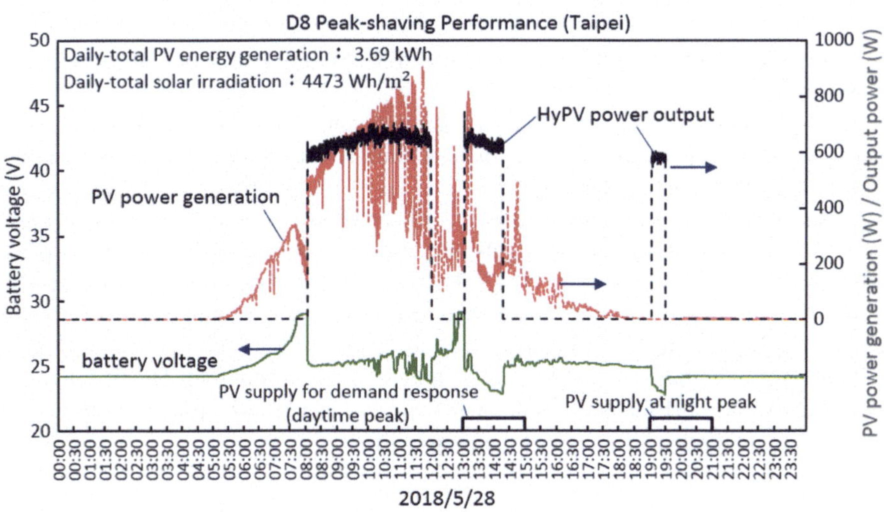


**Fig. 3.5** Peak-shaving performance of HyPV D8 (2018/5/28)

**Table 3.2** Continuous performance of D8

| 2018 (D8) | Solar irradiation, Wh/m$^2$ | PV energy generation, Wh | Total HyPV output, Wh | Specific PV generation, kWh/kWp | HyPV output in peak hours, Wh | HyPV output time in peak hours, min | Ratio of HyPV output in peak hours to total HyPV output |
|---|---|---|---|---|---|---|---|
| 5/12 | 6,423 | 4,690 | 3,916 | 4.69 | 1,141 | 120 | 0.29 |
| 5/13 | 4,543 | 3,460 | 3,584 | 3.46 | 1,102 | 120 | 0.31 |
| 5/14 | 5,068 | 3,740 | 3,428 | 3.74 | 1,277 | 120 | 0.37 |
| 5/15 | 3,470 | 2,530 | 2,542 | 2.53 | 1,027 | 120 | 0.40 |
| 5/17 | 5,142 | 4,220 | 3,542 | 4.22 | 1,315 | 120 | 0.37 |
| 5/18 | 6,210 | 4,870 | 4,672 | 4.87 | 1,340 | 120 | 0.29 |
| 5/20 | 5,787 | 4,910 | 4,851 | 4.91 | 1,474 | 136 | 0.30 |
| 5/21 | 4,396 | 3,750 | 3,476 | 3.75 | 1,179 | 118 | 0.34 |
| 5/23 | 3,518 | 2,920 | 2,778 | 2.92 | 1,113 | 109 | 0.40 |
| 5/24 | 5,749 | 4,660 | 4,490 | 4.66 | 1,506 | 156 | 0.34 |
| 5/25 | 4,688 | 3,840 | 3,757 | 3.84 | 907 | 89 | 0.24 |
| 5/26 | 5,741 | 4,750 | 4,610 | 4.75 | 1,503 | 145 | 0.33 |
| 5/27 | 6,336 | 4,970 | 4,819 | 4.97 | 1,711 | 156 | 0.36 |
| 5/28 | 4,472 | 3,690 | 3,565 | 3.69 | 1,046 | 102 | 0.29 |
| 5/29 | 4,557 | 3,780 | 3,654 | 3.78 | 809 | 81 | 0.22 |
| Average | 5,073 | 4,052 | 3,846 | 4.05 | 1,230 | 121 | 0.32 |

total HyPV output is 0.32. That is, about one third of total HyPV output energy is in peak hours.

A long-term monitoring of D3 shows that the peak-shaving performance is satisfactory with the ratio of HyPV output in peak hours to the total HyPV output reaching 50.9% (Fig. 3.6). That is, about half the total HyPV output happens in peak hours.

In summer (June and July), the peak power shortage in grid is most serious. Table 3.3 shows that, in June and July, the average ratio of HyPV output in peak hours to the total HyPV output is 0.75. This means that three-fourth of total HyPV output happens in peak hours. The average time of HyPV output in peaks is 203 min which covers most of the two peaks. In June, when the peak power supply in grid is the highest, the ratio of HyPV output in peak hours to the total HyPV output is 0.83 and reaches 1.0 for about half of the month. This ratio varies with the solar irradiation and the load demand of the house.

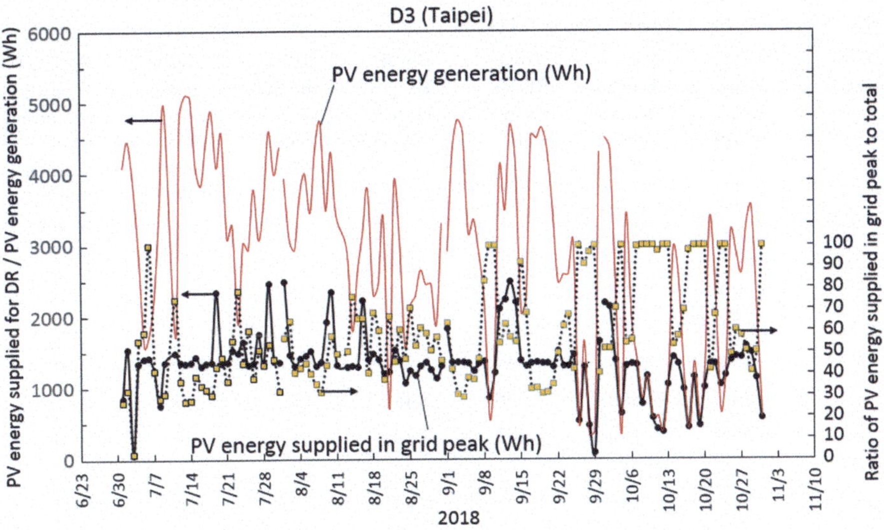


**Fig. 3.6** Peak-shaving performance of HyPV D3 (2018)

### 3.3.2 *Smart Hot Water Supply Performance*

The electric water heater in HyPV is a preheater to store excess solar PV energy when the battery is fully charged. A backup water heater is thus needed for stable hot water supply (Fig. 3.1). In the present study, the preheater is modified using a smart water temperature control. This will turn the preheater into the main hot water supply system all day long and also eliminates the backup water heater.

A SCR controller is used to regulate the electric heating element in the water tank according to the control scheme as follows:

(1) For PV excess energy storage during daytime, the maximum water temperature is set at 70 °C. It is powered by solar PV energy only.
(2) At sunset, the electric water heater is turned on in full capacity by grid power to boost the water temperature to reach 70 °C only one time. This usually can meet the hot water demand at night.
(3) At rest of the day, the heater is powered by grid and the water temperature is maintained at 45–50 °C using grid power.

Figure 3.7 shows the temperature variation of hot water supply of D3 in continuous four days. The thermal storage in daytime is activated when the air conditioners (load) are not used or at low load (April 19 and 20). At sunset around 18:00, the electric heater was fully boosted by grid power to quickly raise the water temperature up to 70 °C. The hot water load occurs at midnight and causes a fast temperature drop. The electric heater in full power boosted the water temperature back to 45 °C again within 15 min.

**Table 3.3** Peak-shaving performance of D3

| 2018 (D3) | Solar irradiation, Wh/m$^2$ | PV Energy generation, Wh | Total HyPV output, Wh | Specific PV generation, kWh/kWp | HyPV output in peak hours, Wh | HyPV output time in peak hours, min | Ratio of HyPV output in peak hours to total output |
|---|---|---|---|---|---|---|---|
| 6/4 | 6,242 | 4,227 | 4,346 | 3.00 | 4,346 | 235 | 1.00 |
| 6/5 | 4,940 | 4,370 | 3,482 | 3.10 | 3,482 | 240 | 1.00 |
| 6/6 | 4,983 | 3,859 | 3,088 | 2.74 | 3,087 | 238 | 1.00 |
| 6/7 | 6,480 | 5,065 | 4,243 | 3.59 | 4,243 | 240 | 1.00 |
| 6/8 | 6,302 | 4,860 | 5,786 | 3.45 | 4,155 | 240 | 0.72 |
| 6/9 | 6,384 | 5,243 | 1,302 | 3.72 | 1,302 | 120 | 1.00 |
| 6/10 | 2,696 | 2,657 | 3,091 | 1.88 | 2,090 | 236 | 0.68 |
| 6/11 | 1,084 | 1,251 | 1,273 | 0.89 | 1,273 | 240 | 1.00 |
| 6/12 | 6,983 | 5,654 | 4,146 | 4.01 | 2,209 | 240 | 0.53 |
| 6/13 | 4,381 | 3,810 | 2,758 | 2.70 | 2,126 | 240 | 0.77 |
| 6/14 | 2,306 | 2,378 | 2,216 | 1.69 | 1,638 | 234 | 0.74 |
| 6/15 | 1,561 | 1,601 | 1,071 | 1.14 | 1,071 | 240 | 1.00 |
| 6/16 | 5,383 | 3,985 | 3,120 | 2.83 | 1,889 | 240 | 0.56 |
| 6/17 | 5,696 | 4,243 | 3,007 | 3.01 | 2,267 | 234 | 0.65 |
| 6/18 | 3,889 | 3,047 | 3,553 | 2.16 | 2,543 | 240 | 0.72 |
| 6/19 | 4,076 | 3,250 | 1,973 | 2.30 | 1,464 | 235 | 0.74 |
| 6/20 | 2,880 | 2,419 | 3,073 | 1.72 | 2,572 | 240 | 0.79 |
| 6/21 | 3,161 | 2,479 | 3,125 | 1.76 | 3,124 | 226 | 1.00 |
| 6/22 | 3,318 | 2,598 | 2,217 | 1.84 | 2,217 | 202 | 1.00 |
| 6/23 | 3,637 | 2,897 | 2,306 | 2.05 | 2,305 | 186 | 1.00 |
| 6/24 | 3,427 | 3,195 | 1,562 | 2.27 | 1,562 | 121 | 1.00 |
| 6/25 | 4,781 | 5,707 | 3,536 | 4.05 | 2,963 | 160 | 0.84 |
| 6/26 | 5,375 | 4,293 | 3,976 | 3.04 | 2,147 | 180 | 0.54 |
| 6/27 | 5,530 | 5,133 | 3,874 | 3.64 | 2,708 | 240 | 0.70 |
| 6/28 | 6,915 | 6,099 | 4,747 | 4.33 | 3,742 | 187 | 0.79 |
| June average | **4,496** | **3,773** | **3,112** | **2.68** | **2,501** | **217** | **0.83** |
| 7/8 | 4,776 | 4,408 | 4,007 | 3.13 | 2,874 | 146 | 0.72 |
| 7/9 | 7,354 | 7,919 | 6,654 | 5.62 | 4,171 | 232 | 0.63 |
| 7/10 | 4,760 | 4,405 | 4,012 | 3.12 | 2,908 | 213 | 0.72 |
| 7/11 | 2,359 | 2,293 | 2,156 | 1.63 | 2,156 | 218 | 1.00 |
| 7/12 | 6,658 | 6,499 | 5,136 | 4.61 | 4,171 | 240 | 0.81 |

(continued)

**Table 3.3** (continued)

| 2018 (D3) | Solar irradiation, Wh/m$^2$ | PV Energy generation, Wh | Total HyPV output, Wh | Specific PV generation, kWh/kWp | HyPV output in peak hours, Wh | HyPV output time in peak hours, min | Ratio of HyPV output in peak hours to total output |
|---|---|---|---|---|---|---|---|
| 7/13 | 7,320 | 8,001 | 7,490 | 5.68 | 4,411 | 240 | 0.59 |
| 7/14 | 7,429 | 7,922 | 7,311 | 5.62 | 3,926 | 201 | 0.54 |
| 7/15 | 6,343 | 6,109 | 5,207 | 4.33 | 4,026 | 209 | 0.77 |
| 7/16 | 6,983 | 6,700 | 5,531 | 4.75 | 4,547 | 235 | 0.82 |
| 7/17 | 7,264 | 7,356 | 6,470 | 5.22 | 4,615 | 238 | 0.71 |
| 7/18 | 7,162 | 7,575 | 6,656 | 5.37 | 4,350 | 222 | 0.65 |
| 7/19 | 6,681 | 6,341 | 5,378 | 4.50 | 4,290 | 219 | 0.80 |
| 7/20 | 7,408 | 7,948 | 6,909 | 5.64 | 4,231 | 220 | 0.61 |
| 7/21 | 4,344 | 4,931 | 4,140 | 3.50 | 3,093 | 240 | 0.75 |
| 7/22 | 4,679 | 4,424 | 3,686 | 3.14 | 3,062 | 224 | 0.83 |
| 7/23 | 2,366 | 2,708 | 2,221 | 1.92 | 2,221 | 155 | 1.00 |
| 7/24 | 3,850 | 4,023 | 3,527 | 2.85 | 0 | 0 | 0.00 |
| 7/25 | 3,838 | 3,944 | 3,702 | 2.80 | 2,136 | 120 | 0.58 |
| 7/26 | 4,887 | 5,026 | 4,917 | 3.57 | 2,321 | 130 | 0.47 |
| 7/27 | 4,355 | 5,629 | 3,276 | 3.99 | 2,211 | 122 | 0.67 |
| 7/28 | 4,874 | 5,728 | 4,260 | 4.06 | 2,571 | 130 | 0.60 |
| 7/29 | 5,490 | 6,755 | 4,209 | 4.79 | 1,812 | 213 | 0.43 |
| 7/30 | 5,673 | 5,364 | 3,425 | 3.80 | 2,040 | 193 | 0.60 |
| 7/31 | 6,400 | 6,351 | 5,153 | 4.50 | 3,556 | 166 | 0.69 |
| July average | **5,552** | **5,765** | **4,810** | **4,088** | **3,154** | **189** | **0.67** |
| Overall average | **5,014** | **4,749** | **3,925** | **3.37** | **2,821** | **203** | **0.75** |

Figure 3.8 shows the continuous temperature variation of hot water supply of D8. The thermal storage at daytime is activated on May 10, 12, 13. The reached highest water temperature depends on the solar PV generation. At sunset around 18:00, the electric heater was fully boosted by grid power to quickly raise the water temperature to 70 °C. The hot water drainage at around 6 L/min was turned on at 17:00, 20:00, 21:00 and 22:00 by a timer control. This causes a rapid temperature drop. However, the smart control brings the temperature back to 45 °C within 15 min.

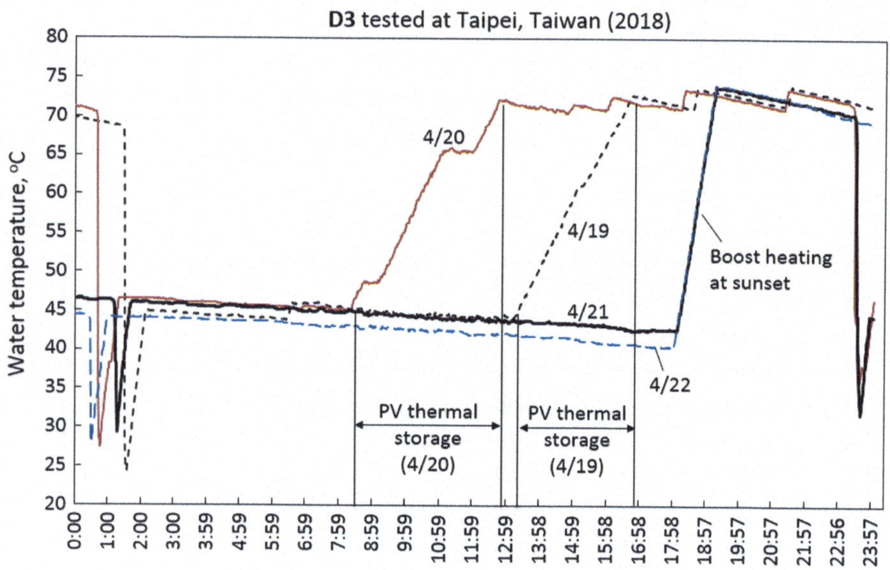


**Fig. 3.7** Smart hot water temperature control of HyPV D3 (2018)

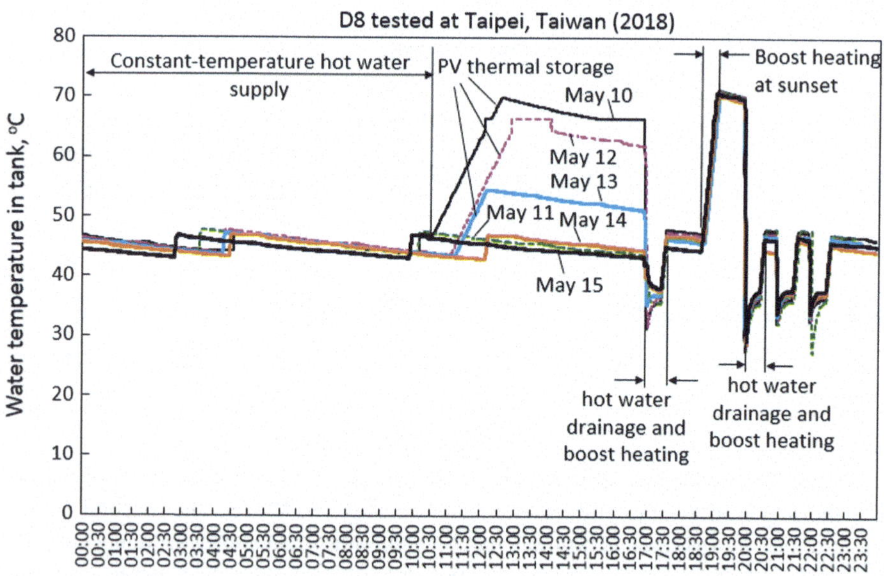


**Fig. 3.8** Smart hot water temperature control of HyPV D8 (2018)

## 3.4 Conclusion

The hybrid solar PV system for self-consumption (HyPV) operates in stand-alone PV mode or grid mode using switching technique [1]. Since no solar PV power is fed into the grid, the HyPV system design match between the load power and the energy storage capacity needs to be optimized in order to reduce the PV generation loss and system cost. Besides battery, an electric water heater is installed to store solar PV energy as heat (when the battery is fully charged) to replace part of expensive battery. The dual energy storage would reduce the system installation cost (Table 3.1). Furthermore, HyPV can add a peak-shaving function to supply solar energy only during peak hours of the grid. Besides, the electric water preheater is further modified using a smart control and turned into the main hot water supply system to eliminate the backup water heater. Two HyPVs were built for field test. The long-term performance shows that the peak-shaving function is satisfactory and the smart hot water supply was also achieved.

## References

1. Hsu, P.C., Huang, B.J., Lin, W.C., Chang, Y.J., Chang, C.J., Li, K., Lee, K.Y.: Effect of switching scheme on the performance of a hybrid solar PV system. Renew. Energy **96**, 520–530 (2016)
2. Bocklisch, T.: Hybrid energy storage systems for renewable energy applications. In: 9th International Renewable Energy Storage Conference, IRES 2015. Energy Proced. **73**, 103–111 (2015). https://doi.org/10.1016/j.egypro.2015.07.582
3. Lee, S.C.: Operation analysis of a photovoltaic lighting system with battery and heater. Sol. Energy **85**, 2144–2153 (2011). https://doi.org/10.1016/j.solener.2011.06.002
4. Thim, F., Rothert, M., Kever, F.: Where is the optimum? Comparison of system topologies for small PV hybrid systems. In: 7th International Conference on PV-Hybrid and Mini-Grid, 10–11 April 2014. Bad Hersfeld, Germany (2014)
5. Huang, B.-J., Hsu, P.-C., Wu, P.-H., Wang, Y.-H., Tang, T.-C., Wang, J.-W., Dong, X.-H., Wu, W.-H., Lee, M.-J., Yeh, J.-F.: Hybrid energy storage of solar PV system for self-consumption. In: International Conference and Workshop REMOO-2017. Energy for Tomorrow, 10–12 May 2017, Venice, Italy (2017)
6. Huang, B.-J., Hsu, P.-C., Wang, Y.-H., Tang, T.-C., Wang, J.-W., Dong, X.-H., Lee, M.-J., Yeh, J.-F., Dong, Z.-M., Wu, M.-H., Sia, S.-J., Li, K., Lee, K.-Y.: Solar home system with dual energy storage. SN Applied Sciences 1:973 (2019). https://doi.org/10.1007/s42452-019-1000-8
7. Hsu, P.-C., Huang, B.-J., Wang, Y.-H., Tang, T.-C., Wang, J.-W., Dong, X.-H., Li, K., Lee, K.-Y.: Solar power sharing between two PV systems in a solar pyramid micro-grid. Int. J. Smart Grid Clean Energy **6**(2), 96–103 (2017)
8. Dong, Z.-M., Hsu, H.-Y., Huang, B.-J., Wu, M.-H., Wu, W.-H., Hsu, P.-C., Li, K., Lee, K.-Y.: Power dispatching control of pyramid solar micro-grid. Int. J. Smart Grid Clean Energy (2019). In print

# Chapter 4
# Influential Factors on Using Reclaimed and Recycled Building Materials

**Zahra Balador, Morten Gjerde and Nigel Isaacs**

**Abstract** When resources are in decline, opportunities to create circular resource flows cannot be ignored. Reuse and recycling of building materials can significantly contribute to these efforts. Reuse and recycling of building materials and consumption of reclaimed and recycled building materials as environmental practices have potential to enhance resource efficiency in the construction industry, leading to a reduction in the amount of waste produced and energy consumed. Counting the use of reclaimed and recycled building materials as a pro-environmental behavior and studying the influential factors is one of the first steps towards establishing this behavior. It is important to have a comprehensive view of the process in which the pro-environmental behavior is generated and it can therefore be useful to study relevant variables influencing pro-environmental behaviors. A current lack of quantitative data linking these issues hinders the effectiveness of solutions. This paper investigates some of the factors influencing use of reclaimed and recycled building materials based on the perceptions of the main stakeholders of the construction industry in New Zealand, discussing these in the context of literature. Results show that price and self-satisfaction are the most influential factors among the factors we examined, and age, gender, and income are predictors of these factors. We also note that environmentalists and regulators are less positive, and producers and consumers are more positive about barriers against the use of reclaimed and recycled building materials. The study results can help direct and focus efforts to divert waste from landfill.

## 4.1 Introduction

Living Planet Report published a report in 2010 that consumption at rates that are 50% faster than Earth can sustain has led to problematic situations for natural resources. Anthropocentric economies can expand forever without any limitation from resources, and research shows that businesses looking to eliminate waste and

Z. Balador (✉) · M. Gjerde · N. Isaacs
School of Architecture, Victoria University of Wellington, Wellington, New Zealand
e-mail: Zahra.balador@vuw.ac.nz



J. Littlewood et al. (eds.), *Sustainability in Energy and Buildings*, Smart Innovation, Systems and Technologies 163, https://doi.org/10.1007/978-981-32-9868-2_4

toxic chemical production must limit their use of natural resources [1]. Treating waste as potential resource presents a valuable opportunity for all industry producers and other stakeholders. A circular resource flow means a system with minimal waste where waste from one industrial activity is treated as input for other activities. This transformation requires preparation, such as infrastructure, technical breakthroughs and regulatory environment [2].

The greatest share of $CO^2$ emissions can be attributed to the production of building materials specifically steel, cement and timber. Almost 70% of the environmental impact of building materials arises through their production. Revising material and resource flows becomes necessary as pressures form growing population levels and the need for buildings continue [3]. Research needs to be robust enough to be trusted in businesses and society. The complexity of this sustainable transformation is because of the large numbers of actors and stakeholders [4]. Studies showed that 33% of waste in the construction process happens in the design stage [5, 6]. Current knowledge is mostly focused on the green purchase of products and not directed toward the building industry. We also note fragmented and inconclusive studies for the use of reclaimed and recycled building materials. The earlier results of studies that examine stakeholders' perceptions are not consistent [7]. Accordingly, there is a need to consult stakeholders to clarify the current situation and opportunities.

## 4.2 Literature Review

Understanding people's perceptions around environmental issues is crucial when it comes to solving environmental problems. Increasing volumes of waste is one of these problems. There are different factors influencing the process of shaping an environmentally friendly behavior, which in the current study relates to the use of reclaimed and recycled building materials. Interrogation of people's perceptions will enable relevant factors, including barriers, to be identified.

### *4.2.1 Stakeholders*

Bahamon notes that construction is one of the most polluting industries; therefore, architects, being chief stakeholders within the construction industry, are a prime candidate for changing the attitude towards recycling. Architects make important decisions about building materials and methods and could include other people's interests in this process, accepting that they can also have good ideas [8].

Addis's opinion is that there is an increasing trend among designers, architects, builders and other stakeholders for using more recycled building materials [9]. Denne suggests that producers should consider raw material use, recycling processes of waste and waste reduction. Consumers should select products which limit waste and know about disposal options. Governments should intervene in waste management because private decisions cannot reduce waste alone, and there is a delay between the production, purchase, and disposal which make it difficult to see the consequences as fast as the decisions are made [10].

### 4.2.2 Influential Factors

Eco-labeling and environmental information on products will increase knowledge and have a positive influence on sustainable consumption behavior. Besides, economic incentives such as discounts and subsidies should be considered, because the price is important even for non-green purchasers [11].

Dursun's study showed that positive influence of personal health and the environment was more significant than economic factors on green buying behavior. In addition, when we talk about buying natural, organic, and recycled products, the quality of life, health and safety are the most important factors. Nevertheless, for simple buying behavior, economic factors were found to be significant. Saving money encourages people to show some of the pro-environmental behaviors such as repairing, reusing and recycling. Immediate costs of recycling can be a barrier (55% of the reasons behind not recycling), but long-term benefits of it is hidden from some stakeholders (especially non-green purchasers) [12]. According to earlier research in this field, price, quality, and convenience are among the most influential factors that affects sustainable consumption behavior [13–15].

For manufacturers to meet expected future market demands, they will need to consider concepts of sustainability in their own practices, including use of green products, in order to be successful. By finding sustainable alternatives, businesses can cut costs as depleting natural resources cause prices to increase and can benefit from sustainable decisions being more appealing to customers and investors. Building strong relationships with stakeholders as members of society plays an important role [16].

Attitudes toward the use of these products are positively affected by the social and environmental image of the company [17, 18], good feelings [19], transparency of information provided by manufacturers [20–23]. Attitudes are negatively influenced by gaps between companies' claims and actual performance, difficult access to information [17, 20] and doubt and dissatisfaction about products and manufacturers [21]. Sometimes green products do not support consumer needs, which lead to poor perceptions. High pricing, low confidence, also high compromise can make it worse [24]. Other studies also support these ideas and have found that providing additional information about the benefits, performance and quality of the environment-friendly products can have positive impacts on green purchase behaviors [17, 25–28]. Auditing perceptions of people to find values of a good product, understanding their present and future lifestyles, knowing how they communicate with media, help them have a positive image [29].

Effective environmental strategies are also opportunities for competitive advantage, since managing conflicting stakeholder interests and being seen to be responsible can increase profitability from sustainable practices and environmental strategies. This can also be a motivation for manufacturers to be more environmentally friendly and innovative [26, 30].

Generally, demographic characteristics such as gender [31], education [28], racial and cultural background lead to different conceptualizations [27, 32], which affect the relationship between consumer value and consumer satisfaction with buying behavior; for example women are more concerned about social issues, and men respond better to emotional communication [33]. Therefore producers and suppliers can communicate rationally rather than emotionally with regard to sex and age of them [18].

## 4.3 Methodology

Similar to other studies [12, 15, 17, 21, 30, 32] that have examined influential factors and perceptions of people around green purchase and incentives of environment-friendly behaviors, a survey questionnaire was determined to be the most appropriate data collection method. A five-point Likert scale was used, with average scores for each demographic classification calculated for each measured item. The scoring scale ranged from strongly agree to strongly disagree. The original questionnaire consisted of more questions in a study to report in a broader area of findings and this paper reports on only a part of these findings.

Freeman defined a stakeholder as "any group or individual who can affect or is affected by the achievement of the organization's objective" (R. E. Freeman, 2010). A stratified sample was the sampling method, and the population of principal stakeholders was divided into industry related subgroups. For the consumer stakeholder group a simple random sampling was deployed. Using email addresses sourced from the websites of different stakeholder organizations, potential respondents were then asked to identify other members of the industry subgroup through a snowball sampling technique. This technique was used in order to make the sample bigger, in case it would be needed as a backup method. Main stakeholders in this study include regulators, consumers, manufacturers and suppliers, NGOs and environmental activists, builders, architects, and designers. Approximately 600 stakeholders participated from three cities of Auckland (43%), Wellington (35%), and Christchurch (22%) in New Zealand in a survey of attitudes and behaviors.

Descriptive statistics, multi-regression, and one-way ANOVA were used in the data analysis, and assumptions of normality, linearity, homoscedasticity were checked before these analyses.

## 4.4 Descriptive Results

The sample consisted of 159 architects, 160 builders, 103 manufacturers, and suppliers, 52 environmentalists and NGOs, 48 regulators and 144 consumers. Table 4.1 shows that how much the sample is over/underrepresented according to the construc-

**Table 4.1** Descriptive statistics of sociodemographic data

| Characteristics | n | Valid | |
|---|---|---|---|
| | | Survey sample (%) | Construction industry |
| *Gender* | | | |
| Men | 429 | 64.8 | 83% [34] |
| Women | 233 | 35.2 | 17% |
| *Age* | | | |
| 18–29 years old | 87 | 13.1 | 21.8% [35] |
| 30–49 years old | 328 | 49.5 | 52.1% |
| 50–64 years old | 186 | 28.1 | 26% |
| 65 years and over | 61 | 9.2 | 4.4% |
| *Education* | | | |
| Secondary school | 48 | 7.3 | 7.7% [36] |
| Vocational training | 194 | 29.3 | 69.2% |
| Bachelor's degree | 197 | 29.8 | 6% |
| Postgraduate degree | 211 | 31.9 | 1.8% |
| None of the above | 12 | 1.8 | 11.7% |
| *Income* | | | |
| Less than $19,999 | 47 | 7.1 | N/A[a] |
| $20,000 to $39,999 | 64 | 9.7 | |
| $40,000 to $79,999 | 216 | 32.7 | |
| $80,000 to $109,999 | 144 | 21.8 | |
| $110,000 or more | 190 | 28.7 | |

[a]The mean value for 2016 was 64640 in the building construction industry [37]

tion industry characteristics in New Zealand. As a summary of representativeness of the sample we can say that men are underrepresented, and women are overrepresented; age distribution is almost close to the main population; education level of secondary school, vocational training and no education are under-represented, and tertiary qualifications are overrepresented.

The descriptive statistics report the frequencies. The first two statements in the questionnaire measure the general knowledge of respondents. One asks people to provide their understanding of how strongly the construction industry pollutes and the second one about the benefits of using reclaimed and recycled building materials. Results showed that respondents are unclear on whether the construction industry is one of the most polluting industries. Most respondents (66%) agree that benefits of using reclaimed and recycled building materials outweigh the drawbacks. We note that despite the fact these respondents represent key stakeholder groups in the building construction industry, they are not aware that construction is one of the most polluting activities in cities.

One of the key reasons why people do not use reclaimed and recycled building materials is poor availability, as 50% of respondents stated that they cannot find these products in nearby suppliers' stores and 30% stated that they have inadequate information on their availability. This issue is also affected by other factors like advertisement, because only 20% of people disagree that they buy these products because they are influenced by advertisements. This fact shows that advertisement have their impact on people purchase behavior like other normal products, and as we know based on the literature green purchasers are more conscious about the transparency of information of products. Therefore, it is not only a matter of advertising, but also access to product information. Results indicate that only 12% of respondents say that needed information are easily available when purchasing a reclaimed and recycled building material. Specifically, when it comes to the use of building materials, knowing specifications of materials is a key fact influencing the purchase behavior because of building codes and standards to be met.

Respondents then answered questions regarding their perceptions of producers and suppliers of reclaimed and recycled building materials. Most of the people (76–79%) believe that public and environmental image of the manufacturer and supplier is important when buying these building materials as green products. In addition, 36% of them believe that these producers do not communicate enough with people to know their needs, and 43% do not have an opinion about this issue. However, 61% of respondents agree that there is a gap between claim and performance of these producers in the building construction industry and only 4% think that there is not a gap. This is also consistent with the literature that usually one of the reasons people do not trust environment-friendly products is because of this gap information [17, 20].

## 4.5 Discussion

### 4.5.1 *Effect of Sociodemographic Data*

Multiple regression was run to predict influential factors from gender, age, education, and income. Results show that "knowledge of impacts of construction industry" and "increasing concern among people" cannot be predicted by variables of demographic data. But age and gender statistically significantly predicted "self-satisfaction", $F(2, 624) = 7.774$, $p = 0.000$, $R2 = 0.024$. Women believe that "self-satisfaction" is more important in a green purchase than men, and age has a contrary relationship with it. Also, age predicted "price", $F(1, 625) = 25.932$, $p = 0.000$, $R2 = 0.040$, and predicted "availability", $F(1, 625) = 9.277$, $p = 0.002$, $R2 = 0.015$. "price" and "availability" have a contrary relationship with age. Interestingly, income predicted (statistically significant) "advertisement", $F(1, 625) = 19.169$, $p = 0.000$, $R2 = 0.030$, and this is contrary relationship. Similarly, age and income predicted "accessibility of information", $F(2, 624) = 8.590$, $p = 0.000$, $R2 = 0.027$ which is a contrary relationship. Income and gender are statistically significant predictors of "knowledge of benefits", $F(2, 624) = 7.542$, $p = .001$, $R2 = 0 .024$. Women think that use of these materials has more benefit, but "Knowledge of benefit" has a contrary relationship with income. As we can see age and gender statistically significantly predicted "public" and "environmental" image of companies, $F(1, 605) = 15.143$, $p = 0.000$, $R2 = 0.024$, $F(1, 605) = 7.918$, $p = 0.005$, $R2 = 0.013$, respectively. "public image" has a contrary relationship with age, and "environmental image" is more important for women. Also, gender statistically significantly predicted "gap between claim and performance", $F(1, 605) = 6.825$, $p = 0.009$, $R2 = 0.011$. Surprisingly, this test shows that men are more suspicious about the gap between claim and performance of producers.

It is clear from the results of the regression that age, gender, and income variables, especially age in most cases added statistically significantly to the prediction of influential factors, $p < 0.005$. This finding is consistent with the literature that confirms the predictability of these influential factors on green purchase behavior based on the sociodemographic data specifically the same three variables of age, gender, and income.

### 4.5.2 *Effect of Stakeholders' Roles*

Influential factors are compared between different groups of stakeholders through one-way ANOVA test as well. Results show that stakeholders have statistically different opinions about this fact that construction industry is one of the most polluting industries, ($F(5,621) = 4.242$, $p = 0.001$), this analysis also revealed that interestingly architects know more about this fact than manufacturers and consumers. There was a statistically significant difference between groups as determined by one-way ANOVA for "increasing concern among people" ($F(5,621) = 5.592$, $p = 0.000$) and a Tukey post hoc test revealed that manufacturers agree more than builders, environmentalists, and regulators, and consumers agree more than environmentalists that there is a growing environmental concern for reuse and recycling. Also, ANOVA showed that perceptions about "availability" of reclaimed and recycled building materials are different among different stakeholders, ($F(5,621) = 4.168$, $p = 0.001$) and based on a Tukey post hoc test, consumers' group has a more positive attitude towards availability than builders, architects, and environmentalists. Accessible "information" is another item that respondents have different opinions about. It is statistically different ($F(5,621) = 7.395$, $p = 0.000$) and a Tukey post hoc test showed that manufacturers think more positively that required information for purchasing the reclaimed and recycled building materials are easily available than regulators, also consumers think more positively about this issue than builders, architects, environmentalists and regulators. Perceptions of respondents about "communication" of producers with consumers are also statistically different among stakeholders, ($F(5,601) = 2.935$, $p = 0.013$) and based on a Tukey post hoc test, consumers and manufacturers agree more than architects that producers of these building materials communicate enough with their consumers. But there was not a significant difference for "self-satisfaction", "price", "advertisement", "public image", "environmental image", "gap of claim and performance", and "knowledge of benefits" among different stakeholders. This may have been in part due to the fact that specialists are more aware and knowledgeable about the environment than lay people.

We noted that environmentalists are not optimistic about environmental issues, which can in part be as a consequence of the nature of their professional interests compared to others. Similarly, regulators are more negative than others in their views about environmental issues. As seen in Fig. 4.1, when the question is about environmental problems experts agree to a greater extent, but when it is about the influential factors they disagree. Comparisons of opinions around the measured items indicate that the general trend of respondents' attitude about different influential factors is the same across different stakeholders groups.

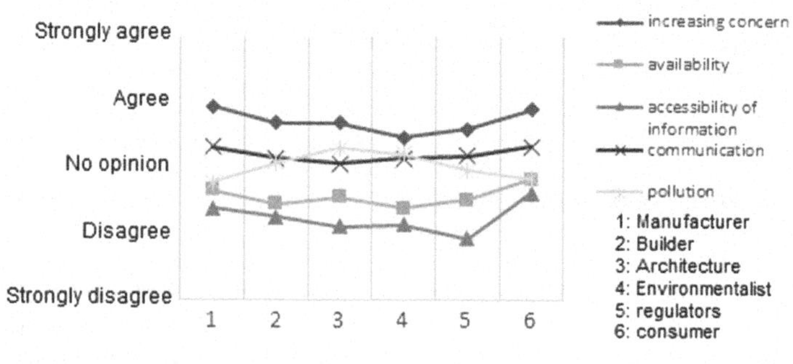


**Fig. 4.1** Different perceptions of stakeholders

## 4.6 Conclusion

Hence studying the perceptions of these stakeholders and investigating their differences and opportunities is one of the first steps towards this sustainable transformation. In the light of the results, we see that although reclaimed and recycled building material is an environment-friendly alternative to conventional materials, still, price, availability, and other normal marketing influential factors play their role in purchasing behavior of consumers. However, price and self-satisfaction are the most influential factors among the studied factors. We should not ignore this fact that green purchasers are more careful, price conscious, not loyal to brands, favor new products and communicate product information [31] than normal consumers. This finding should be considered by producers, manufacturers, and suppliers since people pay attention to the image these producers show to the society and they can see the gap between their claim and performance. Especially for building materials, because this is the architect who decides on material choice at the end of the day and architects are more aware and knowledgeable about the required specifications of materials than ordinary people. Examination of prediction of influential factors and investigating sociodemographic data of stakeholders offers much more insight into the fact that age, gender, and income are three strong predictors, which can be helpful for producers to plan future strategies.

## References


1. Parkes, C., Borland, H.: Strategic HRM: transforming its responsibilities toward ecological sustainability—the greatest global challenge facing organizations. Thunderbird Int. Bus. Rev. **54**(6), 811–824 (2012)
2. Dominish, E., et al.: Australian Opportunities in a Circular Economy for Metals: Findings of the Wealth from Waste Cluster. The Wealth from Waste Cluster (2017)
3. Iacovidou, E., Purnell, P.: Mining the physical infrastructure: opportunities, barriers and interventions in promoting structural components reuse. Sci. Total Environ. **557**, 791–807 (2016)
4. Markard, J., Raven, R., Truffer, B.: Sustainability transitions: an emerging field of research and its prospects. Res. Policy **41**(6), 955–967 (2012)
5. Osmani, M., Glass, J., Price, A.: Architect and contractor attitudes to waste minimisation. Proc. Inst. Civ. Engineers—Waste Resour. Manag. **159**(2), 65–72 (2006)
6. Poon, C., Jaillon, L.: A guide for minimizing construction and demolition waste at the design stage. The Hong Kong Polytechnic University, Dept. of Civil and Structural Engineering (2002)
7. Prasnikar, J., et al.: An integral approach to corporate environmentalism and its application to a country in transition. Zbornik radova Ekonomskog fakulteta u Rijeci, časopis za ekonomsku teoriju i praksu. Proc. Rij. Fac. Econ., J. Econ. Bus. **30**(1), 89–113 (2012)
8. Bahamón, A.: Rematerial: From Waste to Architecture. WW Norton and Company Incorporated (2010)
9. Addis, B.: Building with Reclaimed Components and Materials: A Design Handbook for Reuse and Recycling. Routledge (2012)
10. Denne, T., Bond-Smith, S.: Economic Factors of Waste Minimisation in New Zealand. Ministry for the Environment, New Zealand (2012)

11. Biswas, A., Roy, M.: Green products: an exploratory study on the consumer behaviour in emerging economies of the East. J. Clean. Prod. **87**, 463–468 (2015)
12. Dursun, I., et al.: Pro-environmental consumption: is it really all about the environment? J. Manag. Mark. Logist. **3**(2) (2016)
13. Freeman, L.: The Greening of America II: Savy Marketers Keep Aboard the Environmental Cause. Advertising Age (November 13, 1990), P-514 (1990)
14. Ottman, J., N.B.: Books, Green marketing: opportunity for innovation. J. Sustain. Prod. Des. **60** (1998)
15. Tilikidou, I., et al.: Pro-environmental purchasing behaviour during the economic crisis. Mark. Intell. Plan. **32**(2), 160–173 (2014)
16. Ministry for the Environment: Simply Sustainable. Ministry for the Environment, New Zealand (2005)
17. Grimmer, M., Bingham, T.: Company environmental performance and consumer purchase intentions. J. Bus. Res. **66**(10), 1945–1953 (2013)
18. Sonnenberg, N.C., Erasmus, A.C., Donoghue, S.: Significance of environmental sustainability issues in consumers' choice of major household appliances in South Africa. Int. J. Consum. Stud. **35**(2), 153–163 (2011)
19. Denne, T. et al.: Recycling: Cost Benefit Analysis. Report prepared for the Ministry for the Environment (New Zealand), Covec, Ltd. (2007)
20. Britzelmaier, B., Burger, S.: Reasons for the low aceptance of ethically sustainable investmens. In: 5th Annual EuroMed Conference of the EuroMed Academy of Business. EuroMed Press (2012)
21. Lemke, F., Luzio, J.P.P.: Exploring green consumers' mind-set toward green product design and life cycle assessment. J. Ind. Ecol. **18**(5), 619–630 (2014)
22. Cheng, B., Ioannou, I., Serafeim, G.: Corporate social responsibility and access to finance. Strateg. Manag. J. **35**(1), 1–23 (2014)
23. Hart, S.L.: A natural-resource-based view of the firm. Acad. Manag. Rev. **20**(4), 986–1014 (1995)
24. Luzio, J.P.P., Lemke, F.: Exploring green consumers' product demands and consumption processes. Eur. Bus. Rev. **25**(3), 281–300 (2013)
25. Heikkurinen, P.: Image differentiation with corporate environmental responsibility. Corp. Soc. Responsib. Environ. Manag. **17**(3), 142 (2010)
26. Rodriguez-Melo, A., Mansouri, S.A.: Stakeholder engagement: defining strategic advantage for sustainable construction. Bus. Strategy Environ. **20**(8), 539–552 (2011)
27. Sandhu, S., et al.: Consumer driven corporate environmentalism: Fact or fiction? Bus. Strategy Environ. **19**(6), 356–366 (2010)
28. Tilikidou, I.: Evolutions in the ecologically conscious consumer behaviour in Greece. EuroMed J. Bus. **8**(1), 17–35 (2013)
29. Bendell, J. and A. Kleanthous, *Deeper luxury: Quality and style when the world matters*. 2008: WWF-UK
30. Driessen, P.H., Hillebrand, B.: Integrating multiple stakeholder issues in new product development: an exploration. J. Prod. Innov. Manag. **30**(2), 364–379 (2013)
31. Shrum, L., McCarty, J.A., Lowrey, T.M.: Buyer characteristics of the green consumer and their implications for advertising strategy. J. Advert. **24**(2), 71–82 (1995)
32. Guerci, M., Longoni, A., Luzzini, D.: Translating stakeholder pressures into environmental performance–the mediating role of green HRM practices. Int. J. Hum. Resour. Manag. **27**(2), 262–289 (2016)
33. Hur, W.M., Woo, J., Kim, Y.: The role of consumer values and socio-demographics in green product satisfaction: the case of hybrid cars. Psychol. Rep. **117**(2), 406–427 (2015)
34. BCITO. Snapshot of Women in Construction (2012). https://bcito.org.nz
35. Figure, N.Z.: Age Distribution of People Working in the Building Construction Industry in New Zealand (2013). [cited 2013]

36. Figure, N.Z.: Highest Qualifications of People Working in the Building Construction Industry in New Zealand (2013) [cited 2013]. https://figure.nz/chart/zNtAhhyNmReAAz4o-ZUCa3NMLpY1tr03o
37. Figure, N.Z.: Average Earnings in the Building Construction Industry in New Zealand. 2018 December 2016 February 26, 2018]. https://figure.nz/chart/LqNIqJaXhjH9w2Y4-KndZEYtAMmuAIcOO

# Chapter 5
# Energy and Economic Analyses for Supporting Early Design Stages: Introducing Uncertainty in Simulations

**Giacomo Chiesa and Elena Fregonara**

**Abstract** The aim of this paper is to highlight the potentialities of a synergic application of energy analysis and economic analysis for supporting design strategies, considering the effects of the territorial location in terms of energy consumptions and prices/market variations. In order to simulate the effects of the specific projects' potential location on energy and economic input data, uncertainty is introduced as a proxy in the conjoint energy and economic analysis. A two-phases methodology is proposed, considering, first, a massive set of dynamic energy simulations by EnergyPlus of a sample office unit by varying, in this first step of the research, its location and related envelope thermal insulation levels. Minimal U-values are assumed for each simulation according to the local climate zone in accordance to DM 26.06.15, while simulated energy needs for space heating and cooling are analyzed by using a devoted python script. Second, a stochastic Global Cost calculation is proposed, considering the Global Cost (EN 15459:2007) the fundamental of the Life Cycle Cost Analysis (ISO 15686: 2008, Part 5). In this first step of the work, focus is posed at the Global Cost calculation considering solely the running costs component, in stochastic terms. The Probability Analysis solved through the Monte Carlo Method is used to represent the possible effects of uncertainty on the "energy-economic items" definition, affecting the NPVs results. The Italian territory is considered including 7978 Italian Municipalities and the related climate zones.

## 5.1 Introduction

In this paper, a multidisciplinary analysis is proposed based on a synergic application of energy analysis and economic analysis, considering the effects of the territorial location in terms of energy consumptions and market prices variations on design strategies, even in the presence of risk and uncertainty. Specifically, a two-phase methodology is proposed and experimented considering the Italian territory, including 7978 Municipalities and the related climate zones. The analysis develops, firstly,

G. Chiesa (✉) · E. Fregonara
Politecnico di Torino - DAD, Viale Mattioli 39, 10125 Turin, Italy
e-mail: giacomo.chiesa@polito.it

 
J. Littlewood et al. (eds.), *Sustainability in Energy and Buildings*, Smart Innovation, Systems and Technologies 163, https://doi.org/10.1007/978-981-32-9868-2_5

a massive set of dynamic energy simulations of a sample office unit by varying its location and related envelope thermal insulation levels. Second, a stochastic Global Cost calculation—fundamental of the Life Cycle Cost Analysis (LCCA)—is proposed. The Probability Analysis solved through the Monte Carlo Method is used to represent the possible effects of uncertainty on the LCCA results.

The paper is articulated as follows: Sect. 5.2 illustrates the literature background, Sect. 5.3 presents the methodological proposal, Sect. 5.4 illustrates the results of the application, Sect. 5.5 discusses the work and Sect. 5.6 concludes.

## 5.2 Domestic Energy Use in Italy

The requirement to reduce the energy needs in the building sector for space heating, cooling and ventilation is an essential aspect of current European regulations and efforts. Buildings are in fact responsible by more than 40% of the total energy consumptions and related greenhouse gas emissions [1], while a large amount of these consumptions is related to guarantee internal comfort conditions [2]. EPBD recast Directive 2010/31/EU, Directive 2009/28/EC, and the recent Directive (EU) 2018/844 focus on the need to increase the energy efficiency of systems, speed-up the usage of renewable sources, and reduce the energy needs acting on the envelope performances. Nevertheless, if large efforts are underlined to limit the heating needs—which are at the basis of local regulations on minimal U-value definition—further efforts are needed to curve the constant increase of cooling energy consumptions [3]. The need to consider both cooling and heating needs during early design stages to optimiz1e technological choices is in fact recognized [4]. Recent parametric studies focus on the energy consumption optimization considering technological and parametric approaches—e.g., [5]. Nevertheless, they base on a limited number of locations and do not include uncertainty aspects in simulations. New approaches including statistical model development based on the elaboration of massive simulation results and random variations constitute a new frontiers of algorithmic analyses—e.g., [6]—needed to be expanded. According to the regulatory framework and the recent literature, the energy sustainability in buildings must be managed considering jointly environmental and economic issues [7]. According to an operative viewpoint, the LCCA approach [8, 9] as described in the Standard ISO 15686-5:2008—Part 5, and to the Global Cost calculation defined in the Standard EN 15459:2007 (and Guidelines accompanying the Commission Delegated Regulation (EU) No 244/2012, following the Directive 2010/31/EU–EPBD recast), is recognized as a suitable approach for evaluating economic sustainability. Furthermore, LLCA applications with risk and uncertainty elements are proposed [10].

## 5.3 Methodology

This analysis aims at developing models to support early design choices including the effect of uncertainty in the simulation database used to model development.

In the first phase, a sample office unit is assumed to generate a large database of expected energy needs for both space cooling and heating when local maximal U-values of closures are assumed in accordance to DM 26.06.15 (Appendix B), 2019–2020 (requalification)—see Table 5.1. Two opposite working open rooms of 89 $m^2$ connected by a corridor are designed in DesignBuidler—see Fig. 5.1—considered being part of a larger office building, last floor with exposed roof. Confined rooms are assumed to be at the same temperature (adiabatic surfaces), while external vertical walls have a window-to-wall ratio of 70% and are north and south facing. Further configurations will be implemented in future. Suggested occupation and templates for office areas were assumed by the program, including simple HVAC definition for heating cooling and ventilation, self-dimensioned in EnergyPlus, including mechanical ventilation (minimum IAQ airflows). For each opaque element, the corresponding U-value from the table was imputed in the EnergyPlus *.idf file for the related climate zone by changing the thickness of the insulation layer ($\lambda = 0.034$ W/mK).

**Table 5.1** Maximal U-vale [kWh/$m^2$K] of closures for each Italian climate zone

| Type of closure—see UNI 82901:1981 | Climate zone | | | | |
|---|---|---|---|---|---|
| | A/B | C | D | E | F |
| Roofing | 0.32 | 0.32 | 0.26 | 0.24 | 0.22 |
| Floor | 0.42 | 0.38 | 0.32 | 0.29 | 0.28 |
| Vertical wall | 0.40 | 0.36 | 0.32 | 0.28 | 0.26 |
| Transparent closure | 3.00 | 2.00 | 1.80 | 1.40 | 1.00 |

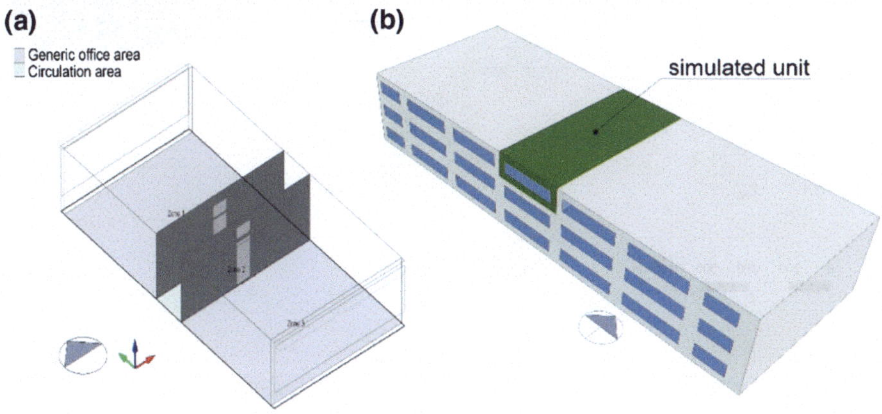


**Fig. 5.1** **a** The sample office unit considered, and **b** its location on a sample building

Other layers are for vertical closures: cement plaster (2 cm, outermost), insulation, concrete block medium density (16 cm), and gypsum plaster (2 cm); and for horizontal roof: asphalt (1 cm), concrete high density (6 cm), insulation, cast concrete (16 cm). Differently, considering that transparent closures are directly related to the type of used window, specific window types are selected from the DesignBuilder database: Clr 3/13 mm Air (U-value 2.761) for zone A/B; Dbl LoE Clr 3/13 mm Air (1.978) for zone C; Dbl LoE Clr 6/13 mm Air (1.772) for zone D; Tpl. LoE (e5 = 1) Clr 3/13 mm Air (1.270) for E; and Tpl LoE (e2–e5 = 1) Clr 3/13 Air (0.993) for F. Internal blinds, medium reflectivity, and external fixed overhang of 0.5 m are assumed as shading systems. A power density of 10 W/m$^2$ was defined [11].

A set of local typical meteorological year (TMY) conditions for each Italian Municipality are generated using the Meteonorm database (radiation period 1991–2010; temperature period 2000–2009). Files are generated using batch mode, while input files are defined through a python script elaborated by authors connected to the GIS shape file of Municipalities released by ISTAT. Climate zones are upgraded to the last classification according to DPR 412–93 and further upgrades and integrations. The entire set of the 7978 considered locations is reported in Fig. 5.2.

In the second phase, the economic analysis is conducted according to the following steps: (1) calculation of the Life Cycle Cost for each climate zone, in terms of Global Cost, incorporating the results of the analysis of the energy consumptions;

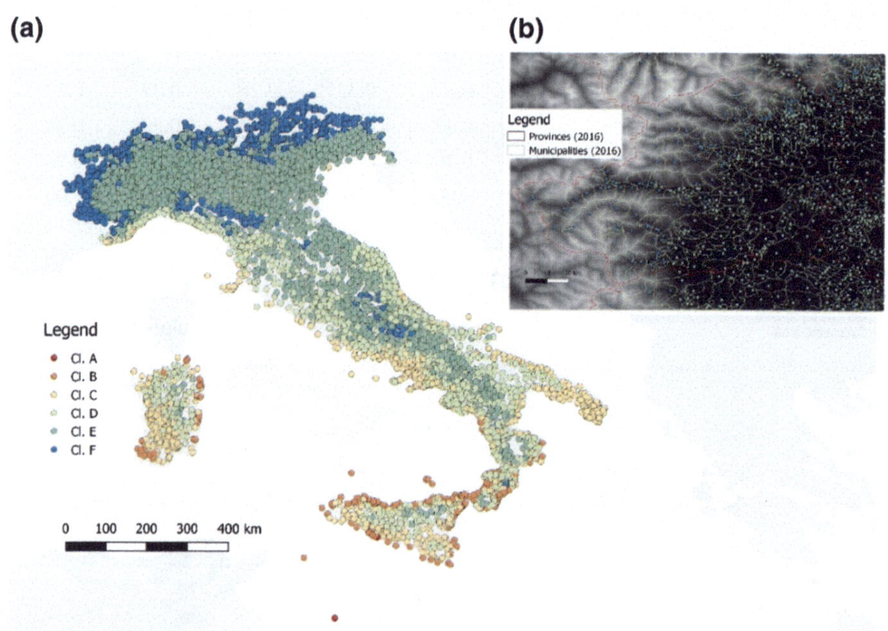


**Fig. 5.2** **a** Geo-localization of the 7978 Municipalities used for sample building simulations classified according to their Italian climate zone {A/B; C; D; E; F}. **b** sample zooming on the Turin Province (NW of Italy)

(2) calculation of the economic performance indices through the Life Cycle Cost Analysis (LCCA) approach, assuming a reference building considered in all the different potential locations; (3) comparison between the project under evaluation and the reference building's indicators, from both an energy and an economic viewpoint.

The step (1) of the economic analysis is developed through the LCCA approach, which results are expressed through quantitative indicators: Net Present Value (NPV), Savings to Investment Ratio (SIR), Adjusted Internal Rate of Return (AIRR), Simple Pay-back Period (SPB), Discounted Pay-back Period (DPB). The Global Cost calculation is based on the Eq. (5.1)

$$C_G(\tau) = C_I + \sum_j \left[ \sum_{i=1}^{\tau} \left( C_{a,i}(j) \cdot R_d(i) \right) - V_{f,\tau}(j) \right] \quad (5.1)$$

where $C_G(\tau)$ represents the global cost, referred to starting year $\tau_0$; $C_I$ represents the initial investment costs; $C_{a,i}$ (j) represents the annual cost during year i of component j, including the annual running costs (energy costs, operational costs, maintenance costs) and periodic replacement costs; $R_d$ (i) represents the discount factor during the year i; $V_{f,\tau}$(j) represents the residual value of the component j at the end of the calculation period, referred to the starting year. In this proposal the Global Cost formula is assumed, and simplified with reference to a building project, according to Eq. (5.2) [12]:

$$C_G = C_I + \sum_{t=0}^{N} \frac{C_o + C_m}{(1+r)^t} \quad (5.2)$$

where $C_G$ represents the Global Cost; $C_I$ represents the initial investment costs; $C_o$ represents the operating and energy costs; $C_m$ represents the maintenance costs, t stands for the year in which the costs arise and N stands for the number of years within the timespan taken into account for the application; r represents the discount rate.

Then, for including uncertainty in the LCCA application, the Probability Analysis is proposed, with the simulation method [13]. This implies the definition of PDFs functional forms, founded on random number generation, both for model input and for model output. Thus, it is necessary to quantify the marginal contribution of each input variable (expressed in stochastic terms) on the output. Also the model output is calculated in terms of stochastic Global Cost, as in Eq. 5.3

$$\hat{C}_G = \hat{C}_I \sum_{t=0}^{N} \frac{\hat{C}_o + \hat{C}_m}{(1+\hat{r})^t} \quad (5.3)$$

where $\hat{C}_G$ is the stochastic Global Cost, $\hat{C}_I$ is the stochastic investment cost, $\hat{C}_o$ is the stochastic operating and energy cost, $\hat{C}_m$ is the stochastic maintenance cost, t is the year in which the cost occurred, N is the number of years of the period considered for the analysis, r̂ is the stochastic discount rate. As a first step, in this study the methodology is related solely to the calculation of the NPV of the $C_o$ item (running costs), this last in stochastic terms.

## 5.4 Simulations and Results

### *5.4.1 Energy Analysis*

Results of energy analyses conducted according to the described methodology are used to support the further Sect. 5.2. Thanks to a python code elaborated for this paper, the district cooling and heating net consumptions for conditioned building area [kWh/m$^2$ y] reported to the correspondent energy source values (natural gas for heating and electricity for cooling) are tabulated for each location by reading the EnergyPlus Output report. Locations are subdivided by climate zones. It can be noted that the cooling energy demand results to be dominant for classes A, B, and partially C, but remains an important voice also in class D, and partially E, such as can be expected considering the medium-to-high internal gains of office spaces and the large WWR—even if shading systems are included. Figure 5.3 synthetizes this consideration by simply regressing the simulated consumptions as function of the local climate zone. Further analyses are planned in the further research steps including uncertainty by considering for example different sources and reference periods for TMY, variations in the system efficiencies, and different building configurations.

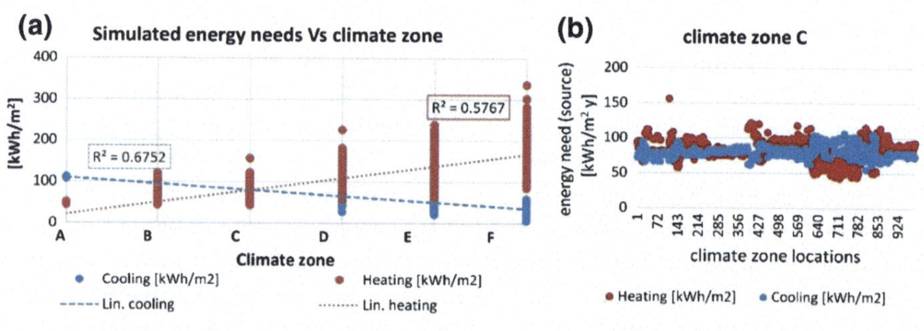


**Fig. 5.3** **a** Heating and cooling energy needs (natural gas and electricity) plotted in accordance to local climate zone and related linear regression lines; **b** distribution of energy needs (sources) for locations in climate zone C

## 5.4.2 Economic Simulation

The methodology illustrated in Sect. 5.3 is applied to the case-study mentioned above. The results of the economic simulations are performed through the software @Risk (by Palisade Corporation, release 7.5). The Net Present Value (NPV) calculation, related to the running costs components, is the focus of the present analysis. As a preliminary step, the results of the first simulation are presented in Table 5.2. The Probability Distribution Functions—PDFs, are calculated considering: the consumption levels for heating and cooling, for the different Municipalities included into the climate zones; the parametric costs distinguishing between energy cost for cooling and natural gas for heating. Cost definition is based on Energy Plus assumptions. Lastly, Discount rate with a range of values defined on the indications of the official documents is assumed. From the Table 5.2 a great variety of the PDFs emerges. This is not surprising given the wide heterogeneity of the Italian territory, and the variety of the conditions even inside each climate zone. It must be considered, particularly for the cooling season, that the Italian climate classification, which is used for maximal U-value and heating period definitions, is based solely on the $HDD_{20}$ (see DPR 412–93, Annex A, and further modifications), although different climatic indexes were recently included in the UNI 10349-3:2016.

As a second step, the Monte Carlo Method is applied for calculating, in stochastic terms, the output values for the NPV for the running costs. Figure 5.4 presents, as example, the results considering the climate zone E with the relative statistics. The following Fig. 5.5 illustrates the simulation results. Specifically, the Tornado graphs represent graphically the effects of the inputs on the output mean, considering all the climate zones. The Tornado graphs presented in Fig. 5.5 show a net prevalence of the energetic factors. In fact, the NPVs are influenced by the energy components in all the climate zones. Cooling is always prevalent, while heating approaches the weight of the cooling solely in relation to the coldest climate zones (E and F). Regarding this last point, it must be considered that the minimum values for the thermal transmittance fixed by the law are established on the basis of the climate zone which refers to local $HDD_{20}$.

The analysis is concluded by comparing the simulation output results (NPVs) for each climate zone, as reported in Table 5.3. From Table 5.3, emerges that the climate zone A + B—corresponding to the warmest areas—is shifted toward the lowest values. In general, the distributions are quite wide, and climate zone F—corresponding to mountain areas—stands out for the shape of the distribution, particularly wide. Cost values related to climate zone E are slightly higher, as it includes the greatest number of Municipalities.

**Table 5.2** Energy consumptions by climate zone and cost items: probability distribution functions (graphics and statistics)

| Energy and cost items | Distribution | Graph | Min | Mean | Max | 5% | 95% |
|---|---|---|---|---|---|---|---|
| Heating [kWh/m2] ClZone A–B | Loglogistic | 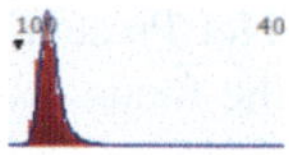 | 113.77 | 142.09 | 394.01 | 127.01 | 162.66 |
| Heating [kWh/m2] ClZone C | Laplace | 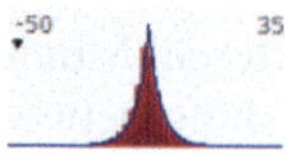 | – 4.066 | 139.66 | 335.95 | 102.38 | 176.89 |
| Heating [kWh/m2] ClZone D | Laplace | 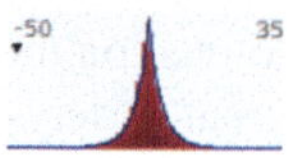 | – 37.39 | 140.95 | 303.78 | 100.71 | 181.15 |
| Heating [kWh/m2] ClZone E | Loglogistic | 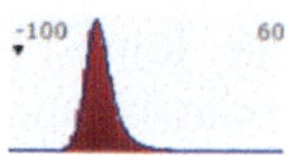 | 18.58 | 118.01 | 582.81 | 68.04 | 185.69 |
| Heating [kWh/m2] ClZone F | Gamma | 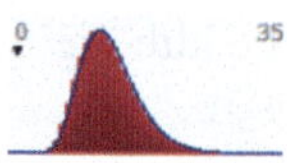 | 39.39 | 120.39 | 332.18 | 72.59 | 182.55 |
| Cooling [kWh/m2] ClZone A-B | Triangular |  | 7.19 | 35.28 | 49.46 | 16.41 | 48.39 |
| Cooling [kWh/m2] ClZone C | ExtValue | 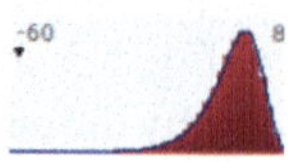 | – 55.99 | 45.89 | 75.23 | 21.39 | 63.02 |
| Cooling [kWh/m2] ClZone D | Triangular |  | 6.63 | 47.83 | 73.79 | 20.24 | 67.94 |
| Cooling [kWh/m2] ClZone E | Laplace | 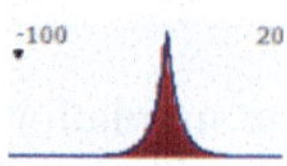 | – 66.16 | 60.92 | 168.68 | 32.61 | 89.19 |
| Cooling [kWh/m2] ClZone F | Weibull | 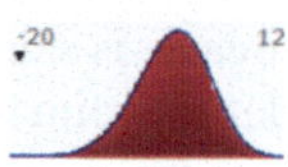 | – 1.11 | 57.41 | 114.17 | 28.36 | 84.04 |
| Heating supply (€/KWh) | Triangular |  | 0.10 | 0.11 | 0.12 | 0.10 | 0.12 |
| Electric power supply (€/KWh) | Triangular |  | 0.20 | 0.25 | 0.30 | 0.22 | 0.28 |
| Discount rate | Triangular |  | 1.3% | 1.7% | 2.5% | 1.3% | 2.2% |

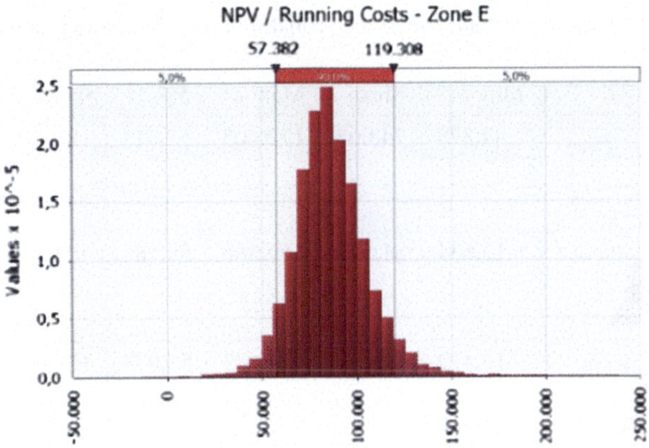


| Statistics | |
|---|---|
| Minimum | -13.462 |
| Maximum | 238.862 |
| Mean | 86.254 |
| Std Dev | 19.270 |
| Variance | 371349766,2 |
| Skewness | 0,573838872 |
| Kurtosis | 5,290754664 |
| Median | 84.841 |
| Mode | 80.278 |
| Left X | 57.382 |
| Left P | 5% |
| Right X | 119.308 |
| Right P | 95% |

**Fig. 5.4** The probability distribution functions and statistics for the climate zone E

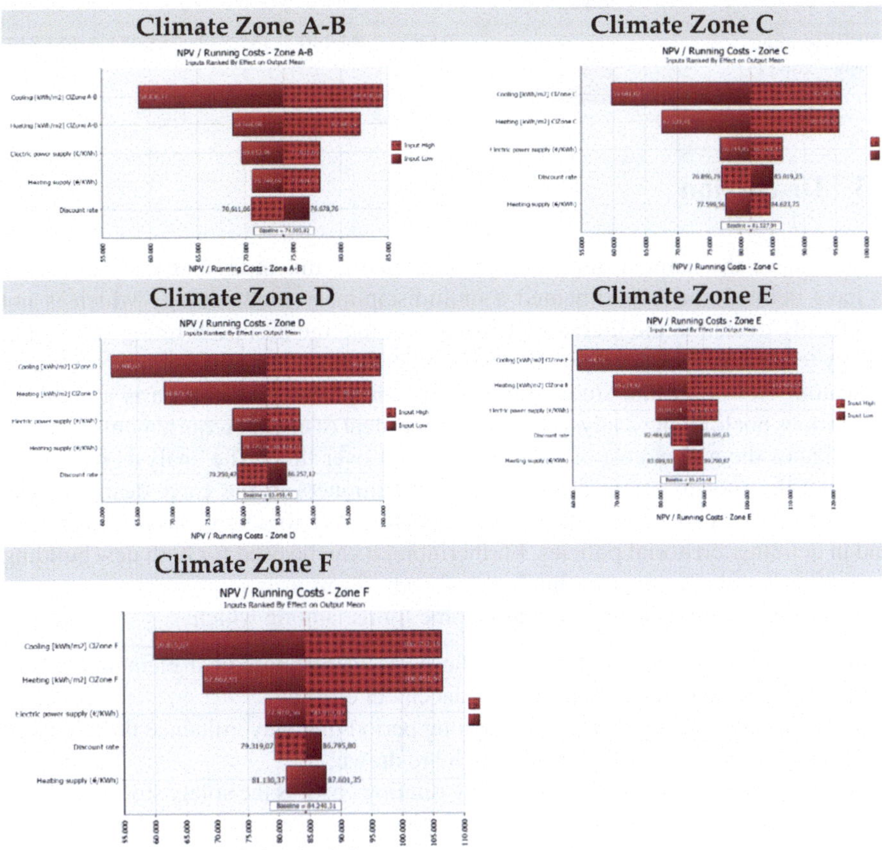


**Fig. 5.5** The inputs ranked by the effects on the output mean, for each climate zone

**Table 5.3** The output values of NPVs for the climate zones: the probability distribution functions and statistics

| Output | Graph | Min | Mean | Max | 5% | 95% |
|---|---|---|---|---|---|---|
| NPV/Running costs—zone A–B | 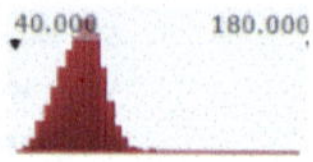  | 44,376 | 74,004 | 172,605 | 57,615 | 88,859 |
| NPV/Running costs—zone C | 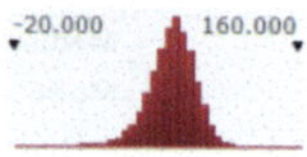  | – 13,634 | 81,528 | 159,388 | 58,188 | 102,058 |
| NPV/Running costs—zone D | 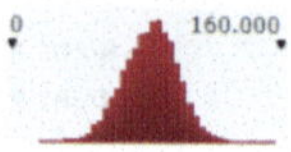  | 19,479 | 83,458 | 154,473 | 58,074 | 106,678 |
| NPV/Running costs—zone E | 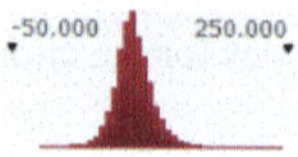  | – 13,462 | 86,254 | 238,862 | 57,382 | 119,308 |
| NPV/Running costs—zone F | 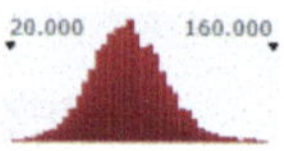  | 24,876 | 84,240 | 155,533 | 55,481 | 114,813 |

## 5.5 Discussion

According to the authors, the research presented in this study has the main merit to have defined and experimented a multidisciplinary methodology which is not commonly treated in the literature. Considering that the best performing solution in energy terms does not correspond to the best performing solution from the economic indicators viewpoint, the study assumes that the preferable solution is a trade-off which does not limit the analysis to a strict fulfilment of energy requirements, but also investigates the global cost of each intervention over time. The analysis represents a supporting tool to orient designers and practitioners in the early design phase, decision makers in the decision processes, public authorities in governance activities and in defining territorial policies. Furthermore, it can be used for both new building construction projects and existing buildings retrofit projects.

Beside the potentials, there are also some limits, among which

- the difficulty in defining local specific costs of technological elements (initial investments, and maintenance and replacement costs);
- the variability of energy prices in the long period that may influence the results of the analyses and the conclusions which are drawn;
- the analysis was conducted considering running costs as the solely stochastic variable. It would be preferable to complete the analysis also with stochastic investment costs and stochastic maintenance cost, aiming at calculating a final stochastic Global Cost.

The last point is demanded to a future research, assuming this study as a first step toward the definition of the territorial-location influence on the Global Cost (considering uncertainty in cost component as a proxy of territorial variations). In the meanwhile, the study of the effects of different design choices and design constraints on the above indicators is on progress. Lastly, as just is underlined in the Table 5.2, the great variation of the PDFs suggests that the present climate zones may be updated according to additional criteria, considering eventually clusters of Municipalities.

## 5.6 Conclusions

In this work, a two-phase methodology was presented. First, a massive set of dynamic energy simulations was considered, by varying the location and related envelope thermal insulation levels of a sample office unit. Second, a stochastic Global Cost calculation was proposed, in order to simulate the effects of the specific projects' location on energy and economic input data. Uncertainty was introduced as a proxy in the conjoint energy and economic analysis. The results show how energy and economic analyses can be fruitfully used in synergy when addressing different possible design strategies.

## References


1. European Commission: Clean Energy for All Europeans. COM(2016) 860 final. European Commission, Brussels (2016)
2. Chiesa, G., Grosso, M., Pearlmutter, D., Ray, S.: Editorial. Advances in adaptive comfort modelling and passive/hybrid cooling of buildings. Energy Build. **148**, 211–217 (2017)
3. Santamouris, M.: Cooling the buildings—past, present and future. Energy Build. **28**, 617–638 (2016)
4. Heiselberg, P. (ed.): Ventilative Cooling Design Guide. IEA EBC Annex 62. Aalborg Un. Press, Aalborg (2018)
5. Košir, M., Gostiša, T., Krist, Z.: Influence of architectural building envelope characteristics on energy performance in Central European climatic conditions. J. Build. Eng. **15**, 278–288 (2018)
6. Chiesa G. et al.: Insulation, building mass and airflows—Provisional and multivariable analysis. Sustain. Mediterr. Constr. **8**, 36–40
7. Gundes, S.: The use of life cycle techniques in the assessment of sustainability. Procedia—Social Behav. Sci. **216**, 916–922 (2016)
8. Langdon, D.: Life cycle costing (LCC) as a contribution to sustainable construction: a common methodology—final methodology (2007). http://ec.europa.eu/enterprise/sectors/construction/studies/life-cycle-costing_en.htm
9. König, H., Kohler, N., Kreissig, J., Lützkendorf, T.: A Life Cycle Approach to Buildings, Principles, Calculations, Design Tools. Detail Green Books, Regensburg (2010)
10. Boussabaine, A., Kirkham, R.: Whole Life-Cycle Costing: Risk and Risk Responses. Blackwell Publishing, Oxford (2004)
11. Menezes, A.C., Cripps, A., Buswell, R.A., Wright, J., Bouchlaghem, D.: Estimating the energy consumption and power demand of small power equipment in office buildings. Energy Build. **75**, 199–209 (2014)

12. Fregonara, E., Lo Verso, V.R.M., Lisa, M., Callegari, G.: Retrofit scenarios and economic sustainability. A case-study in the Italian context. Energy Procedia **111**, 245–255 (2017)
13. Fregonara, E., Ferrando, D.G., Pattono, S.: Economic-environmental sustainability in building projects: introducing risk and uncertainty in LCCE and LCCA. Sustainability **10**(6), 1901 (2018)

# Chapter 6
# Using Evidence Accumulation-Based Clustering and Symbolic Transformation to Group Multiple Buildings Based on Electricity Usage Patterns

**Kehua Li, Zhenjun Ma, Duane Robinson and Jun Ma**

**Abstract** This paper presents a cluster analysis-based strategy to group multiple buildings based on their electricity usage patterns. In this strategy, an evidence accumulation-based clustering algorithm was first used to cluster the daily electricity usage profiles of all buildings over the course of a whole year. The whole year electricity usage time series data of all buildings were then transformed into symbolic representations based on the clustering result of daily electricity usage profiles and further clustered using an agglomerative hierarchical clustering algorithm to group the buildings. The one-year hourly electricity usage time series data collected from 40 buildings on a campus were used to evaluate the performance of this strategy. The result showed that this strategy can group the buildings effectively so that the electricity usage patterns of the buildings in the same group had high similarity to each other while that among different groups were remarkably different. The results from this study can be potentially used to assist in the planning and operation of building energy systems.

## 6.1 Introduction

Grouping building energy usage patterns has been considered as a helpful tool to enhance planning and operation of building energy systems, management of local energy generation and the development of tariff strategies [1]. Over the last decade, many studies have focused on the development of effective strategies for grouping building energy usage patterns [2–6]. Cluster analysis, as a powerful tool which can group objects to ensure that the differences among the objects in the same group are much smaller than the differences among the groups [7], has been used for this purpose in both grouping daily energy usage profiles (DEUPs) and grouping buildings based on their electricity usage patterns. McLoughlin et al. [8] developed

K. Li · Z. Ma (✉) · D. Robinson
Sustainable Buildings Research Centre, University of Wollongong, Wollongong 2522, Australia
e-mail: zhenjun@uow.edu.au

J. Ma
SMART Infrastructure Facility, University of Wollongong, Wollongong 2522, Australia

 
J. Littlewood et al. (eds.), *Sustainability in Energy and Buildings*, Smart Innovation, Systems and Technologies 163, https://doi.org/10.1007/978-981-32-9868-2_6

a strategy based on cluster analysis to determine the representative daily electricity usage profiles of residential buildings on a national scale. Self-organizing map (SOM) technique was used as the clustering technique after a comparison with *k*-means and *k*-medoids methods. Cluster analysis was conducted with daily profiles over a 6-month period in order to detect the variations of electricity usage in different time scales. A number of typical electricity usage profiles which represented common patterns of residential building electricity usage were presented. Miller et al. [9] developed a strategy based on Symbolic Aggregate approXimation (SAX) and *k*-means to characterise the hidden information from building and sub-systems operating data to facilitate energy savings.

In previous studies on grouping buildings based on electricity usage patterns, the similarity among buildings was calculated based on the percentage of typical DEUPs in each building. For instance, Tsekouras et al. [10] proposed a two-stage method to classify electricity usage patterns. In the first stage, *k*-means and other eight clustering algorithms were used to identify typical DEUPs of all electricity customers. The electricity customers were then clustered based on their major typical DEUPs. Ma et al. [6] developed a strategy based on hierarchical clustering to identify representative daily heating load profiles of multiple buildings in a campus, and the percentages of different typical daily heating load profiles in each building were then used to classify the buildings into groups. However, the time distribution of different types of DEUPs, which is an important aspect of energy usage behaviour, was not considered in these studies.

To address this gap, a cluster analysis-based strategy for grouping multiple buildings based on electricity usage patterns was proposed in this paper. Different from the previous studies, the variation of electricity usage in each building in a whole year was considered for grouping multiple buildings. An evidence accumulation-based clustering (EAC) algorithm, which is more robust than other conventional clustering algorithms, was used to identify clusters in the DEUPs of multiple buildings. Symbolic transformation and agglomerative hierarchical clustering (AHC) were then used to group the buildings based on the clustering result of the DEUPs. The one-year hourly electricity usage data collected from 40 buildings of a university campus were used to evaluate the performance of the proposed cluster analysis-based strategy.

## 6.2 Description of the Proposed Strategy

### *6.2.1 Outline of the Proposed Strategy*

The outline of the proposed cluster analysis-based strategy for grouping multiple buildings based on their electricity usage patterns is demonstrated in Fig. 6.1. The strategy consists of four steps, which are data collection, data pre-processing, clustering of DEUPs, and grouping of buildings. In the first step, building electricity usage data were collected from building automation systems. The collected data were then

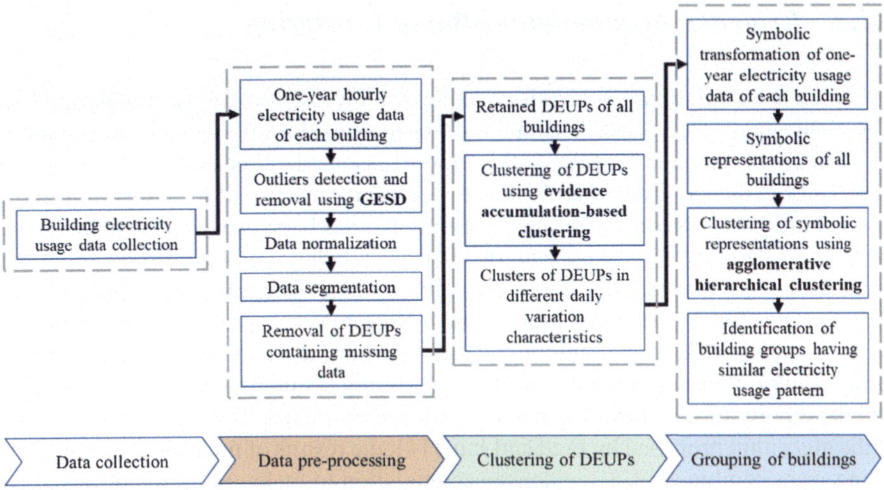


**Fig. 6.1** Outline of the proposed cluster analysis-based strategy

preprocessed in the second step, where initially, the electricity usage time series data from each building in the same year were selected from the raw data. The outliers in the selected data were then detected and removed using generalised extreme studentised deviation (GESD) test method. As different buildings have different magnitudes of the electricity usage, the time series data of each building were standardised to zero mean and one standard deviation to avoid their influencing on the clustering of DEUPs. The one-year time series data of each building were then segmented into 24-h profiles in order to form DEUPs.

In the third step, the retained DEUPs of all buildings were used as the input data in which each DEUP was treated as a data point. Cluster analysis of the DEUPs was then conducted using an EAC algorithm (to be introduced in Sect. 6.2.2). After the identification of the clusters of the DEUPs, the buildings were grouped in the last step. In this step, the one-year electricity usage time series data of each building were first transformed into symbolic representations based on the clustering result of DEUPs (to be explained in Sect. 6.2.3). Then, the symbolic representations of all buildings were clustered using an AHC algorithm using Ward's method with N-gram distance [11]. Compare with other commonly used linkage criteria such as single linkage, complete linkage and group average linkage, Ward's method showed a better performance in terms of clustering accuracy and computational cost [12] and therefore was used in this study. Building groups, in which the buildings had a similar electricity usage pattern during the whole year, and the buildings having a unique electricity usage pattern were identified from the clustering result.

### 6.2.2 Evidence Accumulation-Based Clustering

Evidence accumulation is a cognitive model widely used in decision-making [13]. The basic idea of evidence accumulation is to consider all available information simultaneously to make decision-making more reasonable. Fred and Jain [14] applied the idea of evidence accumulation to enhance the robustness of the clustering result of *k*-means. *k*-means is a clustering algorithm that has been proved helpful in identifying building electricity usage patterns. However, using *k*-means method, the same input data and the same parameters typically produce different clustering results in different runs for the random initialization of clusters [7]. In practice, to use *k*-means method, the input data are commonly clustered multiple times and the optimal clustering result is then determined as the final clustering result, which is time consuming and the quality of the clustering result is still not promised. To overcome this issue, in the algorithm proposed by Fred and Jain [14], the results of multiple *k*-means clustering were combined using evidence accumulation to make use of the information discovered form each clustering. The combined result was further clustered using Minimum Spanning Tree (MST) and the clustering result was then considered as the final clustering result. The test of this algorithm showed that the combined clustering algorithm can make clustering result more robust compared to *k*-means. In this paper, an EAC algorithm which is similar to the one used by Fred and Jain [14] was used. However, Partitioning Around Medoids (PAM) and AHC using Ward's method were employed instead of *k*-means and MST for clustering analysis since PAM and AHC have similar advantages to *k*-means and MST but they perform better with noises and outliers in the input data [6, 7].

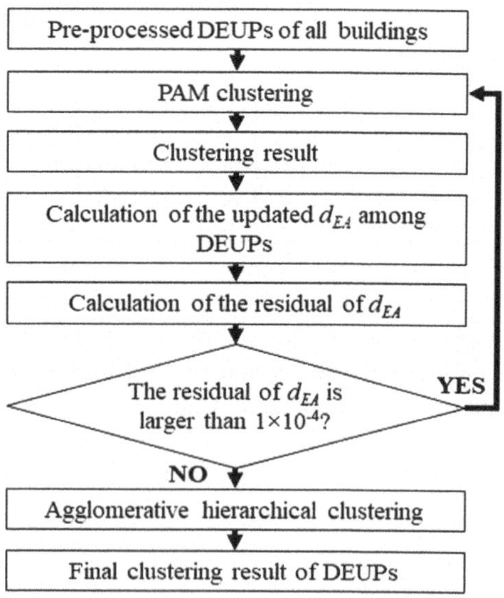


**Fig. 6.2** Flowchart of an EAC algorithm

The EAC algorithm used in this paper is shown in Fig. 6.2. In this algorithm, the preprocessed DEUPs of all buildings were first clustered using PAM. The clustering results were then used to calculate and update the evidence accumulation-based dissimilarity, $d_{EA}$, between each pair of the DEUPs. These steps will be repeated until the residual of $d_{EA}$ was less than $1 \times 10^{-4}$. Once the residual matched the threshold, the final updated $d_{EA}$ between each pair of the DEUPs was further clustered using AHC to identify the clusters in DEUPs. The $d_{EA}$ was calculated according to Eq. (6.1).

$$d_{EA}(x, y) = \frac{vote(x, y)}{G} \tag{6.1}$$

where $x$ and $y$ are two DEUPs, $G$ is the number of iterations conducted in the whole process and *vote* is the frequency of grouping $x$ and $y$ in the same cluster in each PAM clustering. For instance, if two DEUPs were clustered in the same group in three times in four PAM clusterings, the $d_{EA}$ would be 3/4. In this paper, the number of clusters in each PAM clustering was set as the square root of $n$ as per suggested in [14], where $n$ is the number of DEUPs in the input data.

### 6.2.3 Symbolic Transformation and Clustering of Symbolic Representations

In this paper, a clustering-based symbolic transformation method proposed by Das et al. [15] was used to capture the one-year electricity usage patterns in multiple buildings. As illustrated in Fig. 6.3, each DEUP in the one-year electricity usage time-series data was first signed with a label based on the clustering result of the DEUPs. The 365 sequent labels of a building, which correspond to the 365 days of a year, were then identified as the symbolic representation of the electricity usage time series data

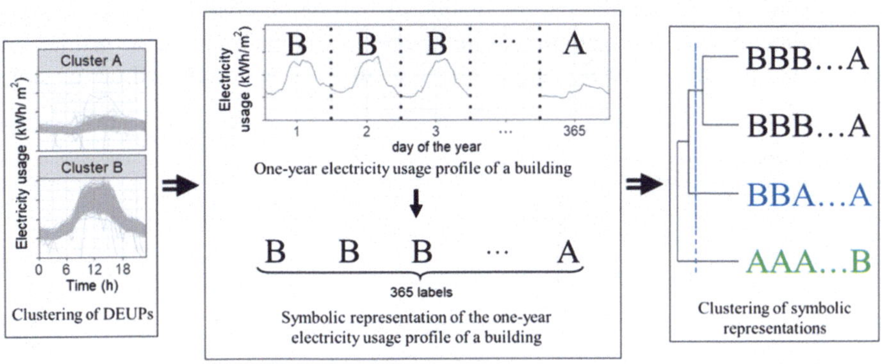


**Fig. 6.3** Illustration of the symbolic transformation and the clustering of symbolic representations

of this building. It is noted that the DEUPs removed during the data preprocessing step were signed with a blank label in the symbolic transformation. After symbolic transformation, the 40 symbolic representations (corresponding to 40 buildings) were further clustered using AHC so that the buildings in the same group had the similar symbolic representations and thus had the similar electricity usage patterns. AHC was used since it performs better than other existing clustering algorithms with high dimensional input data, for example, the symbolic representations in this study which have 365 dimensions (i.e. 365 labels).

## 6.3 Test of the Proposed Strategy

Forty buildings at the University of Wollongong, Australia were used in the test of the proposed strategy. The data of the electricity usage per hour in each building were retrieved in 2015 and used for the test and evaluation of the performance of the proposed strategy. The detailed information of these 40 buildings can be found in [16].

### *6.3.1 Clustering of DEUPs*

As described in Sect. 6.2, the outliers were first detected and removed from the one-year electricity usage time series data of each building using GESD test. Each of the one-year electricity usage time series data was then standardised and segmented into the DEUPs. The DEUPs which contain the missing data were not considered in the following steps. A total of 14,108 DEUPs remained for all 40 buildings when data preprocessing was completed.

All the DEUPs were then clustered using the EAC algorithm described in Sect. 6.2.2. In each PAM clustering, the desired number of clusters was $\sqrt{14{,}108} \approx 119$. A total of 221 iterations were conducted before the residual of $d_{EA}$ became smaller than the threshold. Figure 6.4 shows the correlation between the residual of $d_{EA}$ and the number of iterations (i.e. G). A smaller residual indicated more stable

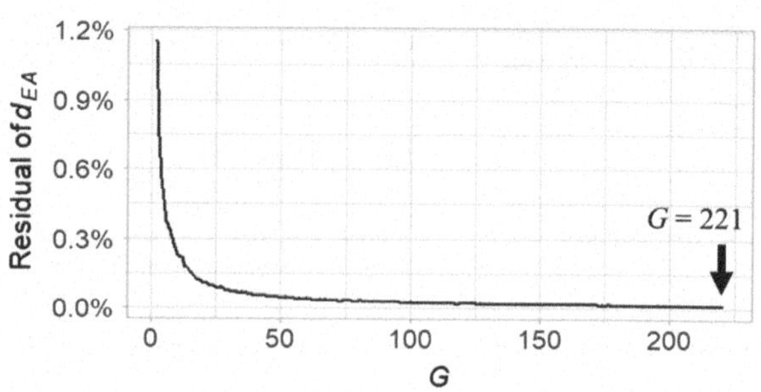


**Fig. 6.4** Change of the residual of $d_{EA}$ in the EAC

input data to the AHC and therefore was more desirable. It can be seen that the change of the residual decreased considerably when $G$ increased from 1 to 50 while it became stable and was close to zero when $G$ was larger than 100. After conduction of 221 times PAM clustering, the $d_{EA}$ between each pair of the DEUPs was calculated and further clustered using AHC. The dendrogram of the AHC is demonstrated in Fig. 6.5. To make each cluster fairly distinctive while avoiding too many clusters of the DEUPs, five clusters were identified by selecting 15 as the threshold to cut the dendrogram.

The variations of the DEUPs identified in each cluster are shown in Fig. 6.6, in which the darker lines indicated the median of the DEUPs in each cluster. In Cluster A, the DEUPs experienced a continuous increase from 8:00 to 18:00. After a peak electricity demand at around 19:00, the electricity usage reduced until 8:00 next day. In Cluster B, the electricity usage during working hours (from 8:00 to 17:00) was

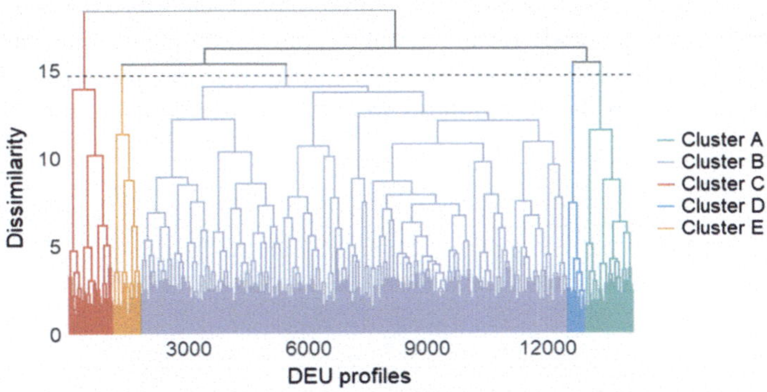


**Fig. 6.5** Dendrogram of the AHC in clustering of DEUPs

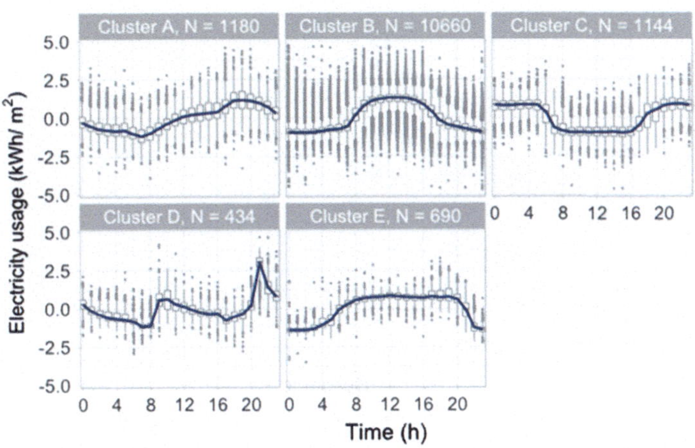


**Fig. 6.6** Clusters identified from DEUPs of all buildings

clearly higher than that during non-working hours. Cluster E also had a clear high electricity usage period while the duration of that period (from 6:00 to 21:00) was longer than the high electricity usage period of Cluster B. In Cluster C, the electricity usage from 6:00 to 18:00 was clearly lower than that during the other periods. The DEUPs in Cluster D had two peaks of electricity demand. The highest peak occurred at 21:00 and the other one occurred at 10:00.

### *6.3.2 Grouping of Buildings*

Through the clustering of the DEUPs, all DEUPs of the 40 buildings were clustered into five groups. In this step, for each building, the one-year electricity usage time series data were transformed into a symbolic representation based on the label of each DEUP. Figure 6.7 demonstrates the symbolic representations using a heatmap, in which each column stands for a symbolic representation of a building and the colours indicated different cluster labels in the symbolic representations. The symbolic representations of the 40 buildings were then clustered using the AHC algorithm.

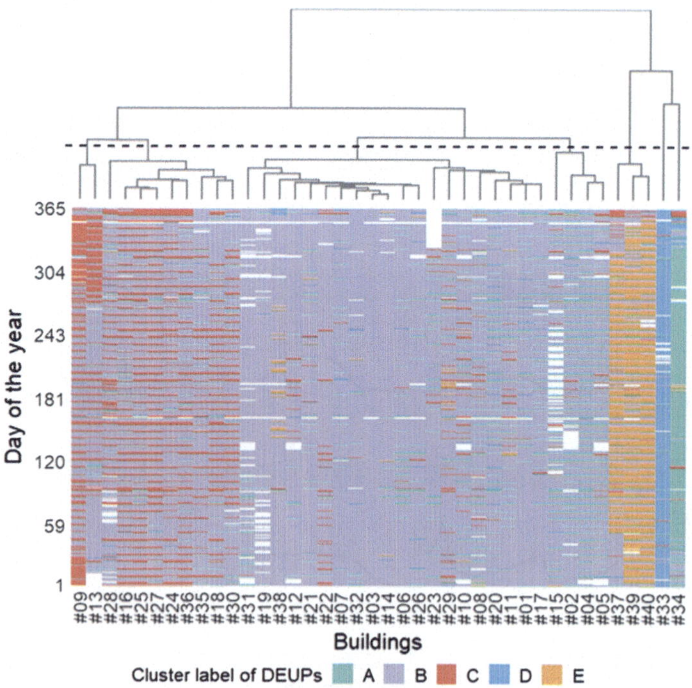


**Fig. 6.7** Clustering of symbolic representations of 40 buildings

In this study, the number of the clusters in the AHC was visually determined with the dendrogram (see Fig. 6.7) since the number of the data points (i.e., buildings) was small. Among the seven clusters (i.e. groups of buildings) identified, some groups consisted of several buildings while some groups only included a single building. For example, building 33 formed an individual group and buildings 37, 39 and 40 were in the same group. It can be seen in Fig. 6.7 that the difference among the symbolic representations in the same group was considerably smaller than the difference among the symbolic representations in different groups.

The difference among building electricity usage patterns in each building group can be clearly seen in Fig. 6.8. For instance, in Building Group 7, most DEUPs during the weekdays belonged to the DEUPs Cluster E, which had a longer high electricity demand period (see Fig. 6.6). Since the buildings in this group consisted of sports centres and buildings used as common space/laboratory (as shown in Fig. 6.9), the longer high electricity demand can be explained by the longer operation hours in these types of buildings. It can be seen from Fig. 6.9 that Building Groups 5 and 6 consisted of student accommodations only while the composition of the DEUPs in the two groups was obviously different (see Fig. 6.8). The information discovered from this step can be potentially used to support the planning and operation of building energy systems at the campus level. For example, the building groups identified using the proposed strategy can be used in building energy benchmarking to improve the accuracy of the benchmarking result.

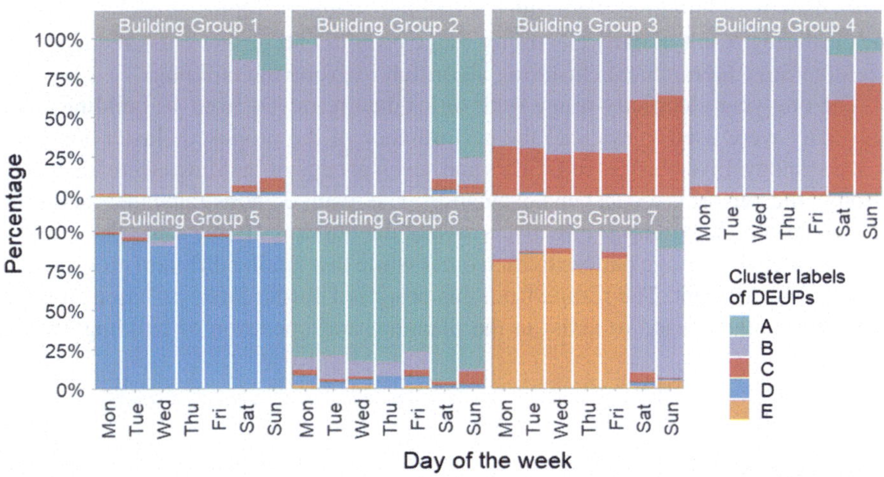


**Fig. 6.8** The percentage of DEUPs into different days of a week in each building group

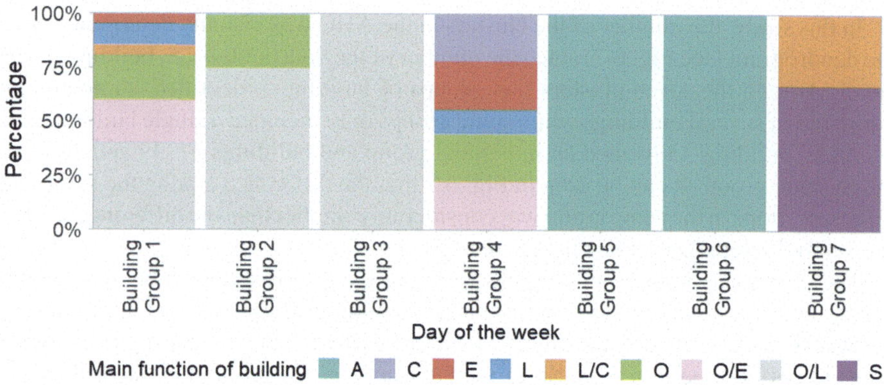


**Fig. 6.9** The functions of buildings in each building group (A: student accommodation; C: common area; E: educational room; L: laboratory; O: office; S: sports centre)

## 6.4 Conclusions

In this paper, a cluster analysis-based strategy to group multiple buildings based on building electricity usage patterns was presented. In this strategy, the daily electricity usage profiles of all buildings in a whole year were first clustered using an evidence accumulation-based clustering algorithm. Then, the whole year electricity usage time series data of all buildings were transformed into symbolic representations based on the clustering result of daily electricity usage profiles and further clustered using an agglomerative hierarchical clustering algorithm to group the buildings.

The one-year electricity usage time series data retrieved from 40 buildings in a university were used to evaluate the performance of the proposed cluster analysis-based strategy. Five clusters of DEUPs and seven groups of buildings were identified through the performance test and evaluation. The result demonstrated that this strategy can group the buildings effectively so that the buildings in the same group had a similar electricity usage pattern to each other while that among different groups were remarkably different. The results from this proposed cluster analysis-based strategy can be potentially used to assist in the planning and operation of building energy systems.

## References

1. Chicco, G.: Overview and performance assessment of the clustering methods for electrical load pattern grouping. Energy **42**(1), 68–80 (2012)
2. Park, J.Y., Yang, X., Miller, C., Arjunan, P., Nagy, Z.: Apples or oranges? Identification of fundamental load shape profiles for benchmarking buildings using a large and diverse dataset. Appl. Energy **236**, 1280–1295 (2019)
3. Tardioli, G., Kerrigan, R., Oates, M., O'Donnell, J., Finn, D.P.: Identification of representative buildings and building groups in urban datasets using a novel pre-processing, classification,

clustering and predictive modelling approach. Energy Build. **140**, 90–106 (2018)
4. Gianniou, P., Liu, X., Heller, A., Nielsen, P.S., Rode, C.: Clustering-based analysis for residential district heating data. Energy Convers. Manag. **165**, 840–850 (2018)
5. Wei, Y., Zhang, X., Shi, Y., Xia, L., Pan, S., Wu, J., Han, M., Zhao, X.: A review of data-driven approaches for prediction and classification of building energy consumption. Renew. Sustain. Energy Rev. **82**, 1027–1047 (2018)
6. Ma, Z., Yan, R., Nord, N.: A variation focused cluster analysis strategy to identify typical daily heating load profiles of higher education buildings. Energy **134**, 90–102 (2017)
7. Tan, P.-N., Steinbach, M., Kumar, V.: Introduction to Data Mining. Pearson Addison Wesley, Boston (2005)
8. McLoughlin, F., Duffy, A., Conlon, M.: A clustering approach to domestic electricity load profile characterisation using smart metering data. Appl. Energy **141**, 190–199 (2015)
9. Miller, C., Nagy, Z., Schlueter, A.: Automated daily pattern filtering of measured building performance data. Autom. Constr. **49**, 1–17 (2015)
10. Tsekouras, G.J., Hatziargyriou, N.D., Dialynas, E.N.: Two-stage pattern recognition of load curves for classification of electricity customers. IEEE Trans. Power Syst. **22**, 1120–1128 (2007)
11. Ma, Z., Yan, R., Li, K., Nord, N.: Building energy performance assessment using volatility change based symbolic transformation and hierarchical clustering. Energy Build. **166**, 284–295 (2018)
12. Murtagh, F.: Expected-time complexity results for hierarchic clustering algorithms which use cluster centres. Inf. Process. Lett. **16**, 237–241 (1983)
13. Gold, J.I., Shadlen, M.N.: The neural basis of decision making. Annu. Rev. Neurosci. **30**, 535–574 (2017)
14. Fred, A.L., Jain, A.K.: Data clustering using evidence accumulation. In: Proceedings 16th International Conference on Pattern Recognition, p. 40276. IEEE, Quebec (2002)
15. Das, G., Lin, K.I., Mannila, H., Renganathan, G., Smyth, P.: Rule discovery from time series. In: Proceedings of the Fourth International Conference on Knowledge Discovery and data mining, pp. 16–22. AAAI, California (1998)
16. Li, K., Ma, Z., Robinson, D., Ma, J.: Identification of typical building daily electricity usage profiles using Gaussian mixture model-based clustering and hierarchical clustering. Appl. Energy **231**, 331–342 (2018)

# Chapter 7
# Laboratory Tests of High-Performance Thermal Insulations

**Zsolt Kovács, Sándor Szanyi, István Budai and Ákos Lakatos**

**Abstract** These days the use of the traditional (conventional) insulation materials requires greater thicknesses (e.g., up to 25 cm) in order to reach significant reduction in the heat loss and in the emission of the greenhouse gases. In some cases the use of these thicknesses cannot be implemented. In the last decade, a new direction was designated in order to decrease the insulating thicknesses furthermore, to increase insulating capability. For this the era of the nano-technological/super/advanced insulation materials started. In this article, thermal conductivity measurements carried out on advanced insulation materials (vacuum ceramic hollow micro-spheres, vacuum insulation panels, and graphite expanded polystyrene) with Holometrix Lambda 2000 type heat flow meter will be presented. These measurements will be completed with scanning electron microscope tests, and in one case with an optical microscope test. The reached thermal conductivity values give promising applicability limit of the materials.

## 7.1 Introduction

In order to reduce the energy demand as well as the emission of green houses gases of buildings thermal insulations materials are used [1–3]. Beside the use of commercial insulation materials (e.g., mineral wool and pure expanded polystyrene), the application of the advanced insulation materials is widespread, too. Building construction is one of the sectors that would highly benefit from high performance insulation materials with enhanced performances. These days the two mostly men-

Z. Kovács
INOX-THERM Kft, Honvéd u. 8. 1. em 2, Budapest 1054, Hungary

S. Szanyi · Á. Lakatos (✉)
Faculty of Engineering, Department of Building Services and Building Engineering, University of Debrecen, Ótemető str. 2-4, Debrecen 4028, Hungary
e-mail: alakatos@eng.unideb.hu

I. Budai
Faculty of Engineering, Department of Engineering Management and Enterprise, University of Debrecen, Debrecen, Hungary



J. Littlewood et al. (eds.), *Sustainability in Energy and Buildings*, Smart Innovation, Systems and Technologies 163, https://doi.org/10.1007/978-981-32-9868-2_7

tioned advanced/high-performance/super insulation materials are the aerogel and the vacuum insulation panels. In the last two decades' great effort was made in order to find the best insulation materials, with lower thermal conductivity (<0.024 W/mK) and with acceptable production cost [4–8]. It was showed in recent papers that aerogel has very low thermal conductivity (<0.018 W/mK) [9–14]. Vacuum Insulation panels (VIP) are also said to be a good insulation material. Newly produced panels should have center of panel thermal conductivity with about 0.004 W/mK. However, its aging (loss of vacuum, in gassing) is their handicap. Furthermore, they are very damageable [15–20]. Moreover, the vacuum contaminant hollow micro-spheres as well as the graphite expanded polystyrene should be taken into account as nanotechnological insulation [7, 21–23]. The main research aim of this paper is to present thermal conductivity measurements results on the advanced insulation materials. Furthermore, we would like to show comparisons between the structures and thermal conductivities.

## 7.2 Materials and Methods

### *7.2.1 The Applied Theory*

As it was previously presented both either from the authors or the researchers all over the world the thermal conductivity of a porous material can be given with as the following:

$$\lambda_T = \lambda_{c,g} + \lambda_r + \lambda_{c,s} + \lambda_{conv} + \lambda_{coupling} + \lambda_{leak} \tag{7.1}$$

where ($\lambda_{c,g}$), ($\lambda_r$), ($\lambda_{c,s}$) and ($\lambda_{conv}$) are the conductive part of the gas filling, and the solid material, the radiation part and the convective part of the gas filling (if there is enough place). If the density of the solid skeleton is increasing the ($\lambda_{conv}$) will decrease, resulting lower effective thermal conductivity. Moreover the $\lambda_{coupling}$ and $\lambda_{leak}$ can be defined as the followings: $\lambda_{coupling}$ = thermal conductivity term accounting for second order effects between the various thermal conductivities in Eq. (7.1), $\lambda_{leak}$ = leakage thermal conductivity. In order to reach a thermal conductivity as low as possible, each of the above thermal contributions have to be minimized. Normally, the leakage thermal conductivity $\lambda_{leak}$, representing an air and moisture leakage driven by a pressure difference, is not considered as insulation materials and solutions are supposed to be without any holes enabling such a thermal leakage transport. The coupling term $\lambda_{coupling}$ can be included to account for second order effects between the various thermal conductivities in Eq. (7.1). This coupling effect can be quite complex and will be neglected in the rest of this article. This $\lambda_T$ is called to total (effective) thermal conductivity [4, 5, 20–22]. Once the convective part of the gas, can be reduced by decreasing the cell size or evacuating the gas. While the gas conductive par can be reduced properly to the Knudsen effect. The heat conduction of the gas happens by microscopic diffusion and collision of particles or quasi-particles (molecule, phonon, electron, etc.). When the linear dimensions of the medium are comparable to the mean free path of the particles, the heat transport from particle to particle is inhibited. The particle will collide with the pore wall with increasing probability and the heat conduction due to intermolecular collision will be less and less probable as the pore size become smaller.

$$\lambda_{(gas+gas/solid)} = \lambda_{(gas,0)}/(1 + 2\beta\,Kn) \tag{7.2}$$

In Eq. 7.2 $\lambda_{gas,0}$ is the thermal conductivity of bulk air at ambient conditions. The Knudsen effect is included in this model though the Knudsen number, Kn = l/d, with l being the gas molecule mean free path and d being the dimension where the gaseous transport takes place. If the pore size within the material is decreased below a certain level, i.e., a pore diameter of the order of 40 nm or below for air the gas thermal conductivity, and thereby also the above-mentioned total thermal conductivity, becomes very low (<0.004 W/mK) [21, 22].

The radiative heat transfer is due to the emittance of electromagnetic radiation in the infrared wavelength region from a material surface. It can be manipulated by the addition of an opacifier or reflective agent. The opacifier will lower the conductivity by increasing the absorption coefficient, according to the Stefan–Boltzmann equation,

$$\lambda_r = (16n^2\sigma T^3)/(3\alpha_r) \tag{7.3}$$

where n is the dimensionless refractive index, σ [W/m$^2$K$^4$] is the Stefan–Boltzmann constant, T [K] is the mean temperature of the materials and $\alpha_r$ is the Rosseland mean absorption coefficient. The Stefan–Boltzmann equation does not consider near-field radiation effects [21, 22].

### *7.2.2 Thermal Conductivity Measurements*

The thermal conductivity measurements were executed after drying the samples in a VentiCell drying instrument to changeless weight at 70 °C. With this device materials can be dried setting different air temperatures (up to 523 K). It works with hot air circulation using an inbuilt ventilator, after the drying of the samples were conditioned at 30 °C. For measuring the thermal conductivity of each polystyrene samples, Lambda 2000 Heat flow meter (HFM) was used. On this equipment is designed to determine the thermal conductivity of insulation materials in accordance with standard ASTM C518 and ISO 8301 protocols. Three different samples from each type were placed in the test section between two plates which are maintained at different temperatures ($T_1$ = 42 °C and $T_2$ = 22 °C, with $T_{mean}$ = 30 °C) during the test. After achieving thermal equilibrium and establishing a uniform temperature gradient throughout the sample, thermal conductivity is determined. To determine the thermal conductivity of a sample, five independent measurements were carried out. The thermal conductivity of analysed material was the mean value of the five measured results. The details of the measurements are clearly written in previous papers [4, 9, 10].

### 7.2.3 *Visualization with Scanning Electron Microscope*

The materials were analysed using a Hitachi TM-3030 scanning electron microscope (SEM) with Bruker Quantax 70 EDS system. The TM3030 is equipped with premium signal detectors which have been incorporated in FE-SEM and VP-SEM and provide unparalleled image quality. The detectors can be effectively operated under low-vacuum conditions and can support SE and BSE image observation without metal coating. The TM3030 can be used to observe BSE images with magnification range: 15–30,000x. Element mapping with X-ray analysis with a wide detection area (30 $mm^2$) was executed on the samples, too.

### 7.2.4 *The Tested High-Performance Insulation Materials*

**Vacuum Insulation Panels**

Different types of Vacuum Insulation Panels and their properties and their aging are well introduced in the literature. They are said to be a very promising insulation materials, with very low center of panel thermal conductivity (~0.004 W/mK). However, they are very vulnerable and their applicability is limited.

In this paper, two different types of sandwich (encapsulated) panels were tested. One of them was with 10 mm thick core material and the other one was with 20 mm thick core material. The core material is packed in a laminate with about 90–110 μm thickness. We knew from the manufacturer, that the pressure inside was 0.5 mbar at the beginning. The main core material was fumed silica. The tested samples were the so-called encapsulated VIPs, with expanded polystyrene substrates. The type "A" is 10 mm thick expanded polystyrene (EPS) on one side of a foil packed 20 mm thick core material, type "B" was with 10 mm EPS on two sides of 10 mm vacuum panel. Three test specimens were used for the measurements from the different types and five tests were carried out on each. The results were the averages of the measured values (15 thermal conductivity value). Furthermore, one sandwich panel was unpacked and the core material was measured with atmospheric pressure inside.

**Ceramic hollow micro-spheres**

The properties and developments of silica micro-spheres were presented by others earlier [7, 21, 22]. These spheres are containing vacuum inside a solid silica shell, so their thermal conductivity must be very low. However, to be an applicable insulating a carrier material is necessary. Usually these powders are mixed with a paint or plaster and applied on walls or on the external surface of piping in 2–5 mm thickness. Furthermore, due to the high thermal conductivity differences among the contaminants (approximately two orders of magnitude) much higher thermal insulation capability will be measured. In this paper, a microsphere contaminated paint embedded into a glass-fiber net was tested with about 5 mm thickness.

**Graphite-enhanced polystyrene**
Graphite-expanded polystyrene is usually not mentioned as advanced or high performance insulation materials, however its manufacturing contains a nano-technological part. In order to reduce the thermal conductivity of the graphite enhanced polystyrene, graphite nanotubes or carbon particles as well as graphite flakes can be added to the polystyrene grains during the manufacturing procedure, since these materials have much lower thermal conductivity (0.03 W/mK) than the pure EPS. The reason for this is mentioned above Eq. 7.3. By adding graphite particles to the polystyrene melt, significant reduce in the thermal radiation can be reached. Graphite has very good infrared radiation absorbing. In this paper a conventional (available on the market) graphite enhanced polystyrene was tested with thickness of 4 cm and compressive strength 100 kPa [23].

## 7.3 Results and Discussion

### *7.3.1 The Images of the Scanning Electron Microscope*

In order to visualize the samples and their containing elements were put under Scanning Electron Microscope tests, while the surface of the graphite enhanced polystyrene was analysed by an optical microscope. The test results as images and element mappings can be found in Figs. 7.1, 7.2 and 7.3.

**Core material of Vacuum Insulation Panel**
From Fig. 7.1a, b one can see the structure of the fumed silica as well as the elements (Si, O and Fe contamination).

**The ceramic hollow micro-spheres**
From Fig. 7.3 one can see the inhomogeneity of the size of the spheres. The diameters vary between 10 and 100 μm. The contain elements are the Zn and the silicon with oxygen.

**Optical microscope view of the graphite enhanced polystyrene**
Due to the nature (cell size) of the polystyrene in this case here only an optical microscope test was carried out. From Fig. 7.3 one can see the cell sizes and distributions in ~20x magnifying. As it is observable the graphite polystyrene has open cell structure.

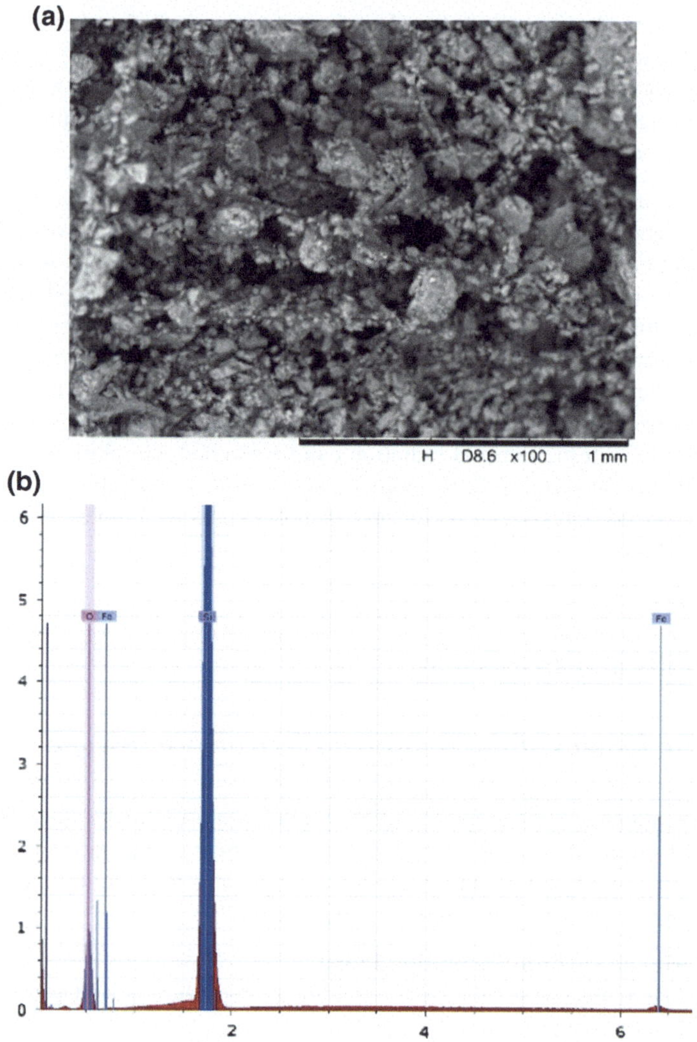


**Fig. 7.1** **a** The SEM image of the core material of the VIP, **b** the element map of the core material of the VIP

### *7.3.2 Thermal Conductivity Measurement Results*

In Table 7.1 the thermal conductivities measured with Holometrix Lambda 2000 equipment is given. In Table 7.1 the thermal conductivities of the graphite enhanced polystyrene as well as the silica hollow microsphere's can be found. The thermal conductivities of the vacuum insulation panels will be presented in Table 7.2.

In Table 7.2 the results of thermal conductivity measurements can be found carried out on vacuum insulation panels. Two different types of panels were tested. Two types

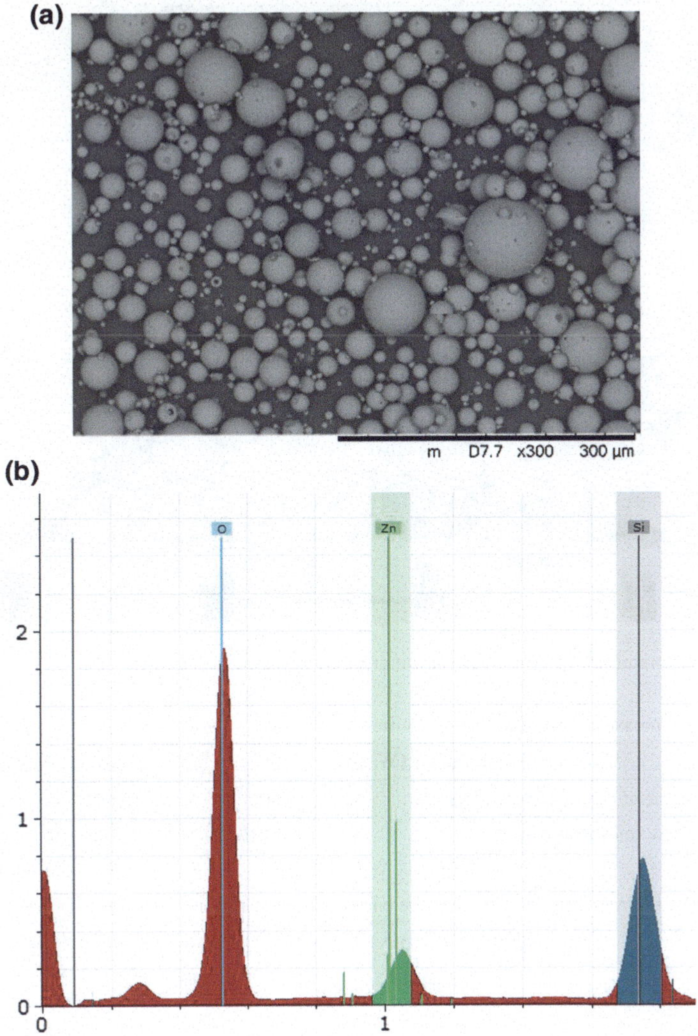


**Fig. 7.2** **a** The SEM image of the core material of the ceramic hollow micro-spheres, **b** the element map of the ceramic hollow micro-spheres

of the VIP's were tested as it was above mentioned ("A" and "B"). From the equivalent thermal conductivities center of panel thermal conductivities were calculated, too.

**The calculation method for the center of panel thermal conductivity**

In order to calculate the center of panel thermal conductivity of the vacuum insulation panel the following calculations should be used. First, from the total measured thermal conductivity of the sandwich panel the thermal conductivity of the EPS carrier should be subtracted by applying the total resistance method. Since, the equivalent thermal conductivity should be reached.

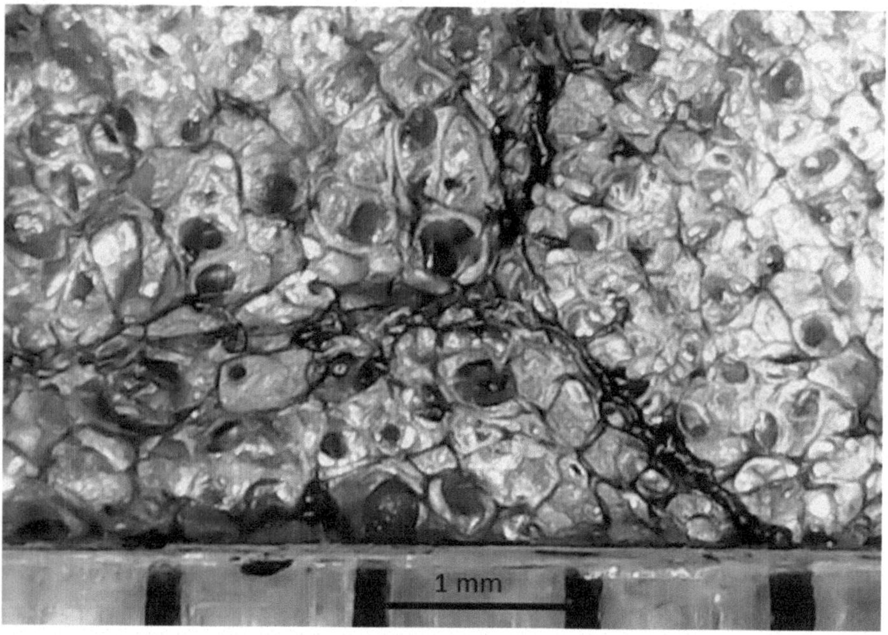


**Fig. 7.3** The optical microscope view of the graphite enhanced EPS (20x)

**Table 7.1** The measured thermal conductivities

| Materials | Thermal conductivity (W/mK) | St. dev. (W/mK) |
|---|---|---|
| Graphite enhanced expanded polystyrene | 0.037 | 5.20E−04 |
| Ceramic hollow microspheres | 0.016 | 1.78E−04 |

**Table 7.2** The thermal conductivities of the encapsulated vacuum insulation panels

| | Type "A" (W/mK) | Type "B" (W/mK) | Core material without layers (W/mK) |
|---|---|---|---|
| Equivalent total thermal conductivity of sandwich panels | 0.0144 | 0.018 | 0.0235 |
| Calculated thermal conductivity without EPS (effective thermal conductivity) | 0.0063 | 0.0089 | – |
| Calculated center of panel thermal conductivity | 0.0047 | 0.0055 | – |

$$\lambda_{eq} = \lambda_{COP} + \psi * d(P/A) \quad (7.4)$$

where $\psi$ is the linear thermal transmittance, P is the semiperimeter, A is the area and d is the thickness of the panel [24–26]. For $\psi$, 0.013 and 0.0127 W/mK were used if the VIPwas 1 cm thick or 2 cm thick respectively, here should be mentioned that thermal conductivity of the EPS layer was 0.035 W/mK [19, 27].

## 7.4 Conclusions

In this paper, thermal conductivity tests in combination with scanning electron microscopy or optical microscope images were presented on the so-called nano-technological/advanced/super insulation materials. Once, graphite enhanced polystyrene, ceramic hollow micro-sphere contaminated paint as well as vacuum insulation panels were tested. The results show promising application possibilities of these materials. It should be emphasized that the use of the vacuum insulation panels is reliable. We got the lowest thermal conductivity values for the VIP sandwich panels.

**Acknowledgements** The publication is supported by the EFOP-3.6.1-16-2016-00022 project. The project is co-financed by the European Union and the European Social Fund. The authors would say special thanks to Dr. Sc. Samuel Brunner from Empa–Swiss Federal Laboratories for Materials Science and Technology, for the consultations regarding the vacuum insulation panels.

## References

1. Kalmár, F.: Energy analysis of building thermal insulation. In: 11th Conference for Building Physics, 26–30 Sept 2002, pp. 103–112. Dresden, Germany (2002)
2. Schiavoni, S., D'Alessandro, F., Bianchi, F., Asdrubali, F.: Insulation materials for the building sector: a review and comparative analysis. Renew. Sustain. Energy Rev. **62**, 988–1011 (2016)
3. Lee, W.L., Yik, F.W.H.: Regulatory and voluntary approaches for enhancing building energy efficiency. Prog. Energy Combust. Sci. **30**, 477–499 (2004)
4. Lakatos, Á.: Thermophysical investigations of nanotechnological insulation materials. AIP Conf. Proc. **1866**, 030003 (2017)
5. Jelle, B.P., Gustavsen, A., Baetens, R.: The path to the high performance thermal building insulation materials and solutions of tomorrow. J. Build. Phys. **34**(2), 99–123 (2010)
6. Jelle, B.P.: Traditional, state-of-the-art and future thermal building insulation materials and solutions – properties, requirements and possibilities. Energy Build. **43**, 2549–2563 (2011)
7. Bozsaky, D.: Thermodynamic tests with nano-ceramic thermal insulation coatings. Pollack Period. **12**(1), 135–145 (2017)
8. Berge, A., Johansson, P.: Literature Review of High Performance Thermal Insulation. Chalmers University of Technology, Sweden (2012)
9. Lakatos, Á.: Investigation of the moisture induced degradation of the thermal properties of aerogel blankets: measurements, calculations, simulations. Energy Build. **139**, 506–516 (2017)
10. Lakatos, Á.: Stability investigations of the thermal insulating performance of aerogel blanket. Energy Build. **185**, 103–111 (2019)

11. Aegerter, M., Leventis, N., Koebel, M. (eds.): Aerogels Handbook, pp. 537–564. Springer, New York (2011)
12. Baldinelli, G., Bianchi, F., endelis, S., Jakovics, A., Morini, G.L., Falcioni, S., Fantucci, S., Serra, V., Navacerrada, M.A., Díaz, C., Libbra, A., Muscio, A., Asdrubali, F.: Thermal conductivity measurement of insulating innovative building materials by hot plate and heat flow meter devices: a round Robin test. Int. J. Therm. Sci. **139**, 25–35 (2019)
13. Stahl, Th, Brunner, S., Zimmermann, M., Wakili, K.G.: Thermo-hygric properties of a newly developed aerogel based insulation rendering for both exterior and interior applications. Energy Build. **44**, 114–117 (2012)
14. Cuce, E., Cuce, P.M., Wood, C.J., Riffat, S.B.: Toward aerogel based thermal superinsulation in buildings: a comprehensive review. Renew. Sustain. Energy Rev. **34**, 273–299 (2014)
15. Baetens, R., Jelle, B.P., Thue, J.V., Tenpierik, M.J., Grynning, S., Uvslokk, S., Gustavsen, A.: Vacuum insulation panels for building applications: a review and beyond. Energy Build. **42**(2), 147–172 (2010)
16. Pons, E., Yrieix, B., Brunner, S.: Evaluation of VIPs after mild artificial aging during 10 years: focus on the core behavior. Energy Build. **162**, 198–207 (2018)
17. Alam, M., Singh, H., Brunner, S., Naziris, C.: Experimental characterisation and evaluation of the thermo-physical properties of expanded perlite—Fumed silica composite for effective vacuum insulation panel (VIP) core. Energy Build. **69**, 442–450 (2014)
18. Brunner, S., Ghazi Wakili, K., Stahl, T., Binder, B.: Vacuum insulation panels for building applications—continuous challenges and developments. Energy Build. **85**, 592–596 (2014)
19. Brunner, S., Erb, M.: Vacuum insulation for building applications product declaration and design values. Technical Report (2012). https://doi.org/10.13140/2.1.1171.8720
20. Lakatos, Á.: Thermal conductivity of insulations approached from a new aspect. J. Therm. Anal. Calorim. **133**(1), 329–335 (2018)
21. Liao, Y., Wu, X., Liu, H., Chen, Y.: Thermal conductivity of powder silica hollow spheres. Thermochim. Acta **526**(1–2), 178–184 (2011)
22. Jelle, B.P., Tilset, B.G., Gao, T., Grandcloas, M., Lovvik, O.M., Bohne, R.A., Mofid, A.S., Ng, S., Sagvolden, E.: High-performance nano insulation materials for energy-efficient buildings. In: Proceedings of TechConnect World Innovation Conference 2017, pp. 289–292, Washington DC, USA, 14–17 May 2017. At: Washington DC, USA (2017)
23. Lakatos, Á., Deák, I., Berardi, U.: Thermal characterization of different graphite polystyrene. Int. Rev. Appl. Sci. Eng. **9**(2), 163–168 (2018)
24. Wakili, K.G., Bundi, R., Binder, B.: Effective thermal conductivity of vacuum insulation panels. Build. Res. Inf. **32**(4), 293–299 (2004)
25. Wakili, K.G., Stahl, Th, Brunner, S.: Effective thermal conductivity of a staggered double layer of vacuum insulation panels. Energy Build. **43**(6), 1241–1246 (2011)
26. Caps, R., Wallaschek, I., Abu-Shawriyeh, I., Brunner, S.: Methods for evaluation of thermal conductivity increase of VIPs. In: 12th International Vacuum Insulation Symposium on IVIS 2015, Nanjing, China, 19–21 Sept 2015. NUAA (2015)
27. Isaia, F., Fantucci, E.S., Capozzoli, A.A., Perino, M.: Thermal bridges in vacuum insulation panels at building scale. In: Proceedings of the Institution of Civil Engineers—Engineering Sustainability, vol. 170, 1 Feb 2017, pp. 47–60. Themed issue on sustainability in energy and buildings—Part 1

# Chapter 8
# A Conceptual Methodology for Estimating Embodied Carbon Emissions of Buildings in Sri Lanka

**Amalka Nawarathna, Zaid Alwan, Barry Gledson and Nirodha Fernando**

**Abstract** The reduction of whole building life cycle carbon requires equal attention on both operational carbon (OC) and embodied carbon (EC). Regardless of that, the Sri Lankan building sector concentrates mainly on reducing OC. The importance of reducing EC is yet to be fully realised. Therefore, there an opportunity exists to encourage research community, practitioners and policy makers associated with Sri Lankan building sector regarding EC reduction. The main drive towards EC reduction is estimation or assessment, as without this, it is impossible to realise the extent to how much is needed to be reduced. Even though, EC estimation process has been well developed and documented over the last years, it is yet a challenging process to Sri Lanka due to many difficulties such as difficulty in setting up a suitable estimation scope, unavailability of an up to date country-specific EC coefficient database for building materials and time consuming and work intensive nature of the estimation procedure due to unavailability of a suitable estimation tool. To deal with these challenges, initially it requires to establish a conceptual EC estimation methodology appropriate to the Sri Lankan building context. Accordingly, this study developed an EC estimation methodology based on the Life Cycle Assessment (LCA) approach; setting up the scope, performing EC coefficient inventory analysis and undertaking EC estimation and interpretation. Subsequently, the developed methodology was verified in terms of its applicability using a proposed commercial building in Sri Lanka. As a further study, it is intended to develop a statistical tool to estimate EC during design stage of commercial buildings in Sri Lanka conforming to this methodology.

A. Nawarathna (✉) · Z. Alwan · B. Gledson
Northumbria University, Newcastle upon Tyne, UK
e-mail: amalka.ranathungage@northumbria.ac.uk

N. Fernando
University of Salford, Salford, UK



J. Littlewood et al. (eds.), *Sustainability in Energy and Buildings*, Smart Innovation, Systems and Technologies 163, https://doi.org/10.1007/978-981-32-9868-2_8

## 8.1 Introduction

The building sector, being responsible for about 28% of global carbon emissions, has been recognised as a major culprit for the high presence of atmospheric carbon [1]. Therefore, buildings represent a critical piece of global carbon reduction targets. The whole life cycle carbon of a building is made up of OC and EC [2]. Until recently, a priority had been given to OC reduction while embodied carbon was considered to be proportionally insignificant [3]. As a result, buildings became more efficient in terms of OC, but enabling EC to gain a larger proportion of whole building life cycle carbon emissions. On account of that, there is now increasing interest around embodied energy (EE) and EC reduction among scholars and various actors related to building and construction industry in developed countries [4]. However, developing countries still do largely focus on OC reduction regardless of fast growing building activities [5].

Rapid economic development has increased building activities in developing countries [6]. As a result, these substantial developments contribute to global building carbon emissions in considerable amounts. Therefore, equal emphasis on both OC and EC reduction is much desirable in developing countries too, in order to minimise whole building carbon emissions.

Sri Lanka is one such developing country where more focus is yet being giving on OC reduction regardless of rapid development of its building sector. Therefore, there is a necessity of encouraging Sri Lankan building sector on EC reduction. The main drive towards carbon emission reduction is estimation or quantification [7]. Even though, EC estimation process has been well developed and documented owing to continuous research and development in this area over the last years, EC estimation is yet a challenging process as it encountered many difficulties. Unavailability of an up-to-date EC coefficient data base for building materials, difficulty in choosing an estimation scope and time consuming and work intensive nature of estimation due to unavailability of an appropriate estimation tool are the main challenges [8]. Therefore, the promotion of EC estimation and reduction in Sri Lankan building sector requires addressing these challenges at first.

Accordingly, this work intended to establish a conceptual methodology so as to better estimate EC of buildings in Sri Lanka by setting up a suitable scope and applying a temporary solution for the unavailability of an up-to-date and country-specific EC coefficient database. This methodology was developed based on the LCA approach and complying to existing EC estimation processes. The developed mechanism was applied in a real- life case to verify its applicability as well.

Eventually, it is intended the proposed methodology to be used in developing a statistical tool to estimate EC during design stage of buildings. While providing the solution for time consuming and work intensive nature of estimation process, it will support to identify the EC reduction opportunities and possible solutions at the earliest possible stage.

## 8.2 EC Estimation

EC is defined as the carbon emissions associated with energy consumption (embodied energy) and chemical processes during the extraction, manufacture, transportation, assembly, replacement and deconstruction of construction materials or products while OC is associated with energy consumption for building heating, cooling, lighting and for other electrical and electronic devices use in the building [2].

Unlike operational carbon which emits only during operational phase of a building, EC emits over all the phases of a building, starting from product stage to disposal stage [9]. Therefore, EC estimation is a quite complex and time-consuming process.

Few standards and guidelines have already been developed to establish a standard EC estimation process. For instances, BS EN 15978:2011- CEN/TR- Sustainability of construction works-Assessment of environmental performance of buildings—Calculation method (2011) and professional guidance developed by [10, 11, 12, 13] can be shown. However, EC estimation of a whole building life cycle is still challenging for most of the countries due to limited available and accessible data. Therefore, depending on the data availability, studies related to EC estimation have used different system boundaries [9].

EC estimation can be conducted at any stage of a building development; early design, design development, detailed estimating, construction and in use [14]. However, unlike the OC, EC cannot be reversed. Once the building is constructed, the opportunity to reduce EC has almost gone [15]. Therefore, even with limited data, greater opportunities for EC reduction exist primarily during the early design or design development stage of a building's life [16].

## 8.3 Challenges for Estimating EC in Sri Lankan Buildings

With the recognition of building sector as a large contributor for the increasing carbon presence in the atmosphere, Sri Lanka has also initiated many carbon reduction measures in the building sector. A Code of Practice for energy efficient buildings developed by Sri Lanka Sustainability Authority [17] and a green building rating system (Green$^{SL}$) developed by the Green Building Council, Sri Lanka [18] are two examples for such initiatives. However, their primary focus is on increasing operational energy efficiency and thereby, reducing OC emissions. The EC share within a building's whole life cycle carbon has, to date, been overlooked by policy makers as well as the research community. Except for limited research, no standards, policies or measures have been developed to reduce EC of buildings in Sri Lanka. [19, 8] have conducted two studies to assess the EC content of carefully selected buildings. They were confined to two different estimation procedures based on the availability and accessibility for data. In addition, [20] has carried out work to assess EE and EC content of widely used building materials in Sri Lanka. [21, 22] have also conducted spate investigations around selected building materials, but their contents have limited only to EE contents.

These circumstances prompted part of the present research team in a previous study [23], to identify what the major reasons for the slow uptake of EC estimation and reduction in Sri Lanka are. Similarly, the research team also attempted to recognise the possibility of estimating EC of Sri Lankan buildings. Both of these studies concluded that EC estimation is quite challenging in Sri Lanka, but it is not impossible if few challenges are addressed. These are (a) difficulty in setting up a suitable estimation scope, (b) unavailability of an up-to-date country-specific database for building material EC coefficients, and (c) the lack of open source EC estimation tools and models [8].

EC coefficients of building materials are the most important input of EC estimation process. Even though, there are many database/inventories developed worldwide, they are geographically specific. The only database which has been developed by Pooliyadda [20] to present the EE and EC coefficients of several widely used building materials in Sri Lanka is also not up to date. Therefore, there is a need of an up to date national database in Sri Lanka.

Even though number of commercial software tools have been developed to facilitate the EC estimation process such as Athena Eco Calculator and Athena Impact Estimator, Building for Environmental and Economic Sustainability (BEES), Environment Agency Carbon Calculator, and Global Emissions Model for integrated Systems (GEMIS), they are either country/regional specific or building type specific [23]. Therefore, there is a limited opportunity for adapting them to Sri Lankan context.

## 8.4 Conceptual Methodology

By establishing a conceptual methodology to estimate EC in Sri Lankan buildings, it intends to address the challenges of difficulty in setting up an applicable estimation scope and unavailability of an up to date country-specific EC coefficient database for building materials while taking the initiative to develop an EC estimation tool in future to address the challenge of time-consuming and work intensive estimation process. The development of the methodology followed the LCA approach defined in ISO 14040 Environmental Management - Life cycle assessment standard. It includes; (1) setting up the scope, (2) performing EC coefficient inventory analysis, (3) undertaking EC estimation and (4) accurately interpreting the results.

### 8.4.1 *Setting up the Scope*

The first step of the EC estimation involves defining the scope or the boundary levels. Figure 8.1 graphically illustrates the arrangement of the scope, with the following subsections which describe each boundary in brief.

**Estimation Stage**. Considering the fact that more EC reduction opportunities lie during the early stage of a building development, the estimation stage of this methodology is confined to the design development stage of a building. According to RIBA Plan of Work [24], the developed design stage is defined as Stage 3 of a building project. At this stage, building specification gets further developed when compared to preparation (Stage 1) and concept design stage (Stage 2) and it details of what products, systems and finishes to use [14]. Therefore, this would be an ideal stage of a building's life cycle to carry out an EC estimation and take necessary measures to reduce EC accordingly.

**Elements of a Building**. RICS [2] identifies the carbon intensive elements that need to be focused on when estimating EC. These are the: foundations; basement retaining walls; ground floor; frame; upper floors; roof; stairs and ramps; external walls; windows and external doors; internal walls and partitions; wall, floor and ceiling finishes; roads, paths and paving; external drainage and external services. Even though it is important to account all the elements and stages of a building life cycle to get a representative EC value of a building, many past studies have been generally confined only to carbon intensive elements or the carbon hotspots depending on the availability and accessibility of material data. Therefore, this methodology also bounds to carbon intensive elements which have been identified as the main substructure

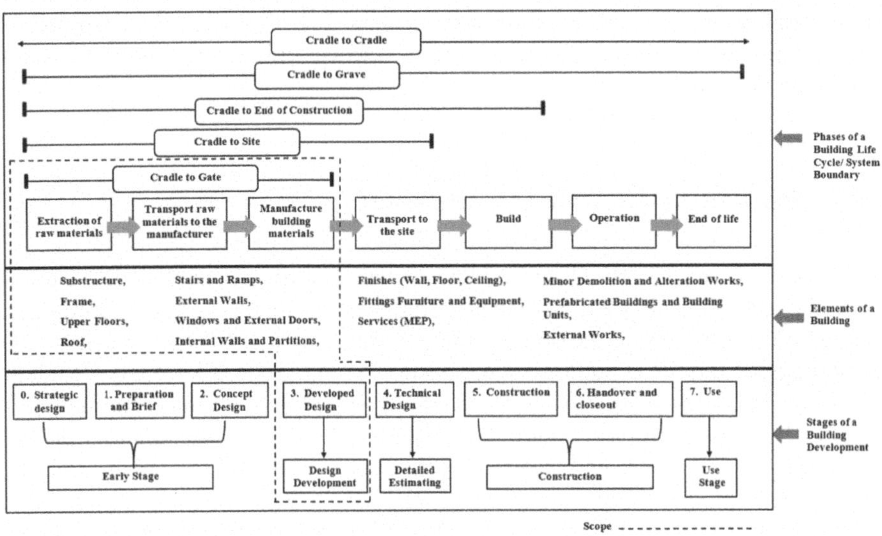


**Fig. 8.1** Setting up the scope

and superstructure elements of a building which consumes large amounts of materials [2]. They are; substructure; frame; upper floors; roof; stairs and ramps; external walls; windows and external doors; internal walls and partitions. These elements are defined in compliance with The Building Cost Information service (BCIS) element classification developed by RICS [25].

**Building Phase/System Boundary**. The estimation boundary of the building life cycle is limited to the product stage of a building, which is called the cradle to gate (raw material extraction to production of building materials) boundary level. This is mainly due to, (a) that the product stage is the second most significant area of carbon emissions from the entire life cycle of a building after the operational carbon emissions [26] and (b) the appropriate EC coefficient database which is used for this methodology has been limited to cradle to gate boundary level.

### 8.4.2 *Identifying EC Coefficients of Building Materials*

EC coefficients of building materials or products are the main input of EC estimation procedure. Therefore, the second step involves identifying suitable EC coefficients of building materials. As a solution for the unavailability of an up to date national EC coefficient database, the Inventory of Carbon and Energy (ICE) version 2.0, 2011 developed by [27] is used. Even though there are number of other databases available such as SimaPro 9.0, Athena LCI, Hutchins UK Building Black Book, EcoInvent 3.3 and AusLCI, they are either country or regional specific and most of them require license to use. ICE is chosen as it is free and open access database which has been drawn on data from around the globe. It specifically mentions that it can be used as 'proxy data' in the absence of country-specific data [27]. This would be a temporary solution until any national database develops, however, this can still provide useful analysis and conclusions.

### 8.4.3 *Estimating EC and Interpretation*

This phase involves EC estimation and interpretation. It was adopted the EC estimation process introduced by RICS [2] in preparation of this methodology. Accordingly, it includes (I) determining the constituent materials required for each element, (II) calculating the weights of building materials used per $m^2$ of each element, (III) identifying the EC coefficients of each material and apply them on material quantities to derive EC content of each material, (IV) adding EC of all materials to establish overall EC per element $m^2$ and multiplying them by total $m^2$ to derive elemental EC, (V) adding EC of all structural and envelope elements to establish the overall EC of a building skeleton. These steps are graphically illustrated with a brief description for each step in Fig. 8.2. The estimation results can be interpreted to provide the overall EC of a building skeleton in $kgCO_2$ or EC $kgCO_2/m^2$.

## 8.5 Case Study

Due to the current growing demand for commercial buildings in Sri Lanka, the applicability of the proposed methodology was verified against a proposed commercial building that is to be located in Delkanda, Colombo, Sri Lanka. Its main purpose is to provide rentable spaces for office and bank activities. It is a four story building in which its footprint covers 500 $m^2$ and extends vertically over 17.15 m with average floor height of 4 m. It provides a gross internal floor area (GIFA) of 1916 $m^2$.

### *8.5.1 Results*

Conforming to the EC estimation methodology explained in Sect. 8.4, EC estimation was carried out to this selected building. To make the data processing and estimating easier, a calculation sheet was developed as in Table 8.1. It clearly presents the elemental and total EC of the building. According to the results gained, it emits 400,639.97 $kgCo_2$ or 209 $kgCo_2/m_2$ GIFA of EC from its substructure and superstructure elements. The largest EC contributor of this building is walls (32%) which include both internal and external followed by substructure (22%), frame (17%), upper floors (16%), roof (8%), stairs, ramps and landings (3%) and windows and external doors (2%).

## 8.6 Discussion

The main purpose of proposing this conceptual EC estimation methodology is to address the prevailing challenges for estimating EC in Sri Lankan buildings as mentioned in the Sect. 8.3. Initially, the developed methodology attempted to deal with the challenges of difficulty in setting up a suitable scope and unavailability of an up-to-date country-specifc EC coefficient database. Later, it is expected this methodology to be used by the present research team to develop a statistical tool to estimate EC during design stage of commercial buildings in Sri Lanka.

Further, this will facilitate to encourage relevant personnels to incorporate EC reduction into decision-making processes to reduce environmental impacts. By developing comprehensive or peer-reviewed datasets covering EC emissions associated with different building types conforming to this methodology, an industrial benchmark can be drawn in order to better inform the decision-making process in terms of the EC performance of a building. Several aspects of EE and EC have already been

**I**

**Determine the constituent materials required for each element**

The first step is to identify all the materials required for each element separately. This can be carried out referring to building specifications.

Ex: External Wall : Bricks, Mortar

**II**

**Calculate weights of building materials per 1m$^2$ of each element**

The second step is to calculate the gross mass (kg) of each material per 1m$^2$ of each element.

Ex:
Bricks: Quantity of bricks per m$^2$ (nr) x mass of a single brick (kg) = Mass per m$^2$ (kg)
Mortar: Volume of mortar per m$^2$ (m$^3$) x Density (kg/m3)= Mass per m$^2$ (kg)

**III**

**Identify the EC coefficients of each material and apply them on material quantities to derive EC content of each material**

In order to derive EC content of each material quantity, this step involves multiplying EC coefficient of materials by material quantities.EC coefficients can be identified as mentioned in Section 4.2

Ex: Bricks: Mass per m$^2$ (kg) x EC coefficient (kg$CO_2$/kg) = EC (kg$CO_2$)
Mortar: Mass per m$^2$ (kg) x EC coefficient (kg$CO_2$/kg)) = EC (kg$CO_2$)

**IV**

**Add EC of all materials to establish overall EC per 1m$^2$ of element and multiply by total m$^2$ to derive elemental EC**

As the fourth step, elemental EC value is derived by multiplying EC per 1m$^2$ of element by total m$^2$ of element.

Ex:
EC of bricks + EC of Mortar = EC per 1m$^2$ of external wall
EC per 1m$^2$ of external wall x Total m$^2$ of external walls = EC of External Walls

**V**

**Add EC of all selected carbon intensive elements to establish the overall EC of a building skeleton**

The last step is to add all the elemental EC together to derive overall EC of building skeleton.

Ex:
EC of substructure + EC of frame + EC of upper floors + EC of walls (external and Internal) + EC of roof + EC of stairs and ramps + EC of windows and external doors = Overall EC of a building skeleton

**Fig. 8.2** Estimation process *Source* [2]

**Table 8.1** Results of the estimation

| Element | Includes | Main building materials | | Quantity | Unit | Mass (kg) | EC factor (Kg $CO_2$/kg) | EC (kg $CO_2$) | Sub total |
|---|---|---|---|---|---|---|---|---|---|
| Substructure | All work below underside of screed, or underside of lowest floor finishes including DPM, excavtaions, foundations | Insitu- mass concrete (Grade 15) | | 19.25 | $m^3$ | 46961 | 0.093 | 4367.373 | **89,847.67** |
| | | Insitu- reinforced concrete (Grade 20) | Concrete steel bar | 65 | $m^3$ | 149028 | 0.155 | 23099.34 | |
| | | | | 5 | t | 5000 | | | |
| | | Insitu- reinforced concrete (Grade 30) | Concrete steel bar | 183 | $m^3$ | 389420 | 0.158 | 61528.36 | |
| | | | | 11.8 | t | 11800 | | | |
| | | Width > 300 mm damp proof membrane- Polythene (0.5 mm thk) | | 406 | $m^2$ | 203 | 4.2 | 852.6 | |
| Frame | Load bearing framwork (beams and columns), main floor and roof beams, ties and roof trusses of framed buildings, stanchion and beam casings for structural or protective purpose | Insitu- reinforced concrete (Grade 25) | Concrete steel bar | 103 | $m^3$ | 259982 | 0.252 | 68035.46 | **68.035.46** |
| | | | | 21 | t | 21000 | | | |
| Upper Floors | Upper floors, service floors, meazzanines, structural screeds, suspended floors over basements, balconies | Water proof —Asphalt-1" | | 128.5 | $m^2$ | 6900 | 0.059 | 407.1 | |
| | | Insitu- reinforced concrete (Grade 25) | Concrete steel bar | 147 | $m^3$ | 385313.5 | 0.164 | 63191.41 | **63,598.51** |
| | | | | 12 | t | | | | |

(continued)

**Table 8.1** (continued)

| Element | Includes | Main building materials | | Quantity | Unit | Mass (kg) | EC factor (Kg $CO_2$/kg) | EC (kg $CO_2$) | Sub total |
|---|---|---|---|---|---|---|---|---|---|
| Roof | Roof structure (beams and trusses in unframed buildings) including eaves and verges, plates and ceiling joists, gable ends: internal walls and chimneys above plate level, parapet walls and balustrades Roof coverings. Specialist roof systems. Roof drainage. Roof lights, skvlighrs, and openings, Roof features | Insitu- Reinforced concrete (Grade 25) | Concrete steel bar | 77 | $m^3$ | 201830.9 | 0.161 | 32494.77 | **33,847.63** |
| | | | | 4 | t | 4000 | | | |
| | | water proof —Asphalt- 1” | | 427 | $m^2$ | 22929.9 | 0.059 | 1352.864 | |
| Walls (external and internal) | External enclosing walls above ground level, External enclosing walls below ground level, Solar/Rain screening, External Sofits, subsidiary walls, balustrades and proprietary balconies, Façade access/cleaning systems Internal walls, partitions, balustardes, room dividers and cublicles | Partition wall | A1 frame-(3” × 1.5” × 1/8”) | 98 | m | 180.32 | 8.24 | 1485.837 | **126,830.24** |
| | | | Aluminium sheet (2 mm thk) | 74 | $m^2$ | 200.54 | 8.24 | 1652.45 | |

(continued)

**Table 8.1** (continued)

| Element | Includes | Main building materials | | Quantity | Unit | Mass (kg) | EC factor (Kg $Co_2$/kg) | EC (kg $Co_2$) | Sub total |
|---|---|---|---|---|---|---|---|---|---|
| | | Brick work | bricks 112.5 mm thick | 246 | $m^2$ | 32472 | 0.23 | 7468.56 | |
| | | | bricks 225 mm thick | 1387 | $m^2$ | 366168 | 0.23 | 84218.64 | |
| | | | mortar (1:5) | 145141 | kg | 169402 | 0.146 | 24732.69 | |
| | | Metal canopy | A1 frame- flat bar (3" 1.5") | 53 | m | 417.11 | 8.24 | 3436.986 | |
| | | | Polycarbonate (10 mm thk) | 53 | $m^2$ | 636 | 6.03 | 3835.08 | |
| Stairs and Ramps | Stair/Ramp structures, finishes, balustardes and handrails, ladders/chutes/slides | Insitu- reinforced concrete (Grade 25) | Concrete steel bar | 17 | $m_3$ | 44560 | 0.195 | 8689.2 | **12,618.43** |
| | | | | 2.1 | t | 2100 | | | |
| | | non slippery ceramic tiles (0.03 × 0.03 m) | | 120 | $m^2$ | 1008 | 0.74 | 745.92 | |
| | | Tile grout | | 2.4 | $m^3$ | 5189 | 0.127 | 659.003 | |
| | | Timber (1.25" × 1.25")- hard wood | | 0.4838 | $m^3$ | 348.723 | 0.86 | 299.9018 | |
| | | Stainless steel (25 mm dia) | | 315 | m | 1222.2 | 1.82 | 2224.404 | |
| Windows and External Doors | Windows including frames, linings, trims, ironmongery, glazing, lintels, dpcs and reveals | A1 frame (1.5" × 1" × 1/8") | | 355.5 | m | 351.945 | 8.24 | 2900.027 | **5,862.02** |
| | | Tempered glass (6 mm thk) | | 119.5 | $m^2$ | 1792.5 | 1.27 | 2276.475 | |
| | External doors, fanlights, frames, linings, trims, ironmongery, glazing, lintels, thresholds, dpcs and reveals | Al (1.5"x l" × 1/8") | | 42 | m | 41.58 | 8.24 | 342.6192 | |
| | | Tempered glass (6 mm thk) | | 18 | $m^2$ | 270 | 1.27 | 342.9 | |
| **Total** | | | | | | | | | **400,639.97** |

started to be incorporated to various decision-making processes related to buildings in some countries. Some Local Authorities, including for instance, Brighton and Hove City Council have already included mandatory EC estimations as part of the building planning process. And also some of the green building rating systems such as BREEAM, LEED and Green Star have already recognised EC measurement and its mitigation as part of minimising building life cycle carbon impacts [9]. Therefore, EC figures obtained using this mechanism can be used to inform such decision-making processes.

## 8.7 Conclusions

EC estimation and reduction is yet a challenging process to Sri Lankan building sector due to many difficulties such as difficulty in setting up a suitable scope, unavailability of an up to date country-specific EC coefficient database and time-consuming nature and work intensive nature of estimation process. Therefore, with the intention of addressing these challenges, this work proposed a conceptual EC estimation methodology appropriate to Sri Lankan building sector. It was developed complying to LCA approach. Depending on the data availability, the scope of the methodology covered cradle to gate as the system boundary, design stage as the estimation stage and substructure and superstructure elements of a building as the elements focused for estimation. EC coefficients of the building materials were extracted from the Inventory of Carbon and Energy (ICE) version 2.0, as a temporary solution for the unavailability of an up-to-date country-specific database. In preparation of EC estimation process, it was adopted the estimation process introduced by RICS. Eventually, the developed methodology was verified for its applicability using a commercial building to be located in Sri Lanka. The results of the estimation indicated that the building emits 400,639.97 $kgCo_2$ or 209 $kgCo_2/m^2$ GIFA of EC from its substructure and superstructure elements. In addition to addressing the prevailing challenges of EC estimation procedure in Sri Lankan buildings, this methodology would encourage building related decision makers and other relevant personnel to incorporate EC reduction into decision-making processes in order to reduce environmental impacts.

## References

1. International Eneregy Agency: Buildings—Tracking Clean Energy Progress. https://www.iea.org/tcep/buildings/
2. RICS: Methodology to calculate embodied carbon of materials—RICS Information Paper (2012). https://doi.org/10.1007/978-94-007-4444-8_4
3. Ibn-Mohammed, T., Greenough, R., Taylor, S., Ozawa-Meida, L., Acquaye, A.: Operational versus embodied emissions in buildings—a review of current trends. Energy Build. **66**, 232–245 (2013). https://doi.org/10.1016/j.enbuild.2013.07.026

4. Nebel, B., Alcorn, A., Wittstock, B.: Life cycle assessment: adopting and adapting overseas LCA data and methodologies for building materials in New Zealand (2009)
5. Ruuska, A.: Role of embodied energy, operational energy and related greenhouse gas emission of buildings in the context of developing tropical countries. In: SB13 Singapore-Realising Sustainability in Tropics, pp. 205–211 (2013)
6. Lucon, O., Ürge-Vorsatz, D., Ahmed, A.Z., Akbari, H., Bertoldi, P.: Buildings (2014)
7. Institute, B.S.: BS EN 15978:2011—Sustainability of construction works—Assessment of environmental performance of buildings—Calculation method. (2011)
8. Nawarthna, A., Alwan, Z., Fernando, N., Gledson, B.: Estimating embodied carbon emissions of buildings in developing countries: a case study from Sri lanka. In: Fourth International SEEDS Conference, pp. 821–831 (2018)
9. International Eneregy Agency: IEA EBC ANNEX 57—Overview of Annex 57 Results (2016)
10. Lockie, S., Berebecki, P., Davies, J., London Paul Dunne, D., Fiske, J., MacDonald, M., Matt Fulford, R., Markus Gaebel, S., Green, A., Steve Hadden, G., Hitchin, R., Craig Jones, C., Ecology Stefaan Martel, C., Jeff Maxted, B., Newman, G., Nsibande, C., Properties Jerry Percy, B., Andy Powell, G., Ralph, P., Langdon Martin Russell-Croucher, D., Nigel Sagar, R., Matthew Saunders, S., Michael Smithing, R., Steele, K., Anna Surgenor, A., Eddy Taylor, U., Graham Watts, R., David Weight, C., Langdon Jim Wiltshire, D.: RICS Professional Guidance, Global Methodology to calculate embodied carbon Alliance for Sustainable Building Products Acknowledgments Methodology to calculate embodied carbon RICS guidance note, Global 1st edn. Green Building Certification (2014)
11. Sturgis, S.: Embodied and whole life carbon assessment for architects (2018). https://doi.org/10.1097/01.prs.0000244906.48448.5d
12. AIA Guide to Building Life Cycle Assessment in Practice Authorship and Acknowledgements (2010)
13. UKGBC: Practical how-to guide: Measuring Embodied Carbon on a Project Background to BRE & UK Green Building Council. (2016)
14. Ashworth, A.P.: S: Cost Studies of Buildings. Routledge, Oxon (20107)
15. Jones, C.: No Title. http://www.betterbuildingspartnership.co.uk/impact-embodied-carbon-real-estate-cls-holdings-plc-explore-issue
16. Low-Carbon Construction Innovation and Growth Team: Low Carbon Construction Executive Summary (2010)
17. Sri Lanka Sustainable Energy Authority: Code of practice for energy efficient buildings in Sri Lanka-2008 (2008)
18. GBCSL: Green Sl ® Rating System for Built Environment (2015)
19. Kumanayake, R., Luo, H., Paulusz, N.: Assessment of material related embodied carbon of an office building in Sri Lanka. Energy Build. **166**, 250–257 (2018). https://doi.org/10.1016/j.enbuild.2018.01.065
20. Pooliyadda, S.P.: Energy Content and Carbon Emission Audit of Building Materials (2000)
21. Jayasinghe, C.: Embodied energy of alternative building materials and their impact on life cycle cost parameters. In: Icsecm, pp. 1–20 (2011)
22. Udawattha, C., Halwatura, R.: Embodied energy of mud concrete block (MCB) versus brick and cement blocks. Energy Build. **126**, 28–35 (2016). https://doi.org/10.1016/j.enbuild.2016.04.059
23. Nawarathna, A., Fernando, N., Alwan, Z.: Reasons for the slow uptake of embodied carbon estimation in the Sri Lankan building sector.**12**, 102–107 (2018)
24. RIBA: RIBA Plan of Work 2013 Overview (2013)
25. Surveyors, R.I. of C.: Elemental Standard Form of Cost Analysis Principles, Instructions, Elements and Definitions 4th (NRM) Edition. BCIS (2012)
26. Government, H.M.: Low Carbon Construction (2010)
27. Hammond, G., Jones, C.: Inventory of Carbon and Energy (ICE) version 1.5, p. 136 (2011). https://doi.org/10.1680/ener.2011.164.4.206

# Chapter 9
# Suitability of Passivhaus Design for Housing Projects in Colombia

**Vincenzo Costanzo, J. E. Carrillo Gómez, Gianpiero Evola and Luigi Marletta**

**Abstract** The aim of this investigation is to study the applicability of the Passivhaus standard to typical housing projects in the Colombian market. Dynamic simulations run in IESVE for a detached house and an apartment block revealed that under the business-as-usual construction scenario slightly cold indoor conditions are perceived in the cold climate of Bogotá ($-0.8 \leq \text{PMV} \leq 0.15$), while pronounced overheating ($1.45 \leq \text{PMV} \leq 1.95$) is predicted in the hot climate of Barranquilla. The implementation of typical passive design measures significantly improves indoor conditions in the cold climate of Bogotá, making these housing designs fulfil the Passivhaus energy requirements. On the other hand, in the hot climate of Barranquilla, these measures are not sufficient to meet the standard prescriptions, and thus further passive strategies should be sought along with the use of highly efficient cooling systems.

## 9.1 Introduction

Since its birth in Germany in the late 1980s, the application of the Passivhaus standard has focused mainly on new constructions located in the cold climates of Central and Northern Europe [1]. The main features promoted by the standard were the envelope superinsulation, the airtightness and the use of mechanical ventilation systems with heat recovery. These technological trademarks are deemed fundamental to achieve energy needs for space heating lower than 15 kWh $m^{-2}$ $y^{-1}$ and primary energy consumption (including heating, domestic hot water and electrical appliances) below 120 kWh·$m^{-2}$ $y^{-1}$. These targets, together with low air infiltration rates (below 0.6 $h^{-1}$ at a pressure difference of 50 Pa) are the pillars of the Passivhaus standard.

So far, its application proved to be successful in achieving the proposed goals, and to represent a solid knowledge base for the implementation of the nearly Zero Energy Building (nZEB) concept introduced in the EU Directive 2010/31 [2]. However, its

V. Costanzo (✉) · G. Evola · L. Marletta
DIEEI, University of Catania, 95125 Catania, Italy
e-mail: vincenzo.costanzo@unict.it

J. E. Carrillo Gómez
Architect, Bogotá, Colombia



J. Littlewood et al. (eds.), *Sustainability in Energy and Buildings*, Smart Innovation, Systems and Technologies 163, https://doi.org/10.1007/978-981-32-9868-2_9

dissemination overseas and for warm climates proved to be questionable because of the strict insulation and airtightness requirements that may easily lead to overheating issues and increased cooling energy demand [3–5].

These aspects were investigated in the framework of the Passive-On project in relation to Southern Europe [6], and led to the introduction of a new limit for the cooling energy demand (15 kWh $m^{-2}$ $y^{-1}$). The Passive-On project also included this last item in the calculation of the total primary energy needs, which should always keep below 120 kWh $m^{-2}$ $y^{-1}$. Furthermore, a less-strict threshold for the air tightness is allowed (1 $h^{-1}$ at 50 Pa) if the outdoor temperature does not drop below 0 °C.

Although some authors investigated the applicability of the Passivhaus standard in warm climates outside Europe [7–9], there is a lack of knowledge about the possibility to implement it within the current trends in the construction market. This research aims at advancing this knowledge by presenting dynamic simulations of typical housing projects located in different climates of Colombia. The comparison between the base case scenario and the passive design solution shows the benefits achievable in terms of thermal comfort and energy consumption, along with the compliance to the standard.

## 9.2 Methodology

### *9.2.1 Case Studies in Bogotá and Barranquilla*

The buildings selected as case studies for assessing the suitability of the Passivhaus Standard in this country are placed in two cities representative of these different climate conditions, namely the coastal city of Barranquilla and the capital city Bogotá.

Barranquilla is defined as a tropical savanna climate according to Köppen-Geiger classification [10], with average outdoor air temperatures constantly above 31 °C and precipitations mainly concentrated from May to October (peak values of around 170 mm). On the other hand, Bogotá is characterized by a subtropical oceanic climate (Cfb climate zone) with average temperatures almost constant throughout the year (around 13 °C) and average precipitations of 67 mm per month. Table 9.1 summarizes the main climate variables of the two cities according to reference winter and summer periods.

For the sake of considering the current trends in the Colombian construction market, two recurrent house typologies are addressed in this study, namely a single-family detached house and an apartment block with several flats per floor. These two typologies have been further distinguished in relation to the different climate and real estate markets of the two cities, taking real projects under current development as references (see Fig. 9.1).

The detached house in Bogotá hosts six people has a net floor area of 145 $m^2$ and the outer walls are made of 15-cm thick clay bricks internally covered with 3 cm of

**Table 9.1** Main climatic variables in Bogotá and Barranquilla—mean values

| | ***Bogotá*** | | ***Barranquilla*** | |
|---|---|---|---|---|
| | Winter (Nov-Feb) | Summer (Jun-Sep) | Winter (Nov-Feb) | Summer (Jun-Sep) |
| Outdoor dry bulb temperature (°C) | 13.1 | 13.2 | 25.2 | 26.5 |
| Relative humidity (%) | 81.8 | 78.2 | 81.1 | 85.5 |
| Daily global horizontal irradiance (Wh $m^{-2}$) | 391.2 | 356.7 | 425.2 | 472.2 |
| Wind speed (m $s^{-1}$) | 1.9 | 2.2 | 1.2 | 1.1 |

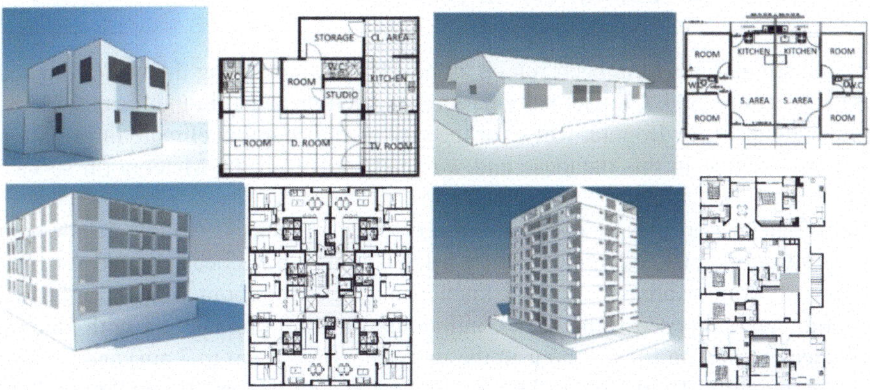


**Fig. 9.1** Case study buildings in Bogotá (left column) and in Barranquilla (right column)

stucco and 3 cm of plaster. The roof is made of slate tiles resting on a 5-cm screed layer and a 10-cm thick reinforced concrete structure. Windows are double-glazed panes (6 mm thickness) filled with an air cavity of 20 mm. On the other hand, the detached house in Barranquilla has a net floor area of about 50 $m^2$ and hosts three people. Outer walls are made of 22-cm thick clay bricks externally covered with 5-cm stucco, while the roof is made of tiles resting on a 5-cm thick timber board. Windows are single-glazed with 6-mm clear panes. For what concerns the apartment block, the one selected in Bogotá has 25 apartments whose net floor area ranges from 50 to around 100 $m^2$. The same construction specifications of the detached house case are used. On the other hand, in Barranquilla the apartment block has ten floors and hosts 45 apartments whose net floor area range from 45 to 120 $m^2$. Outer walls are made of 20-cm thick clay bricks internally covered with 5-cm stucco and plaster layers, while the roof is made of slate tiles resting on 5-cm thick screed layer and on a 30 cm concrete slab. Windows are single glazed with a clear pane of 12 mm. The resulting U-values are summarized in Table 9.2.

**Table 9.2** U-values of the envelope components

| Scenario | Detached house ($W\ m^{-2}\ K^{-1}$) | Apartment block ($W\ m^{-2}\ K^{-1}$) |
|---|---|---|
| ***Bogotá*** | | |
| Outer walls | 1.68 | |
| Roof | 2.31 | |
| Windows | 2.81 | |
| ***Barranquilla*** | | |
| Outer walls | 1.82 | 1.61 |
| Roof | 2.21 | 1.22 |
| Windows | 5.26 | 5.30 |

### 9.2.2 *Dynamic Simulations in IESVE*

Thermal simulations have been run on an hourly time step using the well-validated IESVE software [11]. Climate information has been provided in the format of.epw weather file [12] for Bogotá, while for the city of Barranquilla climate information was not available in this database and was gathered from the Meteonorm service [13].

Different user profiles for occupancy and electric appliances—which include both discontinuous electric equipment and artificial lights—have been created for bedrooms and common areas such as the kitchen and the living room. More specifically, bedrooms are supposed to be used continuously from 9 pm to 7 am with electric appliances turned on for one hour in the morning (from 6 am to 7 am) and for five hours in the night (from 6 pm to 11 pm). As for the common areas, people use them for one hour in the morning (from 7 am to 8 am) and when coming back from work (from 6 pm to 9 pm). Electric appliances are used according to this user profile, while for the remaining of the day they are supposed to be used at 20% of their capacity.

Internal gains include the sensible heat released by people with sedentary activity (1 met, corresponding to 90 W per person), the artificial lights (6 $W\ m^{-2}$ in the bedrooms and 8 $W\ m^{-2}$ in the common areas) and other electric equipment (10 $W\ m^{-2}$ for every room). Finally, air infiltration rate is set constant and equal to the Passivhaus threshold of 0.6 $h^{-1}$, in the absence of reliable experimental measurements and local standard specifications.

In order to investigate the different behavior of the buildings both under the operation of an HVAC system and without it (i.e., in free-running conditions), two sets of simulations were run

1. *Thermal comfort simulations*: the buildings are supposed to be operated in free-running conditions, i.e., without any mechanical system in use. Comfort conditions are appraised by means of the PMV and PPD values defined by Fanger's theory [14].
2. *Energy simulations*: thermal loads are calculated based on a heating set point temperature of 16 °C in Bogotá and a cooling set point temperature of 24 °C in

Barranquilla, only during the occupancy hours. These set point temperatures are coherent with local climate features (i.e., Bogotá does not need any cooling while Barranquilla does not need any heating) and local people thermal preferences, as inferred from previous housing projects and reported in Ref. [15].

The integration of the thermal loads over one year returns the yearly energy demand for space heating and cooling. In this way, the comparison among the different design options is not dependent on the mechanical systems employed, and is thus focused only on the impact of passive measures.

Finally, Primary Energy consumption is calculated by adding the electricity demand for lights, appliances and Domestic Hot Water (DHW) production (assuming a hot water consumption of 40 l per person per day) to these figures and considering national thermo-electric conversion coefficients.

## 9.3 Results and Discussion

This section discusses the results of the simulations for the base case and the passive design scenarios for both the detached house and a typical middle floor of the apartment block. This allows excluding boundary effects that are likely to influence the thermal behavior of flats located at the ground and top floors in the apartment block.

### *9.3.1 Base Case Scenario*

**Bogotá**. The thermal comfort conditions achieved in the detached house are shown on the left-hand side of Fig. 9.2. Here, a red polygon identifies the average Predicted Mean Vote (PMV) and Percentage of People Dissatisfied (PPD) achieved throughout the year in free-running conditions in the different rooms.

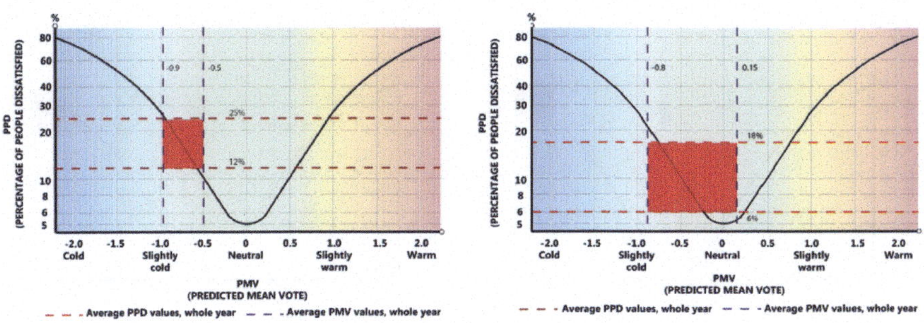


**Fig. 9.2** Thermal comfort in Bogotá: detached house (left) and apartment block (right)

As one can observe, the detached house ensures rather good comfort conditions, the thermal sensation experienced by the occupants being close to neutrality or just slightly cold ($-0.9 \leq \text{PMV} \leq -0.5$) with a PPD ranging from 12 to 25%.

Similar thermal conditions are predicted for a middle floor of the apartment block (right-hand side of Fig. 9.2): here, PMV ranges from −0.8 to 0.15 and PPD is comprised between 6 and 18%. Since the construction technique is the same in both buildings, the slightly different results are imputable to the lower amount of surfaces exposed outdoors, as reported in Sect. 9.2.1.

In terms of energy needs, Table 9.3 indicates that the heating demand of the detached house amounts to 31.8 kWh $m^{-2}$, while for the intermediate floor of the apartment block this figure is 15.6 kWh $m^{-2}$. On the other hand, the higher electricity needs for the equipment in the typical floor than in the detached house (48.3 kWh $m^{-2}$ against 19.3 kWh $m^{-2}$) leads to a higher Primary Energy (PE) consumption in the apartment block, which amounts to 106.6 kWh $m^{-2}$ against 82.8 kWh $m^{-2}$ predicted for the detached house.

**Barranquilla**. The climate conditions experienced in Barranquilla are very different from those in Bogotá, and this is clearly reflected in the indoor thermal conditions predicted by the simulations in the detached house and the apartment block.

As shown in Fig. 9.3, warm conditions are predicted in both typologies for most of the year: in the detached house PMV values range from 1.45 to 1.82 with PPD value comprised between 45 and 80%, while for the typical floor of the apartment block the corresponding ranges are $1.60 \leq \text{PMV} \leq 1.95$ and $42\% \leq \text{PPD} \leq 70\%$.

Interestingly, the cooling energy demand of both housing typologies are rather similar (52.3 kWh $m^{-2}$ for the detached house against 47.5 kWh $m^{-2}$ of the apartment

**Table 9.3** Yearly energy demand and PE consumption for the buildings in Bogotá

| | Detached house (kWh $m^{-2}$) | Typical floor (kWh $m^{-2}$) |
|---|---|---|
| Heating demand | 31.8 | 15.6 |
| Electric equipment demand | 19.3 | 48.3 |
| PE consumption | 82.8 | 106.6 |

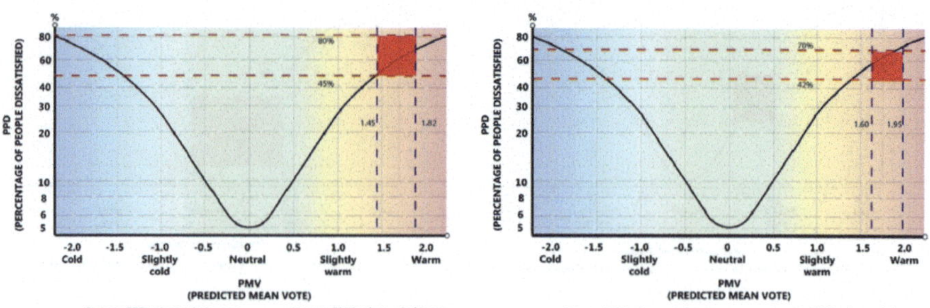


**Fig. 9.3** Thermal comfort in Barranquilla: detached house (left) and apartment block (right)

**Table 9.4** Yearly energy demand and PE consumption for the Barranquilla study buildings

| | Detached house (kWh m$^{-2}$) | Apartment (kWh m$^{-2}$) |
|---|---|---|
| Cooling demand | 52.3 | 47.5 |
| Electric equipment demand | 31.4 | 63.6 |
| PE consumption | 130.3 | 185.5 |

block, see Table 9.4). This can be explained considering that the lower amount of surfaces exposed outdoors reduces the heat losses through the envelope when the outdoor temperature is low (mainly at night), thus causing overheating.

The resulting PE consumption is 130.3 and 185.5 kWh m$^{-2}$ for the detached house and the typical floor in order.

### 9.3.2 *Passive Design Scenario*

From the results of the simulations concerning the base case scenario, it is clear that passive design solutions should be differentiated between the two cities, aiming at improving the thermal resistance of the envelope in Bogotá while also implementing some passive cooling measures in Barranquilla. Therefore, the proposed retrofit scenarios are:

- ***Bogotá***: application of a 10-cm thick polyurethane board on the outer side of the outside walls (with a resulting U-value of 0.21 W m$^{-2}$ K$^{-1}$, see Fig. 9.4 for details) and use of double-glazed windows with argon filling (U = 2.61 W m$^{-2}$ K$^{-1}$) in both the housing typologies;
- ***Barranquilla***: in the detached house, the existing clay bricks are replaced by two layers of 20-cm thick vermiculite bricks with insulating properties ($\lambda$ = 0.27 W m$^{-1}$ K$^{-1}$) separated by an air gap of 8 cm and externally covered with a polyurethane board of 5 cm (U = 0.22 W m$^{-2}$ K$^{-1}$). Existing windows are replaced with double-glazed ones with argon filling (U = 2.61 W m$^{-2}$ K$^{-1}$), to

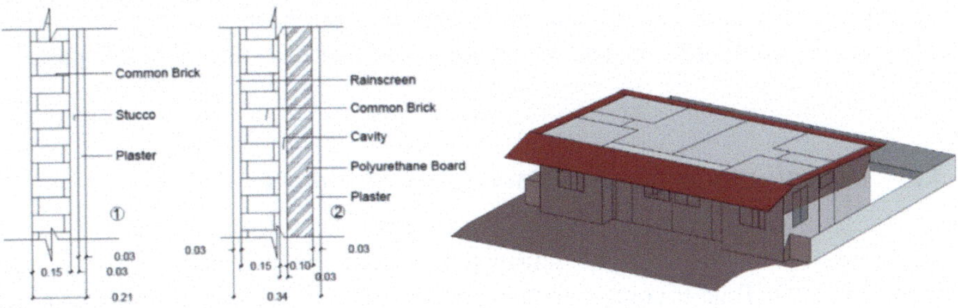


**Fig. 9.4** External walls in the base case and in the passive design for buildings in Bogotà (on the left) and the addition of shading devices in Barranquilla (on the right)

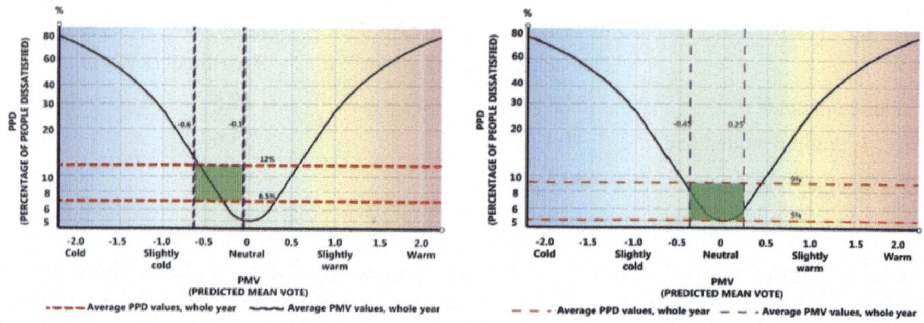


**Fig. 9.5** Thermal comfort in Bogotá: detached house (left) and apartment block (right). (on the left) and the addition of shading devices in Barranquilla (on the right)

be shaded by a continuous overhang on each side of the house (see Fig. 9.4). The windows will be opened at night from 10 pm to 6 am whenever the outdoor air temperature is at least 2 °C lower than indoors (night ventilation strategy). In the retrofit scenario of the apartment block, the outer walls are kept the same as in the base case, except for the application of an exterior 10-cm thick polyurethane board ($U = 0.19\ W\ m^{-2}\ K^{-1}$). Moreover, the shading devices will be placed only above the windows rather than continuously on each side of the building.

**Bogotá**. The implementation of the passive measures described above brings noticeable improvement in terms of thermal comfort. In fact, if looking at Fig. 9.5 the comfort polygon in the detached house is delimited by $-0.6 \leq PMV \leq -0.1$ and PPD between 6.5 and 12%, while in the case of the typical floor the boundaries are $-0.45 \leq PMV \leq 0.25$ and PPD between 5 and 9% respectively. This means that in Bogotá comfortable thermal conditions would be perceived (i.e. close to thermal neutrality) by the vast majority of the occupants.

**Barranquilla**. The retrofit design solutions discussed above can improve indoor comfort conditions by lowering both PMV and PPD values to the range $0.9 \leq PMV \leq 1.45$ and $31\% \leq PPD \leq 47\%$ in the detached house and to $1.2 \leq PMV \leq 1.5$ and $40\% \leq PPD \leq 52\%$ in the intermediate floor of the apartment building (see Fig. 9.6). However, the indoor environment would be still perceived as slightly warm, with a share of people dissatisfied as high as 50%. This calls for the need of further passive and low-energy cooling strategies, which are not further investigated in this paper.

### *9.3.3 Compliance with the Passivhaus Standard*

As reported in Table 9.5, both the detached house and the intermediate floor of the apartment block in Bogotá comply with the standard prescriptions in terms of heating energy demand, which amounts to 8.7 and 4.2 kWh $m^{-2}$ in order, against a threshold value of 15 kWh $m^{-2}$.

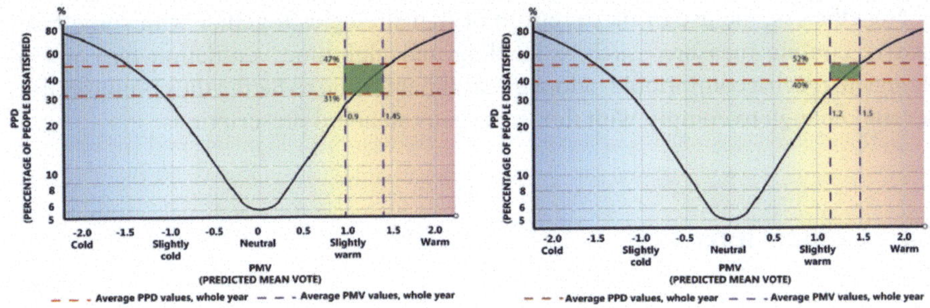


Fig. 9.6 Thermal comfort in Barranquilla: detached house (left) and apartment block (right)

**Table 9.5** Compliance with the Passivhaus standard of the passive design solutions

| | Detached house ($kWh\ m^{-2}$) | Apartment ($kWh\ m^{-2}$) | Passivhaus standard ($kWh\ m^{-2}$) |
|---|---|---|---|
| ***Bogotá*** | | | |
| Heating energy demand | 8.7 | 4.2 | $\leq 15$ |
| Primary energy consumption | 48 | 99.4 | $\leq 120$ |
| ***Barranquilla*** | | | |
| Cooling energy demand | 27.8 | 34 | $\leq 15$ |
| Primary energy consumption | 96.1 | 172.3 | $\leq 120$ |

Primary Energy is not a concern as well, since the threshold value of 120 $kWh\ m^{-2}$ is largely complied with both by the detached house (PE = 48 $kWh\ m^{-2}$) and the apartment block (PE = 99.4 $kWh\ m^{-2}$). It is worth noticing that these outcomes depend also on the use of a heating set point temperature of 16 °C, which may be considered too low by residents in European and North American countries.

As far as Barranquilla is considered, the threshold for the cooling energy demand (15 $kWh\ m^{-2}$) is largely exceeded both by the detached house (27.8 $kWh\ m^{-2}$) and by the apartment (34 $kWh\ m^{-2}$). For what concerns the Primary Energy consumption, only the detached house complies with the Passivhaus criterion (96.1 $kWh\ m^{-2}$ against 120 $kWh\ m^{-2}$), although this is mainly due to the lower amount of electric appliances installed in comparison to the apartment block, that encompasses four apartments per floor.

It thus emerges how the implementation of typical passive design measures, such as improving the thermal resistance of the envelope, works well in improving indoor thermal comfort and in reducing the heating energy demand down to the prescription of the Passivhaus standard as demonstrated for the housing projects in Bogotá.

Nonetheless, the use of these retrofit design strategies together with some basic passive cooling measures such as shading and nighttime ventilation may not be enough to reach a comfortable indoor environment while reducing the cooling energy demand enough to comply with the Passivhaus standard prescriptions.

## 9.4 Conclusions

The present paper proposed an analysis of typical housing projects in Colombia, namely a detached house and an apartment block, in order to assess their thermal and energy performance and to ascertain if they are suitable for an upgrade to the Passivhaus standard.

Dynamic simulations carried out for the cold climate of the capital city (Bogotá) and the hot climate of a coastal city (Barranquilla) revealed that, under the business-as-usual construction scenario, in Bogotá both the detached house and a typical middle floor of an apartment block experience slightly cold or close to neutrality thermal conditions ($-0.9 \leq$ PMV $\leq 0.15$ and $6\% \leq$ PPD $\leq 25\%$). On the other hand, in Barranquilla overheating is quite pronounced in both the housing typologies, being $1.45 \leq$ PMV $\leq 1.95$ and $42\% \leq$ PPD $\leq 80\%$.

When implementing basic passive design measures, the housing projects would fulfill the Passivhaus standard criteria in Bogotá but not in Barranquilla. In fact, although comfort conditions are significantly improved as demonstrated from the calculation of Fanger's comfort indexes ($0.9 \leq$ PMV $\leq 1.5$ and $31\% \leq$ PPD $\leq 52\%$), for the hot climate conditions of Barranquilla the cooling energy demand and PE consumption would be still higher than the standard thresholds for both the housing typologies. On the one hand, this implies the use of additional passive measures and highly efficient cooling systems to fulfill the standard requirements, but on the other hand it may suggest a revision of the standard criteria for hot climates by—for example—prescribing the use of reflective coatings and of very performing electric heat pumps.

The outcomes of this simulation effort, suggesting improvements in the thermal resistance values of the envelope and greater attention to design of shading devices, may be of help for the Colombian building sector in both retrofit and new constructions towards a decarbonisation of the building stock.

## References


1. Passive House Institute: http://passivehouse.com/02_informations/01_whatisapassivehouse/01_whatisapassivehouse.htm. Accessed 5 March 2019
2. Directive 2010/31/EU of the European Parliament and of the Council of 19 May 2010 on the energy performance of buildings
3. Consoli, A., Costanzo, V., Evola, G., Marletta, L.: Refurbishing an existing apartment block in Mediterranean climate: towards the Passivhaus standard. Energy Procedia **111**, 397–406 (2017)
4. Costanzo, V., Fabbri, K., Piraccini, S.: Stressing the passive behavior of a Passivhaus: an evidence-based scenario analysis for a Mediterranean case study. Build. Environ. **142**, 265–277 (2018)
5. Piderit, M.B., Vivanco, F., Van Moeseke, G., Attia, S.: Net Zero Buildings—a framework for an integrated policy in Chile. Sustainability **11**, 1494 (2019)
6. Ford, B., Schiano-Phan, R., Zhong, D.: The Passivhaus standard in European warm climates: design guidelines for comfortable low energy homes. School of the built environment, University of Nottingham (2007)
7. Badescu, V., Rotar, N., Udrea, I.: Considerations concerning the feasibility of the German Passivhaus concept in Southern Hemisphere. Energy Effic. **8**, 919–949 (2015)
8. Schnieders, J., Feist, W., Rongen, L.: Passive Houses for different climate zones. Energy Build. **105**, 71–87 (2015)
9. Attia, S., Zawaydeh, S.: Strategic decision making for zero energy buildings in Jordan. In: Proceedings of the ASHRAE International Conference on Energy and Indoor Environment for Hot Climates, February 24–26, Doha (2014)
10. Kottek, M., Grieser, J., Beck, C., Rudolf, B., Rubel, F.: World map of the Köppen-Geiger climate classification updated. Meteorol. Z. **15**(3), 259–263 (2006)
11. IES, Integrated Environmental Solutions: https://www.iesve.com/. Accessed 5 March 2019
12. Wilcox, S., Marion, W.: User's manual for TMY3 Data Sets. Technical Report NREL/TP-581-43156 (2008)
13. Meteonorm: https://meteonorm.com/en/. Accessed 5 march 2019
14. Fanger, P.O.: Thermal Comfort. Danish technical Press, Copenhagen (1970)
15. Natarajan, S., Rodriguez, J., Vellei, M.: A field study of indoor thermal comfort in the subtropical highland climate of Bogota, Colombia. J. Build. Eng. **4**, 237–246 (2015)

# Chapter 10
# Earth–Air Heat Exchanger Potential Under Future Climate Change Scenarios in Nine North American Cities

**A. Zajch, W. Gough and G. Chiesa**

**Abstract** Earth–air heat exchanger (EAHE) potential was evaluated for nine locations representing a range of North American climates to determine the impact of future climate scenarios. This was motivated by the intrinsic relationship between climate conditions and EAHE potential. Future weather files were created using general circulation model (GCM) outputs for three representative concentration pathways (RCP) scenarios and two separate timeframes. These were used in a climate-based approach for estimating EAHE potential based on their capacity to provide either heating or cooling. The results demonstrated, cooling capacity of a hypothetical EAHE system decreased and heating capacity increased due to changes in demand. The largest effects were observed for later time periods and more drastic changes to radiative forcing. From the chosen locations, heating dominated sites are most susceptible to unfavorable changes in cooling capacity due to the synchronous timing of changes in cooling capacity relative to peak demand. This work highlights the importance of conducting a climate change impact assessment on EAHE potential in these climates to ensure system effectiveness into the future. Further work is required to explore the impact of depth and efficiency in addition to the resolution of climate changes models used.

## 10.1 Introduction

A building's heating and cooling are inherently determined by its climate. The evolution of regional climates presents an important challenge for sustainable buildings. For example, in North America, increased cooling and decreased heating needs are projected to grow energy use in cooling dominated zones [1]. Understanding climate

A. Zajch (✉) · W. Gough
Department of Physical and Environmental Sciences, University of Toronto Scarborough, 1265 Military Trail, Toronto, ON M1C 1A4, Canada
e-mail: andrew.zajch@utoronto.ca

G. Chiesa
Department of Architecture and Design, Politecnico Di Torino, Viale Pier Andrea Mattioli 39, 10125 Turin, Italy



J. Littlewood et al. (eds.), *Sustainability in Energy and Buildings*, Smart Innovation, Systems and Technologies 163, https://doi.org/10.1007/978-981-32-9868-2_10

sensitivity can help in the design of resistant systems which will remain feasible despite evolving conditions.

Earth–air heat exchangers (EAHEs) can provide conditioned air directly to a building using the near subsurface as a thermal reservoir. The heating and cooling capability of a system is dependent on its length, diameter, operational regimes (air speed), depth and surrounding environment [2]. EAHEs are also intrinsically related to a location's climate as they rely on the natural temperature difference between the air passing through the system and the subsurface environment. This is especially the case for open systems, the topic of this work, since they employ ambient air as the working fluid. Typically, cooler climates are better suited for cooling and warmer climates observe effective heating [3–5]. As a result, there is an interesting trade-off between heating and cooling needs and potential determined by the climate. Contrasting the heating and cooling potential for the warmest and coldest days in three Mexican climates identified the climate with the largest annual temperature amplitude as the least susceptible to unfavorable heating or cooling [5]. In another example, [6] demonstrated the enhanced usable days for a hot–dry climate compared to a cold climate due to its ability to provide greater cooling and pre-cooling while having comparable heating potential. Exploring eight bioclimatic zones throughout Brazil [7] showed the highest changes in heating and cooling degree hours produced by the EAHE occurred for the coldest and warmest zones, respectively. The inherent relationship between climate and EAHE potential motivated the exploration of EAHE potential, and therefore feasibility, under future climate scenario conditions. The objective of this study was to estimate the degree and range of climate change impacts across North American climates to reveal the sensitivity of EAHEs to possible future climate conditions. This knowledge is essential for designers in the present to ensure systems remain usable into the future.

## 10.2 Methodology

### *10.2.1 Generating Future Weather Files*

To estimate the potential of EAHE systems under future climate conditions, current climate conditions need to be augmented using projected changes of relevant atmospheric variables such as temperature and incoming solar radiation. The current climate conditions were described using weather files representative of typical behavior at the selected locations, hereafter referred to as sites. These weather files form a baseline for EAHE potential that can be contrasted with future conditions. *Meteonorm* was used to generate weather files for nine sites encompassing the observed range of climate conditions across North America [8] (Fig. 10.1). The nine chosen sites therefore provided sufficient representation of both the size and seasonal contrast between heating and cooling demands in North America to provide a representative sample of climates.

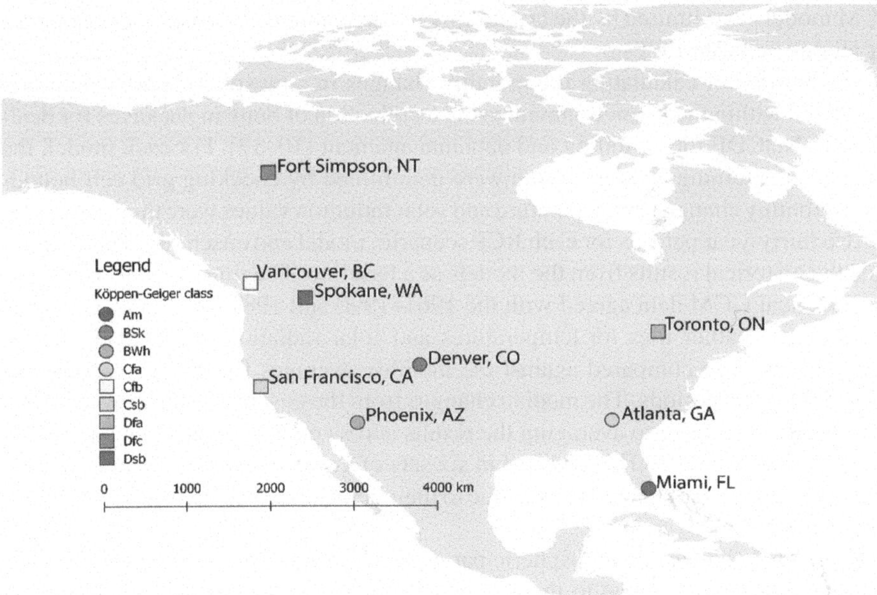


**Fig. 10.1** A map showing the 9 sites investigated and their climate classification

CMIP5 (Coupled Model Intercomparison Project Phase 5) multi-model ensemble outputs for the monthly historical and projected mean, minimum and maximum daily temperatures as well as downward shortwave incoming radiation were downloaded in NetCDF format [9]. CMIP5 describes a collaborative project to assemble climate models outputs and scenarios utilizing a consistent structure. Four GCMs (General Circulation Model) with three ensemble runs (r1i1p1, r2i1p1, and r3i1p1) were used to encompass the inherent variability across and within the models, respectively (Table 10.1). Results for three representative concentration pathways (RCPs), indicative of changes in radiative forcing due to changes in atmospheric composition, were employed to highlight the range of projected future climates. The selection of

**Table 10.1** CMIP5 multi-model ensemble models used for future weather file creation

| Institute ID | Modeling group | Model version used |
|---|---|---|
| CCCMA | Canadian centre for climate modeling and analysis | CanESM2 |
| IPSL | Institut Pierre-Simon laplace | IPSL-CM5A-LR |
| MIROC | Atmosphere and Ocean Research Institute (The University of Tokyo), National Institute for Environmental Studies, and Japan Agency for Marine-Earth Science and Technology | MIROC5 |
| MPI-M | Max-Planck Institut für Meteorologie (Max-Planck Institute for Meteorology) | MPI-ESM-LR |

GCM models was limited by the availability of the required variables and ensembles for all three chosen RCPs.

The process of calculating the monthly changes in atmospheric parameters was conducted within the R Environment using a collection of built-in packages for dealing with NetCDF files, plotting and data management [10–13]. For each model, the grid cells containing a given station were determined by checking grid cell boundaries. Monthly changes in temperature and solar radiation values were then calculated for two thirty-year periods for each RCP scenario, model and ensemble. The changes used the historical results from the models as a baseline. The time periods used from the historical GCM data agreed with the 1961–1990 and 1981–1990 windows used to generate weather files for temperatures and solar radiation, respectively. These time periods were compared against the monthly averages for the 2021–2050 and 2061–2090 time periods. The median changes from the ensembles for a given model were determined prior to averaging the results across models for a given parameter, timescale and scenario. This resulted in six sets of average monthly changes of four atmospheric parameters for each site due to their being three scenarios and two future time periods.

Calculated changes in atmospheric parameters relevant for estimating the EAHE potential were then applied to the weather files representing current conditions by using the "morphing" methodology [14]. The relative changes in temperature and solar radiation for each site modeled by the GCM outputs were used to either shift, scale or apply both changes to the weather files. A scaling factor was used for solar radiation to preserve the diurnal cycle while a scaling factor in addition to a shift was used for the dry bulb temperature to modify both the diurnal amplitude and mean annual temperatures, respectively (e.g., [15]). The scaling factor for the dry bulb temperature was derived from the change in diurnal temperature range described by the maximum and minimum daily temperatures [14]. Monthly shifting and scaling resulted in the creation of new 'future' weather files representative of conditions under future climate conditions at a given site.

### *10.2.2 Climate-Based Approach*

The climate-based approach for estimating EAHE potential provides a parametric approach driven by the climate data at a hypothetical location [16]. While this method may not have the sophistication of dynamical building simulations it does allow for a rapid regional assessment of EAHE potential without discerning system configuration. Instead a single tunable efficiency value is assigned to describe the system's performance. Empirical analysis of EAHE performance has shown this value ranges between 0–1 but is often >0.5 [17, 18]. The climate-based approach uses the efficiency ($\varepsilon$) parameter as a tunable value that can be used with knowledge of the inlet air ($T_{in}$) and surrounding ground ($T_g$) temperatures to determine the outlet air temperature of the system ($T_{out}$) (Eq. 10.1) [16].

$$T_{out} = T_{in} - \varepsilon(T_{in} - T_g) \tag{10.1}$$

The hourly inlet air temperature was derived from the site's weather file. The surrounding ground temperature at a given depth and time was calculated using an analytical description of the ground temperature at depth (Eq. 10.2) [16, 19]. The annual temperature mean ($T_{g,mean}$) and amplitude ($T_{g,amplitude}$) as well as the phase shift ($T_{g,phase}$) of the annual signal was generated using the *CalcSoilSurfTemp* built-in package in EnergyPlus [20]. The depth (z), thermal diffusivity of the subsurface (α) were tuned and set, respectively, for the simulations. A depth of 2.5 m, efficiency of 0.5 and a thermal diffusivity of 6.45 $\times 10^{-7}$ $m/s^2$ was used for the simulations. The ground temperature is computed for each hour in a year (t, seconds) over the duration of the entire annual cycle ($t_0$). Once the hourly outlet temperature throughout the annual cycle was calculated, the heating and cooling potential of the system was evaluated.

$$T_g = T_{g,mean} - (T_{g,amplitude})\left(e^{-z\sqrt{\frac{\pi}{t_0\alpha}}}\right)\cos\left(\frac{2\pi}{t_0}\left(t - T_{g,phase} - \frac{z}{2}\sqrt{\frac{t_0}{\pi\alpha}}\right)\right) \tag{10.2}$$

The heating or cooling produced by the system is determined by comparing the heating and cooling degree hours at the inlet and outlet of the system. A baseline of 18 °C was used for both heating (< 18 °C) and cooling (>18 °C) following the Canadian reference temperature for degree days which also corresponds to an approximation of the ASHRAE baseline (65 °F) [21]. These were calculated for every hour within the weather file. Hourly heating and cooling degree hours at the inlet ($DH_{in}$) and outlet ($DH_{out}$) are summed and then compared to generate a metric of relative capacity (Eq. 10.3). A pseudo-operational control was used in the calculation to exclude the degree hour change from the annual sum if it was unfavorable. Conditioning was deemed unfavorable if the temperature change of the air was counterproductive to heating or cooling. For example, if the calculated outlet air temperature is warmer than the inlet but cooling was required, the cooling degree hour change is set to zero instead of a negative value. Therefore, this metric described the relative reduction in degree hours by the system, applicable for both heating and cooling. EAHE capacity was compared for future scenario results ($DH_{future}\%$) against baseline historical results ($DH_{base}\%$) to highlight the change in heating and cooling potential (Eq. 10.4).

$$DH\% = \frac{DH_{out} - DH_{in}}{DH_{in}} \tag{10.3}$$

$$\Delta DH\% = DH_{future}\% - DH_{base}\% \tag{10.4}$$

## 10.3 Results and Discussion

Changes to heating and cooling degree hours at the locations, under future climate scenarios and time periods, demonstrated larger changes to cooler and warmer climates, respectively. Heating and cooling demand peaked in the winter and summer seasons, respectively (Fig. 10.2). Miami and Phoenix were cooling dominated while Atlanta provided a case with more balanced heating and cooling. The remaining sites predominantly required heating throughout the year. Regardless of the climate regime, cooling degree hours increased while the heating degree hours decreased at all nine sites. The monthly distribution of changes also varied across the sites, although they followed a similar pattern to the seasonal heating or cooling need (Fig. 10.3). It is also important to note that scenario RCP8.5b represented the largest estimated change in degree hours at a given site. For example, a site such as Miami with cooling required throughout the year had changes to cooling degree hours throughout the year, peaking in the summer with the cooling demand. The largest changes to cooling degree hours were observed in Phoenix which also observed the greatest cooling need. Similarly, Fort Simpson showed the highest change in heating degree hours. Since the EAHE directly siphons the ambient air, the changes to degree hours at the sites also represent alteration of the inlet air temperature of the system.

Under future climate change scenarios, EAHE cooling capacity decreased while heating capacity typically increased (Table 10.2). In terms of capacities, the observed behavior was dependent on the heating/cooling demand. Cooler climates with a larger heating load, such as Denver, Toronto, Vancouver, Spokane, and San Francisco, witnessed a convergence of heating and cooling capacity. For the other warmer climates, the contrasts between heating and cooling capacities increased for Miami and Phoenix and decreased for Atlanta. The magnitude of the changes HDH% and CDH%, (see Table 10.2) had a predictable pattern for the scenarios and time periods

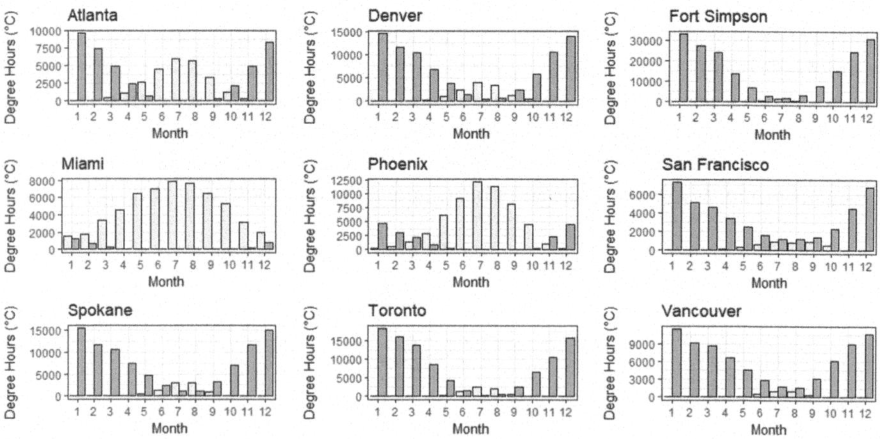


**Fig. 10.2** Historical baseline cooling and heating degree hours (°C) calculated for the 9 sites

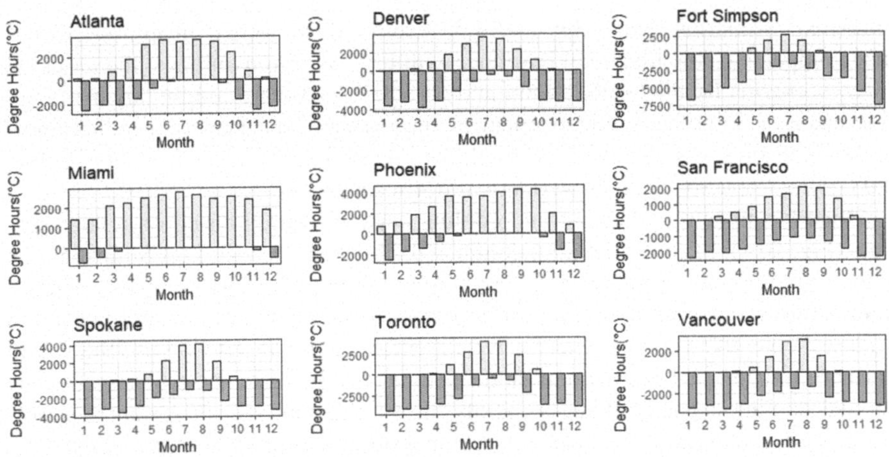


**Fig. 10.3** Monthly difference between future scenario and baseline cooling (light gray) and heating (dark grey) degree hours (°C) for the RCP8.5 scenario 2061–2090 (b) timeframe

**Table 10.2** EAHE capacity (%, Eq. 10.3) for the nine sites for baseline and future scenario conditions (a:2021–2050, b:2061–2090). It was evident that sites with a strong contrast in heating and cooling demand likewise displayed a larger contrast between heating and cooling capacities

| Site | Mode (%) | Baseline | RCP2.6 | | RCP4.5 | | RCP8.5 | |
|---|---|---|---|---|---|---|---|---|
| | | | a | b | a | b | a | b |
| Atlanta | HDH | 36 | 43 | 44 | 44 | 47 | 45 | 53 |
| | CDH | 53 | 44 | 44 | 43 | 41 | 42 | 35 |
| Denver | HDH | 18 | 21 | 21 | 21 | 22 | 21 | 25 |
| | CDH | 81 | 71 | 71 | 70 | 66 | 69 | 58 |
| Fort Simpson | HDH | 17 | 18 | 18 | 18 | 18 | 18 | 19 |
| | CDH | 100 | 100 | 100 | 100 | 100 | 100 | 98 |
| Miami | HDH | 86 | 93 | 93 | 93 | 94 | 93 | 96 |
| | CDH | 21 | 18 | 18 | 18 | 17 | 18 | 16 |
| Phoenix | HDH | 45 | 55 | 55 | 57 | 63 | 59 | 75 |
| | CDH | 43 | 39 | 39 | 38 | 36 | 38 | 33 |
| San Francisco | HDH | 20 | 25 | 26 | 25 | 28 | 26 | 35 |
| | CDH | 85 | 70 | 69 | 70 | 62 | 67 | 51 |
| Spokane | HDH | 18 | 21 | 21 | 21 | 22 | 21 | 26 |
| | CDH | 88 | 79 | 79 | 78 | 74 | 77 | 65 |
| Toronto | HDH | 17 | 20 | 20 | 20 | 21 | 20 | 24 |
| | CDH | 94 | 83 | 83 | 83 | 77 | 80 | 64 |
| Vancouver | HDH | 15 | 18 | 19 | 19 | 20 | 17 | 24 |
| | CDH | 98 | 89 | 88 | 88 | 81 | 86 | 68 |

simulated. Scenarios 2.6a, 2.6b, and 4.5a exhibited the lowest magnitude difference compared to the historical baseline, with little variation amongst the cases. In order, scenarios RCP8.5a, RCP4.5b, and RCP8.5b had increasingly larger magnitude changes. Ultimately, independent of the scenario and period, a reduction of cooling capacity was observed for all climate types. However, the monthly heating and cooling capacity changes were not uniform across sites.

Monthly impacts on heating and cooling capacity varied based on a site's climate (Fig. 10.4). In cooler, heating dominated, climates (Denver, Fort Simpson, San Francisco, Spokane, Toronto, and Vancouver) cooling capacity changes occurred from summer and, increased, into fall months. Heating capacity changes peaked in August–September (late summer-early fall) often extending into winter. Sites with warmer climates (Atlanta, Miami, and Phoenix) on the other hand observed heating and cooling capacity changes mainly during the winter–spring months. Cooling capacity changes at these sites were also observed into summer months. The magnitude of monthly changes to capacity also showed a climate bias. While the ΔCDH% in Miami peaked at roughly 10%, other sites with significant cooling capacity changes showed capacity decreases >25%, reaching near 50% for peak changes under RCP8.5b scenarios. Consistent with the observed changes for Miami, the coldest site Fort Simpson exhibited the lowest observed ΔHDH% at 0–5%. Moreover, sharp spikes (100% change) were also noticed during typically during transitional seasons at the sites. These indicated a complete removal or addition of a heating or cooling need. At Miami and Phoenix, the spikes demonstrated a loss of heating needs that were typically met by the EAHE. For cooler sites, the spikes indicated an addition of cooling needs that could be completely addressed by the EAHE. Despite these monthly increases to cooling capacity at cool climate sites, a reduction of cooling capacity on an annual scale was consistently observed across all sites.

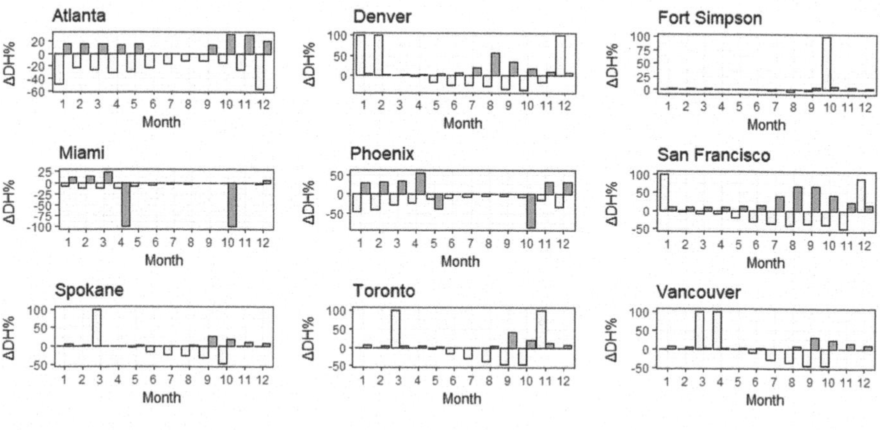


**Fig. 10.4** Monthly difference in cooling and heating capacities (CDH% and HDH%) between future scenarios and baseline results (ΔCDH%, light grey, and ΔHDH%, dark grey) for the RCP8.5 scenario 2061–2090 (b) timeframe

Distinguishing between heating and cooling dominated sites, it was evident that due to the timing of capacity changes and demand periods warmer climates appear more resistant to climate changes. For warmer climates (Atlanta, Miami and Phoenix), cooling capacity decreased during off-peak cooling demand periods (winter-spring) whereas in cooler climates these changes were more coincident with peak demand (summer–fall). Conversely, heating capacity increased during peak (winter–spring) and off-peak (summer-fall) for warmer and cooler climates, respectively. This behavior for cold climate sites can be attributed to the cycle of favorable soil–air temperature differences used by EAHEs. When comparing the change in HDH and ΔHDH% for cooler climates, HDH changes observed a peak in the winter, with similar spring and fall values, while the ΔHDH%, peaked only in the fall. Due to the dampening and phase shift of the surface temperature signal at depth, the annual surface maximum temperature experienced during the summer is delayed at depth. As a result, the ground's maximum temperatures are shifted into the fall instead of the summer, creating warmer ground temperatures less favorable for cooling and more beneficial for heating. On the other hand, the timing of beneficial and unfavorable changes coinciding with peak and off-peak demand in warmer climates demonstrated an apparent resistance to projected climate changes.

Cooler climates with a mix of heating and cooling needs (Denver, San Francisco, Spokane, Toronto, and Vancouver) appeared most susceptible to the impacts of future climate changes. While all sites showed unfavorable changes to cooling capacity on an annual scale, the timing of these negative changes for cooler climate sites enhances the impact on the effectiveness of EAHEs for building conditioning. EAHES in these cooler climates will experience reduced effectiveness in providing cooling during peak demand while heating capacity increases will occur during periods when needs are minimal. Warmer climates on the other hand observed impacts during off-peak periods when cooling needs were lower implying a less noticeable effect on cooling. EAHEs at cooler sites are therefore vulnerable to future climate changes. If an EAHE system was capable of meeting most cooling needs for current climate conditions, future climate conditions may require additional cooling capability to be increased by altering the system or installing other forms of cooling. Climate impacts should therefore be considered by designers in these climates to ensure that EAHEs do not underwhelm or fail in meeting growing cooling demands.

## 10.4 Conclusion

The results demonstrated a potential reduced feasibility for cooler climates due to the reduction of cooling capacity coincident with increasing peak cooling demands. Ultimately, these results implied a regional variation in climate impacts that needs to be explicitly defined for designers looking to employ EAHEs. However, the results were only generated for one set of system conditions. Future work will consist of the analysis for varying soil thermal diffusivities, efficiencies, and depths. Other limitations stem from the assumptions and resolution of the approach. The resolution of the GCM data does not necessarily encompass local scale climate effects that may be important for a given site. Further work is required to identify the difference in EAHE potential that can be described if more robust downscaling is used.

**Acknowledgements** We would like to acknowledge the World Climate Research Programme's Working Group on Coupled Modelling, which is responsible for CMIP, as well as thank the climate modeling groups (see Table 10.1) for producing and making available their model output. Also, we acknowledge the U.S. Department of Energy's Program for Climate Model Diagnosis and Intercomparison which provides coordinating support and led development of software infrastructure in partnership with the Global Organization for Earth System Science Portals for CMIP.

## References

1. Wang, H., Chen, Q.: Impact of climate change heating and cooling energy use in buildings in the United States. Energy Build. **82**, 428–436 (2014)
2. Bordoloi, N., Sharma, A., Nautiyal, H., Goel, V.: An intense review on the latest advancements of earth air heat exchangers. Renew. Sustain. Energy Rev. **89**, 261–280 (2018)
3. Zajch, A.M., Gough, W.A.: Employing the climate based approach for determining earth-air heat exchanger feasibility in the context of canadian climates. In: Grand Renewable Energy 2018, International Conference and Exhibition. Yokohama, Japan (2018)
4. Lee, K.H., Strand, R.K.: The cooling and heating potential of an earth tube system in buildings. Energy Build. **40**, 486–494 (2008)
5. Xamán, J., Hernández-López, I., Alvarado-Juárez, R., Hernández-Pérez, I., Álvarez, G., Chávez, Y.: Pseudo transient numerical study of an earth-to-air heat exchanger for different climates of México. Energy Build. **99**, 273–283 (2015)
6. Fazlikhani, F., Goudarzi, H., Solgi, E.: Numerical analysis of the efficiency of earth to air heat exchange systems in cold and hot-arid climates. Energy Convers. Manag. **148**, 78–89 (2017)
7. Alves, A.B.M., Schmid, A.L.: Cooling and heating potential of underground soil according to depth and soil surface treatment in the Brazilian climatic regions. Energy Build. **90**, 41–50 (2015)
8. Meteotest: Meteonorm handbook part I. Meteotest, Bern (2017)
9. ESGF CMIP5 data search: https://esgf-node.llnl.gov/search/cmip5/. Accessed 13 Dec 2018
10. R Core Team: R: A language and environment for statistical computing. R Foundation for Statistical Computing, Vienna, Austria. https://www.R-project.org/ (2018)
11. Pierce, D.: ncdf4: Interface to unidata netCDF (version 4 or earlier) format data files. R package version 1.16. https://CRAN.R-project.org/package=ncdf4 (2017)
12. Wickham, H.: ggplot: Elegant Graphics for Data Analysis. Spring, New York (2016)
13. Wickham, H., François, R., Henry, L., Müller, M.: dplyr: A grammar of data manipulation. R package version 0.7.6. https://CRAN.R-project.org/package=dplyr (2018)
14. Belcher, S., Hacker, J., Powell, D.: Constructing design weather data for future climates. Build. Serv. Eng. Res. Technol. **26**, 49–61 (2005)
15. Shen, P., Lukes, J.R.: Impact of global warming on performance of ground source heat pumps in US climate zones. Energy Convers. Manag. **101**, 632–643 (2015)
16. Chiesa, G.: EAHX – Earth-to-air heat exchanger: Simplified method and KPI for early building design phases. Build. Environ. **144**, 142–158 (2018)
17. Chiesa, G., Simonetti, M., Grosso, M.: A 3-field earth-heat-exchange system for a school building in Imola, Italy: Monitoring results. Renew. Energy **62**, 563–570 (2014)
18. Pfafferott, J.: Evaluation of earth-to-air heat exchangers with a standardised method to calculate energy efficiency. Energy Build. **35**, 971–983 (2003)

19. Labs, K.: Earth Coupling. In: Cook, J. (ed.) Passive Cooling, p. 207. MIT, Boston (1989)
20. U.S. Department of Energy: Auxiliary programs. EnergyPlus™ Version 8.9.0 Documentation. 23 March 2018
21. Government of Canada. http://climate.weather.gc.ca/glossary_e.html (2018). Last accessed 24 Sep 2018

# Chapter 11
# Developing a Didactic Thermal Chamber for Building Envelope Material Testing

**Bechara Nehme, Fadi Moucharrafie, Tilda Akiki, Rida Nuwayhid, Paul Abi Khattar Zgheib and Barbar Zeghondy**

**Abstract** In this paper, we present a didactic thermal chamber. This chamber has been developed for educating architecture students about building envelopes and their effects on resident comfort. The chamber is equipped with an instrumentation system capable of recording temperature data for different geometric points on the chamber. The chamber was successfully used to compare the thermal transmittance behavior of building blocks using different compositions. Results showed that adding polystyrene aggregate in the traditional cavity concrete block increases its thermal resistivity by 50%. With the developed apparatus, students are able to understand and visualize the thermal behavior of buildings.

## 11.1 Introduction

Building envelopes are the protective barrier between the external climate and the internal desired design thermal comfort within buildings in accommodated spaces. The better the heat flow through the building envelope is controlled, the more effectively is the desired thermal comfort level achieved, thus resulting in less

B. Nehme (✉) · T. Akiki
School of Engineering, Department of Electrical, Telecommunications and Computer Engineering, Holy Spirit University of Kaslik (USEK), Jounieh, Mount Lebanon, Lebanon
e-mail: becharanehme@usek.edu.ib

F. Moucharrafie · P. A. K. Zgheib
School of Architecture and Design, Architecture Department, Holy Spirit University of Kaslik (USEK), Jounieh, Mount Lebanon, Lebanon
e-mail: fadimoucharrafie@usek.edu.lb

R. Nuwayhid
Maroun Semaan Faculty of Engineering & Architecture, Department of Mechanical Engineering, American University of Beirut (AUB), Beirut, Lebanon

B. Zeghondy
Department of Mechanical Engineering, Holy Spirit University of Kaslik (USEK), Jounieh, Mount Lebanon, Lebanon



J. Littlewood et al. (eds.), *Sustainability in Energy and Buildings*, Smart Innovation, Systems and Technologies 163, https://doi.org/10.1007/978-981-32-9868-2_11

energy usage from the energy source, specifically when utilizing nonrenewable energies [1].

The importance of developing a didactic apparatus in conjunction with the course is motivating to students as it helps them visualize theory while also eliminating any prior misconceptions they may have acquired [2].

Phenomenologically, heat flows from spaces of higher temperatures to spaces of lower temperatures. As such, if it happens that one's accommodation is in a cold climate area, then care has to be given to retaining the heat that is generated inside the accommodated space as long as possible before losing it to the outside, otherwise it would be necessary to re-heat the space to achieve the desired thermal comfort [3, 4].

On the other hand, if it happens that the accommodated space is in a hot climate area, it would be imperative to minimize heat gains entering the occupied space, otherwise the impinged heat has to be rejected to the outside through using, for example, an air-conditioning system to achieve the desired thermal comfort level.

To achieve desired thermal comfort within occupied spaces using less artificially generated energy, building designers must create a building envelope that efficiently intercepts heat flow at a rate that achieves thermal comfort as long as possible within the occupied space.

Hence, to achieve an effective interception of the heat flow, the building envelope should be well insulated, and thus efficiently resisting the heat flow.

Building materials to be used in the construction/assembly of the building envelope should accordingly have a high thermal resistance value to obtain an effective interception of the heat flow.

The rate of heat flow (Q) is given by the driving potential due to a temperature difference ($\Delta$T) divided by the resistance to the flow, this can be stated as:

$$Q = \frac{\Delta T}{R}; R = \frac{1}{U}$$

Where R is the thermal resistance in $m^2K/W$ and U is the thermal transmittance coefficient or U-value in $W/m^2K$. The higher the resistance value the better the insulation, in this sense [5].

## 11.2 Theoretical Study of the Thermal Transmittance Coefficient (U-Value)

### *11.2.1 Building Practice in Lebanon*

One of the main used construction blocks in wall construction of buildings in Lebanon is the so-called cavity concrete block (CCB), also known as the concrete masonry unit (CMU) [6].

It would be wise to study its thermal behavior to allow a better understanding of its effective operation in resisting heat flow in building construction practice.

Practical testing as well as theoretical studies were applied on the Lebanese CCB/CMU blockwork to allow a better understanding of their thermal behavior.

### 11.2.2 The 15 cm CMU Used in the Lebanese Building Construction Practice

The 15 cm CMU consists of concrete and air cavities, two rows of air cavities surrounded by concrete. The first row consists of four cavities which possess oval form each of 8 cm length and 6.2 cm width. The second row consists of five cavities, three oval-shaped cavities with equivalent dimension to the above-mentioned cavities and two smaller but circular cavities each of a 3.1 cm diameter. All the above cavities have equal heights; each height is equal to 18.5 cm.

Density of the concrete of the studied 15 cm CMU is equal to 2116 kg/m$^3$.

Material conductivity: Concrete's conductivity is equal to 1.2 W/mK [7] (refer to Table 1), while still air conductivity at 1 atmosphere pressure is 0.0262 W/mK [8] (Fig. 11.1).

**Method of Calculation of the thermal transmittance coefficient (U-Value) of the 15 cm CMU.** The blockwork was divided into two layers in series namely, layer 1 and layer 2 (Fig. 11.2). Layer 1 consists of four oval-shaped cavities possessing the following dimensions:

$$\text{Length} = 8\,\text{cm},\ \text{Width} = 6.2\,\text{cm},\ \text{Height} = 18.5\,\text{cm}.$$

While layer 2 consists of five cavities, three oval-shaped cavities possessing the same dimensions as those in layer 1 and two smaller cavities each with the following dimensions: Diameter = 3.1 cm → Radius = 1.55 cm, Height = 18.5 cm

**Assumption.** To better define the fraction of the face area of the heat flow paths, the oval shaped cavities are converted to equivalent rectangle with the same area. Retaining the oval length (8 cm) for the long side of the rectangle, the width of the rectangle is determined and is equal to 4.96 cm. The circle-shaped cavities were also converted to equivalent rectangles with the same cross-sectional area of the circle resulting in the following dimensions: W = 1.521 cm, L = 4.96 cm.

**Calculation.**

*Layer 1:*

- Non-bridged layers

The thermal resistance is given by $R = \frac{e}{\lambda}$, where e is the thickness in m and λ is the thermal conductivity in W/(m.k).

$$R_{C1} = \frac{0.0182\,\text{m}}{1.2\,\text{W/mk}} = 0.015\,\text{m}^2\text{k/W}, \quad R_{C'} = \frac{0.0072\,\text{m}}{1.2\,\text{W/mk}} = 0.006\,\text{m}^2\text{k/W}$$

**Table 11.1** Shows experiment results on heat flow through different types of blockworks used in the Lebanese construction practice (The thermal VS the regular CMU)

| Time of the day | External wall temp. in °C | Internal wall temp. in °C | Interior chamber air temp. in °C | Solar intensity in W/m2 | Ambient air temp. in °C | Percentage of heat crossing to imping-ing H1/H (%) | Heat retarded in (%) |
|---|---|---|---|---|---|---|---|
| Thermal concrete masonry unit 15 cm thickness | | | | | | | |
| Blockwork with additional Polystyrene aggregate (2 mm diameter balls). Aggregates weight Proportion to cement and relative density. Cement: 1 unit (e.g. 50 kg); Crushed lime stone in powder size 1 unit (e.g. 50 kg); Gravel (crushed lime stone with gravel grain size between 4 mm to 7 mm) 4 units (e.g. 200 kg); Water according to cement to water ratio 3 units to 1 unit simultaneously; Polystyrene added to achieve with the other aggregates a volume of 1m3 and a mixture density of 1150 kg/m$^3$ | | | | | | | |
| 11:15 am | 30 | 19.5 | 20.5 | 585 | 22.7 | 65 | 35 |
| 12:00 pm | 42 | 21 | 21 | 620 | 23.7 | 50 | 50 |
| 12:30 pm | 42 | 23 | 22 | 620 | 23.5 | 55 | 45 |
| 13:00 pm | 42 | 24.5 | 23 | 565 | 23 | 58 | 42 |
| 13:30 pm | 40.5 | 26 | 24 | 490 | 23.1 | 64 | 36 |
| 14:00 pm | 43.5 | 27.5 | 25 | 390 | 22.7 | 63 | 37 |
| Regular concrete masonry unit 15 cm thickness. Cement: 1 unit (e.g. 50 kg); Crushed lime stone in powder size 1 unit (e.g. 50 kg); Gravel (crushed lime stone with gravel grain size between 4 mm to 7 mm) 4 units (e.g. 200 kg); Water according to cement to water ratio 3 units to 1 unit simultaneously. Mixture density 2116 kg/m$^3$ | | | | | | | |
| 11:15am | 19.5 | 18.5 | 18.5 | 140 | 18.1 | 95 | 5 |
| 12:00 pm | 27 | 19 | 19 | 250 | 19 | 70 | 30 |
| 12:30 pm | 25 | 20 | 19.25 | 320 | 19.2 | 80 | 20 |
| 13:00 pm | 28.5 | 21 | 20 | 310 | 19.1 | 74 | 26 |
| 13:30 pm | 24.5 | 21.5 | 20.5 | 235 | 19.05 | 88 | 12 |
| 14:00 pm | 25.5 | 22 | 21 | 170 | 18.85 | 86 | 14 |

Please note that margin of reading error is approximately 5%

- Bridged layer

Fraction of face area of cavities: (8 cm × 18.5 cm) × 4 = 592 cm$^2$
Block face area: 20 cm × 40 cm = 800 cm$^2$
$\frac{592\,\text{cm}^2}{800\,\text{cm}^2} = 0.74$ Face area fraction of air cavities $= f_{air}$.
$\frac{800\,\text{cm}^2}{800\,\text{cm}^2} = 1 \rightarrow 1 - 0.74 = 0.26$ Face area fraction of concrete $= f_C$.

$$R_C = \frac{0.0496\,\text{m}}{1.2\,\text{W/mk}} = 0.041\,\text{m}^2\text{k/W}, \quad R_{air} = \frac{0.0496\,\text{m}}{0.0262\,\text{W/mk}} = 1.893\,\text{m}^2\text{k/W}$$

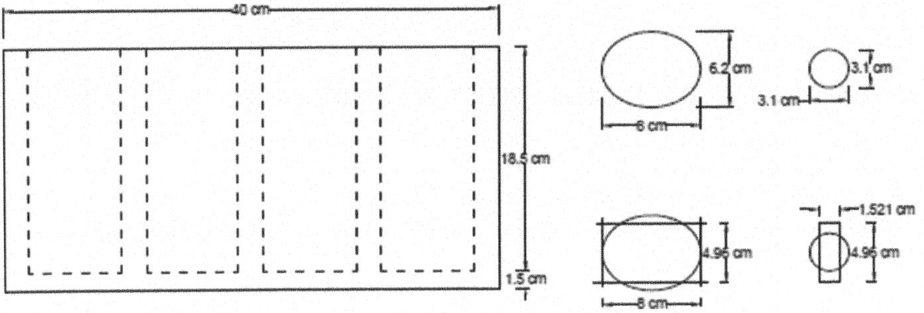


**Fig. 11.1** 15 cm CMU, $0.8 \times (6.2 \times 8) = 39.68\ \text{cm}^2 \rightarrow$ oval, $8 \times 4.96 = 39.68\ \text{cm}^2 \rightarrow$ rectangle

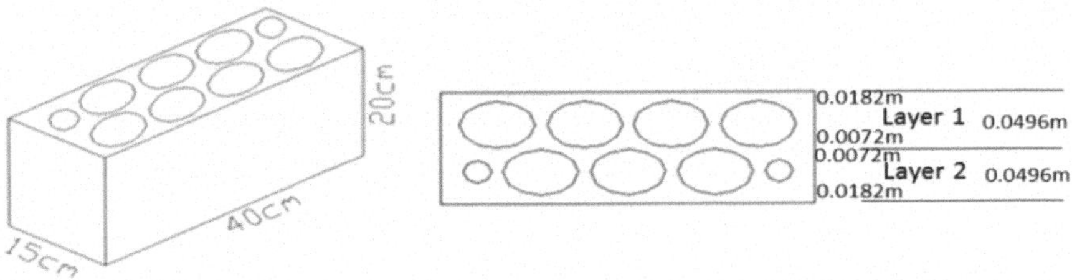


**Fig. 11.2** 15 cm CMU views

In this layer-bridged layer, heat flows in parallel through the concrete and air in the cavities [9].

$$\frac{1}{R_{Equ\ Bridged\ Layer\ 1}} = \frac{f_{air}}{R_{air}} + \frac{f_C}{R_C} = \frac{0.74}{1.893} + \frac{0.26}{0.041} = 6.731\ \text{W/m}^2\text{k}$$

$$R_{Equ\ Bridged\ Layer\ 1} = \frac{1}{6.731} = 0.149\ \text{m}^2\text{k/W}$$

$$R_{Total\ Layer\ 1} = R_{C1} + R_{C'} + R_{Equ\ Bridged\ Layer\ 1} = 0.015 + 0.006 + 0.149 = 0.17\ \text{m}^2\text{k/W}$$

*Layer 2:*

- Non-Bridged Layer

$$R_{C2} = \frac{0.0182\ \text{m}}{1.2\ \text{W/mk}} = 0.015\ \text{m}^2\text{k/W}, \quad R_{C''} = \frac{0.0072\ \text{m}}{1.2\ \text{W/mk}} = 0.006\ \text{m}^2\text{k/W}$$

- Bridged Layer

Fraction of face area of air cavities

$$3 \times (8\,\text{cm} \times 18.5\,\text{cm}) = 444\,\text{cm}^2$$

$$2 \times (1.521\,\text{cm} \times 18.5\,\text{cm}) = 56.277\,\text{cm}^2$$

Total face area of cavities: $444\,\text{cm}^2 + 56.277\,\text{cm}^2 = 500.277\,\text{cm}^2$
Face area fraction of cavities of the heat flow path: $\frac{500.277\,\text{cm}^2}{800\,\text{cm}^2} = 0.63 = f_{air}$
Face area fraction of concrete of the heat flow path: $1 - 0.63\,air = 0.37 = f_c$

$$R_{C2} = \frac{0.0496\,\text{m}}{1.2\,\text{W/mk}} = 0.041\,\text{m}^2\text{k/W}, \quad \text{R}_{air} = \frac{0.0496\,\text{m}}{0.0262\,\text{W/mk}} = 1.893\,\text{m}^2\text{k/W}$$

$$\frac{1}{R_{Equ\,Bridged\,Layer\,2}} = \frac{f_{air}}{R_{air}} + \frac{f_C}{R_C} = \frac{0.63}{1.893} + \frac{0.37}{0.041} = 9.357\,\text{W/m}^2\text{k}$$

$$R_{Equ\,Bridged\,Layer\,2} = \frac{1}{9.357} = 0.107\,\text{m}^2\text{k/W}$$

$$\begin{aligned} R_{TotalLayer2} = R_{C2} + R_{C''} + R_{Equ\,Bridged\,Layer\,2} &= 0.107 + 0.015 + 0.006 \\ &= 0.128\,\text{m}^2\text{k/W} \end{aligned}$$

$$R_{Total\,block\,15\,\text{cm}\,CMU} = R_{Total\,layer\,1} + R_{Total\,layer\,2} = 0.17 + 0.128 = 0.30\,\text{m}^2\text{k/W}$$

$$U_{Total\,blockwork\,15\,\text{cm}\,CMU} = \frac{1}{R_{Total\,blockwork\,15\,\text{cm}\,CMU}} = \frac{1}{0.30} = 3.33\,\text{W/m}^2\text{k}$$

### *11.2.3 Thermal Concrete Masonry Unit*

The thermal concrete block possesses polystyrene aggregate and has a density of $1150\,\text{kg/m}^3$ and a conductivity equal to 0.60 W/mk [7]. The polystyrene balls used in this mix have a diameter of 2 mm. The total weight obtained is 7.70 kg.

Using the same calculation method, we can get the values of the resistance and U-value of the thermal concrete masonry unit

$$\begin{aligned} U_{Total\,block\,work\,thermal\,15\,\text{cm}\,CMU} = \frac{1}{R_{Total\,block\,work\,thermal\,15\,\text{cm}\,CMU}} &= \frac{1}{0.577} \\ &= 1.73\,\text{W/m}^2\text{k} \end{aligned}$$

**Conclusion**. Theoretically the resistance performance to heat flow path of the thermal CMU compared to the regular CMU is $\left(\frac{0.577\,\text{m}^2\,\text{K/W}}{0.30\,\text{m}^2\,\text{K/W}} = 1.92 \cong 2\right)$ almost two times better. Compare the above results to the below obtained practical results of the tested CMUs in the thermal chamber. (see conclusion in Sect. 4.1)

## 11.3 Controlled Experiment to Measure the Heat Flow Through a Thermal Chamber

From the previous section it was found that the U-values of the regular block was 3.33 $W/m^2K$ while that for the thermal block was 1.73 $W/m^2K$. Thus, theoretically one would expect the heat flow through the thermal block to be reduced by at least 52% (i.e., 1.73/3.33) if not more.

In order to obtain practical evaluations of the heat flows associated with building blocks, a thermal chamber was designed, built, and operated.

The Thermal Chamber is made from pinewood and glass fibers insulation, it consists of six parts, namely three side walls, a top cover, a base and a non-erected free side (see Fig. 11.3). The free side is left to construct the test walls, which in this experiment were made from two different types of concrete masonry units. The two types of concrete masonry units have the same configuration, dimensions, and volume but differ only in their ingredients which are closely described in Table 11.1.

$$R_{Total} = R_{Is} + R_{Wood} + R_{Glass-Fiber} + R_{Os} = 0.125 + 0.143 + 0.833 + 0.078 = 1.179 \mathrm{m^2k/W}$$

This resistance is sufficiently large to allow the assumption that the heat loss through these walls is relatively small. As such, Fig. 11.3 below a sketch of the experimental thermal chamber which allows the execution of a controlled experiment to detect the heat flow through the one variable namely the concrete masonry unit (CMU) walls and to observe their thermal behavior.

### 11.3.1 Experimental Methods

The experiment of measuring the heat impinging to the heat crossing through two different types of blockwork walls but of the same thickness namely 15 cm, was executed on two different days of the month of November 2003 at the 34th Earth Latitude North at the American University in Beirut Lebanon.

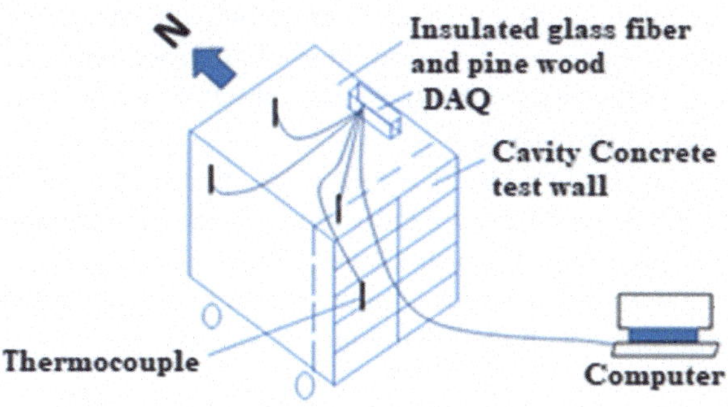


**Fig. 11.3** Mobile thermal chamber

Thermal Concrete Masonry Unit was used on a mostly sunny day of the above-mentioned November, while the regular concrete masonry unit was used in a more or less cloudy day of the same month.

Experiment on the thermal chamber, using two different types of blockworks with the same configuration and volume but with different densities, namely 2116 and 1150 kg/m$^3$. Both CMUs have the same aggregates, but the lighter blockwork contained polystyrene balls of 2 mm diameter as an additional aggregate. Putting the above mixes in similar casting forms gave each blockwork equivalent volume but two different densities (as mentioned above).

In order to demonstrate the theoretical knowledge of building envelopes the thermal behavioral data of the one cubic meter chamber observed and carefully read taking into account some of the used assumptions.

Readings of "long time intervals" should be elaborated because of the acquired high thermal capacity of the chamber. This requires lengthy observations/investigations which include the thermal behavior of the chamber during the nights' periods. For that, an automatic recording instrumentation system was chosen and installed: an NI cDAQ 9191, a NI 9213 thermocouple module. 7 K-type thermocouples were used to measure the temperatures at seven selected points.

Solar intensity was measured using a pyranometer (epply type) placed on the vertical surface of the test section.

## *11.3.2 Data Acquisition System and Thermocouples*

K-type thermocouples have been chosen because their range responds to the desired application. 10 cm thermocouple heads were chosen with a shielded cable for noise rejection and outdoor environment protection.

The seven thermocouples were installed at the following points: External ambient exposed to sunlight (E1), external ambient out of direct sunlight (E2), external surface of the insulated chamber wall (E3), internal surface of the insulated chamber wall (I1), external surface of test wall (E4), internal surface of test wall (I2), internal ambient temperature (I3). The latter is our main concern since we are studying the effect of building envelopes on resident thermal comfort.

The seven thermocouples are then connected to the DAQ. The NI 9213 module has 16 differential Channels for thermocouple acquisition with a speed up to 75S/s. The module has a resolution of 24 bits, one internal cold-junction and one internal autozero channel. It has an internal preprocessing circuit and accepts J, K, T, E, N, B, R, S thermocouple types. The NI 9213 thermocouple module was found to fit the needs of this project.

The NI cDAQ 9191 is a one slot ethernet and Wi-Fi chassis compatible with the NI 9213. Its ethernet bandwidth is much higher than that required for this project, even though it has been chosen for this project since the thermal chamber and the computer will be spaced approximately 30 m away from each other. The ethernet connection with its 100 m range is suitable.

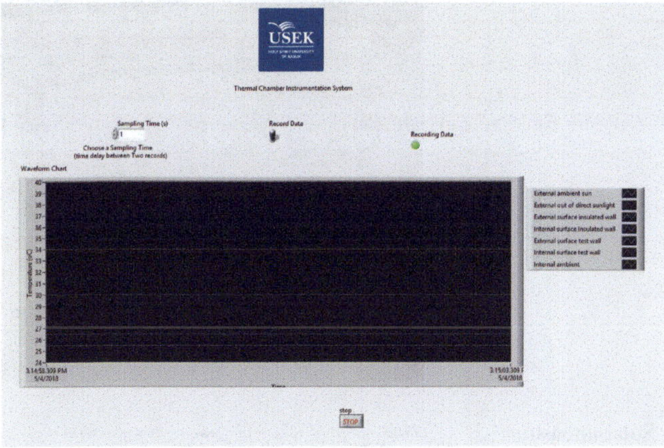


**Fig. 11.4** Windows application

### *11.3.3 Windows Application*

A windows application is deployed so that students and instructors can easily visualize and record the temperatures of the chamber. An application based on LabVIEW called "Thermal Chamber Instrumentation System" and its appropriate VI has been developed. The measured temperatures are displayed in a chart and they are saved in an excel file for future analysis. The user can control the system to start measurement and can control the sampling rate. Figure 11.4 shows the front panel of the application.

## 11.4 Thermal Chamber Experimental Results

The chamber was deployed as already noted to test the two 15 cm MCU. Keeping in mind that the tests were on separate days (since there is only one chamber), the ambient conditions are not quite identical and hence some margin of error is expected in comparisons (Fig. 11.5).

Figure 11.6 shows the solar flux intensities and ambient temperatures during the two experiments performed on separate days on the two types of blocks.

It is to be noted that the solar fluxes were quite higher when testing the thermal block (a sunnier day) as were the ambient temperatures. For this reason, when calculating heat flows normalization will be used later.

Figure 11.7 shows the obtained results for both experiments. The blue-colored data in the diagram are for the CMU with the additional polystyrene aggregate while the red colored data are for the regular CMU.

**Fig. 11.5** External view of the chamber

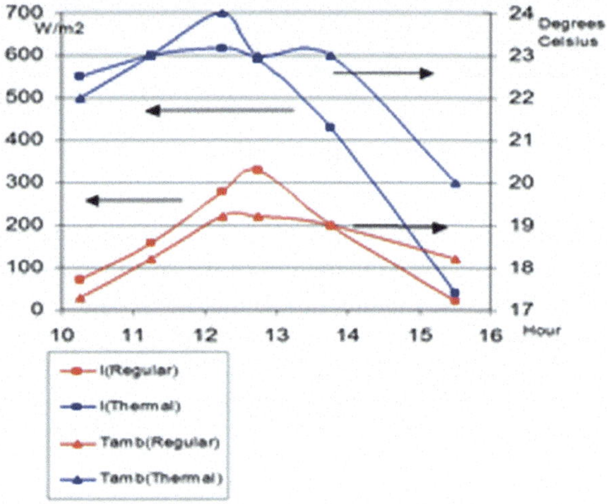


**Fig. 11.6** Solar intensities (I) on outer surface of CMU

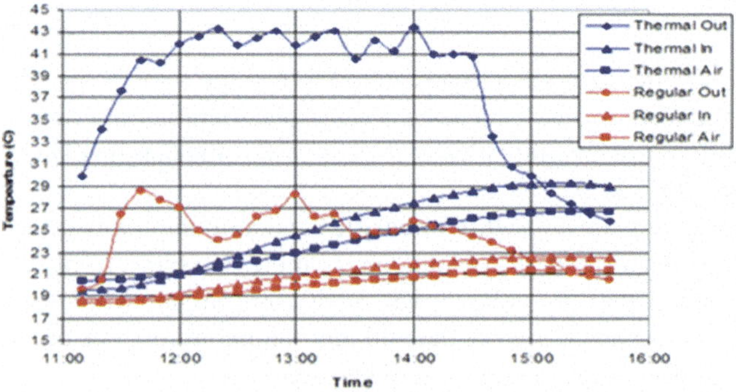


**Fig. 11.7** Experimental results for the two MCU. Thermal out = outer surface temperature of the CMU, Thermal in = interior surface temperature of the CMU, Thermal or regular air = interior air temperature inside the thermal chamber for each type

The thermal blocks were tested on a much sunnier day and hence the resulting temperatures are higher than for the regular block. However, what is relevant and most striking in Fig. 11.7 is the large difference between the outer and inner temperatures for the thermal block relative to that for the regular block. For example, at 12 noon, ΔT across the thermal block is approximately 21 °C while only 8 °C for the regular block.

Another way of comparing the two blocks, given that the solar input is not the same is to define the ratio of "Heat crossing the block—to the heat impinging on the block". Here called the Heat Flow Ratio (HFR). Figure 11.8 clarifies this.

Figure 11.9 shows the heat flow ratio during the test period. The relative heat flow into the regular block is consistently higher than for the thermal block. In fact, on average it can be estimated that the heat crossing through the regular 15 cm CMU is approximately 1.4 times more than the thermal 15 cm CMU with additional polystyrene aggregate.

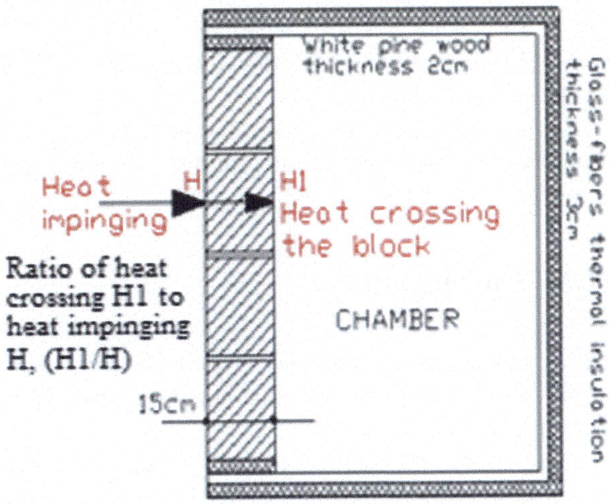


**Fig. 11.8** Section through the Thermal Chamber showing the tested wall and the wooden structure of the chamber with its insulation

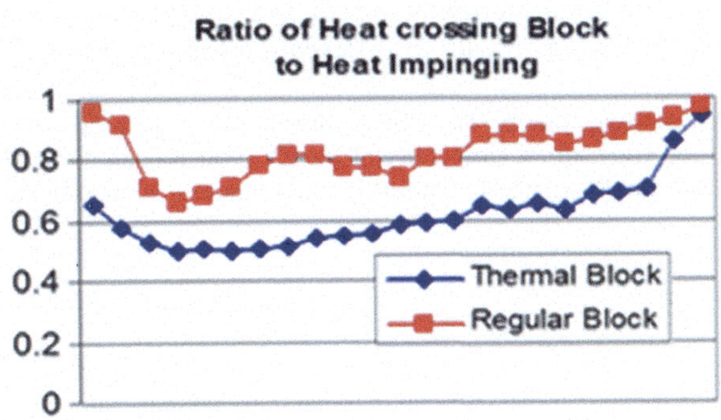


**Fig. 11.9** Heat Flow ratio (HFR) for the two CMU

As a result of the theory and experimentation, a table is developed comparing the thermal features of the commonly used 15 cm block. This is shown in Table 11.1.

## 11.5 Discussion

The calculation and experimentation showed better results when using CMU with Polystyrene. Best performance of the CMU with Polystyrene aggregate was when heat was retarded up to 50% (100% – 50% = 50%) meaning ratio of heat crossing to heat impinging was 50%, while worst performance was when retarding ratio counted only up to 35% (100% – 65% = 35%).

Best performance of regular CMU as it retarded 30% of the impinging heat (100% – 70% = 30%), worst performance of the regular CMU was when it retarded only 5% of the impinging heat (100% – 95% = 5%).

The average insulation performance of the thermal CMU with polystyrene aggregate is 35% + 50% + 45% + 42% + 36% + 37%/5% + 30% + 30% + 26% + 12% + 14% = 2, twice better than the regular CMU with the same thickness namely 15 cm.

As future works, different insulation materials would be tested using the thermal chamber. Also the effect of integrating solar panels on buildings could be studied.

## 11.6 Conclusion

The designed and built chamber is primarily for didactic use where students from the architecture department can understand the effect of building envelopes on resident thermal comfort.

Compared to the theoretical results obtained in calculating the R-values and U-values of the thermal and regular CMUs, the results of the tested CMUs in the thermal chamber show equivalent performance, meaning that the thermal chamber showed accurate results proving that it is a useful tool for students to study the thermal performance/behavior of different building materials.

**Acknowledgements** The Authors would like to thank the **HCR (Higher Center of Research)** of the **Holy Spirit University of Kaslik (USEK)** and the **CNRS-L (The National Council for Scientific Research–Lebanon)** for their financial support to the project.

## References


1. Rajasekar, E., Ashok, K., Ramachandraiah, A.: Thermal performance of building envelops. In: Encyclopedia of Sustainable Technologies, Anonymous. Elsevier, pp. 169–188 (2017)
2. Nehme, B., Akiki, T.: Implementing a didactic photovoltaic energy laboratory for developing countries. In: 2016 3rd International Conference on Renewable Energies for Developing Countries (REDEC), pp. 1–4 29 Sept 2016
3. Suresh, S., Srikanth, M., Robert, B.: Passive building energy savings: a review of building envelope component. Renew. Sustain. Energy Rev. **15**(8), 3617–3631 (2011)
4. Mirrahimi, S., Mohamed, M.F., Haw, L.C., Ibrahim, N.L.N., Yusoff, W.F.M., Aflaki, A.: The effect of building envelope on the thermal comfort and energy saving for high-rise buildings in hot–humid climate. Renew. Sustain. Energy Rev. **53**, 1508–1519 (2016). http://www.sciencedirect.com/science/article/pii/S1364032115010254
5. Gaspar, K., Casals, M., Gangolells, M.: A comparison of standardized calculation methods for in situ measurements of façades U-value. Energy Build. **130**, 592–599 (2016). http://www.sciencedirect.com/science/article/pii/S0378778816307824
6. Nehme, B., Jelwan, J., Youssef, R., Zeghondy, B.: Infrared thermography for assessing thermal bridges and humidity in lebanese building components. In: 2018 4th International Conference on Renewable Energies for Developing Countries (REDEC), pp. 1–6 (2018)
7. Nuwahid, R., Moucharrafie, F.: Results obtained from the conductivity tests at steady state on tiles (dimensions: Height × Length × width = 3 cm × 30 cm × 30 cm consecutively) of the same mix (ingredients) and density of the above given values in Table 1. Executed at the mechanical engineering laboratories at the American University of Beirut, using the Hilton Conductivity Machine
8. Engineering ToolBox. Air —Thermal Conductivity [Online]. https://www.engineeringtoolbox.com/air-properties-viscosity-conductivity-heat-capacity-d_1509.html
9. Li, R.S.: Optimization of thermal via design parameters based on an analytical thermal resistance model. In: ITherm'98. Sixth Intersociety Conference on Thermal and Thermomechanical Phenomena in Electronic Systems (Cat. No.98CH36208), pp. 475–480 (1998)

# Chapter 12
# The Relationship Between the Form of Enclosed Residential Areas and Microclimate in Severe Cold Area of China

**Tingkai Yan, Hong Jin and Hua Zhao**

**Abstract** This study selected a typical closed residential area in Harbin to explore the impact of residential layout on microclimate in severe cold regions. The relationship between living patterns and microclimates is studied by comparing the temperature and wind speed of enclosed buildings with different building densities and floor area ratios. Using PHOENICS, simulation method studies the relationship between settlement morphology and microclimate by comparing the temperature and wind speed of enclosed buildings with different building densities and floor area ratios. Results show that the building density and the floor area ratio have a strong correlation with microclimate. The influence of the plot ratio on the microclimate is greater than the building density, and the effect on the temperature is greater than the impact on the wind speed. Building density is negatively correlated with wind speed and positively correlated with temperature. Floor area ratio is negatively correlated with wind speed, while the relationship between floor area ratio and temperature is complexly and strongly influenced by shadows between buildings.

## 12.1 Introduction

In recent years, China's severe cold regions have experienced rapid economic development and accelerated urbanization. There is a big difference in the spatial form between new residential areas and old urban areas [1]. The climatic conditions and economic development in China's severe cold areas are affecting the settlement

T. Yan (✉) · H. Jin (✉) · H. Zhao
School of Architecture, Harbin Institute of Technology, 150001 Harbin, China
e-mail: 759597330@qq.com

H. Jin
e-mail: 7662572@qq.com

T. Yan · H. Jin
Key Laboratory of Cold Region Urban and Rural Human Settlement Environment Science and Technology, Ministry of Industry and Information Technology, 150001 Harbin, China

Heilongjiang Cold Region Architectural Science Key Laboratory, 150001 Harbin, China



J. Littlewood et al. (eds.), *Sustainability in Energy and Buildings*, Smart Innovation, Systems and Technologies 163, https://doi.org/10.1007/978-981-32-9868-2_12

spatial layout. The increasing urbanization rate in Northeast China has had a significant impact on urban microclimate in recent years [2]. The high-density population and high-intensity human activities will cause urban environmental problems such as large consumption of energy generation with harmful gases and particulate pollutants, the impact of urban spatial form is most serious for urban microclimate [3]. The city spatial pattern directly affects the reflectivity of the urban underlying surface facing the solar radiation or the long ground wave net radiance. In addition, different urban underlying surfaces such as asphalt pavement, cement pavement, Dutch brick pavement, and grassland have different effects on the urban thermal environment [4]. In the layout of urban buildings, the height and combination of buildings are considered to be factors affecting the urban microclimate, and the impact on the urban wind environment is more obvious [5]. The sky visibility factor will cause the changes of short-wave and long-wave radiation between buildings, and then make the space's microclimate change [6]. In addition, different building layouts and urban structures affect the absorption of solar radiation by the underlying surface, resulting in changes in the long-wave radiation emission of the underlying surface, which affects changes in the temperature and wind conditions. The urban form under the high-density development model has a more obvious impact on the urban climate [7]. The building layout and the underlying surface material directly affect the urban thermal environment, which causes the urban heat island. The building layout also affects the urban wind environment. Finally, different building combinations have a great impact on urban microclimate [8].

## 12.2 Simulated Background

Harbin belongs to the temperate continental monsoon climate. In summer, it is affected by the Pacific subtropical air mass; the dominant wind direction is the southwest wind. In winter, under the influence of polar continental air mass, the dominant wind direction is the southerly wind. In winter, the lowest temperature reaches −30.9 °C. Winter begins in November and ends in February of the following year. As shown in Fig. 12.1, the average temperature in winter is below −5 °C, and the max-

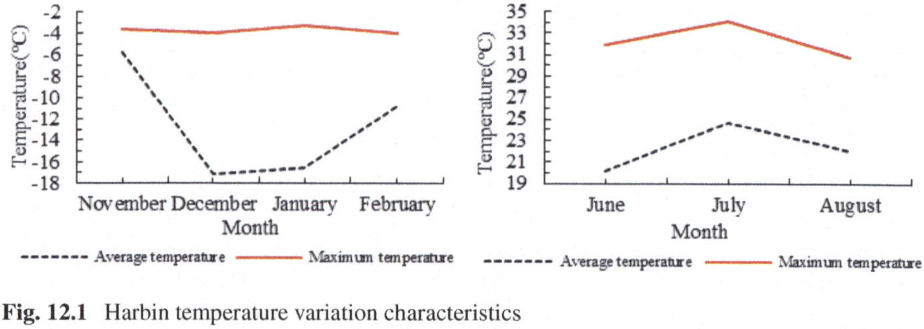


**Fig. 12.1** Harbin temperature variation characteristics

imum temperature is −3.9 °C. The winter is extremely cold and long. The average summer temperature is 22.3 °C, and the highest average temperature is in July, the average temperature is 24.7 °C, the highest temperature is 34.1 °C; the number of days with temperature exceeding 30 °C is 18 days. The number of hot days is fewer, and the overall temperature is stable and relatively cool [9].

## 12.3 Research Object and Boundary Condition

### *12.3.1 Research Object*

The study selected the representative enclosed residential area in Nangang District of Harbin and divided the spatial form of 2 km × 2 km into 16 (500 m × 500 m) spatial slices, and selected the spatial slice with a building density of 20-35%, such as 1-3, 3-3, 4-1, 4-4. As shown in Figs. 12.2 and 12.3, a spatial slice with a building density of 15% is added for comparison, as shown in Fig. 12.4.

**Fig. 12.2** Schematic diagram of the study object

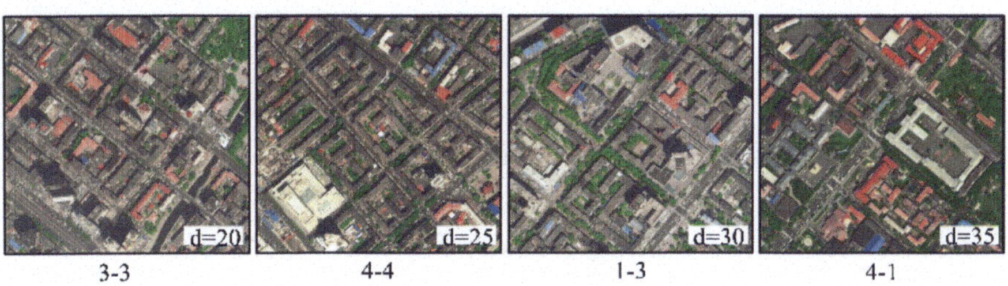


**Fig. 12.3** Actual spatial form of the study object

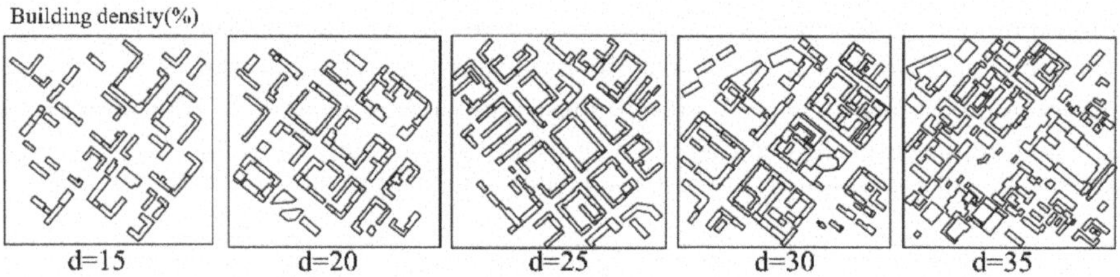


**Fig. 12.4** Building density research model diagram

**Table 12.1** Planning and floor area ratio control plan

| | Multistory building | Medium high-rise buildings | High-rise buildings | Super high-rise |
|---|---|---|---|---|
| Levels | Below 6 level | 7–11 level | 12–18 level | 19 level or more |
| Floor area ratio | 0.8–1.2 | 1.5–2.0 | 1.8–2.5 | 2.4–4.5 |

In the floor area ratio simulation study, the spatial pattern with a building density of d = 25% is used as the basic layout scheme. Combined with the current urban planning law for the definition of different building heights and the control detailed planning of different residential land, as shown in Table 12.1, five groups of case ratios of 1, 1.5, 2.5, 3.5, and 4.5 were selected for research.

### *12.3.2 Calculation Area*

In the initial stage of the simulation, by expanding the range of the simulated mesh, the 4x region of the model range is selected as the model simulation buffer to eliminate the boundary influence and improve the simulation result accuracy. The simulation center is refined with a more detailed grid as shown in Fig. 12.5. The boundary conditions as shown in Table 12.2. Based on the obtained meteorological information, the wind and temperature field on July 29, 2017 are used as boundary conditions for simulation. The specific model parameters are shown in Table 12.3.

## 12.4 Simulation Results Analysis

### *12.4.1 Analysis of Building Density Simulation Results*

**Wind Field Simulation Analysis.** By comparing the simulation results of five groups, as shown in Fig. 12.6, the building layout has a great influence on the urban

**Fig. 12.5** Schematic diagram of mesh grid

**Table 12.2** PHOENICS boundary conditions table

| External temperature (°C) | External pressure (pa) | Direct solar radiation (W/m$^2$) | Wind speed (m/s) | Wind direction | Diffuse solar radiation (W/m$^2$) |
|---|---|---|---|---|---|
| 23 | 101325.0 | 400 | 8 | E-S-E | 300 |

**Table 12.3** Material property setting table

| Attributes | Building | Road | Hard paving |
|---|---|---|---|
| Density (kg/m$^3$) | 2150.00 | 2000.00 | 2000.00 |
| Specific heat capacity (J/kg K) | 744.19 | 500.00 | 500.00 |
| Thermal conductivity (W/m K) | 1.16 | 1.16 | 1.15 |
| Emissivity | 0.93 | 0.95 | 0.95 |
| Solar absorption coefficient | 0.65 | 0.65 | 0.65 |
| Convective heat transfer coefficient (W/m °C) | 30.00 | 10.00 | 10.00 |

wind environment. Results show that in densely built areas, wind speed has decreased significantly. Comparing with case 1–5, with the increase of the building density, the wind speed in the central area decreases. When the building density is 20%, the average wind speed in the central area is more than 2.5 m/s; the area of the wind speed less than 1.75 m/s decreases; the average wind speed of the windward street is more than 3 m/s, and the maximum wind speed can reach 5.2 m/s; when the building

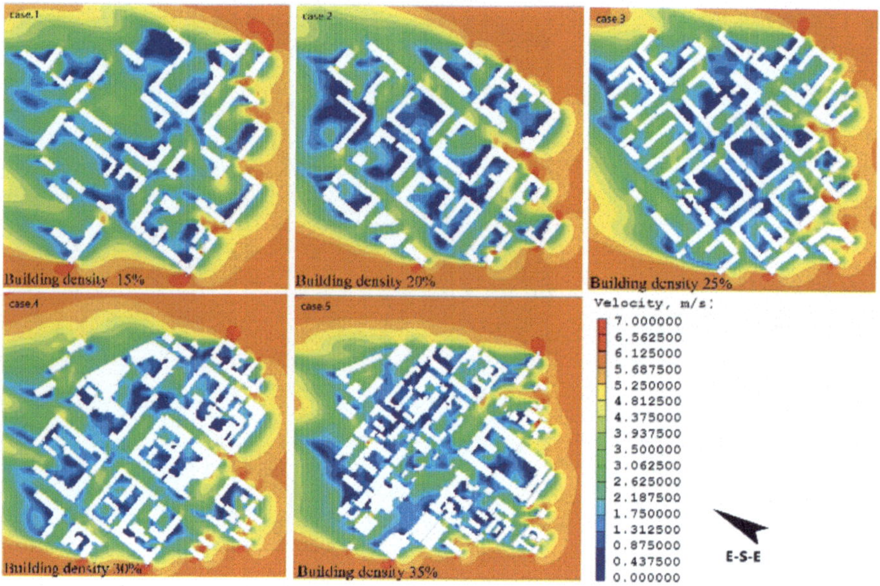


**Fig. 12.6** Building density simulation wind field distribution map

density is more than 30%, the wind speed in the central area decreased significantly, and the average wind speed is less than 2.5 m/s.

By comparing the proportion of wind speed as shown in Table 12.4, the proportion of the building density in the area with wind speed less than 2 m/s is 41.35%, and the proportion of building area is the largest. When the building density is 20%, the proportion of area with V > 5 m/s is 23.83%, and the proportion is the largest. Among the area proportion of 2 m/s < V < 5 m/s, 20, 25, and 35% of the building density are similar. In conclusion, when the building density is 25%, the wind environment of pedestrian space is the most suitable.

**Building Density Temperature Simulation Analysis**. Figure 12.7 shows that the building density has a great influence on the temperature distribution, and the space temperature is higher in the dense buildings. The smaller the enclosed space, the higher the internal temperature such as in case 5. The main reason is that the building surface of the enclosed space is affected by solar radiation and heat transfer, which result in the temperature changes of enclosed space. The larger the ratio of the

**Table 12.4** Composition density and wind speed area ratio

| Building density (%) | 15 | 20 | 25 | 30 | 35 |
|---|---|---|---|---|---|
| 0 m/s ≤ V ≤ 2 m/s area ratio (%) | 25.32 | 31.11 | 41.35 | 28.56 | 35.96 |
| 2 m/s ≤ V ≤ 5 m/s area ratio (%) | 66.45 | 45.06 | 53.42 | 65.54 | 52.47 |
| V ≥ 5 m/s area ratio (%) | 8.23 | 23.83 | 5.23 | 5.90 | 11.57 |

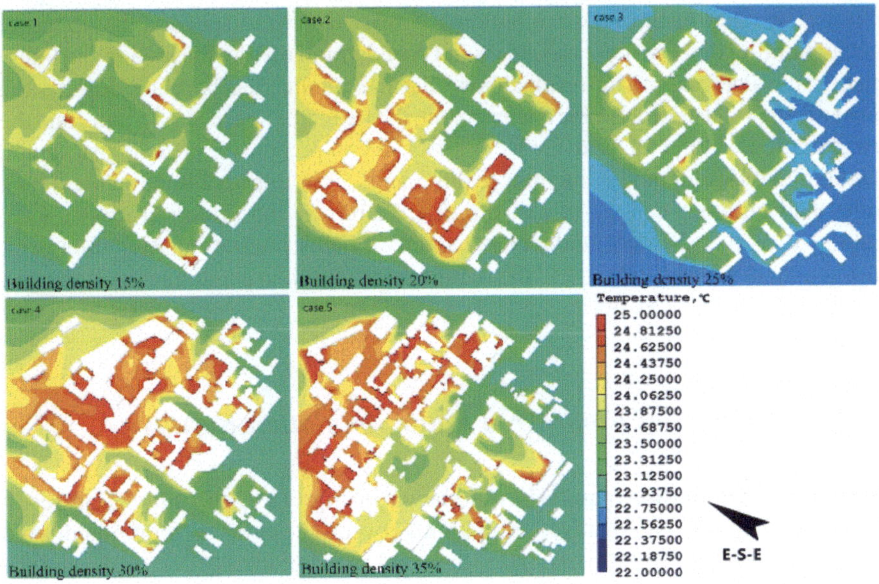


**Fig. 12.7** Building density simulation temperature field distribution map

building surface area to enclosed space, the more obvious the increase of internal temperature is. Therefore, the enclosed inner space area's temperature is smaller than that of the larger space area, and the temperature of the enclosed inner space is higher. The results of case 1–5 show that when the building density reaches 25%, the temperature of the simulated area is the lowest and the building density reaches 35%; the area of T (>25 °C) is larger and widely distributed.

**Building Density and Microclimate Correlation**. Comparing the building density with wind speed and average temperature trend as shown in Fig. 12.8, the study found that with the increase of the building density, the temperature rises and the wind speed decreases. And the building density increases from 15 to 20% and the average temperature rises faster; when the density increases to 20%, the temperature rises

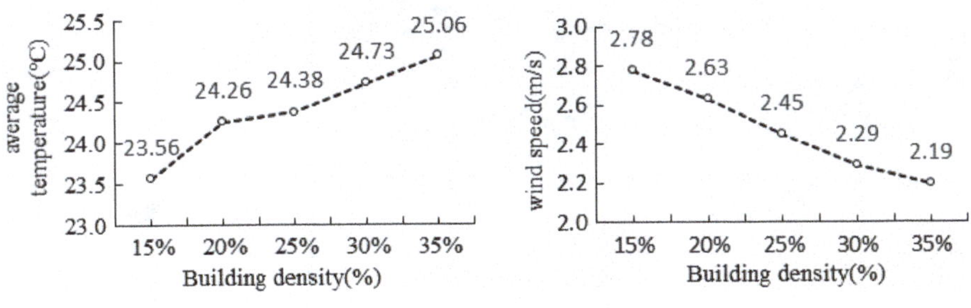


**Fig. 12.8** Relationship between building density and average temperature/wind speed

slowly. The relationship between building density and wind speed is the opposite. As the building density increases, the wind speed gradually decreases and the linear relationship decreases. The results show that the smaller the building density, the better the regional microclimate environment, the better the space temperature and the better ventilation effect. Due to the factors such as comprehensive urban construction, planning and design, and economic requirements, it is impossible to create a plan with small building density in real projects; therefore, the preferred index of the building density is 20–25% and it can guarantee a better microclimate environment, meanwhile, it is also conducive to the requirements of urban economic development.

### 12.4.2 Analysis of Floor Area Ratio Simulation Results

**Wind Field Simulation Analysis**. As shown in Fig. 12.9, the simulation result displays that there is a negative correlation between the floor area ratio and wind speed. As the floor area ratio increases, the wind speed gradually decreases, and the area of the simulated center breeze area gradually increases. Comparing case 1 and case 2, when the floor area ratio changes from 1 to 1.5, the wind speed decreases more obviously the wind speed v $\leq$ 2 m/s area increases, and the regional ventilation capacity decreases. Comparing case 2 and case 3, when the floor area ratio increases from 1.5 to 2.5, the wind speed changes greatly, and the deceleration rate is faster.

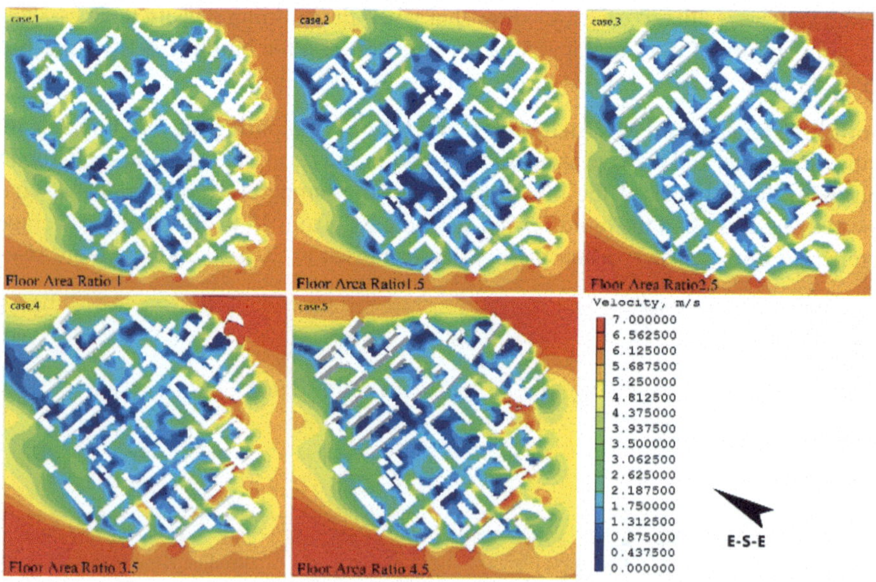


Fig. 12.9 Floor area simulation wind field distribution map

Compared with case 4 and case 5, when the floor area ratio increases from 3.5 to 4.5, the wind speed does not change significantly, and the area of wind speed $v \leq 2$ m/s increases accordingly. However, the area of the wind speed $v \geq 5$ m/s does not change much. Therefore, the results show that the change in the floor area ratio directly affects the wind environment of the building.

By comparing the wind speed ratio, as shown in Table 12.5, the wind speed $V \leq 2$ m/s area ratio, the plot ratio of 1.5 accounted for the largest, 26.78%. In the area ratio of 2 m/s $\leq V \leq$ 5 m/s, the ratios of floor area 1.5, 3.5, and 4.5 are similar. The comprehensive comparison shows that when the floor area ratio is 1.5, the wind speed is mostly 0 m/s $\leq V \leq$ 2 m/s, the pedestrian space wind environment is the most suitable.

**Temperature Field Simulation Analysis**. In order to study the influence of the floor area ratio's change on the central city's thermal environment, see Fig. 12.10. The results show that there is a positive correlation between the floor area ratio and the temperature, that is, as the floor area ratio increases, the regional temperature

**Table 12.5** Ratio of floor area to wind speed area

| Floor area ratio | 1 | 1.5 | 2.5 | 3.5 | 4.5 |
|---|---|---|---|---|---|
| 0 m/s $\leq V \leq$ 2 m/s area ratio (%) | 18.86 | 26.78 | 22.47 | 23.65 | 19.99 |
| 2 m/s $\leq V \leq$ 5 m/s area ratio (%) | 71.02 | 61.67 | 66.74 | 62.69 | 62.94 |
| $V \geq$ 5 m/s area ratio (%) | 10.12 | 11.55 | 10.79 | 13.66 | 17.07 |

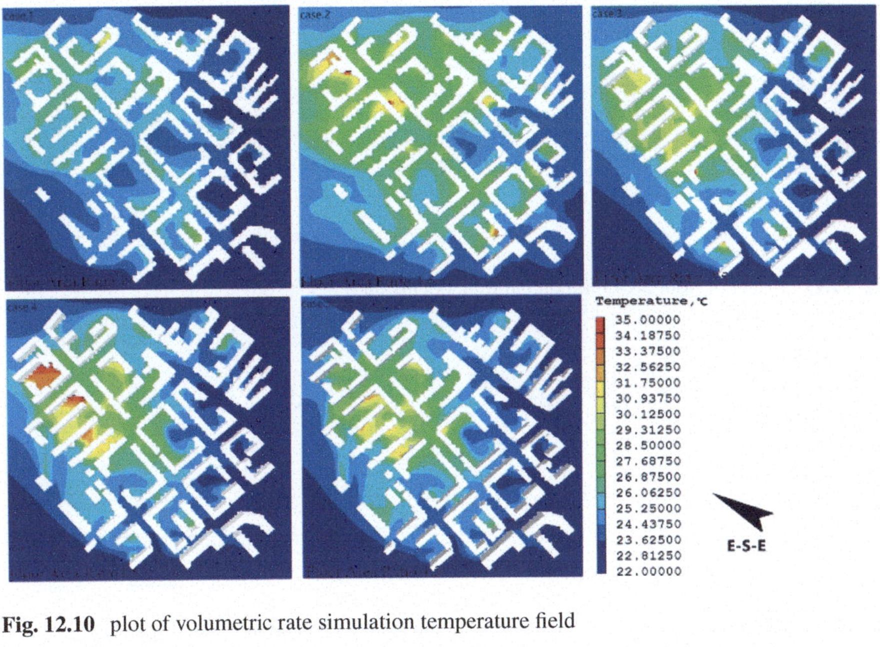


**Fig. 12.10** plot of volumetric rate simulation temperature field

gradually increases, and the high-temperature region gradually increases. Comparing case 1 and case 2, when the floor area ratio is from 1 to 1.5, the temperature rise is more obvious, and the area of temperature T ≥ 30 °C increases. Comparing case 2 and case 3, when the floor area ratio is from 1.5 to 2.5, the temperature change is small, the temperature rises slowly, and the area of T ≥ 30 °C increases, but the maximum temperature decreases; comparing with case 3 and case 4 is obtained. When the floor area ratio is from 2.5 to 3.5, the temperature changes obviously, and the area of T ≥ 35 °C appears. Comparing case 4 and case 5, the results show that when the floor area ratio is >4.5, the high-temperature area decreases while the maximum temperature decreases.

Therefore, the change of the floor area ratio affects the temperature of the air around the building. As the floor area ratio increases, the temperature rises first and then decreases. When the floor area ratio increases from 3.5 to 4.5, the shadow in the space basically covers the ground layer, as shown in Fig. 12.11, the blockage between buildings is more serious, the temperature drops, and the maximum temperature decreases. The floor area ratio increases from 1 to 3.5 and the temperature rises rapidly. When the floor area ratio is >3.5, as the floor area ratio continues to increase, the temperature tends to decrease.

**Floor Area Ratio and Microclimate Correlation.** By comparing the floor area ratio with the wind speed and the average temperature as shown in Fig. 12.12, it is found that when the floor area ratio is between 1 and 2.5, the temperature rises rapidly. When the floor area ratio reaches 2.5, the temperature changes slowly. When the floor area ratio reaches 3.5, the temperature drops. There is a negative correlation between the floor area ratio and wind speed. As the floor area ratio increases, the

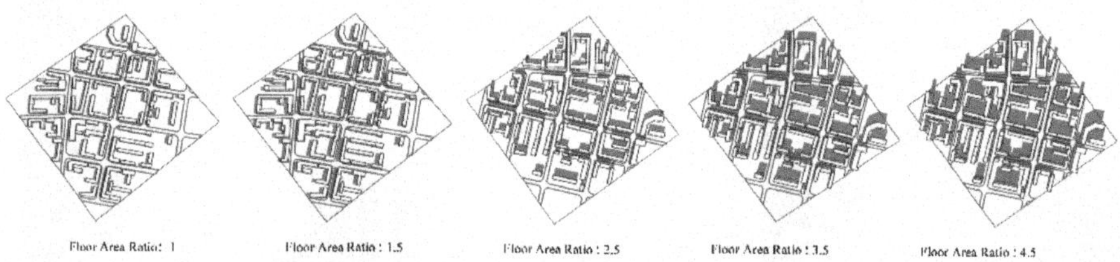


**Fig. 12.11** Effect of floor area ratio on building occlusion

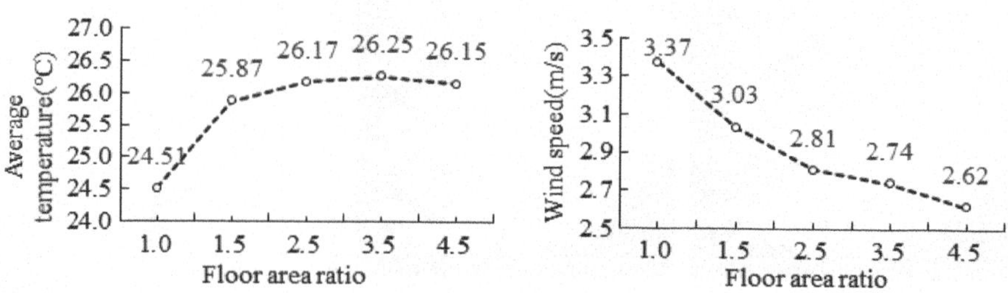


**Fig. 12.12** Relationship between floor area ratio and average temperature/wind speed

wind speed decreases gradually. When the floor area ratio is 1–2.5, the wind speed drops significantly.

## 12.5 Conclusion

Through the investigation, the paper summarizes the typical enclosed residential area in Harbin and establishes a spatial model. By comparing the air temperature and wind speed of different building densities and floor area ratios, it is concluded that there is a close relationship between the building density, floor area ratio, and microclimate.

In the spatial form of the enclosed residential area, the effect of plot ratio on microclimate is more than the density of the building. The influence on temperature is about 3 °C, the effect on wind speed is 0.75 m/s, and the effect on temperature is greater than wind speed.

There is a positive correlation between building density and temperature. As the density increases, the air temperature gradually increases. When the building density is between 20 and 25%, the temperature tends to be stable. The building density is negatively correlated with wind speed. As the building density increases, the wind speed decreases, and when the building density is 25%, the space wind environment at the pedestrian height is most suitable.

The relationship between the floor area ratio and the temperature is complicated. As the floor area ratio increases, the temperature gradually increases. When the floor area ratio reaches 2.5, the temperature tends to be stable. When the floor area ratio exceeds 3.5, the space temperature decreases due to the excessive shadow blocking area; the floor area ratio is negatively correlated with the wind speed, and the wind speed gradually decreases as the floor area ratio increases.

## References


1. Wowo, D., et al.: Study on the relationship between urban morphology and urban microclimate. Archit. J. 16–21(2012)
2. Yumeng, J., et al.: Effects of openings on the wind-sound environment in traditional residential streets in a severe cold city of China. Environ. Plan. B Urban Anal. City Sci. (2018). https://doi.org/10.1177/2399808318805490
3. Bereitschaft, B., et al.: Urban form, air pollution, and $CO_2$ emissions in large U.S. metropolitan areas. Prof. Geogr. **65**(4), 612–635 (2013)
4. Yang, Y., Zou, Z., et al.: Comparative study on the thermal environment effect of six urban underlying surfaces. Acta Sci. Nat. Univ. Pekin. (2017)
5. Elnabawi, M.H., Hamza, N., et al.: Numerical modelling evaluation for the microclimate of an outdoor urban form in Cairo, Egypt. HBRC J. **11**(2), 246–251 (2015)
6. He, X., et al.: Influence of sky view factor on outdoor thermal environment and physiological equivalent temperature. Int. J. Biometeorol. **59**(3), 285–297 (2015)
7. Chun, B., Guldmann, J.M.: Spatial statistical analysis and simulation of the urban heat Island in high-density central cities. Landsc. Urban Plan. **125**, 76–88 (2014)

8. Erell, E.: The application of urban climate research in the design of cities. Adv. Build. Energy Res. **2**(1), 27 (2008)
9. Civil Building Thermal Design Code (GB50176-97). National Standards of People's Republic of China (1997)

# Chapter 13
# The Successful Introduction of Energy Efficiency in Higher Education Institution Buildings

**Dirk V. H. K. Franco, Marijke Maes, Lieven Vanstraelen, Miquel Casas and Marleen Schepers**

**Abstract** The conditioning of buildings equals 40% of the energy consumption and is responsible for 36% of the greenhouse gas emissions (within the EU). Given that a complete renewal of the existing building stock would take about 100 years, investing in energy efficiency (EE) in existing buildings is crucial to reduce energy consumption and greenhouse gas emissions. In addition, besides energy reduction and energy flexibility also the transition towards sustainable and renewable sources is important. In this study, we report the EE potential and the financial scenario, obtained from Energy Quick Scans (EQS) for a higher education institution. The introduction of an Energy Performance Contract (EPC) with the help of an Energy Service Company (ESCO) is also discussed.

## 13.1 Introduction

Since the beginning of the industrial revolution, the global economy has extracted and used 0.5 trillion tonnes of oil equivalent of fossil fuels and has led to 1.2 trillion tonnes of $CO_2$ emissions [1]. The international climate debate started more than 25 years ago. The optimists notice an important evolution in the fact that not only the EC countries have committed themselves to conducting a climate policy, that collects us collectively from the two degrees average heating [2–4]. But the debate is ongoing and the society is asking for or taking themselves more initiatives. The SDGs are often taken as a frame of reference. Moreover, it is often possible to see the interactions

D. V. H. K. Franco (✉) · M. Maes
PXL, Central Administration, Building A, Elfde Liniestraat 25, 3500 Hasselt, Belgium
e-mail: dirk.franco@pxl.be; dirk.franco@uhasselt.be

D. V. H. K. Franco
Faculty of Science, UHasselt, Universitaire Campus, 3590 Diepenbeek, Belgium

L. Vanstraelen · M. Casas
Energinvest, Energinvest 107 rue Joseph Coosemansstraat, 1030 Brussels, Belgium

M. Schepers
PXL Tech, Universitaire Campus, 3590 Diepenbeek, Belgium



J. Littlewood et al. (eds.), *Sustainability in Energy and Buildings*, Smart Innovation, Systems and Technologies 163, https://doi.org/10.1007/978-981-32-9868-2_13

between climate, environment and wellbeing, for certain interventions [5, 6]. Energy is SDG nr 7 '*Ensuring access to affordable, reliable and modern energy for all has come one step closer due to recent progress in electrification, particularly in LDCs, and improvements in industrial energy efficiency. However, national priorities and policy ambitions still need to be strengthened to put the world on track to meet the energy targets for 2030*' [5].

The targets for the EC are

- a 40% cut in greenhouse gas emissions compared to 1990 levels
- at least a 27% share of renewable energy consumption
- indicative target for improvement in energy efficiency at EU level of at least 27% (compared to projections), to be reviewed by 2020 with an EU level of 30% in mind
- support the completion of the internal energy market by achieving the existing electricity interconnection target of 10% by 2020, with a view to reaching 15% by 2030 [7, 8].

## 13.2 Background and Related Work

### *13.2.1 Buildings, Energy and Climate/Environment*

The overall energy consumption for buildings equals 40% and they are also one of the most significant sources of greenhouse gas emissions at 36% (for EU) [7]. In addition, as the energy performance of buildings is poor and as the average life span of a building is over 50 years [8] a complete renewal of the existing building stock would take about 100 years. Investing in building refurbishment is crucial to reduce energy consumption and greenhouse gas emissions and relevant for the achievement of the energy and climate objectives of the European Union (EU) for 2020 and 2050 [7, 9, 10]. In Belgium, 22.9% of the $CO_2$ emissions is caused by buildings and the production of heat and electricity [11]. The introduction of an EPC (an output-driven approach rather than an input-driven approach) with an ESCO (a public or private organization that provides integrated energy solutions to its customers) and/or a facilitator, is more and more introduced [12–14].

### *13.2.2 Role of Higher Educational Institutions (HEI)*

All stakeholders must play a crucial role in the (energy) transition. [15–18]. The quadruple model, a strong interaction between government, knowledge institutes, industry and society catalyze the transition towards a truly sustainable society. The

University college PXL is active in all domains of sustainable development (research, education, campus greening…) [15].

### 13.2.3 Output-Driven Instead of Input-Driven

Often the Trias Energetica is applied, however, the chronological procedure (design, installation, maintenance, repair, replacement) is not very effective in view of energy management, as there is no concrete feedback [12–14]. It is clear that, although there is a more or less linear relation between the 'E-level of buildings' and the real energy consumption, on individual dwelling level, the E-level as such cannot be used for the prediction of the real energy consumption [19, 20], as, apart from the building concept, also the occupants behaviour has an impact [21, 22] and can lead to quite differing energy consumptions between dwellings with the same E-level. An output-driven approach is possible by means of an ESCO. An ESCO focuses on improving energy efficiency or energy savings in the existing buildings. Typical for ESCO is the provision of performance contracts, where not only a contractual guarantee is given to the customer on the estimated energy savings but also comfort or energy delivery can be guaranteed via a performance contract [23, 24]. In addition to the new technologies, new business models will be developed [25, 26]; (see Fig. 13.1).

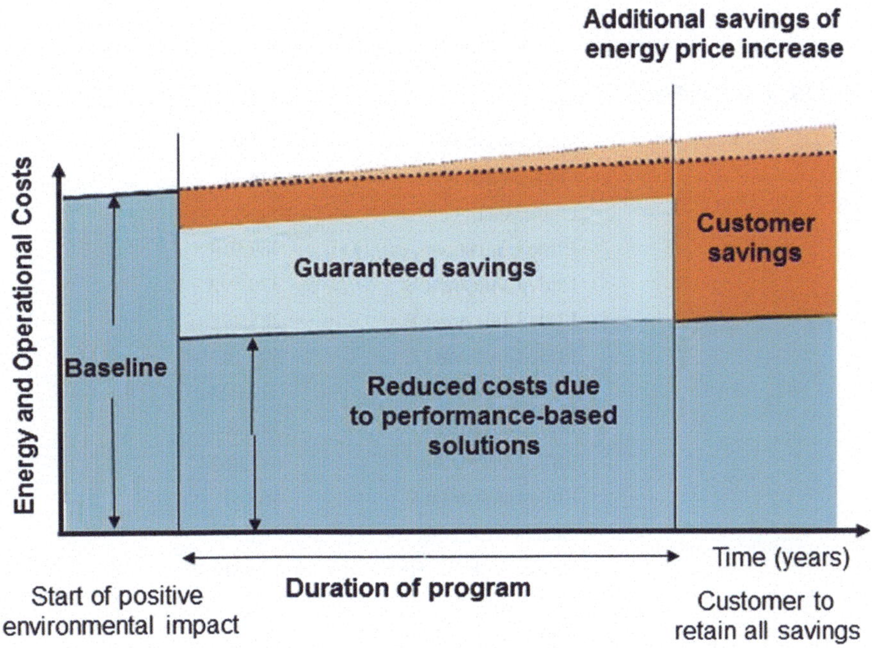


**Fig. 13.1** EPC concept [23]

### 13.2.4 Methods for Energy Quickscans

In this study, the first results of a campaign of Energetic Quickscans (EQS) in 12 PXL buildings are reported (see Table 13.1). The evaluations of the investment scenarios were realized on the basis of matrices developed by Energinvest. These matrices are standard tools in Excel and allow an energy auditor to evaluate the appropriate energy savings with a level of accuracy that is equal to that of a Quickscan (± 20%). The approach is similar to an energy audit that an ESCO would perform in the framework of a public contract for an EPC project. The investment matrix aims for PXL to:

(1) Identify the investment measures that an ESCO could propose in the context of a tender;
(2) Evaluate their technical and financial potential and return;
(3) Validate the economic pertinence of an EPC;
(4) Admit to assign a correct scope of an EPC contract (required investments, expected energy savings, financial montage);
(5) Facilitate the competition between ESCOs during the audit phase of an EPC assignment (the data of the investment matrices can be communicated to the ESCOs in the specifications)
(6) Validate the pertinence of the measures proposed by the ESCOs during the EPC assignment.

The EPC concept is illustrated in Fig. 13.1 and Table 13.2 and the role of the different partners is clarified in Fig. 13.2.

**Table 13.1** PXL buildings

| Site | Consumption (kWh/y) |
|---|---|
| Guffenslaan P | 402.428 |
| Guffenslaan N | 474.313 |
| Guffenslaan O | 74.875 |
| Elfde Liniestraat A | 469.914 |
| Elfde Liniestraat B | 376.392 |
| Elfde Liniestraat B | 285.422 |
| Elfde Liniestraat C | 592.143 |
| Elfde Liniestraat D | 377.923 |
| Elfde Liniestraat D | 348.466 |
| Elfde Liniestraat F | 617.862 |
| Elfde Liniestraat E | 46.167 |
| Qcanal M | 42.652 |
| Diepenbeek H CHF1 | 1.060130 |
| Vilderstraat T 1 | 1,923.368 |
| Vilderstraat T 2 | 114.034 |
| Vilderstraat T 3 | 17.185 |

**Table 13.2** Different scopes for an EPC contracting

| Scope of the EPC project | Actions |
|---|---|
| EPC maintained | Recommissioning<br>quick wins |
| EPC techniques | Measures in the fields of production, distribution and emission of heating and sanitary hot water with a payback time (PBT) of maximum 12 years<br>Renovation/optimization of stoves and fireplaces<br>Renovation of sanitary hot water installations Place insulating reflectors behind the radiators<br>Relamping and relighting |
| EPC building envelop | Improvement of the building shell (PBT maximum 20 y)<br>Replacing old windows and doors, overhauling roofs<br>Isolation walls |

## *13.2.5 Investment*

**Time value money**

As stated above, energy savings result in economic savings. The latter has been calculated by taking into account the time value or opportunity cost of money. One Euro today has a higher value than one Euro in the future, as today's Euro can be invested in, for instance, stocks and shares that bring along some return on investment.

Therefore, future cash flows have been recalculated to their 'present' value (NPV) according to the following equation [27, 28]:

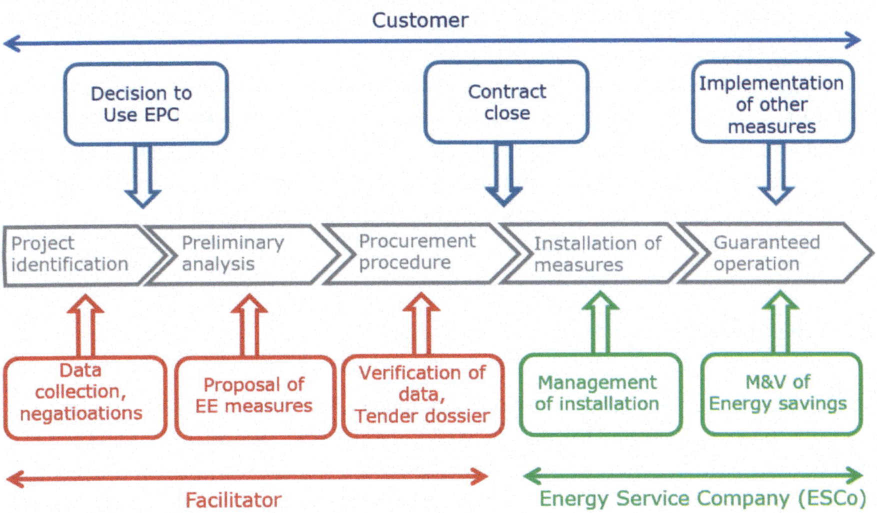


**Fig. 13.2** Different steps and roles in an EPC [35]

$$NPV = \frac{CF_t}{(1+d)^t} \tag{13.1}$$

With

NPV Net present value of the future cash flow
$CF_t$ cash flow in year t
d discount rate reflecting the time value of money

Here, we used a social discount rate of 4% as advised by the Flemish Government for public investments [29].

**Total Cost of Ownership (TCO)**

In order to estimate the profitability of the EPC project in TCO [30–32] and to know the impact on the budget, financial analyses were done through simulations. In addition to the direct expenditures and benefits of the energy component of the EPC project, expense savings and opportunity costs will be generated by comparing the EPC project with a Business As Usual (BAU) reference scenario.

### *13.2.6 Social Innovation*

In order to successfully implement and execute the new trends with all stakeholders, social efforts are necessary. The concept of an output-driven approach (EPC) with the support of an ESCO is a new way of thinking. It is important that we transform the possible resistance of co-workers and users of the buildings into ownership. In this way, an intrinsic motivation and enthusiasm will arise for working with an EPC with the help of an ESCO/facilitator. Figure 13.3 shows the interaction needed to come to good energy performance of building [15].

The interactions between the physical environment, knowledge, attitudes and behaviours [21, 22, 32] are crucial for EE projects in HEI. The PXL will, therefore, undertake an energy literacy survey (a self-administered online questionnaire [33, 34]) on energy-related attitudes and the effects of EE projects on their health and well-being for students and all users/stakeholders of the building.

## 13.3 Results

Based on the data collected in the investment matrix, we have 2 investment scenarios: *Base case* 'heating/Sanitair Hot Water (SHW) and lighting' and an investment volume of €2,610,903 which generates annual savings of €242,569 € with a simple payback time of 11.5 years and an *Enhanced case* 'heating/SHW and lighting', an investment volume of €4,145,854, which generates annual savings from €296,581 with a simple payback time of 15.2 years.

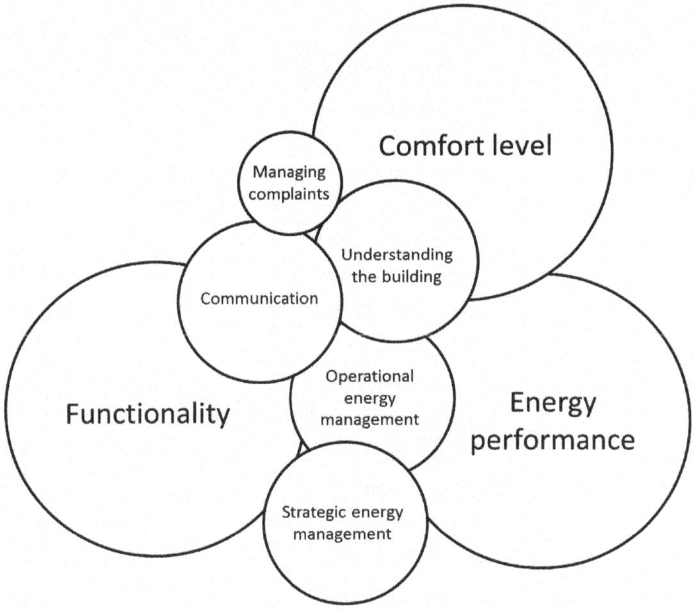


**Fig. 13.3** Different types of interactions in order to obtain good energy performance [15]

### *13.3.1 Base Case*

These EQS highlight the potential on heating, ventilation, cooling, lighting and insulation on the outer shell. The influence of the lighting measures (reduced internal heat release) on the efficiency of the boiler room measures was also taken into account. The total energy consumption for heating/SHW and lighting is estimated at 9.9 million kWh, while the potential energy savings are estimated at 3.07 million kWh or 31%. These energy savings provide a financial saving of €226,720/y for one investment of €2.6 million, with a simple payback time of 11.5 years. The energy consumption decreases with 42.4 kWh/m$^2$/y (94.5 kWh/m$^2$/y compared to 136.9 kWh/m$^2$/y). The investment amount is 36.1 €/m$^2$ (Table 13.3).

### *13.3.2 Enhanced Case*

The total energy consumption for heating/SHW and lighting, of course, remains 9.9 million kWh, while the potential energy savings are estimated at 3.9 million kWh (compared to 3.07 million kWh in the base case) or 39.12%.

These energy savings result in a financial saving of €271,940/year (Base case: €226,720/year) for an investment of €4,1 million (base case: €2,6 million), with a simple payback time of 15.2 years (base case: 11.5 years). Energy consumption

**Table 13.3** Base and enhanced case PXL buildings

| Topic | Base case | Enhanced case |
|---|---|---|
| Consumption total (kWh/y) | 9.891.356 | 9.891.356 |
| Consumption total (kWh/m$^2$/y) | 136.9 | 136.9 |
| Consumption total (€/y) | 778.044 | 778.044 |
| Consumption total (€/m$^2$/y) | 10.8 | 10.8 |
| Total investment (€) | 2.610903 | 4.145.854 |
| Total investment (€/m$^2$) | 36.1 | 57.4 |
| Savings (kWh/y) | 3.329.616 | 4.156.002 |
| Savings (kWh/m$^2$/y) | 46.1 | 57.5 |
| Savings (kWh/y) | 3.065465 | 3.869.488 |
| Savings incl rel (kWh/m$^2$/y) | 42.4 | 53.6 |
| Savings (€/y) | 242.569 | 296.581 |
| Savings (€/m$^2$/y) | 3.4 | 4.1 |
| Savings (€/y) | 226.720 | 271.940 |
| Savings (€/m$^2$/y) | 3.1 | 3.8 |
| Savings total (%) | 33.66 | 42.02 |
| Savings total incl effect rel (%) | 30.99 | 39.12 |
| Payback time (y) | 11.5 | 15.2 |
| Reduction $CO_2$ (tonne/y) | 606 | 762 |

decreases by 53.6 kWh/m$^2$/y (83.3 kWh/m$^2$/y compared to 136.9 kWh/m$^2$/y). The investment amount is 57.4 €/m$^2$. Tables 13.4 and 13.5 below show the EPC/BAU income and expenses that are included in the simulations (amounts including VAT).

**Table 13.4** EPC/BAU benefits

| EPC savings | Base case | Enhanced case |
|---|---|---|
| Energy benefits/y | €274.331 | €329.047 |
| BAU savings | | |
| Reparations and maintenance apart from the maintenance contract | | €26.500/y during 25 y |
| Maintenance contract<br>Expenses for internal maintenance and exploitation | | €142.464/y<br>0.5 fte (€67.000/y) |

**Table 13.5** EPC/BAU costs

| EPC costs | Base case | Enhanced case |
|---|---|---|
| Investments<br>Facilitator | €2.831.196<br>€181.500 | €4.472.083<br>€181.500 |
| EPC exploitation and maintenance costs | €480.000/y | €480.000/y |
| BAU avoided expenses during this EPC period | €95.000/y or 75% of the estimated EPC investment of the base case scenario (NPV) | |

## 13.4 Discussion

### *13.4.1 Base Case*

Gas (heating/SHW) at 35.5% has an energy saving in kWh for financial savings from €153,890. The investment amounts to €1.8 million, representing 68% of the total investment. The investment per $m^2$ amounts to 24.4 €/$m^2$. The investment return is 8.7 Euro cent/y per Euro invested (simple payback time of 11.5 y).

Electricity (lighting): 18.77% energy saving in kWh for financial savings from €72,830. The investment amounts to €848,488 or 32% of the total investment. The investment per $m^2$ amounts to 11.7 €/$m^2$. The return on investment is almost the same as 8.6 Eurocents/year per invested Euro (simple payback time of 11.7 years). If we look ino the detail of all 12 sites with regard to gas, we notice that five sites represent 70% of the investments (and 67% of the financial savings (Elfde Linie A, B, C, F and Diepenbeek H).

Seven sites represent 87% of the investments (and 88% of the financial savings) (the already mentioned buildings and Guffenslaan N and Vilderstraat T). The remaining 5 sites represent only 13% of the investments. The savings percentages per site vary considerably, from 3% to 82%, with an average of 39%. The simple payback time is a maximum of 16 years, with an average of 10 years. As far as light is concerned, 3 sites represent 74% of the investments (and 68% of the financial savings). The remaining 7 sites represent only 26% of the investments. The savings percentages per site vary, from 8% to 29%, with an average of 19%. The payback time is a maximum of 16 years, with an average of 12 years. Here too, the savings rate for the 3 sites rises barely (up to 22% compared to 19% for all sites). The payback time even rises to 13 years.

### *13.4.2 Enhanced Case*

Gas (heating/SHW): 45.94% energy saving in kWh for financial savings from €199,110. The investment amounts to €3.1 million, representing 75% of the total investment. The investment per $m^2$ amounts to 43.1 €/$m^2$. The investment return is 6.4 Euro cent/year per Euro invested (simple payback time is 15.6 y).

Electricity (lighting): 20.66% energy saving in kWh for financial savings from €80,280. The investment amounts to 1,033,038 € or 25% of the total investment. The investment per $m^2$ amounts to €14.3/$m^2$. The return on investment is slightly lower.

The conclusions for the different buildings are largely similar to those of the base case scenario, in particular with regard to the risks and benefits in an EPC project.

These fluctuations of potential and profitability demonstrate

(1) the importance of being able to avoid risk arising from investment concentration
(2) benefits through the pooling of buildings in an EPC project.

After all, from a budgetary perspective, for example, limiting the scope of the EPC project to the seven most important buildings makes it possible to limit the investment budget, but it hardly affects the performance of the project. The savings percentage only rises from 39% to 42%. The payback time remains identical: 10 years. Comparable projects (public buildings, $m^2$ and similar investment possibility) can be used as a benchmark. In project 1, an energy reduction of 30% (kWh) (HVAC and relighting, contract 12 y) was possible. In project 2, 36.2% energy savings (HVAC, relighting and isolation, 15 y) or 43.8% (HVAC, relighting and isolation (20 y and higher investment (+ 44%)) are possible. In project 3, the energy savings differ from 24.4% (kWh) (only HVAC (12 y)) or 20,8% on gas and electricity (HVAC and relighting) (15 y). Independent of the investment scenario and funding, the EPC project proves to be profitable with positive NPVs over the lifetime of the project (25 years). The comparison of the scenarios shows that the profitability of the project decreases with the increase in the investment volume. The financing options (own resources, bank loan) is fundamentally dependent on the budgetary choices of the PXL. As of course, the financing with own resources has a strong impact on the budgetary situation of the PXL during the first 2 years, it yields more budgetary benefits in the following years.

## 13.5 Conclusion

The analysis and comparisons with reference cases show that the energy savings are realistic and that the ESCO market must be able to realize these objectives. The choice of the specific scenario will ultimately be determined by the PXL ambition vision. Important parameters are the objectives in terms of costs, energy and $CO_2$ reductions and the financial limitations of PXL. The energy efficiency also yields Non-Energy Benefits (NEB) as better image and improved comfort (more stable temperatures, which can stimulate less absenteeism, more creativity for teachers, students and all occupants of the building). This project facilitates also further knowledge exchange. In a first step, the PXL-Tech campus (as a living lab) can be the ideal pilot, which can then be rolled out in other high-energy buildings of other PXL departments. The interactions between the physical environment and students' knowledge, attitudes and behaviours are crucial for EE projects in HEI. The link between sustainability and well-being offers the opportunity to shift the attention of students more to corporate social responsibility in combination with the Sustainable Development Goals. This is very important as these initiatives deal with the involvement and behaviour of future generations—'decision-makers'.

**Acknowledgements** The views expressed in this article are solely the responsibility of the authors. The authors are very grateful to Christine Schoeters, Laura Franco and Viviane Mebis in assisting to finalize an earlier version of the article.

## References

1. Fouquet, R.: Historical energy transitions: Speed, prices and system transformation. Energy Res. Soc. Sci. **22**, 7–12 (2016)
2. Bienstman, M.: Op Eigen kracht, De bevrijding van Gas, Olie en kernenergie. België: Borgerhof and Lambrechts (2017)
3. European Environment Agency (23–24/10/2014). Accessed from http://www.consilium.europa.eu/uedocs/cms_data/docs/pressdata/en/ec/145356.pdf
4. Bruyninckx, H. (2014). European Environment Agency 'Signals 2014', Well-being and the environment. Consulted on January 10th, 2019. Accessed from http://www.eea.europa.eu/publications/signals-2014/download
5. The 2030 Agenda for Sustainable Development Goals, 2016 Consulted on January 10th, 2019 Accessed from https://sustainabledevelopment.un.org/?menu=1300
6. Green building &and the Sustainable Development Goal http://www.worldgbc.org/green-building-sustainable-development-goals Last Accessed 1 Jan 2018
7. Technology Roadmap, Energy-efficient Buildings: Heating and Cooling Equipment, OECD, IEA, Paris (2010)
8. S.N.: Directive 2002/91/EC of the European parliament and of the council on the energy performance of the buildings. Off. J. Eur. Communities **4**, 65–71 (2003)
9. Lützkendorf, T., Balouktsi, M.: From energy demand calculation to life cycle environmental performance assessment for buildings: status and trends.In: Green and Technology (2018)
10. Azvedo, I., Fouquet, D., Glachant, J-M., Kaderják, P., Kotek, P., Meeus, L.,... von der Fehr, N-H.: How to refurbisch all buildings by 2050, Final Report. Think, 1–64 (2012). Accessed from: https://www.eui.eu/projects/think/documents/thinktopic/thinktopic72012.pdf
11. LNE, 2019 https://www.lne.be/ecocampus Last Accessed 10 Jan 2019 https://www.duurzaambedrijfsleven.nl/future-leadership/24021/expertpanel-de-sdgs-zijn-voor-bedrijven-het-nieuwe-winnen Webpagina last Accessed 10 Jan 2019
12. Langlois, P., Hansen, S.J.: World ESCO Outlook. The Fairmont Press, Lilburn (2012)
13. Bertoldi, P., Boza-Kiss, B., Panev, S., Labanca, N.: ESCO Market Report 2013. European Commision, Luxembourg (2014)
14. Bleyl, J.W.: ESCO market development: a role for facilitators to play. including national perspectives of task 16 experts IEA DSM task 16 discussion paper April. Accessed from http://www.ieadsm.org/wp/files/Tasks/Task%2016%20-%20Competitive%20Energy%20Services%20(Energy%20Contracting,%20ESCo%20Services)/Publications/Facilitators_Task16_Discussion%20paper_incl.%20national%20subchapter_140505.pdf
15. Franco, D.V., De Vocht. A., Martens, H., Thewys, T., Vanheusden, B., Schepers, M., Segers, J.P.: Sustainable education an essential contribution at quadruple helix interaction towards a sustainable paradigm shift. In: World Environmental Education Congress (2017)

16. Disterheft, A., Caeiro, S., Azeiteiro, U.M., Leal Filho, W.: Sustainability Science and Education for Sustainable Development in Universities: A Way for Transition. In: Caeiro, S., Filho, W., Jabbour, C., Azeiteiro, U. (eds.) Sustainability Assessment Tools in Higher Education Institutions. Springer, Cham (2013)
17. Franco, D., De Langhe, R., Venken, J.: Energy Efficiency Services in Buildings: A Tool for Energy Transition Central Europe Towards Sustainable Building, p. 215 (2016)
18. Experpanel: 'De SDG's zijn voor bedrijven het nieuwe winnen' [website]. (z.j.). Consulted on January 10th, 2019. https://www.duurzaambedrijfsleven.nl/future-leadership/24021/expertpanel-de-sdgs-zijn-voor-bedrijven-het-nieuwe-winnen
19. Taranu, V., Verbeeck, G.: A closer look into the European Energy Performance Certificates under the lenses of behavioural insights – a comparative analysis. Energ. Effi. **11**(7), 1745–1761 (2018). https://doi.org/10.1007/s12053-017-9576-6
20. Verbeeck, G.: Lessons from reality: examples of near zero energy dwellings in practice (2014). https://doi.org/10.13140/2.1.1121.8888
21. Berardi, U.: Stakeholders' influence on the adoption of energy-saving technologies in Italian homes. Energy Policy **60**, 520–530 (2013)
22. Albino, V., Berardi, U.: Green buildings and organizational changes in Italian case studies 21(6), 387–400 (2012)
23. About EPC Consulted on January 10th, 2019. Accessed from http://www.eu-esco.org/index.php?id=21
24. Energy Performance Contracting. Consulted on January 10th, 2019. https://www.belesco.be/solutions/energy-performance-contracting
25. Hannon, M.J., Foxon, T.J., Gale, W.F.: 'Demand pull' government policies to support product-service system activity: the case of energy service companies (ESCos) in the UK. J. Clean. Prod. 900–915 (2015)
26. Bocken, N., Boons, F., Baldassarre, B.: Sustainable business model experimentation by understanding ecologies of business models. J. Clean. Prod. **208C**, 1498–1512 (2018). https://doi.org/10.1016/j.jclepro.2018.10.159
27. Berk, J., DeMarzo, P.: Corporate Finance, 3rd edn. Pearson Education (2014)
28. Kuppens, T., Van Dael, M., Vanreppelen, K., Thewys, T., Yperman, J., Carleer, R., Schreurs, S., Van Passel, S.: Techno-economic assessment of fast pyrolysis for the valorization of short rotation coppice cultivated for phytoextraction. J. Clean. Prod. **88**, 336–344 (2014)
29. Milieueconomie. Consulted on January 10th, 2019 https://www.lne.be/milieueconomie
30. Smart Energy Performance Contracting. Consulted on January 10th, 2019. https://www.energinvest.be/services/smart-energy-performance-contracting
31. Vanstraelen, L., Marchand, G., Casas, M., Creupelandt, V., Steyaert, E.: Increasing capacities in Cities for innovating financing in energy efficiency. In: A Policy Framework for Sustainable Real Estate in the European Union: Multidisciplinary Approaches to an Evolving System (2015). https://doi.org/10.1007/978-3-319-94565-1_5
32. Franco, D.V.H.K., Kuppens, T., Beckers, D., Cruyplandt, E.: Energy efficiency in school buildings? how to use in a successful way the triple bottom line framework? In: Kaparaju, P., Howlett, R., Littlewood, J., Ekanyake, C., Vlacic, L. (eds.) Sustainability in Energy and Buildings 2018. KES-SEB 2018. Smart Innovation, Systems and Technologies, vol 131. Springer, Cham (2019)
33. Cotton, D., Miller, W., Winter, J., Bailey, I., Sterling, S.: Developing students' energy literacy in higher education. Int. J. Sustain. High. Educ. **16**(4), 456–473 (2015)
34. Cotton, D.R.E., Shiel, C., Do Paco, A.: Energy saving on campus: A comparison of students' attitudes and behaviours in the UK and Portugal. J. Clean. Prod. **129**, 586–595 (2016)
35. Facilitators Guideline Consulted on April 10th, 2019. https://guarantee-project.eu/ie/wp-content/uploads/sites/6/2013/11/EPC-Facilitator-Guidelines.pdf

# Chapter 14
# LCA Integration in the Construction Industry: A Case Study of a Sustainable Building in Aveiro University

**Kamar Aljundi, Fernanda Rodrigues, Armando Pinto and Ana Dias**

**Abstract** This work aims to apply Life Cycle Assessment (LCA) methodology to evaluate the potential environmental impacts of three structural solutions (mixed, steel and reinforced concrete) of a building in Aveiro University (Portugal), comparing two life spans for the building: 50 and 100 years. Designing a 100-year life span building requires structural materials with higher resistance and higher environmental impacts in construction phase than a 50-year design. Therefore, a detailed assessment of materials and environmental impacts for those two different life spans was performed. Data were collected on the amounts of materials consumed in each alternative and on the environmental impacts associated with the production of each material. The work concludes that the concrete alternative is environmentally more sustainable than the other two structural solutions. A sensitivity analysis can show whether the 100-year design could compensate for the 50-year design, particularly when the first one needs maintenance once, while the second one requires maintenance twice. Moreover, the sensitivity analysis helped finding out the reasons for uncertainty of LCA results, highlighting the need for a treatment and more sufficient maintenance actions avoiding using materials with high environmental impacts.

## 14.1 Introduction

The construction industry is contributing largely to natural resources depletion and to the serious increase in the biocapacity of the planet. For example, the actual ecological footprint was estimated in 1.6 planets and it is still increasing, a value that is not sustainable in the long term [1]. The embodied energy in building construction could be nearly 15–20% of the whole building life cycle energy consumption, which

K. Aljundi (✉) · A. Pinto
National Laboratory for Civil Engineering, Lisbon, Portugal
e-mail: kaljundi@lnec.pt

F. Rodrigues
RISCO, Civil Engineering Department, University of Aveiro, Aveiro, Portugal

A. Dias
CESAM, Department of Environment and Planning, University of Aveiro, Aveiro, Portugal



J. Littlewood et al. (eds.), *Sustainability in Energy and Buildings*, Smart Innovation, Systems and Technologies 163, https://doi.org/10.1007/978-981-32-9868-2_14

can be higher with the building energy efficiency increase, namely in mild climates like Mediterranean climate. Moreover, the building sector contributes to 38% of summer smog and 20% of heavy metal emissions [2]. With the continuous increase in the global environmental crisis and the correspond growth of warnings about the climate changes, sustainability could be the last claim to reduce those negative effects. However, the construction sector is contributing largely to these effects, which could be minimized by applying sustainability measures in the building sector, such as (i) selecting sustainable materials; (ii) creating sustainable sites and work places; (iii) taking care of the water efficiency using suitable equipment and resilience strategies; (iv) analysing and controlling the energy need; (v) enhancing the quality of the indoor environment and (vi) innovating the design process. Therefore, acting in the design phase is essential to decrease the impacts of all phases of a building from the beginning. Life cycle assessment (LCA) is a tool that can be used for this purpose, supporting the choice of more sustainable materials and solutions with less environmental impacts [3]. For instance, energy savings could approach 50% and the decrease of carbon dioxide emissions could reach 30% if construction materials are carefully selected [2, 4].

LCA is recognized as an innovative method, which helps applying and improving sustainability concept in the construction industry throughout all stages of the building life cycle, considering raw materials' extraction, construction, use and maintenance/final disposal or demolition. To help classify the different kinds of LCA studies in the construction sector, several authors have used a simple classification distinguishing Building Material and Component Combination (BMCC) LCA and Whole Process Construction (WPC) LCA [5, 6].

The first objective of this work is to quantify and to analyse the environmental impacts of three most common structural solutions in Portugal in order to identify the most sustainable one. For that purpose, a real built solution, which is a mixed structure consisting basically of steel structure with mixed slabs and reinforced concrete foundations, is compared with two equivalent alternatives: a steel structure alternative and a 100% reinforced concrete one. Moreover, since the structural materials must have a long life with a sufficient quality that enables them to keep the building stable and safe, a design scheme of 50 years' life span and a design scheme of 100 years' life span are also assessed and compared. This study aims to verify how the 100-year life span design can compensate for the 50-year life span design, using materials with higher environmental impacts and reducing the required actions of maintenance.

## 14.2 Methodology

The LCA study follows the definitions and procedures described in the international standards ISO 14040 and ISO 14044 [7, 8], and includes the four mandatory phases:

1. Goal and scope definition: The goal of this study states the intended application, the reasons for carrying out the study. The scope of the LCA study is to ensure that the breadth, the depth and the detail of the study are compatible and sufficient to address the stated goal.
2. Inventory analysis: Involves data collection and calculation procedures to quantify relevant inputs and outputs of a product system.
3. Life cycle impact assessment: Aims to evaluate the significance of potential environmental impacts using the results of the life-cycle inventory analysis.
4. Interpretation phase: It is the phase where the results are discussed and explained.

### 14.2.1 Case Study

The case study is the most recent building of the Art and Communication Department of Aveiro University with a plan area of 1600 $m^2$. It aims to provide a creative educational environment for its students and professors throughout its halls, auditorium and classrooms. It was selected since it intends to be an example of a scholar sustainable building. The underground floor is composed of backstage rooms situated below the auditorium. The ground floor has two distinct parts: an auditorium with all its supporting areas, such as the toilets, halls and circulations; and a scholar part with 7 classrooms and laboratories. The first floor consists of the rest of the auditorium and their supporting rooms and circulations, which vertically occupies the same area as on the ground floor. Like the ground floor, there is also a scholar part with 2 classrooms, 1 studio, 2 laboratories and 2 offices, with an open zone between them. The second and the last floor are only dedicated to scholar purposes. It has a big room for the researchers and the Ph.D. students and 12 classrooms for design, 1 students' room and offices. The building is a mixed superstructure of steel and concrete. Concrete piles foundations majorly compose its foundations. The basement floor has a concrete foundation slab and concrete bearing walls. The superstructure beams and columns are mainly in steel and there are 20 steel bracing systems. The slabs in the first and second floors are collaborative of steel and concrete. Finally, the building has three metallic staircases supported by steel beams and columns.

### 14.2.2 Goal and Scope Definition

The goal of the present study is to compare the environmental impacts of three types of structures (mixed structure, only steel structure, only reinforced structure). Those three alternatives are pre-designed using the Eurocode and studied according to two life spans, 50 and 100 years, considering the maintenance requirements and the design needs for each life span as it is recommended in E464 LNEC [9].

In fact, the concrete structure could have carbonation and sea chlorine penetration ions which will affect its reinforcement and its stability. Therefore, it is assumed that the covering layer should be substituted after 50 years considering the worst hypothesis. With that maintenance action, the durable age of the reinforced concrete structure is being extended for more than 50 years. It is also assumed that this new concrete will be the same as that of the construction phase.

Since the life cycle of all alternatives is 100 years, the alternatives with 50-years life span, will be maintained twice, while the alternatives with 100 years will be maintained once only. Thus, in alternatives with 50-years life span, the concrete will be maintained twice after the first and the second 50-years of its 100 years life cycle.

On the other hand, in alternatives with 100 years life span, the concrete will be only substituted once at the end of its life cycle.

According to [9], alternatives with 100 years life span will be designed with a concrete which its structural class is higher than that of alternatives with only 50 years life span.

Finally, those six alternatives are analysed in a cradle-to-cradle LCA approach for a 100 years life cycle considering two life span designs, using SimaPro (see Table 14.1).

The three alternatives were modelled using Autodesk Revit as a Building Information Modelling tool [10] to provide the study with the quantities of each material in each alternative (see Figs. 14.1, 14.2 and 14.3).

**Table 14.1** Description of the six alternatives analysed by LCA using SimaPro

| Alternative | Description | Life span design | Alternatives |
|---|---|---|---|
| 1 | Real case study (mixed) | 50 years life span | Mixed.50 |
| | | 100 years life span | Mixed.100 |
| 2 | Only steel structure | 50 years life span | Steel.50 |
| | | 100 years life span | Steel.100 |
| 3 | Only reinforced concrete | 50 years life span | Concrete.50 |
| | | 100 years life span | Concrete.100 |

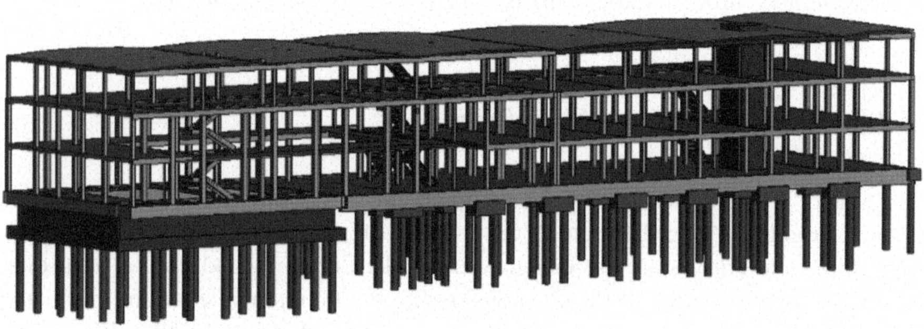

**Fig. 14.1** 3D Revit model of alternative 1

Fig. 14.2 3D Revit model of alternative 2

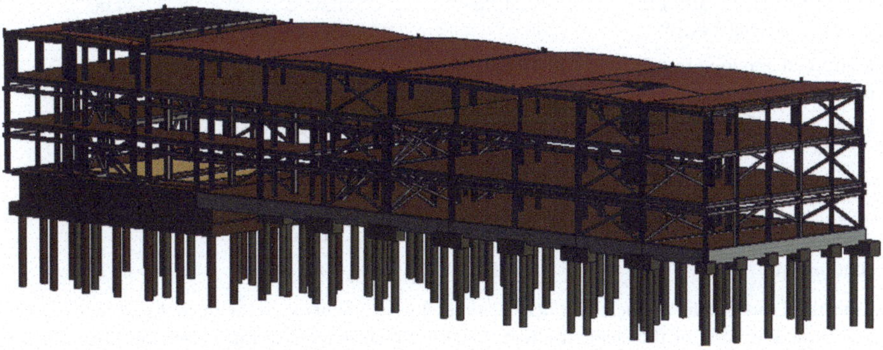

Fig. 14.3 3D Revit model of alternative 3

The functional unit (reference unit in relation to which the results are expressed) in all the alternatives is the building.

### 14.2.3 Inventory Analysis

The case-study building was modelled using Autodesk Revit as a BIM methodology tool, considering its real and constructed design as it was provided by the contractor. The same software provided this study with materials' quantities which were compared with the contractor's bill of quantities, as it is presented in Table 14.2.

Table 14.2 validates the Revit model created, since the difference between the values of the real bill of quantities and quantities extracted automatically in Revit is less than 5% in all the fields, except painting which is even less than 5.5%.

Data on the production of each material were taken from Ecoinvent database (version 3.3) embedded in SimaPro [11]. The materials used in the structure are summarized in Table 14.3, which also provides a short description of each material

Table 14.2 Structural materials' quantities of the real case study

| Elements and their functional units | Bill of quantities | Revit quantities | Difference (%) |
|---|---|---|---|
| Steel columns and beams (kg)[a] | 389,482 | 379,437 | 2.5 |
| Reinforced concrete piles (m) | 1059 | 1030 | 2.82 |
| Reinforced concrete elevator box ($m^3$) | 17.81 | 17.91 | −0.55 |
| Reinforced concrete in columns ($m^3$) | 3.86 | 3.68 | 4.66 |
| Reinforced concrete bearing walls ($m^3$) | 177.97 | 187.31 | −4.98 |
| Reinforced concrete in collaborative slabs ($m^3$)[b] | 328.97 | 345.93 | −4.90 |
| Haircol 59S steel part of the collaborative slabs (kg)[b] | 21959 | 23092 | −4.90 |
| Reinforced concrete in solid slabs ($m^3$) | 389.59 | 407 | −4.27 |
| Formwork ($m^2$)[c] | N.A. | Foundations: 2458 | N.A. |
| | | Beams: 1231 | |
| | | Slabs: 1275 | |
| | | Columns: 41 | |
| Painting ($m^2$) | 7617 | 8048 | −5.35 |

[a]In the BIM model, the connections between the steel elements were neglected and their weight was assumed as 12% of the steel structure weight, which is the real ratio that could be obtained from the contractor's bill of quantities of the case-study building
[b]In the contractor's bill of quantities, there is only the area of collaborative slabs, so its steel weight was indirectly calculated from that area value of 2530.5 $m^2$
[c]In the contractor's bill of quantities, there is only the area of collaborative slabs, so its steel weight was indirectly calculated from that area value of 2530.5 $m^2$. In the contractor's bill of quantities, there is no value for formwork because it is included in the reinforced concrete articles. Therefore, it is impossible to do a validation for this value

and the main characteristics of the processes found in Ecoinvent are embedded in SimaPro.

### *14.2.4 Life Cycle Impact Assessment*

The impact categories considered in this study are (i) The global warming potential (kg $CO_2$ eq); (ii) Ozone depletion potential (kg CFC eq); (iii) Acidification Potential Total (kg $SO_2$ eq); (iv) Eutrophication Potential Total (kg NOx eq) and (v) Smog Potential (kg $O_3$ eq). To calculate the total impacts of each alternative, the impacts' indicators for each functional unit of each construction material must be multiplied by the quantities provided in the inventory stage. The sum of all environmental

**Table 14.3** Inventory analysis of materials database

| Material | Description | Ecoinvent SimaPro classification |
|---|---|---|
| Concrete in collaborative slabs | C20/25, XC3 + XS1 | C20 with a density of 2335 kg/m$^3$ |
| Concrete in solid slabs | C35/45, XC4 + XS3 + XA2 | C35 with a density of 2315 kg/m$^3$ |
| Concrete in foundations | C35/45, XC4 + XS3 + XA2 | C35 with a density of 2315 kg/m$^3$ |
| Concrete for 100 years design | C45/50, XC4 + XS3 + XA2 | C50 with a density of 2300 kg/m$^3$ |
| Steel | Hot rolled steel | Steel, hot rolled, low alloyed steel |
| Wood | Plywood | Plywood, outdoor use |
| Painting | Epoxy with zinc | Only zinc coating |

impacts of each material of each building gives the global environmental impacts of the alternative. Therefore, the global functional unit of this LCA is one building. In this study, the midpoint approach is used, since it allows predicting the potential environmental impacts of the previous categories which are significantly affecting humanity. In this approach, the impact assessment method used is TRACI, which is the Tool for the Reduction and Assessment of Chemical and other Environmental Impacts developed by the U.S. Environmental Protection Agency (EPA) and is primarily used in the United States [12].

## 14.3 Results

The environmental impacts of each category obtained for each alternative (functional unit) are depicted in the Figs. 14.4, 14.5, 14.6, 14.7 and 14.8.

The comparison of the six alternatives shows that the reinforced concrete alternative is environmentally more sustainable than the mixed and the steel structure, for all impact categories analysed. Moreover, the 100-year design has less environmental impacts than the 50-year design, highlighting that materials with high environmental impacts and only with maintenance required once per life cycle could compensate the use of materials with lower environmental impacts and twice the maintenance needs during the life span.

In order to compare the results obtained in the present study with other studies' results and validate them, it is essential to divide them by the constructed area (m$^2$), which is equal to 4704 m$^2$. Therefore, the results per square metre are depicted in Table 14.4.

Ngo et al. [13] showed that the global warming impacts per m$^2$ are almost 470 kg $CO_2$ eq for a concrete building and 780 kg $CO_2$ eq for a steel-framed building. In

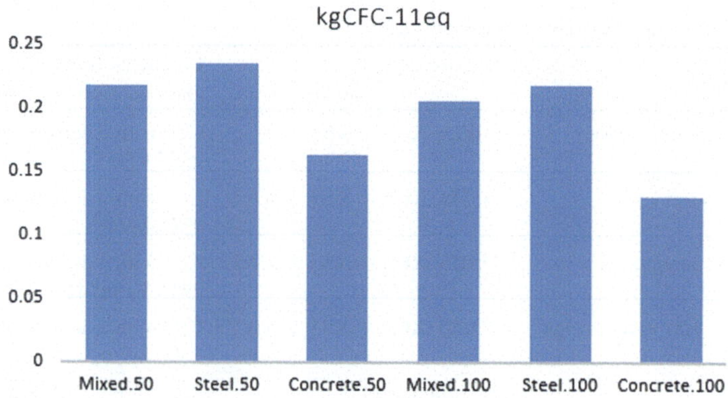


**Fig. 14.4** The ozone depletion impact category calculated in SimaPro

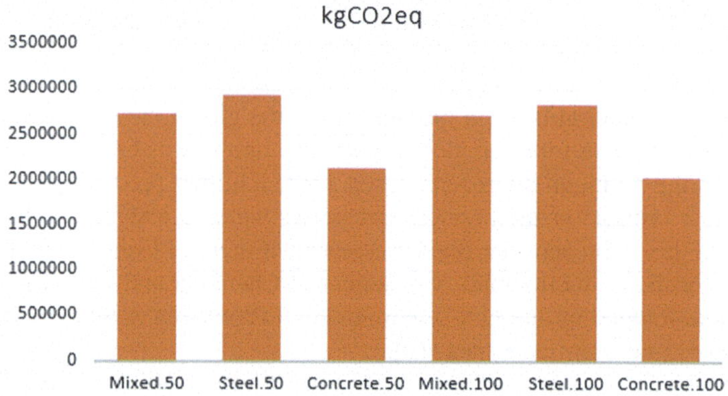


**Fig. 14.5** The global warming impact category calculated in SimaPro

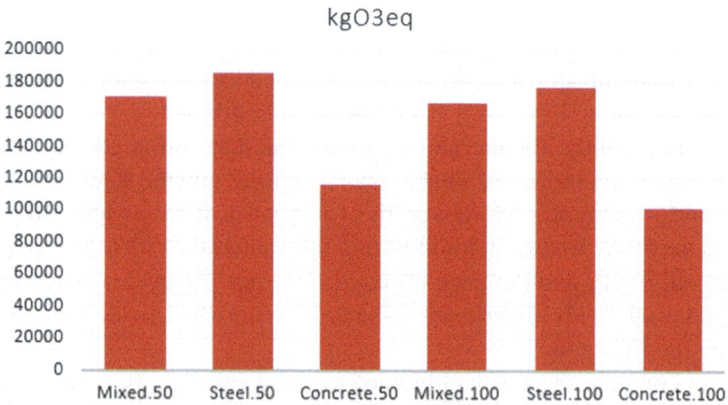


**Fig. 14.6** The smog impact category calculated in SimaPro

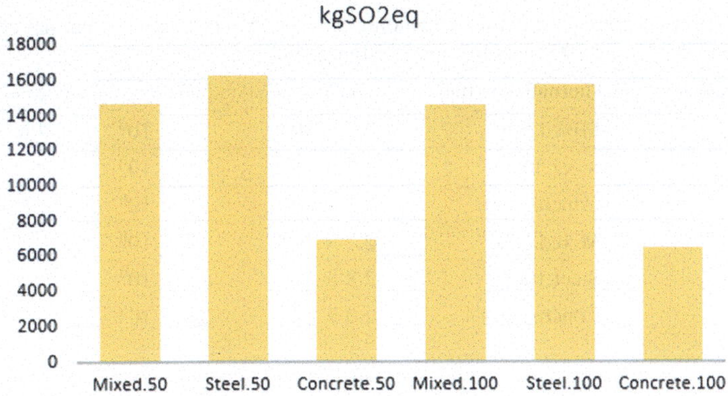


**Fig. 14.7** The acidification impact category calculated in SimaPro

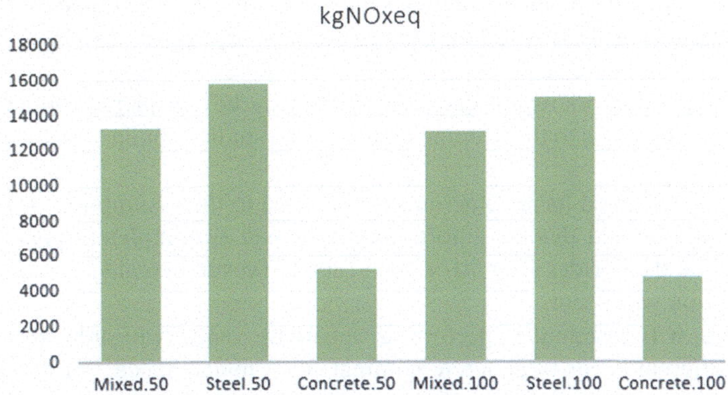


**Fig. 14.8** The eutrophication impact category calculated in SimaPro

**Table 14.4** LCA impacts of each alternative per $m^2$

| Impact Cat. | Mixed.50 | Steel.50 | Concrete.50 | Mixed.100 | Steel.100 | Concrete.100 |
|---|---|---|---|---|---|---|
| kg CFC-11 eq/$m^2$ | $4.7 \times 10^{-5}$ | $5.0 \times 10^{-5}$ | $3.5 \times 10^{-5}$ | $4.4 \times 10^{-5}$ | $4.7 \times 10^{-5}$ | $2.8 \times 10^{-5}$ |
| kg $CO_2$ eq/$m^2$ | $5.8 \times 10^2$ | $6.3 \times 10^2$ | $4.5 \times 10^2$ | $5.8 \times 10^2$ | $6.0 \times 10^2$ | $4.3 \times 10^2$ |
| kg $O_3$ eq/$m^2$ | $3.6 \times 10^1$ | $4.0 \times 10^1$ | $2.5 \times 10^1$ | $3.5 \times 10^1$ | $3.8 \times 10^1$ | $2.2 \times 10^1$ |
| kg $SO_2$ eq/$m^2$ | 3.1 | 3.5 | 1.5 | 3.1 | 3.4 | 1.4 |
| kg NOx eq/$m^2$ | 2.8 | 3.4 | 1.1 | 2.8 | 3.2 | 1.0 |

**Table 14.5** Global warming emissions when only 50% or 25% of concrete nominal cover needs maintenance in the 50 and 100 years life span design period

| Calculation tool | Alternative name | 100% | 50% | 25% |
|---|---|---|---|---|
| SimaPro | (Mixed.50) | $2.7 \times 10^6$ | $2.7 \times 10^6$ | $2.6 \times 10^6$ |
| | (Steel.50) | $3.0 \times 10^6$ | $2.9 \times 10^6$ | $2.6 \times 10^6$ |
| | (Concrete.50) | $2.1 \times 10^6$ | $1.9 \times 10^6$ | $1.8 \times 10^6$ |
| | (Mixed.100) | $2.7 \times 10^6$ | $2.7 \times 10^6$ | $2.7 \times 10^6$ |
| | (Steel.100) | $2.8 \times 10^6$ | $2.8 \times 10^6$ | $2.7 \times 10^6$ |
| | (Concrete.100) | $2.0 \times 10^6$ | $2.0 \times 10^6$ | $1.9 \times 10^6$ |

this study, the concrete alternative is responsible for 454 and 433 kg $CO_2$ eq per $m^2$, which are very close to [13] results. Moreover, in this study, the steel structure has 626 kg $CO_2$ eq per $m^2$, a value which has the same order of magnitude as that of the other studies, such as [13]—780 kg $CO_2$ eq per $m^2$—and [14]—640 kg $CO_2$ eq per $m^2$.

Concluding, this work results can, in a certain way, be considered validated, since they are generally close to the average results of the studies obtained in the literature review.

The results obtained have some uncertainty, due to the assumption of the worst situation that assumed that the entire nominal cover is completely penetrated by carbonations and chlorides after 50 or 100 years. However, in reality, it is unsure that this assumption will occur.

Therefore, it is essential to perform a sensitivity analysis in order to know the effects of different needs of concrete nominal cover maintenance. For that purpose, the global warming impacts of the six alternatives are calculated, assuming that, half and quarter of the concrete quantity is maintained. Then, those impacts are compared with the impacts of the original assumption when 100% of the concrete nominal cover is maintained. Table 14.5 presents the entire global warming emissions released from the three alternatives in a 50-year life span design period.

Comparing the 50-year design with the 100-year design highlights that the concrete is dominating the global warming impacts in Alternatives 1 and 3. In fact, decreasing the maintenance quantity of concrete by 50 and 75% made the impacts of the 100-year design higher than the impacts of the 50-year design in these alternatives. In Alternative 2, where the steel is dominating its impacts, the change in the maintained quantity of concrete was small and, thus, the 50-year design is still worse than the 100-year design option (see Fig. 14.9).

## 14.4 Conclusions

This work highlights the need to achieve more sustainable buildings, since they largely contribute to significant environmental impacts, such as global warming

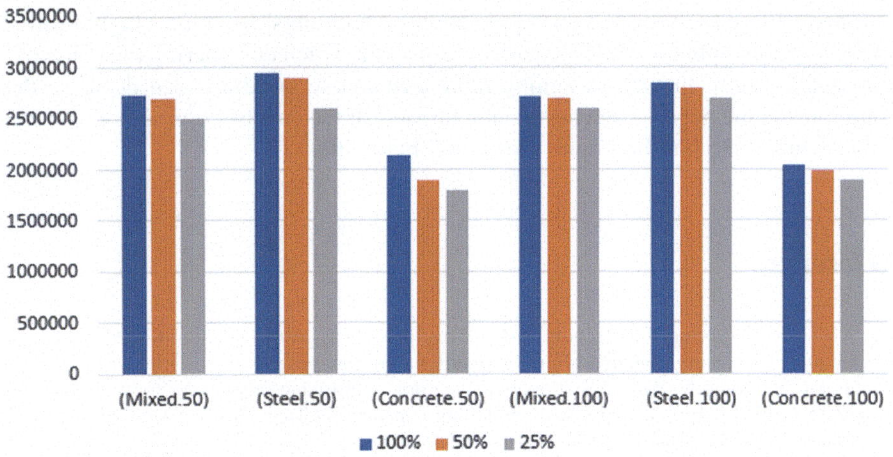


**Fig. 14.9** Sensitivity analysis of the different scenario of concrete maintenance considering the global warming impacts only

and ozone depletion, in addition to their significant impacts on the limited natural resources and the biocapacity of the Earth. Therefore, these requirements culminate with the need of choosing more sustainable and green materials in the construction sector, which could be achieved using LCA, since LCA is a methodology that can successfully predict the environmental impacts of materials and processes.

The results obtained in this study demonstrate that the concrete alternative, which is the most common structural solution used in Portugal, has less environmental impacts than the other two structural solutions. On the other hand, this case study is based on a structure that is environmentally less sustainable than the concrete one (the most common). A fact that highlights the necessity to display LCA analysis since the early design when it is possible to decide whether the structural solution could be the most sustainable.

In addition, it highlights the vital impact of maintenance, since maintaining once in a 100-year design and using materials with higher environmental impacts could have a better environmental performance than maintaining twice in a 50-year design and using materials with lower environmental impacts. Moreover, it is important to highlight the need of studying different maintenance methods for the concrete structure, since it is a fundamental material in the construction sector and its impacts cannot be neglected in the context of more durable and sustainable buildings.

On the other hand, LCA application in the construction sector has various obstacles, such as its complexity, since it is needed to consider the enormous number of materials and components used in the building. In addition, those databases are often not representative of the real processes, which leads to uncertainty in the LCA results that requires cooperation from the technical community to help create a national database.

Finally, these LCA studies and applications at buildings' level have high positive impacts on the replication of urban areas, especially when considering the construction densification. In fact, designing more sustainable buildings can decrease the pollution and the negative environmental impacts of the construction sector, and can consequently help building more green and smart cities.

## References


1. WWF (2016). Living planet report. Risk and resilience in a new era. Gland, Switzerland. http://awsassets.panda.org/downloads/lpr_2016_full_report_low_res.pdf
2. Huang, B., Xing, K. Pullen, S.: Energy and carbon performance evaluation for buildings and urban precincts: review and a new modelling concept. J. Clean. Prod. (Elsevier Ltd.) 1–12 (2015). https://doi.org/10.1016/j.jclepro.2015.12.008
3. Ortiz, O., Castells, F., Sonnemann, G.: Sustainability in the construction industry: a review of recent developments based on LCA. Constr. Build. Mater. (Elsevier Ltd.) **23**(1), 28–39 (2009)
4. Pearce, A.R., Hastak, M., Vanegas, J.A.: A decision support system for construction materials selection using sustainability as a criterion. In: Proceedings of the NCSBCS Conference on Building Codes and Standards, pp. 1–4 (1995)
5. Ferreira, J., Duarte Pinheiro, M., De Brito, J.: Economic and environmental savings of structural buildings refurbishment with demolition and reconstruction—a Portuguese benchmarking. J. Build. Eng. (Elsevier) **3**, 114–126 (2015)
6. Kulahcioglu, T., Dang, J., Toklu, C.: A 3D analyzer for BIM-enabled life cycle assessment of the whole process of construction. HVAC&R Res. **18**(1–2), 37–41 (2012)
7. International Organization of Standards ISO (1997) Environmental management—Life cycle assessment—Principles and framework
8. International Organization of Standards ISO (2006) Environmental management—Life cycle assessment—Requirements and guidelines
9. Laboratório Nacional de Engenharia Civil (LNEC) (2007) Especificação E464 Metodologia prescritiva para uma vida útil de projeto de 50 e de 100 anos face às ações ambientais
10. Autodesk Revit (2016) https://www.autodesk.com
11. Eco-invent database. http://www.ecoinvent.org/database/database.html
12. PE International: Handbook for Life Cycle Assessment (LCA) Using the GaBi Education, pp. 1–66 (2010)
13. Ngo, T., Mirza, A., Gammampila, R., Aye, L., Crawford, R.: Life cycle energy of steel and concrete framed commercial buildings, pp. 1–10 (2009)
14. Kaziolas, N.D., Zygomalas, I., Baniotopoulos, C.C., Eleftherios, S.G.: Life cycle assessment of a steel-framed residential building. In: The Fourteenth International Conference on Civil, Structural and Environmental Engineering (2013). https://doi.org/10.4203/ccp.102.152

# Chapter 15
# Standard-Based Analysis of Measurement Uncertainty for the Determination of Thermal Conductivity of Super Insulating Materials

**Chiara Cucchi, Alice Lorenzati, Sebastian Treml, Christoph Sprengard and Marco Perino**

**Abstract** The assessment of the thermal performance of super insulating materials (SIM), characterized by thermal conductivities down to 0.002 W/(m K), is strongly influenced by even small absolute measurement uncertainties. Moreover, current standard measurement devices are constructed mostly for the so-called conventional insulation materials with values down to 0.015 W/(m K) and subsequently higher thicknesses of the investigated panels compared to SIM. For these reasons, this paper describes methods to determine the standard-based measurement uncertainty for the determination of thermal conductivity of SIM. A sensitivity study shows that temperature difference and thickness determination are the most influential parameters. Therefore, it is recommendable to use adequate temperature differences above 15 K and to take special care about the method for the determination of the panel thickness that may depend also on the flatness and torsion of the investigated panel.

## 15.1 Introduction

Over the last 10 years, the so-called super insulating materials (SIM), having significantly lower values of thermal conductivity compared to conventional insulation materials (e.g., mineral wool and polystyrene rigid foam), began to gain market shares in the construction sector [1]. The most important SIM are vacuum insulation panels (VIPs), reaching around 0.002–0.007 W/(m K) of thermal conductivity $\lambda$, and advanced porous materials (APM) with values of $\lambda$ around 0.015–0.020 W/(m K) [1].

C. Cucchi (✉) · S. Treml · C. Sprengard
FIW Forschungsinstitut für Wärmeschutz e.V., Lochhamer Schlag 4, 82166 Gräfelfing, Germany
e-mail: cucchi@fiw-muenchen.de

A. Lorenzati · M. Perino
Politecnico di Torino, corso Duca degli Abruzzi 24, 10129 Turin, Italy

 
J. Littlewood et al. (eds.), *Sustainability in Energy and Buildings*, Smart Innovation, Systems and Technologies 163, https://doi.org/10.1007/978-981-32-9868-2_15

The more advanced applications of VIPs and APM in the building sector, characterized by very different boundary conditions, now require intense investigation of the thermal conductivity measurement uncertainty obtainable with standard measuring equipment. In case of thermal insulation materials, the thermal conductivity is used to design hygrothermal functional building elements and to predict future energy savings among decades after the construction or thermal renovation of a building, thus influencing also the related operating costs of the building.

The determination of thermal conductivity can be performed both with steady-state methods and transient methods. The steady-state methods have a lower measurement uncertainty and therefore can still be seen as the reference method. Steady-state methods are the guarded hot plate (GHP) and the heat flow metre (HFM) method, according to EN 12664 [2], EN 12667 [3], ISO 8302 [4] and ISO 8301 [5].

For conventional insulation materials with values of thermal conductivity down to 0.029 W/(m K), a typical measurement uncertainty equal to 2% for GHP and 3% for HFM can be reached [3]. It is doubtful how the combined measurement uncertainty will evolve for significantly lower values of $\lambda$ reached by SIM.

This paper describes investigations of the combined standard-based measurement uncertainty for the determination of the thermal conductivity by the use of the HFM method. For this purpose, a sensitivity study was carried out based on a systematic variation of the relevant parameters and test conditions. Furthermore, the determination of the measurement uncertainty of selected parameters under real lab conditions is discussed.

## 15.2 Calculation of Measurement Uncertainty

The uncertainty analysis was investigated considering both Type A and Type B evaluations (based respectively on a statistical approach or other scientific and relevant information available).

### *15.2.1 Uncertainty Type A and Type B*

The Guide to the expression of uncertainty in measurement (GUM) [6] defines two different categories of uncertainty, considering both random and systematic errors: Type A (statistical analysis of a series of observations) and Type B (use of non-statistical available information). The associated uncertainty of each measurement is obtained combining Type A and Type B components. The combined standard uncertainty is always a Type B uncertainty.

Type A uncertainty is determined through a set of experimental measurements. The uncertainty is indeed obtained from a Probability Density Function (PDF or $p(x)$ [7]) derived from an observed frequency distribution. On the contrary, Type B uncertainty is derived from a beforehand assumed PDF (subjective probability).

In other words, the PDF allows calculating the probability that the variable has a value in a specified interval. The standard uncertainty is indeed equal to the standard deviation of the PDF of the observed phenomenon (where $\mu_x$ is the mean of the measurements, defined as the best estimate) [Eq. (15.1)]:

$$u(x) = \sigma = \sqrt{\sigma^2} = \sqrt{\int (x - \mu_x)^2 p(x)dx} \tag{15.1}$$

### 15.2.2 Combined Measurement Uncertainty

The measurand $Y$ is usually not directly measured but is expressed as a function $f$ of other measured parameters $X_i$. In this case, the uncertainties of the various parameters will propagate through the function $f$ to an uncertainty in $Y$. When these quantities are uncorrelated, the uncertainty of $y$ (estimate of the measurand $Y$) is calculated through the combined standard uncertainty, $u_c(y)$, which is statistically equivalent to a standard deviation. It is obtained combining the uncertainty contributions of each input quantity [Eq. (15.2)]:

$$u_c(y) = \sqrt{\sum_{i=1}^{N} \left(\frac{\partial f}{\partial x_i}\right)^2 \cdot u(x_i)^2} \tag{15.2}$$

where

- $u(x_i)$ is the Type A or Type B standard uncertainty of each parameter;
- $\frac{\partial f}{\partial x_i}$ is the sensitivity coefficient of the input quantity $x_i$;
- $\left(\frac{\partial f}{\partial x_i}\right)^2 \cdot u(x_i)^2$ is the uncertainty contribution of each input quantity $x_i$ on the output uncertainty $u_c(y)$.

## 15.3 Sensitivity Analysis for Standard-Based Measurement Uncertainty of HFM

To achieve general conclusions about the applicability of the current standards to SIM, a theoretical study was performed, considering the standard reference Type B uncertainties provided by EN 1946-3 [8] (which contains minimum requirements on measurement uncertainty for the experimental assessment of thermal conductivity, in accordance with EN 12667 [3] and EN 12664 [2]). Consequently, the recommendations and conclusions are based on theoretical assumptions and may be strongly affected by real practical applications.

This investigation was performed both in case of GHP [9] and HFM [8] apparatuses, but for the sake of brevity, only the results related to the HFM method are shown, since for the HFM method the measurement uncertainty was more sensitive for the investigated parameters than that of GHP.

### 15.3.1 Model for Calculation of Measurement Uncertainty for HFM

The assessment of the thermal conductivity (λ [W/(m K)]) with the HFM method is obtainable through Eq. (15.3):

$$\lambda = \frac{f_{cal}(\vartheta) \cdot Q \cdot t}{\Delta\vartheta} = \frac{\varphi \cdot t}{\Delta\vartheta} \tag{15.3}$$

where $f_{cal}(\vartheta)$ [W/(m$^2$ μV)] is the HFM plates calibration factor, $Q$ [μV] is the electric signal from the transducer, $t$ [m] is the sample thickness, $\Delta\vartheta$ [K] is the temperature difference between the HFM plates and $\varphi$ [W/m$^2$] is the specific heat flux crossing the sample. Since each plate has its own temperature, the calibration factor should be calculated for each plate's actual temperature, obtaining two different values of thermal conductivity to be averaged.

In order to consider also the influence of the apparatus' geometrical aspects and quality, the standard EN 1946-3 [8] proposes two additional uncertainty contributions: $\Delta\lambda_E$ (edge heat losses [−]) and $\Delta\lambda_O$ (imperfect contact [−]).

Moreover, if the calibration factor and the measured electrical signal are unknown, the specific heat flux uncertainty depends on the following errors: $\Delta\lambda_K$ (specimen calibration accuracy [−]]), $\Delta\lambda_L$ (maximum permissible non-linearity of the calibration [−]) and $\Delta\lambda_g$ (maximum allowable calibrating drift [−]).

All these parameters can be included in the formula of the thermal conductivity by adding them as factors of 1.0 and considering a defined relative error in the calculation. The extended equation of the thermal conductivity is provided by Eq. (15.4):

$$\lambda = \frac{\left(\varphi \cdot \Delta\lambda_K \cdot \Delta\lambda_L \cdot \Delta\lambda_g\right) \cdot t}{\Delta\vartheta} \cdot \Delta\lambda_E \cdot \Delta\lambda_O \tag{15.4}$$

The thermal conductivity uncertainty is calculated through the combined standard uncertainty provided in Eq. (15.2) applied to Eq. (15.4).

### 15.3.2 *Influence of Thickness and Temperature Difference, Assuming Maximum Probable Errors According to EN 1946-3:1999*

To check if the thermal properties of SIM can be properly evaluated with HFM, a sensitivity analysis was performed. According to EN 1946-3 [8], three different equipment configurations with different sizes of the metering area can be identified: $150^2$ mm$^2$ (equipment A), $200^2$ mm$^2$ (equipment B) and $300^2$ mm$^2$ (equipment C). For each equipment typology, the standard [8] provides also the maximum probable relative errors (Table 15.1). Consequently, the absolute errors of all the parameters were quantified basing on a set of data, composed by minimum thickness (depending on the equipment type: equipment A, minimum specimen thickness $t_{\text{min}} = 0.015$ m; equipment B, $t_{\text{min}} = 0.025$ m; equipment C, $t_{\text{min}} = 0.03$ m), minimum temperature difference (10 K for all equipment), minimum measurable thermal conductivity (0.015 W/(m K) for all equipment) and minimum calculated heat flux (equipment A, 3.0 W/m$^2$; equipment B, 1.1 W/m$^2$; equipment C, 1.5 W/m$^2$) (Table 15.1).

Based on the absolute errors, the combined measurement uncertainty $u_c(\lambda)$ was calculated considering different specimen thicknesses (0.01, 0.02, 0.04 and 0.08 m), temperature differences (5, 10, 15 and 20 K) and specimen thermal conductivities between 0.002 W/(m K) (fiber-based VIP), 0.004 W/(m K) (fumed silica-based VIP), 0.008 W/(m K) (aged VIP) and 0.016–0.020 W/(m K) (range of the APM). The results related to samples with a thickness equal to 10 mm and 40 mm are shown in Fig. 15.1a and b, respectively.

The temperature difference and the specimen thickness were found to be very influential on the obtained measurement uncertainty. For a specimen thickness of 10 mm (Fig. 15.1a), equipment A provides the most accurate results for a thermal conductivity above 0.004 W/(m K). However, equipment B enables also accurate results with an uncertainty below 2.0% for a temperature difference above 15 K over the full range of the observed thermal conductivity and therefore can be seen as the most versatile. For a specimen with 40 mm thickness (Fig. 15.1b), the measurement uncertainty increases significantly, especially for very low values of thermal conductivity. However, for a temperature difference of 15 K, a threshold value of less than 2% measurement uncertainty is achieved down to 0.013 W/(m K) for equipment A, down to 0.005 W/(m K) for equipment B and down to 0.007 W/(m K) for equipment C. Also in this case, equipment B can be identified as the most versatile equipment over the range of the observed thermal conductivity. For these reasons, in Sect. 15.3.3, the results for equipment B are discussed more deeply.

**Table 15.1** HFM—maximum relative error [8] and absolute errors for the calculation of the $u_c(\lambda_{HFM})$, equipment A, B and C [8]

| Measurand | Errors for calculation | Symbol | Maximum relative error | Absolute errors | | | |
|---|---|---|---|---|---|---|---|
| | | | [%] | Unit | Equipment A | Equipment B | Equipment C |
| $\lambda$ | Edge heat loss | $u(\Delta\lambda_E)$ | 0.5 | [W/(m K)] | 0.000075 | 0.000075 | 0.000075 |
| $\lambda$ | Imperfect contact | $u(\Delta\lambda_O)$ | 0.5 | [W/(m K)] | 0.000075 | 0.000075 | 0.000075 |
| $t$ | Specimen thickness | $u(t)$ | 0.5 | [m] | 0.000075 | 0.000125 | 0.00015 |
| $\Delta\vartheta$ | Temperature difference | $u(\Delta\vartheta)$ | 1.0 | [K] | 0.1 | 0.1 | 0.1 |
| $\varphi$ | Calibration accuracy of the specimen | $u(\Delta\lambda_K)$ | 1.5 | [W/m$^2$] | 0.045 | 0.016 | 0.023 |
| $\varphi$ | Maximum non-linearity of the calibration | $u(\Delta\lambda_L)$ | 1.0 | [W/m$^2$] | 0.03 | 0.01 | 0.02 |
| $\varphi$ | Maximum calibrating drift | $u(\Delta\lambda_q)$ | 1.0 | [W/m$^2$] | 0.03 | 0.01 | 0.02 |

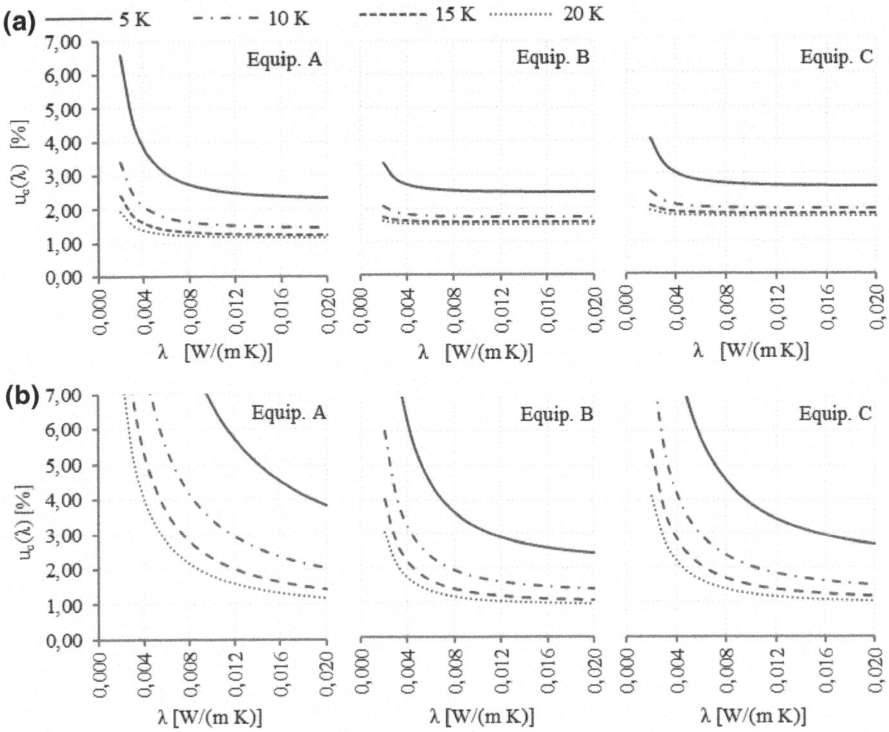


**Fig. 15.1** HFM—combined relative uncertainty as a function of thermal conductivity and temperature difference, equipment A, B and C according to [8]. **a** $t = 10$ mm; **b** $t = 40$ mm

## *15.3.3 Sensitivity of Increasing Error of Single Parameters on the Combined Uncertainty of Thermal Conductivity*

Figure 15.2a shows the isolines for selected levels of uncertainty as a function of temperature difference $\Delta\vartheta$ and specimen thickness $t$ (equipment B), considering the maximum relative error as shown in Table 15.1. The spread between the curves is quite wide, but with temperature differences lower than 10 K or for specimen thicknesses lower than 10 mm, the distance between the isolines becomes smaller, thus indicating a high sensitivity of $u_c(\lambda)$ in this range of values. The curves, referred to uncertainty $u_c(\lambda) = 1\%$, were obtained only in case of $\lambda > 0.008$ W/(m K), $\Delta\vartheta > 20$ K and $t > 0.02$ m. The minimum observable $u_c(\lambda)$, in case of samples with λ equal to 0.004 W/(m K) and 0.002 W/(m K) (SIM), is < 2.0% with $\Delta\vartheta > 10$ K. The measurement uncertainties of lower thermal conductivities are more influenced by the temperature difference. Considering a fixed thickness, high variations in temperature difference are necessary for uncertainty reduction (especially for the lowest thermal conductivities and desired uncertainty value). Otherwise, for a defined value of $\Delta\vartheta$, a smaller variation in thickness is required for the uncertainty improvement.

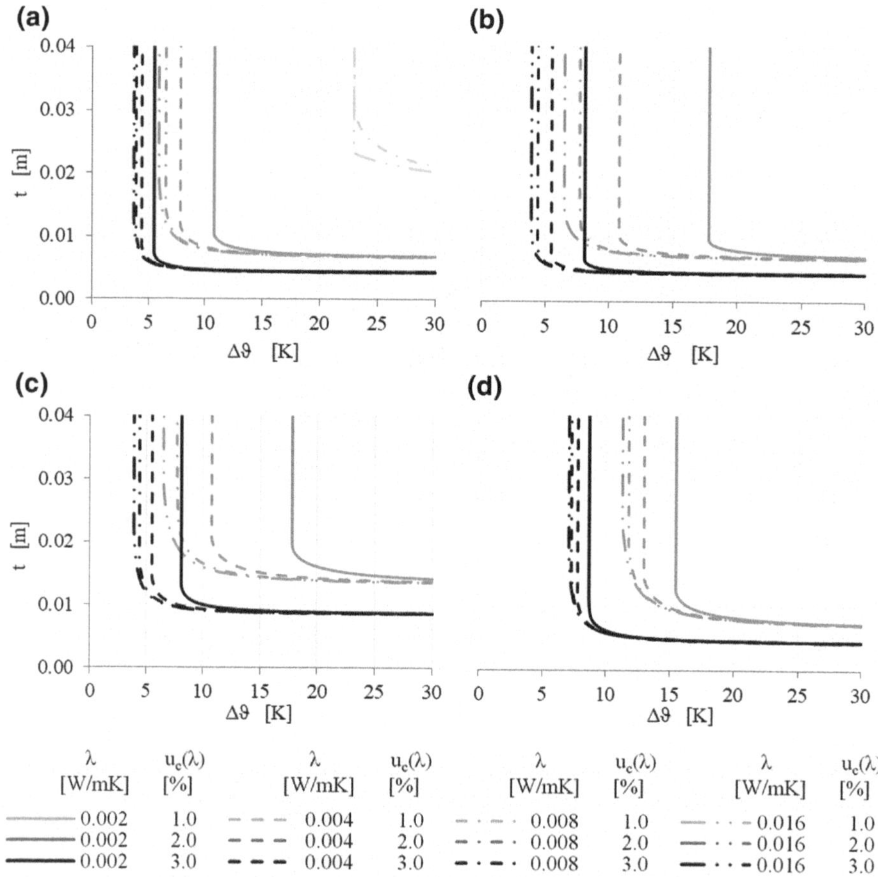


**Fig. 15.2** Uncertainty isolines for different values of λ, as a function of the temperature difference $\Delta\vartheta$ and the specimen thickness $t$, equipment B. **a** Assuming the maximum uncertainty according to [8]; **b** doubling the heat flux error; **c** doubling the thickness error; **d** doubling the $\Delta\vartheta$ error

To identify the most influencing parameters involved in the measurement uncertainty budget, a sensitivity analysis was carried out, considering variable uncertainties of the temperature difference and the sample thickness.

Figure 15.2b–d shows the variation of the combined measurement uncertainty $u_c(\lambda)$ isolines if the single measurement uncertainty of the heat flux (Fig. 15.2b), the thickness (Fig. 15.2c) or the temperature difference (Fig. 15.2d) are increased by a factor of 2 compared to the values given in Table 15.1 (Fig. 15.2a).

The effects of a doubled value of heat flux uncertainty cause a right shifting and a broader spread of all the curves (Fig. 15.2b), compared to the standard case (Fig. 15.2a), depending on the thermal conductivity. For instance, fiberglass VIPs (0.002 W/(m K) with a thickness of 10 mm could reach an uncertainty equal to 2% with a testing temperature difference higher than around 18 K (instead of approximately 11 K required in the standard case). Otherwise, for thermal conductivities higher than 0.002 W/(m K), it is possible to compensate for the increase in the heat flux uncertainty by the rise of $\Delta\vartheta$.

The increase in the thickness uncertainty (Fig. 15.2c) has the same effects of the heat flux increased uncertainty (right shifting and a more significant spread), but it also moves the isolines upwards. This effect is hard to compensate with an increase in the temperature difference. For any values of $\Delta\vartheta$, specimens with a thickness of 10 mm could reach uncertainty values not lower than 3%, while $u_c(\lambda) < 2\%$ is available only for $t > 15$ mm and requires adequate high temperature difference (for each considered thermal conductivity). Doubling the uncertainty of the temperature difference $\Delta\vartheta$ (Fig. 15.2d) simply shifts the isolines to the right, thus increasing $u_c(\lambda)$ for any observed value of temperature difference and specimen thickness.

## 15.4 Uncertainty of Selected Parameters Under Real Lab Conditions

The combined uncertainty of thermal conductivity is influenced by different measurands as shown in the sensitivity analysis. The most important parameters are the thickness and the temperature difference (since they strongly affect the heat flux through the specimen). To check if the previously assumed measurement uncertainty given in the relevant standard is achievable in principle for SIM, investigations to determine the measurement uncertainty of these parameters under real lab conditions were performed.

### *15.4.1 Uncertainty of Thickness Determination*

The thickness determination for SIM requires special attention. Due to the conformation of the panel, different variables must be taken into account. In fact, VIPs and APM are oftentimes not perfect, rigid and plane.

To observe the uncertainty of thickness determination, three exemplary VIP specimens, differing in thickness and flatness, were tested. Specimen I is 10 mm thick and visible flat, specimen II is also 10 mm thick but has a visible curvature and specimen III is 30 mm thick and visible flat. On these exemplary VIPs, the measurement uncertainty (Type A) was determined for different setups of measurement equipment (*ME*, 1 = measuring table, 2 = calliper), position of the measurement point (*Pos*, C = centre of panel, E = edge of panel), number of measurement points ($N_{MP}$) and number of operators (NOP) (Fig. 15.3). Based on the achieved results (Table 15.2), some recommendations and experiences concerning repeatability and reproducibility can be derived. To respect as much as possible the repeatability of the measure,

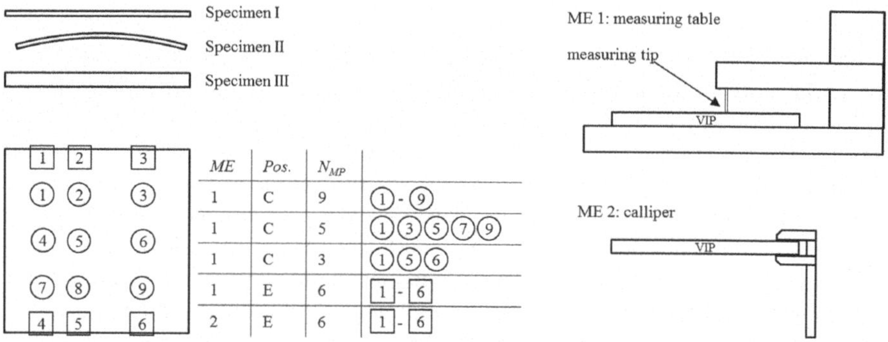


**Fig. 15.3** Principle sketch of the specimen (top, left), the distribution of the measurement positions (bottom, left) and the measurement equipment *ME* 1 (measuring table) and *ME* 2 (calliper)

the thickness was measured in the same conditions and only one parameter at a time was changed.

It is obvious that the measurement uncertainty decreases with an increasing number of measuring points ($N_{MP}$ = [3, 5, 9]). However, the expected measurement uncertainty lower than 0.5% (according to EN 1946-3 [8], Table 15.1) can also be met with 3 measuring points for all tested specimens. The specimen II shows a significantly higher measurement uncertainty (due to curvature) for 3 and 5 measuring points compared to specimen I. These differences are eliminated when measured at 9 measuring points. To check for repeatability and reproducibility, the thickness measurement using 9 measurement points was repeated five times by the same operator and five times (each time) by a different operator ($N_{OP}$ = [1, 5]). The results indicate both a higher mean measurement uncertainty and also higher spreading of the measurement uncertainty in the case of 5 five operators. The main reason for this is due to the individual hand force that is necessary to guarantee proper contact of the panel with the measuring table.

The last set of observations shall indicate whether there is an influence of the used measurement equipment and the position of the measurement. When measured with the measuring table (*ME* 1) in the centre of the panel, the lowest measurement uncertainties were obtained. A repetition of the measurement with the same equipment, considering only measurement positions at the edge of the panel, led to an increase in uncertainty for all the tested specimen. This effect can be explained with more irregularities of the panels at the edge. The use of a standard calliper enables only measurements at the edge of the panels. Thereby lower uncertainty determined on the edge for specimen II may depend on the more curved specimen; therefore, the use of the measuring table to measure the thickness at the edge of the specimen is less accurate because it is difficult to apply a constant pressure to avoid the influence of the curvature.

**Table 15.2** Uncertainty of the thickness determination for specimens I, II and III with different measurement equipment (*ME*), position of measurement (*Pos*), number of measurement points ($N_{MP}$) and number of operators ($N_{OP}$)

| *ME* | *Pos* | $N_{MP}$ | $N_{OP}$ | Specimen I | | Specimen II | | Specimen III | |
|---|---|---|---|---|---|---|---|---|---|
| – | – | – | – | $u(t)$ [mm] | $u(t)$ [%] | $u(t)$ [mm] | $u(t)$ [%] | $u(t)$ [mm] | $u(t)$ [%] |
| 1 | C | 3 | 1 | 0.044 | 0.41 | 0.052 | 0.48 | 0.037 | 0.13 |
| 1 | C | 5 | 1 | 0.023 | 0.22 | 0.033 | 0.30 | 0.026 | 0.09 |
| 1 | C | 9 | 1 | 0.021 | 0.20 | 0.021 | 0.19 | 0.018 | 0.06 |
| 1 | C | 9 | 1[a] | $\mu = 0.065$<br>$\sigma = 0.004$ | – | $\mu = 0.071$<br>$\sigma = 0.009$ | – | $\mu = 0.058$<br>$\sigma = 0.005$ | – |
| 1 | C | 9 | 5[a] | $\mu = 0.067$<br>$\sigma = 0.011$ | – | $\mu = 0.084$<br>$\sigma = 0.018$ | – | $\mu = 0.076$<br>$\sigma = 0.015$ | – |
| 1 | C | 6 | 1 | 0.030 | 0.28 | 0.017 | 0.16 | 0.022 | 0.08 |
| 1 | E | 6 | 1 | 0.047 | 0.45 | 0.053 | 0.49 | 0.041 | 0.14 |
| 2 | E | 6 | 1 | 0.058 | 0.58 | 0.020 | 0.20 | 0.041 | 0.14 |

[a] $\mu$ = mean value and $\sigma$ = standard deviation of measurement uncertainty based on 5 repetitions obtained on the same specimen by 1 and 5 operators

### 15.4.2 Uncertainty of Temperature Difference Determination

The temperature difference between the hot and the cold plate is measured by a variable number of thermocouples. Thus, the combined uncertainty of the temperature difference (Eq. 15.5), (Type B), depends on the thermocouple uncertainty $(u(\vartheta_{h/c}))$ and the number of thermocouples used $(n)$.

$$u(\Delta\vartheta) = \sqrt{\frac{\sqrt{u(\vartheta_{h/c})^2 \cdot n}^2}{n} + \frac{\sqrt{u(\vartheta_{h/c})^2 \cdot n}^2}{n}} \quad (15.5)$$

To observe the influence of the number of thermocouples on the measurement uncertainty of the temperature difference, a total number of 10 and 20 thermocouples were assumed. The combined relative uncertainty (Eq. 15.2) was applied using both the error provided by the standard ISO 8302 [4] and the error derived from an extended calibration already used in the laboratory for thermocouples T and N. The absolute error for the thermocouples is constant.

Both the results (Fig. 15.4) show that an increase in the number of thermocouples decreases the combined relative uncertainty. However, the results obtained with the error from the standard meet the limit of 1% imposed by the standard only for some type of thermocouples and for temperature differences higher than 22 °C (Fig. 15.4a).

Thermocouples of type T are suitable for low temperatures (−40 to 40 °C), while type N thermocouples are suitable for high temperatures (0 to 375 °C) (Fig. 15.4b). However, also for calibrated thermocouples of type T, the number of thermocouples

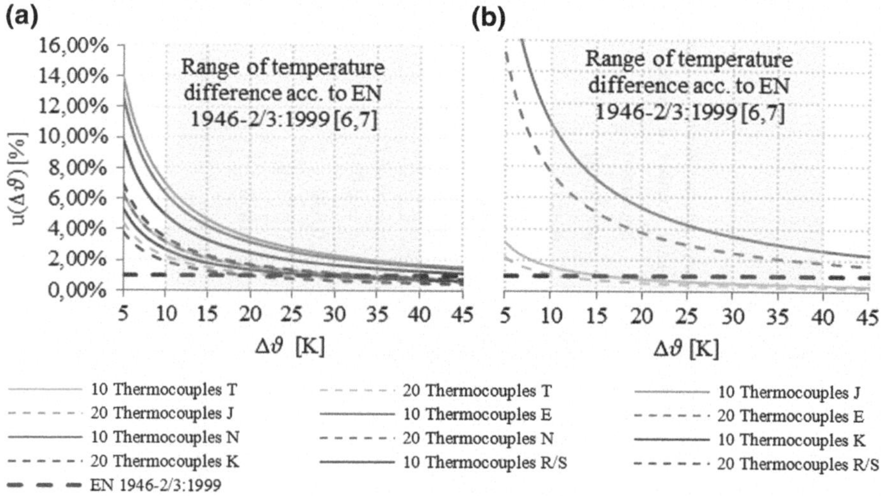


**Fig. 15.4** Combined relative uncertainty of the temperature difference between 550 K with 10 and 20 thermocouples. **a** Thermocouples with special error [4]; **b** calibrated thermocouples

**Table 15.3** Combined relative uncertainty for 10 and 20 calibrated thermocouples T at the temperature difference between 10 and 16 K

| n° Thermocouples | $\Delta\vartheta$ [K] | 10 | 11 | 12 | 13 | 14 | 15 | 16 |
|---|---|---|---|---|---|---|---|---|
| 10 | $u(\Delta\vartheta)$ [%] | 1.58 | 1.44 | 1.32 | 1.22 | 1.13 | 1.05 | 0.99 |
| 20 | $u(\Delta\vartheta)$ [%] | 1.12 | 1.02 | 0.93 | 0.86 | 0.80 | 0.75 | 0.70 |

has to be increased to respect the standard limit of 1% for temperature differences less than 15 K (Table 15.3).

## 15.5 Recommendations and Outlook

Temperature difference and thickness are very influential parameters on the measurement uncertainty of thermal conductivity. The increase in the thickness determination uncertainty is difficult to compensate by an increase in temperature difference (especially for very low values of thermal conductivity). In any case, a minimum temperature difference of 15 K is recommendable.

Upon the determination of the uncertainty for selected parameters under real lab conditions, the following distinct recommendations can be derived from the conducted work. For the determination of the panel thickness, the influence of the user shall be minimized by using measurement equipment that guarantees a controlled pressure force on the specimen. In the case of curved panels, the increase in the number of measurement points helps to minimize the measurement uncertainty as well as the choice of measurement equipment that is not influenced by the support situation on the measuring table.

Concerning the choice of thermocouples, the results confirm that the thermocouples of type T are suitable for low temperatures while type N thermocouples are suitable for high temperatures. For temperature differences lower than 15 K, an increase in the number of thermocouples helps to keep the measurement uncertainty in the specifications of the standards.

The results obtained from the performed theoretical analysis, highlight that the combined relative uncertainty based on data from real sensors is lower than the uncertainty based on the exemplary measurement uncertainties from the standard EN 1946-3 [8]. Therefore, the calculation of the uncertainty under real lab conditions should be emphasized and studied further.

**Acknowledgements** This work was developed in the framework of the IEA-EBC Annex 65—Subtask 2 [1]. The authors are very grateful to all the partners involved in the project.

## References


1. Holm, A., Sprengard, C.: International Energy Agency, IEA EBC Annex 65: Long-term performance of super-insulating materials in building components and systems. Subtask 2: Characterisation of materials & components—Laboratory scale. IEA Energy Technology Network (2018). Under revision
2. EN 12664:2001. Thermal performance of building materials and products—Determination of thermal resistance by means of guarded hot plate and heat flow meter methods—Dry and moist products of medium and low thermal resistance
3. EN ISO 12667:2001. Thermal performance of building materials and products. Determination of thermal resistance using guarded hot plate and heat flow meter methods. Products of high and medium thermal resistance
4. ISO 8302:1991. Thermal insulation; determination of steady-state thermal resistance and related properties; guarded hot plate apparatus
5. ISO 8301:1991. Thermal insulation; determination of steady-state thermal resistance and related properties; heat flow meter apparatus
6. BIPM—Bureau International des Poids et Mesures (2008). JCGM 100:2008. Evaluation of measurement data—Guide to the expression of uncertainty in measurement (GUM), https://www.bipm.org/en/publications/guides/. Last accessed 7 February 2019
7. ISO 3534-1:2006. Statistics—Vocabulary and symbols—Part 1: general statistical terms and terms used in probability
8. EN 1946-3:1999. Thermal performance of building products and components—Specific criteria for the assessment of laboratories measuring heat transfer properties—Part 3: measurements by heat flow meter method
9. EN 1946-2:1999. Thermal performance of building products and components—Specific criteria for the assessment of laboratories measuring heat transfer properties—Part 2: Measurements by guarded hot plate method

# Chapter 16
# Field Experimental Study on Energy Performance of Aerogel Glazings with Hollow Silica: Preliminary Results in Mid-Season Conditions

**C. Buratti, E. Moretti, E. Belloni, F. Merli, V. Piermatti and T. Ihara**

**Abstract** In the last decades noticeable research efforts focused on aerogels, due to their thermal and acoustic insulation properties and potential applications in energy efficient translucent windows. In this work, thermal-energy and lighting performance of innovative double glazing units is evaluated through in-field experimental campaigns. Two identical rooms (Test and Reference) and a weather station are installed on the roof of a building in the Campus of Engineering (University of Perugia, Italy). An aerogel glazing system (AGS), consisted of aerogel granules mixed with opaque hollow silica in the interspace of a double glazing, is mounted in the window frame of the Test Room, whereas a standard double glazing system (SGS) with air in interspace is mounted in the Reference Room. The main quantities that influence the thermo-hygrometric and lighting conditions are monitored; preliminary results (June 2018) show that the indoor air temperature in the Test Room, especially the peak values, is about 6–7 °C lower than the one in the Reference Room. This trend is due to the presence of silica dust, which contributes to the reduction of the solar factor. At the same time, the maximum illuminance values close to the window are about 5000–8000 lux with SGS and about 2000–2500 lux with AGS; the Useful Daylight Illuminance (UDI) calculated during the working hours (8 a.m–6 p.m.) shows for AGS values 57% higher than SGS, highlighting the aerogel ability to diffuse light and to reduce glare.

## 16.1 Introduction

In the last ten years research noticeable efforts focused on silica aerogels, due to their thermal and acoustic insulation properties. Silica aerogels can be either transparent (monolithic form) or translucent (granular form). Advanced glazing solutions with

C. Buratti (✉) · E. Moretti · E. Belloni · F. Merli · V. Piermatti
Department of Engineering, University of Perugia, Via G. Duranti 67, 06125 Perugia, Italy
e-mail: cinzia.buratti@unipg.it

T. Ihara
Finishing Materials Group, Construction Material Engineering Department, Takenaka Research & Development Institute, Takenaka Corporation, Tokyo, Japan

 
J. Littlewood et al. (eds.), *Sustainability in Energy and Buildings*, Smart Innovation, Systems and Technologies 163, https://doi.org/10.1007/978-981-32-9868-2_16

aerogel in the gap were investigated and some translucent solutions are now available on the market [1–5].

Glazing units with granular aerogel were largely investigated in the literature. Thermal, optical, and acoustic performance on real prototypes were measured in [6–11], while the possible application in building was assessed by numerical simulations in [12–15]. In general, aerogel windows show very good thermal, optical, and acoustic performance, depending on the particle size of the aerogel granules [6, 7, 10]. The numerical analysis for several reference buildings in different climate conditions shows that aerogel glazing systems have a significant impact on the building energy performance, depending on the window to wall ratio, the building location, and the aerogel sample. In particular the new solutions allow to reduce the annual energy consumption when compared with conventional double glazing, especially in cold climates [12–15].

Starting from previous findings about thermal, energy, acoustic and lighting properties of aerogel window system through in-lab characterization or numerical analysis, the main original contribution of this research is their in-field experimental investigation at pilot scale. Nowadays in situ performance investigation of new aerogel glazing systems is still missing in research literature at our best of knowledge [16].

Thermal and lighting performance of innovative glazing systems with mixed granular aerogel was studied by means of an in-field experimental campaign. The innovative prototype consists of aerogel granules in the interspace (24 mm) between two float glasses; large granules are mixed with opaque hollow silica, in order to reduce the solar factor of the glazing and to improve energy performance in inter-mediate climate conditions.

The study is carried out by monitoring two boxes (Test and Reference Rooms) identical in terms of sizes, construction materials, and orientation. They have south-oriented windows: the aerogel glazing is mounted in the window frame of the Test Room, whereas a standard double glazing system with the same total thickness, but with air in interspace, is mounted in the Reference Room. In order to evaluate the influence of the innovative solution in comparison with standard glazing systems, the in situ performance of the aerogel window is studied in free-floating regime.

The paper focusses on the preliminary experimental results at the end of spring season (June 2018), in order to investigate behavior (thermal and daylighting performance) of the aerogel granules mixed with hollow silica in comparison to the conventional double glazing.

## 16.2 Methodology

### *16.2.1 Experimental Field C.A.S.E.T.T.E.: Development and Description*

The experimental field at pilot scale C.A.S.E.T.T.E. (Coupled Advanced Systems for Experimentation on Translucent and Thermal-insulated Envelopes) is composed of two boxes and an external weather station installed on the roof of the Department of Engineering (University of Perugia, Italy). The rooms (1.60 m × 2 m × 2 m height) are identical in terms of size, construction materials, and orientation, and have south-oriented windows: the double glazing with aerogel and hollow silica in interspace is mounted in the window frame of the Test Room, whereas a standard double glazing system with air in interspace is mounted in the Reference Room. The positioning of the two rooms allows to avoid the influence of obstacles and shadings that can affect the responses of the two boxes.

C.A.S.E.T.T.E. were built using sandwich panels consisting of insulating material into two metal layers. The panels of the vertical walls and of the floor are 50 mm thick, while the ones of the roof are 40 mm thick. The panel of the walls has a transmittance of 0.43 W/m$^2$ K, while the one of the roof is 0.53 W/m$^2$ K, evaluated in compliance with EN UNI 14509 standard [17]. POLIISO PLUS panels were installed inside the boxes, glued to the walls and to the ceiling; they consist of a rigid closed-cell polyiso foam, expanded between two multilayer metallized paper supports, with a thermal conductivity at 10 °C equal to 0.023 W/m K, and in particular a thermal resistance of 3.30 m$^2$ K/W (for a total thickness of 80 mm).

The total transmittances of the envelope elements are reported in Table 16.1. A photo of the experimental field is shown in Fig. 16.1. The south wall surfaces are equipped with a two-wing window; each glass has 470×1380 mm dimensions; the north façades are equipped with a door with 1000×1800 mm dimensions. The window frame is made of aluminum with thermal break, thick 70 mm, fixed with metal bars to the walls both inside and outside.

**Table 16.1** U-values of the main rooms components

| | Total transmittance (W/m$^2$ K) |
|---|---|
| Door/Floor | 0.375 |
| Wall | 0.167 |
| Roof | 0.182 |

Fig. 16.1 The experimental field C.A.S.E.T.T.E. with a particular of the two windows

### 16.2.2 Glazing Systems

The aerogel glazing system (AEG) installed in the Test Room is composed of a tempered float glass 6 mm thick, aerogel granules mixed with hollow silica in the interspace (24 mm thickness), and a second tempered float glass (6 mm), for a total thickness of 36 mm. A cross section of the sample is shown in Fig. 16.2.

The total thermal transmittance of the only single glazing pane is about 5.7 $W/m^2$ K, as certified by the manufacturer. The aerogel mixture (granules + hollow silica) was supplied by Takenaka Corporation and it was directly used for the manually filling process. The interspace of the system is delimited by a metal (aluminum) spacer. The final sealing is obtained by a specific two-components sealant. The total thermal transmittance of the aerogel glazing system is 0.7 $W/m^2$ K and it

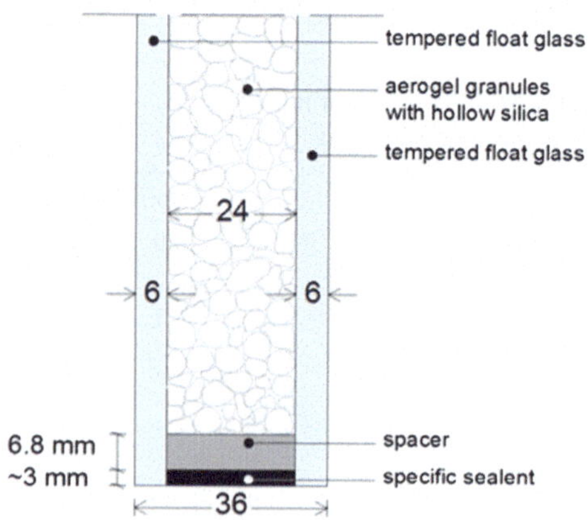


Fig. 16.2 Cross section of the aerogel glazing system (AGS)

was calculated starting from thermal resistance measured with a Guarded Hot Plate facility, in compliance with EN 674 [18].

The standard glazing system (SGS) is composed of a tempered float glass 6 mm thick, interspace of 15 mm thickness, and a second tempered float glass (6 mm); the total thickness is 27 mm and the total transmittance is 2.7 W/m$^2$ K, as attested by the manufacturer.

### *16.2.3 Measurement Equipment*

In the experimental field both an external weather station for the weather data analysis and internal probes for the evaluation of the inside parameters are installed.

The weather station is able to collect and store outdoor air temperature, relative humidity, global solar radiation on the horizontal plan, wind velocity and direction (Fig. 16.3a).

Inside each room, thanks to a multi-acquisition system Babuc and two delta-loggers, the main quantities that influence the thermo-hygrometric conditions are measured: air temperature, globe temperature, relative humidity, air velocity, illuminance in two different points (E1, 0.5 m from the window and E2, 1.35 m from the window), incoming solar radiation entering through the glazing systems, and surface temperatures (walls, floors, and glazing), as shown in Fig. 16.3b, c.

Thanks to the black globe temperature ($T_g$) and to the air temperature ($T_a$), it is possible to calculate the mean radiant temperature ($T_{mr}$) by means of the equations suggested in UNI EN ISO 7726 [19].

(a) (b) (c)

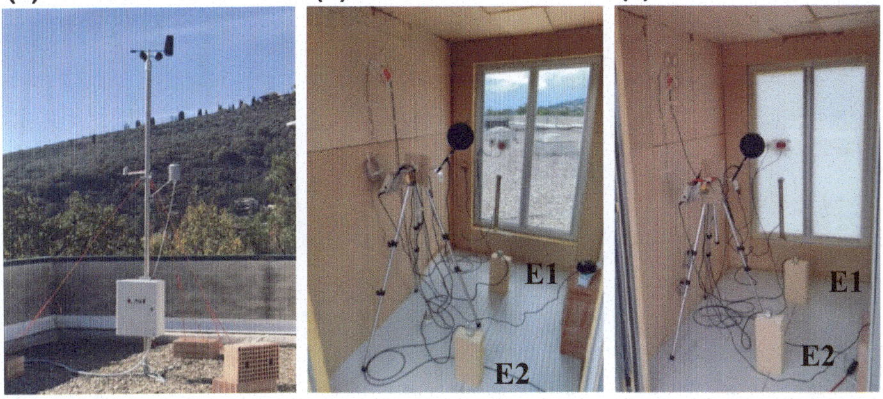


**Fig. 16.3** Weather station with its sensors (**a**); probes installed in the Reference Room (**b**) and in the test room (**c**)

## 16.3 Results and Discussion

### *16.3.1 Thermal Performance*

The experimental campaigns were carried out in different climatic conditions by monitoring the rooms during the various seasons of the year. Preliminary data collected in June 2018 are analyzed and represented in graphs, in order to compare the behavior of the Test Room with the innovative glazing system with the Reference Room, in which the standard glazing system is installed.

In Fig. 16.4a comparison between the temperatures of the two rooms is shown: air temperatures ($T_a$), surface temperatures of the glazings ($T_{sur}$), and mean radiant temperatures ($T_{mr}$); the outside air temperature is also reported. The outside air temperatures in June, in central Italy, reach medium-high values during the day (about 25 °C but with peak of 30 °C), and medium–low values in the night (in 12–15 °C range). The indoor air temperature in the Test Room, especially the peak values, is lower than the one in the Reference Room (differences of about 6–7° C). This trend is due to the presence of silica dust, which contributes to the reduction of the solar factor. During the night the two chambers show comparable values of temperature; however, the Test Room is a little bit hotter than the Reference Room, in compliance with its lower value of transmittance.

The trend of the surface temperatures of the SGS is close to the air temperature; furthermore the surface temperatures of the AGS are 7–8 °C lower than the ones of

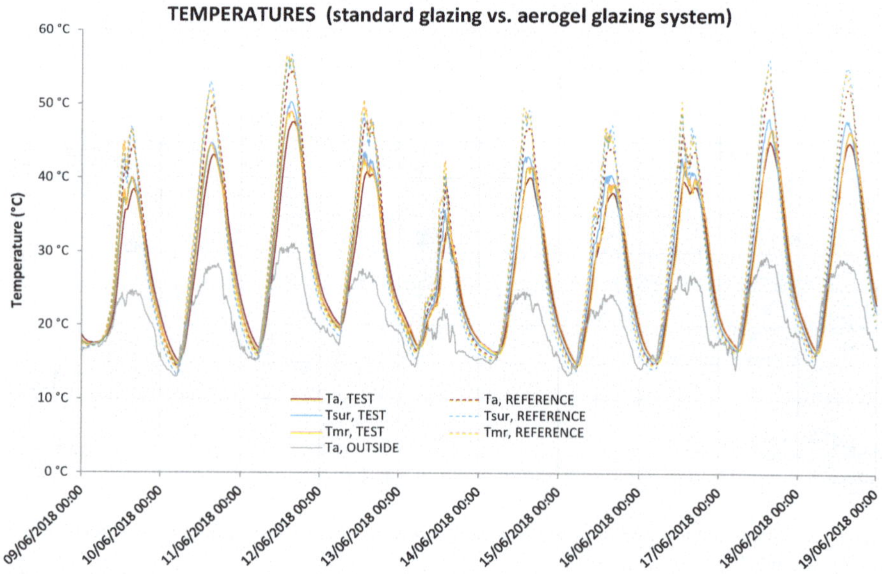


**Fig. 16.4** Comparison between the temperatures in the Test and Reference Rooms

the SGS during the sunny hours, while during the night they are 1–2 °C higher than the ones of the Reference Room.

For the sake of completeness, a typical sunny day is isolated in order to highlight the behavior of the temperatures in the rooms (Fig. 16.5): during the hottest hours of the day the mean radiant temperature is a little bit higher than the air temperature, but the increase is the same in the Test and Reference Room.

Also the west wall and the floor temperatures of the Reference Room are always higher than those of the Test Room. During the warmest hours of the day, the difference of the floor temperatures is about of 7–8 °C, while the surface temperatures of the wall differ for more than 10–11 °C; during the night, the surface temperatures are about the same.

A summing-up section using the indoor air temperature as indicator is reported in Fig. 16.6. Air temperature inside the rooms is plotted vs. the outside temperatures: blue points represent data in the Test Room, red points data in the Reference Room. Regression lines are also showed and a significant value of $R^2$ is found (in the 75 ÷ 81% range). When the outside air temperature is higher than 17 °C, in the Test Room with aerogel, indoor air temperatures are lower than the ones with SGS. As the outside temperature increases, the difference of the measured indoor temperature between Reference and Test rooms also increases. For example, when the outside temperature of the air is 30 °C the temperatures in the Test Room are 14% lower than the ones in the Reference Room with standard glazing system.

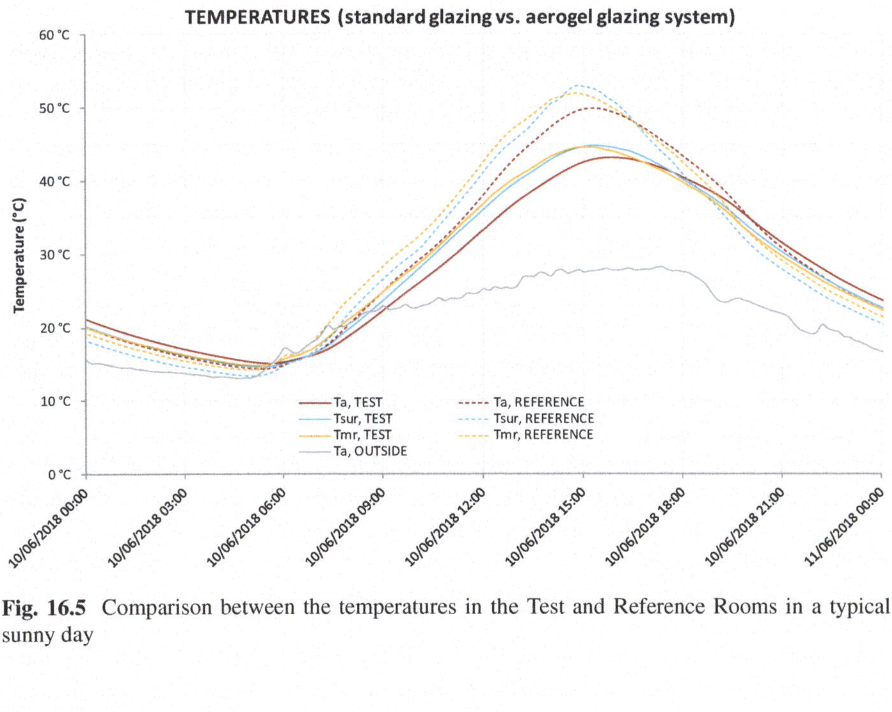


**Fig. 16.5** Comparison between the temperatures in the Test and Reference Rooms in a typical sunny day

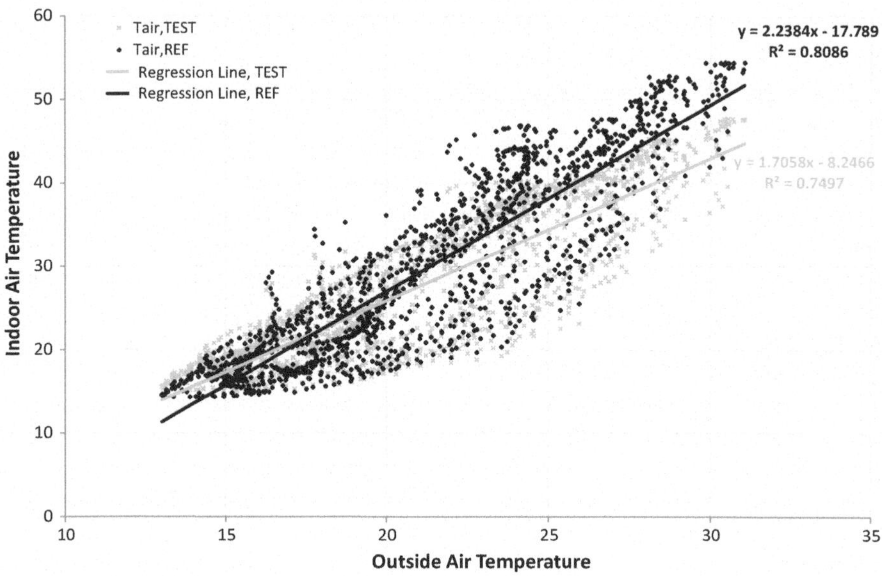


**Fig. 16.6** Indoor air temperature trends versus outside temperatures of the different glazing systems

## *16.3.2 Lighting Performance*

The illuminance was measured inside the rooms in two different points, respectively 50 and 135 cm from the window (E1 and E2), at a height of about 40 cm from the floor, in order to avoid the shading of the parapet of the window.

For all the days the values of the illuminance in the Reference Room are higher than those of the Test Room; this trend is a consequence of the diffusing behavior of the granular aerogel. The illuminance peak values close to the window (E1) are about 5000–8000 lux with the conventional glazing and about 2000–2500 lux with the aerogel glazing. Lower differences are observed in the point E2, far from the window: peaks of 1700–2500 lux are found in the Reference Room and of 1000–1500 lux in the Test Room. Figure 16.7a represents in details the trends in a typical day of June: the difference of illuminance between the two rooms in E1 is about 3500 lux and the one in E2 is about 700 lux. This behavior shows the aerogel ability to diffuse the light and to reduce glare.

By a probe sited vertically and close to the glazing system, the global radiation entering the window is measured; data are reported in Fig. 16.7 together with outside solar radiation on the horizontal plan and outside air temperature. The solar radiation through the glazing with aerogel is always lower than the one of the standard glazing system; in a typical day, the maximum peak value is about 100 $W/m^2$ lower (Fig. 16.7b).

In order to assess the daylight quality, the illuminance data were used to calculate the Useful Daylight Illuminance (UDI) [20–23]. UDI indicates the levels of day-

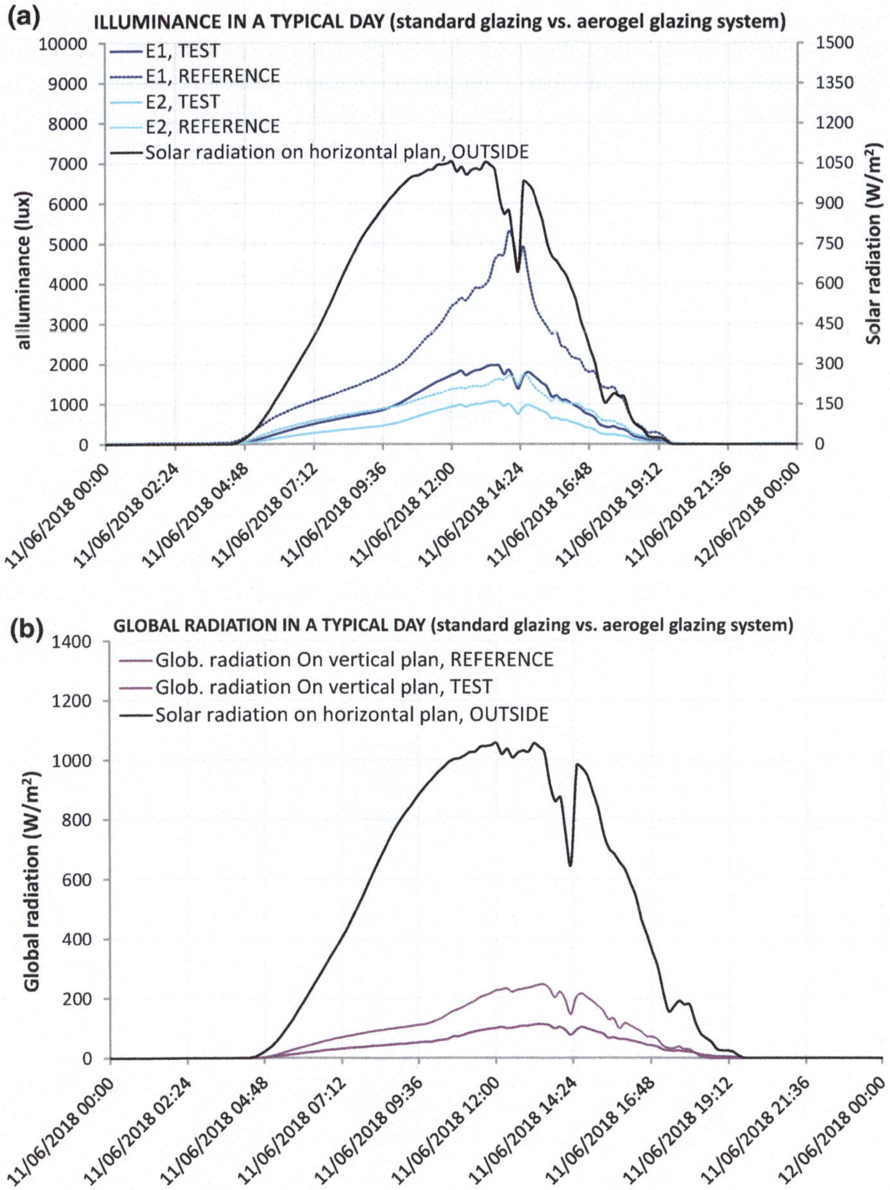


**Fig. 16.7** Comparison of the illuminances (**a**) and the global radiation (**b**) in a typical day

lighting associated with occupant discomfort (i.e. glare) and unwanted solar gain based on the horizontal illuminance level [24, 25]. The conventional approach produces a single percentage number (daylighting factor) for each examined point in the space. It informs not only on useful levels of daylighting illuminance, but also on the unwanted solar gain, and the period in which illuminance is lower than a fixed useful value.

The values of the measured illuminances (every 10 min.) were post-processed in order to evaluate the differences between the two glazing systems in the two different reference points (E1 and E2). An occupancy period of 8 a.m –6 p.m. was considered. The chosen pre-established ranges are the ones generally used in the Literature [21–25].

- UDI Fell-short, less than 100 lux;
- UDI Achieved, illuminance values in the 100–2000 lux range, where the lighting comfort conditions are reached;
- UDI Exceeded, for illuminances above 2000 lux, which are considered excessive and may cause visual and/or thermal discomfort.

Figure 16.8 shows the frequency distribution in time of each illuminance range. The graph is divided into two parts (E1 and E2 reference points): each column of the graph shows the UDI percentages of each illuminance range. With the black and white colors the percentages with an illuminance less than 100 lux and above 2000

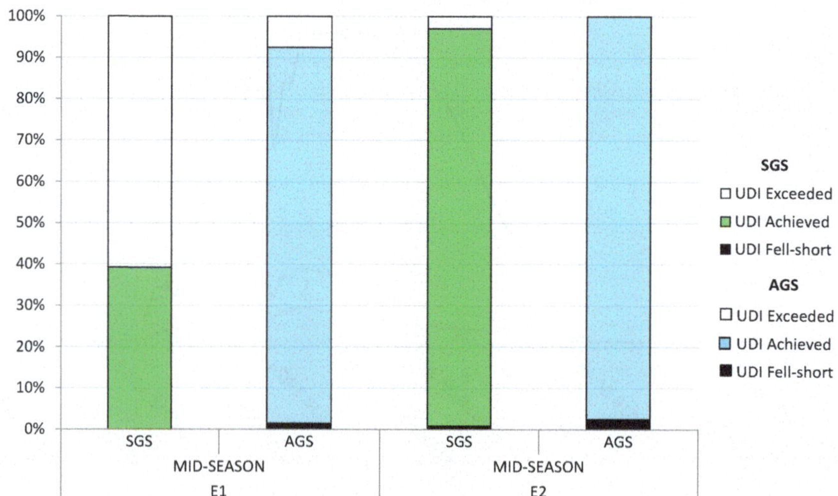


| | SGS (MID-SEASON E1) | AGS (MID-SEASON E1) | SGS (MID-SEASON E2) | AGS (MID-SEASON E2) |
|---|---|---|---|---|
| UDI Exceeded | 61 | 8 | 3 | 0 |
| UDI Achieved | 39 | 91 | 96 | 98 |
| UDI Fell-short | 0 | 1 | 1 | 2 |

**Fig. 16.8** Useful daylight Illuminance (UDI) analysis: comparison between the two transparent solutions in the June experimental campaign

lux are shown, respectively. Green and light blue colors indicate the UDI Achieved for the SGS and AGS, respectively.

In E1 (close to the window), results show that the UDI Achieved of the innovative glazing is higher than the Reference Room one: it is 57% higher than SGS for the considered season. In E2 (far from the window), very low differences between the two glazing solutions are reached in terms of UDI and the achieved is nearly 100% for both the glazing solutions.

## 16.4 Conclusions

Thermal-energy and lighting performance of innovative double glazing units with aerogel is evaluated through experimental campaigns at pilot scale in two rooms (Test and Reference) installed on the roof of a building. An innovative sample (AGS) with a mixture of aerogel granules and opaque hollow silica in the interspace of a double glazing is mounted in the Test Room and a standard double glazing system (SGS) with air in interspace in the Reference Room.

The main thermal and lighting parameters are monitored and the preliminary results (June 2018) are discussed: the indoor air temperature in the Test Room (peak values) is about 6–7 °C lower than the one in the Reference Room, while during the night the two chambers show comparable values of temperature. In absence of solar radiation, the Test Room is a little bit hotter than the Reference Room, due to its lower transmittance value. The same behavior is observed for the west wall and the floor surface temperatures and for the glass surface temperature of SGS and AGS. Furthermore, when the outside air temperature is higher than 17 °C, in the Test Room with aerogel indoor air temperatures are lower than the ones with SGS and the difference increases as the outside temperature increases, resulting in a value 14% lower when outside temperature is 30 °C.

The illuminance in the Reference Room is always higher than those of the Test Room, with peak values close to the window of about 5000–8000 lux and 2000–2500 lux with the conventional and the aerogel window respectively. A similar behavior is observed for the global radiation entering the window (the maximum peak value is about 100 $W/m^2$ lower for AGS).

Finally data of the Useful Daylight Illuminance (UDI) attests the good behavior of the aerogel window, confirming the ability of this innovative material to diffuse the light and to reduce glare risk.

## References


1. Baetens, R., Jelle, B.P., Gustavsen, A.: Aerogel insulation for building applications: a state-of-the-art review. Energy Build. **43**, 761–769 (2011)
2. Riffat, S.B., Qiu, G.: A review of state-of-the-art aerogel applications in buildings. Int. J. Low-Carbon Technol. **8**(1), 1–6 (2012)
3. Cuce, E., Cuce, P.M., Wood, C.J., Riffat, S.B.: Toward aerogel based thermal superinsulation in buildings: a comprehensive review. Renew. Sustain. Energy Rev. **34**, 273–299 (2014)
4. Berardi, U.: Aerogel-enhanced solutions for building energy retrofits: insights from a case study. Energy Build. **159**, 370–381 (2018)
5. Gao, T., Ihara, T., Grynning, S., Jelle, B.P., Gunnarshaug, Lien A.: Perspective of aerogel glazings in energy efficient buildings. Build. Environ. **95**, 405–413 (2016)
6. Gao, T., Jelle, B.P., Ihara, T., Gustavsen, A.: Insulating glazing units with silica aerogel granules: the impact of particle size. Appl. Energy **128**, 27–34 (2014)
7. Lv, Y., Wu, H., Liu, Y., Huang, Y., Xu, T., Zhou, X., Huang, R.: Quantitative research on the influence of particle size and filling thickness on aerogel glazing performance. Energy Build. **174**, 190–198 (2018)
8. Ihara, T., Grynning, S., Gao, T., Gustavsen, A., Jelle, B.P.: Impact of convection on thermal performance of aerogel granulate glazing systems. Energy Build. **88**, 165–173 (2015)
9. Huang, Y., Niu, J.: Energy and visual performance of the silica aerogel glazing system in commercial buildings of Hong Kong. Constr. Build. Mater. **94**, 57–72 (2015)
10. Buratti, C., Merli, F., Moretti, E.: Aerogel-based materials for building applications: influence of granule size on thermal and acoustic performance. Energy Build. **152**, 472–482 (2017)
11. Huang, Y., Niu, J.: Application of super-insulating translucent silica aerogel glazing system on commercial building envelope of humid subtropical climates: Impact on space cooling load. Energy **83**, 316–325 (2015)
12. Ihara, T., Gao, T., Grynning, S., Jelle, B.P., Gustavsen, A.: Aerogel granulate glazing facades and their application potential from an energy saving perspective. Appl. Energy **142**, 179–191 (2015)
13. Garnier, C., Muneer, T., McCauley, L.: Super insulated aerogel windows: impact on daylighting and thermal performance. Build. Environ. **94**(1), 231–238 (2015)
14. Lolli, N., Andresen, I.: Aerogel versus argon insulation in windows: a greenhouse gas emissions analysis. Build. Environ. **101**, 64–76 (2016)
15. Mujeebu, M.A., Ashraf, N., Alsuwayigh, A.H.: Effect of nano vacuum insulation panel and nanogel glazing on the energy performance of office building. Appl. Energy **173**, 141–151 (2016)
16. Cotana, F., Pisello, A.L. Moretti, E., Buratti, C.: Multipurpose characterization of glazing systems with silica aerogel: in-field experimental analysis of thermal-energy, lighting and acoustic performance. Build. Environ. **81**, 92–102 (2014)
17. UNI EN 14509:2013. Self-supporting double skin metal faced insulating panels—Factory made products—Specifications
18. EN 674, Glass in building—Determination of thermal transmittance (U value)—Guarded hot plate method (2011)
19. UNI EN ISO 7726:2002. Ergonomics of the thermal environment—Instruments for measuring physical quantities
20. Ramos, G., Ghisi, E.: Analysis of daylight calculated using the EnergyPlus programme. Renew. Sustain. Energy Rev. **14**, 1948–1958 (2010)
21. Zhang, A., Bokel, R., van den Dobbelsteen, A., Sun, Y., Huang, Q., Zhang, Q.: Optimization of thermal and daylight performance of school buildings based on a multi-objective genetic algorithm in the cold climate of China. Energy and Build. **139**, 371–384 (2017)
22. Mardaljevic, J., Heschong, L., Lee, E.: Daylight metrics and energy savings. Light. Res. Technol. **41**(3), 261–283 (2009)
23. Cantin, F., Dubois, M.C.: Daylighting metrics based on illuminance, distribution, glare and directivity. Light. Res. Technol. **43**(3), 291–307 (2011)

24. Nabil, A., Mardaljevic, J.: Useful daylight illuminance: a new paradigm for assessing daylight in buildings. Light. Res. Technol. **37**(1), 41–57 (2005)
25. Nabil, A., Mardaljevic, J.: Useful daylight illuminances: A replacement for daylight factors. Energy Build. **38**, 905–913 (2006)

# Chapter 17
# 'Zukunftsquartier'—On the Path to Plus Energy Neighbourhoods in Vienna

**Jens Leibold, Simon Schneider, Momir Tabakovic, Thomas Zelger, Daniel Bell, Petra Schöfmann and Nadja Bartlmä**

**Abstract** This paper presents an approach to define and implement a 'Zukunftsquartier' (future neighbourhood) in the context of the densely populated city environment of Vienna, which is in line with the national energy targets 2050. The 'Zukunftsquartier' project explores the feasibility of plus energy neighbourhood concepts at four prospective project sites in Vienna. The case studies evaluate the potential of demand side management, innovative renewable energy systems including photovoltaic and near-surface geothermal energy by hourly energy balancing and are compared for the Austrian building code and 'passive house' construction standards. Due to the high floor space index of urban projects, all investigated concepts failed to achieve a positive energy balance, except theoretical variants with unfeasibly high PV utilization of virtually the entire roof and façade surfaces. To offset the unintended effect of plus energy being harder to achieve in a dense urban context, we propose a correction factor for the target energy balance of 'plus energy' buildings and neighbourhoods based on the floor space index. Together with a second energy balance adjustment, reflecting the prospective renewable energy system (RES) of Austria 2050, most ambitious variants (utilizing ground heat and moderate PV surfaces) achieved 'plus energy' standard for dense urban areas and life cycle costs compared to conventional realizations within 30 years.

J. Leibold (✉) · S. Schneider · M. Tabakovic · T. Zelger · D. Bell
University of Applied Sciences Technikum Vienna, Giefinggasse 6, 1210 Vienna, Austria
e-mail: leibold@technikum-wien.at

P. Schöfmann
UIV Urban Innovation Vienna GmbH, Operngasse 17-21, 1040 Vienna, Austria
e-mail: schoefmann@urbaninnovation.at

N. Bartlmä
IBR & I, Institute of Building Research & Innovation ZT GmbH, Wipplingerstraße 23/3, 1010 Vienna, Austria
e-mail: nadja.bartlmae@building-research.at



J. Littlewood et al. (eds.), *Sustainability in Energy and Buildings*, Smart Innovation, Systems and Technologies 163, https://doi.org/10.1007/978-981-32-9868-2_17

## 17.1 Introduction

With the Smart City Vienna Framework Strategy, published in 2014, Vienna committed to a path of decarbonisation. Developing sustainable, secure and affordable strategies to supply (new) urban districts with energy is one of the many challenges in this context. The City Administration of Vienna aims at realizing an innovative showcase of modern city quarters as part of its governmental agreement (2015). The intended exploratory project ought to make a valuable contribution and, with the help of a competent consortium in the field of research–planning–implementation, to substantially advance the preparation of such a showcase city quarter with new knowledge and experience.

With the support of the City of Vienna and numerous developers, at least five urban mixed quarters, which will be developed over the next 2–5 years and whose energy supply has not yet been decided, are being investigated in this exploratory project. One task of the project tries to answer the question of adequate system boundaries and indicators for positive energy neighbourhoods. Furthermore, the consortium develops and evaluates early-stage concepts and options for the neighbourhoods to determine the most promising ones for realization and detailed planning in the next step.

So far, for four quarters, preliminary draughts of energy concepts based on the local energy situation and the requirements of stakeholders/users have been developed.

## 17.2 Aim

This paper presents an approach to define and implement a 'Zukunftsquartier' (future neighbourhood) in the context of the densely populated city environment of Vienna, which is in line with the national energy targets 2050. Therefore, a proposal is presented on how a compensation between low-density and highly dense urban areas, in terms of 'effort sharing' can be achieved. This approach is essential as a push towards the development of high-density urban plus energy districts and can be observed on both, international and national levels [6]. Development of plus energy neighbourhoods, or even districts, is not easy to achieve in terms of technical feasibility and marginal costs [4]. Through the analysis, modelling and simulation of the considered neighbourhoods, including their technical and economic conditions, and the subsequential derivation of recommendations for action (e.g. for the planning process, for technology combinations and for stakeholder integration), the project aims to provide insights into the broader applicability of the plus energy neighbourhood concept.

## 17.3 Status Quo Plus Energy Quarters

There is no uniform definition for system boundaries and calculation method for plus energy quarters. Common sense is that a plus energy quarter produces more energy than it consumes throughout the year. How to reach the 'plus' (efficiency in energy demand, increase in renewable energy production or use of smart controls is open as well as the considered main indicators, like operating, final or primary energy or $CO_2$-emissions. In general, definitions are based on those applied for plus energy buildings, supplemented by general electricity demand of the quarter, like outdoor lighting in the neighbourhood. Considered energy services commonly include heating, cooling, ventilation and hot water demands plus the electricity like user electricity, lighting and auxiliary requirements. The extended view at neighbourhood or district level can increase the ratio of internal consumption and thus are improving the profitability. Reasons are mixed usage effects and the circumstance that energy can be exchanged between the buildings; however, there is no positive effect on the annual energy balance. Generally, the way to a plus energy neighbourhood is similar to the challenges of plus energy buildings. The neighbourhoods that own energy supply capabilities are limited by the available plot size (for solar and ambient energy), or more accurately by the ratio of the conditioned space to the available plot size (also known as 'floor space index' or FSI). This is the predominant factor for the on-site renewable energy supply (RES) potential of any building or quarter.

### *17.3.1 Plus Energy Districts in Urban Context*

In Austria, there are so far no realized projects for plus energy quarters in dense urban areas that have achieved the desired goals. In Europe, while the number of projects in the implementation or planning stage is quite high (compare [3]), a few projects are in operation (for example, Hunziker Areal, Zurich, Switzerland, Fleuraye, Carquefou/Nantes, France). To guarantee the actual implementation of the concepts from the planning to the operation stage, the economic feasibility is crucial. In Austria, for example, the concept of the quarter Reininghaus Süd was initially planned as a plus energy network [8]. In the realization phase, the PV system was saved and therefore no plus energy was reached.

## 17.4 Methodology and Assumptions

For the preliminary calculation of the envisaged quarters, the energy demand as well as the on-site potential for renewable energies are determined in dynamic simulations, described under Sect. 17.4.1. The following variations based on Fig. 17.1 were calculated. The main considerations of the project are variations of the building stan-

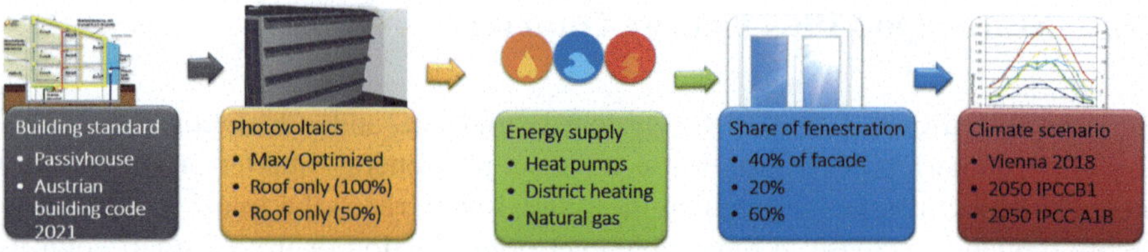


**Fig. 17.1** Main variants which are considered for calculations

dard, different photovoltaics allocations and energy supply systems. Three different shares of fenestration and various climate scenarios for 2050 are additional subjects.

### *17.4.1 Energy Demand Calculation and Potential for Renewables Energies*

**Simulation Method, Zoning and Usage Profiles**

For the calculation of the heating and cooling demand, a simplified dynamic calculation of the neighbourhood was implemented in a single-zone model, in which all relevant heat flows (into the system positive, from the system negative) in hourly resolution are incorporated. By rearranging Eq. 17.1, the respective capacity can be determined.

$$C^* \, dT/dt = \dot{Q}_T + \dot{Q}_v + \dot{Q}_s + \dot{Q}_I + \dot{Q}_H + \dot{Q}_C \tag{17.1}$$

$C$ Effective storage capacity in $Wh/m^2_{NFA}K$
$dT$ Temperature difference between two timesteps in K
$dt$ Time difference in h
$\dot{Q}_T$ Transmission heat flux in $W/m^2$
$\dot{Q}_V$ Ventilation heat flow in $W/m^2$
$\dot{Q}_S$ Solar gains due to transparent components in $W/m^2$
$\dot{Q}_I$ Internal gains by persons, equipment, lighting, etc. in $W/m^2$
$\dot{Q}_H$ Heating capacity in $W/m^2$
$\dot{Q}_C$ Cooling capacity in $W/m^2$.

The following assumptions were made:

- Transmission heat flux according to PHPP results in Table 17.1 outlined characteristic values, depending on building energy standard.
- Increased summer air exchange via windows for natural cooling was not taken into account.
- A heavy construction method was assumed with $C = 204\,Wh/m^2_{NFA}K$.

**Table 17.1** Used characteristic values for building envelope

| Characteristic values | Passive house | OIB 2021 | |
|---|---|---|---|
| Exterior wall | 0.12 | 0.18 | $W/m^2\,K$ |
| Roof | 0.1 | 0.12 | $W/m^2\,K$ |
| Ceiling | 0.12 | 0.19 | $W/m^2K$ |
| Basement ceiling | 0.15 | 0.19 | $W/m^2\,K$ |
| Windows | 0.83 | 0.93 | $W/m^2\,K$ |
| Thermal bridges | 0.03 | 0.06 | $W/m^2_{NFA}\,K$ |

- Distinction between heating and cooling periods is determined monthly, depending on the quality of the building envelope, resulting in three possible building states, either heating, cooling or 'freerunning'.
- Assumptions for the internal heat are differentiated into winter and summer case according to ÖNORM B 8110-5: The internal loads in cooling mode are assumed to be twice as high as in the heating mode, with linear interpolation in the transitional period.
- The solar gains are assumed for the simplified calculation with a fixed shading factor of 0.75, resulting in a total solar transmission rate of 0.39 including g-value, frame section, etc.

There are two setpoint room temperatures considered, one each for heating and cooling:

- Minimum target room temperature: This must not be under-(heating case) or exceeded (cooling case) even in the case of DSM. Without DSM, this represents the target room temperature.
- Maximum and minimum target room temperature: Use of the building storage mass for the DSM, this represents the maximum (heating mode) or minimum (cooling mode) room temperature.

**Photovoltaics**

The assessment for the potential of the usable solar energy for PV was performed for four exploitation strategies of different size and cost with PV sites. The variants are the maximum technical potential, maximum roof utilization, half roof utilization and an optimized case. For the roof allocation, a 15° inclination, east/west orientation was carried out for each case. In addition, maintenance corridors and distances of 0.6 m to the roof edge are provided between the module rows. The bifunctional PV canopies (sun protection) are mounted at an angle of 30°. In the case of façades, the considered window area proportions (20, 40, 60%) are deducted from the yields. For the optimized variant, special areas, if any, are also taken into account and removed (keyword: conflict of use). In addition, no PV modules were planned for façades with annual irradiation of <500 kWh/m$^2$. The 3D model in PV sites takes into account shading of surrounding buildings. The PV yields can be displayed in hourly resolution, which is an important prerequisite in combination with the dynamic energy demand calculation for determining the own consumption rate. The simulations are carried out with a highly efficient state–of-the-art solar panel from LG with 320 W power (LG 320N1C G4), which is characterized by a typical size of 1.6 m by 1 m.

**Energy Supply**
One conventional and one plus energy neighbourhood energy concept was developed for each use case by taking the local renewable potentials into account. Plus energy variants for the considered neighbourhoods are implemented by using on-site PV, waste heat and geothermal potentials via heat pumps. The potential for the use of near-surface geothermal energy by geothermal probes was determined for the rough concepts with a withdrawal capacity of 35 W/m. The maximum probe length is assumed to be 150 m if no restrictions exist on-site. The annual coefficient of performance of the heat pump was assumed to be 5 for heating, 7.5 for cooling and 3 for domestic hot-water demand. Distances between the probes are 8 m without regeneration or 5 m with regeneration.

### *17.4.2 System Boundaries*

Current system boundaries serve many functions but fail to connect directly to local and national climate goals. Hence, three extensions to the common system boundary of the primary energy balance for building operation are proposed [9].

**System Boundary Extension 1: Regional Renewable Peak Shaving**
The Austrian renewable energy strategy [7] aiming at 100% renewable energy supply by the year 2050, requires a fivefold capacity increase of wind power. However, how to integrate this volatile energy supply in the future energy system? An increased energy flexibility in buildings and districts [5] in combination with other storage technologies could be the solution. Alham et al. [1] and [10] show that it is both technically and economically feasible to dispatch wind power generation in accordance with building demand side response. Therefore, the system boundary of the plus energy neighbourhood is extended to include possible peak shaving of regional wind power due to demand response potentials of the buildings.

**System Boundary Extension 2: National RES User Credit**
On the basis of the 'renewable Austria 2050' scenario, the renewable energy from large-scale wind parks, water power stations and biomass will first be allocated to energy uses, which are difficult to supply locally: Industry, public transport and large-scale power2hydrogen or power2gas. The remaining RES from large-scale power plants can be nationally allocated to all inhabitants as an 'individual renewable credit', which can be taken into account for primary energy balancing of a building: The cumulative RES credit of all building inhabitants' counts towards its PEB.

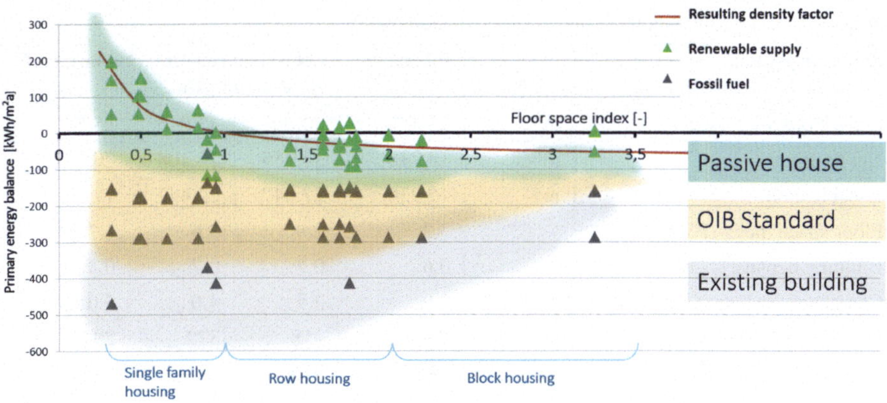


**Fig. 17.2** Graphical dependence of the '*primary energy credit*' determined by energy supply system, building density and building standard

**System Boundary Extension 3: Density Factor**

Contrarily to the energy demand, the local RES potential is approximately proportional to the plot size. Thus, the lower the floor space index of a building is, the easier it will achieve plus energy standard. In Fig. 17.2, the maximum primary energy balance achievable for projects of different floor space index, as compiled by the research project *SC_Microquartiere* [2] is shown. The red coloured curve shows the proposed density factor, depending on the floor space index.

Conversely, it is virtually impossible to achieve conventional plus energy standard at a certain higher floor space index—there is not enough on-site renewable energy potential for the useable floor area. This leads to the effect that the more efficient a building is in terms of land use, the more difficult, if not impossible, it will be to achieve a plus. Paradoxically, the classical NZEB as well as plus energy standard, which aim to improve energy efficiency and use of renewables on-site, indirectly promote less efficient use of the finite resource that is buildable land. Therefore, depending on the floor space index, a density factor is introduced as extension 3.

## *17.4.3 Investment Costs Difference*

For the project, no full life cycle assessment is carried out, but a simplified differential cost analysis. Following are the assumptions to determine the resulting costs for the variants:

- The imputed real interest rate of the investment was assumed with 2%.
- Total costs within the first 30 years (excluding the development of income).
- Subsidies not taken into account.
- Residual values were not considered.

Table 17.2 Structural data from the considered quarters

| Zukunftsquartier | | Pilzgasse | Ottakringerjeben | An der Kuhtrift | Marx Hub |
|---|---|---|---|---|---|
| Gross floor area (GFA) | $m^2$ | 23,435 | 40,069 | 33,010 | 25,740 |
| Net floor area | $m^2$ | 18,748 | 32,055 | 26,408 | 20,592 |
| Useable area | $m^2$ | 15,936 | 27,247 | 22,447 | 17,503 |
| Floor space index (FSI) | – | 3.2 | 2.8 | 3.6 | 1.7 |
| Building density | – | 0.6 | 0.7 | 0.7 | 0.3 |
| Characteristic length (lc value) | m | 3.9 | 2.9 | 4.0 | 5.0 |
| Households (fictitious) | number | 268 | 458 | 377 | 294 |
| Persons (fictitious living) | number | 536 | 916 | 755 | 588 |

The cost calculation of the building services represents a conservative estimate because

- Maximum service life of 30 years is not identical with the actual service life of individual components (higher realistic service life of components such as geothermal probes are not taken into account).
- The average increase in energy prices and maintenance costs is assumed to be in line with the inflation rate.

## 17.5 Case Study

All the quarters under consideration comprise at least three building complexes with a total floor area of above 20,000 $m^2$ and are characterized by high mixtures of usage.

Table 17.2 summarizes the structural data of the considered neighbourhoods. Unless otherwise stated, the reference unit for all specific key figures is the net floor area (NFA), which is conditioned and ventilated. The net floor area is presumed with 80% of the gross floor area (GFA). The usable area was assumed to be 85% of the net floor area.

## 17.6 Results

The proposed system boundaries are applied to four plus energy quarters in Vienna, while Extension 1 (wind peak shaving) is used in general. The results for highly efficient passive house building standard are shown in Fig. 17.3. As can be seen, PV installation size has by far the biggest impact on the achievable primary energy balance of the variants. All four quarters could achieve the conventional plus energy standard, PEB balance > 0 only under the assumption of utilizing most of the building

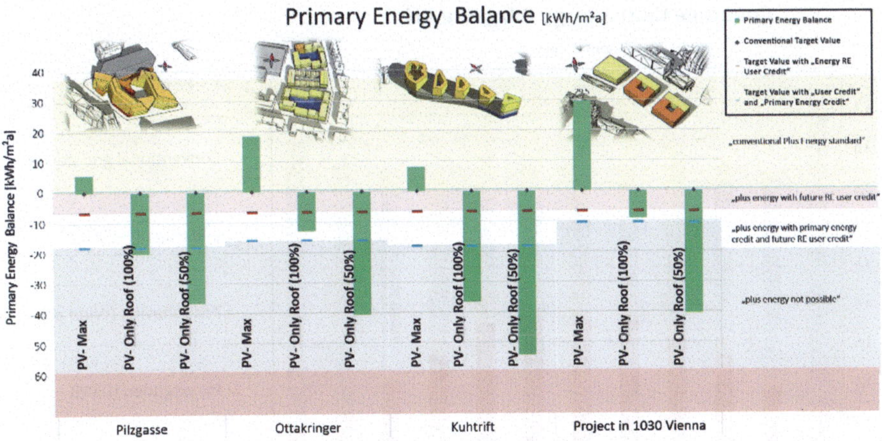


**Fig. 17.3** Primary energy balance for four Plus Energy Quarters. Each Quarter is visualized in three variants: (1) PV max: All building surfaces with an insolation >500 kWh/m$^2$a are utilized for PV generation, (2) PV Roof only (100%) PV utilization on the entire roof area and (3) PV Roof only (50%): Only half of the roof

surfaces for PV power generation. Although technically possible, this is economically unfeasible. Apart from the extensive PV Max strategy, the more moderate variants all require adaptations to the classical primary energy balancing method to be plus energy feasible. The use of system boundary Extension 2 is marked with a red line and Extension 3 (density factor) in blue. Results show that a realistic roof PV allocation of 50% is not adequate. Therefore, for each site, an optimized variant (PEQ) was determined with optimized roof allocation and a share of PV façades depending on the resulting energy deficit.

The monthly results for the power supply are shown in Fig. 17.4. The PEQ variant clearly shows that the renewables (solar and wind surplus) in combination with DSM measures can cover well over 50% of the electrical energy requirement in the winter period. In very unfavourable climatic months (cold, almost no wind and solar energy), such as in December 2015, just about 25% can be generated from the considered renewable sources. The PV surpluses in the summer half-year can cover a significant proportion of the future e-mobility needs of living, working and educational persons in the neighbourhood.

As shown in Fig. 17.5, the exemplary additional costs of a plus energy project area are mainly caused by the PV system, the ventilation system and the highly efficient heat/cooling distribution and storage system. Due to the partly less complex equipment standard of the reference variant (multi-split system attics, fixed shading, etc.), the differential costs are relatively moderate and are well below 10% of the planned construction costs. Maintenance and financing costs increase production costs by approx. 80%. The energy costs savings (on the right side) resulting in a total 'profit' over 30 years.

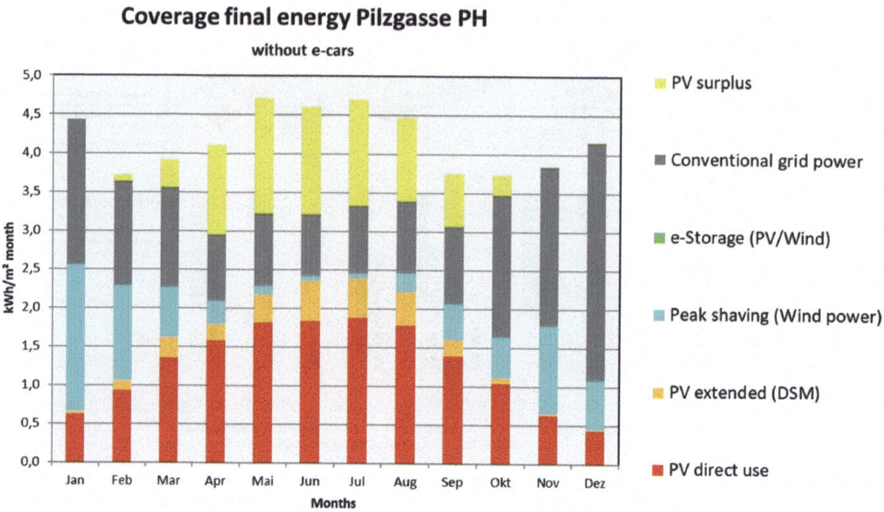


**Fig. 17.4** Monthly coverage of final energy for the case area 'Pilzgasse'

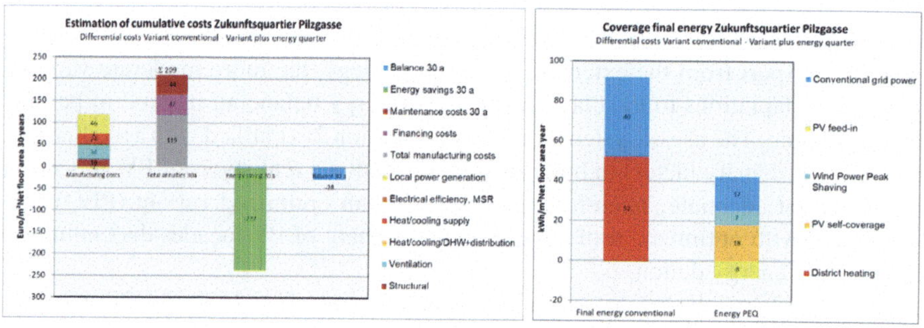


**Fig. 17.5** Estimation of additional costs, considering annuities (left side) and energy savings. On the right side division of achievable energy savings

## 17.7 Conclusion

The results show that plus energy concepts with reduced PV areas in the façades are feasible for all considered quarters under consideration of consistent system boundaries. The proportion of façade-integrated PV is decisive for cost-effectiveness because investment costs are higher and the energy yield, compared to roof systems is lower. Due to the predominant mix of uses, high own consumption rates of the PV yield between 60 and 70% are achievable for all considered neighbourhoods, which are important for economic reasons. Depending on the future expansion rate of e-mobility, PV surpluses in summer can be largely absorbed. For the neighbourhood Kuhtrift (big car parking planned), in an estimation with relevant e-car share and low loading capacities, almost 100% own consumption was achieved. The estimation of differential costs shows large differences. In three quarters, there are not only moderate additional costs but also one with high additional burdens.

The main influencing factors are

- Reference standard of the conventional variant
- Predominant mixture of usage
- Availability of waste heat, otherwise, expenses for the active regeneration of the geothermal probes, like PVT collectors are challenging.

**Acknowledgements** The authors thankfully acknowledge the Austrian Research Promotion Agency (FFG) and the Austrian Federal Ministry of Mobility, Innovation and Technology for enabling this publication by funding the research project 'Zukunftsquartier' in the research programme 'Stadt der Zukunft'.

## References

1. Alham, M.H., Elshahed, M., Ibrahim, D.K., Abo El Zahab, E.E.D.: A dynamic economic emission dispatch considering wind power uncertainty incorporating energy storage system and demand side management. Renew. Energy. **96**, 800–811 (2016). https://doi.org/10.1016/j.renene.2016.05.012
2. Fellner, M., Zelger, T., Leibold, J., Huemer-Kals, V., Kleboth, A., Granzow, I., Fleischhacker, A.: Smart City MIKROQUARTIERE. Vienna (2018)
3. Gollner, C., Hinterberger, R., Noll, M., Meyer, S., Schwarz, H-G.: Booklet of positive energy districts in Europe. Preview (2019)
4. Iturriaga, E., Aldasoro, U., Terés-Zubiaga, J., Campos-Celador, A.: Optimal renovation of buildings towards the nearly Zero Energy Building standard. Energy. **160**, 1101–1114 (2018). https://doi.org/10.1016/j.energy.2018.07.023
5. Jensen, S.Ø., Marszal-Pomianowska, A., Lollini, R., Pasut, W., Knotzer, A., Engelmann, P., Reynders, G.: IEA EBC annex 67 energy flexible buildings. Energy Build. **155**, 25–34 (2017). https://doi.org/10.1016/j.enbuild.2017.08.044
6. Koutra, S., Becue, V., Gallas, M.-A., Ioakimidis, C.S.: Towards the development of a net-zero energy district evaluation approach: a review of sustainable approaches and assessment tools. Sustain. Cities Soc. **39**, 784–800 (2018). https://doi.org/10.1016/j.scs.2018.03.011
7. Österreich, E.E.: Energiewende 2013 – 2030 – 2050 (2015)
8. Partoll, M.: +ERS Plusenergieverbund Reininghaus Süd, Endbericht (2016)
9. Schneider, S., Bartlmä, N., Leibold, J., Schöfmann, P., Tabakovic, M., Zelger, T.: New system boundaries! Abolishing the efficiency paradigm. RealCorp Paper (2019)
10. Wu, J., Zhang, B., Jiang, Y., Bie, P., Li, H.: Chance-constrained stochastic congestion management of power systems considering uncertainty of wind power and demand side response. Int. J. Electr. Power Energy Syst. **107**, 703–714 (2019). https://doi.org/10.1016/j.ijepes.2018.12.026

# Chapter 18
# Electrical Devices Identification Driven by Features and Based on Machine Learning

**Andrea Tundis, Ali Faizan and Max Mühlhäuser**

**Abstract** The analysis of energy data of electrical devices in Smart Homes (SHs) represents an important factor for the decision-making process of energy management both from the consumer perspective by saving money and also in terms of energy redistribution and $CO_2$ emissions reduction, by knowing how the energy demand of a building is composed in the Smart Grid (SG). A proactive monitoring and control mechanism motivates the need to face with the identification of appliances. In this context, the paper proposes a model for the automatic identification of electrical devices driven by 19 features that are formalized through a mathematical notation. On the basis of such proposed features, three different classifiers are trained and experimented, by evaluating their accuracy, for the identification of 33 types of appliances.

## 18.1 Introduction

In the last decade, the traditional power systems, which consist of mechanical and electrical parts, have been evolving toward more sophisticated equipment, by integrating and interconnecting software-centered devices for the management and control of their functionalities and usage. This change is leading to a highly interconnected and intelligent infrastructure known as Smart Grid (SG) [1]. Thanks to the presence of the software, the SG can monitor the network and take decisions in the energy distribution by implementing a dynamic load management on the basis of available resources, electricity usage, weather conditions, and so on [2]. Furthermore, consumers can have benefits in terms of saving money by optimizing the

A. Tundis (✉) · M. Mühlhäuser
Technische Universität Darmstadt, Darmstadt, Germany
e-mail: tundis@tk.tu-darmstadt.de

M. Mühlhäuser
e-mail: max@tk.tu-darmstadt.de

A. Faizan
Software AG, Darmstadt, Germany
e-mail: ali.faizan@softwareag.com



J. Littlewood et al. (eds.), *Sustainability in Energy and Buildings*, Smart Innovation, Systems and Technologies 163, https://doi.org/10.1007/978-981-32-9868-2_18

scheduling of their devices usage in specific hours of the day or week according to the electricity costs and so on. Moreover, as an actor can be producer and consumer (socalled prosumer) of electricity in a SG, such flexibility enables a different way to distribute the energy and to deal with unexpected emergency situations, resulting from faults and failures in the network [3].

In this scenario, an important role is played by Smart Homes (SHs). They are equipped with electrical devices that can be controlled and monitored remotely not only to achieve economic benefits by saving electricity but also by contributing in the reduction of $CO_2$ emission in the environment [4, 5]. They provide a new perspective toward the usage of the energy in everyday life and, in particular, in the relationship between energy utilities and consumers. Typically, the traditional homes have devices that work locally and manually, usually by switching them on/off by pushing a button, with limited control, which can make difficult the energy management. A SH, instead, represents the convergence of energy efficient, controllable electrical appliances, and real-time access to energy usage data. This combination of device management and smart grid enables to proactively manage energy use in ways that are convenient, cost-effective, and good for the environment. This can be realized by creating user profiles through the analysis of their habits, based on daily, weekly, or seasonal use of the devices.

However, behind the advantages of a more intelligent energy grid management, one of the main challenges for enabling such a proactive control relies on the automatic recognition, identification, and classification of the electrical appliances. This in turn requires to face with several factors [6], such as: (i) *power consumption extraction*, that is the process of measuring the energy from different devices in order to identify recurrent consumption patterns; (ii) *multimode functionality*, this means that some devices can have multiple operation modes which can be misleading for their identification due to such a complex behavior; (iii) *parallel usage*, this is an important factor that has to be faced, since typically more than one device is in operation at the same time; (iv) *similar characteristics*, because many devices can present similarities in the way they use the energy (e.g., consumption, charging time); (v) *external effects*, because the data could be spoiled by external and random factors, which are not predictable, such as temperature, communication failures, human influences, and so on.

In this context, this work proposes a model for the automatic identification of electrical devices after they are plugged in the electrical socket. The model is based on a set of 19 features which are able to characterize different electrical appliances and distinguish them from others. They are derived by analyzing three main aspects: (i) *power consumption*: related to the electricity being consumed by a device for a certain period of time, (ii) *working schedule*: which includes the hours of the days and the time duration when the device is turned on/off, and (iii) *location*: that represents the place where an electrical device is connected on the basis of the electrical socket within the house. Then machine learning techniques are used to experiment the model through different classifiers, by using a dataset of 33 types of appliances [7]. The goodness of the proposed model is evaluated in terms of its accuracy, for the identification and classification of electrical devices, respect to the existing approaches discussed in Sect. 18.2.

The rest of the paper is structured as follows. In Sect. 18.2 the related works are presented. The reference model and the considered aspects are discussed in Sect. 18.3. Section 18.4 presents and elaborates the proposed features. In Sect. 18.5 the results are discussed, whereas Sect. 18.6 concludes the paper.

## 18.2 Related Work

This section discusses the most relevant research efforts and related solutions, which have been proposed for supporting the identification of electrical devices.

In particular, in [8], a middle layer to connect sockets and devices, which is centered on Measurement and Actuation Units (MAUs), is presented. The MAUs monitors and analyzes the electrical power consumption of any connected device individually by providing fined grain analysis. The main information for the classification is based on temporal behavior of the appliances, power consumption, shape of the power consumption, and level of noise.

In [6] instead, different energy measurements, such as active power, reactive power, phase shift, root mean square voltage, and current, by collecting data of each device in an isolated way, are considered. This approach aims to provide a plug and play tool to create energy awareness on the basis of real-time energy consumption of electrical devices. Additionally, multimode functionality, parallel usage of devices and external effects are also tackled. The authors discussed the difficulties in support the identification of devices which have a multimode operation compared to those with a single operation mode by resulting in a extensive training for deriving an appropriate classification model.

In [9], an approach based on Non Intrusive Appliance Load Monitoring (NIALM) at meter level, to detect whether the device is switched on or off is discussed. When a change occurs in the overall electrical power signal of the house, the change is analyzed and compared to the already-known patterns available in a database. Another centralized approach, for monitoring power signal, exploits the ZigBee device which is attached to the main electrical unit [10]. It is used to identify in real time the appliance which contributes to each spike of energy. Another research effort, based on a centralized approach, is described in [11], in which the authors used custom data collector and, in particular, a power interface oscilloscope and a computer as hardware. It allows to detect electrical noise to classify electrical device in home by exploiting the electrical noise as additional parameters. Whereas, time-series measurements, which represent electrical signatures of different electrical devices, are used in [12] for their identification.

Two main approaches, to face with automatic identification of electrical devices, emerged from the above-related works. One is based on the employment of additional monitoring devices either distributed [8, 12] or centralized [9–11] which results expensive in terms of money for their installation and hardly scalable; the second one, that does not exploit any additional devices, is centered on energy measurements [6], but it lacks in the categorization and formalization of the adopted features. Our work stands out from the previous ones because (i) a set of features that characterize electrical devices are proposed and formalized (ii) a model, based on their combination, is used to identify and recognize devices when they are plugged into the circuit without additional monitoring devices, (iii) better performances in term of higher accuracy are reached.

## 18.3 Electrical Devices Identification Model

In this Section, the proposed *Electrical Device Identification Model (EDIM)* for the identification of electrical devices is described. It is inspired by the following questions: (i) HOW MUCH does a device consume? (ii) "WHEN does a device consume?" (iii) WHERE is a device used? By facing with these questions, three main related *feature classes* have been identified, namely, *Energy and Power Consumption*, *Temporal Usage*, and *Appliance Location*, as depicted in Fig. 18.1.

*Energy and Power Consumption*. This feature class focuses basically on the measurement of power and energy at various levels and at specific points of time. The features belonging to this class aim to extract information related to the electricity consumption of a device in order to characterize it, by answering the question "HOW MUCH does a device consume?" In this class, the following consumption-related features are identified: *Daily Power Consumption*, *Max Power*, *Power Deviation*, *Average Power*, *Average Active Power*, *Lower Activity Power (or MinPower)*, *Energy Consumption*, *Average Peak Value*, *Power Dense Location*, and *Standby Devices*.
*Temporal Usage*. This feature class focuses on the use of a device, mainly from a temporal point of view, by considering the question "WHEN does a device consume?" The features that fall into this class try to extract information, regardless of the amount of energy consumed, with the aim of identifying temporal usage patterns such daily, weekly, and seasonal related to a single device as well as sequence-parallel relationships between multiple devices (eg. the dryer after the washing machine, or the decoder along with the television). In this class, the following time-related features are identified: *On-Off Time*, *Active Time*, *Average Active Time*, *Active Duration*, *Most often Usage Time*, *Devices Used in Sequence*, and *Devices Used in Parallel*.
*Appliance Location*. The place of use of a device is another important indicator, since some electrical devices are often used in the same place (e.g., the hairdryer in the

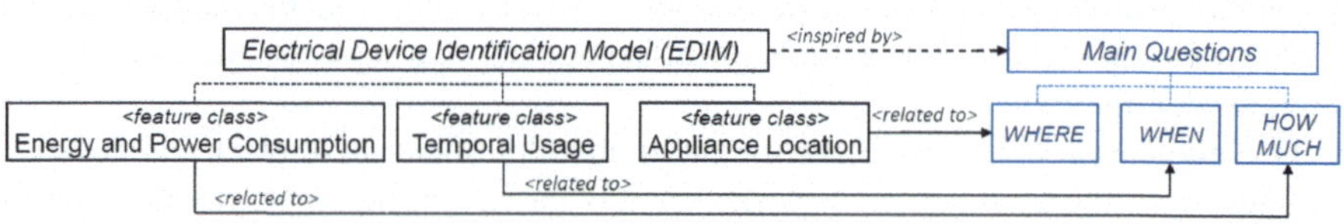


**Fig. 18.1** The electrical device identification model and related questions

bathroom, the kettle in the kitchen). Some of them are movable and others are not. As a consequence, some devices can be used in more than one location inside a house and they can be active not more than in one location at time. This feature class is driven by the question "WHERE is a device used?" Considering this aspect, in this class, the following location-related features are identified: *Place of Use*, *Sequence of Usage Location*.

In the next section, the abovementioned features, which have been proposed in each feature class, are elaborated.

## 18.4 Features Description

After providing an overview of the main *feature classes* and having identified and proposed for each of them a set of features, this section gives, for each single feature, a more detailed description along with a possible formalization.

*On-Off Time*. This feature is used to know, in which instant of time $i$ of a day $D$, a device $j$ is turned On/Off. The function $Status_j(i)$ is used to check the working status of $j$ at the time $i$ [s] $\in D$, in order to identify when a change occurs, based on the previous instant of time $i-1$, where $\lambda$=5[W] is the On-threshold.

$$T_j^{On-Off}(i) = \begin{cases} Off, & when\ \text{Status}_j(i-1) > \lambda\ \ and\ \text{Status}_j(i) = 0 \\ On, & when\ \text{Status}_j(i-1) = 0\ \ and\ \text{Status}_j(i) > \lambda \end{cases} \tag{18.1}$$

*Daily Power Consumption*. It extracts the total power $P_j^{Tot}$ used by a device $j$ within a day $D$ by summing the power values $p_j(i)$ measured each second $i$.

$$P_j^{Tot} = \sum_{i \in D} p_j(i), \text{ where } i \text{ corresponds to a second [s]} \tag{18.2}$$

*Max Power*. This feature is used to calculate the maximum power value $p$ used by a device $j$ within a specific day $D$.

$$P_j^{Max} = max\{p_j(i)\}, \text{ where } i[s] \in D \tag{18.3}$$

*Power Deviation*. This feature deals with the power deviation which is computed as the sum of the difference between the *Max Power* of a device $j$ within a reference day $D$ and its power consumption at every time unit $i[s] \in D$. Only when $j$ is in operation $Status_j(i) > \lambda = 5[W] = On$.

$$P_j^{Dev} = \sum_{i \in D} (P_j^{Max} - p_j(i)) \Leftrightarrow Status_j(i) = On \tag{18.4}$$

*Average Power.* Given a device $j$ this feature calculates the average power used from it, related to a day $D$ by considering both active and non-active operation time $i \in D$.

$$P_j^{Avg} = \frac{\sum p_j(i)}{count(i)}, \forall\, i \in D \tag{18.5}$$

*Average Active Power.* Given a device $j$ this feature calculates the average power used from it, related to a day $D$ by considering only the active operation time $i \in D$, such that $Status_j(i) > \lambda = 5[W] = On$.

$$P_j^{AvgAct} = \frac{\sum p_j(i)}{count(i)}, \forall\, i \in D \text{ such that } Status_j(i) = On \tag{18.6}$$

*Lower Activity Power (Min Power).* This feature is used to calculate the minimum power value $p$ used by a device $j$ within a specific day $D$, by considering only the active operation time $i \in D$, such that $Status_j(i) > \lambda = 5[W] = On$.

$$P_j^{Min} = min\{p_j(i)\}, \text{ where } i[s] \in D \text{ such that } Status_j(i) = On \tag{18.7}$$

*Energy Consumption.* Given a time period $D$ (e.g., a day, a week, and a month) divided into a set of $n$ sub-periods $\{d_1, d_2, ..., d_n\} \subset D$. This feature is used to calculate the energy consumption of a device $j$ in a specific subperiod $b \in D$.

$$EC_j^b = \sum_i^D p_j(i) \Leftrightarrow\; i \in b \subset D \tag{18.8}$$

*Average Peak Value.* Given a reference period of time $D_j$ (e.g., a day, a week, and a month) as a disjoint list of $K = \{1, 2, ..., k\}$ time intervals $I_j$=$\{sp_j(h_0, h_1), sp_j(h_2, h_3), ..., sp_j(h_{k-1}, h_k)\}$ in which a device $j$ is actively used. The average peak value of a device $j$ $APV_j$ calculates the average of all the peak values within the considered period of time $D_j$. Here, $peak(sp_j(h', h'')) = max\{p_j(i)\}$ with $h' < i < h''$ is the max value of energy consumed from the the device $j$ in the time interval $[h',\; h'']$.

$$APV_j(D_j) = \frac{\sum peak(sp_j(h', h''))}{k},\; \forall\, sp_j(h', h'')) \in\; I_j \tag{18.9}$$

*Power Dense Location.* Given a location $l$ and a set of $n$ devices J=$\{j_1, j_2, ..., j_n\}$. This feature provides the amount of power consumed in $l$ from all the devices in J in an arbitrary period of time D, if the total power consumed is more than a reference threshold $PD_{threshold}$.

$$PD_D^l = \sum_j^J \sum_{i=0}^T p_j^l(i) > PD_{threshold} \tag{18.10}$$

*Active Time*. This feature counts the number of times that a device $j$ is turned on in an arbitrary day D. For example, a dishwasher is typically turned on once or twice a day.

$$T_j^{Act}(D) = count(T_j^{On-Off}(i)) \Leftrightarrow T_j^{On-Off}(i) = On, \ \forall\, i \in D \tag{18.11}$$

*Active Duration*. Given a device $j$, its active duration in an arbitrary day $D$ represents the overall time $i$ in which $j$ is active, that is $Status_j(i) = On$.

$$T_j^{ActiveDur}(D) = count_j(i) \ \Leftrightarrow \ Status_j(i) > \lambda = 5[W] = On, \ \text{with } i \in D \tag{18.12}$$

*Average Active Time*. This feature calculates the active average duration of a device $j$ within an arbitrary day D.

$$T_j^{AvgAct}(D) = \frac{T_j^{ActiveDur}(D)}{T_j^{Act}(D)} \tag{18.13}$$

*Sequence of Usage Location*. Given a set of locations $L = \{l_1, ..., l_z, ..., l_k\}$, that represent specific places (e.g., a kitchen, a bathroom, a bedroom, a living room, and so one) in a given house $h$. This feature computes the list of locations in a house $h$, where a device $j$ is chronologically used $p_j^{l_z}(i_w) > \lambda = 5[W] = On$ in an arbitrary day $D$.

$$So\bar{U}L_j^h(D) =< l_1(i_1), ..., l_z(i_w), l_{z+1}(i_{w+1}), ..., l_k(i_w) >_j^h, \text{with } i_1 ... i_w \in D \tag{18.14}$$

$$\forall \ < i_w, i_{w+1} > \text{with } i_w \ < i_{w+1} \text{ and } p_j^{l_z}(i_w)\ \lambda \text{ and } p_j^{l_{z+1}}(i_{w+1}) > \lambda$$

*Most Often Usage Time*. Given a reference period of time $D_j$ (e.g., a day, a week, and a month) as disjoint list of $K = \{1, 2, ..., k\}$ time intervals $I_j$=$\{sp_j(h_0, h_1), sp_j(h_2, h_3), ..., sp_j(h_{k-1}, h_k)\}$ in which a device $j$ is actively used. The most often usage time of a device $j$ indicates the longest interval of time $\Delta h_j$=$< h^{'}, h^{''} >_j$ such that $sp_j(h^{'}, h^{''}) \in I_j$ and $h^{'}, h^{''} \in K$, in which the device $j$ is used.

$$\Delta h_j \ = max < h^{'}, h^{''} >_j = max\{(h^{''} - h^{'})_j\} = max\{sp_j(h^{'}, h^{''})\} \tag{18.15}$$

*Devices Used in Sequence*. Given two instants of time $i_1$ and $i_2$ with $i_2 \geq i_1$. A device $j_2$ works in sequence after $j_1$ $seq((j_2, j_1))$, when $j_1$ stops working at time $i_1$, that is, $Status_{j1}(i_1 - 1) = On$ and $Status_{j1}(i_1) = Off$ and the device $j_2$ starts working in a subsequent instant of time $i_2$, that is, $Status_{j2}(i_2) = Off$ and $Status_{j2}(i_2 + 1) = On$. So this feature $S\bar{E}Q_J(D)$ return all the possible couples of devices $j_1, j_2 \in J$, which work in sequence in a reference period D.

$$S\bar{E}Q_J(D) = \{seq(j_1, j_2)\}_D \ \forall\, j_1, \ j_2 \in J \text{ set of devices} \tag{18.16}$$

*Devices Used in Parallel*. Given two instants of time $i_1$ and $i_2$ with $i_1 > i_2$, $i_1 > (i_2 + 1) - \Delta t$ and $\Delta t \geq 1$. The device $j_2$ works in parallel with $j_1$ $par((j_2, j_1))$, when $j_1$ stops working at time $i_1$, that is, $Status_{j1}(i_1 - 1) = On$ and $Status_{j1}(i_1) = Off$ and the device $j_2$ starts working in a instant of time $i_2$, that is, $Status_{j2}(i_2) = Off$ and $Status_{j2}(i_2 + 1) = On$. This feature $P\bar{A}R_J(D)$ returns all the possible couples of devices $j_1, j_2 \in J$, which work in parallel at least for a threshold $\Delta t$ in a reference period D.

$$P\bar{A}R_J(D) = \{par(j_1, j_2)\}_D \; \forall j_1,\; j_2 \in J \text{ set of devices} \tag{18.17}$$

*Standby Devices*. Given a set of $n$ devices J=$\{j_1, j_2, ..., j_n\}$, this feature calculates a subset of devices $SB_{Dev} = \{j_1, j_2, ..., j_k\} \subset J$ which are neither Off nor On, rather those which present a standby mode, that is a power consumption $0 < p_j(i) < \lambda = 5[W]$ for at least an uninterrupted period of time $\delta t$.

$$SB_{Dev} = < j_1, j_2, ..., j_k >, \text{if } \exists\, p_j(i) \text{ such that } 0 < p_j(i) < \lambda, \text{ and } continuous(i) > \delta t \tag{18.18}$$

*Place of Use*. Given a set of locations $L = \{l_1, l_2, ..., l_k\}$, that represent specific places (e.g., a kitchen, a bathroom, a bedroom, and a living room) in a given house $h$. This feature allows to know in which location $z \in L$ the device $j$ was used $p_j^z(i) > \lambda = 5[W] = On$. That is, it was On for at least one time unit $i$ in an arbitrary day D.

$$z_j^h(i) \Leftarrow p_j^z(i) > \lambda, \text{ with } z \in L \text{ and } i \in D \tag{18.19}$$

## 18.5 Experiments and Results Discussion

The dataset used to evaluate the proposal consists of a collection of traces related to the daily use of different electrical devices, which is publicly available [7]. Each entry of the dataset contains basic data such as the identifier of a device, the time unit with a granularity of a second, which is used to collect the data of each device, the amount of energy consumed in a time unit, and so on. Starting from such raw data, additional information is extracted by using the features, that have been proposed in Sect. 18.4. It is worth noting that the useful information extracted through the features and, which is used to train the classifiers, is based on traces/records that have a duration of 24 h, and not on single measurement instants (i.e., real time). Both the basic and extracted information is used to train and test different classifiers. As the dataset is labeled, the case in consideration falls under a supervised learning problem, as a consequence, only supervised machine learning techniques have been considered and compared. In particular, Random Forest, Bagging, and LogitBoost algorithms have been selected and experimented, as they showed the best performance in literature.

The classifiers have been trained with the same *rational*, that is 80% of the data has been used for the training, whereas 20% of the data to validate them. A first result

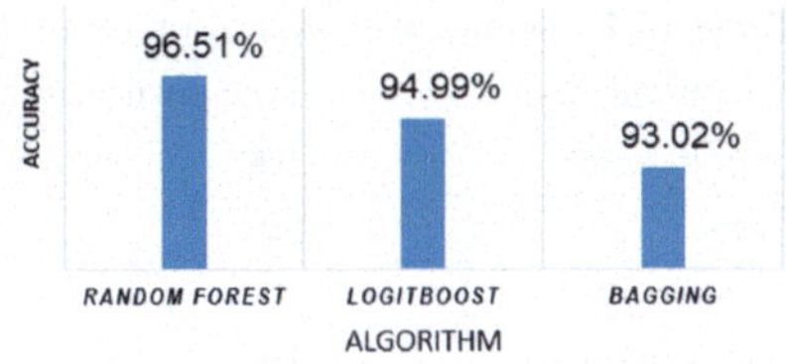


**Fig. 18.2** Accuracy of the different classifiers

is shown in Fig. 18.2 in terms of accuracy of the different classifiers with fivefold cross validation. A general observation is that all the classifiers presented an accuracy higher than 90%, by providing an indication of goodness of the selected features. Indeed, they show a certain degree of independence from a specific classifier and, as a consequence, they are well suited to describe appliance types and distinguishing them from others. Moreover, among them the best performance is reached by the Random Forest algorithm with a classification accuracy of 96.51%, that is why the rest of the analysis is specifically based on its use for the subsequent evaluation. The details are reported in Table 18.1 by showing the result values for *true* and *false positives* as well as the *precision*, which measures the proportion of actual negatives that are correctly identified as such, and the *recall*, which measures the proportion of actual positives that are correctly identified as such, for each kind of device.

In particular, as we can see, our implementation reaches at least 80% of accuracy and almost all the devices are always classified correctly, which can be seen from the true positive ratio which is equal to 100%. Whereas for other devices the false positive ratio is, however, very low. Additionally, both the results obtained by calculating the precision and the recall values comply with the observed values related to the true positive rate. Only for one device, and in particular for the Water Kettle a lower level of classification is shown. This is associated with the limited number of instances available during the training phase of the classifier that made very hard the training phase of the classifier for this typology of device. In general, some devices present electrical characteristics which are easier to recognize and which require few instances for training the classifier, others require instead a greater number of instances. However, more than 60% of the devices are correctly identified and classified, as we can see from the true positive rate and precision, without errors.

In summary, we globally obtained better performances in terms of higher level of accuracy compared with other related works by only using 19 features respect to huge amount of features, for example, 517 features used in [8], which require both more computational resources and computing time.

## 18.6 Conclusion

The paper focused on the automatic identification of electrical devices by proposing a model based on 19 distinct features. The information extracted by applying such features has been used to train three classifiers, which showed a high level of accuracy

**Table 18.1** Accuracy classification for each appliance provided by the Random Forest

| Electrical device | True positive/recall (%) | False positive (%) | Precision (%) |
|---|---|---|---|
| Alarm clock | 100 | 0 | 100 |
| Iron | 80 | 0.2 | 65 |
| Xmas lights | 99 | 0 | 100 |
| Wash machine | 100 | 0.2 | 98.3 |
| TV-LCD | 100 | 0 | 100 |
| Water fountain | 100 | 0 | 100 |
| Projector | 97 | 0 | 100 |
| Lamp | 88 | 0.2 | 85 |
| Stove | 100 | 0 | 100 |
| Dishwasher | 100 | 0 | 100 |
| Laptop | 96 | 0 | 100 |
| Water kettle | 57 | 0.3 | 58 |
| Amplifier | 100 | 0 | 100 |
| Coffee | 100 | 0 | 100 |
| Tv-rec | 96 | 0 | 100 |
| Ethernet | 95 | 0 | 100 |
| Dryer | 100 | 0 | 100 |
| Microwave | 100 | 0.1 | 95 |
| Monitor-CRT | 92 | 0.2 | 93 |
| Monitor-TFT | 100 | 0 | 100 |
| Desktop PC | 100 | 0 | 100 |
| Toaster | 100 | 0.1 | 95 |
| Refrigerator | 100 | 0 | 100 |
| DVD | 99 | 0.1 | 94.1 |
| Coffee machine | 100 | 0 | 100 |
| Freezer | 100 | 0 | 100 |
| Media-centre | 99 | 0 | 100 |
| PlayStation | 87 | 0 | 100 |
| Vacuum cleaner | 100 | 0 | 100 |
| Printer | 100 | 0.1 | 98 |
| Router | 100 | 0 | 100 |
| USB harddrive | 100 | 0 | 100 |
| TV-CRT | 100 | 0 | 100 |
| Weighted average (%) | 96.51 | 0.04 | 96.4 |

as symptom of goodness of the proposed features. Further details related to the Random Forest classifier, that provided the highest accuracy equal to 96.51%, have been discussed, as it outperformed respect to the related works. An ongoing work aims to analyze the importance of each feature and to use such model to enhance the energy management between SHs and SGs.

**Acknowledgements** This work was partially performed in the context of the TAKEDOWN, an EU Horizon 2020 Research and Innovation Programme, Grant Agreement no 700688.

## References

1. Karnouskos, S.: Cyber-physical systems in the smartgrid. In: 2011 9th IEEE International Conference on Industrial Informatics, pp. 20–23 (2011)
2. Eurostat (2018). https://www.statista.com/statistics/418078/electricity-prices-for-households-in-germany. Accessed 28 Feb 2019
3. Zafar, R., Mahmood, A., Razzaq, S., Ali, W., Naeem, U., Shehzad, K.: Prosumer based energy management and sharing in smart grid. Renew. Sustain. Energy Rev. **82**, 1675–1684 (2018)
4. Smart Home: The Human Side of the Smart Grid. http://www.smartgrids-cre.fr/media/documents/1003-CapG-SmartHome.pdf. Accessed 28 Feb 2019 (2010)
5. Alam, M.R., Reaz, M.B.I., Ali, M.A.M.: A review of smart homespast, present, and future. IEEE Trans. Syst. Man Cybernet. Part C (Appl. Rev.) **42**(6), 1190–1203 (2012)
6. Abeykoon, V., Kankanamdurage, N., Senevirathna, A., Ranaweera, P., Udawalpola, R.: Real time identification of electrical devices through power consumption pattern detection. Pervasive Comput. **10**(1), 40–48 (2016)
7. Tracebase (2017). http://www.tracebase.org. Accessed 28 Feb 2019
8. Reinhardt, A., Baumann, P., Burgstahler, D., Hollick, M., Chonov, H., Werner, M., Steinmetz, R.: On the accuracy of appliance identification based on distributed load metering data. In: Sustainable Internet and ICT for Sustainability (2012)
9. Hart, G.W.: Residential energy monitoring and computerized surveillance via utility power flows. Technol. Soc. Mag. **8**(2), 12–16 (1989)
10. Ruzzelli, A.G., Nicolas, C., Schoofs, A., O'Hare, G.M.P.: Real-time recognition and profiling of appliances through a single electricity sensor. In: 7th Annual IEEE Communications Society Conference on Sensor, Mesh and Ad Hoc Communications and Networks (SECON), pp. 1–9 (2010)
11. Patel, S.N., Robertson, T., Kientz, J.A., Reynolds, M.S., Abowd, G.D.: At the flick of a switch: detecting and classifying unique electrical events on the residential power line (nominated for the best paper award). In: UbiComp (2007)
12. Ridi, A., Gisler, C., Hennebert, J.: Automatic identification of electrical appliances using smart plugs. In: 8th International Workshop on Systems, Signal Processing and their Applications (WoSSPA), pp. 301–305 (2013)

# Chapter 19
# Maslow in the Mud. Contrast Between Qualitative and Quantitative Assessment of Thermal Performance in Historic Buildings

**Marcin Mateusz Kołakowski**

> *... not everything that can be counted counts,*
> *and not everything that counts can be counted.*
> —William Bruce Cameron [1]

**Abstract** This paper argues that the notion of comfort is first and foremost related to subjective choices and individual value systems. The article presents results from research on the perception and measurements of the thermal qualities of heritage buildings in Lincolnshire, UK. The qualitative and quantitative results identified a strong contrast between different methodologies. Inhabitants describe as comfortable houses which would not be considered comfortable if a standard positivist approach was used. The conflict presented will be discussed in the context of sustainable strategies in architecture.

## 19.1 Introduction: Who Owns Sustainability?

Sustainable design—arguably the most important challenge of architecture today—seems to be dominated methodologically by qualitative research and searching for technological solutions. Yet, the core of the problem seems to lie in human preferences and decisions—and those can be understood only from psychological perspectives. Researchers engaging with sustainability who go as far as to recognise the value of psychology sometimes make a 'methodological gesture' by referring to well-being and Maslow's pyramid [2]. It seems easy to use. It orders 'needs' using a structure that is hierarchical and linear: physiological needs → security → social needs → esteem needs → self-actualisation… Lupo [3] even claims that by following Maslow's theory we could understand the essence of well-being and achieve it

M. M. Kołakowski (✉)
University of Lincoln, Lincoln LN6 7TS, UK
e-mail: mkolakowski@lincoln.ac.uk



J. Littlewood et al. (eds.), *Sustainability in Energy and Buildings*, Smart Innovation, Systems and Technologies 163, https://doi.org/10.1007/978-981-32-9868-2_19

on a global scale by simply satisfying basic needs: the right quantity of food and safe, dry and warm houses…

However, Maslow's system has been contested by many key figures of psychological schools such as Schaller et al. [4]. Erich Fromm, who developed his own psychological theory of needs, pointed out that 'This list is a somewhat unsystematic enumeration, and regrettably Maslow did not try to analyse the common origin of such needs in the nature of Man' [5]. Basing on his own psychoanalytical research, Fromm claimed that commonly people are ready to sacrifice Maslowian basic needs and comfort in order to fulfil psychological needs which he defined as (1) relatedness—a need to relate to other living beings, (2) transcendence—a need to belong to something larger than oneself, (3) rootedness—being able to define one's ideological framework, (4) sense of identity and (5) the need for orientation and meaning [6].

Notably, Fromm's system sheds light on recent attempts to develop a human-centred approach in understanding the relationship between sustainable strategies in architecture and the set of values embraced by its users. This issue can be particularly clearly identified in the conflict between sustainable principles and the historical value of buildings.

Fouseki and Cassar [7] propose that the driving question for sustainable architecture—apart from technological aspects—should also be 'what does this building mean for those who use it?' Architecture is a medium of adding meaning to a building, which brings relevance to the theory of Fromm but also of Bourdieu [8] who argues that the social dimension includes 'cultural capital'—a cultural identity which is mediated through features of buildings. Lynch [9] and Ingold [10] argue that a sense of meaning is often created by occupants' interaction with the building. This is particularly visible in the case of inhabitants who consciously decide to live in historic buildings.

Even Tweed and Sutherland [11], who initially tried to draw on Maslow's hierarchy, eventually came to the conclusion that utilitarian needs in architecture can be overrun by cultural appeal. These authors attempted to reconcile sustainability with other paradigms and concluded that heritage values in the built environment straddle the three pillars of sustainability—the economic, environmental and social dimensions. In their article entitled '*Irrational homeowners? How aesthetics and heritage values influence thermal retrofit decisions in the United Kingdom*', Sunikka-Blank and Galvin [12] found that homeowners develop their own sophisticated strategies for balancing between the need for retrofit and respecting heritage. They concluded that sustainability must consider cultural and heritage issues which cannot easily be quantified.

Unlike Sunikka-Blank and Galvin, who researched retrofitted houses and their owners, this paper presents results of 18-month-long research on a unique group of buildings and their inhabitants who seem content to compromise the conventional sense of comfort because they see value in living in Lincolnshire Mud & Stud cottages—some of which were built over 400 years ago.

This research addresses a far greater issue—the sustainable agenda relating to the modernisation of historical building stock. This is a major global challenge considering that only in the UK, according to the National Statistics Survey, 21% dwellings were built before 1919 and 85% before 1990, which is when thermal standards were first introduced [13].

## 19.2 Research Subject and Methodology

Vernacular style known as 'Mud and Stud' (M&S) is claimed by local enthusiasts to be unique only to Lincolnshire in the UK [14]. According to renowned British building historian Maurice Barley, M&S cottages are a unique type of buildings which were built for the poorest part of village society between 1400 and 1850 [15]. Barley claims that M&S is one of the least known British building techniques. Robert alarmed that Lincolnshire's vernacular tradition has been almost lost and has already been destroyed in great part [16]. An example of this type of buildings could be the 'Black Swan' cottage in Conings by (Fig. 19.1) which appeared on prints from that period proving that the quality of this architecture was appreciated locally.

However, since 1990s several researchers and enthusiasts have started to take a closer look at M&S heritage. Naomi Field created the first rigorous reports [17, 18]. Building historian Rodney Cousins created a list of around 400 remaining M&S buildings and published a monograph dedicated to this heritage. Cousins organised hundreds of talks dedicated to this Lincolnshire style contributing to its popularisation [14]. Architect David Glew looked at the possibility of adapting M&S for new build

**Fig. 19.1** Mud and stud cottage 'Black Swan' in Coningsby, Lincolnshire, UK, 1894

[19, 20]. M&S buildings have slowly become local pride and started to be popularised thanks to the local group East Midlands Earth Structure Society (EMESS).

The M&S construction—a method which Hugo Houben and Hubert Guillaud classify as 'cob on post' [21]—developed in Lincolnshire where good quality timber was scarce. Local builders developed a way of creating solid walls by combining timber substructure together with a thin earth cover. Built economically, M&S cottages proved to be a very efficient way of using the materials available. Today they are valued for their architectural charm—central chimney, white walls and thick thatched roofing.

The research on M&S buildings conducted between 2015 and 2018 by the author of this paper, Dr. Magdalena Baborska-Narożny and Ian Keeling, focused on both the thermal performance and the narratives associated with those buildings. Questionnaires with invitations to participate in the research and a return envelope were sent to all available 88 addresses of M&S buildings from a catalogue created by EMESS. 23 questionnaires were returned with answers. For a further 'monitoring phase' of research, 12 cottages were chosen based on the criteria of best-preserved features and accessibility. Selected cottages were all houses that inhabitants voluntarily decided to live in and were well aware of the heritage associated with M&S. In the selected cottages, thermal and humidity sensors were installed which collected data over the period of 18 months. Three sensors were typically installed in each house: in the living room, kitchen and the bedroom. Thermal imaging was completed. About one-hour-long semi-structured interviews were conducted with the inhabitants. Qualitative and quantitative data were compared and cross-referenced [22]. This mixed methodology offered very contrasting results between the quantitative and qualitative data obtained.

## 19.3 Contrast Between Qualitative and Quantitative Research

Qualitative research based on interviews painted an overall very positive picture of M&S cottages. A common theme in all semi-structured interviews was the very positive emotions towards the cottages. When asked: 'Are you happy with this house?', all interviewees answered 'Yes' and offered additional explanations. One-third of interlocutors praised their cottage's thermal properties claiming that the houses are '*cosy and warm in winter*'. When asked 'Does the building you live in meet your expectations regarding thermal performance?' one of the typical answers was: '*Oh, yes. It's fantastic. I don't know why more people don't build out of mud*'. It is worth noticing that those answers seem to echo opinions presented in publications about earth buildings. Rodney Cousins wrote, 'these cottages are known for being cosy and warm in winter but cold in summer' [14].

However, those opinions stand in contrast to results of sensor monitoring. It must be noted, however, that readings from the sensors installed varied greatly. This is understandable taking into consideration the fact that M&S cottages were constructed over 400 years ago, and they have been readapted in manifold ways and various heating systems have been installed in them.

For the purpose of this paper, the contrast will be illustrated by an example of Cottage A (all names were anonymised). This building may be seen as representative since it still bears all the original features. The main central fireplace for coal and wood is still the main source of space heating. 1.5 tonnes of coal are used during the winter—‘*one basket a day*’. Sporadically, a 2 KW oil electric heater is used additionally. An electric immersion heater is installed for heating water and shower. A coal stove and an electric cooker are in use in the kitchen. Inhabitants of Cottage A are enthusiastic about the quality of their home. In their opinion it is warm and provides a good environment to live in. It could be illustrated by statements such as: ‘*It is warm in winter. It works really well*’.

Results derived from thermal imaging and sensor monitoring paint a somewhat different picture. Living room in Cottage A was the warmest place because this is where the fireplace was located. According to the interviewees, this is also the place where the whole family spends most of the time. The kitchen and the bedroom were colder and the temperature there rarely exceeded 15 °C and sometimes dropped below 10 °C (see Fig. 19.2). The results presented below illustrate the sensor readings taken during the coldest month of February. The limitation of space does not allow presenting other months but even in April on several occasions, the bedroom temperature fell to as low as 10 °C, whilst temperature in the kitchen rarely exceeded 15 °C. The average temperature (Fig. 19.3) does not reach the commonly accepted standards, such as those recommended by WHO (see below).

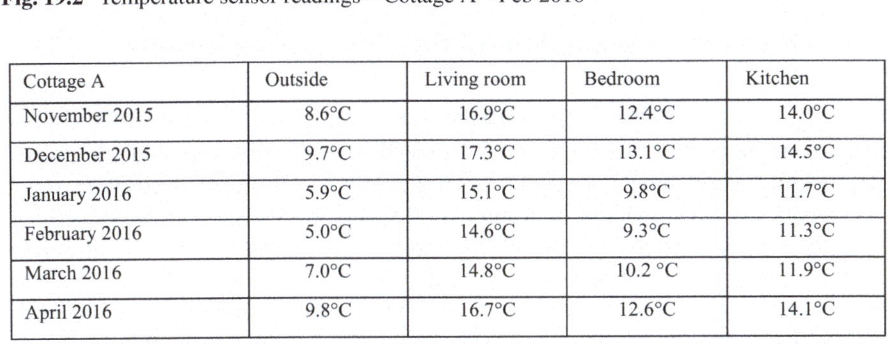


**Fig. 19.2** Temperature sensor readings—Cottage A—Feb 2016

| Cottage A | Outside | Living room | Bedroom | Kitchen |
|---|---|---|---|---|
| November 2015 | 8.6°C | 16.9°C | 12.4°C | 14.0°C |
| December 2015 | 9.7°C | 17.3°C | 13.1°C | 14.5°C |
| January 2016 | 5.9°C | 15.1°C | 9.8°C | 11.7°C |
| February 2016 | 5.0°C | 14.6°C | 9.3°C | 11.3°C |
| March 2016 | 7.0°C | 14.8°C | 10.2 °C | 11.9°C |
| April 2016 | 9.8°C | 16.7°C | 12.6°C | 14.1°C |

**Fig. 19.3** Average temperature (sensor readings)—Cottage A

**Fig. 19.4** Thermal imaging from inside of one of M&S cottages

The results from thermal monitoring show that temperature often reached a level far below what could be expected from habitable buildings or homes described as 'warm and cosy'. In fact, the results obtained suggest conditions that are below the standard of indoor temperature recommended by The World Health Organization (WHO) which is 18 and 21 °C if babies or elderly people live in the house [23]. WHO recommendations echoed research by Kenneth Collins on the effect of low temperature on health: 'At temperature below 16 °C, resistance to respiratory infections may be diminished. Both low and high relative humidities promote respiratory illnesses. At temperatures below 12 °C, cold extremities and slight lowering of core temperature can induce short-term increases in blood pressure' [24]. Thermal imaging demonstrated that in certain places, on cold days, the temperature of M&S walls sometimes dropped to freezing point (Fig. 19.4).

From the point of view of this research, the most important question is, however, why do inhabitants of M&S cottages claim that their homes are warm and comfortable? The answer could have an objective and subjective nature.

## 19.4 Objective Explanations of the Contrasting Results

The objective results obtained from sensor monitoring were in accord with the theoretical modelling of the performance of earth walls which depends on their clay/straw mix, specific weight and additives. In the case of M&S walls, the material consists of a straw-earth mix with weight of around 1600 kg/m$^3$. It offers thermal conductivity ($\lambda$) in the range of 0.8 W/mK, which for M&S walls (usually 250–300 mm thick) translates into U-value of around 2 W/Km$^2$. This means that the thermal performance of M&S walls is times times worse than the currently required UK standards of 0.3 W/Km$^2$. Any slight changes to the straw-earth mix or the wall thickness would not significantly change the fact that without considerable alteration those walls cannot offer the required insulation.

The thermal transmittance (U-value) does not describe all the properties of earth constructions. Another important factor which could positively influence the internal microclimate of earth constructions is the high thermal mass [25]. Once warmed up, due to thermal inertia, the wall '*keeps the warmth*'. Walls radiate the heat and this could give a pleasant impression that despite the temperature fluctuations outdoors the house '*keeps its warmth*'. This is also why earth walls could give an impression that they '*keep cold longer*' on hot summer days and 'hold heat' during the cold winter days.

Second, due to its molecular structure, an unfired earth wall '*as if automatically*' absorbs and releases water molecules offering natural regulation of humidity. This means that whilst mould could develop on conventional walls in lower temperature, earth walls are more resistant to this problem. The mechanism behind the phenomenon was described in detail by Minke [25]. The monitoring of humidity at M&S houses, as predicted, identified relative humidity at the level of around 55% which is the recommended level from the health point of view. This result supports theoretical literature about earth architecture [25, 26].

Those positive aspects mitigate but do not change the fact that objectively the monitored houses were below expected standards of modern architecture in terms of thermal performance.

## 19.5 Subjective Explanations of the Contrasting Results

The interviews shed light on how problems with performance have been overrun by other positive values of the house. This could be illustrated by an interview with two people A and B living together in one of the researched houses:

*A: If you weren't careful, your bills would be phenomenal. Because if you wanted to keep it at 22 degrees, you'd have to have the heating on ALL the time. I don't know how you feel… you're probably not used to…*

> *B: It's quite hot when the fire is going.*
>
> *A: At night-time I get the fire going because that's the time when I like to be warm.*
>
> *B: I think I like the way with the central fire. The whole way the house is laid out. Just how it works.*

The analyses of answers from the returned questionnaires identified two groups: the first, smaller group, is enthusiastic about the cottages, and the second, larger group, is aware of the thermal issues but still enjoys living in M&S. When asked a question about the positive aspects of living in an M&S cottage, 17% of respondents repeated the claim about the positive thermal properties of M&S houses ('*They are well insulated and warm in winter*'). Other responses pointed to the general atmosphere ('*It has an atmosphere which makes you feel good*') (21%), aesthetics (47%), connection to history (39%) and unique character (31%). Answers related to 'negative aspects' identified the awareness of problems of low thermal performance (52%), high cost of living and maintenance (52%) and restrictions connected with living in a historical building (39%).

What is worth noting is that even those respondents who mentioned low thermal performance as the most problematic still highly praise the way of living in an M&S cottage. The questionnaire also asked respondents to agree or disagree with a set of statements. The answering system followed the Likert scale (strongly agree, agree, no answer/don't know, disagree and strongly disagree). When asked about the statement "*The building I live in meets my expectations regarding thermal comfort*," twice as many respondents agreed or strongly agreed than disagreed or strongly disagreed. When asked about the statement '*The building I live in is affordable with regards to bills*', 2.6 times more respondents agreed or strongly agreed than disagreed or strongly disagreed.

The seeming paradox of the contrast between the results could be explained by 'the forgiveness factor' also identified by other researches. Adrian Leaman and Bill Bordass noticed that overall the satisfaction with buildings described as 'green' could sometimes be greater than the satisfaction with individual elements. They believed that the green ethos is part of a 'psychological transaction'. 'They trade-off good things against bad, reach compromises and put up with shortcomings within reason' [26]. Leaman and Bordass called users who are ready to adapt and cope 'satisfiers'. It is a direct reference to the concept of Herbert A. Simon—a renowned economist, political scientist and cognitive psychologist—who used the concept of 'satisficing' to describe decision makers under circumstances in which an optimal solution cannot be determined [27]. As he put it 'A "satisficing" path is a path that will permit satisfaction at some specified level of all its needs' [28]. The 'forgiveness factor' related to green buildings has also been identified by Gou et al. and Khoshbakht et al. [29, 30] as well as Max Deuble and Richard de Dear [31].

Questionnaires and interviews with M&S inhabitants also provided strong indication that the factor which diminishes the problem of lack of comfort in this case is the positive attitudes towards heritage architecture, which is related to needs defined by Fromm, namely relatedness, transcendence, rootedness and sense of identity. Unlike Leaman and Bordass, who identified forgiveness related to green buildings, what played a decisive role in this case was the paradigm related to historical architecture, which was investigated through a set of questions. Respondents were asked to agree or disagree with a series of statements according to the Likert scale (Figs. 19.5, 19.6 and 19.7). The results presented below identified strong tendencies towards favouring historical values in architecture over conventionally understood comfort.

During the semi-structured interviews, inhabitants of the 12 selected M&S cottages were asked a direct question related to forgiveness: 'Do you believe that the current internal climate you experience in your home would be acceptable if experienced within a modern equivalent?' In the case of this question, even those inhabitants who were very enthusiastic about their buildings admitted that they would expect better thermal quality if they were to live in a new build. All of them admitted that they would not mind even introducing technological solutions as long as they do not destroy the character of the cottage.

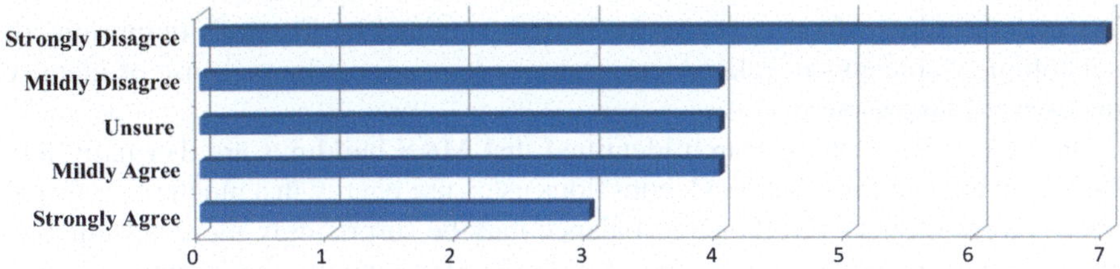


**Fig. 19.5** Reaction to statement 'If historical architecture does not respond to current needs, it should be changed, retrofitted or modernised'

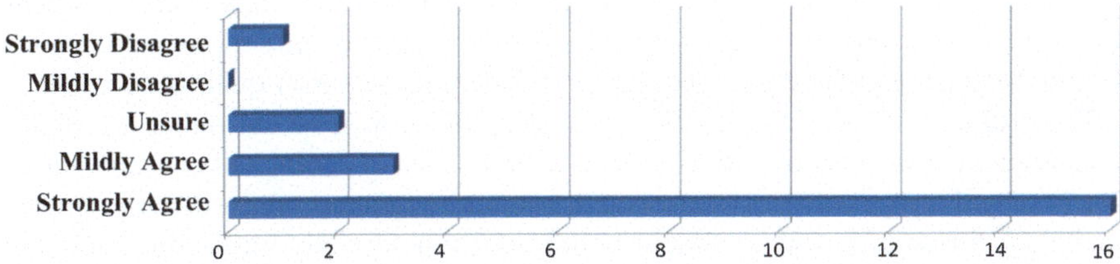


**Fig. 19.6** Reaction to statement: 'Living in historical buildings is particularly attractive because of their specific qualities and character'

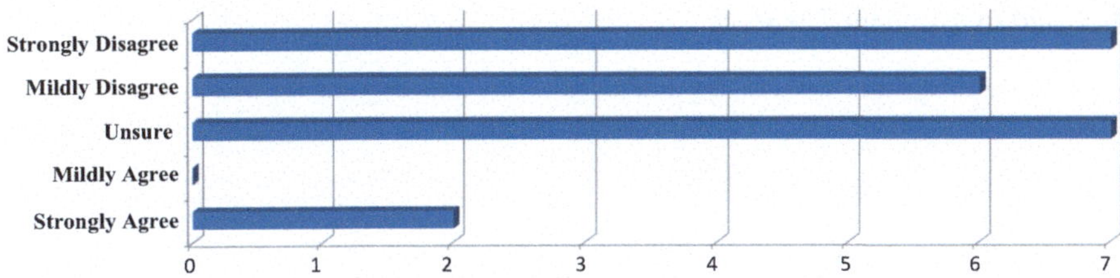


**Fig. 19.7** Reaction to statement: 'Historical architecture is interesting for historians and ethnographers but in order to meet current needs it would be better to invest in new high-quality architecture'.

## 19.6 Conclusions in Relation to Sustainability

The question concerning reconciliation of individual and global needs in the context of sustainability is far from simple and poses methodological challenges. The research presented in this paper focused on heritage houses demonstrates how the positivist, quantitative assumptions according to which the performance of buildings can easily be translated into well-being of inhabitants may be very deceiving.

Those who decided to live in M&S cottages see their homes as far more than shelters to live in. Their cottages are a way of belonging to a certain idea, group and tradition. Living in those houses is full of meaning and the inhabitants often feel that they are part of something greater than individual interest: '*It is a privilege of being a guardian of heritage*' stated one respondent during a research interview.

This understanding of comfort may be better explained by Fromm's definitions of psychological concepts of relatedness, transcendence, rootedness, sense of identity and the need for meaning.

On the one hand, the research identified that M&S buildings are thermally less efficient, whilst on the other hand, conscious users use energy in a much more frugal way. As a result, the overall energy balance may be surprisingly more sustainable. This kind of user behaviour must be a key factor in sustainable assessment.

The awareness of various paradigms should not prevent searching for solutions to improve the properties of historic buildings, but this should be done in a way that respects the value system of the inhabitants. It is important to note that users are not against improving the thermal performance of their dwellings even by means of high-tech solutions. What they are vehemently against is the callous indifference towards heritage and the dominance of quantitative factors over qualitative values.

The problem of sustainability is not the lack of know-how but rather a psychological deficiency—lack of will to engage with the idea of ecological consciousness. This is why the psychological perspective cannot be ignored whilst discussing sustainability. Smart cities or intelligent buildings are potentially interesting tools but the social and psychological sciences need to be incorporated into the assessment process and research. If the process of introducing the sustainability ethos is to succeed, it is crucial to recognise the fact that data, knowledge, know-how, information and intelligence are powerless where there is lack of meaning and wisdom, as T. S. Eliot said in his poem *The Rock*.

*Where is the wisdom we have lost in knowledge?*
*Where is the knowledge we have lost in information?*

## References


1. Cameron, W.: Informal Sociology: A Casual Introduction to Sociological Thinking. Random House, New York (1963)
2. Maslow, A.: Theory of human motivation. Psychol. Rev. **50**(4), 370–396 (1943)
3. Lupo, R.: A Measure of Net Well-Being that Incorporates the Effect of Housing Environmental Impacts. BRE, London (2014)
4. Schaller, M., Neuberg, S., Griskevicius, V., Kenrick, D.: Pyramid power: a reply to commentaries. Perspect. Psychol. Sci. **5**, 335–337 (2010)
5. Fromm, E.: Anatomy of Human Deconstructiveness. Holt Reinehard and Winston, New York (1973)
6. Fromm, E.: The Sane Society. Routledge, London (2002)
7. Fouseki, K., Cassar, M.: Energy efficiency in heritage buildings—future challenges and research needs. Hist. Environ.: Policy & Pract. **5**(2), 95–100 (2014)
8. Bourdieu, P.: Distinction: A Social Critique of the Judgement of Taste. Routledge, London (1984)
9. Lynch, K.: The Image of the City. The MIT Press, Cambridge, MA, USA (1960)
10. Ingold, T.: The perception of the environment: essays in livelihood. In: Dwelling and Skill, Routledge, London (2000)

11. Tweed, C., Sutherland, M.: Built cultural heritage and sustainable urban development. Landscape Urban Plann. **83**(1), 62–69 (2007)
12. Sunikka-Blank, M., Galvin, R.: Irrational homeowners? How aesthetics and heritage values influence thermal retrofit decisions in the United Kingdom. Energy Res. Soc. Sci. **11**, 97–108 (2015)
13. Department for Communities and Local Government: English Housing Survey Housing stock Report. DC&LG Publication London (2008)
14. Cousins, R.: Lincolnshire Buildings in the Mud and Stud Tradition. Heritage Trust of Lincolnshire, Sleaford (2000)
15. Barley, M.: Lincolnshire and the Fens, p. 28. B T Batford Ltd., London (1952)
16. Roberts, D.L.: The Vernacular buildings of Lincolnshire. Archaeological Journal **131**, 298–308 (1974)
17. Field, N.: Withern cottage. In: Field, N., White, A. (eds.) A Prospect of Lincolnshire, pp. 92–5. Lincoln (1984)
18. Field, N.: Mainly Mud and Stud. In: Vernacular Architecture Group, Spring Conference: Lincoln (2016)
19. Glew, D.: New Mud-and-Stud Construction in England and the Problem of Thermal Insulation. In: Proceedings of Terra 2012 Terra 2012: XIth International Conference on the Study and Conservation of Earthen Architectural Heritage (22–27 April 2012), Pontifical Catholic University of Peru, Lima (2012)
20. Glew, D.: Mud-and-Stud construction: compliance with modern planning and building regulations. In: English Heritage, Terra 2000 8th International Conference on The Study and Conservation of Earthen Architecture, pp. 139–143. Torquay, UK (2000)
21. Houben, H., Guillard, H.: Earth Construction: A Comprehensive Guide, Intermediate Technology Publication, p. 188. London (1994)
22. Kołakowski, M.: How to measure well-being in architecture: the benefits of using mixed-method research. Architectus **49**(1), 15–32 (2017)
23. World Health Organization: The Effects of the Indoor Housing Climate on the Health of the Elderly (September). Graz (1982)
24. Collins, K.: Low indoor temperatures and morbidity in the elderly. Age Ageing **15**(4), 212–220 (1986)
25. Minke, G.: Building with Earth: Design and Technology of a Sustainable Architecture, pp. 15–17. Birkhäuser, Berlin (2006)
26. Leaman, A., Bordass, B.: Are users more tolerant of 'green buildings'? Build. Res. Inf. **635**(6), 662–673 (2007)
27. Simon, H.A.: Rational choice and the structure of the environment. Psychol. Rev. **63**(2), 129–138 (1956)
28. Ibid. p. 136
29. Gou, Z., Prasad, D., Lau, S., Siu, Y.: Are green buildings more satisfactory and comfortable? Habitat Int. **39**, 156–161 (2013)
30. Khoshbakht, M., et al.: Are green buildings more satisfactory? Rev. Glob. Evid. Habitat Int. **74**, 57–65 (2018)
31. Deuble, M., de Dear R.: Green occupants for green buildings: the missing link? In: Proceedings of Conference: Adapting to Change: New Thinking on Comfort, Cumberland Lodge, Windsor, UK, 9–11 April 2010. Network for Comfort and Energy Use in Buildings, London (2012)

# Chapter 20
# Hidden Building Performance Evaluation Sources: What Can Trip Advisor and Other Informal User-Generated Data Tell Us?

**Julie Godefroy**

**Abstract** There are well-known benefits to carry out Building Performance Evaluation (BPE), but also well-known barriers, including lack of support from clients which limits the ability to carry out extensive BPE exercises. Focusing on user feedback evaluation, this paper presents a methodology to expand current approaches by using data that is publically available and requires little resource to analyse: the systematic analysis of reviews from a travel or visitor website (in this case, Trip Advisor). Two case studies are presented (one hotel, one leisure centre). The first used the proposed method alongside more formal BPE methods, which helped to interrogate findings and validate the method in principle. The second uses it as the main BPE method. The analysis draws out which design and operational issues are mentioned most often by users and in what way (i.e. positively or negatively). It can then investigate these issues in more detail and over time, comparing the early stages of occupation with later, more 'settled', stages. Both case studies indicate significant potential to generate valuable feedback on building briefing, design, delivery and operations, both in relation to specific buildings and as a way to better understand user needs and expectations and inform future projects. Formal and extensive methods are invaluable, and should not be replaced by less formal ones which have inherent limitations; however, built environment professionals risk missing valuable opportunities to learn and improve building performance by concentrating on formal methods alone, particularly in a world where user-generated content grows in quantity, availability and prominence.

## 20.1 Introduction

The benefits of Building Performance Evaluation (BPE) are well documented, but for a number of reasons it is not commonly carried out in the UK. In addition to well-established, extensive and robust BPE methods, this paper makes the case for more flexible approaches that would allow feedback to be gathered more often

J. Godefroy (✉)
Julie Godefroy Sustainability Ltd., London, UK
e-mail: julie@juliegodefroysustainability.co.uk



J. Littlewood et al. (eds.), *Sustainability in Energy and Buildings*, Smart Innovation, Systems and Technologies 163, https://doi.org/10.1007/978-981-32-9868-2_20

by designers, based on the non-intrusive use of informal and publically available user feedback from sources such as popular review websites. It focuses on post-occupancy evaluation (POE) of user feedback rather than other aspects of BPE such as energy consumption. The proposed POE methodology is presented in Sect. 20.3; the POE method is then applied to two case studies presented in Sects. 20.4 and 20.5. Section 20.6 discusses the potential benefits and limits of using this approach more widely.

## 20.2 Context: Why New Approaches Are Needed

### *20.2.1 Current Approaches*

The benefits of BPE are well documented, including improving the performance of individual buildings, better responding to user needs, and gathering lessons for designers and the wider built environment industry to learn and gradually improve. In the UK, the Usable Buildings Trust [1] and PROBE studies [2] were instrumental in developing methodologies, carrying out BPE, and disseminating lessons. For a recent review, see for example the RIBA [3].

A range of methods is available depending on the intended purpose and available resources, including site visits, informal or structured interviews with occupants and FM teams, and occupants surveys. Standardised surveys have the benefit of offering benchmark against other projects, while bespoke surveys may better meet the objectives of a particular project. Recent years have also seen new opportunities through widespread technological development of monitoring devices and consumer wearables.

A number of standardised surveys are now well-established in offices and, to a lesser extent, educational environments, notably the Building Use Studies (BUS), the CBE Berkeley method and the Leesman Index. They cover similar issues including overall satisfaction, facilities management, thermal comfort, perceived air quality, light and views out, acoustics and design features such as storage and layout.

Methodologies are less developed in other environments, particularly those with large numbers of 'transient' users who are not permanently based in that building, e.g. transport hubs, retail and hospitality, or those with vulnerable populations or young children whose feedback can be difficult to gather reliably. Methods tend to be more bespoke but examples are available in a range of sectors [3].

### *20.2.2 Limitations of Current Approaches*

Despite the well-known benefits and methodologies available, BPE is far from being routinely carried out in the UK. There are multiple reasons for this, including [3]

- Client's lack of interest and/or reluctance to allocate additional fees for something that they think designers should carry out anyway
- Worries from clients, designers and building management teams about the possible findings that might emerge from BPE
- Perceptions that BPE is necessarily a resource-intensive and lengthy process.

There are, however, growing pressures on the UK designers to carry out BPE routinely:

- Growing concerns about the performance of buildings including built quality, energy consumption, and user satisfaction, comfort and health: the construction industry is sometimes described as 'the next Volkswagen scandal', and there have been calls for more regulatory focus on operational performance [4].
- Professional bodies: Chartered Institution for Building Services Engineers (CIBSE) has emphasised the importance of BPE for a number of years, and the Royal Institute of British Architects (RIBA)wish to give it more emphasis [5].
- Market and technological trends: the growth in consumer devices generating data on the quality of people's environment (e.g. air pollution), combined with sites or apps based on user-generated content (e.g. Rate my Building) may, at least in some cases, create significant pressure on building owners, operators and designers to understand and improve building performance.

In this context, it is thought that there is a professional interest and duty for designers to increase their efforts to gather feedback, both through convincing clients about the needs and benefits of BPE and through gathering feedback in other ways if needed.

## 20.3 Research Question and Proposed POE Methodology

### *20.3.1 Rationale*

The proposed method is the use of reviews posted on public websites, such as visitor feedback sites. A background search has found a small number of similar analyses with promising results [6], but they typically covered overall visitor experience rather than focusing on BPE insights about design, construction and Facilities Management (FM) issues. The rationale behind the proposed methodology is simple:

- Barring formal POE methods, this ought to be better than not attempting to gather and analyse any feedback at all.
- Valuable observations on some aspects of building performance can be made without intrusive methods, e.g. through a walk-around [7]. It could even be argued that designers are not presenting clients with a 'fair' offer if they only mention BPE as an extensive service, without offering any feedback otherwise.
- The opinion of building users on how buildings perform ought to be given consideration even if it was expressed outside of formal BPE processes (Fig. 20.1).

**Fig. 20.1** Simple observations on building performance and user interventions to improve comfort, both available from the outside or public areas: (left) Highly glazed façade, all blinds are down; (right) Plug-in electric heater installed by receptionist in front of busy entrance doors

As the method relies only on few resources and public information, it circumvents two of the main barriers to BPE, i.e. the need for extensive resources and client approval.

### *20.3.2 Proposed POE Methodology*

The proposed method is as follows: for a given building, the reviews left on publically available review websites which are analysed as a way to assess occupant feedback. In the case of this paper, Trip Advisor was used as the main source, but others may also be (e.g. Google reviews, Booking.com). Wherever possible, this should be accompanied by a site visit, at the very least of the outside and publically accessible areas. The reviews are analysed as systematically as possible:

- Issues mentioned in the reviews are categorised as follows: (1) Design issues, broken down into key themes and/or building areas; where possible, this should differentiate between the brief (e.g. whether a particular facility is provided) and the design response (e.g. people's satisfaction with that facility); (2) FM issues, broken down into cleanliness, comfort and technical issues; (3) General operational issues (e.g. price and quality of service).
- The analysis records show how often each issue is mentioned, whether it is in a positive or negative way and relevant detailed comments.
- The analysis separately records reviews in the initial period after opening (here, 6 months) and those in the later stage. This intends to identify whether feedback has changed after the initial handover period, for example, due to users getting more familiar with the building or to the common 'teething' problems of a new building being sorted through fine-tuning and remediation.

The application of the method in two case studies is described in the following sections. In the first, the proposed POE method was used alongside more formal and established methods, which allowed an initial validation of the method; in the second, it was used as the main method of investigation. While the key elements of feedback on both buildings are presented, the focus of this paper is not on the technical analysis of these specific buildings, but to test the usefulness and reliability of the proposed method.

## 20.4 Case Study 1

### *20.4.1 Background*

The idea of using a review website was first developed as part of a BPE funded by Innovate UK (then Technology Strategy Board). The building is a high-end boutique hotel, opened in September 2012. The analysis was carried out in 2012–2013.

Established POE methods were used to gather feedback from permanent users (i.e. staff), including interviews and BUS surveys. In terms of gathering feedback from customers, a log of complaints was available but this only provided an angle biased to negative aspects. The hotel operator, whilst supportive of the BPE process and interested in the findings, had reservations about subjecting their customers to surveys. Furthermore, having been developed largely on offices and schools, the BUS surveys were considered poorly suited to a hotel, with some questions that were irrelevant or of little value and without coverage of some themes of interest to hotels. Feedback from hotel guests was therefore gathered through a variety of sources which provided different angles and which could be checked against each other:

- Customer feedback received by hotel staff, made available via staff interviews
- Reception desk logs
- Customer feedback on two popular review websites. These were used by the hotel management team themselves to track and respond to feedback. The hotel provided assurances that they did not post reviews themselves to skew the ratings, a concern often mentioned in relation to such websites.

### *20.4.2 Analysis and Findings*

Customer reviews on both review websites showed high rankings for the hotel. This was supported by high occupancy levels and a high proportion of repeat visits. The analysis of customer reviews was carried out over 2 periods: the first 6 months of operation of the hotel, which included 48 reviews, a period of 3 months approximately 1 year later, which included 40 reviews. The analysis is illustrated in Fig. 20.2.

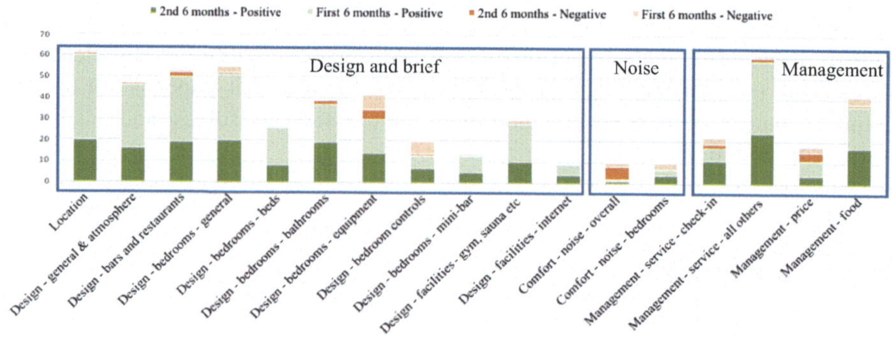


**Fig. 20.2** Case study 1—hotel: Categorisation of reviews available from public website

The feedback is in large majority positive, with only 12% of the mentions being negative. Design and brief issues represent 71% of the total, FM and comfort issues 4% (all about noise) and general management issues 26% of the total.

**Issues mentioned most often**. The issues most commented on by guests were as follows:

- Location
- Quality of service
- General quality, design and atmosphere of the bedrooms (with additional comments on specific aspects of the bedrooms, e.g. beds and bathrooms)
- Bars and restaurants
- General quality, design and atmosphere of the hotel as a whole.

The first two are driven by the hotel operator and highlight the limit of influence of designers on overall user feedback.

**Feedback on brief and design**. The recurrent positive feedback on location and on the provision of ancillary facilities (bars, spa, etc.) reflects the client's brief. The feedback on design issues was in large majority positive, in particular on beds and bathrooms ('huge beds', 'rain showers' and 'huge baths' were highly appreciated). This is an example of useful feedback for designers, as the value attributed to luxury water features poses a clear challenge to sustainability and water consumption objectives. The 'modern feel', 'stylish' design and lighting scheme were also commented on positively.

**Feedback on FM and comfort**. Heating, Ventilation and Air Conditioning (HVAC) systems and associated FM issues were rarely, if at all, mentioned. It is probably the case that, had their performance been unsatisfactory, for example with insufficient, excessive, or noisy ventilation, this would have been identified in guest feedback.

**Handover versus settled period**. In the first few months after opening, the reviews repeatedly included negative feedback on a small number of issues. Such occurrences significantly reduced over time, and correlated well with measures by the hotel operator:

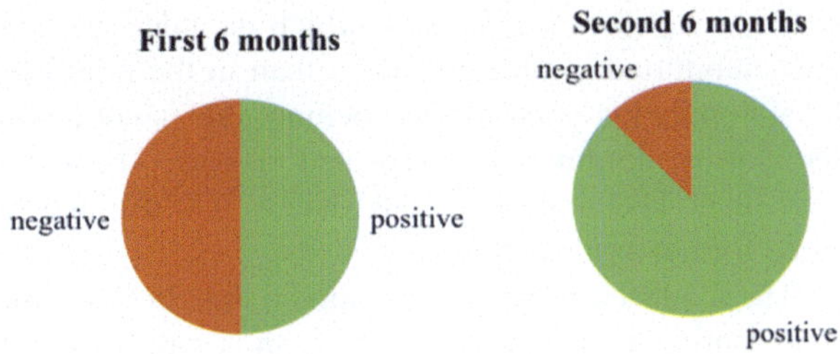


**Fig. 20.3** Case study 1—hotel: Changes to reviews on bedroom controls over time, illustrating the benefits of fine-turning and user training

- Noise: Works were carried out to improve the acoustic performance of the façade and reduce noise disturbance from the bars into the neighbouring spaces.
- Bedroom controls: The FM team carried out a fine-tuning programme and staff became more familiar with the room controls, thereby improving their guidance to guests. The resulting improvement in customer feedback is shown in Fig. 20.3.

**Reliability of findings**. Overall, the analysis correlated well with measures known to have been taken in the building's design and operation, and with other forms of feedback (e.g. staff feedback and reception logs). The FM team noted it was also in line with feedback on another travel review website. This led to the conclusion that the method could be a useful compromise towards formal POE methods, and it wastherefore, applied more recently in another building, as described in Case Study 2.

## 20.5 Case Study 2

### *20.5.1 Background*

This second case study is a leisure centre, opened in 2013. A similar approach to Case Study 1 was applied, with reviews analysed in late 2018 and a site visit early 2019 with the architect. A notable difference with Case Study 1 is that the building is not subjected to formal BPE. The site visit, however, gave the opportunity to verify early findings through simple observations and discussions with the architect and FM team; this happened post-analysis, in order to avoid bias when analysing the reviews.

### *20.5.2 Analysis and Findings*

At the time of the analysis, the building's overall review score was average (3.0), with reviews evenly spread from 1 (Terrible) to 5 (Excellent). The analysis of reviews was carried out over two periods: the first 6 months of operation, which included 46 reviews, and the past 2 years, which included 40 reviews. Far fewer reviews have been submitted in recent months than in the initial period, probably reflecting the local population's interest at opening; this is not a national attraction and visits are likely to be dominated by repeated visitors. The average score barely has changed over time. The analysis of the reviews is illustrated in see Fig. 20.4, following the same format as Case Study 1.

The feedback is much more mixed than in Case Study 1, with negative mentions representing approximately 55% of the total. It should be noted that the building is very busy: as a popular and needed facility, there may be a level of 'forgiveness' which on balance explains the overall average ratings. Design and brief issues represent 42% of the total, FM management issues 24% and general management issues 34%. Compared to Case Study 1, design issues are much less prominent and FM much more, probably because of perceived inadequacy in this case study—see details below. General management issues represent a similar share.

**Issues mentioned most often**. The issues mentioned most often are as follows:

- Service: staff friendliness and helpfulness
- Design of the pools: more in the initial phase but still relatively frequent
- Overall management: timetable, bookings and related levels of 'busy-ness'
- Overall building design: very prominent initially, much less recently
- FM: cleanliness.

As in Case Study 1, two of these dominant issues (staff and overall management) are outside the control of designers.

**Feedback on brief and design**. Feedback is in majority positive but there are strong caveats, especially in the first 6 months. Feedback on the design of the main facilities was usually positive (e.g. light levels and views, diving boards, slides, 'modern' feel). The design of external areas, the size of the building and pool and the provision of outdoor facilities were often commented on negatively, as was the building's quick deterioration, e.g. rusty slides and broken lockers. Many of the issues

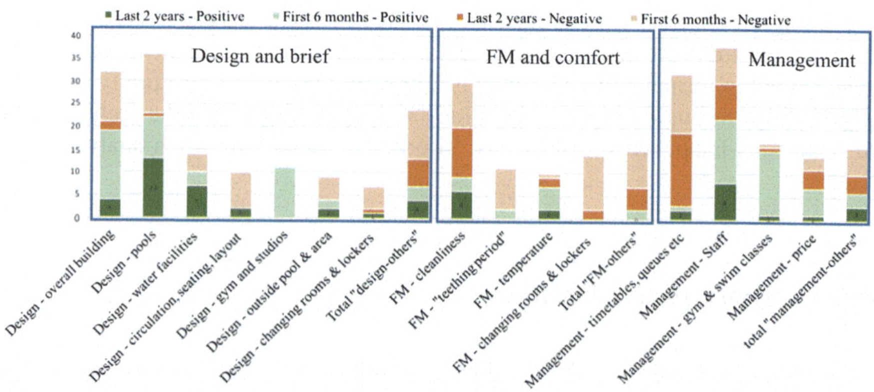


**Fig. 20.4** Case study 2—Leisure Centre: Categorisation of reviews available from public website

receiving negative feedback are not about the design response itself, but about the brief or as-built quality.

**Feedback on FM and comfort**. FM features highly in reviews, especially cleanliness, with reviews in large majority negative although slightly less in recent months. Broken lockers are a recurring issue, and recent reviews may indicate a general downhill trend on maintenance, e.g. broken windows and missing ceiling tiles.

The reviews do not identify recurring comfort issues, with very few comments other than isolated ones, e.g. 'light and airy' gym area. The pool water temperature is mentioned relatively often, and overwhelmingly positive. The noise was not mentioned, which is surprising given how busy the building is—this may reflect people's expectations of acoustics in leisure and sports buildings.

**Handover versus settled period**. Early reviews regularly mention 'teething problems', most acknowledging it is common in new buildings. This level of forgiveness and understanding of the building delivery process in the general public is interesting.

**Reliability of findings**. There was a good correlation between this analysis and the feedback from the project architect and FM team, in particular:

- Size of the building and pool: Following an early consultation which led to high expectations, significant budget cuts had to be made and subsequent changes were not well received by the local population. Negative comments often displayed discontent and/or mistrust about the client (i.e. the council).
- Early degradation: Budget cuts and a contractor-led process led to changes such as materials selection. The visit confirmed visible rust, to an extent which makes it surprising that reviews do not mention it more often.
- Design of external areas: Due to budget constraints, the council carried out the design through their in-house team, rather than as part of the overall project.
- Cleanliness and insufficient maintenance: this matched impressions on site.

## 20.6 Discussion

### *20.6.1 Discussion and Caveats on the Usefulness of the Method*

There are caveats and limitations to this method, in particular:

- As a non-standardised method, it offers opportunities for bias, particularly if the person carrying out the study was involved in the original design. This is addressed here by being as systematic and data-based as possible. In Case Study 1, the analysis was subjected to peer-review; in Case Study 2, the author of the study had no previous involvement with the building.
- When making observations from site visits, some caution is needed to avoid drawing conclusions from a snapshot in time.

- While no evidence of fake reviews was found in the case studies, the potential should be borne in mind, with some judgment and caution in the analysis.
- People going through the effort of posting a review may be particularly enthusiastic or frustrated by their experience rather than expressing a representative sample of overall opinions. This is not seen as a significant problem in the context of this study: the method does not claim to provide an absolute assessment of the building, but to identify key areas of success or concern.
- Strictly speaking, the analysis is limited to what people mention, compared to formal surveys which systematically gather feedback on set topics. However, one can use the lessons available from decades of POE to identify recurrent parameters important to users (e.g. noise, thermal comfort, FM quality and responsiveness) so that, if such a parameter is barely mentioned in reviews, it may be assumed to perform well enough in that building not to be noticed or worth mentioning by users.

Within these limitations and points of caution, it is considered that the two case studies highlight significant potential for the proposed method to generate useful feedback through analysis by an experienced professional. The findings correlated reasonably well with other methods, and it was possible to identify:

- Overall trends as well as more specific issues.
- Separate feedback on issues related to the brief, design, FM and broader operations which designers have no control over.
- Trends over time, with insights from comparing the 'settled' stages with the initial period of occupation, which typically still has issues to resolve.

Beyond identifying issues specific to individual buildings, one valuable output is in highlighting the consequences of decisions such as value-engineering and inadequate commissioning and FM, these issues are repeatedly found in BPE [8, 9], and this method allows some (albeit imperfect) quantified illustration: users do notice.

The method also provides a POE process for buildings with 'transient' users, which are often not addressed by formal POE methods. Visitor review sites only apply to some building types but similar user-generated feedback sites are increasingly available in other sectors, for example, 'Rate my Building' used on campuses in Australia, 'Rate my Apartment' used for the US university accommodation and Care Homes UK.

### *20.6.2 Conclusions and Further Areas of Development*

Possible areas for development should seek to draw trends and further standardise the method by applying the method more widely to

- More buildings, including somewhere more extensive BPE is carried out, allowing to further check its validity and limitations. For example, it would be useful to see whether there are trends between first impressions (e.g. reported friendliness of reception staff and queues) and overall satisfaction: how influential are first impressions on the overall review?

- Several buildings of the same type, as this could help identify trends in what matters to users in that sector and therefore inform future project briefs.
- The same building, over time: because the method is not disruptive and requires few resources, it can be used regularly to track changes, for example, in management regimes or in seasons.
- Wider building types, not subject to visitor/travel reviews but where user-generated feedback sites are becoming more common (see Sect. 20.6.1).

The method has limitations and caution should be exercised. Ideally, it should be used alongside other tools, such as site visits and interviews. However, it does not *rely* on this, i.e. if used cautiously, it could prove valuable even in the absence of other forms of BPE. It should not be a replacement of formal BPE methods, but an option if nothing else can be done, or a routine first step to exercise our professional duty, identify key issues and pique a client's or occupier's interest before a more extensive exercise.

**Acknowledgements** Case Study 1 was carried out while at Hoare Lea through Innovate UK (TSB) funding. Case Study 2 is partly being carried out as consultant for Wilkinson Eyre Architects.

## References

1. Usable Buildings Trust. http://www.usablebuildings.co.uk/
2. Available from https://www.cibse.org/building-services/building-services-case-studies/probe-post-occupancy-studies
3. Riba, S., Hay, S., Bradbury, S., Dixon, D., Martindale, K., Samuel, F., Tait, A.: Pathways to POE, Value of Architects, University of Reading, RIBA (2016)
4. Committee on Climate Change, UK Housing: Fit for the Future? February 2019
5. Godefroy, J.: Coming together—CIBSE and RIBA collaborate on sustainability overlay. CIBSE J. (2018)
6. Cuau, C., Pim, H.: Invisible Insights for the British Museum: Learning from TripAdvisor Reviews, Museum-iD (date unknown). https://museum-id.com/invisible-insights-learning-from-tripadvisor-reviews/
7. Bordass, B., Leaman, A., Cohen, R., Walking the Tightrope (2001)
8. Bordass, B., Bunn, B., Cohen, R., Ruyssevelt, P., Standeven, M., Leaman, A.: The PROBE project: technical lessons from PROBE 2, CIBSE National Conference (1999)
9. Innovate UK, Building Performance Evaluation Programme, Findings from Non-domestic Projects (2016)

# Chapter 21
# Use of an Object-Oriented System for Optimizing Life Cycle Embodied Energy and Life Cycle Material Cost of Shopping Centres

**K. K. Weththasinghe, André Stephan, Valerie Francis and Piyush Tiwari**

**Abstract** Shopping centres are an integral part and a critical component of urban cities in most economies. Typically, the shorter refurbishment cycle and frequent tenant replacements in shopping centres cause excessive use of building materials over its service life. This drastic use of resources, consequently, increase life cycle embodied energy (LCEE) and life cycle material cost (LCMC) of shopping centres. Therefore, careful selection of materials is vital to reduce the negative environmental impacts and material costs. Current research on the implications of material choices on LCEE and LCMC of shopping centres are insubstantial and decisions makers are left with limited information to make better selections. Therefore, selection of energy efficient, cost-effective and environmentally responsive materials and assemblies has been a critical process for the professionals who are involved in decision-making. This paper proposes the use of object-oriented programming (OOP) to develop a mathematical model to generate combinations of building assemblies with minimum LCEE and LCMC of shopping centres through material selection. The model is based for sub-regional shopping centres in Australia, yet can be applied for any similar property type with modifications to databases and model architecture. However, scope of this paper is limited to the development of model architecture with detailed explanations on databases and computing core development. Even though, the detailed presentation of development of OOP structure provides proper insight to the mathematical core for future application.

## 21.1 Introduction

Shopping centres are the largest component of the retail property sector and an important aspect of modern cities [1]. They are inevitably one of the major energy and resource users in the built environment over the life cycle [2, 3]. The extended opening hours, high ceilings and excessive use of illuminations and heating and cooling cause the heaps in operational energy use over life cycle while shorter refurbishment

K. K. Weththasinghe (✉) · A. Stephan · V. Francis · P. Tiwari
The University of Melbourne, Parkville, VIC 3010, Australia
e-mail: kweththasing@student.unimelb.edu.au



J. Littlewood et al. (eds.), *Sustainability in Energy and Buildings*, Smart Innovation, Systems and Technologies 163, https://doi.org/10.1007/978-981-32-9868-2_21

cycle and frequent tenant replacements cause the excessive embodied energy and material use [3, 4] in shopping centres. The necessity to maintain trendy aesthetics in shopping centres require incessant maintenance and upgrades [5]. Additionally, tenant leases cause frequent material replacements depending on the lease periods and tenant replacements [6, 7]. Hence, materials used in shopping centres serve a shorter life before becoming obsolete [8] resulting in increased life cycle embodied energy (LCEE) in shopping centres when compared to other residential and commercial buildings [9]. Furthermore, this sustained use and disposal of building materials over the life cycle ultimately result in an elevated life cycle material cost (LCMC) [10]. Therefore, material selection for shopping centres needs to be more thorough and observant to mitigate negative environmental impacts of increased LCEE and to reduce LCMC. Hence, this paper aims to propose an approach to identify combinations of building materials and assemblies with optimum LCEE and LCMC for shopping centre development in Australia.

## 21.2 Literature Review

Shopping centres are major energy and resource users over their life cycle [11]. Being community places, shopping centres use large amounts of energy to maintain the good and comfort visual [12]. They require continuous maintenance and refurbishments over the years to attract and sustain foot traffic and tenants [13] (refurbishment: any 'remodeling, refashioning and general renovation of a building, site, product or infrastructure' [14]). Tenants of shopping centres are often required to refurbish the premises during or at the end of their leases [9, 15]. According to the *Retail Leases Act 2003*, a tenant entering a new lease for a retail shop has the right to a minimum tenancy period of up to 5 years (which can be renewed at the end of lease). Hence refurbishments can occur during the 5-year lease period and when it is terminated and beginning of a new tenant. This is comparable to the findings by [9] who stated that refurbishment frequency of retail shops is every 2–10 years. Empirical findings through interviews with sub-regional shopping centre managers in Australia also solidifies the findings of literature on refurbishment frequency.

The shorter refurbishment cycle and frequent tenant replacements inevitably cause excessive building material usage in shopping centres [16]. Building materials are replaced long before reaching the end of their expected service lives due to economic, functional or social obsolescence during refurbishments [8]. Hence, the recurrent embodied energy (the energy required for the maintenance and replacement of building materials or systems during the building useful life [17]) becomes vital in shopping centres, notably compared to other commercial and residential building types [9]. Consequently, LCEE (combination of initial and recurrent embodied energy and demolition energy [18]) of shopping centres become immense [9, 11]. Hence, careful attention needs to be given to material selection process of shopping centres [19] to reduce LCEE.

Since the evolving concern on sustainability in built environment has already made a significant impact in shopping centre development and redevelopments in terms of operation energy use, next is to address the embodied energy issue [16]. However, it is important to maintain a trade-off between minimization of operational and embodied energy since the extreme measures to minimize one can lead to a significant escalation in the other. Therefore, selection of environmentally responsive building materials is challenging not only due to prior reasons but also due to financial constraints [20]. Typically, initial cost of materials and assemblies can make up to 20–30% of the total project cost [21]. Also, environmentally responsible, innovative building materials are alleged to be more expensive than conventional building materials [8]. Additionally, frequent material replacements in shopping centres cause an enormous cost over the life cycle [22]. Therefore, minimizing LCMC (total of capital, maintenance and refurbishment and disposal or recycling costs) of shopping centres also becomes vital.

Research has identified cost as one of the main barriers for the selection of materials with improved environmental performances [23]. Also, literature has acknowledged that material selection has a close relationship between embodied energy and cost of buildings [14, 15, 24]. However, the findings of a few research can only be used within a limited boundary due to the lack of comprehensiveness of the approach used to generate life cycle inventory data used and the scope of research [25]. Furthermore, many researches in the area of embodied energy analysis and cost of materials are concentrated around residential and commercial office buildings [23, 25]. Only a limited number of studies are engaged in retail property and embodied energy [9, 12] but none of them are focused on LCEE and LCMC in shopping centres. Hence, the knowledge on availability of building materials with lower embodied energy with minimum LCMC (with minimum negative impact to life cycle operational energy use in the building) for shopping centre construction is lacking.

Several models have been developed to assist in the process of selection of building materials, engaging different optimization methods [15], analytic hierarchy processes [24] and computer-aided software tools [26]. However, the existing models are not precisely focused on the unique nature of shopping centres, and their exceptional material replacement frequencies. Hence, a mathematical model which incorporates the unique features of shopping centres is required to facilitate the selection of combinations of materials and assemblies with minimum LCEE and LCMC. This research niche is therefore addressed in this paper.

Accordingly, the research problem identified can be expressed in the form of a programming problem as follows.

Minimize $f(x)$; subject to $g(x)$ and $h(x)$, where $(x)$ is a vector of $n$ real value design variables which will be alternative building material and assembly combinations and $f$ is the cost function also known as the objective function. $g(x)$ and $h(x)$ are the inequality and equality constraints. The study aims to achieve three objective functions as

$$Minimize\ \text{LCEE}\,(x) \tag{21.1}$$

$$Minimize \text{ LCMC}(x) \quad (21.2)$$

$$Minimize \text{ LCEE} \,\&\, LCMC\,(x) \quad (21.3)$$

The study follows a case study research method with a mathematical modelling process to achieve set forth aim and objectives. The mathematical model is developed for an archetypal single-story sub-regional shopping centre in Australia. Sub-regional shopping centre is selected as the base type for the model since it dominates Australian shopping centre industry in terms of investments [27], GLA (gross lettable area) and with a three-year pipeline of planned developments and redevelopments till 2021 [21, 28]. Case studies are carried out to observe and identify the most representative single-story sub-regional shopping centre which is selected as the base case for archetypal centre design (Since more than 75% of sub-regional shopping centres in Australia are single story buildings the archetypal is designed following the majority [21, 28]). Three different scenarios with different tenant mixes and GLAR (gross lettable area retail) are tested. The results are analyzed to examine the relationships between material selection and LCEE and LCMC and to identify combinations of building materials and assemblies with optimum LCEE and LCMC for sub-regional shopping centres in Australia.

This paper presents the knowledge on the development of mathematical model and databases which can be used as an archetype to resolve similar problems in the built environment.

## 21.3 Object-Oriented Programming (OOP) as a Method

Development of any mathematical model requires automation of complex systems and rigorous calculation processes [29]. LCEE and LCMC calculation of sub-regional shopping centres is a complex task since they are massive building projects with a wide range of building elements and assemblies used. The process of calculations requires different data sets of materials, assemblies, bills of quantities of different types of shops and many other and demanding matrix calculations (further explained in 2.1). Therefore, the model developed needs to accomplish the requirements and specifications to achieve the set forth objectives. One of the main requirements of the model is to assess and compare different scenarios of sub-regional shopping centres at a reduced run time. Further, the model architecture needs to be resilient yet flexible. OOP provides solutions to all these requirements.

OOP is a programming paradigm organized around "objects" and data [29, 30]. An object includes a set of data with associated behaviours. Objects with different attributes and methods are classified under different classes. A "*class*" is basically a blueprint of an object. In other words, the classes describe the objects with attributes, methods and variables. Different types of objects can have different classes [30]. The study defines four class modules as *Materials*, *Assemblies*, *Shop*, and *ShoppingCen-*

*tre*. Each class instance can have attributes attached to it for maintaining its state. Class instances can also have methods (defined by its class) for modifying its state [31]. For instance, *Shop* class has parameters as *length*, *width*, *height*, etc. and the method get_quantity_gfa. Since OOP provides a flexible architecture, the attributes or the parameters of classes can be modified without changing the calculation algorithms. Hence, different instances of a class can be assessed based on their attributes. Furthermore, additions of new methods to a class often do not affect other methods in the class. Thus, updates can be performed with minor impact on the entire model, because of the modular structure [32]. Consequently, OOP is selected as the programming paradigm to develop the model architecture.

To develop the mathematical model, a software programming is required. This study uses Python™ 3.7 as the programming language since it is free and opensource [33]. It also provides a large array of libraries including classes and functions which are designed specially to tackle specific aspects of programming [31]. Different Python-based modules are available for free for database generation, to develop matrices and other numerical operations and to generate graphs and charts. OOP in Python is proven to be an efficient method for real-world applications [29]. Hence, the programming language chosen to develop the model is Python.

### 21.3.1 Model Architecture

A general software tool consists of three main compartments as graphical user interface (GUI), computing core and the databases. However, at this stage the research is limited to developing the computing core and the databases only. The computing core is the model architecture containing all the classes with defined attributes and methods for all related calculations to quantify LCEE and LCMC of the shopping centre for different assembly combinations. The data required to execute the methods are mostly extracted from the databases. The databases are the data sets of materials, assemblies, shops and shopping centres which provide attribute values to the classes.

### 21.3.2 Classes and Databases

To understand the development of classes, the reader should be familiar with the hierarchy of different levels. Figure 21.1 demonstrates the sequence of classes in the computing core.

*Materials* class imports data from materials database. Materials database provides material details under fields of material_id (unique for each material), material_name, material_type, material_unit (m, $m^2$, $m^3$, t, etc.) material_eec, material_unit_price, etc. Material instances are created at *Materials* class using the database. Assemblies database is created to store different assemblies which can be constructed using the materials objects. Assemblies database delivers datasets of assemblies under different

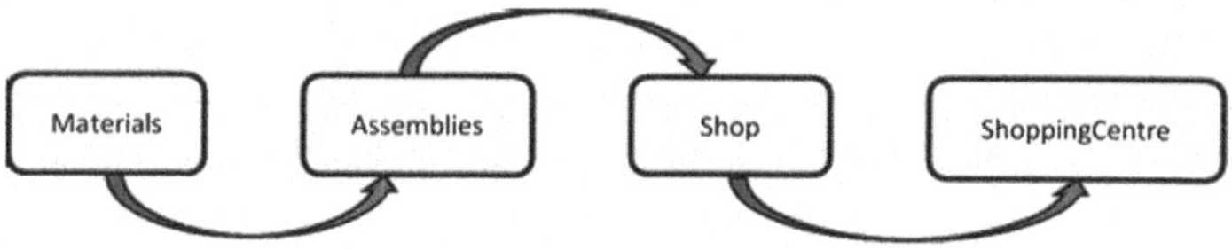


**Fig. 21.1** Sequence of levels of classes and databases

fields such as assembly_id, assembly_type (assemblies are categorized under 13 assembly_types as foundation, column, internal_wall, etc. based on AIQS building elements and sub-elements [34]), assembly_unit (m, $m^2$, $m^3$), etc. Further it contains details of quantities of each material in a unit of assembly and respective material_id which will be used to determine assembly embodied energy coefficient (eec) and unit price. For instance, consider M7 and M30 are respective material_id of 12 mm rebars and 20Mpa concrete (Table 21.1).

To build 1 m of assembly_id BM01, following quantities of M7 and M30 are required.

$$\mathrm{M7{-}0.001\,(t)\ and\ M30{-}0.09(m^3)}$$

Since materials instances delivers material_eec and material_unit_price, to determine assembly_eec and assembly_unit_price instances of *Assemblies* class access the material objects created at *Materials* class. This process is methodologically formed in *Assemblies* class to create assembly objects with required steps and calculations. *Shop* class is the largest class in the model with several methods to perform required calculations to gradually determine shop_lcee, shop_lcmc and develop shop_boq. Unlike previous classes *Shop* class import data from two different databases: shops catalogue database and shops database. Shops catalogue database provides data on different types of shops in the shopping centre under 16 categories (e.g. clothing, food_supplies, health_&_beauty, etc. [27, 35]). Each shop type is given a unique id same as previous and characteristics are defined in different fields. Also, default assemblies for each assembly_type are defined for each shop_type (Table 21.2).

Shops database defines the geometry of different shops in the archetypal shopping centre. Every shop has a shop_type as defined in shops catalogue database and attribute values of length, width, height and span. This format of database allows

**Table 21.1** Sample assembly

| Assembly_id | Assembly_type | Assembly_name | Assembly_unit |
|---|---|---|---|
| BM01 | beam | 300 * 300 concrete beam | m |

**Table 21.2** Sample shop_type

| Shop_type_id | Beam | ceiling_finish | Column | Door |
|---|---|---|---|---|
| CL_01_RF_5 | BM04 | CF01 | CL04 | DR01 |

the researcher to develop parametric shop designs following the shoe-box concept. Working with parametric designs provides flexibility to the shopping centre design and thus allow for future modifications to the model. The geometries defined for each shop are used in *Shop* class to generate BOQ (bills of quantities) with automated calculations of quantities of building assembly types. *ShoppingCentre* class defines the three different scenarios of archetypal shopping centre (with different tenant mixes).

Constraints to the model are also provided as databases. Assembly compatibility matrix (assembly compatibility with each other is demonstrated with TRUE and FALSE operators) and Assembly shop compatibility matrix (compatibility of assemblies with each shop type) are the two main constraint databases. The model uses these classes and databases to gradually determine the combinations of assemblies for each assembly type with minimum LCEE and LCMC for sub-regional shopping centres in Australia. The following section provides a demonstration of how these classes are utilized to determine the combinations of assemblies with minimum LCEE.

### *21.3.3 Assessment of LCEE*

Aim of this study is to identify combinations of building assemblies with minimum LCEE and LCMC for sub-regional shopping centre construction in Australia using OOP model. Figure 21.2 demonstrates the flowchart to assess the LCEE.

In Fig. 21.2,

| | |
|---|---|
| $EEC_A$ | Embodied energy coefficient of an assembly |
| $IEE_S$ | Initial embodied energy of a shop |
| $REE_S$ | Recurrent embodied energy of a shop |

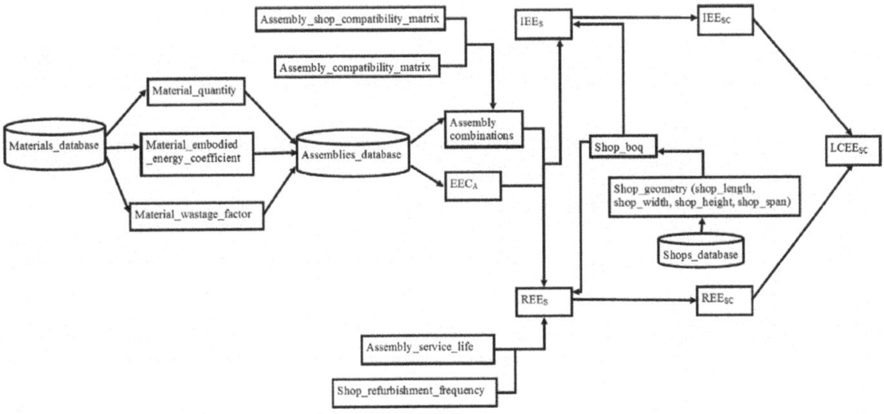


**Fig. 21.2** LCEE assessment flow chart

$IEE_{SC}$ Initial embodied energy of shopping centre
$REE_{SC}$ Recurrent embodied energy of shopping centre
$LCEE_{SC}$ Life cycle embodied energy of shopping centre.

At *Shop* level assembly combinations are generated for different shop types using assemblies database, shop database, shops_catalogue and compatibility matrices. Shops of different assembly combinations are then combined to develop alternatives of the three shopping centre scenarios. The LCEE of every alternative shopping centre is quantified following the above process. LCEE is defined as the combination of IEE (initial embodied energy) and REE (recurrent embodied energy) of the shopping centre. The study neglects the impacts of demolition energy towards LCEE since literature analysis finds it negligible when life cycle approach is considered [3, 14, 19, 30]. As shown in Fig. 21.2 the IEE of the centre is the sum of IEE of all the shops in the centre which in turn, is the sum of the IEE of its assemblies, itself the sum of the IEE of constituting materials. Following equations denote IEE calculation at different levels.

$$IEE_{sc} = \sum_{s}\sum_{a}\sum_{m} EEC_m \times Q_{m,a,s} \tag{21.4}$$

$IEE_{sc}$ IEE of shopping centre
$EEC_m$ Embodied energy coefficient of material m
$Q_{m,a,s}$ Quantity of material m in assembly a in shop s.

The non-material energy inputs at the construction stage are added as per Eq. 21.5. This includes inputs associated with the financial, communication, marketing and other service sectors, which are considered as further sideway inputs [25]. $Q_{M,A,S}$ is also modified by incorporating the wastage factor of material.

$$IEE_{sc} = \sum_{s}\sum_{a}\sum_{m}(EEC_m \times Q_{m,a,s} \times WF_m) + \left[TERREBS - \sum_{m=1}^{M}(TER_m)\right] \times C_{sc} \tag{21.5}$$

$WF_m$ Material wastage factor
$TERREBS$ Total energy requirement of the retail building sector (GJ per currency unit)
$TER_m$ Total energy requirement of the input-output pathway representing material m (GJ per currency unit)
$C_{sc}$ Total cost of the shopping centre.

The REE of shopping centres can be calculated using the following Eq. 21.6 developed by [36].

$$REE_{sc} = \sum_{s}\sum_{a}\sum_{m}[RR \times (EEC_m \times Q_{m,a,s} \times WF_m) + (TERREBS - TER_m - NATER_m \times C_{m,a,s}) \tag{21.6}$$

$REE_{sc}$ Recurrent embodied energy of shopping centre

$RR$ Replacement rate
$NATER_m$ The total energy requirement of all input–output pathways not associated with the installation or production process of material m being replaced (GJ per currency unit)
$C_{m,a,s}$ Cost of material m in assembly a in shop s.

In Eq. 21.7, RR can be characterized in the following manner.

$$RR = \left\{\left\lceil \frac{POA_{a,s}}{SL_{m,a,s}} - 1\right\rceil \Leftrightarrow \left\lceil \frac{POA_{a,s}}{SL_{m,a,s}} - 1\right\rceil \leq RF_s\right\}; \left\{RF \Leftrightarrow \left\lceil \frac{POA_{a,s}}{SL_{m,a,s}} - 1\right\rceil \geq RF_s\right\} \quad (21.7)$$

$POA_{a,s}$ Period of analysis of assembly a in shop s
$SL_{m,a,s}$ Service life of the material m in assembly a in shop s
$RF_s$ Refurbishment frequency of shop s.

However, RR can also be represented as the refurbishment frequency (RF) of the shops in shopping centre.

A similar sequence of calculations is used to generate LCMC values of different assembly combinations for archetypal shopping centre scenarios. However, unlike recurrent embodied energy, recurrent material cost is discounted to present value to account for time value of money. Once the calculations are fully performed, the combinations of assemblies responsible for minimum LCEE and minimum LCMC independently are recorded achieving objective functions 1 and 2. Then the model executes the optimization process to identify combinations which lead to minimum trade-off of LCEE and LCMC achieving the third objective function. However, this paper does not deliver particulars of the optimization process.

## 21.4 Discussion and Conclusion

This paper aimed in discussing the process of developing a mathematical model using OOP to investigate the combinations of building assemblies with optimal LCEE and LCMC for sub-regional shopping centre construction in Australia. The research problem is detailed in the beginning to provide the reader a clear understanding of the system and what is required of the model. Application of OOP to resolve the research problem is demonstrated in Sect. 21.3. Assessment of LCEE of the archetypal sub-regional shopping centre is exhibited as a sample. The research by Stephan and Crawford [36] and Stephan and Stephan [37] are pioneering in the area of life cycle energy assessments in residential buildings, which are used as fundamentals for the model. The automated calculations of building elements make preparation of bills of quantities (BOQ) of the shops an effortless process. Defining parametric shop designs based on basic building geometries (length, width, height and span) to prepare BOQ allows future expansions to the model regarding design upgrades (changes to shopping centre layout, shop designs). The developed system is primarily focused on shopping centres which is currently an under-researched area. Yet, it can be used to assess life cycle energy and cost impacts of other property types with some modifications and alterations to databases and programming. Furthermore, the model can be used in future scenarios of more advanced materials and assemblies with updates to *Materials* and *Assemblies* databases.

However, the model developed, and the results are subject to limitations due to uncertainty and variability in the data. Notably, the LCEE quantification process used in the model experience several limitations due to use of hybrid embodied energy coefficients [17], calculation of the non-material energy inputs (following the equations developed by Crawford [25]) and the algorithms used to calculate material replacement rate. Additionally, the model performs LCEE and LCMC calculations for materials and assemblies in the databases developed based on empirical data and research data for innovative materials and assemblies. Even though assembly-assembly compatibility matrix and shop-assembly compatibility matrix are used to address design, structural and material constraints the results are subject to limitations in reliability and uncertainty for future scenarios. Nevertheless, the model is comparatively comprehensive and provides reliable results, overcoming the drawbacks of previous life cycle energy assessment models. Hence, this approach can be applied in similar optimization problems in the built environment. Even though this paper does not review the optimization process, the detailed presentation of development of OOP structure provides proper insight to the mathematical core for future applications.

## References

1. LaSalle, J.L.: Australian Shopping Centre Investment Review & Outlook (2019)
2. Juaidi, A., AlFaris, F., Montoya, F.G., Manzano-Agugliaro, F.: Energy benchmarking for shopping centers in Gulf Coast region. Energy Policy **91**, 247–255 (2016)
3. Haase, M., Skeie, K.S., Woods, R.: The key drivers for energy retrofitting of European shopping centres. Energy Procedia **78**, 2298–2303 (2015)
4. Braslavsky, J.H., Wall, J.R., Reedman, L.J.: Optimal distributed energy resources and the cost of reduced greenhouse gas emissions in a large retail shopping centre. Appl. Energy **155**, 120–130 (2015)
5. Coleman, P.: Shopping Environments: Evolution, Planning and Design: Taylor and Francis, Jordan Hill (2007)
6. Scott, K.M.: Shopping Center Tenant Coordination (2006)
7. Wakefield, K.L., Baker, J.: Excitement at the mall: determinants and effects on shopping response. J. Retail. **74**(4), 515–539 (1998)
8. Holtzhausen, H.J.: Embodied Energy Impact Arch. Dec. **II**, 377–385 (2007)
9. Fieldson, R., Rai, D.: An assessment of carbon emissions from retail fit-out in the United Kingdom. J. Retail Leisure Prop. **8**(4), 243–258 (2009)
10. Lewry, A.J., Suttie, E.: Ecoshopping: energy efficient & cost competitive retrofitting solutions for retail buildings. Romanian J. Civil Eng. **6**(1), 87–100 (2015)
11. Climateworks Australia.: Low Carbon Growth Plan for Australia Retail Sector Summary Report (2011)
12. Thompson, B.: Green retail: Retailer strategies for surviving the sustainability storm. J. Retail Leisure Prop. **6**(4), 281–286 (2007)

13. Kocaili, B.E.: Evolution of Shopping Malls: Recent Trends and the Question of Regeneration (2010)
14. Bansal, D., Singh, R., Sawhney, R.L.: Effect of construction materials on embodied energy and cost of buildings—A case study of residential houses in India up to 60 $m^2$ of plinth area. Energy Build. **69**, 260–266 (2014)
15. Castro-Lacouture, D., Sefair, J.A., Flórez, L., Medaglia, A.L.: Optimization model for the selection of materials using a LEED-based green building rating system in Colombia. Build. Environ. **44**(6), 1162–1170 (2009)
16. Yudelson, J.: Sustainable retail development. [electronic resource]: new success strategies. Springer; [New York]: International Council of Shopping Centers, c2009. Dordrecht; New York (2009)
17. Crawford, R.H., Bontinck, P.-A., Stephan, A., Wiedmann, T., Yu, M.: Hybrid life cycle inventory methods—a review. J. Clean. Prod. **172**, 1273–1288 (2018)
18. Ramesh, T., Prakash, R., Shukla, K.K.: Life cycle energy analysis of buildings: An overview. Energy Build. **42**(10), 1592–1600 (2010)
19. Máté, K.: Remediating Shopping Centres for Sustainability (2013)
20. Inyim, P., Zhu, Y., Orabi, W.: Analysis of time, cost, and environmental impact relationships at the building-material level. J. Manage. Eng. **4**, 4016005 (2016)
21. Shopping Centre Council of Australia.: Shopping Centre Redevelopment (2017)
22. Aktas, G.G.: Sustainable Approaches in Shopping Center Public Interiors: Lighting and Finishing Materials, pp. 183–187 (2011)
23. Jiao, Y., Lloyd, C.R., Wakes, S.J.: The relationship between total embodied energy and cost of commercial buildings. Energy Build. **52**, 20–27 (2012)
24. Akadiri, P.O.: Understanding barriers affecting the selection of sustainable materials in building projects. J. Build. Eng. **4**, 86–93 (2015)
25. Crawford, R.H.: Life Cycle Assessment in the Built Environment. Spon Press, Abingdon, Oxon (2011)
26. Seo, S., Tucker, S.N.: Ambrose MD. Selection of Sustainable Building Material Using LCADesign Tool (2007)
27. CBRE Research.: Sub-Regional Shopping Centres: A case of middle child syndrome. Report. CBRE RESEARCH, Australia (2018)
28. Jones Lang LaSalle.: Australian Shopping Centre Investment Review & Outlook (2017)
29. Lott, S.F., Fatouhi, D.: Mastering Object-Oriented Python: Grasp the Intricacies of Object-Oriented Programming in Python in Order to Efficiency Build Powerful Real-World Applications. Packt Publishing (2014)
30. Tokoro, M., Pareschi, R.: Object-Oriented Programming: 8th European Conference, ECOOP '94, Bologna, Italy, 4–8 July 1994: proceedings. Springer-Verlag (1994)
31. Phillips, D.: Python 3 object oriented programming. [electronic resource]: harness the power of Python 3 objects. Packt Open Source, Birmingham, U.K. (2010)
32. Stanger, N., Pascoe, R.: Exploiting the Advantages of Object Oriented Programming in the Implementation of a Database Design Environment, p. 525 (1997)
33. Python Software Foundation.: Python Language Reference, Version 2.7. Python Software Foundation Wilmington, DE (2010)
34. The Australian Institute of Quantity Surveyors.: Australian Cost Management Manual. Australian Institute of Quantity Surveyors, 2000–2008, Canberra, A.C.T. (2000)
35. International Council of Shopping Centres.: Dictionary of Shopping Centre Terms. International Council of Shopping Centers, pp. 63–64 (2005)
36. Stephan, A.: Crawford RH. The Relationship Between House Size and Life Cycle Energy Demand: Implications for Energy Efficiency Regulations for Buildings. Energy. 1158 (2016)
37. Stephan, A., Stephan, L.: Reducing the total life cycle energy demand of recent residential buildings in Lebanon. Energy. **74**, 618–637 (2014)

# Chapter 22
# Hygrothermal Characterization of High-Performance Aerogel-Based Internal Plaster

**Stefano Fantucci, Elisa Fenoglio, Valentina Serra, Marco Perino, Marco Dutto and Valentina Marino**

**Abstract** The development of novel high-performance thermal insulating products represents a key action within the deep energy retrofit strategies on the existing building stock. In the framework of the Horizon 2020 project Wall-ACE, a highly efficient thermal insulating plaster, based on silica aerogels, has been developed. In this paper, the development process aimed at achieving a thermal conductivity lower than 0.030 W/mK is presented. Moreover, the hygrothermal characterization process aimed at assessing the data for the dynamic heat and moisture transfer (HMT) simulations, according to EN 15026:2008, is described.

## 22.1 Introduction

Around 83% of the buildings in EU countries were built before 1991 when poor energy-saving measures were adopted. Around half of them were built before the 1960s, of which, a large part is so far non-insulated.

The thermal insulation of the building envelopes represents, therefore, one of the most effective energy-saving strategies. However, historical and technological constraints, as well as the need to save internal space, can represent a limitation for an extensive application of thermal insulation. For this reason, the development of suitable envelope retrofit solutions that can overcome the present-day barrier for a deep energy retrofit represents an essential step towards the next targets of 40% reduction of the emissions by 2030 (Paris agreement).

In the framework of the EU Horizon 2020 project Wall-ACE, a highly efficient thermal insulating plaster based on silica aerogels has been developed for the internal wall insulation.

S. Fantucci (✉) · E. Fenoglio · V. Serra · M. Perino
Department of Energy, Politecnico di Torino, TEBE Research Group, C.so Duca Degli Abruzzi 24, 10129 Turin, Italy
e-mail: stefano.fantucci@polito.it

M. Dutto · V. Marino
Vimark srl, Strada Spartafino 2, 12100 Peveragno, CN, Italy



J. Littlewood et al. (eds.), *Sustainability in Energy and Buildings*, Smart Innovation, Systems and Technologies 163, https://doi.org/10.1007/978-981-32-9868-2_22

According to the national legislation, all the retrofit solutions must be verified as far as the absence of surface and interstitial condensation is concerned, this assessment can be performed using a simplified steady-state method based on EN ISO 13788:2012 standard (Glaser method). Nevertheless, several studies highlighted that the application of this simplified procedure often determines an overestimation of the condensation risk [1–3]. For this reason, especially when internal insulation is adopted; dynamic coupled heat and moisture transfer (HMT) simulations, accordingly to EN 15026 standard, are always preferable. Moreover, HMT analyses can be a valid support for designers to

- Assess the actual advantages achievable by adopting advanced thermal insulating layers in terms of energy saving as well as Indoor Environmental Quality (IEQ);
- Optimize the application of the insulating materials with the aim of achieving the best performance.

To perform reliable HMT simulations, a number of material's hygrothermal properties is needed. These properties have to be determined through a series of laboratory tests following measurement protocols suggested by various national/international standards.

Unfortunately, most of the materials that are under development and sometimes, even those that are already marketed, are not accompanied by such data. Indeed, only a few studies report a complete and exhaustive hygrothermal characterization [4–7] while most of the studies are mainly focused on thermal characterization.

In this framework, the aims of this paper are as follows:

- To present the development process of a high-performance aerogel-based thermal insulating plaster with a thermal conductivity below 0.030 W/mK;
- To describe the experimental protocol and the methodologies to be adopted for the material hygrothermal characterization;
- To provide guidelines and raise the awareness of manufacturers and designers that extensive dataset of physical properties should be always provided; thus, an example of datasheet containing the minimum information to fully simulate the hygrothermal behaviour of materials is also shown.

## 22.2 Development of a High-Performance Insulating Plaster

One of the goals of the H2020 project Wall-ACE is to develop an internal thermal insulating plaster characterized by a thermal conductivity $\lambda$ lower than 30 mW/mK (project target value) adopting the Kwark® granular aerogel as aggregate. To achieve this goal, an optimization process was pursued, starting from a series of mixtures in which a different aerogel and other mineral lightweight aggregates ratio was analysed (samples A-B-C). Since the $\lambda$-value achieved by the first set of samples was not considered satisfactory (0.06–0.09 W/mK, Fig. 22.1), in a second phase a com-

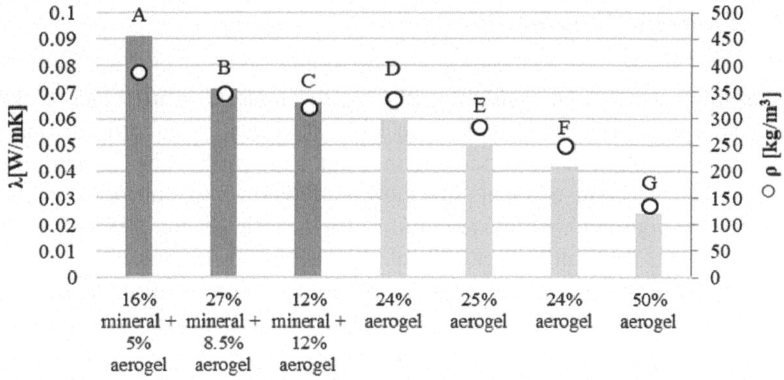


**Fig. 22.1** Comparison of the thermal conductivity and density of the different thermal insulating plasters developed

plete replacement of the mineral lightweight aggregate with a full aerogel aggregate formulation was done. Further, formulations were then developed by increasing the aerogel content to 24–25% in mass (D-E-F). The difference in the density as well as in the thermal conductivity of the three mixtures depends on the different aerogel particle size and distribution. Although the use of 24–25 m% of aerogel as lightweight aggregate determines an important reduction of the thermal conductivity (down to 0.042 W/mK, Fig. 22.1) the results achieved did not fulfil the project target value of 0.030 W/mK. For this reason, a new formulation (sample G) was developed, starting from the same particle size of sample F, which had shown a better performance if compared with the previous mixtures (D-E). Limiting the aerogel content to ~50% aerogel in mass (a good compromise considering the main aspects involved, as thermal properties, mechanical properties and cost) it was possible to reach the project target value, achieving the lowest thermal conductivity of ~0.024 W/mK (Fig. 22.1) and the lowest dry-bulk density (~139 kg/m$^3$).

## 22.3 Experimental Characterization Methods

Several standardized methods must be adopted to characterize the hygrothermal properties of insulating plasters; thus, in the following paragraphs, a summary of the methodologies is reported for each physical property to be measured.

The presented experimental tests allow to obtain the minimum required data to simulate the combined effect of the heat and the water vapour transmission. Nevertheless, to allow the effect of liquid water transport to be analysed, an additional liquid water absorption test, according to EN 1015-18:2004 [11], should be performed.

### *22.3.1 Thermal Conductivity*

To be classified as thermal insulating plaster, the material must reach a thermal conductivity lower than 0.2 W/mK. The EN 12667:2001 standard [8] has to be followed to measure the thermal performance since the product has a medium–high thermal resistance. The instruments that can be adopted for this purpose are the heat flux meter or the guarded hot plate apparatus. The heat flux meter (HFM, Fig. 22.2a) presents two plates that can be heated/cooled at different temperatures.

The sample size has to be in accordance with the size of the instrument adopted and its measurement area. Moreover, the standard reports the maximum and minimum thickness that the sample must have, according to the overall size and to the thermal conductivity value. In particular, the sample must have a thickness $\leq$8 times the side size (a prismatic sample of 40 $\times$ 40 $\times$ 5 cm has been adopted). In case of rigid products, as the thermal plasters, the samples must have smooth, plane and parallel surfaces. Furthermore, due to the irregular surfaces of the specimens and to reduce the contact resistance between the plates and the sample, contact sheets (e.g. rubber mats) must be placed at the interface. The surface temperature for the thermal conductivity calculation should be measured by means of external thermocouples placed between the rubber mats and the plaster specimen.

**Temperature-dependent thermal conductivity**.
To measure the thermal conductivity of the material, sufficient high heat flux through the samples must be generated. This can be achieved, imposing a temperature difference between the HFM plates in the range of 10 K to 50 K. In order to determine if the temperature can affect the thermal conductivity, the $\lambda$-measurements can be performed using at least two different average temperatures between the plates (e.g. 10 °C and 40 °C). From the heat fluxes and the temperatures, measured by the HFM, the equivalent thermal conductivity of the specimens (Eq. 22.1) can thus be determined as follows:

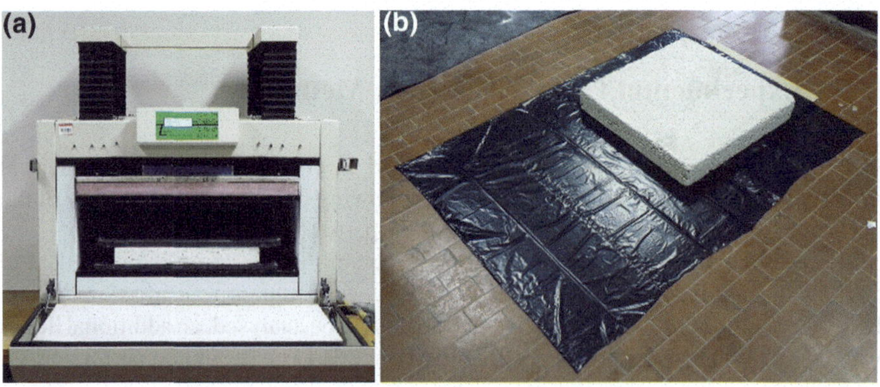


**Fig. 22.2** **a** HFM for thermal conductivity measurement; **b** sealing operation of a plaster sample

$$\lambda = \left(\frac{s \cdot \phi}{\Delta\theta}\right) \quad (22.1)$$

λ: equivalent thermal conductivity [W/mK], s: sample thickness [m], ϕ: specific heat flux [W/m$^2$], Δθ temperature difference between the plates [°K].

**Moisture-dependent thermal conductivity**.
The moisture content can affect the thermal performance of the plasters. Therefore, the thermal conductivity measurement must be repeated starting from the dried specimens and then, step by step, moving to various cases in which the sample is tested at its equilibrium moisture content with increasing relative humidity.

For the λ measurement in dry conditions, the samples are previously dried at 105 °C until the constant mass is reached. For the measurement performed on the moist material, the sample has to be conditioned in an environment at constant RH level.

Since the measurement can require several hours, before starting the test, in both dry and moist condition, the specimen must be sealed in a vapour-tight envelope to avoid any variation of the moisture content due to the migration of water vapour from the environment to the sample or vice versa.

## 22.3.2 *Specific Heat Capacity*

The specific heat capacity measurements were performed according to the methodology reported in [9] through the adoption of a heat flux meter (Fig. 22.2a).

The HFM measures the heat required to increase the temperature of the samples for a preset temperature difference. A step-up of the temperature between an initial set point (T1, e.g. 15 °C) to the final set point (T2, e.g. 25 °C) is applied to both the HFM plates. The HFM thus measures the heat fluxes exchanged with the lower and the upper plate during this transient phase (Fig. 22.3). The total thermal energy stored in the sample is finally obtained as the integral of the net heat fluxes:

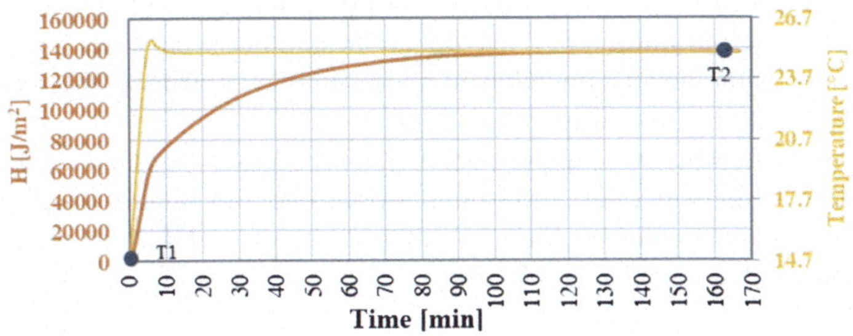


**Fig. 22.3** The heat per unit of the area required by the sample to increase the temperature from T1 to T2

$$H = \sum_{i=1}^{N} [\phi_{upper} + \phi_{lower}]\tau \left[\frac{J}{m^2}\right] \quad (22.2)$$

From the H value, the specific heat, $c_p$ [J/(kg K)], can be calculated by means of the following Eq. (22.3):

$$c_p = \frac{H}{\Delta T \cdot s \cdot \rho} \quad \left[\frac{J}{kg \cdot K}\right] \quad (22.3)$$

where: Δt is the difference between the first and the second set point temperature [°C]; s is the sample thickness [m]; ρ is the sample density [kg/m$^3$].

The specimens have to be preliminarily conditioned in an oven and then sealed in an envelope in the same way as for the λ measurement. Moreover, even for this test, the rubber mats should be adopted and placed at the interface between the instrument plates and the sample. In such a case, it is necessary that the heat capacity of the rubber sheets has to be subtracted from the value obtained for the assembly (sample + rubbers).

### *22.3.3 Water Vapour Diffusion Resistance Factor*

The water vapour permeability test can be performed according to the wet cup test UNI EN ISO 12572:2016 [12] that allows characterizing buildings products and materials under isothermal conditions.

The samples (dimensions of 12 × 12 × 3 cm) were preliminarily stored in a controlled environment (23 ± 1 °C of temperature and 50 ± 5% of relative humidity) and have to be kept in such conditions for a time period long enough to have a variation of less than 5% of weight over five consecutive weighing.

For the test, a saturated solution of water and potassium nitrate ($KNO_3$) was prepared. The aqueous solution allows reaching a relative humidity (RH) of ~94% inside the cup. The solution is then placed in the cup with a minimum depth of 15 mm. The sample must be sealed on the cup with an appropriate hydrophobic material (e.g. silicon, bee wax, aluminium tape), leaving an air gap between water and specimen of 15 ± 5 mm. Then the specimen has to be stored in a controlled environment (23 °C and 50% of relative humidity) (Fig. 22.4).

To determine the water vapour permeability, it is necessary to calculate the average mass flow rate of the water vapour through the sample; this quantity is given by the ratio between the mass change (Δm1, 2) and the time interval during which the mass change occurred that is given as follows:

$$\Delta \dot{m}_{1,2} = \frac{m_2 - m_1}{t_2 - t_1} \left[\frac{kg}{s}\right] \quad (22.4)$$

**Fig. 22.4** The cup filled with the saturated solution, the tested specimen and the sealing operation

The water vapour permeability δ of the material is calculated as follows:

$$\delta = d\frac{G}{A \cdot \Delta p}\left[\frac{kg}{m \cdot s \cdot Pa}\right] \tag{22.5}$$

where G is the water vapour flow rate through the sample (the mean of five successive determination of $\Delta\dot{m}_{1,2}$), d is the sample thickness [m], A is the area of the specimen [$m^2$] and Δp is the water vapour pressure difference across specimen [Pa].

Finally, the water vapour diffusion resistance factor (μ) can be determined through Eq. 22.6:

$$\mu = \frac{\delta_{air}}{\delta} \tag{22.6}$$

where $\delta_{air}$ is the water vapour permeability of air.

### 22.3.4 Hygroscopic Sorption/Desorption

To simulate the effect of the moisture storage in building components, the hygroscopic sorption properties must be determined according to UNI EN ISO 12571:2013 [10].

The samples must be exposed to an increasing or decreasing humidity level with a constant temperature of 23 °C. When the equilibrium is reached at each humidity level, the moisture content (mass by mass, u) can be assessed. By repeatedly increasing or decreasing (step by step) the relative humidity, different values of u can be assessed and, finally, the adsorption or desorption isotherm curve can be obtained by plotting the equilibrium moisture content value against the relative humidity.

The specimen has to be representative of the material and, if the density is less than 300 kg/m$^3$, shall have an exposed area of almost 100 × 100 mm.

The adsorption curve measurement can be performed using a climatic chamber. For this measurement, a preliminary drying phase of the specimens is required. The

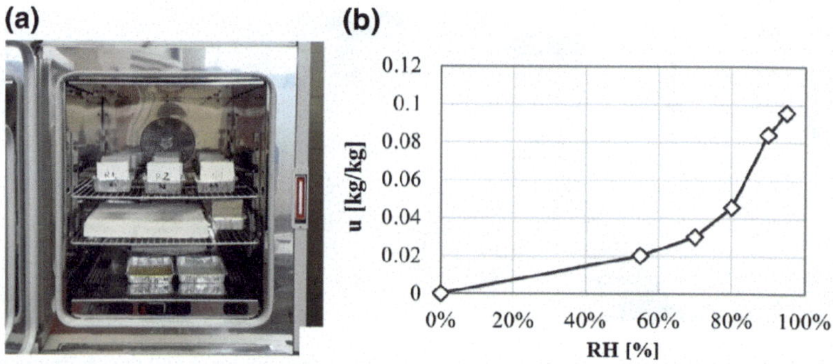


**Fig. 22.5** **a** Sample conditioned in the climatic chamber; **b** an example of adsorption curve for the thermal plaster (F)

duration of this phase has to be prolonged until a constant mass is reached (according to ISO 12570). Then the specimen must be placed in the climatic chamber, and the humidity level shall be set at the lower level selected for the test. A minimum of four RH levels (varying between 30 and 95%) is required. The mass is considered constant when the change in mass between three consecutive weighings is less than 0.1% of the overall sample mass. When the constant mass is reached, it is thus possible to switch to the next and higher RH level. In the case of desorption curve, the first measurement point has to start with the higher relative humidity value (i.e. 95% or sample in free water saturation condition) and then switching to a lower value until 35% of RH is reached. The equilibrium moisture content for every relative humidity level (u) is calculated according to Eq. (22.7):

$$u = \frac{m - m_0}{m_0}\left[\frac{kg}{kg}\right] \tag{22.7}$$

After the collection of at least four points at different RH value, the adsorption/desorption curve can be drawn plotting u value against RH, as shown in Fig. 22.5b.

## 22.4 The Hygrothermal Properties Datasheet

The previously reported standards allow obtaining an exhaustive overview of the hygrothermal properties of thermal insulating plaster. Generally, all the heat and moisture transfer simulation software (e.g. WUFI® [13], Delphin [14]) requires the definition of all these properties for an accurate hygrothermal simulation. In Fig. 22.6, an example of technical datasheet is reported with the aim of simplifying the filling phase on the software schedule and also to have a summarized, but a complete overview, of the material properties.

| Plaster F | | | | | |
|---|---|---|---|---|---|
|  | General information | | | | |
| | Binders | | Mineral, < 0.5% organic | | |
| | Aggregates | | Aerogel | | |
| | Other compound | | (e.g. fibres) | | |
| | Water ratio | | 1:1 | | |
| | Application | | Internal | | |
| | Other | | | | |
| Technical data | | | | | |
| | Value | | u.m. | | |
| $\rho_{dry}$ | 337 | | kg/m³ | | |
| $\lambda_{10,dry}$ | 0.060 | | W/mK | | |
| cp | 988 | | J/kgK | | |
| μ | 8 | | - | | |
| $C_m$ | n.a. | | kg/(m²min$^{0.5}$) | | |
| Thermal conductivity as function of temperature [a] | | | | | |
| $T_{average}$ [°C] | 10 | 25 | 40 | … | |
| λ [W/mK] | 0.060 | 0.061 | 0.062 | .. | |
| Thermal conductivity as function of moisture content [a] | | | | | |
| m.c. [kg/m³] | 0 | … | … | 240 | |
| λ [W/mK] | 0.060 | … | … | 0.176 | |
| Adsorption curve [b] | | | | | |
| R.H. [%] | 55% | 70% | 80% | 90% | 95% |
| u [kg/kg] | 0.0203 | 0.030 | 0.046 | 0.084 | 0.096 |
| (a) At least 2 point are required; | | | (b) At least 4 point are required | | |

**Fig. 22.6** Example of a datasheet for material hygrothermal simulation

## 22.5 Conclusions

The study shows the development process of a new high-performance aerogel-based internal plaster and a synthesis of the tests to be performed to obtain a comprehensive hygrothermal characterization of the materials, that is useful to perform heat and moisture transfer simulations.

Increasing the aerogel content to a 50% (in mass) allows reaching the project target value ($\lambda \leq 0.03$ W/mK), with thermal conductivity of ~0.024 W/mK.

Furthermore, moisture and temperature-dependent thermal conductivity, specific heat capacity, hygroscopic sorption and water vapour diffusion resistance factor were identified, as the required properties, to perform accurate heat and moisture transfer simulations. For each identified hygrothermal property, the associated test procedure that was adopted for the characterization of the newly developed products was described.

Finally, an example of datasheet containing the minimum required information to fully simulate the hygrothermal behaviour of building insulating materials is reported.

**Acknowledgements** The research project Wall-ACE has received funding from the EU Horizon 2020 research and innovation programme under the Grant Agreement No. 723574. The authors wish to thank the project partner Vimark srl, which provided the specimens to be tested in the PoliTO labs and ENERSENS that provided the Kwark® aerogel for the development of the aerogel-based insulating plaster.

## References

1. Cascione, V., Marra, E., Zirkelbach, D., Liuzzi, S., Stefanizzi, P.: Hygrothermal analysis of technical solutions for insulating the opaque building envelope. In: 72nd Conference of the Italian Thermal Machines Engineering Association, ATI2017, 6–8 September 2017, Lecce, Italy
2. Mumovic, D., Ridley, I., Oreszczyn, T., Davies, M.: Condensation risk: comparison of steady-state and transient methods. Building Serv. Eng. Res. Technol. **273** (2006)
3. Fantucci, S., Fenoglio, E., Serra, V., Perino, M.: Coupled Heat and Moisture Transfer Simulations on Building Components Retrofitted with a Newly Developed Aerogel-Based Coating, Building simulation 2019 (in preparation)
4. Maia, J., Ramos, N.M.M., De Freitas, V.P., Sousa, Â.: Laboratory tests and potential of thermal insulation plasters. In: 6th International Building Physics Conference, IBPC 2015
5. Liuzzi, S., Rubino, C., Stefanizzi, P., Petrella, A., Boghetich, A., Casavola, C., Pappalettera, G.: Hygrothermal properties of clayey plasters with olive fibers. Constr. Build. Mater. **158**, 24–32 (2018)
6. Pavlík, Z., Forta, J., Pavlíková, M., Pokornya, J., Trník a, A., Cerny, R.: Modified lime-cement plasters with enhanced thermal and hygric storage capacity for moderation of interior climate. Energy Build. **126**, 113–127 (2016)
7. Mazhoud, B., Collet, F., Pretot, S., Chamoin, J.: Hygric and thermal properties of hemp-lime plasters. Build. Environ. **96**, 206–216 (2016)
8. EN 12667:2001—Thermal performance of building materials and products—Determination of thermal resistance by means of guarded hot plate and heat flow meter methods—Products of high and medium thermal resistance
9. Tleoubaev, A., Brzezinski, A.: Thermal Diffusivity and Volumetric Specific Heat Measurements Using Heat Flow Meter Instruments for Thermal Conductivity 29/Thermal Expansion 17 Conference
10. EN ISO 12571:2013—Hygrothermal performance of building materials and products—Determination of hygroscopic sorption properties
11. UNI EN 1015–18:2004—Methods of test for mortar for masonry—Determination of water absorption coefficient due to capillary action of hardened mortar
12. ISO 12572:2016—Hygrothermal performance of building materials and products—Determination of water vapour transmission properties—Cup method
13. WUFI: https://wufi.de/en/. Last accessed 01 Mar 2019
14. Delphin, http://bauklimatik-dresden.de/delphin/index.php. Last accessed 01 Mar 2019

# Chapter 23
# Combining Conservation and Visitors' Fruition for Sustainable Building Heritage Use: Application to a Hypogeum

**Benedetta Gregorini, Michele Lucesoli, Gabriele Bernardini, Enrico Quagliarini and Marco D'Orazio**

**Abstract** The exploitation of Building Heritage generally leads to sustainability issues in terms of environmental preservation and tourist enjoyment. When these requirements are not jointly respected, occupancy issues can provoke degradation phenomena on indoor environment (i.e., building materials and surfaces with artistic and historical value) or conditions of discomfort during visitors' fruition. Hence, our research defines a combined strategy to solve at the same time both the issues: guaranteeing the conservation of Building Heritage (and its artefacts) while ensuring optimal visitors' fruition tasks. The Building Heritage conservation is pursued by a monitoring campaign of ideal (undisturbed) indoor conditions and by the evaluation of the human presence effect considering thermal loads as main driver. The visitors' fruition is analyzed by assessing individuals' behavioral patterns in terms of attention given to the hosted artifacts (where and how the visitors' attention is posed?), through a wearable eye tracking system. The strategy is applied to a hypogeum environment characterized by high reliefs on walls and vaults. This scenario is considered since its isolated hygrothermal conditions are strongly influenced by human presence. Results showed that the environmental preservation is reached when considering the fruition model proposed by the stakeholder. Furthermore, the eye tracking analysis revealed high-level of visitors' engagement towards significant spaces only when exposed to adequate lighting conditions and/or in a good conservation state.

## 23.1 Introduction

Developing strategies and solutions to guarantee sustainable use and exploitation of the Building Heritage is becoming an important and challenging issue. Such strategies are addressed to contribute to environmental protection of the Heritage while leading to positive socio-economic impacts, especially in case of tourist attractions. The desired outcome of a sustainable touristic use is that resources will be managed in such

B. Gregorini (✉) · M. Lucesoli · G. Bernardini · E. Quagliarini · M. D'Orazio
Department of Construction, Civil Engineering and Architecture, Università Politecnica Delle Marche, via Brecce Bianche, 60131 Ancona, Italy
e-mail: b.gregorini@pm.univpm.it

 
J. Littlewood et al. (eds.), *Sustainability in Energy and Buildings*, Smart Innovation, Systems and Technologies 163, https://doi.org/10.1007/978-981-32-9868-2_23

a way that economic, social, and aesthetic needs can be fulfilled while maintaining cultural integrity and touristic enjoyment [1]. In this way, sustainable strategies have to: (1) guarantee the conservation of the goods/building; (2) guaranteeing the use and exploitation of the buildings by allowing people to have an effective experience of them.

Regarding the Built Environment preservation, it can be generally pursued by reducing and contrasting critical conditions in building heritage use such as the ones due to visitors' access and activities [2]. In fact, tourists' massive and/or permanent presence inside the architectural spaces may cause the variation of indoor environment parameters. In particular, hygroscopic buildings materials, such as wood and stones surfaces (which can generally have, in such scenarios, artistic and historical values), can suffer water absorption and desorption and consequential biological deterioration phenomena from visitors' presence-induced hygrometric fluctuations, while, inadequate air temperature and heat transfer may cause expansion, drying up and increasing of the fragility of surface layers of the materials [3].

Nowadays, current solutions for the environmental protection of such elements are mostly technological ones [4, 5] and they can be mainly distinguished between supply side management (include building electrical system retrofits, such as HVAC implant [6], and the use of renewable energy) and demand side management, aimed at reducing building heating and cooling demand through the use of energy efficient equipment and low energy technologies, by essentially providing building fabric insulation and windows retrofit [7, 8]. However, if on one hand these solutions are an effective opportunity for environmental protection on common existing buildings, they could not be applied to historical architectures needing to preserve their key testimonial knowledge into the society [9].

Anyway, while considering the impact of visitors' presence in the Building Heritage environment, defining solutions starting from users' behaviors and impact in the environment plays a key role in the environmental protection and the conservation of the building itself. Research studies have pointed out that changes in behavioral variables, especially heating temperature settings, may have stronger effects on environmental protection than technological solutions [10]. In particular, guidelines for visitors' management (in terms of staff education and public awareness of preventive conservation strategies) have been developed in the last years in relation to museums, with the goal to develop sustainable Building Heritage uses and exploitation while maintaining ideal conditions for the hosted goods and chattels [2].

Such visitor's management actions should be combined to key point two related strategies in order to define the best presence timing in respect to the changes of environmental conditions [11]. In this sense, the definition of user's behaviors in such scenarios can play a key role in managing visitors' flows and tour for Building Heritage. Recent studies have pointed out that eye tracking technology can evaluate human perception of hosted goods and chattels, since they can detects where the visitor focuses his/her attention (where is looking at) [12, 13]. The application of such technology to museum environments may help the stakeholders to evaluate fruition time and behavioral/motion patterns in relation to visitors' issues [14]. The fruition patterns are investigated through wearable eye tracking system, letting the visitor to move freely in the environment [14], or through fixed eye tracking system with a visitor seated in front of a painting [15].

This study wants to combine key points 1 and 2 to propose a first step towards sustainable solutions in a particular kind of building heritage: an underground built environment in which preservation issues of building stone surfaces are significant because of its historic and artistical value, while exploitation should be guaranteed [16]. According to previous researches [17], due to the isolation from the outdoor, such spaces are close to saturation with a nearly constant temperature. These specific microclimatic parameters are susceptible to induced alterability in occupancy conditions [18]: the hygrothermal loads due to visitors' presence can alter indoor climate conditions leading to conservation problems because of increasing biocolonization, biodegradation and salt crystallization that cause material loss in walls and artefacts [19].

To this aim, the occupants' (visitors') influence on the environmental parameters (i.e., air temperature and relative humidity) is firstly defined. Then, eye tracking technologies are employed to assess the main drivers guiding the users' attention during the fruition of the hypogeum. According to the case study, a fruition model effectiveness is evaluated and suggestions on how to improve visitors' experience (by, e.g., organizing visitors' paths, by also using technological systems, like lighting systems) are provided.

## 23.2 Phases, Materials, and Methods

This work is divided in three phases. In the first phase, the visitors-indoor environment parameters relationships for Building Heritage conservation purposes are considered: a monitoring system is defined and the methods to test the visitors' influence on environmental parameters are presented (according to methods of Sect. 23.2.1). The second phase defines and discusses both the technology and the calculation method to test the visitor's fruition patterns, by an evaluation of how long the visitors' look at the artistical and historical valued elements (according to methods of Sect. 23.2.2). The third phase concerns the application to the case study, which is an historical hypogeum (Osimo, Italy). After defining the fruition models according to the stakeholder, in Sect. 23.3, the model is tested in the hypogeum environment by using the methods defined in phases 1 and 2, so as to understand whether the human presence disturbs indoor conditions and if the proposed model guarantees tourists' enjoyment and engagement.

### 23.2.1 Visitors' Influence on Environmental Parameters: Systems and Methods

The monitoring system structure is settled according to previous researches combining requirements for environmental protection of building heritage spaces to the sustainability and low impact of wireless networks [20].

The requirements and application methods for the indoor environment characterization are addressed by international regulations for cultural heritage monitoring [21–24] by considering: the measurement, at least, of environmental temperature and relative humidity; the application of the system by guaranteeing a complete spatial and temporal description of the indoor conditions (both for undisturbed and visiting conditions). Moreover, recent studies show that the application of Wireless Network to Building Heritage spaces, being non-invasive, mimetic, and long lasting, is particularly well suited for long-term monitoring and online diagnosis of the conservation state [25, 26].

The monitoring system proposed by [20] can be adopted since it accomplishes such issues. From a general point of view, this system is controlled by a central elaboration unit that storages and processes monitoring data. Thanks to the wireless connection, it can be placed in the staff office, away from visitors' access. The base stations (National Instrument MyRIO-1900—http://www.ni.com/it-it/support/model.myrio-1900.html, last access: March 2018) collect raw data of temperature and relative humidity and monitoring devices, provided with combined temperature and relative humidity sensors (Sensirion SHT-31D—https://www.adafruit.com/product/2857, last access: March 2018), directly measures these environmental parameters. The monitoring sensors are positioned at 1.50 m high from the ground alongside tourist attraction points, where visitors stop more frequently and for a longer time. Monitoring devices and base stations are linked through flat cables connection: these guarantee a low impact on the heritage environment. The application of the system in the chosen scenario is described in Sect. 23.2.3.

Methods for testing the influence of the visitor's presence are defined according to conservation strategies that involve the users' behaviors. In addition to the two control algorithms presented in [20], that set the acceptability range for temperature and relative humidity under undisturbed conditions, a third algorithm for visiting conditions is defined According to conservation strategies that involves human behaviors model, the proposed visiting model should not cause unacceptable alterations in one of the critical environmental parameters in comparison to the undisturbed conditions [27]. The chosen critical environmental parameter is air temperature because of sensible heat load due to visitors' presence. In addition, air temperature is a reliable variable, easily measured, and it gives a clear response to change in underground environment [28]. The proposed algorithm imposes that a visiting use can be considered sustainable only when the thermal variation is limited within the natural fluctuation of the environment. Equation 23.1 shows the analytical calculation of the algorithm where $T_{vu}$ is the registered temperature in a visiting use (°C), $T_{min}$ is the minimum temperature (°C) and $T_{max}$ corresponds to the maximum temperature (°C), both evaluated under undisturbed conditions:

$$T_{min} < T_{vu} < T_{max} + \varepsilon_T \tag{23.1}$$

For the upper bound, the instrumental error of temperature sensors $\varepsilon_T$ is added only when it is comparable to the temperature range. It has no significance in the lower bound, since thermal variation is only due to added human thermal loads. Moisture loads are not considered since relative humidity is still at saturation point [17].

### 23.2.2 Visitors' Behavior and Fruition: Systems and Methods

The eye tracking system tested in this study is the Tobii Pro Glasses 2 (https://www.tobiipro.com/siteassets/tobii-pro/brochures/tobii-pro-glasses-2-brochure.pdf/?v=6, last access: March 2018), with improved Pupil Centre Corneal Reflection (PCCR) technique. This type of technology is chosen because, being a wearable eye tracker with four eye cameras and a full HD wide angle scene camera placed in the head unit (glasses), the user is allowed to unrestrictedly move around the spaces. Moreover, the recoding unit saves the eye tracking data, in terms of fixation location, to an SD card and it communicates via wireless with the controller software, by allowing a simple use inside each part of the building. The gaze sampling frequency is 50 Hz and, according to previous studies, it can be considered adequate to visiting use purpose [12]. The eye tracking system was used in combination with both its Controller Software and its Analysis Software Tobii Pro Lab. The Controller Software, installed on the central unit of the monitoring system, allows the calibration of the tool, the start and stop recording and the real time monitoring of the data, while the Tobii Pro Lab software provides tools for post-analysis of recorded data (i.e., fixation data export).

The fixation of one object (or one space section) is a measure of visitor's engagement (how long the visitor stares at the object) [13] and is calculated as in Eq. (23.2):

$$Object\ Perception\ (\%) = \frac{\sum Gaze\ Event\ Duration}{Maximum\ Gaze\ Event\ Duration} \tag{23.2}$$

where the sum of Fixation Duration (ms) is the sum of fixation duration for a single object and the Maximum Fixation Duration (ms) is the maximum fixation duration among all objects. 560 ms was taken as a lower bound for the minimum duration of a fixation: according to previous studies, this is the minimum time for a significative visual fixation [29]. This equation allows to determine the engagement of visitors towards the different space areas.

### 23.2.3 Case Study Application

Palazzo Campana underground built environment was chosen for the testing campaign of the proposed model for a sustainable use of the Building Heritage. The hypogeum environment was dug in building subsoil, mainly composed of sandstone. Thanks to this type of stone, in the main corridor and in the circular hall, the walls, vaults and dome were engraved by artistic-valued high reliefs. Through the years, the hypogeum has been visited by numerous and non-regulated groups of visitors comported rapid thermal fluctuation causing a severe state of degradation.

The suitable fruition model was determined in accordance with Palazzo Campana stakeholder. The chosen fruition model was set on a group of 10 people for a 10 min maximum visiting time.

Before the tests started, the monitoring system was set in the hypogeum environment and it measured air temperature and relative humidity under undisturbed conditions (i.e., people were not allowed to enter, the lighting system was switched off) for a significant period of time. The system was applied according to its definition requirements shown in Sect. 23.2.1. As shown in Fig. 23.2, four monitoring sensors were placed in the main corridor alongside the main high reliefs. The monitoring devices were linked to two base stations and the central unit was sheltered from the direct visitors' fruition and placed in the staff office, near above the entrance of the hypogeum. Sampling time was set up to 15 min, also according to current building heritage monitoring regulations. For undisturbed conditions, the system defined the maximum and minimum temperature ($T_{max}$ and $T_{min}$), according to Eq. (23.1).

After the week, the test for the visiting conditions was performed using the fruition model suggested by the stakeholder. The visiting test was repeated for three times, spacing each visit with a break of 10 min. This helped the hypogeum to re-establish the undisturbed conditions. The system set up was maintained unaltered apart from the sampling time: it was reduced to 30 s to obtain a more detailed trend of the thermal loads due to human presence. The control algorithm presented in Eq. (23.1) was run by the system evaluating the registered temperature $T_{vu}$ during the visit use. At the end of each visit, all the 30 people filled out a survey. The question was "was the time adequate to appreciate the environment (e.g., spatial development, figures and details)?" with a likelihood answer (in a scale from 1 to 5). This tool permitted to investigate whether the time of visit was adequate or not.

The eye tracking test was performed in order to evaluate the fruition patterns inside the hypogeum environment by following these steps: (1) Outside the hypogeum, the visitor wore the eye tracking glasses and the calibration on pupil movement process started; then, the visitor could enter the hypogeum; (2) At the starting point of view (see Fig. 23.1), the eye tracking system started recording and the visitor was allowed to move freely inside the corridor of the hypogeum; (3) At the end of the visit, the visitor reached the ending point of view (see Fig. 23.1) and the eye tracking system stop recording.

The test was performed 10 times with 10 different visitors: only one visitor per time was allowed to enter in order to focus visitor's attention on the scene. The

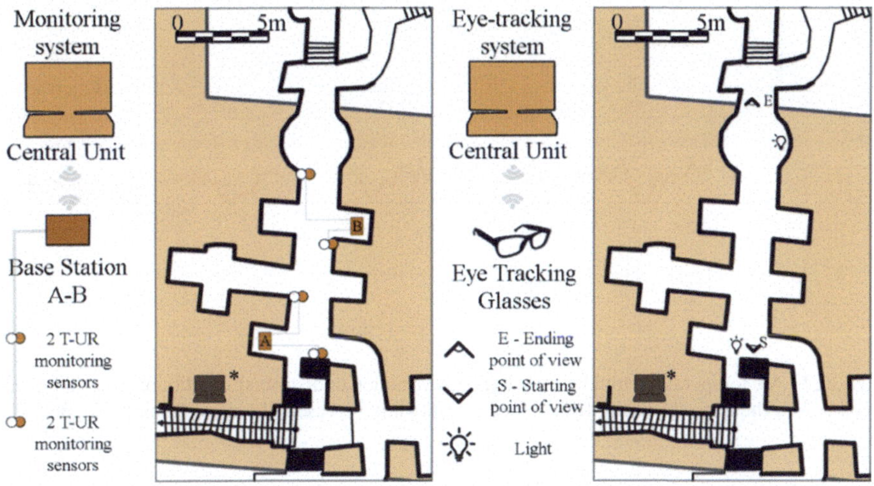


**Fig. 23.1** Case study application: left: monitoring system main structure and its application to in the hypogeum environment; right: eye tracking system main structure and its application

hypogeum environment was divided according to these space sections: walls high reliefs, vaults high reliefs and recess for both the main corridor and the circular hall. The lighting system already set in the environment was not modified but, according to initial setup experiments performed in different parts of the hypogeum, the eye trackers seems not to succeed in measuring data in case of light intensity lower than 3 lux.

## 23.3 Results

Figure 23.2 shows the air temperature of the main corridor under undisturbed conditions: it was mainly constant over the week and it varied between 15.92 and 16.18 °C. The vertical bars indicate the instrumental error ($\varepsilon_T = 0.3$ °C) and it was added since it is comparable to the natural variation. Relative humidity was constantly at saturation point.

As shown in Fig. 23.3, during the tested visiting conditions: the air temperature varied in between the range set in accordance with Eq. 23.1, during all the three visits. Moreover, the results of the survey expressed an average likelihood level equal to 4.36 and it confirmed that the visiting time was adequate for the touristic enjoyment.

From the eye tracking results, it is evident that the object perception, and therefore the fruition patterns of visitors, was primarily influenced by the light intensity and the position of the lighting fixtures (Fig. 23.4). The object perception in terms of time was longer where the light intensity was higher: the average object perception in the circular hall was 30% with a light intensity at the ground equal to 60 lux, while in the

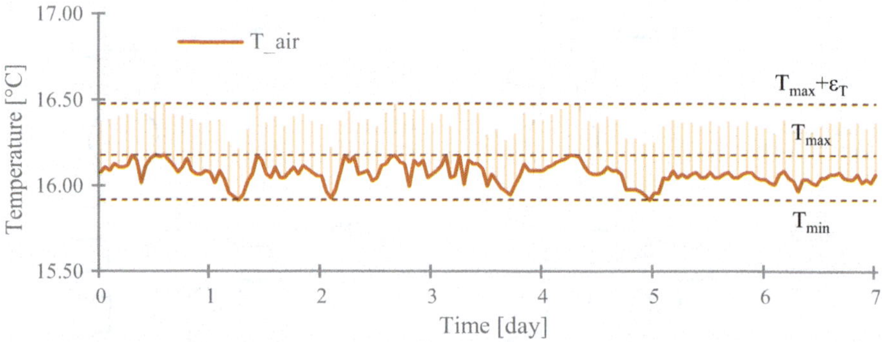


**Fig. 23.2** Air temperature trend in the hypogeum under undisturbed conditions. Results of the week before the tests are showed since they significantly represent the whole monitoring period

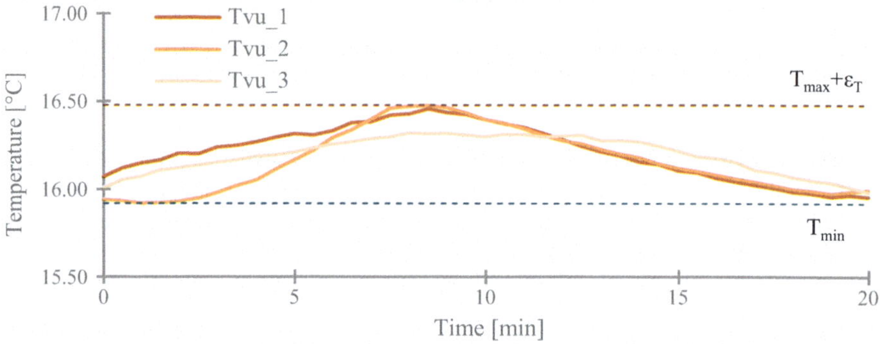


**Fig. 23.3** Air temperature trend of the hypogeum under visiting use conditions, by showing the limits of undisturbed conditions, which are never reached during the visiting time

corridor, the average object perception was about 20% with 30 lux of light intensity. Moreover, the visitors' perception was encouraged when the lighting fixture created strong effects of light and shadows, enhancing the depth of the relief. In fact, the object perception was 90% for the high relief (b) while, for high relief (a), it was only 65%.

Indeed, where constant lighting conditions of the surfaces (both vaults and walls) occurred, the conservation state of the figures seems to influence the visitors' perception. In fact, in the main corridor, the more observed space section was the high relief (c), with 86%, while the less observed figure was the high relief (d) with an object perception of 5%. This was probably due to the strong deterioration phenomena that affect these figures. In general, the recesses fixation was lower than 20%.

**Fig. 23.4** Fixation patterns by including the views of high reliefs

## 23.4 Discussion

Hypogeum environment are mainly characterized by specific microclimatic conditions: by the isolation with the outdoor weather, temperature is kept constant over the time and relative humidity is nearly to saturation [17]. This work confirms this trend for both the physical quantities through the monitoring of the undisturbed conditions (see Sect. 23.2.1). Moreover, according to previous studies [2, 18, 28], visitors presence can alter the indoor environment due to hygrothermal load induced by human activities: the monitored temperature during the visited conditions increased when the visitors were inside the hypogeum and it decreased when the environment was empty (see Sect. 23.3). Relative humidity was permanently at saturation. Regarding the preservation of the Building Heritage environment, it is generally pursued by reducing and contrasting critical conditions [2]. From the result obtained in this work, it is evident that the visiting conditions proposed by the stakeholder can limit the increase of temperature within the natural fluctuation bounds.

Furthermore, as for other literature works [12–14], the wearable eye tracking system allowed to evaluate fruition time and behavioral/motion patterns in relation to visitors' issues [14]. In the hypogeum environment the lighting conditions strongly influenced the object perception, while, the object conservation state of was the second main driver of the fixation.

Future works could extend this strategy to other case scenario of Building Heritage (e.g., museum, exhibition spaces, churches) that need an optimization of tourist fluxes in terms of both environmental condition and fruition patterns.

## 23.5 Conclusions

This work provides a sustainable strategy for touristic purposes that both guarantees the Heritage conservation and its touristic exploitation. A visitors' fruition model, proposed according to the stakeholder, is tested in the considered case study. The sustainability aims in terms of environmental preservation are reached by evaluating the effect of thermal load due to visitors' occupancy in respect to the undisturbed indoor environmental conditions. About exploitation tasks, the eye tracking analysis shows a general high level of engagement of people towards the high reliefs, while lighting system seems to play a key role in visitors' fruition patterns.

## References


1. Pedersen, A.: World Heritage Manuals 1—Managing Tourism at World Heritage Sites: A Practical Manual for World Heritage Site Manager. UNESCO World Heritage Centre, Paris (2002)
2. Lucchi, E.: Review of preventive conservation in museum buildings. J. Cult. Herit. **29**, 180–193 (2018)
3. Litti, G., Audenaert, A., Braet, J.: Energy retrofitting in architectural heritage, possible risks due to the missing of a specific legislative and methodological protocol. In: Proceedings of the European Conference on Sustainability, Energy and Environment (2013)
4. Ma, Z., Cooper, P., Daly, D., Ledo, L.: Existing building retrofits: methodology and state-of-the-art. Energy Build. **55**, 889–902 (2012)
5. Cabeza, L.F., de Gracia, A., Pisello, A.L.: Integration of renewable technologies in historical and heritage buildings: a review. Energy Build. **177**, 96–111 (2018)
6. Bonacina, C., Baggio, P., Cappelletti, F., Romagnoni, P., Stevan, A.G.: The Scrovegni Chapel: the results of over 20 years of indoor climate monitoring. Energy Build. **95**, 144–152 (2015)
7. Sciurpi, F., Carletti, C., Cellai, G., Pierangioli, L.: Environmental monitoring and microclimatic control strategies in "la Specola" museum of Florence. Energy Build. **95**, 190–201 (2015)
8. Roberti, F., Oberegger, U.F., Lucchi, E., Troi, A.: Energy retrofit and conservation of a historic building using multi-objective optimization and an analytic hierarchy process. Energy Build. **138**, 1–10 (2017)
9. Mazzarella, L.: Energy retrofit of historic and existing buildings. The legislative and regulatory point of view. Energy Build. **95**, 23–31 (2015)
10. Lidelöw, S., Örn, T., Luciani, A., Rizzo, A.: Energy-efficiency measures for heritage buildings: a literature review. Sustain. Cities Soc. **45**, 231–242 (2019)
11. Lobo, H.A.S.: Tourist carrying capacity of Santana cave (PETAR-SP, Brazil): a new method based on a critical atmospheric parameter. Tour. Manag. Perspect. **16**, 67–75 (2015)
12. Cantoni, V., Merlano, L., Nugrahaningsih, N., Porta, M.: Eye tracking for cultural heritage. In: Proceedings of the 17th International Conference on Computer Systems and Technologies 2016—CompSysTech'16, pp. 307–314 (2016)
13. Eckstein, M.K., Guerra-Carrillo, B., Miller Singley, A.T., Bunge, S.A.: Beyond eye gaze: what else can eyetracking reveal about cognition and cognitive development? Dev. Cogn. Neurosci. **25**, 69–91 (2017)
14. Mokatren, M., Kuflik, T., Shimshoni, I.: Exploring the potential of a mobile eye tracker as an intuitive indoor pointing device: a case study in cultural heritage. Futur. Gener. Comput. Syst. **81**, 528–541 (2018)
15. Bauer, D., Stofer, K.: Capturing visitors' gazes : three eye tracking studies in museums. (2015)

16. Galeazzi, C.: The typological tree of artificial cavities: a contribution by the commission of the Italian Speleological Society. Opera Ipogea—J. Speleol. Artif. Cavities. **1**, 9–18 (2013)
17. Scatigno, C., Gaudenzi, S., Sammartino, M.P., Visco, G.: A microclimate study on hypogea environments of ancient roman building. Sci. Total Environ. **566–567**, 298–305 (2016)
18. Hoyos, M., Canaveras, J.C., Sanchez-Moral, S., Sanz-Rubio, E., Soler, V.: Microclimatic characterization of a karstic cave: human impact on microenvironmental parameters of a prehistoric rock art cave (Candamo Cave). Environ. Geol. **33**, 231–242 (1998)
19. Vereecken, E., Roels, S.: Review of mould prediction models and their influence on mould risk evaluation. Build. Environ. **51**, 296–310 (2012)
20. Stazi, F., Gregorini, B., Gianangeli, A., Bernardini, G., Quagliarini, E.: Design of a smart system for indoor climate control in historic underground built environment. Energy Procedia. **134**, 518–527 (2017)
21. UNI 10829: Works of art of historical importance—Ambient conditions for the conservation—Measurement and analysis (in Italian) (1999)
22. UNI EN 15757: Conservation of Cultural Property—Specifications for temperature and relative humidity to limit climate-induced mechanical damage in organic hygroscopic materials (2010)
23. UNI EN 15758: Conservation of Cultural Property—Procedures and instruments for measuring temperatures of the air and the surfaces of objects (2010)
24. UNI EN 16242: Conservation of Cultural Heritage—Procedures and instruments for measuring humidity in the air and moisture exchanges between air and cultural property (2013)
25. Mecocci, A., Abrardo, A.: Monitoring architectural heritage by wireless sensors networks: San Gimignano—a case study. Sensors **14**, 770–778 (2014)
26. Martínez-Garrido, M.I., Fort, R.: Experimental assessment of a wireless communications platform for the built and natural heritage. Measurement **82**, 188–201 (2016)
27. Camuffo, D.: Microclimate for Cultural Heritage. Elsevier, Amsterdam (1998)
28. Calaforra, J.M., Fernández-Cortés, A., Sánchez-Martos, F., Gisbert, J., Pulido-Bosch, A.: Environmental control for determining human impact and permanent visitor capacity in a potential show cave before tourist use. Environ. Conserv. **30**, 160–167 (2003)
29. Land, M.F., Hayhoe, M.: In what ways do eye movements contribute to everyday activities? Vision. Res. **41**, 3559–3565 (2001)

# Chapter 24
# Energy Savings and Summer Thermal Comfort for Retrofitted Buildings: A Complex Balance

**Gianpiero Evola, Luigi Marletta and Federica Avola**

**Abstract** In the last years many buildings in Europe have undergone energy retrofit, in order to improve their performance and reduce the energy needs. In most cases, retrofit solutions were based on the application of insulating materials either to the outside walls or to the roof. Such a practice, which undoubtedly improves the winter performance, can also induce non-negligible drawbacks in summer and in other warm months, mostly due to solar gains. The aim of this paper is to study the consequences of a typical retrofit solution, aiming to get the "nZEB" label, on summer thermal comfort. The study is based on dynamic simulations carried out with EnergyPlus on a public residential building, located in the city of Catania, Southern Italy. The analysis is repeated for various values of the insulation thickness, to be applied to the outer face of the external walls. Besides, nighttime ventilation is scheduled in summer at various opening rates of the windows, with the aim to reduce overheating. The results of the simulations may help designers to find the right balance between insulation needs and summer thermal comfort for the retrofit of existing residential buildings.

## 24.1 Introduction

The recent European Directives on Energy Performance of Buildings, implemented in Italy by the Ministerial Decree DM 26/06/2015 [1], have promoted energy efficiency in buildings with the aim to make nearly Zero Energy Buildings (nZEB) compulsory starting from 2021. An nZEB is a building characterized by low energy demand for space heating, space cooling and domestic hot water, which should be to a large extent covered by renewable sources. In Italy, it is possible to label a building as an nZEB if it complies with a series of requisites concerning the thermal performance of its envelope, and if its non-renewable primary energy demand is not higher than for a suitably defined *reference building* [1].

G. Evola (✉) · L. Marletta · F. Avola
University of Catania, DIEEI, 95125 Catania, Italy
e-mail: gevola@unict.it



J. Littlewood et al. (eds.), *Sustainability in Energy and Buildings*, Smart Innovation, Systems and Technologies 163, https://doi.org/10.1007/978-981-32-9868-2_24

In case of energy retrofit of an existing building, it is not always easy to comply with the nZEB standard. Indeed, the insulation of the envelope (walls, roofs, double-glazing windows) allows to reduce heat losses and to improve thermal performance in winter; however, the excessive rate of insulation can easily lead to indoor overheating in summer, especially in hot and humid climatic conditions, like in Southern Italy. This happens because a high thermal resistance of the envelope does not allow releasing heat to outdoors in the nighttime, when the outdoor air temperature is usually lower than the indoor temperature. In this case, as already highlighted in the literature, overheating can produce indoor thermal discomfort and increase the energy demand of the space cooling system [2–4].

In this framework, the main goal of this paper is to show how typical retrofit solutions for the building envelope can have significant drawbacks in terms of summer thermal comfort. This is demonstrated through dynamic energy simulations carried out with EnergyPlus for an existing residential building in Southern Italy. The analysis includes different values of the insulation thickness, and investigates the effectiveness of nighttime ventilation to overcome this problem [5].

The outcomes of this study may help designers find the right balance between insulation and summer thermal comfort for existing buildings to be retrofitted. In fact, this has proven to be a quite difficult task in Southern Europe, as underlined by several ongoing researches [6].

## 24.2 Methodology

As a first step, the energy performance of the selected building has been evaluated through the quasi-stationary calculation model introduced by the UNI Standard 11300-1 [7], which is implemented in several commercial software tools. This step is necessary to understand if the proposed energy retrofit solutions are able to get the nZEB label. Since in this paper attention is only paid to the building performance—and not to the energy systems—the parameters considered at this stage are

- The overall heat transfer coefficient of the envelope, $H'_T$ (in W $m^{-2}$ $K^{-1}$);
- The annual final energy demand for space heating, $EP_{H,nd}$ (in kWh $m^{-2}$);
- The annual final energy demand for space cooling, $EP_{C,nd}$ (in kWh $m^{-2}$);

Then, summer thermal comfort is investigated. In this paper the discussion is based on the values of the operative temperature resulting from dynamic simulations in EnergyPlus 8.6.0. These values are compared to the temperature threshold introduced by the adaptive comfort theory: this was developed by Nicol and Humphreys, and postulates that occupants can react to uncomfortable thermal conditions by adapting their dressing code, by opening the windows and by using shading systems to control solar heat gains [8]. Such adaptive behavior makes occupants accept higher operative temperatures than those suggested by the Fanger's comfort model.

Further information on thermal comfort can be provided by other indicators, such as the *Intensity of Thermal Discomfort* (ITD) and the *Frequency of Discomfort* (FD).

The ITD is defined as the time integral, over the occupancy period P, of the positive difference between the current operative temperature (derived from the simulations) and the upper threshold for comfort [9]:

$$\mathrm{ITD} = \int_P \left(\mathrm{T_{op}}(\tau) - \mathrm{T_{lim}}\right)^+ d\tau. \tag{1}$$

The ITD is an effective way to quantify the uncomfortable thermal sensation due to overheating: indeed, it measures with a single number the intensity and the duration of the thermal discomfort perceived by the occupants. The value of the threshold temperature ($T_{lim}$) corresponds to the upper limit of Category II, as described in the EN Standard 15251:2007 [10]. Category II is suitable when referring to the retrofit of existing buildings. The threshold value is not constant in time and must be calculated on a daily basis as a function of the running mean outdoor air temperature.

On the other hand, the FD represents the length of the period in which the operative temperature overcomes the threshold of Category II, and it is defined as the ratio of the hours of discomfort to the total hours of occupancy. In order to evaluate these indicators, the building has to be simulated in free-running conditions.

## 24.3 Case Study

The case study is a public residential building erected in the 1950s and located in the city of Catania, in Southern Italy (37°28′N, 15°3′E). Here, the climate is warm in winter, as witnessed by the low Heating Degree Days (HDD = 833 °C day), defined relative to a base outdoor temperature of 12 °C. On the other hand, in summer the climate is hot and humid, with peak daily outdoor temperatures frequently exceeding 35 °C. The city is close to the sea and experiences a high diurnal temperature variation (11 °C on average), which is a positive feature to exploit nighttime ventilation.

The building is made up of seven residential levels, while the ground floor hosts commercial activities, and is not addressed in this analysis. Figure 24.1 shows the plan of a typical floor, with the codes used to identify the four apartments (from U1 to U4). The building is 28-m tall and is shaded only on its western side by another tall building. As concerns the outer shell, both the outside walls and the roof are not insulated, as reported in Tables 24.1 and 24.2. The corresponding thermal transmittance is $U = 1.17\ \mathrm{W\ m^{-2}\ K^{-1}}$ for the walls and $U = 1.36\ \mathrm{W\ m^{-2}\ K^{-1}}$ for the roof.

The windows have 4 mm single glazing and aluminum frames without thermal break: hence, they have poor performance, as their overall thermal transmittance is $U = 6\ \mathrm{W\ m^{-2}\ K^{-1}}$. Roller blinds are used to shade all the windows.

A retrofit proposal has then been conceived in order to reach the "nZEB" standard. The retrofit solutions applied to the envelope are

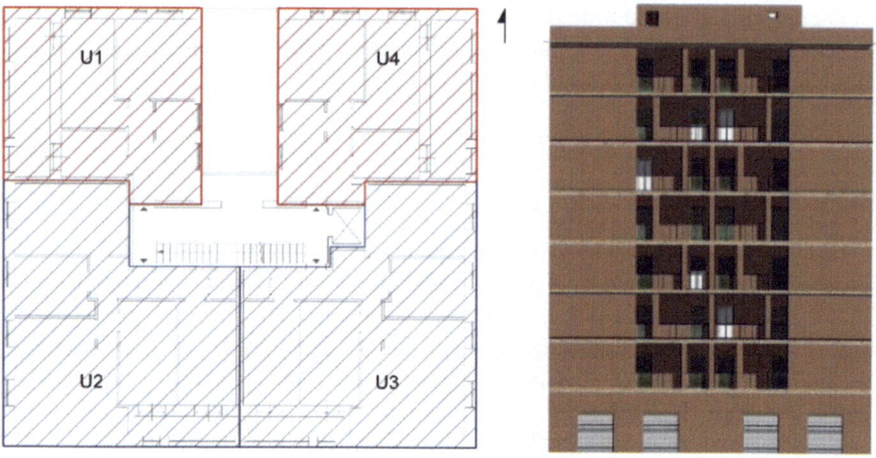


**Fig. 24.1** Building plan for a typical floor and south-facing façade

**Table 24.1** Composition of the external walls (from indoors to outdoors)

| Material | Thickness (mm) | Conductivity (W $m^{-1}$ $K^{-1}$) | Density (kg $m^{-3}$) |
|---|---|---|---|
| Inner gypsum plaster | 20 | 0.70 | 1400 |
| Concrete blocks | 120 | 0.55 | 890 |
| Air gap | 170 | R = 0.15 $m^2$ K/W | 1.2 |
| Concrete blocks | 120 | 0.55 | 890 |
| Outer cement plaster | 30 | 0.90 | 1800 |

**Table 24.2** Composition of the roof (from indoors to outdoors)

| Material | Thickness (mm) | Conductivity (W $m^{-1}$ $K^{-1}$) | Density (kg $m^{-3}$) |
|---|---|---|---|
| Inner gypsum plaster | 20 | 0.70 | 1400 |
| Concrete and hollow blocks | 260 | 0.74 | 1150 |
| Cement screed—bottom layer | 100 | 0.66 | 1200 |
| Cement screed—top layer | 50 | 1.16 | 2000 |
| Cement mortar | 10 | 1.40 | 2000 |
| Tiles | 10 | 3.00 | 2000 |

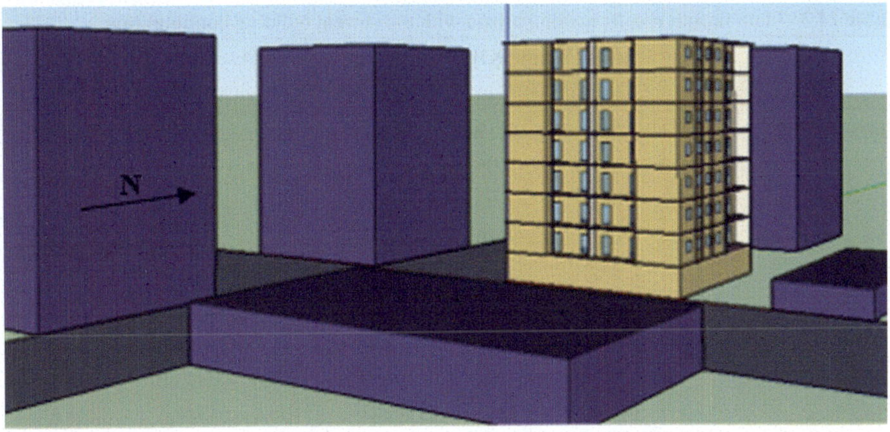


**Fig. 24.2** 3D model of the building and the surrounding area

- outside walls insulation (100 mm of wood fiber applied to the outer face);
- roof insulation (50 mm of polystyrene applied on top of the cement screed);
- installation of double-glazing windows (4-20-4 mm) with argon filling and PVC frames in place of the old windows;
- external venetian blinds to shade the windows and reduce solar heat gains.

Consequently, the new thermal transmittance would be U = 0.36 W $m^{-2}$ $K^{-1}$ for the walls, U = 0.45 W $m^{-2}$ $K^{-1}$ for the roof and U = 1.63 W $m^{-2}$ $K^{-1}$ for the windows. The retrofit solution also includes the insulation of the partition walls between the apartments and the staircase with 40 mm of wood fiber (U = 0.62 W $m^{-2}$ $K^{-1}$).

As concerns natural ventilation and infiltration, a constant air change rate as high as 0.5 $h^{-1}$ is considered, as suggested by the national standard dealing with the thermal performance of buildings [7]. No mechanical ventilation is available in the dwellings. The 3D model of the building was completed by the adjacent buildings in order to account for their shading effect (Fig. 24.2).

## 24.4 Results

Table 24.3 shows the values of the main parameters resulting from the quasi-steady calculation model. Thanks to the proposed retrofit solution, the envelope performance improves, both in winter and in summer: in particular, the energy demand for space heating decreases by more than 70% in many dwellings, and especially in the top floor. On the other hand, the energy demand for space cooling shows a smaller reduction, which in the intermediate floor keeps around 25% and 50%, the lowest values pertaining to the smallest units (U1 and U4).

**Table 24.3** Main retrofit results in comparison with the current building configuration

| | | $H'_T$ [W/(m$^2$ K)] | | $EP_{H,nd}$ (k Wh/m$^2$) | | $EP_{C,nd}$ (k Wh/m$^2$) | |
|---|---|---|---|---|---|---|---|
| | | Retrofit | Current | Retrofit | Current | Retrofit | Current |
| First floor | U1 | 0.56 | 1.19 | 30.5 | 78.7 | 18.0 | 21.8 |
| | U2 | 0.55 | 1.06 | 24.0 | 55.6 | 10.7 | 16.3 |
| | U3 | 0.57 | 1.06 | 20.5 | 48.7 | 13.6 | 30.2 |
| | U4 | 0.57 | 1.19 | 27.1 | 74.2 | 21.2 | 32.4 |
| Intermediate floor | U1 | 0.63 | 1.85 | 18.1 | 81.1 | 20.3 | 24.7 |
| | U2 | 0.57 | 1.54 | 6.9 | 37.4 | 12.5 | 21.5 |
| | U3 | 0.57 | 1.53 | 5.6 | 32.7 | 14.8 | 32.2 |
| | U4 | 0.63 | 1.85 | 15.5 | 77.6 | 23.0 | 32.2 |
| Top floor | U1 | 0.56 | 1.62 | 28.6 | 109.6 | 23.6 | 53.0 |
| | U2 | 0.52 | 1.46 | 19.8 | 73.5 | 13.8 | 40.8 |
| | U3 | 0.52 | 1.45 | 18.5 | 72.1 | 14.2 | 44.2 |
| | U4 | 0.53 | 1.64 | 24.4 | 107.0 | 25.0 | 56.5 |

In any case, the proposed retrofit solution allows to comply with DM 26/6/2015 and to reach the nZEB standard, both for the single apartments and at a building scale.

Furthermore, the results of the dynamic simulations show that the comfort conditions in summer undergo significant changes after retrofit.

Figure 24.3 shows the hourly values of the operative temperature in August, both for the current building configuration and after retrofit, in a room of the intermediate floor facing south (apartment U2). The trends are compared with the thresholds of the adaptive model, referring to Category II and III, in order to understand if comfort conditions are satisfied. From the graph, it is clear that the retrofit causes an increase in the operative temperature, which frequently exceeds 33 °C and constantly overcomes the threshold of Category II. On the contrary, in the current building the temperature is frequently below the thresholds, and the lower peak values occur.

As concerns the other floors, whose results are not shown in Fig. 24.3, the thermal conditions are less severe. Indeed, in the top floor the operative temperature for the rooms facing south currently reaches 33 °C, while in the first floor the peak temperature is around 32 °C. However, the retrofit solution modifies this trend: indeed, the insulation of the roof reduces the peak operative temperature in the top floor to 31 °C, while in the first and in the intermediate floor this increases by around 0.5 °C.

Finally, with the aim to have a wider view of the entire season, it is useful to look at the ITD values (see Fig. 24.4) and the FD values (see Table 24.4). What emerges from these results is that the apartments facing south (U2, U3) are much more uncomfortable than the other ones (U1, U4). As an example, in the top floor the seasonal ITD value is, respectively, around 3000 and 3700 °C h in U2 and U3, while it keeps around 1500 °C h in U1 and U4. Moreover, in the first floor discomfort

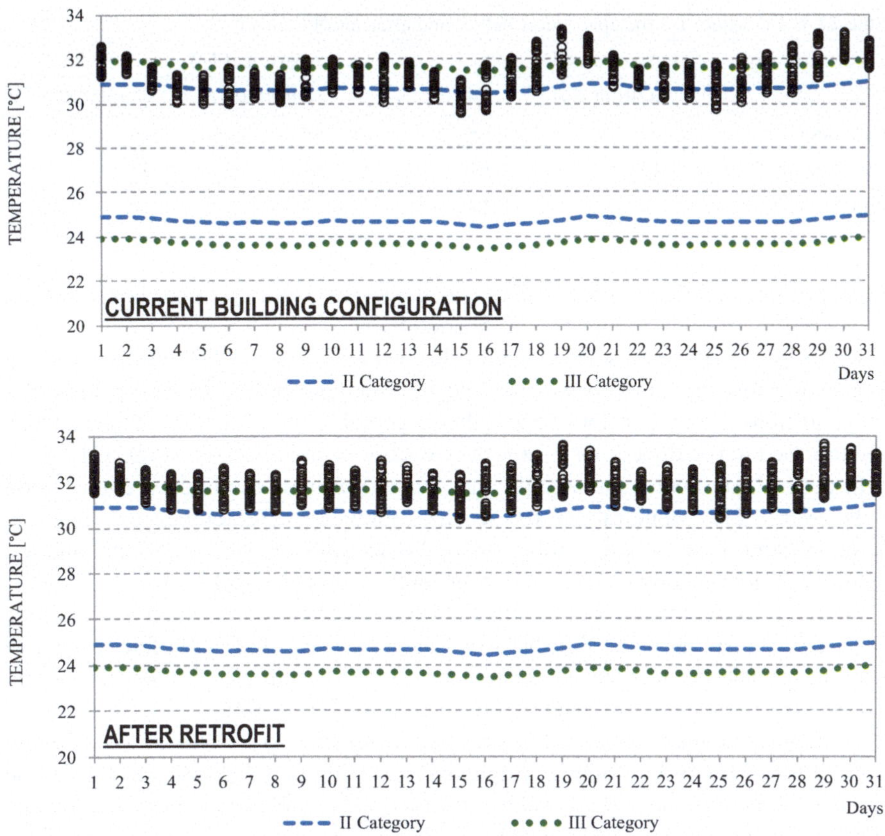


**Fig. 24.3** Operative temperature trend in August for a room facing South in the intermediate floor (each dot corresponds to an hourly value)

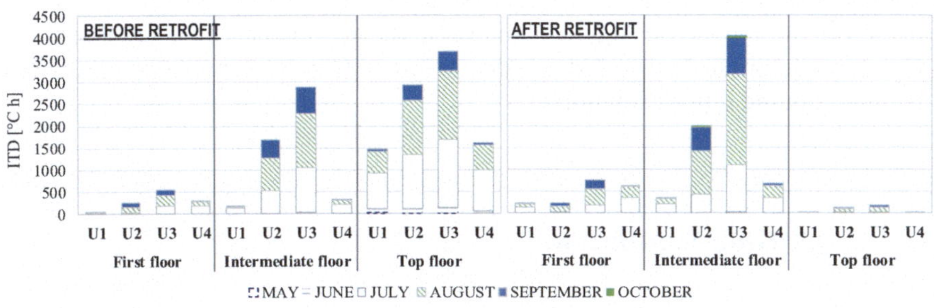


**Fig. 24.4** ITD values for the apartments before and after retrofit

**Table 24.4** FD values for the apartments before and after retrofit (July)

| | Apartment U1 | | Apartment U3 | |
|---|---|---|---|---|
| | Retrofit (%) | Current (%) | Retrofit (%) | Current (%) |
| First floor | 12 | 6 | 14 | 16 |
| Intermediate floor | 14 | 10 | 51 | 47 |
| Top floor | 0 | 43 | 5 | 63 |

is less pronounced than for other floors, as the room in the first floor benefits from the possibility of releasing heat to the ground floor.

According to the results, the retrofit solution cuts down the ITD value in the top floor, while this increases in the first and intermediate floors. However, the comfort conditions in the first floor keep rather acceptable (i.e., low ITD values occur), whereas the intermediate floor is now very uncomfortable (ITD = 4000 °C h).

In general, summer months (July and August) provide the highest contribution to the overall ITD value (about 40% each), followed by September and June. On the other hand, in May and October thermal discomfort is only occasional and not intense, as witnessed by the very low contribution to the overall ITD.

Similar information comes from the FD values. The period in which the operative temperature overcomes the threshold of Category II is often negligible, except in July and August. For this reason, the results reported in Table 24.4 refer only to July and to two significant cases, namely the apartments U1 and U3.

In the intermediate floor, the FD increases from 10% to 14% in U1 (North) and from 47% to 51% in U3 (South), respectively, due to the proposed retrofit. On the contrary, in the top floor the FD values show a drastic reduction in both apartments, approaching a maximum 5% after retrofit. In conclusion

- The rooms facing south show the worst summer comfort conditions;
- During hot months (August and July) retrofit causes overheating especially in the intermediate floors, without ventilation strategy;
- Roof insulation improves thermal comfort in the apartments of the top floor;
- Comfort conditions in the mild months (May and October) are not compromised by the proposed retrofit solution.

## 24.5 Nighttime Ventilation

The second part of this study evaluates the effectiveness of nighttime ventilation as a strategy to dissipate internal gains and avoid overheating. In fact, occupants can open windows during the night, when the outdoor temperature is low.

Hence, a new series of simulations was carried out by introducing a suitable air flow rate from outdoors at nighttime. In particular, two ventilation schemes were considered: single-sided ventilation and cross ventilation. Cross ventilation occurs when at least two windows—placed on different façades—are open, either in the same room or in different rooms of the same apartment; in the latter case, however, indoor doors should be open to allow this ventilation scheme. On the other hand, single-sided ventilation postulates that all open windows are on the same façade, or that indoor doors are closed, thus avoiding cross ventilation.

In both cases, a preliminary analysis has allowed to find out the opening percentage needed to introduce a mean air change rate of 2 $h^{-1}$, which is regarded as a suitable compromise to avoid draught risks. Indeed, higher rates of incoming air would induce high and annoying air velocity indoors (i.e., above 0.5 m/s). To this aim, the results suggest that it is necessary to open 5% of the windows in cross ventilation and 75% in single-sided ventilation.

The new hourly operative temperature trend is shown in Fig. 24.5, which refers to a room located in the intermediate floor and facing south, during August and after the retrofit solution. Similar results can be obtained in single-sided ventilation (but with a much larger open area). In comparison with the results reported in Fig. 24.3, nighttime ventilation has improved the trend of the operative temperature: indeed, temperatures are on average 2 °C lower and almost always below the threshold of Category II. Accordingly, also the ITD and the FD benefit from air ventilation strategies and get negligible (see Tables 24.5 and 24.6).

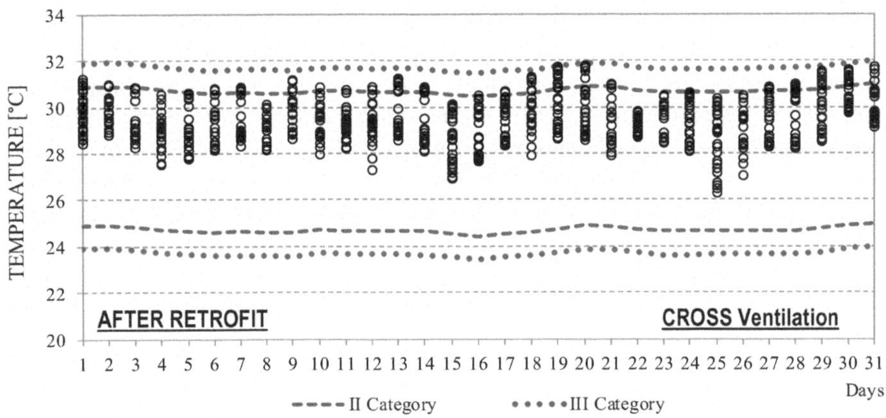


**Fig. 24.5** Operative temperature trend with cross ventilation (August—South)

**Table 24.5** Impact of the ventilation strategies on FD values in August (intermediate floor)

| | Apartment U2 (%) | Apartment U3 (%) |
|---|---|---|
| No nighttime ventilation | 57 | 89 |
| Cross ventilation | 1 | 4 |
| Single-Sided ventilation | 1 | 5 |

**Table 24.6** Impact of the ventilation strategies on the ITD (intermediate floor)

| | Apartment U2 | Apartment U3 |
|---|---|---|
| No nighttime ventilation | 2011.10 | 4043.32 |
| Cross ventilation | 24.93 | 92.46 |
| Single-Sided ventilation | 38.77 | 139.50 |

Thanks to nighttime ventilation, the severely uncomfortable conditions occurring after retrofit in the intermediate floors—especially in the rooms facing south—can be corrected, and the duration and intensity of discomfort sensation are minimized.

## 24.6 Insulation Thickness

The proposed retrofit solution includes the insulation of the outside walls with 10-cm wood fiber panels. However, the choice of the thickness is a delicate matter: using thicker panels would improve winter performance but, at the same time, it is likely to show non-negligible drawbacks in summer. For this reason, other simulations were performed by adopting a higher thickness for the wood fiber (14 cm); in this case, the thermal transmittance of the walls would be $U = 0.28\ W\ m^{-2}\ K^{-1}$.

The results of the calculations and the comparison with the previous results are reported in Table 24.7 (only in relation to U1 and U2, for the sake of brevity). Here, it is clear that those parameters that are related to winter performance ($H'_T$, $EP_{H,nd}$) benefit from thicker insulation in all floors; on the other hand, the parameters describing energy demand and thermal comfort in summer ($EP_{C,nd}$, ITD and FD) worsen in all cases. This is particularly relevant in the intermediate floor: if looking at the dwelling facing south (U2), the ITD increases by around 20% if compared with the lower insulation thickness. As a matter of fact, introducing higher thickness of insulation causes an increase in the operative temperature by around 0.1 °C throughout summer. However, comfort conditions are not compromised in May and October: the FD and the ITD would be still negligible.

## 24.7 Conclusions

This study addresses the retrofit of a public residential building located in Southern Italy and built in the 1950s. The proposed retrofit solution aims to comply with the requirements introduced by DM 26/06/2015, and to reach the nZEB standard. To this aim, the retrofit actions envisage the insulation of roofs and outside/internal walls, and the introduction of double-glazing windows.

Dynamic simulations performed in EnergyPlus allowed investigating internal indoor thermal comfort in summer, both in the current configuration and after retrofit.

**Table 24.7** Effect of the insulation thickness on winter and summer performance

| | | $H'_T$ [W/(m$^2$ K)] | | $EP_{H,nd}$ (k Wh/m$^2$) | | $EP_{C,nd}$ (kWh/m$^2$) | | ITD (°C h) | |
|---|---|---|---|---|---|---|---|---|---|
| | | 10 cm | 14 cm | 10 cm | 14 cm | 10 cm | 14 cm | 10 cm | 14 cm |
| First floor | U1 | 0.56 | 0.55 | 30.5 | 29.1 | 18.0 | 18.1 | – | – |
| | U2 | 0.55 | 0.54 | 24.0 | 23.6 | 10.7 | 10.6 | 9.4 | 10.3 |
| Intermediate floor | U1 | 0.63 | 0.58 | 18.1 | 15.7 | 20.3 | 21.1 | – | – |
| | U2 | 0.57 | 0.54 | 6.9 | 6.7 | 12.5 | 12.6 | 124.6 | 140.1 |
| Top floor | U1 | 0.56 | 0.50 | 28.6 | 23.4 | 23.6 | 24.4 | – | – |
| | U2 | 0.52 | 0.49 | 19.8 | 18.1 | 13.8 | 13.8 | – | – |

In fact, the insulation of the envelope may compromise the summer performance of buildings, especially in hot regions like Southern Italy.

As expected, the simulations show that the proposed retrofit solution affects the comfort conditions in summer. In fact, the envelope insulation allows reducing the operative temperature in the top floor, while also considerably increasing it in the other floors. The indoor operative temperature is often above the comfort thresholds introduced by the adaptive model in the rooms facing south, especially in the intermediate floors, where it even reaches 34 °C. On the other hand, much more comfortable conditions occur in the rooms facing north, east and west. In May and October, comfortable indoor conditions are often obtained, both before and after retrofit.

Starting from these results, nighttime ventilation strategies are set up in order to improve thermal comfort in the intermediate floors. New dynamic simulations are carried out introducing a suitable opening profile of the windows: as a result, the performance of the intermediate floors is improved, and the Frequency of Discomfort is now below 5% even in the hottest months.

Further conclusions regard the insulation thickness. Using thicker insulating materials allows reducing the thermal transmittance from 0.36 W $m^{-2}$ $K^{-1}$ to 0.28 W $m^{-2}$ $K^{-1}$, but in summer this implies slightly higher cooling demand and more pronounced thermal discomfort.

**Acknowledgements** The work described in this paper has been carried out within a research activity financed by ENEA (Agenzia nazionale per le nuove tecnologie, l'energia e lo sviluppo economico sostenibile), dealing with "Energy retrofit of existing public buildings to reach the nearly Zero Energy standard" (PAR 2017 and PAR 2018).

## References

1. Decreto Interministeriale del 26 giugno 2015, "Applicazione delle metodologie di calcolo delle prestazioni energetiche e definizione delle prescrizioni e dei requisiti minimi degli edifici"
2. Stazi, F., Bonfigli, C., Tomassoni, E., Di Perna, C., Munafò, P.: The effect of high thermal insulation on high thermal mass: is the dynamic behaviour of traditional envelopes in mediterranean climates still possible? Energy Build. **88**, 367–383 (2015)
3. Stazi, F., Tomassoni, E., Bonfigli, C., Di Perna, C.: Energy, comfort and environmental assessment of different building envelope techniques in a Mediterranean climate with a hot dry summer. Appl. Energy **134**, 176–196 (2014)
4. Evola, G., Marletta, L., Costanzo, V., Caruso, G.: Different strategies for improving summer thermal comfort in heavyweight traditional buildings. Energy Procedia **78**, 3228–3233 (2015)
5. Evola, G., Marletta, L., Sicurella, F., Tanasiev, V.: Combining thermal inertia, insulation and ventilation strategies for improving summer thermal comfort. In: Proceedings of 34th AIVC Conference, Athens (Greece), 25–26 September 2013 (2013)
6. Attia, S., Eleftheriou, P., Xeni, F.: Overview and future challenges of nearly zero energy buildings (nZEB) design in Southern Europe. Energy Build. **155**, 439–458 (2017)
7. UNI/TS 11300-1. Prestazioni energetiche degli edifici—Parte 1: Determinazione del fabbisogno di energia termica dell'edificio per la climatizzazione estiva ed invernale (2014), in Italian

8. Nicol, J.F., Humphreys, M.A.: Adaptive thermal comfort and sustainable thermal standards for buildings. Energy Build. **34**, 563–572 (2002)
9. Sicurella, F., Evola, G.: A statistical approach for the evaluation of thermal and visual comfort in free-running buildings. Energy Build. **47**, 402–410 (2012)
10. EN Standard 15251. Indoor environmental input parameters for design and assessment of energy performance of buildings addressing indoor air quality, thermal environment, lighting and acoustics (2007)

# Chapter 25
# Building Energy Simulation of Traditional Listed Dwellings in the UK: Data Sourcing for a Base-Case Model

**Michela Menconi, Noel Painting and Poorang Piroozfar**

**Abstract** The need for improving energy efficiency and reducing carbon emissions has made retrofitting existing homes a priority today. A research project has been designed with one of its aims to propose a framework to intervene in traditional listed dwellings (TLDs) to reduce their environmental impact in England, with a special focus on South-East region. Selected case studies in the City of Brighton and Hove, have been modelled and simulated in their status quo using Dynamic Energy Simulation (DES). The models, calibrated using monitored energy and indoor conditions data, are then to be used to simulate the effect of permissible retrofit interventions. DES requires accurate sourcing of multiple input data, to ensure that the models created, closely resemble the real case study dwellings in their energy performance and thermal behaviour. This process can be extremely challenging in the case of simulation of TLDs, where most of the envelope's construction is unknown and intrusive tests are not usually permitted. The data sourcing process is even more complex in the case of dwellings in use, because of the variability of occupancy profiles and patterns of use over time. Providing a brief overview of the methodology adopted in this study, this paper describes, in detail, the approach devised to ensure that the most credible datasets are collected from different sources for generating models that accurately represent the real case study dwellings in their status quo and can be used in the following stages of the analysis to asses potential retrofit interventions.

M. Menconi (✉) · N. Painting · P. Piroozfar
School of Environment and Technology, University of Brighton, Brighton BN2 4GJ, UK
e-mail: m.menconi3@brighton.ac.uk

P. Piroozfar
Digital Construction Lab, University of Brighton, Brighton BN2 4GJ, UK

 
J. Littlewood et al. (eds.), *Sustainability in Energy and Buildings*, Smart Innovation, Systems and Technologies 163, https://doi.org/10.1007/978-981-32-9868-2_25

## 25.1 Introduction

### *25.1.1 Research Background*

The application of Building Energy Simulation (BES) for the analysis of buildings energy performance is an established strategy to assist in the design process of energy efficient buildings [1–4] as well as in the choice of suitable retrofit interventions [5–8]. Simulation has several advantages compared to physical modelling (including mock-up tests and field tests): it is less time consuming and more cost-effective; importantly, in the case of buildings in use, it is non-intrusive [9]; factors like outside weather can be controlled and changed in order to account for the effects of occupancy alone as well as those of one specific retrofit measure or of a combination of more than one [10]. Simulation has been preferred to mathematical modelling [11] as well because it is a more developed method that allows for the testing of different retrofit scenarios with a more user-friendly approach requiring less coding skills. Nevertheless, when it comes to traditional buildings, there are still concerns with regards to the proper implementation of the models for BES [12–15]. Primary concerns are raised around deploying energy simulation for traditional buildings because the processes and synergies that characterize this part of the stock are not always captured by models. The uncertainties around construction materials and thermal behaviours of the traditional stock require skilled modellers with tacit knowledge of local and indigenous building materials and construction methods. Furthermore, when BES is conducted on dwellings in use, the accuracy of the outcome is strictly linked to the "human factor" [13]: both the expertise of the modeller and, the occupants' behaviour and the building's pattern of use.

### *25.1.2 Aim of This Paper*

This paper describes, in detail, the process of creation of a base-case model for a pilot case study (CS) of a traditional listed dwelling (TLD). This work is part of an ongoing research that aims to devise effective and responsive retrofit strategies to improve the energy performance of TLDs in South-East England. The research project uses BES to assess the benefits of potential interventions on nine selected CSs representative of 19th C listed dwellings in Brighton, UK. This paper aims to provide a comprehensive report of all the data collected, their sources and the assumptions made to ensure that the datasets will help generate a model that closely resembles the actual energy performance and thermal behaviour of the real world case.

### 25.1.3 Overview of the Research Methodology

The research is articulated around successive stages of dynamic energy simulation. Once the model is created, the first dynamic energy simulation is run in building's existing condition and the data output at this stage is used for calibration with actual monitored data. The calibrated models are then normalized to generate a baseline performance scenario to be used in the following stages of research to assess the benefits of potential interventions on the selected CSs. Figure 25.1 illustrates the methodology and methods utilized in this research for the creation of realistic dynamic energy models.

IES-VE has been chosen as a suitable energy simulation application for this research, because it is already validated against a number of global as well as regional standards [16]; it has been considered the most appropriate application for energy analysis in several precedent studies (e.g., [17–19] to name but a few); it was also deemed suitable to evaluate the energy performance of traditionally constructed dwellings [20–22]; with its parametric capabilities, it allows for simulation of multiple case scenarios to be applied to the same model, hence the credibility of comparative analysis of the interventions is deemed to be high; it is an application developed in the UK and its use is widespread in the country as well as around the world; it offers a user-friendly interface; and it does not require extensive coding skills.

In order to create accurate models of the selected CSs, a mixed methods approach was developed, for which multiple methods were used to gather, collate and analyze a wide range of input and monitored data. The pilot CS model, produced according to the procedure reported in this paper, was then calibrated using metered data for energy consumption and indoor conditions, in order to ensure that it accurately reproduces the real CS and represents its actual energy performance and thermal behaviour.

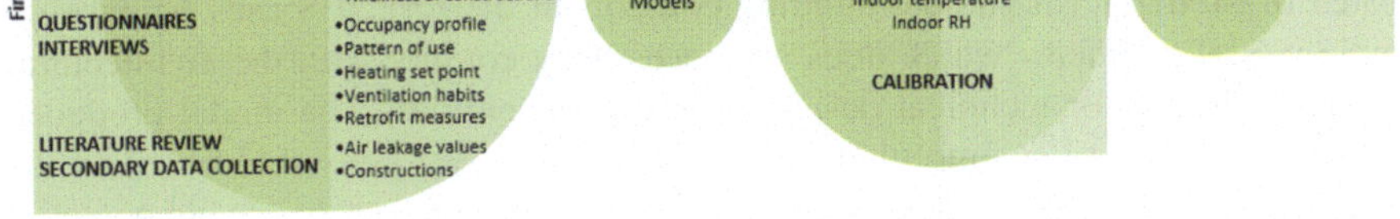


**Fig. 25.1** The research design from the genesis of models to the production of a base-case performance scenario

### 25.1.4 The Paper Structure

This paper starts with setting the scene for the study in an introduction section which includes the research background, the aim and an overview of the research methodology. It will then set out to explain the steps taken for developing the base-case model which will then be used as a basis for a parametric study of introduced permissible measures and interventions to improve the environmental impact of TLDs.

The model specifications include geometry, orientation, site location and weather data, openings, construction materials, and carries on to heating and domestic hot water systems (DHWS), complemented by sources of heat gain and mechanical ventilation rates.

The paper concludes by providing some insight into the data sourcing process to ensure that the models devised for BES closely resemble the actual energy performance and thermal behaviour of the real buildings they would represent in this study.

## 25.2 Model Creation

### 25.2.1 Geometry

The pilot CS is a 1st floor converted flat located in Brunswick Place (City of Brighton and Hove) one of the earliest Regency developments in town. It was built during the 1st half of C19th and is a grade II listed dwelling as a part of the List Entry Number 1204771. Its layout is typical of similarly converted flats in South-East England, with a living area containing kitchen and dining in the front, and a bedroom area in the rear of the flat; its total usable floor area is approximately 70 $m^2$. Its main elevation faces West, overlooking Brunswick Place, and is therefore exposed to the prevailing winds in this area, although partially sheltered by the offsetting buildings along the western side of the place.

In order to generate an energy model of the CS dwelling, first of all, a floor plan was generated in AutoCAD using the drawings and measures taken during the walk-in measured survey. The dxf drawing was then imported to IES and a 3D model was created in ModelIT (the model building component of IES). IES uses a relatively simple model geometry which does not necessarily comply with the architectural drawing protocols or technical design drawing conventions. The model produced in IES was, therefore, a simplified version of the drawings generated based on the surveys. This is crucial to reduce unnecessary model complexity and avoid potential simulation errors as a result of an excessive level of details, which may confuse the application during simulation runs with very little to no impact on the outcomes of simulations. The adjacent volumes were also modelled, to take into account the thermal conditions at the boundaries and to include the shading effects in solar gain calculations. The temperature difference between the adjacent building volumes and the CS was assumed to be null or negligible (adiabatic conditions) as the surrounding flats are also occupied and heated [23, 24] throughout the year.

### 25.2.2 Orientation, Site Location and Weather Data

Activating the APlocate application within ModelIT, data related to the site location and weather was inputted. APlocate uses site data that contain values for latitude, longitude and altitude of different geographical locations throughout the world, taken from standard tables published by CIBSE and ASHRAE.

ASHRAE [25] suggests the use of hourly weather data that corresponds to the same time period as the energy use data to which the model will be calibrated. However, if the savings are to be "normalized" to represent a typical year, the Guidance considers it also acceptable to use a typical year of local weather data or of data from a site that is near the building under investigation (usually an airport).

Brighton and Hove, UK, is a non-ASHRAE location. Therefore, initially, the weather dataset relative to Shoreham Airport was used as the nearest geographical location included in the basic set of simulation weather files. During the following calibration phases, it was decided to adopt an average Brighton weather file provided by Meteo-Norm to allow for a more accurate comparison between simulated and monitored energy consumption, temperature and relative humidity (RH) data and to aid this way in the calibration process.

A further stage of the calibration was finally performed, using the specific Brighton weather file recorded for 2017—to which most of the monitored data pertain—acquired from IES, to increase the level of confidence in the model created.

### 25.2.3 Openings, Shading and Insulation Studies

When the geometry was completed, the types of openings, their exposure, and their use profiles were assigned using the IES module MacroFlo. Each window was modelled using an ad hoc Opening-Type-Properties-Database [26, 27] based on the data collected during the visual and measured survey. It stores information concerning the window's geometry, the leakage characteristics, the degree and timing of window opening and, when appropriate, its dependence on room temperature and/or RH. A profile has therefore been created for each window to define its pattern of use according to the information obtained from the interviewees.

It is worth noting that the coefficients that describe the air leakage characteristics, i.e. Crack Flow Coefficient (CFC) and Crack Length (CL), have been given a value of zero for all the external openings in the MacroFlo Database. The reason for this is that, calculating the infiltration rates using such values requires, beside data about the maximum openable area of each external opening (for which the values were extrapolated from the measured survey), the detailed information about cracks in the building envelope, as well as wind pressure coefficient data for the building surfaces; such data was not possible to acquire at this stage with the instruments available for this research. Although the use of ad hoc accurate data for CFC and CL could potentially generate very accurate results, it also implies the risk of misleading outputs when detailed knowledge of such data is impracticable to gain. Air leakage rates were therefore considered null during the MacroFlo analysis that was only used to calculate natural ventilation rates, using the ad hoc user-defined profiles for each opening. Infiltration rates were instead inputted as uniform values for each room Template in ApacheSim (see Sect. 2.4, Windows). Values of CFC and CL have only been inputted in MacroFlo for the internal closed doors, to take into account air circulating inside the flats between those rooms whose doors are generally permanently closed. These values have been taken from those suggested by IES in the MacroFlo User Guides [26, 27] in the measure of 1.3 l/smPa $^{0.6}$ for the CFC and of 100% for the CL around the opening perimeter.

When the Openings Database was completed, the right orientation was set for the geometry created and the module SunCast was run to perform shading and solar insolation studies. The impacts of air movement (as generated by MacroFlo) and solar shading calculations (as generated by SunCast) were then quantified in terms of heat gains and energy consumption to be used in ApacheSim, the thermal simulation engine in the IES.

## 25.2.4 Construction Templates

IES software calculates the envelope's U-value based on thickness, conductivity, density, heat capacity and resistance of each building element, (referred to as "construction" in the IES terminology). Using the Building Template Manager within ApacheSim [28, 29], Construction Templates were therefore created, to assign constructions and performance characteristics to each surface in the model. Each construction defines the thermal properties of a building element such as wall, ceiling/roof, floor/ground floor, window or door. It consists of layers of different materials, together with thermal properties of the materials, surface properties and other data used in the thermal analysis. Because of the private ownership of the dwelling, it was not possible to use any type of invasive technique, e.g. core sampling, to gain a clear understanding of the components of each construction. Therefore, specific building elements were created based on the assumptions made from the measured thickness of such elements (whenever this measure was possible to take), the visual and tactile inspection as well as the literature review, secondary data collection and conversations with experts about the typical and prevailing construction methods and materials used in the area at that time.

**Walls**. The external walls in the pilot CS are made of rendered brickwork, as confirmed in the List Entry Summary, which talks of "stucco over bricks" [30]. The size of bricks varied slightly throughout the centuries in this area.

To make appropriate assumptions concerning constructions and materials of the envelope, the thickness of each wall was measured, and tactile inspection of the inside was carried out, aimed at establishing the eventual presence of a traditional finish—such as plaster on lath and battens—or of modern dry lining. Given its overall measure and the tactile inspection, the front wall was assumed to be made of imperial size bricks (227 mm × 115 mm), therefore the thickness of the brickwork was estimated to be 464 mm, hence a two brick wall.

The external face of the main elevation is finished in traditional lime-based stucco as it originally was. The IES library does not include such material. Therefore, for the modelling of this layer, a new lime plaster material was created, for which the thermophysical properties of traditional lime plaster were assigned, as suggested by the International Organization for Standardization -ISO- [31, 32]. Such values are also in the range of those proposed by previous research [33–35] given the variability of thermal conductivity, heat capacity and vapour resistivity of this material, as a consequence of its moisture content.

The internal finishing of the front wall was assumed to be lath and plaster. This was a common method for interior finishing as it allowed for a smooth surface; it was frequently used on ornamental or unusual surfaces, like rounded walls, such as those found in the front elevation of the pilot CS. Therefore, the materials build-up of the front elevation has been modelled as: 20 mm of stucco, brickwork, 40 mm of cavity (the thickness of the vertical timber battens), 6 mm of wood (oak essence for the lath) and 15 mm of lime plaster (this thickness takes into account the increased thickness of the plaster forced into the gaps between the lath and the thickness of the layer on top of the lath).

Brick masonry was assumed to be the material used also in party walls (in adiabatic conditions with the adjacent dwellings) and the rear walls. Therefore, given the overall thickness of 250 mm and the tactile inspection (excluding the presence of plaster on lath or dry lining), the rear elevation was estimated to be a single brick wall (brickwork with 227 mm of thickness). The rear elevation was finished in lime-based render externally and in lime plaster internally.

The internal partitions were assumed to be made of brickwork (finished in lime plaster) as well, given their measure and tactile inspection: respectively one-and-a-half-brick walls for the main staircase and half-a-brick walls for the other partitions.

**Internal floors-ceilings**. The internal floor structures are made of timber joists as they were originally built; they were likely finished with floorboards, but these have been replaced with chipboard flooring and carpets and they have been modelled accordingly. The ceilings instead, being generally less modified and still decorated in stucco, have been assumed to be finished in lath and plaster. The measured overall thickness has guided the definition of the composition of the layers of the construction, which were modelled, from inside to outside, as: carpet, chipboard flooring, cavity (220 mm for the timber joists), oak (6 mm for the lath), lime plaster (15 mm for the plaster on lath).

**Windows.** Most of the windows are likely to be the original timber sash ones, as determined through visual inspections, and confirmed during the interviews. The frame materials, type and thickness of glazing as well as type of shading device used (and their profile of use) were modelled in the ApacheSim Constructions Database, with all their other characteristics, affecting the ventilation rates, having been detailed in MacroFlo (see Sect. 2.3).

All the windows are single glazed (with a 6 mm-thick glass) and some of them have shading systems (when present, they were added to the windows Constructions Database and given a pattern of use according to the information produced by the occupants). The values assigned, in the Apache Constructions Database, to the resistance of the shading devices, their shading coefficient and shortwave radiant fractions have been taken from previous research [36], from CIBSE Guide A [32], based on Wood et al. [37] and from IES Apache-Tables [38].

Infiltration rates are certainly a variable to take into account when modelling old sash windows and their influence on energy consumption and indoor conditions have been thoroughly investigated by previous research [37, 39, 40]. The air leakage (resulted from windows and fabrics) was considered in this study as a uniform rate in each Room Template Database in ApacheSim. It was not possible to perform a blower door test to measure the actual air leakage rates of the CSs investigated because of the invasiveness of such test in private dwellings, which were all inhabited during the entire period of this study. Therefore, each room in the dwelling was assigned a specific air leakage value, according to the figures suggested by CIBSE [32]. The CIBSE Guide A ([32], Table 4.24) provides, in fact, empirical values for air infiltration rates for rooms in flats (levels 1–5) on normally exposed sites in winter: they range from 1.40 Air Changes per Hour (ACH) for leaky buildings to 0.25 ACH for extremely airtight buildings. For each room a value has been assigned within the range proposed by CIBSE, taking into account the results of the visual survey (where the windows were assessed in their state of conservation, observing the overall condition for all of them), the thermal imaging survey (that highlighted moderate rates of air leakage around the openings) and in accordance with existing literature concerning similar properties of the same period and area [36, 41].

Therefore, although the air leakage value assigned to the model before the calibration was 1.4 ACH (as suggested by CIBSE for old leaky buildings), during the calibration, such value was fine-tuned (within the range suggested by CIBSE) to achieve acceptable differences between the simulated and monitored data. The process concluded assuming, for the pilot CS, a value of the air leakage in the area of 0.5 ACH, which allowed for a successful calibration of energy and indoor conditions data and is also in line with previous research on similar dwellings [36, 41].

### *25.2.5 Heating and Domestic Hot Water Systems*

Heating and DHWS were inputted in the Apache Systems Tab within the Thermal Template Tab. As part of the visual survey undertaken, data concerning the fuel(s) used, boiler and DHWS tank (wherever present) nameplate and size, as well as type and thickness of any insulation eventually found around the hot water tank, was collected. Then, data concerning the seasonal efficiency and DHWS delivery efficiency was gathered online from the respective producers and inputted in the Apache Systems Tab for each system. The pilot CS, as most of the CS dwellings investigated, has central heating with a combi-boiler.

As the research is set in the UK, the UK National Calculation Methodology (NCM) System Data Wizard in IES was used to aid in the definition of the characteristics of the heating system, adopting some of the default values proposed given a certain boiler efficiency inputted. On the Apache Systems dialog box [42], a set of NCM system types are available for selection. This utility allows describing the characteristics of the heating systems using the method implemented in the BRE Simplified Building Energy Model (SBEM). The system specifications entered here are interpreted into Apache Systems, where they are used for sizing central plant and calculating fuel consumption and carbon emissions. Therefore, in the Apache Systems Tab, only the Boiler Seasonal Efficiency (BSE) was manually inputted according to the value given by the producer. The UK NCM wizard then assigned, by default, a value to the Heating Delivery Efficiency (HDE) depending on the type of system (which, in the pilot CS, was set to: "Central heating using water—radiators"). The software finally calculated the value of the Seasonal Coefficient of Performance (SCoP) given these other two.[1]

The pattern of use of the Heating system and temperature setpoint(s) were also set creating specific profiles in the Thermal Template for each room in the dwelling investigated, in accordance with the data acquired through questionnaires and interviews with the participants.

DHW consumption was estimated based on the number of occupants, using the formula proposed by Energy Saving Trust [43] for the average use (in l/day) of DHW in UK dwellings, as follows:

$$DHW = 40 + 28\,N\,l/day$$

where N is the number of occupants. The value obtained this way was then divided by 24 to obtain the l/h consumed in the dwelling, as required by the software. The value finally found corresponds to the total estimated amount of hot water consumed in the dwelling each hour when the DHWS is in use. This total was then assigned to only one room Thermal Template (the main bathroom in this case), assuming that all the other room Templates have no DHWS when running the simulations.

Questions concerning the hot water usage were included in the questionnaires and interviews to help understand whether the pattern of use of the DHWS presented any variance that may lead to a use in excess (or much below) of the average use given by the formula, because of anomalies in the behavioural patterns (e.g. occupants that take showers somewhere else during the week or that make very little use of hot water). However, this was not the case for the pilot CS.

[1]SCoP is a parameter used in the Apache Systems tab to describe the efficiency of the heating system. This value is linked to the boiler seasonal efficiency (BSE) as per following formula: SCoP = BSE * HDE [42].

### 25.2.6 Heat Gains

Internal heat gains for each room Template [42] were also assigned, using the information acquired through the visual surveys (appliances, lighting fittings) and questionnaire surveys and interviews (occupancy profile, pattern of use) and adopting the values suggested for the heat gains by CIBSE Guide A [32], as detailed below.

**People.** CIBSE Guide A ([32], Table 6.3) provides average heat emission rates per person (male or average for a mixture of men, women and children) depending on the activity and dry bulb temperature of the room. The Guide also suggests that the value of heat gain from a female body can be calculated multiplying by 0.85 the value given for the male body. In each room Template, the actual occupancy pattern, as evinced by the questionnaire survey and interviews, was taken into account to make a decision concerning the values of sensible and latent heat gains (assuming the rooms to be at 20 °C), adapting them from Table 6.3 of the CIBSE Guide A [32].

**Lighting.** The dwelling investigated mainly uses Fluorescent and Tungsten lamps (the occupants showing, however, a tendency to opt for more efficient lamps when in need of renovation).

CIBSE Guide A [32] suggests considering the energy consumption of each lighting equipment as equal to the value of the sensible heat gain from it. Values suggested by CIBSE ([32], Table 6.2) have been used for heat gain generated by lighting in the area of 8–12 $W/m^2$ based on fluorescent lamps. The Guide suggests the upper value for older installations and halved values for LED lighting. Therefore, the value used in each room Template for energy consumption and heat gain generated by lighting fittings was 12 $W/m^2$; the pattern of use of the lighting system was modelled according to the information produced by the occupants.

**Appliances.** The values suggested by CIBSE [32] for energy consumption and heat gain of typical domestic and office equipment [32, Tables 6.6, 6.15 and 6.16] and of electric and gas cooking equipment [32, Tables 6.18 and 6.20] were taken into account, to make a decision about the appropriate data to input for heat gains from appliances. Such values, in CIBSE tables, depend on the appliance rating, the temperature of wash (for washing machines), the type of fuel and the presence or otherwise the lack of a hood (for cooking appliances). Therefore, heat gains and energy consumption for each appliance were taken from the CIBSE Guide and inputted according to the information provided by the visual survey (type of appliances and their rating) and confirmed or complemented by the questionnaire survey and interviews. Profiles of use were also created for each appliance, according to the information provided by the participants with the questionnaire/interviews and concerning their occupancy profile as well as the frequency and length of use of the appliances.

### *25.2.7 Mechanical Ventilation Rates*

Where extract fans were found, (in this case in the bathroom), the input value of 15 l/s, as suggested by CIBSE Guide A for intermittent fans [32, Table 4.2a] was used in the Room Template Data in ApacheSim, in accordance with the occupation profile of the room, as deducted from the interviews with the occupants.

## 25.3 Conclusions

Performance gap is a growing area of research in BES. Its importance should by no means be understated, it is however very context specific and may vary from one study to another. It could have been argued, in the current research project, that as the interventions are introduced on a case by case basis and their impacts are measured accordingly, model calibration may have not been of such priority or significance in this study. This is correct but only to some extents. To be able to develop the scope and applicability of this research, and increase its validity and reliability through facilitation of cross comparison internally (between the findings for different CSs in this research) and externally (between the findings of this research and those of other studies), it was crucial to ensure that models represent the real performance of the actual CSs as closely as possible, hence the paramount importance of this stage of the study.

The paper provides an insight into the data sourcing process used for the creation of realistic energy models of TLDs. It aims to address a gap frequently pointed out by previous research that considers BES of traditional dwellings extremely challenging because of the lack of detailed knowledge about their materials, constructions, and user behaviour. As a consequence, a wide number of assumptions were necessary to ensure that the created models are accurate representations of the real world case.

The data sourcing process described was finally validated in the subsequent stage of calibration, when iterative simulations were run of the pilot CS model and their output was assessed against measured data concerning energy usage as well as indoor temperature and RH. The input data were checked and, when required, fine-tuned, to allow for a good correspondence between simulated and monitored data. When calibration achieved the expected outputs, the chosen input values were considered realistic for the specific CS. The same data collection and validation process was then repeated also for all the selected CSs investigated. To date, all the models are fully calibrated and ready to be used in the following stages of research, to assess the benefits of potential retrofit interventions on the selected dwellings, both as individual and combined measures.

## References


1. Fesanghary, M., Asadi, S., Geem, Z.W.: Design of low-emission and energy-efficient residential buildings using a multi-objective optimization algorithm. Build. Environ. **49**, 245 (2012)
2. Garber, R.: Optimisation stories: the impact of building information modelling on contemporary design practice. Architect. Design **79**, 6–13 (2009)
3. Nguyen, A.-T., Reiter, S., Rigo, P.: A review on simulation-based optimization methods applied to building performance analysis. Appl. Energy **113**, 1043–1058 (2014)
4. Wang, W., Rivard, H., Zmeureanu, R.: An object-oriented framework for simulation-based green building design optimization with genetic algorithms. Adv. Eng. Inform. **19**, 5–23 (2005)
5. Ascione, F., Bianco, N., De Masi, R.F., De'Rossi, F., Vanoli, G.P.: Energy retrofit of an educational building in the ancient center of Benevento. Feasibility study of energy savings and respect of the historical value. Energy Build. **95**, 172–183 (2015)
6. Kolaitis, D.I., Malliotakis, E., Kontogeorgos, D.A., Mandilaras, I., Karsourinis, D.I., Founti, M.A.: Comparative assessment of internal and external thermal insulation systems for energy efficient retrofitting of residential buildings. Energy Build. **64**, 123–131 (2013)
7. Pernigotto, G., Penna, P., Cappelletti, F., Gasparella, A.: Extensive Utilization Of Dynamic Simulation For Sensitivity Analysis And Optimization Design Of Refurbishment Measures. International High Performance Buildings Conference. Purdue (2012)
8. Stazi, F., Veglio, A., Di Perna, C., Munafo, P.: Experimental comparison between 3 different traditional wall constructions and dynamic simulations to identify optimal thermal insulation strategies. Energy Build. **60**, 429–441 (2013)
9. Yang, Z., Becerik-Gerber, B.: A model calibration framework for simultaneous multi-level building energy simulation. Appl. Energy **149**, 415–431 (2015)
10. Wei, S., Wang, W., Jones, R., De Wilde, P.: Using building performance simulation to save residential space heating energy: A pilot testing. 8th Windsor Conference: Counting the Cost of Comfort in a changing world, 10–13 April 2014 Cumberland Lodge, Windsor, UK. London: Network for Comfort and Energy Use in Buildings (2014)
11. Diakaki, C., Grigoroudis, E., Kolokotsa, D.: Performance study of a multi-objective mathematical programming modelling approach for energy decision-making in buildings. Energy **59**, 534–542 (2013)
12. Barnham, B., Heath, N., Pearson, G.: Historic Scotland Technical Paper 3: Energy Modelling Analysis of a Scottish Tenement Flat. Historic Scotland, Edinburgh (2008)
13. Heath, N., Pearson, G., Barnham, B., Atkins, R.: Historic Scotland Technical Paper 8: Energy modelling of the Garden Bothy, Dumfries House. Historic Scotland, Edinburgh (2010)
14. Ingram, V., Jenkins, D.: Historic Scotland Technical Paper 18: Evaluating energy modelling for traditionally constructed dwellings. Historic Scotland, Edinburgh (2013)
15. STBA: Responsible Retrofit of Traditional Buildings [Online]. Available: http://www.sdfoundation.org.uk/downloads/RESPONSIBLE-RETROFIT_FINAL_20_SEPT_2012.pdf, last accessed 2017/7/20. Sustainable Traditional Buildings Alliance (2012a)
16. IES: ModelIT: Model builder User Guide. https://www.iesve.com/support/userguides?page=5, last accessed 2017/1/11. IES (2016)
17. McNally, Y.: An investigation into energy saving via retrofit compared to replacement housing. Ph.D., Ulster University (2014)
18. Memon, S.: Analysing the potential of retrofitting ultra-low heat loss triple vacuum glazed windows to an existing UK solid wall dwelling. International Journal of Renewable Energy Development (IJRED) **3**, 161–174 (2014)
19. Pomponi, F. & Piroozfar, P.: Double skin façade (DSF) technologies for UK office refurbishments: a systemic matchmaking practice (2015)
20. Flores, J.A.M.: The investigation of energy efficiency measures in traditional buildings in the Oporto World Heritage Site. Oxford Brookes University, PhD (2013)
21. Ingram, V.: Energy performance of traditionally constructed dwellings in Scotland. PhD, Heriot-Watt University (2013)

22. Moran, F.: Benchmarking the energy use of historic dwellings in Bath and the role for retrofit and LZC technologies to reduce CO2 emissions. Doctor of Philosophy, University of Bath (2013)
23. IES: Construct DXF: DXF Drawing Requirements [Online]. Available: https://www.iesve.com/support/userguides?page=5, last accessed 2017/1/6. IES (2015c)
24. IES: Integrated Environmental Solutions [Online]. Available: https://www.iesve.com/, last accessed 2016/10/13. IES (2018a)
25. ASHRAE: ASHRAE Guideline 14-2002 for Measurement of Energy and Demand Savings. American Society of Heating, Refrigerating and Air-conditioning Engineers, Atlanta, GA (2002)
26. IES: MacroFlo Calculation Methods [Online]. Available: http://www.iesve.com/content/downloadasset_9259, last accessed 2017/4/15. IES (2015d)
27. IES: MacroFlo User Guide [Online]. Available: https://www.iesve.com/support/userguides?page=7, last accessed 2016/1/10. IES (2015e)
28. IES: Tabular Building Template Manager (BTM) User Guide [Online]. Available: https://www.iesve.com/support/userguides?page=2, last accessed 2017/1/18. IES (2015f)
29. IES: BTM User Guide—Building Template Manager [Online]. Available: https://www.iesve.com/support/userguides?page=2, last accessed 2017/1/11. IES (2017)
30. Historic England: Listed Buildings. https://historicengland.org.uk/listing/what-is-designation/listed-buildings/. Last accessed 20 Nov 2018
31. BS EN ISO 10456: Building materials and products. Hygrothermal properties. Tabulated design values and procedures for determining declared and design thermal values (2007)
32. CIBSE: CIBSE guide A: Environmental Design. Chartered Institution of Building Services Engineers, London (2015)
33. Černy, R., Kunca, A., Tydlitat, V., Drchalova, J., Rovnanikova, P.: Effect of pozzolanic admixtures on mechanical, thermal and hygric properties of lime plasters. Constr. Build. Mater. **20**, 849–857 (2006)
34. Theodoridou, M., Kyriakou, L., Ioannou, I.: PCM-enhanced Lime Plasters for Vernacular and Contemporary Architecture. Energy Procedia **97**, 539–545 (2016)
35. Vejmelkova, E., Keppert, M., Kersner, Z., Rovnikova, P., Černy, R.: Mechanical, fracture-mechanical, hydric, thermal, and durability properties of lime–metakaolin plasters for renovation of historical buildings. Constr. Build. Mater. **31**, 22–28 (2012)
36. IES: Historic Scotland Technical Paper 5: Energy modelling of a mid19th century villa—Baseline performance and improvement options. Edinburgh: Historic Scotland (2009)
37. Wood, C., Bordass, B., Baker, P.: Research into the thermal performance of traditional windows: timber sash windows. English Heritage (2009)
38. IES: Apache-Tables User Guide. https://www.iesve.com/support/userguides?page=7, last accessed 2017/1/2. IES (2015b)
39. Baker, P.: Historic Scotland Technical Paper 1: thermal performance of traditional windows. Historic Scotland, Edinburgh (2008)
40. Pickles, D.: Energy Efficiency and Historic Buildings: Draught-proofing Windows and Doors. Historic England, London (2016)
41. Porritt, S., Shao, L., Cropper, P., Goodier, C.: Adapting dwellings for heat waves. Sustain. Cities Soc. **1**, 81–90 (2011)
42. IES: Tabular Room Data User Guide [Online]. Available: https://www.iesve.com/support/userguides?page=9, last accessed 2017/1/14. IES (2015 g)
43. EST: Measurement of domestic hot water consumption in dwellings. Energy Saving Trust (2008)

# Chapter 26
# Building Insulating Materials from Agricultural By-Products: A Review

**Santi Maria Cascone, Stefano Cascone and Matteo Vitale**

**Abstract** Construction is one of the sectors with the higher $CO_2$ emissions. To evaluate the building environmental impacts, it is necessary to consider all the life cycle phases: from the extraction of raw materials to their disposal. In the last years, the improvement in building energy efficiency has increased the energy savings during the operational phase, without reducing the impacts in the material production. For this purpose, several researches showed that natural materials are comparable to commercial synthetic products in terms of thermal and acoustic performance. In particular, agricultural by-products, used as building thermal insulation, allow positive impact on $CO_2$ emissions. The general aim of this review paper is to find solutions to minimize the energy consumption during building construction, in both raw material extraction and transportation. In particular, this paper carried out a literature review on the use of agricultural waste materials as thermal insulators. The construction technique and the binder type to create panels were assessed, providing a critical analysis on the future perspective in this sector.

## 26.1 Introduction

The problem of the energy supply and the growing demand for natural resources makes it necessary to implement solutions aimed at minimizing the environmental impacts. Furthermore, cities continue to expand their peripheries to accommodate increases in population migrating from rural areas to urban spaces and, consequently, the upward trend in energy demand will continue in the future [1]. In developed countries, the energy consumption for residential and commercial buildings is about 40%

S. M. Cascone · S. Cascone · M. Vitale (✉)
Department of Civil Engineering and Architecture, University of Catania, Via Santa Sofia 64, 95123 Catania, Italy
e-mail: matteo.vitale@unict.it

S. M. Cascone
e-mail: santi.cascone@darc.unict.it

S. Cascone
e-mail: stefano.cascone@unict.it



J. Littlewood et al. (eds.), *Sustainability in Energy and Buildings*, Smart Innovation, Systems and Technologies 163, https://doi.org/10.1007/978-981-32-9868-2_26

of the total energy used [2]. This consumption could be reduced up to 80% through the thermal insulation of existing building envelope as well as new buildings [3]. To this end, it is necessary to use local raw materials deriving from recycling and to optimize their production processes. In addition, local materials reduce the transportation cost. Finally, an analysis on the sustainability of buildings and construction materials should also consider the disposal phase. By comparing different construction techniques, it is possible to observe that a higher level of recyclability can be achieved using natural materials [4].

In this context, it is necessary to improve building energy performance and to examine sustainable solutions for the built environment. In recent years, several researchers developed technological solutions using natural materials to increase both thermal performance and sustainability in buildings. Lachheb et al. [5] examined the thermal and environmental benefits of the spent coffee grounds, through annual simulations for a typical Moroccan house located in Marrakech. The results indicated that the cooling and heating loads of the building can be reduced up to 20% if the natural material is used instead of the conventional one. From an environmental point of view, the proposed material reduced up to 1500 kg of $CO_2$ per year. Another study, carried out by Cascone et al. [6] assessed the performance of a technological solution consisting timber walls, combining the platform frame system with a filling in compressed straw as an insulating material. The proposed system allowed to reach a lower thermal transmittance, around by 12%, a higher periodic thermal transmittance, around by 12%, as well as the increase thermal lag of about half an hour in comparison with the XLAM stratigraphy.

The objective of this research is to analyse the recent applications of natural insulating materials from agricultural wastes and to compare the different production techniques. The research carries out a literature review of more than scientific papers. For each study analysed, the laboratory results and the manufacturing process are described.

## 26.2 Thermal Insulators

Insulating materials aim to ensure adequate performance and to allow significant energy savings. However, thermal performance is not the only requirement to be assessed. Other parameters such as sound insulation, fire resistance and vapor permeability should also be evaluated. Furthermore, the impacts on both environment and human health by means of Life Cycle Analysis (LCA) should be considered.

Insulating materials derived from petrochemical products, such as polystyrene, or from natural materials using high quantity of energy resources for the manufacturing process, such as glass wool and rock wool, cause harmful effects on the environment. Improving the energy efficiency and the production process could offer a great reduction in $CO_2$ emissions and could decrease the exploitation of non-renewable resources.

It is possible to distinguish two types of insulating products, depending on the raw materials and construction techniques used. The first type is the result of technological innovation. This category includes VIPs (Vacuum insulation Panels), GFP (Gas Filled Panels) and aerogels, which combine lightness, low thickness and high performance [7]. The second category includes natural materials. Further research is needed to improve the manufacturing process of these materials. These materials are based on natural fibers and they are becoming more and more known and commercially widespread [8]. This is due to their easy availability, the low cost, and the lower environmental impact.

### *26.2.1 Natural Insulating Materials*

**Corn cob**. World production of maize was more than 600 million tonnes in 2003, slightly more than rice and wheat. In the international production stage the corn cob is treated as an agricultural waste that needs to be burned, with consequent negative environmental impacts. Instead, these wastes could be transformed into panels to be used in construction as thermal and acoustic insulators [9]. In some ancient buildings located in the north of Portugal, these scraps were previously used as filler materials to isolate the infill walls of the houses. Experimental works show significant similarities in terms of microstructure and chemical composition between a corn cob and extruded polystyrene (XPS). Further tests have evaluated, in addition to the thermal insulation, the acoustic insulation of a panel produced with crushed waste of corn cob. The results show that the acoustic performances are comparable with products commercially used such as glass wool or expanded polystyrene. Good thermal insulation properties, combined with acoustic performance, make the panel suitable for insulating walls or ceilings. Experimental panels were created using urea formaldehyde resin binder, pressed and dried in temperature controlled ovens [10]. A similar procedure involves pressing with a hot plate at high temperature without binder [11]. Other panels were made by tying the corn cobs with wood glue. This process, using natural polymerization, has obtained satisfactory thermal characteristics.

**Cork**. Cork is a natural material that has been used since ancient times and is nowadays mainly used for the bottle stoppers production. During the manufacturing process, over 75% of it becomes waste product. In addition, a large amount of cork comes from the cleaning of forests and pruning activities. This material could be shredded and recycled to obtain cork granules. The production of cork granules is a sustainable solution that recycles a waste product while substantially maintaining the characteristics of the original material. This latter can be used in the building sector as an insulating panel consisting of agglomerates. A great advantage of this material is the possibility to self-binding thanks to the resins possessed inside. The process of pressing at high temperatures allows the fusion of the naturally contained resins in the bark, which act as a natural glue to aggregate the granules and form the panel [12]. The roasting does not alter the characteristics of the cork, on the contrary it improves it, because it allows the granule to expand, incorporating air and improving its insulation characteristics. It has also been shown that cork and plaster are mutually compatible and that new building materials can be made by mixing these components in different fractions of volume. The results of the experimental tests carried out show unsatisfactory mechanical strength values, making the addition of reinforcing fibers necessary. The result of thermal conductivity tests show good thermal insulation properties (0.18 W/mK) compared to a normal plasterboard (0.30 W/mK). Concerning the acoustical insulation characteristics, this composite is not a sound-absorbing material but a reflecting one, and it needs some kind of perforations to behave as an absorbing construction material for sound and noise [13].

**Durian**. Durian (Durio zibethinus) is a fruit produced mainly in the countries of Southeast Asia. The peel is a waste product of the processing industry and can be used for various purposes including the construction of insulating panels for the building industry. The manufacturing process involves shredding the peel and mixing it with the resin in a humidity controlled rotating drum. Subsequently the mixture is pressed at high temperature and left to dry for 24 h. The results obtained from the tests on the experimental samples are satisfactory for the use of the material as insulation for the building industry [14].

**Coconut**. Coconuts typically grow in coastal areas of tropical countries. The high content of lignin present in coconuts allows the use of resins held by the material as a natural binder. The panel can therefore be obtained naturally and without the addition of chemical binders. Experiments were carried out on crushed coconut peels and pressed at 108 °C. The resistance results obtained are comparable to those of MDF wood; these characteristics of resistance are maintained even after immersion in water, resulting overall better than wood [15]. The coconut fibers, after being dried, have also been analyzed in panels created with the addition of chemical and natural adhesives. A study has tested the possibility of creating a high temperature pressed panel with the addition of Urea Formaldehyde and two-component polyurethane resin based on castor oil. The adhesive with the best thermal characteristics turned out to be polyurethane. The latter, besides having a natural origin and respecting the environment, fills the spaces between the fibers better, giving better strength and durability characteristics [16]. Overall, the thermal properties (thermal conductivity of 0.054 W/mK) of the material make it suitable for application as an insulating for walls and ceilings, representing a viable alternative as a sustainable building product.

**Hemp**. Hemp-based building materials are used in non-load-bearing walls, such as finishing plasters and floor or roof insulation. The product is commonly marketed in non-structural blocks composed of a mixture of hemp, lime and cement. A variant introduced by an Italian company plans to create a panel with a binder based on royal jelly, creating panels without chemical glues and completely natural [17]. Hemp chipboard is a material rich in micropores, which allow continuous microcondensation and evaporation processes, thus providing the product with a high thermal, acoustic, and hygrometric insulation. Further characteristics are: good thermal inertia, ability to accumulate heat and release it slowly, good transpiration, absence of toxic fumes in case of fire, low consumption of energy possessed and recyclability at the end of life. The excellent properties of the material make it suitable also for the composition of mortars and plasters, using lime as a binder.

**Walnut shell**. The shells of the walnuts are a waste material available in large quantities in different parts of the world. China is the main producer in the world and is the first interested in finding a use of these waste, other than incineration or landfill. A good proposal to use these by-products is the creation of MDF wood panels with addition of crushed walnut shells [18]. These panels use a urea formaldehyde resin as a glue, a product known for the release in the environment of volatile substances dangerous for human health. It has been shown that the addition of nut shells in MDF panels improves the internal quality of the air through a lower emission of polluting organic substances. The thermal properties of the material are given by the good porosity and have shown a thermal conductivity of 0.076 W/mK.

**Peanut shell**. Most of the national peanut production, estimated at 900,000 tons per year, is processed in several cities in the central area of Argentina. From the processing of peanuts derive different products including shells, which are separated in the selection and processing plants. Usually the shells are disposed of by incineration or used as fuels for boilers. Incineration of the shells causes significant environmental impacts, such as the production of large amounts of $CO_2$ and micro-particles. Peanut shells are an abundant waste material with no market value, but with potential use as a building material [19]. A study has shown that the aggregation of this material with plaster creates a product with good thermal insulation characteristics (0.14 W/mK) [20]. This mixture improves the performance of the plaster, allowing you to obtain panels with better performance or can be applied as wall cladding. In addition to improving the performance of plaster, peanut shells can be used to make panels [21]. The panel manufacturing process is based on hot pressing using a polyester based adhesive. The thermal conductivity of 0.043 W/mK allows its application in building as thermal insulation [22].

**Bamboo**. Thanks to its rapid growth, Bamboo is one of the best sources of lignocellulosic material in tropical and sub-tropical areas. The uses in construction of this material are very common as composite panels. Furthermore, a use of the waste resulting from planing has been studied. The sawdust of bamboo, believed to be a by-product, can be bonded with urea formaldehyde resin to form highly resistant corrugated panels. The panel molding process is performed with a hot press, then two sheets of MDF are glued [23].

**Rice**. Rice hulls are a potential insulating material for construction deriving from the processing of rice. The latter is the primary source of food for half of the world's population producing large quantities of by-products. Testing on bulk material showed, besides an advantageous thermal resistivity (0.05 W/mK), good fire resistance, corrosion resistance and water absorption. Regarding the production of insulating materials, rice straw was used to make panels by hot pressing at high frequency [24]. This procedure allows to reduce the pressing times and increase the final resistance of the panel, maintaining a conductivity of 0.052 W/mK [25].

**Pineapple**. An alternative resource from agricultural residues are pineapple leaves. Traditionally, after the collection of the products, the residues (stem, pineapple leaves, etc.) are destined to incineration, causing environmental problems of pollution. The industrial use of agricultural residues prevents air pollution due to the combustion of residues, which has negative effects on air quality, human health and the environment, and is also economically beneficial for farmers. The production of chipboard panels based on pineapple leaves has the idea of creating a construction material with low thermal conductivity. The particle board can be produced by spraying the natural rubber latex pretreated on pineapple fiber with panels density ranging from 178–232 kg/m$^3$. The adhesive used allows, in addition to reducing production costs, to preserve the environment from pollutants typical of chemical glues. As for the thermal properties, it has been seen that the K of the pineapple (0.039 W/mK) is comparable with the glass wool and polystyrene foams, making this product suitable for use as a thermal insulator [26]. The material coming from agricultural waste is composed of more than 80% cellulose and can also be used as fibrous polyester reinforcement, also allowing an improvement of the thermal performance characteristics [27].

**Opuntia**. A plant that is widespread in the South of Italy is the Opuntia ficus indica. The results obtained on thermal insulation systems that recycle the waste material of the Opuntia ficus indica pruning coming from very widespread and lush plantations in Sicily are satisfactory. In particular, this material difficult to dispose of was treated to obtain an insulating, in the form of a panel or in loose grains, which demonstrated competitive thermal insulation values (0.071). The same material has also been evaluated as bulk filling material, with much better conductivity values (0.057 W/mK). The thermal conductivity of Opuntia ficus indica granules are comparable to that of vermiculite, less than those of expanded clay, pumice, granulated foam glass and cellulose granules [28].

**Sunflower**. From the agricultural scraps of sunflower oil extraction an element composed mainly of ligninocellulosic fibers and proteins is obtained. During the pressing of this by-product the proteins and the fibers act respectively as binders and as reinforcing fibers, obtaining a completely natural panel free of external adhesives. A possible application of this panel is to thermally insulate walls and ceilings since the thermal conductivity (0.038 W/mK) is similar to other commercial insulators such as polystyrene or rock wool [29].

**Sugarcane**. Bagasse is a waste product from sugar cane processing produced in large quantities and mainly intended for combustion. The sugar cane is a material made of lignocellulosic fibers and represents a valid alternative to commercial synthetic insulators. Studies on the thermal conductivity of some fibrous natural materials show that sugar cane has excellent thermal performances (0.046 W/mK), declaring it an excellent product for insulation [30]. The bagasse normally contains—depending on the variety of cane, maturity and method of collection—residual sugars. The latter may cause problems in the production of resin bonded panels as they may not be chemically compatible and may interfere with the bond. The possibility of making the panel without adding external products, by means of a hot pressing process, was therefore evaluated. Regarding the mechanical characteristics, the panel exploits the properties of the lignocellulosic material of which it is composed obtaining acceptable results [31].

**Coffee**. Coffee is one of the most frequently consumed beverages in the world. A product obtained during the roasting process of coffee beans is the chaff. Recent experimental studies have evaluated the use in construction of this waste material. The manufacturing process of the panel was analyzed in three variants: hot-pressed, cold-pressed and with the addition of external glues. From a comparative analysis of the thermal and acoustic characteristics, it results that the panel produced with cold pressing is the best. The latter represents the best production process for sustainable panels because of a lower environmental impact [32]. Another by-product of the coffee industry is spent coffee grounds. Considering the enormous quantities of this product spread throughout the world, there have been many proposals to reuse this material. One of these is the thermal improvement of the plasters. The integration of coffee grounds reduces the conductivity of the plaster up to 0.31 W/mK, ensuring a potential buildings energy containment of about 20% [5].

### *26.2.2 Discussion*

This literature review showed that the manufacturing of the insulating panels highly affects the thermal and environmental performance. Table 26.1 summarizes the binder used in the production process for some agro-waste products. In particular, the most widespread binder for wooden panels is urea formaldehyde, causing emissions harmful to human health and high environmental pollution. Some natural materials allow to take advantage of the glues already included inside by means of the pressing during the production process. Cold pressing requires the least amount of energy. However, in some materials, it could not release the natural binders contained. Therefore, the hot pressing is the most used technique because it dissolves the natural resins and favors the bond (see Table 26.1). When this natural bond is not possible, an external glue is added to improve the aggregation between the granules. Another characteristic is the possibility to enhance the thermal performance of lime and plaster, obtaining natural and highly performing blocks or panels. In addition, a great advantage of insulation made from natural fibers is the possibility of recycling them. Future development and application of natural insulating materials from agricultural waste would allow to comply with all the requirements of the circular economy, contributing to the environmental sustainability in building sector.

## 26.3 Conclusion and Research Perspectives

Products derived from agriculture can be used as insulation material for construction because they are considered carbon-negative materials. The literature review showed that the most used production process is the hot pressing. In addition, some agricultural by-products could be combined with various natural binders instead of urea formaldehyde. Currently, there are many crops worldwide producing a large amount

**Table 26.1** Binder used for some agro-waste products

| Binder / Agro-waste | Epoxy resin | Urea formaldehyde resin | Cold pressed | Hot pressed | Hot pressed with binder | Lime | Plaster | Rubber latex | Polyester binder | Royal jelly | Polyurethane glue | References |
|---|---|---|---|---|---|---|---|---|---|---|---|---|
| Corn cob | | x | | | x | | | | | | x | [9–11] |
| Cork | | | | x | | | x | | | | | [12, 13] |
| Durian | x | | | | x | | | | | | | [14] |
| Coconut | x | | | x | x | | | | | | x | [15, 16] |
| Hemp | | | x | | | x | | | | x | | [17] |
| Walnut shell | | x | | | | | | | | | | [18] |
| Peanut shell | | | | | x | | x | | x | | | [19, 20] |
| Bamboo | | x | | | x | | | | | | | [23] |
| Rice | | | | x | | x | | | | | | [24, 25] |
| Pineapple | | | | | | | | x | x | | | [26, 27] |
| Opuntia | x | | | | | | | | | | | [28] |
| Sunflower | | | | x | | | | | | | | [29] |
| Sugarcane | | | | x | | | | | | | | [31] |
| Coffee | | | x | x | | | | | | | x | [5, 32] |

of waste. For example, the citrus processing industry, widespread in the Mediterranean area, represents a potential supplier of raw materials for natural building applications.

## References


1. Pérez-Lombard, L., Ortiz, J., Pout, C.: A review on buildings energy consumption information. Energy Build. (2008)
2. EURIMA: Cost-effective Climate Protection in the Building Stock of the New EU Member States. Beyond the EU Energy Performance of Buildings Directive. (2005)
3. Lechtenböhmer, S., Schüring, A.: The potential for large-scale savings from insulating residential buildings in the EU. Energy Effic. (2011)
4. Vefago, L.H.M.C., Avellaneda, J.: Recycling concepts and the index of recyclability for building materials. Resour. Conserv. Recycl. (2013)
5. Lachheb, A., Allouhi, A., El Marhoune, M., Saadani, R., Kousksou, T., Jamil, A., Rahmoune, M., Oussouaddi, O.: Thermal insulation improvement in construction materials by adding spent coffee grounds: An experimental and simulation study. J. Clean. Prod. (2019)
6. Cascone, S., Catania, F., Gagliano, A., Sciuto, G.: Energy performance and environmental and economic assessment of the platform frame system with compressed straw. Energy Build. (2018)
7. Schiavoni, S., D'Alessandro, F., Bianchi, F., Asdrubali, F.: Insulation materials for the building sector: A review and comparative analysis. Renew. Sustain. Energy Rev. (2016)
8. Panyakaew, S., Fotios, S.: 321: agricultural waste materials as thermal insulation for dwellings in thailand: preliminary results. PLEAA 2008–25th Conf. Passiv. Low Energy Archit. (2008)
9. Pinto, J., Paiva, A., Pereira, S., Bentes, I., Sá, A.B.: Possible applications of corncob as a raw insulation material. Insul. Mater. Context Sustain. (2016)
10. Akinyemi, A.B., Afolayan, J.O., Ogunji Oluwatobi, E.: Some properties of composite corn cob and sawdust particle boards. Constr. Build. Mater. (2016)
11. Sampathrajan, A., Vijayaraghavan, N.C., Swaminathan, K.R.: Mechanical and thermal properties of particle boards made from farm residues. Bioresour. Technol. (1992)
12. Gil, L.: Cork composites: A review. Mater. (Basel). (2009)
13. Hernández-Olivares, F., Bollati, M.R., Del Rio, M., Parga-Landa, B.: Development of cork-gypsum composites for building applications. Constr. Build. Mater. (1999)
14. Khedari J, Charoenvai, S., Hirunlabh, J.: New insulating particleboards from durian peel and coconut coir. Build. Environ. (2003)
15. Van Dam, J.E.G., Van Den Oever, M.J.A., Keijsers, E.R.P.: Production process for high density high performance binderless boards from whole coconut husk. Ind. Crops Prod. (2004)
16. Fiorelli, J., Curtolo, D.D., Barrero, N.G., Savastano, H., de Jesus Agnolon Pallone, E.M., Johnson, R.: Particulate composite based on coconut fiber and castor oil polyurethane adhesive: An eco-efficient product. Ind. Crops Prod. (2012)
17. Legambiente, R., dell'Osservatorio, R-: 100 Materiali per una nuova edilizia. (2016)
18. da Silva, C.F., Stefanowski, B., Maskell, D., Ormondroyd, G.A., Ansell, M.P., Dengel, A.C., Ball, R.J.: Improvement of indoor air quality by MDF panels containing walnut shells. Build. Environ. (2017)
19. Pelozo, G., Cristóbal, A.A. lbert.: Use of wastes from the peanut industry in the manufacture of building materials. Int. J. Sustain. Dev. Plan. 662–671 (2018)
20. Lamrani, M., Laaroussi, N., Khabbazi, A., Khalfaoui, M., Garoum, M., Feiz, A.: Experimental study of thermal properties of a new ecological building material based on peanut shells and plaster. Case Stud. Constr. Mater. (2017)
21. Arup: The urban bio-loop. Milan (2017)

22. Kokoboard: Kokoboard Thermal Resistant. http://www.kokoboard.com/th/test-analysis/kokoboard-thermal-resistant/. (2011)
23. Yang, F., Fei, B., Wu, Z., Peng, L., Yu, Y.: Selected properties of corrugated particleboards made from bamboo waste (Phyllostachys edulis) laminated with medium-density fiberboard panels. BioResources (2014)
24. Wei, K., Lv, C., Chen, M., Zhou, X., Dai, Z., Shen, D.: Development and performance evaluation of a new thermal insulation material from rice straw using high frequency hot-pressing. Energy Build. (2015)
25. Chabannes, M., Garcia-Diaz, E., Clerc, L., Bénézet, J.C.: Studying the hardening and mechanical performances of rice husk and hemp-based building materials cured under natural and accelerated carbonation. Constr. Build. Mater. (2015)
26. Tangjuank: Thermal insulation and physical properties of particleboards from pineapple leaves. Int. J. Phys. Sci. (2011)
27. Idicula, M., Boudenne, A., Umadevi, L., Ibos, L., Candau, Y., Thomas, S.: Thermophysical properties of natural fibre reinforced polyester composites. Compos. Sci. Technol. (2006)
28. De Vecchi, A., Colajanni, S.: Isolamento termico: dal riciclo all'innovazione. (2016)
29. Evon, P., Vinet, J., Labonne, L., Rigal, L.: Influence of thermo-pressing conditions on the mechanical properties of biodegradable fiberboards made from a deoiled sunflower cake. Ind. Crops Prod. (2015)
30. Manohar, K.: Experimental investigation of building thermal insulation from agricultural by-products. Br. J. Appl. Sci. Technol. (2014)
31. Widyorini, R., Xu, J., Umemura, K., Kawai, S.: Manufacture and properties of binderless particleboard from bagasse I: Effects of raw material type, storage methods, and manufacturing process. J. Wood Sci. (2005)
32. Ricciardi, P., Torchia, F., Belloni, E., Lascaro, E., Buratti, C.: Environmental characterisation of coffee chaff, a new recycled material for building applications. Constr. Build. Mater. (2017)

# Chapter 27
# Energy Consumption and Retrofitting Potential of Latvian Unclassified Buildings

**Anatolijs Borodinecs, Aleksandrs Geikins and Aleksejs Prozuments**

**Abstract** Most of the urban housing stock in Latvia as well as in major part of EU cities consists of apartment multistory buildings. The improvements in energy efficiency of these buildings are the key priorities in many countries. However, unclassified buildings have a significant potential for the application of innovative energy efficient measures. This study analyzes the energy consumption of such buildings as military buildings, police departments, and fire stations, with particular attention on retrofitting options for fire station. Similar to apartment buildings these type of buildings are connected to centralized district heating systems. In the scope of this paper energy consumption of unclassified buildings were analyzed based on statistic data. In addition, the impact of renovation measures on energy consumption was analyzed. The fire station was selected as majority of such buildings have a similar floor plan layout and usage profile.

## 27.1 Introduction

Most of the buildings in Latvia has been built between 1945 and 1990s. Since the 1990s, refurbishment of existing bundling's is one of the top priorities for the European Union. Latvia located in cold climate; thus reduction of energy consumption is a key priority. The average length of heating season is 203 day with average temperature 0 °C [1]. Moreover, climate change and the limited amount of fossil fuels make the energy efficiency topic actual for all countries. Studies, on deep retrofitting [2, 3] have shown that heat consumption of apartment buildings varies from 160 up to 250 kWh/m$^2$ and can be significantly reduced by the application of simple retrofitting measures such as thermal insulation. In some cases, the heat consumption for heating and hot water preparation can exceed 300 kWh/m$^2$. Even buildings built after the 1990s consume unnecessarily large amounts of energy in comparison to already well-known, cost-efficient solutions which can be easily applied in retrofitting or new

A. Borodinecs (✉) · A. Geikins · A. Prozuments
Institute of Heat, Gas and Water Technology, Riga Technical University, Kipsalas street 6A, Riga, Latvia
e-mail: anatolijs.borodinecs@rtu.lv



J. Littlewood et al. (eds.), *Sustainability in Energy and Buildings*, Smart Innovation, Systems and Technologies 163, https://doi.org/10.1007/978-981-32-9868-2_27

building construction [2]. Installation of new ventilation systems with heat recovery [3, 4] as well as extra thermal insulation [5] significantly improve energy efficiency of existing buildings. Nowadays a combination of passive and active methods allows significant improvements [6] in newly constructed buildings energy efficiency level with relatively short payback period [7].

Nowadays in Latvia, special attention is paid to retrofitting of multi apartment buildings. While other types of building, such as unclassified buildings, are not adequately addressed through the government support and promotion of implementation of energy efficient measures. According to Latvian law definitions, unclassified buildings include prisons, security force, police, and fire station buildings also barracks, warehouses, and other similar buildings. There is no statistical information on the total share of unclassified buildings in Latvia. It could be roughly estimated that the total number is less than 1%. Nevertheless, such a small share should not be neglected. Firstly, military sector is the large energy consumers in many countries, especially in heating dominated countries [8, 9]. Secondly, retrofitting of such buildings allow to improve thermal comfort and indoor air quality.

Energy savings in unclassified buildings could be achieved in both in cold climate and hot climate by application of thermal insulation and solar energy [10–12]. Such buildings have completely different heat gain and heat loss balance and usage profile in comparison to traditional public and apartment buildings. Also, the hot water consumption differs from apartment buildings and the hourly usage profile is different.

Military buildings have a potential for significant improvement of energy efficiency [13, 14]. Implementation of energy efficient measures can ensure energy reduction of almost 65% in military buildings depending on the climate zone [12, 15]. While the thermal insulation technical solutions of the external building envelope is similar as to those used in typical public buildings, the improvement of energy efficiency of ventilation and hot water systems should take into consideration specifics of unclassified building usage profile. Improvements of ventilation systems with heat recovery allow to provide not only energy efficiency but optimal indoor air quality and stability of indoor air parameters [16].

## 27.2 Methods

The method is based on real measured and monitored data on unclassified buildings thermal energy consumption. The authors have analyzed heat energy consumption military buildings, police departments, and fire stations. Totally data on 24 military buildings, 27 police departments, and 40 fire stations. All analyzed buildings are connected to district heating systems. Thus, data was provided by these buildings' maintenance organizations for the time period 2016–2019.

For theoretical evaluation of energy consumption in fire station the IDA Indoor Climate and Energy (IDA-ICE) 4.8 chosen. The main advantage of the software is a dynamic energy simulation, which includes calculation of heat flows and heat gains, maintained temperatures, sources of heat losses and energy costs to maintain a comfortable temperature. In addition, it takes into account the human comfort level and metabolic rate.

IDA-ICE was validated according to ISO 13791: 2012 "Thermal performance of buildings - Calculation of internal temperatures of a room in summer without mechanical cooling - General criteria and validation procedures guidelines" [5] and [6]. Climate data is taken from ASHRAE 2013 which differs from ASHRAE 2017 data only by 2.5% [17]. The wind profile from ASHRAE 1993. The accuracy of IDA-ICE simulations was multiply tested [8, 10].

### *27.2.1 Energy Consumption of Unclassified Buildings*

The major part of unclassified buildings has been constructed before the 1990s. The typical soviet construction projects were used for mass construction. Mainly brick external walls and unheated attics with very minimal attic slab thermal insulation was used due to low energy prices and limited availability of thermal insulation. One-pipe heating systems and natural ventilation are the most common technical solutions in all types of unclassified buildings except some very specific buildings as garages, ammunition rooms, indoor shooting ranges, etc. In addition to initial poor technical conditions, unclassified buildings have not undergone proper energy management or energy audits due to the data privacy and limited access to such buildings. According to the data of Latvian Ministry of Economy, the average total annual energy consumption for military buildings is 212 kWh/m$^2$ for buildings built before the 1990s. The measured data for unclassified buildings with variable construction dates are collected and showed in Fig. 27.1 and 27.2. The measurement process was

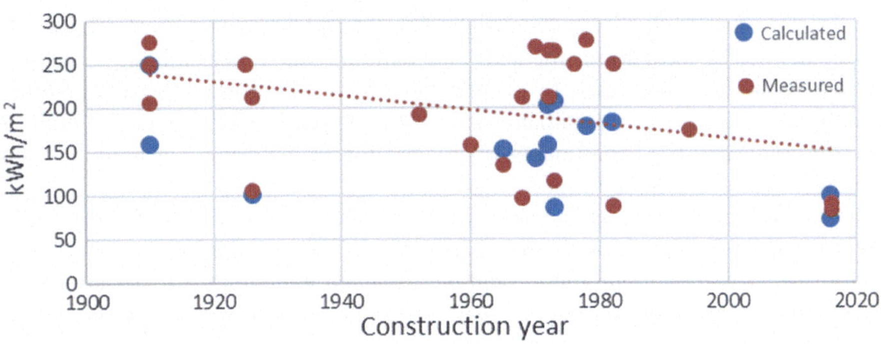


**Fig. 27.1** Calculated and measured total annual energy consumption for military buildings with various construction year; calculation period: 2014–2016 (adopted from source: [18])

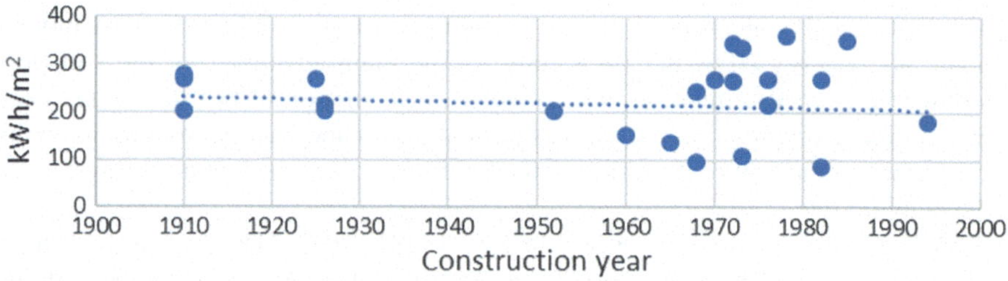


**Fig. 27.2** Calculated total annual energy consumption for military buildings, calculation period: 2011–2014 (adopted from source: [18])

performed in two different periods. The first time the measurements were done in the time period of 2011–2014 while the second time in 2014–2016. Also, for some of the buildings, the calculation of theoretical energy consumption was done according to local regulations.

Figure 27.1 shows that the average energy consumption slightly reduces as the buildings construction date increases. Total average annual measured energy consumption for military buildings is 230 kWh/m$^2$ for the measurements performed in years 2011–2014 (Fig. 27.2). Calculated values are obtained according to Latvian official calculation procedure Regulation No. 348 "Methodology for Calculating the Energy Performance of a Building". This procedure is mainly based on EN ISO 13790:2009 data. Real measure data were provided by the Ministry of Economics.

The analysis for the same group of buildings were repeated three years later to see if there are any changes as during the last year's active information campaigns and EU directives on energy efficiency have started. However, the results showed although there is a slight reduction in energy consumption it is negligible.

The measured energy consumption performed both times is significantly higher than theoretically estimated values. Average theoretical energy consumption estimated by energy auditors for analyzed buildings is 153 kWh/m$^2$ which is by 39% lower than measured during the years 2014–2016. Such difference can occur due to improper definition of such initial set up values as hot water consumption, indoor temperature, supply air exchange, airtightness of building envelope, etc. These values are defined by local norms for apartment buildings and office buildings. The data for unclassified buildings aren't strictly defined by local norms and energy auditors usually take into consideration data for civil buildings. Similar situation is noticed also for police and fire departments where specifics of hot water consumptions and ventilation rates should be taken into account.

The measured average annual energy consumption for police departments is 252 kWh/m$^2$ while for fire departments 317 kWh/m$^2$. The increased energy consumption for fire department buildings could be explained with more strict requirements for ventilation rates and introduction of new technological devices like dedicated ventilation system from firefighting truck exhaust pipes (Fig. 27.3).

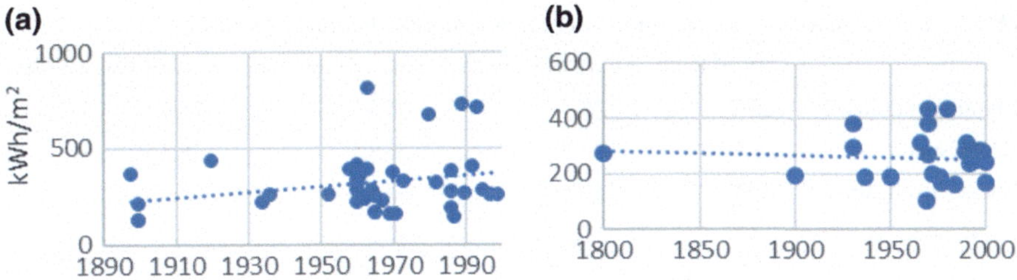


**Fig. 27.3** Measured average total annual energy consumption for fire (**a**) and police (**b**) departments, measurement period: 2011–2014 (adopted from source: [18])

The measured energy consumption of before mentioned buildings significantly exceeds the energy consumption of apartment buildings. Typical Latvian multi apartment annually consumes 190 kWh/m$^2$ for heating and hot water preparation [1].

## 27.3 Fire Department Energy Analyses

In the scope of this study, the impact of renovation measures was evaluated for fire stations. Majority of fire stations have a similar floor layout and occupancy specific. For this purposes, a typical Latvian fire station was chosen (Fig. 27.2). As a first step, energy consumption of non-renovated building was analyzed.

As shown in Fig. 27.4, fire stations were built according to individual projects. However, a typology of the analyzed depots indicates to common construction principles. Namely, the ground floor is used for parking firefighting vehicles whereas the second floor is used for staff needs and administration.

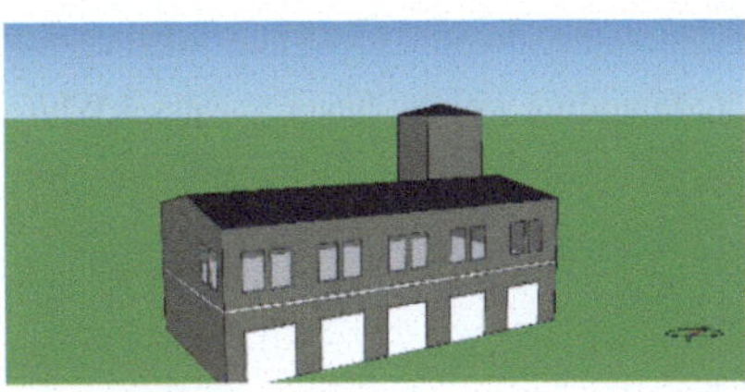

| | |
|---|---|
| Model floor area | 552.0 m$^2$ |
| Model volume | 2531.9 m$^3$ |
| Model ground area | 276.0 m$^2$ |
| Model envelope area | 1245.8 m$^2$ |
| Window/Envelope | 4.0 % |

(a) used model

(b) typical Latvian fire station

**Fig. 27.4** Examples of Latvian fire stations and used model

**Table 27.1** Calculated energy consumption of non-renovated building, $kWh/m^2$

| | City of Riga | City of Liepaja | City of Gulbene | City of Daugavpils |
|---|---|---|---|---|
| Lighting, facility | 25.7 | 25.71 | 25.71 | 25.71 |
| HVAC aux | 28.2 | 25.71 | 28.15 | 28.15 |
| Energy for heating | 337.9 | 319.8 | 385.1 | 367.0 |
| Equipment, tenant | 11.3 | 11.3 | 11.3 | 11.3 |

Table 27.1 presents data on energy consumption of non-renovated buildings in different Latvian cities. Human activity level in garage assumed as 2.0 MET and 1.4 CLO. 1.2 MET and 0.7 CLO were assumed for the office part. 2.0 MET corresponds to standing and medium activity which is relevant to some moderate vehicles maintenance work [19].

The non-renovation scenario includes air infiltration ($q_{50}$) 4.5 $m^3/h\,m^2$ and average U-value of external building envelope 0.90 $W/(m^2\,K)$. Mechanical exhaust ventilation without recovery.

As it can be seen the obtained results are closed to measured data. The difference is less than 10%.

The retrofitting package includes windows' replacement and application of additional thermal insulation. Air infiltration for the retrofitted building is assumed as 1.5 $m^3/h\,m^2$ and average U-value of external building envelope 0.29 $W/(m^2\,K)$. Mechanical exhaust ventilation $n^{-1} = 2$ without recovery (Table 27.2).

Extra scenario takes into consideration exhaust air heat recovery with temperature efficiency of 85% (Table 27.3).

As it can be seen the implementation of exhaust air heat recovery is the most efficient solution to increase the energy efficiency of the fire station. However, more precise study on optimal ventilation rates and optimization of HVAC regimes should be done.

**Table 27.2** Calculated energy consumption of renovated building without heat recovery, $kWh/m^2$

| | City of Riga | City of Liepaja | City of Gulbene | City of Daugavpils |
|---|---|---|---|---|
| Energy for heating | 235.7 | 220.8 | 272.9 | 259.7 |
| Reduction, % | 30 | 31 | 29 | 29 |

**Table 27.3** Calculated energy consumption of renovated building with heat recovery of 85%, $kWh/m^2$

| | City of Riga | City of Liepaja | City of Gulbene | City of Daugavpils |
|---|---|---|---|---|
| Energy for heating | 55.9 | 46 | 73.9 | 68.4 |
| Reduction, % | 83 | 86 | 81 | 81 |

## 27.4 Conclusions

The paper presents the results of calculated and measured data of energy consumption for unclassified buildings that were built before 1990ies. The official data for all unclassified buildings shows that an average measured total annual energy consumptions is 212 kWh/m$^2$ while for separately measured buildings it was about 230 kWh/m$^2$. The calculation of energy consumption performed according to local regulations gave much lower results—153 kWh/m$^2$ which is by 39% lower than measured.

Specifically performed measurements showed that the average annual energy consumption for police departments is 252 kWh/m$^2$ while for fire stations 317 kWh/m$^2$. It must be noted that the energy consumption for fired departments increases for newer buildings due to the installation and use of mechanical supply/exhaust ventilation.

Energy consumption for fire departments is higher than for police departments, which can be explained by architectural peculiarities of this type of buildings. Annual differences in heat energy consumption can be explained not only with different climate conditions, which correlates between fire and police departments. Extra energy is required for extra ventilation of garages and repair shop.

Dynamic energy simulation for typical fire station has shown good correlation with measured data. Based on dynamic energy simulation theoretical energy savings was evaluated. The typical retrofitting package allows 30% reduction of thermal energy for space heating. Installation of exhaust air heat recovery systems allows energy savings of up to 83%. However, a more precise study on operation hours and optimal air change rate should be done.

**Acknowledgements** This study was supported by the European Regional Development Fund project Nr.1.1.1.1/16/A/048 "Nearly zero energy solutions for unclassified buildings".

## References

1. Borodinecs, A., Zemitis, J., Sorokins, J., Baranova, D.V., Sovetnikov, D.O.: Renovation need for apartment buildings in Latvia. Mag. Civ. Eng. **68** (2016)
2. Borodinecs, A., et al.: Specifics of multi-apartment building deep complex retrofitting. In: CESB 2016—Central Europe Towards Sustainable Building 2016: Innovations for Sustainable Future (2016)
3. Korniyenko, S.: Complex analysis of energy efficiency in operated high-rise residential building: Case study. In: E3S Web of Conference (2018). https://doi.org/10.1051/e3sconf/20183302005
4. Pukhkal, V., Vatin, N., Murgul, V.: Central ventilation system with heat recovery as one of the measures to upgrade energy efficiency of historic buildings. Appl. Mech. Mater. (2014). 10.4028/www.scientific.net/amm.633-634.1077
5. Lakatos, Á.: Thermal conductivity of insulations approached from a new aspect. J. Therm. Anal. Calorim., 1–7 (2017). https://doi.org/10.1007/s10973-017-6686-5

6. Petrichenko, M.R., Nemova, D.V., Kotov, E.V., Tarasova, D.S., Sergeev, V.V.: Ventilated facade integrated with the HVAC system for cold climate. Mag. Civ. Eng. (2018). https://doi.org/10.18720/mce.77.5
7. Gorshkov, A.S., Vatin, N.I., Rymkevich, P.P., Kydrevich, O.O.: Payback period of investments in energy saving. Mag. Civ. Eng. (2018). https://doi.org/10.18720/mce.78.5
8. Stavropoulos, G., Skodras, G.: Energy Savings at Military Installations: A Case Study
9. Booth, S., Barnett, J., Burman, K., Hambrick, J., Westby, R.: Net Zero Energy Military Installations: A Guide to Assessment and Planning. NREL Technical Report (2010)
10. Suárez-García, A., et al.: Estimation of photovoltaic potential for electricity selfsufficiency: a study case of military facilities in northwest Spain. J. Renew. Sustain. Energy (2017). https://doi.org/10.1063/1.4995687
11. Caprio, D.M., Soulek, A.B.: MILCON Energy Efficiency and Sustainability Study of Five Types of Army Buildings (2011)
12. Langner, R., Deru, M., Zhivov, A., Liesen, R., Herron, D.: Extremely low-energy design for army buildings: tactical equipment maintenance facility. ASHRAE Trans. (2012)
13. Korniyenko, S.: Evaluation of thermal performance of residential building envelope. Procedia Eng. (2015). https://doi.org/10.1016/j.proeng.2015.08.140
14. Barsi, D., Costa, C., Satta, F., Zunino, P., Sergeev, V.: Feasibility of mini combined cycles for naval applications. In: MATEC Web of Conferences, vol. 245 (2018). https://doi.org/10.1051/matecconf/201824507008
15. Liesen, R., Ellis, P., Zhivov, A., Herron, D.: Extremely low energy design for army buildings: barracks. ASHRAE Trans. **118**, 767–789 (2012)
16. Bilous, I.Y., Deshko, V.I., Sukhodub, I.O.: Building inside air temperature parametric study. Mag. Civ. Eng. (2016). https://doi.org/10.5862/mce.68.7
17. ASHARE. ASHRAE climatic design conditions 2009/2013/2017. www.ashrae-meteo.info/
18. Latvian Ministry of Economy: Data on buildings energy consumption own by government. In: Buildings owned, held and used by public authorities with a total floor area over 250 m2 on 09.07.2018. In accordance with Article 5 (5) of Directive 2010/27/ EU of the European Parliament and of the Council on energy efficiency (2018). https://www.em.gov.lv/lv/nozares_politika/majokli/eku_energoefektivitate/no_direktivas_2012_27_es_par_energoefektivitati_izrietosas_prasibas/. Accessed 2 May 2019
19. Ryńska, E.D., Koźmińska, U., Zinowiec-Cieplik, K., Rucińska, J., Szybińska-Matusiak, B.: Design solutions for nZEB retrofit buildings. Design Solutions for nZEB Retrofit Buildings (2018). https://doi.org/10.4018/978-1-5225-4105-9

# Chapter 28
# Towards a User-Centered and Condition-Based Approach in Building Operation and Maintenance

**Gabriele Bernardini and Elisa Di Giuseppe**

**Abstract** Ensuring a sustainable performance to buildings is a key topic that cannot overlook occupancy conditions and users' behavior. In fact, individuals' actions (man-man and man-built environment interactions) highly affect the overall building efficiency as well as the possibility to ensure the designed level of performance. Such occupancy issues are connected not only to energy consumptions but also to other common building and facility Operation and Management (O&M) tasks, especially in relation to building components and technological systems maintenance (e.g., elevators, doors, flooring, devices, etc.) and building use (e.g., cleaning, visitors' flows, room occupancy, etc.). Monitoring and understanding users' behavioral patterns and their effect on building components through smart and integrated cognitive systems can optimize predictive and corrective actions, by linking a "user-centered" to a "condition-based" approach. This paper critically reviews results from previous works on conditioned-based O&M and proposes improvements to the approach, based on a user-centered point of view. A general framework is proposed by combining monitoring tasks (spaces use, occupants' actions, and flows) with occupants' awareness/engagement, through management and communication platforms. Framework data can be used to derive occupancy profiles (including models of interactions with building devices) and inputs for condition-based analyses, in order to allow designers and Building and Facilities managers to improve actions planning.

## 28.1 Introduction

The design of sustainable built environments is a priority goal that should include the assessment of the impact of users' behavior on building conditions during the whole building life cycle, also considering several possible use scenarios and situations (i.e., normal fruition versus emergency) [1–3]. Occupants adopt specific behaviors depending on surrounding environmental drivers, over time and space, and interact with building devices to reach or restore optimal conditions. In this way, their actions

G. Bernardini (✉) · E. Di Giuseppe
Università Politecnica delle Marche, via Brecce Bianche, 60131 Ancona, Italy
e-mail: g.bernardini@univpm.it



J. Littlewood et al. (eds.), *Sustainability in Energy and Buildings*, Smart Innovation, Systems and Technologies 163, https://doi.org/10.1007/978-981-32-9868-2_28

modify the building conditions themselves by affecting the designed performances: significant differences between expected and effective conditions can appear depending on behavioral issues, according to studies on Post-Occupancy Evaluations (POE) and real-world case study analyses [3–5]. "Corrective" actions are needed to restore or improve the designed performance, thus inducing additional social, physical, psychological, environmental impacts and management costs, especially in existing buildings and during the whole life cycle [6–9]. In this context, public buildings and spaces (i.e., offices, high occupancy density buildings, and transportation hubs) represent the most critical scenarios [4, 10]. Significant examples of action-effect relations among "human desire-interaction with building components-affected building performance" are (but not limited to): (i) maintaining comfort conditions-modifications of building devices status, for instance of heating and cooling systems-energy consumption [1, 4]; (ii) remaining in or moving towards safe spaces-emergency and evacuation response, such as pre-movement and movement actions-building emergency safety [7, 10]; (iii) reaching a journey destination-wayfinding strategies and route selection- visitors' flow, crowd, queueing, and travel time management [10].

Designing sustainable architectural spaces means developing and implementing methods and solutions to optimize these processes while improving users' conditions and limiting individual's actions negative impact. Monitoring and understanding users' behavioral patterns and their effect on building components through smart and integrated cognitive systems can optimize predictive and corrective actions. At this aim, a combination of building monitoring and "user-centered" ("behavioral-based") standpoints can be used [7, 8, 11, 12]. These approaches are typically based on:

1. *analyzing occupants' behaviors* in relation to the surrounding conditions, during the building use and while interacting with devices, systems, and spaces. Real-world scenarios and POE techniques can be used to derive related input data;
2. *modeling evidence-based users' behavioral patterns*, in relation to the considered drivers and additional surrounding conditions (including social ones);
3. *implementing obtained models in simulation software or remote control and interaction systems and validating them* through additional experimental data;
4. *applying them to assess building performance* versus behavioral patterns and then *proposing solutions to improve performance* while increasing users' satisfaction (e.g., comfort, safety, and fruition). Simulation software is mainly used in the design phase, while a remote control and interactions systems are exploited in real scenarios, through case studies, during building operation.

Such processes have been codified and widely applied to building emergency safety [2, 7] and building operations tasks in relation to energy consumption optimization [8, 13]. However, the substantial impact of occupants' behavior on building performances is generally shared by all Building Operations and Maintenance (O&M) issues, including systems repair and building use [14–17]. However, aside from common occupancy issues [16], a general operational framework for buildings "user-centered" O&M seems not to be properly addressed so far. Hence, this work tries to overcome this lack by proposing a novel framework for buildings O&M that takes advantage of occupants' awareness/engagement, through management and communication platforms. The paper provides a critical analysis of existing literature on building O&M (Sects. 28.2), identifying definitions and key challenges to propose a framework for optimizing building O&M (Sect. 28.3), and finally discusses how the results triangulate research gaps and challenges (Sect. 28.4).

## 28.2 Methodology

Building O&M studies selected for this review are published in the last five years and retrieved in the Scopus database, according to the keywords defined in Table 28.1. Over 1490 publications have been retrieved (Table 28.1). The review is performed mainly addressing review papers for main concepts and research works for in-depth analysis and application topics, primarily published in scientific journals.

In recent decades, great attention was given to users' impact on building O&M, as demonstrated by the increasing number of publications in the field (Fig. 28.1). Considering Scopus database, "Engineering" Subject Area, and specific keywords ("building", "maintenance", "operation", "condition-based"), from 1995 to date, the number of publications has more than quintupled reaching over than 420 only in

**Table 28.1** Summary of search results on Building O&M studies in Scopus database, 2014–18

| Keywords, searched within title, abstract, publication keywords | No. of publications (journal) | Main journals (number of publications) |
|---|---|---|
| Operation AND building AND occupant | 455 (297) | Energy and Buildings (85); Building and Environment (44); Applied Energy (20) |
| Operation AND building AND occupant AND/OR behavior | 146 (100) | Energy and Buildings (43); Building and Environment (15); Applied Energy (11) |
| Maintenance AND building AND user OR occupant | 463 (235) | Energy and Buildings (13); Automation in Construction (11); Facilities (10) |
| Maintenance AND building AND condition AND based | 330 (151) | Construction and Building Materials (11); Automation in Construction (7); J. of Perf. of Constructed Facilities (6) |
| Maintenance AND building AND user AND condition | 59 (27) | Facilities (3); Automation in Construction (2); IEEE Sensors Journal (2) |
| Maintenance AND building AND user OR occupant AND/OR behavior | 43 (27) | Energy and Buildings (5); Sensors Switzerland (3); A Z ITU Journal of The Faculty of Architecture (1) |

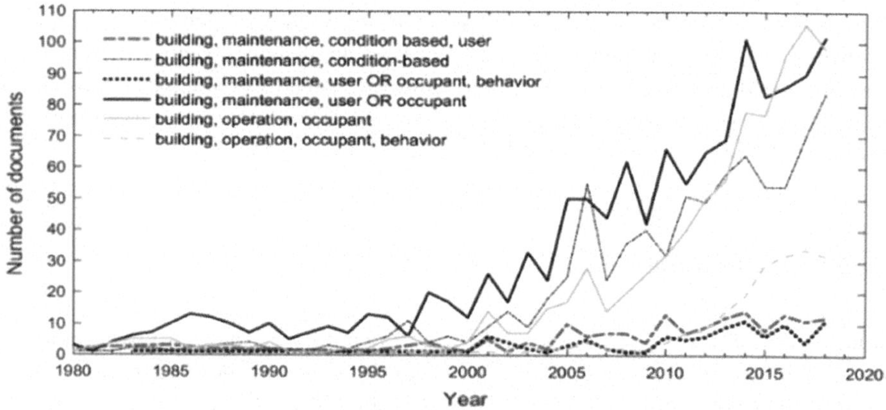


**Fig. 28.1** Trend of publications on Building O&M in Scopus database, "Engineering" Area, period 1980–2019, according to keywords in Table 28.1 (www.scopus.com; last access: 6/3/2019)

2018, while those especially focused on a behavioral standpoint (previous keywords plus "user", or "occupant", or "behavior") mainly concerns energy consumption and grew during the last five years, reaching 188.

## 28.3 Findings

### *28.3.1 Building Operation Literature Overview*

"*Once a building is constructed and operational*" [3], the performances connected to the building life cycle are included within the Building Operations issues [17]. Ensuring good Building Operations means optimizing the building exploitation by means of technological (on building services) and management (on building planning use and users' actions) strategies [16]. Main investigated topics concern [3, 4, 6, 10, 15, 16, 18]: utilities and related consumptions; building services functioning such as heating, lighting, cooling, pumping, elevators, and safety systems; effects on the Indoor Environmental Quality (IEQ); effects on the continuous operation of the building in terms of systems safety. Recent works tried to include additional building use and fruition tasks, such as building access towards security tasks and spaces use planning, by including operational strategies in relation to occupants' flows and movement [10, 18, 19]. Collected data generally consider, at least, operational status (e.g., time-dependent consumption) and environmental conditions measurement [3, 6, 18]. Occupancy conditions detection over time and space, such as duration, number of occupants, and users' actions, is also provided when a "user-centered" approach is considered [4, 8, 19, 20]. POE is used to provide data in several scenarios, define gaps between designed and real performance, develop simulators for the design phase, and provide automated systems to optimize the use of resources or even to interact with people [1, 4, 13, 15, 16].

Building Automation Systems (BAS) and Internet of Things (IoT) technologies are applied in several case studies [18, 21], focusing on data collection about building energy and occupants' comfort [4, 12]. Applications addressed to individuals' movement seems to be mainly limited to: transportation hub activities such as boarding; elevators operation towards occupants' flows; systems for occupants' evacuation support [7, 10]. Attempts to define a general management system considering several building performances have been performed, usually preferring BIM-based solutions because of their capabilities in database management, analysis and sharing among stakeholders [19, 22, 23]. Main limitations deal with [18, 24]: traditional use of BASs only to control indoor climate; limited databases; lack of a general framework and unique BAS to avoid the overlapping of functioning and data collection by different monitoring and interaction systems/networks; partial real time applications including interactions with users.

### *28.3.2 Building Maintenance Literature Overview*

Maintenance strategies are complementary solutions to ensure a proper building exploitation and performance during the life cycle since they avoid or correct failures by restoring and/or keeping up existing functions or elements [14]. From a regulation perspective, "EN 13306:2017. Maintenance - Maintenance terminology" is the reference standard for the definitions in this field. Two main categories of maintenance are outlined, depending on when the fault of the building element (damage of the component or a break in a certain operation) is detected [14, 18, 25, 26]:

- before the fault, *planned (preventive) maintenance*, to evaluate and mitigate the degradation of the element, to prevent or reduce the probability of fault. *Time-Based Maintenance* (TBM) does not consider the actual condition of an item and could lead to over-maintenance (restoration at the scheduled time also in the absence of degradation). *Condition-Based Maintenance* (CBM) occurs after the direct observation of degradation through inspection by professionals or remote-control systems, such as BAS, and could include the application of tests and systems metrics monitoring at scheduled time intervals, on request or according to a continuous detection approach. CBM could adopt a *Predictive Maintenance* approach to optimize actions, based on the assumption that "*most abnormalities do not occur instantaneously*" but are due to degradation [25]. Monitoring allows detecting abnormalities that could lead to probable faults, also according to prediction models' outcomes[1];

[1] *Proactive Maintenance* tries to improve predictions by aggregating data from different systems, to trace dependencies among them. *Opportunistic Maintenance* combines maintenance activities in the same moment, to reduce the intervention costs.

- after the fault, *Corrective Maintenance* (CM) to restore required functions without delay or within a reasonable delay depending on building use and maintenance rules.

The abnormality/fault should be detected as soon as possible to improve the effectiveness of actions implementation. Hence, CBM-based strategies improve this process [14, 25]. BAS are essential for real time monitoring and prediction systems, as confirmed by several works on building services, such as lighting, heating, and cooling [16, 18, 26, 27]. Detection problems can occur to devices which are generally not directly monitored, such as doors, flooring, and façade elements. Some studies then suggest the importance of end users' and stakeholders' engagement by using BASs in combination to BIM-based or mobile application-based networks to provide and collect information on the building status [21, 23, 28], as well as to define the users' satisfaction level [29].

### *28.3.3 Key Challenges in O&M: Uncertainties and Human Factors*

The review highlights Key Challenges (KC) to be investigated [1, 3, 14, 18, 30]:

- **KC1**. Uncertainties and lack of data about existing buildings and building components/services/devices, which increase problems in O&M actions, BAS implementation, and systems compatibility;
- **KC2**. Expected versus real performance of the elements, even in case of similar elements under similar building conditions, because of construction-related factors, such as old and new/restored components compatibility issues and on-site works;
- **KC3**. Expected versus real performance of the elements (focusing on maintenance due to degradation) because of occupancy, users' actions and particular building operational conditions, for instance, use of spaces during the time, differences in hosted activities. Combining CBS and UCS could support this challenge;
- **KC4**. Forecasting costs and actions scheduling due to: (a) KC1-related factors; (b) incoming innovative techniques that could be applied in the future; (c) changes in expected use/fruition; (d) macroeconomic scenario and budget uncertainties.[2]
- **KC5**. Improving the dialog among users, stakeholders, building managers, and maintenance professionals to have adequate and immediate feedback on the building devices status (i.e., breaks/faults) and to make people aware of their behavior's impact. UCS could support this challenge.

[2]Points (b) and (c) can lead to delays in actions performing ("Waiting option") [14].

### *28.3.4 A Novel Framework for Optimizing Building O&M*

A novel framework is proposed to this aim (Fig. 28.2), combining *Conditioned-Based Systems* (CBS) and *User-Centered Systems* (UCS). In this framework, *creating database*, *simulating to plan* and implementing *actions* are considered as the main actions and performed through *monitoring* (i.e., *Occupancy & users' behavior monitoring* and *Building monitoring*) and *users' engagement* techniques.

CBS monitors environmental quality conditions, performance and durability of building devices and components, by onboard sensors [10, 15, 18]. Devices connected to building utilities can be directly controlled (e.g., cooling, heating, and elevators). Devices like doors and windows can be monitored by using on-device sensors, as contact sensors: people interact with them by changing their status, and so affecting operations (e.g., energy) and maintenance (e.g., fire doors use and interventions on degradable parts, like hinges and handles) [8, 10]. "Hard -to-be-monitored" devices like flooring, ceilings, and walls, should be monitored through occupancy detection systems (for example flooring degradation in public buildings also depends on vis-

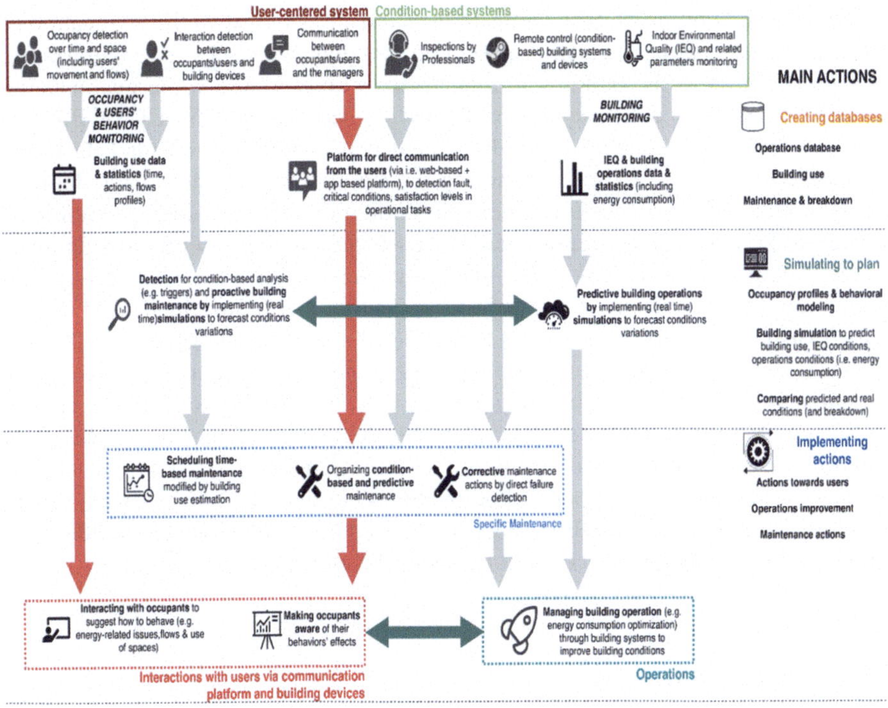


**Fig. 28.2** Proposed O&M framework, including: involved monitoring systems (continuous line boxes); adopted techniques (capitalized terms); main actions and related tasks (description and icon); "one" and "two" ways (green) dependencies (red ones relate to a high users' engagement); main system outputs (dashed line boxes)

itors' flows), and by users' feedbacks [7, 8, 10, 21]. CBS also retrieve data from professionals' inspections.

UCS adopts sensors of different types [8, 12, 18, 20]: on-room, like passive infrared sensors for presence detection; on-device, for door/windows/blinds contact sensors; electric input power sensors; individual such as RFID on badges (suitable for controlled access spaces) smartphone application, wi-fi heatmaps services (especially for wide public spaces with a significant presence of occupants during the time). Then, *creating database* uses UCS outputs to derive occupancy profiles and evidence effective trends in spaces use over time depending on performed activities [10]. Since occupancy profiles can also vary over the building lifecycle, evolutionary fruition and operation models can be derived to forecast future scenarios, especially in case of building adaptation [19]. Collected occupancy profiles and actions patterns provide consistent data for *building simulation* models, mainly concerning energy, visitors' movement, and flows, use of elevators, safety and space fruition for crowd management [7, 18, 27]. Simulation tasks can be so applied within the BAS for planning purpose or real time *predictive operations*: short terms prediction can apply actions aimed at timely improving users' satisfaction and building performances [7, 12].

The combination of UCS and CBS in the same model ensures the application of *proactive approaches*, especially in building maintenance issues like optimization of visitors' flows in the building to reduce stress on elevators while guaranteeing adequate travel time and safe conditions in crowded spaces [26]. UCS-CBS data could be then used to compare predicted and real conditions and for Life Cycle Cost estimation [1, 3, 7]. Then, monitoring and simulation data allow better defining O&M outputs for *implementing actions* depending on effective arising conditions, to manage [15, 26, 30]: operations and TBM scheduling aimed at minimizing interferences with occupancy tasks and improving users' satisfaction; data-driven and simulation-driven predictive (or proactive) operations and CBM; timely CM. In this sense, *users' engagement* in respect to the O&M status can be encouraged by the UCS, via BAS-connected *platform for direct communication*, such as web-based communication platform with apps interface [24, 29], by: (1) collecting feedback on users' satisfaction, information on elements degradation/fault (i.e., processes in buildings, e.g., cleaning, space fruition; "hard-to-be monitored" components); (2) interacting with them to support "good practices" in building use by making them aware of benefits for themselves (e.g., thermal comfort; flows inside buildings to comfortably/quickly move), community and stakeholders.

## 28.4 Discussion

The outcoming proposed framework can respond to gaps in knowledge and practice evidenced by Key Challenges. Table 28.2 discusses how combining CBM and UCS could support a more effective approach to improve O&M during the building lifecycle, thanks to its compliance with Sect. 28.1 general literature criteria, main

**Table 28.2** Compliance of the framework with respect to the main Key Challenges (KC) defined in Sect. 28.3.3, in terms of framework solutions and related main tasks and blocks in Fig. 28.2

| KC | Framework solutions oriented to: Operation [O]; Maintenance [M]; both [O&M] | Actions (eventual essential "framework tasks")/references |
|---|---|---|
| KC1 | Knowledge-based approach to the building by means of structured databases (BIM) and analytics (as data mining techniques on available data) [O&M] | Creating database (-)/ [22, 23] |
| KC2 | Providing direct devices monitoring in BAS to move from TBM to CBM [M]; shared databases to trace trends in elements compatibility and condition-based degradation [O&M]; developing Predictive Models [O&M] | Creating database ("Remote control"); Simulating to plan ("Proactive building maintenance", "Predictive building operations and simulations")/ [14, 18, 25] |
| KC3 | Implementation of UCS in BAS to include analysis of users' actions [O&M]; simulations and scheduling of occupancy according to a probabilistic approach [O]; moving towards predictive and proactive standpoints by integrating detection systems and simulation tools [O&M] | Creating database ("Building use"); Simulating to plan ("Proactive building maintenance", "Predictive building operations and simulations")/ [10, 18, 20, 30] |
| KC4 | Managing O&M actions depending on available and expected budget through simulation systems [O&M]; combining simulations and scheduling aspects with Life Cycle Cost analysis in different implementation scenarios [O&M] | Simulating to plan (-); Implementing actions (from the users' standpoint: "Building use statistics")/ [3, 9, 10, 21] |
| KC5 | Promoting users-stakeholders-professionals connections by means of a unique communication platform and related shared knowledge-based databases [O&M] | Creating database ("Platform for direct communication"); Actions implementation ("Interactions with users")/ [22, 24, 29] |

scientific references, and fundamental key challenges (see Sect. 28.3.3). *Occupancy & users' behavior* and *Building monitoring* techniques provide analysis of users' behaviors in respect to surrounding conditions to identify main drivers (for instance: heating system use against IEQ; wayfinding choices against estimated traveling time in case of queueing).

Then, CBS and UCS are based on concepts of modularity, easy implementation and maintenance (use of wireless connection and low energy devices) required by literature [7, 10, 18]. Future works should apply the framework to real case studies, by taking advantages of existing techniques for monitoring, simulation, and communication with users. Implementation and capability demonstration should start from maintenance tasks, which seem to be underestimated in comparison to operational safety and energy-related topics in current state of art.

## 28.5 Conclusions

Sustainable buildings need to optimize available resources by guaranteeing users' satisfaction. Developing systems able to jointly monitor the built environment ("condition-based" approach) and understand occupants' actions ("user-centered" approach) could lead to substantial optimization of costs and impacts due to O&M during the building use phase, which is linked together. This paper reports an overview of main definitions and key challenges in the field and proposes a general framework to effectively connect "condition-based" to "user-centered" approaches in buildings O&M. For the "user-centered" standpoint, the framework highlights the need for defining unified systems to derive users' actions and occupancy profiles, by including the modeling of interactions with building devices. Combining such issues to condition-based analyses allows building designers and managers to improve actions planning, towards a proactive O&M.

## References


1. Hong, T., Yan, D., D'Oca, S., Chen, C.: Ten questions concerning occupant behavior in buildings: the big picture. Build. Environ. **114**, 518–530 (2017)
2. Kobes, M., Helsloot, I., de Vries, B., Post, J.G.: Building safety and human behaviour in fire: a literature review. Fire Saf. J. **45**, 1–11 (2010)
3. de Wilde, P.: Ten questions concerning building performance analysis. Build. Environ. (2019)
4. Stazi, F., Naspi, F., D'Orazio, M.: A literature review on driving factors and contextual events influencing occupants' behaviours in buildings (2017)
5. Manahasa, O., Özsoy, A.: Do architects' and users' reality coincide? A post occupancy evaluation in a university lecture hall. A/Z ITU J. Fac. Archit. **13**, 119–133 (2016)
6. O'Brien, W., Gaetani, I., Carlucci, S., Hoes, P.-J., Hensen, J.L.M.: On occupant-centric building performance metrics. Build. Environ. **122**, 373–385 (2017)
7. Bernardini, G.: Fire Safety of Historical Buildings. Traditional Versus Innovative "Behavioural Design" Solutions by Using Wayfinding Systems. Springer International Publishing (2017)
8. Naylor, S., Gillott, M., Lau, T.: A review of occupant-centric building control strategies to reduce building energy use. Renew. Sustain. Energy Rev. **96**, 1–10 (2018)
9. Baldoni, E., Coderoni, S., D'Orazio, M., Di Giuseppe, E., Esposti, R.: The role of economic and policy variables in energy-efficient retrofitting assessment. A stochastic life cycle costing methodology. Energy Policy **129**, 1207–1219 (2019)
10. Dong, B., Yan, D., Li, Z., Jin, Y., Feng, X., Fontenot, H.: Modeling occupancy and behavior for better building design and operation—a critical review. Build. Sim. **11**, 899–921 (2018)
11. Ahmadi-Karvigh, S., Becerik-Gerber, B., Soibelman, L.: Intelligent adaptive automation: A framework for an activity-driven and user-centered building automation. Energy Build. **188–189**, 184–199 (2019)
12. Pereira, P.F., Ramos, N.M.M.: Detection of occupant actions in buildings through change point analysis of in-situ measurements. Energy Build. **173**, 365–377 (2018)
13. Annex 66: Definition and Simulation of Occupant Behavior in Buildings: Technical Report: Studying Occupant Behavior in Buildings: Methods and Challenges (2017)
14. Lind, H., Muyingo, H.: Building maintenance strategies: planning under uncertainty. Prop. Manag. **30**, 14–28 (2012)
15. Yoshino, H., Hong, T., Nord, N.: IEA EBC annex 53: total energy use in buildings—analysis and evaluation methods. Energy Build. **152**, 124–136 (2017)

16. Borgstein, E.H., Lamberts, R., Hensen, J.L.M.: Mapping failures in energy and environmental performance of buildings. Energy Build. **158**, 476–485 (2018)
17. Suweero, K., Moungnoi, W., Charoenngam, C.: Outsourcing decision factors of building operation and maintenance services in the commercial sector. Prop. Manag. **35**, 254–274 (2017)
18. Burak Gunay, H., Shen, W., Newsham, G.: Data analytics to improve building performance: a critical review. Autom. Constr. **97**, 96–109 (2019)
19. Simeone, D., Coraglia, U.M., Cursi, S., Fioravanti, A.: Behavioural simulation for built heritage use planning. In: 34th International Conference on Education and Research in Computer Aided Architectural Design in Europe, Oulu, pp. 503–510 (2016)
20. Chen, Z., Jiang, C., Xie, L.: Building occupancy estimation and detection: a review. Energy Build. **169**, 260–270 (2018)
21. Tang, S., Shelden, D.R., Eastman, C.M., Pishdad-Bozorgi, P., Gao, X.: A review of building information modeling (BIM) and the internet of things (IoT) devices integration: present status and future trends. Autom. Constr. **101**, 127–139 (2019)
22. Tan, A.Z.T., Zaman, A., Sutrisna, M.: Enabling an effective knowledge and information flow between the phases of building construction and facilities management. Facilities **36**, 151–170 (2018)
23. Gao, X., Pishdad-Bozorgi, P.: BIM-enabled facilities operation and maintenance: a review. Adv. Eng. Inform. **39**, 227–247 (2019)
24. Vellei, M., Natarajan, S., Biri, B., Padget, J., Walker, I.: The effect of real-time context-aware feedback on occupants' heating behaviour and thermal adaptation. Energy Build. **123**, 179–191 (2016)
25. Shin, J.-H., Jun, H.-B.: On condition based maintenance policy. J. Comput. Des. Eng. **2**, 119–127 (2015)
26. Nowakowski, T., Tubis, A., Werbińska-Wojciechowska, S.: Evolution of technical systems maintenance approaches—review and a case study. In: Intelligent Systems in Production Engineering and Maintenance. ISPEM 2018, pp. 161–174 (2019)
27. Cauchi, N., Macek, K., Abate, A.: Model-based predictive maintenance in building automation systems with user discomfort. Energy **138**, 306–315 (2017)
28. Cha, H.-S., Kim, J., Kim, D.-H., Shin, J., Lee, K.-H.: Mobile application tool for individual maintenance users on high-rise residential buildings in South Korea. In: MATEC Web of Conference, vol. 167, p. 01002 (2018)
29. Pontan, D., Surjokusumo, S., Johan, J., Hasyim, C., Setiawan, M.I., Ahmar, A.S., Harmanto, D.: Effect of the building maintenance and resource management through user satisfaction of maintenance. Int. J. Eng. Technol. **7**, 462–465 (2018)
30. Silva, A., de Brito, J.: Do we need a buildings' inspection, diagnosis and service life prediction software? J. Build. Eng. **22**, 335–348 (2019)

# Chapter 29
# Towards a Near-Zero Energy Landmark Building Using Building Integrated Photovoltaics: The Case of the Van Unnik Building at Utrecht Science Park

**Wilfried van Sark and Eelke Bontekoe**

**Abstract** We assess the feasibility of renovation of a 22-story high-rise building from the 1960s to realize a near-zero energy building by cladding all usable parts of facades and roof using building integrated photovoltaic (BIPV) components. With the present building electricity demand, which includes all energy demand of the building except for heating, it is not possible to generate all demand by BIPV: an annual self-sufficiency ratio of 0.666 or 0.756 is found, using two different roof designs, and 60% coverage of all facades by highly efficient (20%) BIPV modules. Analysis of energy yield on different typical days in summer and winter reveals that the building is self-sufficient for many hours of the day. As on such days, self-consumption is relatively low, which leads to considerable feed-in of surplus electricity to the grid, application of local storage would increase self-sufficiency considerably. Furthermore, it is imperative to lower the electricity demand of the building to reach high self-sufficiency ratios.

## 29.1 Introduction

### *29.1.1 Near-Zero Energy Buildings*

As buildings consume 40% of the European energy supply all new buildings in the EU built from 2020 onwards should be nearly zero energy buildings (NZEBs). This means that the energy demand of the buildings should ideally be met by local generation, at least on an annual basis. Such a building is self-sufficient [1]. The European Performance of Buildings Directive 2010/31/EU is the key market driver for this in order for the built environment to contribute to the substantial lowering of greenhouse gas emissions [2]. The 2016 "Winter Package" communicated by the European Commission stated that also existing buildings should be renovated to

W. van Sark (✉) · E. Bontekoe
Copernicus Institute of Sustainable Development, Utrecht University, Princetonlaan 8A, 3584 Utrecht, The Netherlands
e-mail: w.g.j.h.m.vansark@uu.nl



J. Littlewood et al. (eds.), *Sustainability in Energy and Buildings*, Smart Innovation, Systems and Technologies 163, https://doi.org/10.1007/978-981-32-9868-2_29

become NZEBs [3], and European Parliament has endorsed this. Photovoltaics will play a crucial role here: PV on roofs and façades have the potential to generate a sufficient amount of electricity to supply all necessary demand, also for heating by means of, e.g., heat pumps, if efficiency measures have been undertaken and if the building has enough surface area [4]. For example, in the Netherlands, the present average electricity consumption of a residential building is about 3500 kWh per year, which one can generate using a 4 kWp PV system, optimally oriented, requiring about 21 $m^2$ of roof space. On an annual basis, such a household is self-sufficient.

With the expected increase in electrification of heating and transport, energy demand could easily double and a roof alone may not generate sufficient energy thus needing other surfaces for energy generation. One can imagine large solar parks around cities to provide the necessary energy, in combination with wind parks, but this may not be possible for orographic reasons and/or societal resistance. As an additional area within the city usually is limited, all roof and façade area should be investigated as to how much energy could be generated. In this respect, we have shown in a case study for the city of Utrecht that suburban areas with typical larger roof space than historical inner city neighborhoods are needed to increase the self-sufficiency of the whole city [5]. This also calls for integrated and esthetically pleasing designs of PV integrated building envelope: Building Integrated PV (BIPV) [4, 6, 7].

Optimally situated roofs lead to lowest levelized cost of electricity (LCoE) in terms of €/kWh and economic payback times (PBT) of some 5–7 years, while LCoE and PBT will increase for surfaces with lower energy yield. Proper time-resolved analysis should be performed in conjunction with time-dependent demand in order to find the best options for optimization of self-consumption (direct use of PV energy) and self-sufficiency on the building level. For example, high demand at 6 pm could be satisfied by means of a west oriented PV façade for high self-consumption.

### *29.1.2 Landmark Building Renovation*

Many of the buildings at the Utrecht University campus have been built over 50 years ago, and plans are made for renovation. One of those buildings is the 22-floor "W.C. Van Unnik building" (Fig. 29.1). It is the tallest building of the campus and bears the university logo which is lit at night and can be seen from large distances. It is considered a landmark in the region. As part of the university strategy to reach an energy neutral campus in 2030, deep renovation of this landmark building to a near-zero energy building using BIPV would bring another landmark status to this building. Also, next to the already installed 1.2 MWp system, that generates 1 GWh per year [8], or about 2% of the university's annual electricity demand [8], and that is dispersed over 8 university buildings, it is highly desirable to make PV visible to students, staff, and campus visitors.

The Van Unnik building is a 22-floor building of 76 m height and $57 \times 27$ $m^2$ ground floor area. The building demand at present (2017) is measured and amounts to 1.28 GWh electricity per year, so a 1.5 MWp optimally oriented PV system would

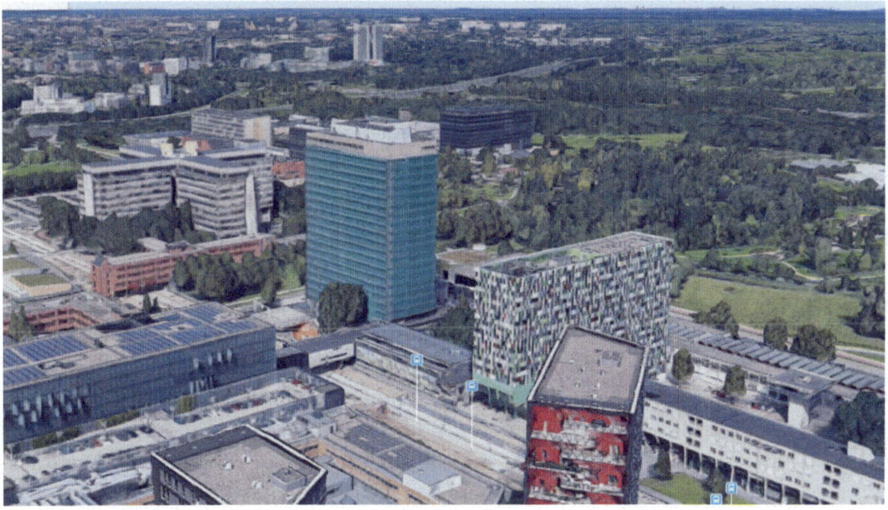

**Fig. 29.1** Aerial view of the Van Unnik building from the South-East side, showing the surrounding buildings as well [9]

suffice to generate that annual demand. Electricity demand thus is 37.7 kWh/m$^2$ total floor area. We note here that about half of the building was not in use due to the planned renovation so that 60–70 kWh/m$^2$ would be a more realistic number. This compares well to a recent benchmark analysis of energy use in buildings in the Netherlands that reported a typical number of 50 kWh/m$^2$ floor area [10]. Note, this excludes heating demand but includes the demand for cooling and ventilation.

This paper investigates in how far a BIPV design would allow the realization of a near-zero energy building. As the building is tall, large façade area would be available for BIPV.

## 29.2 Calculation Methods

### *29.2.1 PV Yield*

A PV yield time series at a certain time resolution can be generated using algorithms available in the open-source Python package PV_LIB [11]. Input data such as tilt angle, azimuth angle, available area, and irradiation is required. We used global horizontal irradiation for 2013 from the Royal Netherlands Meteorological Institute (KNMI) in De Bilt, The Netherlands. Note, 2013 is chosen as annual solar irradiance is closest to the typical meteorological year [1]. The irradiation measurement interval was 10 min. To obtain a 5-min data set linear interpolation was done. Global tilted orientation was calculated using the transposition algorithms in PV_LIB. To illustrate seasonal effects, we will show typical summer and winter day data for clear sky, completely overcast, and mixed clear/cloudy conditions. To calculate PV energy yield, we used a single PV module efficiency of 20%, corresponding to using 320 Wp modules of 1.6 $m^2$ size, and multiplied this with a performance ratio (PR) value of 0.85 to account for temperature and other system losses, which is consistent with well-performing PV systems in the Netherlands [12]. The actual design of a PV system covering façades and roof of a building is facilitated by the availability of 3D drawings of buildings, such as in SketchUp [13]. The design tool developed in the PVSITES project [14] then allows to calculate solar irradiation on all surfaces. We have used this in designing the PV cladding.

The Van Unnik building is a 22-floor building of 76 m height and 57 × 27 $m^2$ ground floor area. It is oriented such that is has a large South and North façade, both having an area of 4332 $m^2$, and a considerably smaller East and West façade of 2052 $m^2$. For every floor, the windows are 1.4 in height, while individual floor height is 3.45. Hence, 60% of the façade is usable for PV modules, excluding windows. The roof area is 1539 $m^2$. The PV energy yield is calculated for N, S, E, and W orientations each at 90° tilt. For the roof, we use two options, either south facing modules at optimum 37° tilt, with proper interrow distance, i.e., a usable fraction of 50%, or a combination of east and west facing modules at 10° tilt.

### 29.2.2 *Self-sufficiency and Self-consumption*

The self-consumption ratio *SCR* is defined as the amount of PV energy $E_{PV,cons}$ that are directly consumed divided by the total amount of PV energy generated $E_{PV,tot}$:

$$SCR = \frac{E_{PV,cons}}{E_{PV,tot}} \tag{29.1}$$

The self-sufficiency ratio is *SSR* defined as the amount of PV energy $E_{PV,cons}$ that are directly consumed divided by the electricity demand of a building $E_{cons}$:

$$SSR = \frac{E_{PV,cons}}{E_{cons}} \tag{29.2}$$

In order to calculate self-sufficiency and self-consumption, electricity demand data at 15-min time resolution was collected using certified kWh meters for the Van Unnik building. Also here, linear interpolation was performed to achieve a 5-min time resolution.

SCR and SSR can be defined for any time interval. In this paper, we use annual and (some) daily ratios.

## 29.3 Results

### *29.3.1 PV System Design, Annual Data Analysis*

We used PVSITES to make a PV system design based on the assumption to fully cover the usable façade area (60% of total façade) and the roof of the building with PV modules tilted south, see for details Fig. 29.2. We have used the Sunpower SPR320E-WHTD module for this, from the database of modules available in PVSites. The dimensions of the building allow to install two rows of 34 modules each in landscape orientation per floor on the south and north facades, and rows of 12 modules on the east and west facades. These are then connected per level to 40 inverters of different capacity and with double MPPT trackers (ABB Trio-7.5-TL-OUTD (east and west) and TRIO-27.6-TL-OUTD (north and south)). For roof and other parts, 28 additional inverters are used.

We find that a total PV capacity of 1.69 MWp is possible, which is calculated to generate 853.1 MWh of solar energy per year. The annual self-sufficiency ratio thus is 0.666. Covering the roof instead with a combination of east and west oriented modules at 10° tilt allows to install 1.84 MWp, which generates 968.0 MWh annually, and self-sufficiency ratio is 0.756. To reach a self-sufficiency ratio equaling one, electricity demand must reduce to about 25 kWh/m$^2$.

The specific yields are 505.9 kWh/kWp and 526.1 kWh/kWp, respectively. A breakdown of these results is shown in Table 29.1. Obviously, the south facing facade shows the largest PV capacity and energy yield, due to its large area, while the specific yield is considerably lower (~60%) than the one for an optimally tilted

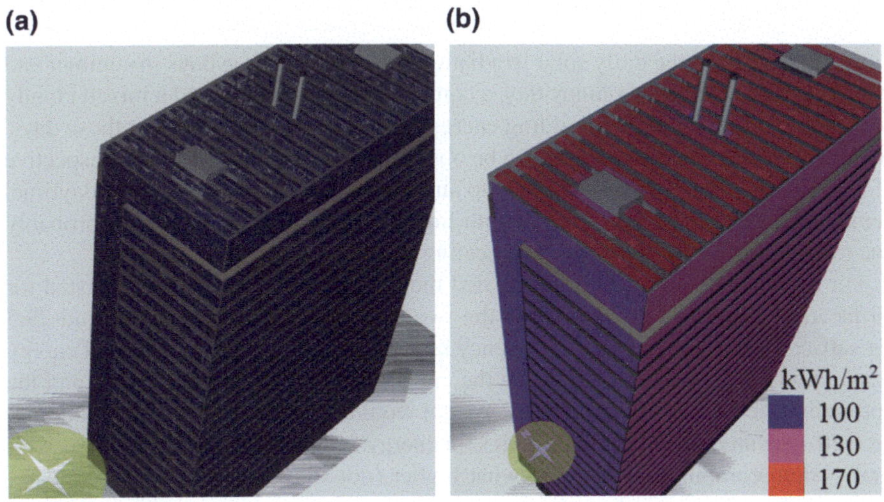


**Fig. 29.2** Design of the PV system for the Van Unnik building, using PVSITES [14]. **a** Aerial view from the south-east, **b** color coded energy generation view

**Table 29.1** Calculated PV capacity, annual yield and specific yield, for the façades and roof

| Surface | PV capacity (kWp) | Annual energy yield (MWh) | Specific yield (kWh/kWp) | Yield per unit area (kWh/m$^2$) |
|---|---|---|---|---|
| South facade | 519.8 | 311.1 | 598.5 | 71.18 |
| North facade | 519.8 | 151.9 | 292.2 | 35.06 |
| East facade | 246.2 | 112.9 | 458.4 | 55.02 |
| West facade | 246.2 | 125.1 | 507.9 | 60.96 |
| Roof (South tilt) | 153.9 | 152.1 | 988.6 | 98.83 |
| Roof (East tilt) | 153.9 | 132.9 | 863.5 | 86.35 |
| Roof (West tilt) | 153.9 | 134.2 | 872.1 | 87.20 |

system-oriented south. A recent detailed study on vertical solar irradiance in Spain shows similar numbers [15]. Although we do not focus on cost-effectiveness here, a combination of systems with high specific yield would be economically of most interest, i.e. a south facade combined with a combined east/west oriented roof system.

Placing PV on the North façade seems not to be an economical solution because of its low specific yield. Alternatively, using the roof of the neighboring Educatorium building (to the North) would be much more cost-effective. Nevertheless, using the North façade has other advantages, such as a higher self-consumption and the visibility of PV. Further on, with declining PV costs and growing demand for renewable energy even using the North façade could soon become economically viable.

## 29.3.2 Daily Data Analysis

Figure 29.3a shows the daily solar irradiation variation for three days in summer and three days in winter, i.e., a sunny day, a completely overcast day and a mixed cloudy day. Figure 29.3b shows the building energy demand for the building for those days. A clear baseload during the night can be observed of about 100–120 kW irrespective of the season, which can be attributed to air circulation equipment. During daytime, demand increases to 220–280 kW, with higher demands in winter, most probably due to higher lighting demands. The heating itself is not included.

For these six days, we have calculated the amount of solar energy harvested for all facades separately as well as for the two roof options, see Fig. 29.4, and 29.5 for self-consumption and self-sufficiency ratios. We also compare the actual energy demand with the net demand in case the energy generated from the facade and the optimally tilted south facing roof PV system would be used directly in the building, i.e., self-consumed. We note that figures for energy from the facades and the east/west roof system combination are somewhat higher (not shown here). It can be clearly observed that in summer the net demand is negative for the clear and cloudy day, and surplus electricity is fed back into the campus grid, see also Table 29.2. Although for the clear winter day self-consumption is high and surplus electricity is fed back

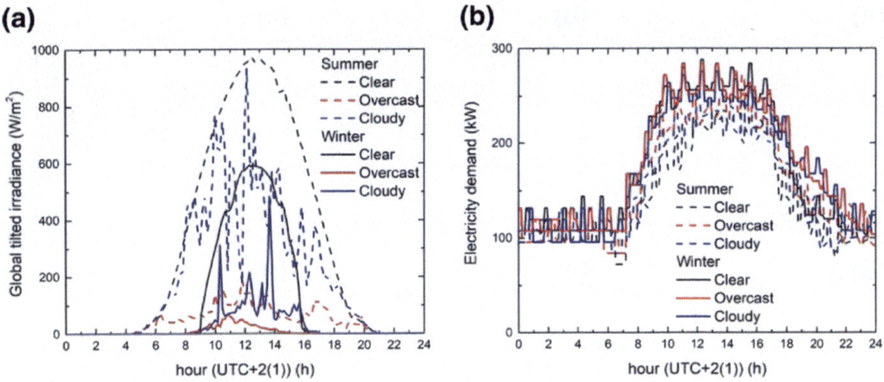


**Fig. 29.3** **a** Solar irradiance for six days in 2017, three in summer and three in winter, for clear, overcast and mixed cloudy conditions, for a south facing system at an optimum tilt. **b** Building electricity demand for the same days. Time step is 5 min

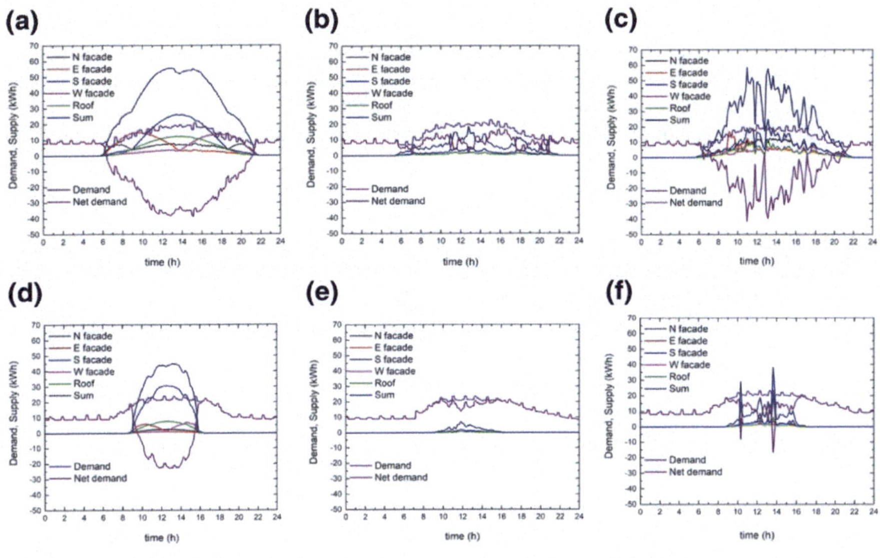


**Fig. 29.4** PV energy from façades and roof using solar irradiance for six days in 2017 (Fig. 29.2), and electricity demand with (net) and without PV. Three days in summer (**a**, **b**, **c**) and three in winter, for clear (**a**, **d**), overcast (**b**, **e**) and mixed cloudy (**c**, **f**) conditions. Time step is 5 min

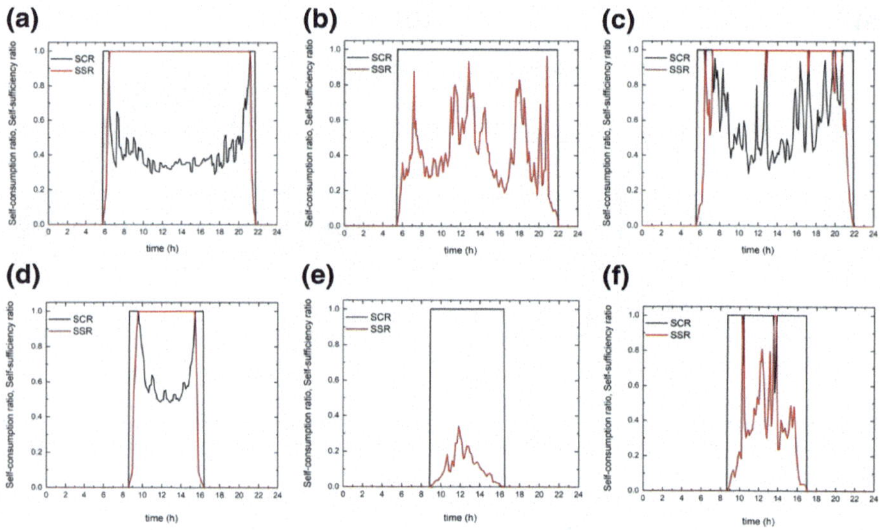


**Fig. 29.5** Self-consumption (black solid line) and self-sufficiency (red solid line) ratios for the Van Unnik building, using data in Fig. 29.4, for three days in summer (**a**, **b**, **c**) and three in winter, for clear (**a**, **d**), overcast (**b**, **e**) and mixed cloudy (**c**, **f**) conditions. We only show results for the case that energy is generated from all facades and the optimally tilted south facing roof PV system. Time step is 5 min

**Table 29.2** Electricity demand, self-consumption, electricity fed back to the grid, and cumulative net demand for six typical days, for the case that energy is generated from all facades and the optimally tilted south facing roof PV system

| Days | Demand (kWh) | Self-consumption (kWh) | To grid (kWh) | Net demand (kWh) |
|---|---|---|---|---|
| Summer, clear | 3474 | 2566 | 4209 | −3301 |
| Summer, overcast | 3804 | 1310 | 0 | 2493 |
| Summer, cloudy | 3662 | 2632 | 2404 | −1374 |
| Winter, clear | 4200 | 1708 | 1159 | 1333 |
| Winter, overcast | 4300 | 215 | 0 | 4084 |
| Winter, cloudy | 4140 | 755 | 40 | 3345 |

into the grid, the net demand is still positive. Even if the net demand is negative, the self-sufficiency ratio does not equal unity, as can be seen in Table 29.3.

The self-consumption and self-sufficiency ratios are shown in Fig. 29.5, which corroborate the findings from Fig. 29.4 in that self-sufficiency equals unity for the moments at which surplus electricity is fed back into the grid. Also, at those times, only a fraction of the PV generated energy is directly self-consumed. On a daily basis, the self-sufficiency ratio is lower to one, obviously due to nighttime demand. As the net demand is negative, surplus electricity could be locally stored in stationary

**Table 29.3** Calculated SCR and SSR for six typical days and two options for the PV system

| Days | Facades and Roof South | | Facades and Roof East/West | |
|---|---|---|---|---|
| | SCR | SSR | SCR | SSR |
| Summer, clear | 0.379 | 0.739 | 0.326 | 0.740 |
| Summer, overcast | 1.000 | 0.344 | 0.998 | 0.395 |
| Summer, cloudy | 0.523 | 0.719 | 0.456 | 0.725 |
| Winter, clear | 0.573 | 0.407 | 0.593 | 0.406 |
| Winter, overcast | 1.000 | 0.051 | 1.000 | 0.058 |
| Winter, cloudy | 0.950 | 0.182 | 0.952 | 0.202 |

battery systems and used later. In that case, the self-sufficiency ratio can equal unity, while still surplus electricity is generated.

### *29.3.3 Discussion*

The results show that the annual cumulative electricity demand is higher than can be generated by PV systems in the configurations used. While the generation capacity of the combined facade and roof system is high, the electricity demand is simply too high. To reach a near-zero energy building, considering only electricity demand, efficiency measures have to be taken such that electricity demand is lowered to about 25 kWh/m$^2$, on annual basis. On a daily basis, electricity import or export will occur, depending on solar irradiance variations.

The suggested design may not be valued in terms of esthetics, while we also do not consider the effect of nearby buildings. This will lead to lower energy yields, which stresses the importance of realizing even lower electricity demands than 25 kWh/m$^2$, perhaps even ambitiously as low as 10 kWh/m$^2$.

## 29.4 Conclusion

We have presented a BIPV design that can be used in the renovation of the landmark Van Unnik building with the purpose of realizing a net-zero energy building. Given the present electricity demand of 37.7 kWh/m$^2$ total floor area where half of the building is utilized, the building cannot be made self-sufficient. An annual self-sufficiency ratio of 0.666 or 0.756 is found, using two different roof designs, while covering 60% of all facades by BIPV modules. Daily self-sufficiency ratios can be unity for sunny days if surplus electricity can be stored locally in batteries. Annual self-sufficiency ratio of unity can be reached if the electricity demand is lowered to about 25 kWh/m$^2$ total floor area.

**Acknowledgements** We would like to thank Frans Tak (UU) for supplying energy demand data, Geert Litjens (UU) for having developed PV_LIB based python codes, the PVSITES project coordinator (Maider Machado) for using their 3D simulation tool prior to public release, and the DEM4BIPV project partners for many stimulating discussions. This work is supported by the "Werkelijk Bouwen aan BIPV" project, which is funded by the European Innovation program for the South of the Netherlands (OPZuid), as part of the European Fund for Regional Development (EFRO) of the European Union, as well as by the KA2 Strategic Partnerships for the higher education program of Erasmus+ under contract 2015-1-NL01-KA203-008882.

## References

1. Litjens, G.B.M.A., Worrell, E., van Sark, W.G.J.H.M.: Influence of demand patterns on the optimal orientation of photovoltaic systems. Sol. Energy **155**, 1001–1014 (2017)
2. European Commission Directive 2010/31/EU of the European Parliament and of the Council of 19 May 2010 on the energy performance of buildings. Off. J. Eur. Union **L 153**, 13–35 (2010)
3. European Commission Proposal for a Directive of the European Parliament and of the Council amending Directive 2010/31/EU on the energy performance of buildings. http://eur-lex.europa.eu/legal-content/EN/TXT/?uri=CELEX:52016PC0765
4. Scognamiglio, A., Røstvik, H.N.: Photovoltaics and zero energy buildings: a new opportunity and challenge for design. Prog. Photovolt. Res. Appl. **21**, 1319–1336 (2013)
5. Litjens, G.B.M.A., Kausika, B.B., Worrell, E., van Sark, W.G.J.H.M.: A spatio-temporal city-scale assessment of residential photovoltaic power integration scenarios. Sol. Energy **174**, 1185–1197 (2018)
6. Heinstein, P., Ballif, C., Perret-Aebi, L.-E.: Building integrated photovoltaics (BIPV): review, potentials. Barriers Myths. Green **3**, 125–156 (2013)
7. Tripathy, M., Sadhu, P.K., Panda, S.K.: A critical review on building integrated photovoltaic products and their applications. Renew. Sustain. Energy Rev. **61**, 451–465 (2016)
8. van Sark, W., de Waal, A., Uithol, J., Dols, N., Houben, F., Kuepers, R., Scherrenburg, M., van Lith, B., Benjamin, F.: Energy performance of a 1.2 MWp photovoltaic system distributed over eight buildings at Utrecht University Campus. In: Proceedings of the 33rd European Photovoltaic Solar Energy Conference and Exhibition, pp. 2284–2287 (2017)
9. Google maps. https://goo.gl/maps/xsCv5q3iJEoeYR8A9. Accessed 10 March 2019
10. Sipma, J.M.: Nieuwe benchmark energieverbruik utiliteitsgebouwen en industriële sectoren. Petten, The Netherlands (2016)
11. Andrews, R.W., Stein, J.S., Hansen, C., Riley, D.: Introduction to the open source PV_LIB for python photovoltaic system modelling package. In: Proceedings of the IEEE 40th Photovoltaic Specialist Conference (PVSC), pp. 170–174 (2014)
12. Kausika, B.B., Moraitis, P., Van Sark, W.: Visualization of operational performance of grid-connected PV systems in selected european countries. Energies **11**, 1330 (2018)
13. Sketchup. www.sketchup.com. Accessed 17 Feb 2018
14. Machado, M., Challet, S., Weiss, I., Román Medina, E., Espeche, J.M., Noris, F., Reijenga, T., Rico, E., Huerta, I., Assoa, Y.B., et al.: Supporting market uptake of building-integrated photovoltaic technologies with the PVSITES project. In: Proceedings of the 33rd European Photovoltaic Solar Energy Conference and Exhibition, pp. 2882–2887 (2017)
15. Díez-Mediavila, M., Rodríguez-Amigo, M.C., Dieste-Velasco, M.I., García-Calderón, T., Alonso-Tristán, C.: The PV potential of vertical façades: A classic approach using experimental data from Burgos. Spain. Sol. Energy **177**, 192–199 (2019)

## Chapter 30
# Internal Insulation of Historic Buildings: A Stochastic Approach to Life Cycle Costing Within RIBuild EU Project

**Elisa Di Giuseppe, Gianluca Maracchini, Andrea Gianangeli, Gabriele Bernardini and Marco D'Orazio**

**Abstract** The application of internal insulation is a widespread and effective solution for energy renovation of historic buildings. However, it entails quite high installation costs and a certain risk of failure due to moisture-related problems. A probabilistic risk assessment of both hygrothermal performance and life cycle costs can be used to address internal insulation issue, in order to support risk management and decision-making. This paper presents the application of a probabilistic approach to Life Cycle Costing developed within the EU project RIBuild (Robust Internal Thermal Insulation of Historic Buildings), to five internal insulations solutions widely used in Italy. The method provides estimates of the range and likelihood of global costs and payback periods, also considering alternative energy and future economic scenarios. The impact of insulation systems service life on global costs is also addressed, in order to highlight the possible connection of the method to a stochastic estimation of insulation systems durability based on hygrothermal and damage assessments.

## 30.1 Introduction

Considering that in today's Europe 30% of all buildings are historic[1] buildings that are expected to last for decades, there is great potential for energy savings, and consequently emission reductions, by their deep renovation. Given the architectural features of the façade of these buildings, the energy retrofit should be properly evaluated considering the need to preserve the cultural value. In this context, the application of internal insulation to the facades is one of the most exploited solutions, given its significant potential for energy savings without compromising the building appearance [1, 2].

[1]Buildings built prior to 1945. They include heritage buildings and other buildings not protected by legislation.

E. Di Giuseppe (✉) · G. Maracchini · A. Gianangeli · G. Bernardini · M. D'Orazio
Università Politecnica delle Marche, 60131 Ancona, Italy
e-mail: e.digiuseppe@univpm.it



J. Littlewood et al. (eds.), *Sustainability in Energy and Buildings*, Smart Innovation, Systems and Technologies 163, https://doi.org/10.1007/978-981-32-9868-2_30

However, the implementation of internal thermal insulation is subject to a certain risk of failure, due to the modification of the hygrothermal performance of the building envelope [3–5], and to high installation costs. The EU project *RIBuild* (Robust Internal Thermal Insulation of Historic Buildings) investigates in-depth how and under what conditions internal insulation can be employed [6]. Next, to the hygrothermal performance, life cycle costs and environmental impacts are important factors to be considered during the decision-making process before installing internal insulation [7, 8].

In the last two decades, Life Cycle Costing (LCC) has become an important decision tool in the building context, both for the development of specific policies and for the single design process. Directive 2010/31/EU introduced the concept of "cost-optimality" of building design solutions [9], and recent Directive 2018/844 encourages "*[...] in relation to buildings undergoing a major renovation, high-efficiency alternative systems, in so far as this is technically, functionally and economically feasible*" [10]. As reported in a comprehensive review of Ferrara et al., a considerable amount of research recently addressed the cost analysis of building design options in Europe [11], mainly referring to standardized LCC methods as those reported in the international standards ISO 15686–5:2008 [12] and EN 15459-1:2017 [13]. However, standard LCC does not fully capture the risk associated with the investment and the calculation is often achieved with notable simplifications related to the cost items and macroeconomic scenarios quantification. In reality, accurate cost analysis relies on the quality of data and long-term forecasts, and data uncertainty is a well-recognized issue associated with LCC methods [8, 14–18]. Ignoring these uncertainties may lead to improper decisions [17].

In this context, as part of *RIBuild* project—work package 5, a "probabilistic" methodology to assess internal insulation affordability, based on an LCC, has been developed, in order to take into account the inherent uncertainties related to the long-term perspective of the building interventions [19]. The method includes a Monte Carlo-based approach to LCC of internal insulation measures and a model to characterize the future macroeconomic scenario for the assessment. Furthermore, the probabilistic approach allows to explicitly consider the uncertainty of the input related to the insulation hygrothermal performance, i.e., the heat transmission loss, its service life, and its maintenance needs. Indeed, a probabilistic analysis of the hygrothermal performance of interior insulation is also developed within *RIBuild* project—work package 4 [3].

In this paper, the stochastic LCC method is used to assess five internal insulations solutions, among those investigated in *RIBuild* project, usually applied for the renovation of historic buildings in Italy. Special attention is given to their economic performance under alternative macroeconomic and building energy scenarios and to the role played by the system service life. Section 30.2 presents the insulation systems under investigation and summarizes the stochastic LCC methodology. Section 30.3 and 30.4 respectively report and discuss the main results, while conclusions and future developments within *RIBuild* and in the general field of historic building renovation are finally drawn in Sect. 30.5.

## 30.2 Methodology

### 30.2.1 The Insulation Systems

Five internal thermal insulation systems typically used in Italy for historic building renovation have been considered in this study, i.e., Expanded Polystyrene (EPS), Calcium Silicate (CaSi), Autoclaved Aerated Concrete (AAC), Cork, and Rockwool (RW). They are applied to an exemplary historic building in Italy in plastered solid bricks masonry, with an overall thickness of about 30 cm and an air-to-air heat transfer coefficient (U-value) of 1.76 W/m$^2$K. The building is supposed to be located in the region Lombardia, belonging to the largest climatic zone of Italy. In Table 30.1, the thermophysical properties of the five insulation systems are reported. The thicknesses of the different internal insulation layers have been computed in order to reach a U-value lower than 0.36 W/m$^2$K, according to the actual Italian law requirements [20], with slight differences due to the commercial insulation thicknesses available in the market.

### 30.2.2 The Stochastic Approach to LCC

**LCC model**. The LCC analysis of the internal insulation systems is based on the procedure described in the European Standard EN 15459-1 [13] that allows computing the Global Costs $\left(GC_{j,0}\right)$, referred to the starting year ($t = 0$), of a specific building design option ($j$) at the end of a determined calculation period ($CP$). The $GC_{j,0}$ formula is adapted in this study by including annual variations of the discount factor and specific price developments rates for human operation and for energy, as follows:

$$GC_{j,0} = CI_j + \sum_{t=1}^{CP}\left[\left(CM_j + CS_{j,t}\right)R_t^{disc}R_t^L + CE_jR_t^{disc}R_t^E\right] - Val_{j,CP} \quad (30.1)$$

where $CI_j$ is the initial investment cost, $CM_j$ the annual maintenance cost, $CS_{j,t}$ the replacement cost, $CE_j$ the annual energy cost, $R_t^{disc}$ the discount factor (based on inflation rate and market interest rate), $R_t^L$ and $R_t^E$ the price development rates (respectively for human operation and for energy), and $Val_{j,CP}$ the residual value of the design option at the end of the CP. In this study, the CP is assumed to be equal to 30 years. Based on the same data inputs, the Payback Period (PP) of each solution is also calculated as the minimum number of years making the cumulative energy saving equalizing the total investment costs.

**Uncertainty propagation**. The stochastic approach to LCC, developed by the authors and described in depth in [19], couples Monte-Carlo (MC) simulations to the model Eq. (30.1), thus requires defining the Probability Density Functions (PDFs)

**Table 30.1** Thermophysical properties of the analyzed internal insulation systems

| Layer | Standard thickness (m) | Density ($kg/m^3$) | Thermal conductivity (W/mK) |
|---|---|---|---|
| *"EPS" insulation system* (U-value = 0.36 $W/m^2K$) | | | |
| Adhesive mortar | 0.006 | 1400.00 | 0.540 |
| **EPS** | 0.080 | 25.00 | 0.035 |
| Adhesive mortar | 0.006 | 1400.00 | 0.540 |
| Plasterboard | 0.0125 | 680.00 | 0.200 |
| Surface rendering | 0.004 | 1200.00 | 0.47 |
| Primer+paint | 0.0002 | 1670.00 | – |
| *"CaSi" insulation system* (U-value = 0.36 $W/m^2K$) | | | |
| Adhesive mortar | 0.006 | 1800.00 | 0.63 |
| **Calcium Silicate** | 0.125 | 290.00 | 0.053 |
| Surface rendering | 0.006 | 1800.00 | 0.63 |
| Primer+paint | 0.0002 | 1670.00 | – |
| *"AAC" insulation system* (U-value = 0.35 $W/m^2K$) | | | |
| Adhesive mortar | 0.006 | 800.00 | 0.18 |
| **AAC** | 0.100 | 90.00 | 0.042 |
| Surface rendering | 0.006 | 800.00 | 0.18 |
| Primer+paint | 0.0002 | 1670.00 | – |
| *"Cork" insulation system* (U-value = 0.35 $W/m^2K$) | | | |
| Adhesive mortar | 0.006 | 1800.00 | 0.60 |
| **Cork** | 0.100 | 150.00 | 0.041 |
| Surface rendering | 0.006 | 1800.00 | 0.60 |
| Primer+paint | 0.0002 | 1670.00 | – |
| *"RW" insulation system* (U-value = 0.33 $W/m^2K$) | | | |
| **Rockwool**[a] | 0.080 | 110.00 | 0.035 |
| Vapor barrier | 0.0002 | 2700.00 | – |
| Plasterboard | 0.025 | 680.00 | 0.200 |
| Surface rendering | 0.004 | 1200.00 | 0.47 |
| Primer+paint | 0.0002 | 1670.00 | – |

[a]Fixed to the wall through a metallic frame

of all LCC variables and parameters. MC method selects values from the input PDFs and inserts them into the output Eq. (30.1) for a proper number of times depending on the envisaged accuracy level. The output parameters distributions are then quantified as a result of the possible variance of the input parameters. In this work, Sobol's sequences have been used as a quasi-random sampling technique in order to generate samples as uniformly as possible from inputs PDFs. Data analysis software "R" has been used for both sample generation and uncertainty propagation [21]. The assessment has been performed in several scenarios to evaluate results robustness. These include:

- 2 building heating scenarios, including natural gas and electricity as energy sources (the most widespread in Italy);
- 4 macroeconomic scenarios [22], i.e.,:
  - *Regular Growth (RG)*, representing an economic situation with a balanced growth path, i.e., moderate growth of Gross Domestic Product (GDP) and inflation rate, and a moderate nominal interest rate;
  - *Intense Growth (IG)*, characterized by more robust growth in terms of GDP, and inflation and interest rates higher than in the RG scenario;
  - *Stagflation (St)*, where the inflation rate is very high;
  - *Deflation (De)*, where the inflation rate is the lowest (near-zero).
- Three replacement scenarios, due to hygrothermal damages (i.e., internal insulation service life assumed to be equal to 10, 20 or 30 years). Replacement costs ($CSs$) are assumed to be equal to investment costs. Periodic maintenance on the system is not foreseen, then maintenance costs ($CM$s) have been always considered equal to zero.

As a result, 120 simulation cases have been obtained from the combination of all the insulation systems, economic and energy scenarios (five insulation systems × two energy scenarios × four economic scenarios × three replacement scenarios).

**Data inputs characterization**. LCC data inputs are grouped into three main categories: design option characteristics (i.e., investment and maintenance costs, service life); building energy performance (energy need and overall efficiency for heating) and energy carrier (national tariffs); macroeconomic scenario (i.e., inflation rate, interest rate, and price development rates). The PDFs of inputs related to the first two categories, obtained as described in [19], are summarized in Table 30.2.

In brief, the normal distributions of pre-renovation and post-renovation energy needs ($Q_H^{pre}$ and $Q_H^{post}$, respectively) have been computed through the annual Heating Degree-Days (HDD) method, considering variable HDD data from 2000 to 2016 of the region Lombardia extracted from the Eurostat database. The statistical distributions of the investment costs are assumed normal and based on producers pricing lists. The wall is assumed clean and ready for the internal insulation installation. Uniform distributions have been associated with both the overall building heating efficiency $\eta_H$ and energy tariffs $EnT$ [19].

**Table 30.2** Input parameters PDFs. The variability on the PDF mean values correspond to the 5 and 95% percentile for the normal distribution (Coefficient of Variability, CoV = 7.5%) and to the min/max values for the uniform distribution

| LCC parameters | PDF | Insulation system (mean values) | | | | | Variability (%) |
|---|---|---|---|---|---|---|---|
| | | EPS | CaSi | AAC | Cork | RW | |
| $Q_H^{pre}$ (kWh/y) | Normal | 96.19 | | | | | ±12 |
| $Q_H^{post}$ (kWh/y) | | 18.64 | 18.10 | 18.60 | 18.17 | 17.32 | ±12 |
| $CI$ (€) | | 44.40 | 220.62 | 90.59 | 103.78 | 62.57 | ±12 |
| $EnT_{gas}$ (€/kWh) | Uniform | 0.075 (taxes included) | | | | | ±15 |
| $EnT_{electricity}$ (€/kWh) | | 0.186 (taxes included) | | | | | ±15 |
| $\eta_{H,gas}$ (-) | | 0.80 | | | | | ±25 |
| $\eta_{H,electricity}$ (-) | | 3.25 | | | | | ±23 |

Finally, concerning the last group of inputs, impacting on the discount factor and the price developments rates of Eq. (30.1), for each macroeconomic scenario, the nominal interest rate (*INT*), the inflation rate (*INF*), and the rate of *GDP*, expressed in real terms, have been forecasted (see [19, 22] for further details).

## 30.3 Results

Figure 30.1a, b shows the Cumulative Density Function (CDF) of Global Costs (*GC*) and of Payback Periods (*PP*) of the five insulation systems related to regular growth macroeconomic scenario; natural gas scenario; a service life equal to 30 years. As can be seen, from a merely economic point of view, under these conditions, the *EPS* insulation system is the best performing solution, followed by the *RW* one, while the *CaSi* option is the worst. Concerning the *GC*s, median values for *EPS* and *RW*

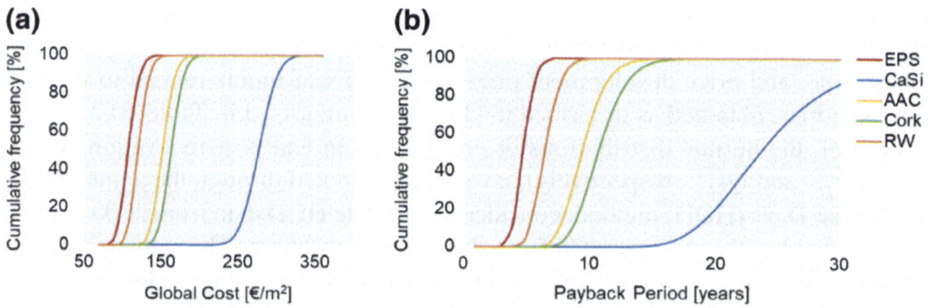


**Fig. 30.1** CDF of the Global Cost (**a**) and Payback Period (**b**) for the different insulation systems, assuming natural gas as the building energy source and regular growth macroeconomic scenario (insulation systems service life = 30 years, calculation period = 30 years)

options are about 107 and 121 €/m$^2$, respectively (Fig. 30.1a). *CaSi* option, instead, is characterized by the highest *GC* mean value, i.e., about 277 €/m$^2$. *AAC* and *Cork* options have intermediate median values, equal to 152 and 164 €/m$^2$, respectively. Concerning the *PP*s, as expected, a similar ranking is obtained (Fig. 30.1b). The lowest *PP* median value is obtained by *EPS* (about 5 years) followed by *RW* (about 7 years). The highest value is reached by *CaSi* (about 23 years), while intermediate values are obtained for *AAC* (about 10 years) and *Cork* (about 11 years) solutions.

Since running costs, such as maintenance and energy costs, are almost the same for all the insulation solutions (the first assumed equal to 0 for all the cases, and the second almost the same due to the similar U-values), the differences between the insulation options obtained in this comparison can be mainly attributed to the different initial investment costs (*CI*s) (see Table 30.2).

In order to evaluate the robustness of the results, and then support designers in the selection of the best performing solution (from a merely economic point of view) under several conditions, Fig. 30.2a compares the *GC*s obtained under different assessment scenarios (different energy sources and macroeconomic scenarios). Despite all 40 cases show considerable *GC*s uncertainty, the ranking of the solutions previously obtained is still confirmed for all the economic and energy source scenar-

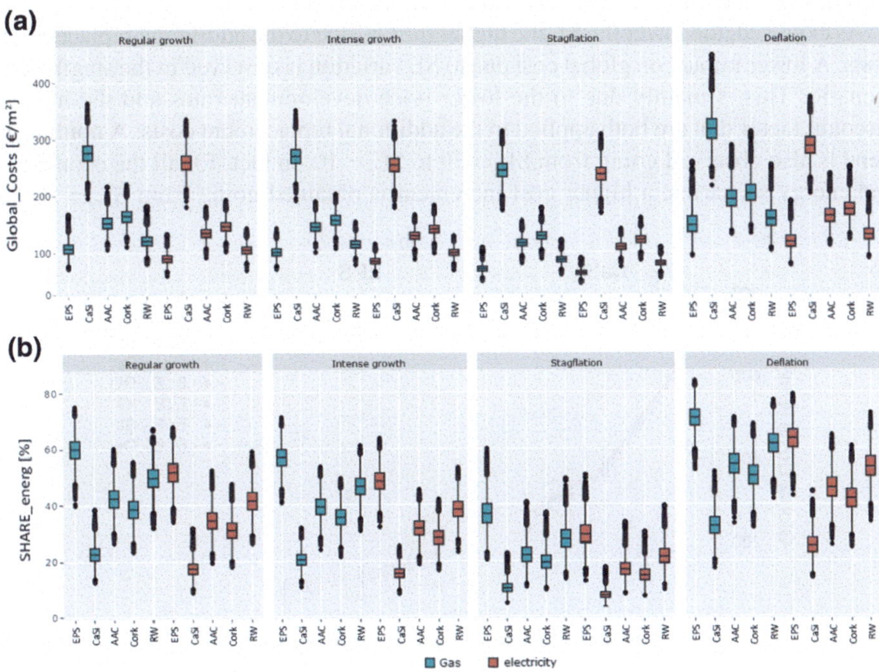


**Fig. 30.2** Box-whiskers plots of the Global Cost (**a**) and cost share of the energy cost (**b**) for each insulation system under different economic scenario and by considering different energy sources (insulation systems service life = 30 years, calculation period = 30 years)

ios. However, some considerations on the impact of the scenarios on the *GC* results can be highlighted.

Concerning energy sources, electricity scenario entails lower costs than natural gas. This is due to the high equipment efficiency, together with the lower electricity tariffs.

Concerning the macroeconomic scenarios, regular growth, and intense growth give rise to similar *GC* values. Highest and lowest *GC*s are instead obtained in the deflation and stagflation scenarios, respectively. In fact, in the stagflation scenario, lower running costs are obtained due to price development rates lower than 1, and a discount factor higher than that of all other macroeconomic scenarios (due to the high inflation rate). In contrast, in the deflation scenario, inflation is the lowest of all scenarios, while discount rates and escalation factors are the highest. This generates higher running costs. This can be clearly gathered from the analysis of the energy costs share on the global costs (Fig. 30.2b). However, the electricity energy source seems to entail lower variations in terms of *GC*s between economic scenarios, than those obtained in the gas energy scenario.

Finally, according to the different "replacement" scenarios, the impact of different Service Life values (SL) on the *GC*s is evaluated. For the sake of brevity, in Fig. 30.3 the mean values obtained for different SLs and related to the best and worst performing solutions, i.e., *EPS* and *CaSi*, have been reported.

As expected, the lower the SL the higher the GC due to the additional replacement costs. A lower impact on global cost due to SL variation is observed in the stagflation scenario. This is mainly due to the lower price development rates and the higher discount factor that are both applied to the additional replacement costs. A nonlinear trend is also observed going from SL = 30 to SL = 10. In fact, for all the economic and energy scenarios, a higher cost increment is obtained going from SL = 20 to

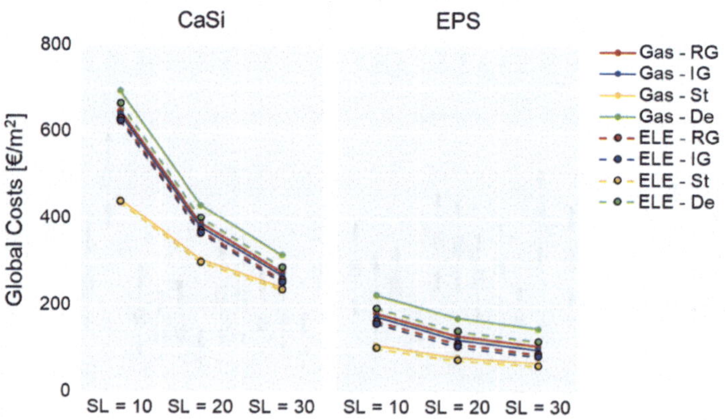


**Fig. 30.3** Mean values of GCs for "*CaSi*" and "*EPS*" design option under different economic and energy scenarios by considering different *SL* values

SL = 10 than that obtained going from SL = 30 to SL = 10. This is mainly due to the different residual value (Val) of the design option at the end of the CP, which is about 50% of the CI in the SL = 20 case and about 0 for the other two "replacement" scenarios.

## 30.4 Discussion

In the example shown, from a merely economic point of view, EPS insulation solution ranks firsts, followed by Rockwool, AAC, Cork, and Calcium Silicate, in all assessment scenarios. This is justified by the lower purchase and installation costs for this system, considering similar running costs for all insulation solutions during the calculation period, due to the same assumed energy and durability performance. This assumption constitutes a limitation of the specific case study, which does not take into account possible factors that could affect the life cycle costs of the renovation measure, such as the systems' repair needs and their real in situ hygrothermal performance, depending on possible moisture-related problems.

Indeed, the stochastic LCC could be coupled to a probabilistic risk assessment of hygrothermal performance, in order to fully capture the risk associated with the investment, effectively supporting decision-making. This means that outputs related to hygrothermal simulations and risk damage assessments (i.e., the probability distribution of the wall heat transmission losses, of the insulation system service life and of maintenance frequency) can be used in the stochastic LCC in order to provide more outstanding and substantial results. Taking into account these factors could even overturn the ranking among the insulation systems obtained in the specific case study, in favor of more expensive solutions, but safer from a hygrothermal point of view. Future applications of the stochastic LCC are foreseen in this direction.

## 30.5 Conclusion

This paper presented a probabilistic life cycle costing assessment of five internal insulations solutions for historic buildings renovation in Italy, based on Monte-Carlo simulations and a stochastic characterization of the macroeconomic scenario. During a calculation period of 30 years, the following costs items have been taken into account: the investment costs, the energy costs, and the replacement costs; while maintenance costs were disregarded. The assessment was performed in two alternative building energy scenarios (gas, electricity), four alternative macroeconomic scenarios, and three replacement scenarios (considering insulation systems service life alternatively equal to 10, 20, 30 years).

Results (global costs and payback periods) for the design options are expressed as probability distributions, rather than a single point estimate, and their robustness assessed in several scenarios. Even if in the specific case study some simplifications have been applied for input data characterization, the stochastic LCC method constitutes an advance over traditional methods, based on deterministic calculations and static economic scenarios.

When coupled to detailed hygrothermal simulations and risk damage assessments to refine related input data, the method constitutes an even more useful decision-making support tool in the field of building renovation.

**Acknowledgements** This work was supported by the European Union's Horizon 2020 research and innovation program under grant agreement N. 637268.

## References

1. de Place Hansen, E.J., Wittchen, K.B.: Energy savings due to internal façade insulation in historic buildings. In: Brostom, T., Nilsen, L., Carlsten, S. (eds.) Conference Report The 3rd International Conference on Energy Efficiency in Historic Buildings. Visby Sweden, pp. 22–31 (2018)
2. Walker, R., Pavía, S.: Thermal performance of a selection of insulation materials suitable for historic buildings. Build. Environ. **94**, 155–165 (2015). https://doi.org/10.1016/j.buildenv.2015.07.033
3. Vereecken, E., Van Gelder, L., Janssen, H., Roels, S.: Interior insulation for wall retrofitting—a probabilistic analysis of energy savings and hygrothermal risks. Energy Build. **89**, 231–244 (2015). https://doi.org/10.1016/j.enbuild.2014.12.031
4. Guizzardi, M., Carmeliet, J., Derome, D.: Risk analysis of biodeterioration of wooden beams embedded in internally insulated masonry walls. Constr. Build. Mater. **99**, 159–168 (2015). https://doi.org/10.1016/j.conbuildmat.2015.08.022
5. Klõšeiko, P., Arumägi, E., Kalamees, T.: Hygrothermal performance of internally insulated brick wall in cold climate: field measurement and model calibration. In: Proceedings of the 2nd CESBP (2013). https://doi.org/10.1081/e-eee2-120046011
6. RIBuild—Robust Internal Thermal Insulation of Historic Buildings. https://ribuild.eu/. Accessed 14 Mar 2019
7. Favi, C., Di Giuseppe, E., D'Orazio, M., et al.: Building retrofit measures and design: a probabilistic approach for LCA. Sustainability **10**, 3655 (2018). https://doi.org/10.3390/su10103655
8. Di Giuseppe, E., Iannaccone, M., Telloni, M., et al.: Probabilistic life cycle costing of existing buildings retrofit interventions towards nZE target: methodology and application example. Energy Build. **144**, 416–432 (2017). https://doi.org/10.1016/j.enbuild.2017.03.055
9. EU: Directive 2010/31/EU of the European Parliament and of the Council of 19 May 2010 on the energy performance of buildings (recast). Off J Eur Union (2010)
10. The European Parliament and the Council of the European Union: Directive (EU) 2018/844 of the European Parliament and of the Council of 30 May 2018 amending Directive 2010/31/EU on the energy performance of buildings and Directive 2012/27/EU on energy efficiency (2018)
11. Ferrara, M., Sirombo, E., Fabrizio, E.: Automated optimization for the integrated design process: the energy, thermal and visual comfort nexus. Energy Build. **168**, 413–427 (2018). https://doi.org/10.1016/j.enbuild.2018.03.039
12. ISO—International Organization for Standardization: ISO 15686-5 Buildings and constructed assets—service life planning—Part5: life-cycle costing (2017)
13. CEN European Committee for Standardization: EN 15459-1:2017. Energy performance of buildings—economic evaluation procedure for energy systems in buildings—Part 1: calculation procedures, Module M1-14 (2017)

14. Moore, T., Morrissey, J.: Lifecycle costing sensitivities for zero energy housing in Melbourne, Australia. Energy Build. **79**, 1–11 (2014). https://doi.org/10.1016/j.enbuild.2014.04.050
15. Ilg, P., Scope, C., Muench, S., Guenther, E.: Uncertainty in life cycle costing for long-range infrastructure. Part I: leveling the playing field to address uncertainties. Int. J. Life Cycle Assess. **22**, 277–292 (2017). https://doi.org/10.1007/s11367-016-1154-1
16. Goh, Y.M., Newnes, L.B., Mileham, A.R., et al.: Uncertainty in through-life costing-review and perspectives. IEEE Trans. Eng. Manag. **57**, 689–701 (2010). https://doi.org/10.1109/TEM.2010.2040745
17. Burhenne, S., Tsvetkova, O., Jacob, D., et al.: Uncertainty quantification for combined building performance and cost-benefit analyses. Build. Environ. **62**, 143–154 (2013). https://doi.org/10.1016/j.buildenv.2013.01.013
18. Fregonara, E., Ferrando, D.G.: How to model uncertain service life and durability of components in life cycle cost analysis applications ? The stochastic approach to the factor method. Sustainability (2018). https://doi.org/10.3390/su10103642
19. Baldoni, E., Coderoni, S., D'Orazio, M., et al.: The role of economic and policy variables in energy-efficient retrofitting assessment. A stochastic life cycle costing methodology. Energy Policy **129**, 1207–1219 (2019). https://doi.org/10.1016/j.enpol.2019.03.018
20. DM 26/06/2015—"Requisiti minimi" Schemi e modalità di riferimento per la compilazione della relazione tecnica di progetto ai fini dell'applicazione delle prescrizioni e dei requisiti minimi di prestazione energetica negli edifici (in Italian) (2015)
21. The R Project for Statistical Computing. https://www.r-project.org/. Accessed 29 May 2018
22. D'Orazio, M., Di, Giuseppe E., Esposti, R., et al.: A probabilistic tool for evaluating the effectiveness of financial measures to support the energy improvements of existing buildings. IOP Conf. Ser. Mater. Sci. Eng. **415**, 012003 (2018). https://doi.org/10.1088/1757-899X/415/1/012003

# Chapter 31
# Process for the Formulation of Natural Mortars Based on the Use of a New Natural Hydraulic Binder

**Santi Maria Cascone, Giuseppe Antonio Longhitano, Renata Rapisarda and Nicoletta Tomasello**

**Abstract** In the so-called "circular economy", i.e., an economy capable of self-healing, waste is not disposed of as in the natural operational evolution of the work but reused. This leads to a reduction in the impact of the building that uses it and an increase, even if ideological, of its value. Respecting the principles of the circular economy, the present study has an aim for the formulation of natural mortars with hydraulic behavior that foresees the reuse of the material coming from the restoration site. At the base of these mortars, there is a new hydraulic binder obtained from the "complete" combination between lime and cocciopesto, a reaction that does not allow the development of residual free lime. Thanks to this feature, and to the lack of its consequences, the obtained mortars can be used to recover the historical buildings also having an archaeological and artistic interest.

## 31.1 Introduction: Objectives and Presentation of the Patent

This study is focused on the formulation of natural mortars with hydraulic behavior, which can be used for the restoration of the walls of historic buildings with archaeological and artistic interest.

The study starts from a question that the authors, and researchers of the sector, have often asked themselves: why, although most of the historic buildings were made with lime and pozzolanic aggregates, reagents on the used lime, their state of conservation today does not appear homogeneous?

But not only another necessity has led the research to the results reported in this paper. In fact, interventions in historical factories, to be compatible with the existing, require natural mortars, i.e., without Portland cement and other artificial industrial substances that could lead to diseases (e.g., ettringite).

S. M. Cascone · R. Rapisarda · N. Tomasello (✉)
University of Catania, 95123 Catania, Italy
e-mail: nicolettatomasello@unict.it

G. A. Longhitano
Catania, Italy



J. Littlewood et al. (eds.), *Sustainability in Energy and Buildings*, Smart Innovation, Systems and Technologies 163, https://doi.org/10.1007/978-981-32-9868-2_31

The mortars to be used in interventions in historic buildings must also guarantee esthetic performances that satisfy the requirements of a restoration site.

And finally from the practical point of view, and in accordance with the principles of the Life Cycle Assessment (LCA), a production process that can be directly done at the site is preferable, according to principles of sustainability and environmental impact close to zero.

Answering to the questions and necessities exposed, the aim of this research concerns the development of a procedure for the formulation of a new natural hydraulic binder:

- based on the "complete" combination between lime and cocciopesto, characteristic that determines the resistance overtime of the historical buildings. From the chemical point of view, the main compounds, such as calcium oxide and silicon dioxide, combined in the "wet mortar phase", are transformed into calcium silicates and will be stable on "complete hardening";
- based on the use of raw materials free of corrective additives and dyes, characteristic that allows end-of-life programming that can provide for the complete recycling of the mixes formulated with this process;
- based on the use of recycled material, an action that allows the reduction of the environmental impact of the work. The reduction in transport associated with the supplying of the material on-site, in fact, contributes to the reduction of $CO_2$—the component most responsible for the increase of urban heat—in the environment.

The reuse of the components of the mix—which derive from the same restoration sites—represents the principles of the so-called "circular economy", defined as an economy capable of "self-regenerating" [1–4]. In the "circular" economy, waste (i.e., what should be disposed of in the current operational evolution of the work) is null, i.e., the waste material is reused by increasing, although ideologically, the value of the work [5]. In the definition of the circular economy, both the quality of the base material and the maintenance over time of the resulting product are fundamental [6, 7].

In order to achieve a "circular" economy, it must be based on the principle of sustainability, defined by Goodland [8] according to three different criteria, such as the "production rule" (which emphasizes the importance of equilibrium between the volume of waste and what can be assimilated from the environment), the "input rule" (which explains how to best use renewable and nonrenewable resources), and the "operating principle" (according to which efficiency improvement has greater importance than capacity).

This procedure, and the restoration mortars produced through its implementation are covered by a patent.

## 31.2 State of the Art and Innovation

In current practice, the natural binder for mortars is derived from the cooked calcium carbonate; suitably worked, this material has been used over the centuries as slaked lime, lime milk or, recently, powdered lime.

As can be seen in the same buildings that have been kept in a good state of conservation up to the present day, was added to the lime-based mortar—due to its basic nature—pozzolanic product (mainly cocciopesto obtained from grinding of cotto through the cooking of blue clays) which included, from the mineralogical point of view, sufficient silicon dioxide.

The combination of silicon dioxide and calcium oxide of which limestone is made generates calcium silicates, which have been defined until today in an imprecise manner, i.e., in an empirical way. The reason why some historical buildings have been preserved and others have deteriorated, although the mortars have a natural hydraulic binder, should be found in these variables, given by the cocciopesto and lime: combined randomly they provided different results, satisfying or not.

The optimal balance of these elements, subject of numerous studies, had not yet been reached before the research presented here. This fact is also demonstrated by the formulation of Portland cements which, although initially starting from the same clay/lime-based compound, with cooking up to and above 1500 °C went so far as to form tricalcium silicates which gave serious problems both on the new buildings and on the historical ones, where the pathology of ettringite degradation is evident just for the combination of the residual lime with the sulfates and finally with the tricalcium aluminates. This type of degradation is mainly caused by tricalcium aluminates present in Portland cement, which, when combined with gypsum (lime and sulfates) also present as a set retarder, degenerates as defined above. The same lime, of basic nature, cannot be used alone as binder: turning when in contact with water in calcium bicarbonate, it causes the formation of gypsum following the reaction with the sulfates, normally present in the acidic rains.

The innovation of the research here exposed is the use of a binder composed of:

- lime containing up to 98% of CaO;
- pozzolanic material reagent on the lime used,

that has hydraulic behavior and does not contain residual free lime (and/or soluble products in general). The mix thus composed is durable over time and characterized by higher resistance to compression and durability.

## 31.3 The Process for Making the Binder and the Formulation of the Natural Mortar

### *31.3.1 Step 1: Considerations Concerning the* Vicat Equation

From the qualitative point of view, the absence of residual free lime improves the natural binder a higher percentage of residual free lime on the mass of the compound decreases the hydraulic capacity of the same mix and even more complex will be the decay pathologies due to the formation of weak acids up to the ettringite. In other words, a quality binder is composed of hydrated lime integrally combined with pozzolanic and/or cocciopesto materials. This binder will be characterized, in the final drying phase, by the absolute absence of free residual elements not integrally combined.

How to obtain such a binder? It is necessary to start from the Vicat equation, according to which the ratio between lime and clay, i.e., the hydraulic index, falls within a range of 0.1–0.5, where 0.5 means the maximum hydraulicity of the compound (*eminently hydraulic lime*). So, it is an uncertain value and its obtainment—although it falls within a defined range—does not prevent the absence of residual free lime, therefore of soluble parts.

The procedure proposed in this research provides that in Vicat's formula, instead of an uncertain value belonging to a range of values, the result of the optimal ratio of clay/lime is equal to 1.

The imposition of the value 1 respects, in part, also the principle of Avogadro, for which in the same volume—apart from the mass—exist and insist the same numbers of atoms "combinable" between elements of different nature. These elements are in our case represented by calcium oxide and silicon dioxide which are transformed into calcium silicates, the only compound that guarantees natural hydraulicity to the mortars.

The unresolved problem before this patent lay in the impossibility to establish what were the optimal quantities of the clay/lime compounds, especially in relation to the mineralogical nature of the compounds, which are variable in the case of clay.

### *31.3.2 Step 2: The Process for the Formulation of Binders to Be Used to Obtain Natural Mortars*

Density, defined as the ratio of mass to volume, is a unique value for each element that makes up lime and clay. If in the pure 98% lime density can be assumed as a fixed value, this does not happen for the clay, composed of variable parts.

If the density can be defined fixed for the lime, and by choice used pure at 98%, even its *apparent density* can be identified as unique; instead, it is variable in the case

| | | | | | | | | | |
|---|---|---|---|---|---|---|---|---|---|
| NATURAL BINDER FUNCTION OF COCCIOPESTO | | | | DATA TO ENTER (COCCIOPESTO | | | EXACT | | |
| COCCIOPESTO | | DENSITY | | MINERALOGICAL CHARACTERISTICS | | % C.PESTO | % DENSITY | DENSITY | |
| Alkaline Metal | Ca | 1.54 | | 8.42 | 8.42 | 9.14 | 0.14 | MEDIA | |
| Not metal | Si | 2.33 | | 61.81 | 61.81 | 67.11 | 1.56 | C.PESTO | |
| Other metal | Al | 2.70 | | 16.38 | 16.38 | 17.79 | 0.48 | 2.65 | |
| Transitional metal | Fe | 7.86 | | 5.49 | 5.49 | 5.96 | 0.47 | | |
| Clay coefficient | | 14.43 | | | | | | | |
| Fire loss | | 0.85% | | 92.1 | 92.1 | | 2.65 | | |
| Other elements | | 7.05% | | | 0.921 | | | | |
| PERCENTAGE CHECK BETWEEN LIME AND COCCIOPESTO AS A FUNCTION OF THE EXAMINED COCCIOPESTO | | | | | | | | | |
| | | DENSITY | Percentages needed to saturate lime-cocciopesto combination | | | | | | |
| CALCIUM | Ca | 1.54 | Fixed index | %variable | 36.73% | | | | |
| Alkaline Metal | Ca | | | | | | | | |
| Not metal | Si | | | | | KG/MC | | | |
| Other metal | Al | 2.65 | Variable index | %variable | 63.27% | | | kg/mc | |
| Transitional metal | Fe | | FROM PROJECT NEW HYDRAULIC NATURAL BIN | | | DATA | NEW BINDER MIX | | |
| | | | | MVA | | TO USE | KG | NEW MVA | Natural |
| | | 0.419 | Lime used | 530 | Hydrated lime | 36.73% | 194.65 | 998,23 | hydraulic |
| | | 0.0419 | Cocciopesto use | 1270 | Cocciopesto | 63.27% | 803.58 | kg/mc | binder |

**Fig. 31.1** Formulary 1, to be used for the construction of the new hydraulic binder. The values shown for the component densities are examples

of clay-based on its granulometric assortment and mineralogical characteristics. As is known, the variants necessary to establish the exact apparent density through the weighing in the laboratory are often complex and inaccurate, as well as strongly dependent on the relative humidity rate of each individual raw material and the operator's hand.

Considering a volume of 1 $m^3$ and comparing the elements from the point of view not of weight but of apparent density, it is possible to establish—the apparent density of the new compound.

On the basis of the above procedure, the use of the formulary 1 illustrated below (Fig. 31.1), which forms part of the patented process, allows—starting from the density of the individual elements used, derived from the specific mineralogical nature of the cocciopesto used each time—to obtain a new hydraulic binder given by the use of hydrated lime (always 98% pure CaO) and cocciopesto perfectly balanced.

In Formulary 1, the values of the density of the Ca and of the cocciopesto components used are reported; all depend on the chosen products (for which values are reported in the relative technical datasheets). If once the components have been established, these density values remain unchanged, it is not the same for their sum: by varying the percentage of the elements, the density of the clay also varies.

The ratio between the density given by the Ca in the hydrated lime and the average density of the main elements of the cocciopesto Ca+Si+Al+Fe will provide the percentage to be used to formulate the new natural hydraulic binder obtained by cooking (therefore with raw materials work at temperatures below 1000 °C) and through first ionization energy.

The percentages obtained to compose the new hydraulic natural binder, in the case in question where the densities are those reported in Formulary 1, are:

– hydrated lime powder 36.73%;
– cocciopesto 63.27%.

The new apparent density of the natural hydraulic binder thus obtained is 998.23 kg. Since the calcium oxide present in the lime and the elements present in the cocciopesto (with pozzolanic characteristics) are completely combined, there will be no residual free lime.

The object of the patent, therefore, defines a "codified procedure" for the formulation of natural mortars with the use of a new hydraulic natural binder obtained by cooking. The heart of the process lies in the possibility—starting from the volumes of aggregates and binders used—to create a mortar that is *universal* as it can be used independently of the context, and free from the diseases that characterize Portland cement.

The process is applicable both on an industrial and artisanal scale, as well as with a very low natural impact, being based on only natural raw materials.

The used cocciopesto will preferably have a granulometric curve between 0 and 500 μ. The aggregates used for the composition of these mortars will be as pure as possible, to minimize the "uncontrolled" soluble parts, which could be the future cause of degradation pathologies for the plaster, both chemical and mechanical.

The binder formulated and composed of hydrated lime and cocciopesto will exclusively use super-ventilated calcium lime CL90 and cocciopesto pozzolanico as a natural hydraulicizing compound: this will ensure, depending on its mineralogical characteristics, enough silicon dioxide, as a guarantee of enough calcium silicates and absence of residual free lime.

Any percentage variation between aggregates and binders, if desired, will still be controllable and will affect only the elastic modulus, allowing a possible variation of stiffness that can be modulated for specific purposes and prescriptions and in conformity with the historical walls.

Essential for the application of the process is the knowledge of the mineralogical characteristics of the selected natural aggregates which, in advance, should be dry mixed with the project volumetric percentages, defined by the calculation matrices shown below.

## 31.4 The Color of the Mortars

In the mix used in the project, colors are selected for "subtractive primaries" (cyan, yellow, and magenta), or colors that are perceived in the contact areas when the additive primary colors are applied. With their use in the appropriate quantities of volume, the desired project mixes are obtained, which return from a close distance single colors while from a distance prevailing color.

The lack of colored artificial pigments in the compound arises from the desire to highlight the aggregates used in the mix (Fig. 31.2, 31.3, 31.4, and 31.5).

| TRANSFORMATION IN PERCENTAGE 100 WORKED BASES | | | | | | | | | | | | | | | | | | |
|---|---|---|---|---|---|---|---|---|---|---|---|---|---|---|---|---|---|---|
| | | | | COCCIOPESTO | COCCIOPESTO | SAND SILK | YELLOW | WHITE LAND | GRANULAR | AZOLE | AZOLE | | | | | NHL | | |
| ENTER DATA | | | | 0-500 | 0-1.3 | 0-3 | 0-500 | 0.1 | 500-800 | 0.5 | 0.3 | NHL | 500-800 | 0-1 | 0-2 | WHITE | VOLUMES | |
| | | | | 0.00 | 3.00 | 0.10 | 0.50 | 1.00 | 1.00 | 0.00 | 1.00 | 0.00 | 0.00 | 0.00 | 0.00 | 0.00 | 0.0660 | % FUNZIONI |
| SAMPLE | PERCENTAGES | | | 0.00% | 45.455 | 1.515 | 7.576 | 15.152 | 15.152 | 0.000 | 15.152 | 0.000 | 0.000 | 0.000 | 0.000 | 0.00% | 100 | |
| DETECTED | 225-240% | % | | | | | | | | | | | | | | | OBTAINED | 225-240% |
| 173 | 0.68 | 0.439 | RED | | 95.45 | 3.11 | 16.89 | 38.18 | 37.12 | 0.00 | 20.30 | 0.00 | | | TO CORRECT | 0.54% | 173.939 | 0.682 |
| 128 | 0.50 | 0.325 | GREEN | | 50.91 | 2.65 | 16.36 | 38.48 | 37.88 | 0.00 | 21.36 | 0.00 | | | | 1.38% | 129.773 | 0.509 |
| 93 | 0.36 | 0.236 | BLUE | | 14.09 | 2.23 | 14.92 | 38.64 | 38.18 | 0.00 | 23.64 | 0.00 | | | | 0.55% | 93.515 | 0.367 |
| 72 | 0.30 | | SHADES | | 8.18 | 0.29 | 2.20 | 20.15 | 19.85 | 0.00 | 22.27 | 0.00 | | | | | 72.939 | 0.304 |
| 147 | 0.61 | | SATURATION | | 80.91 | 1.33 | 5.23 | 36.36 | 19.55 | 0.00 | 3.64 | 0.00 | | | | | 147.015 | 0.613 |
| 146 | 0.61 | | SHADES | | 51.36 | 2.52 | 0.00 | 36.21 | 35.45 | 0.00 | 20.61 | 0.00 | | | | | 146.152 | 0.609 |

**Fig. 31.2** Formulary 2, to be used for the certain identification of the volumes and of the colors

| FORMULAS IN VOLUME AND PERCENTAGES | | | | | | | | | | | | | |
|---|---|---|---|---|---|---|---|---|---|---|---|---|---|
| IDENTITY | E x Kg | 0 | | N.1 | c. cliente | €0,00 | | | | | | 67% | 33% |
| Petrographic description | | Inert lava | | | | cal. Cristall | cal. Oolitica | l. Cristallina | | | Mortar binders | EN 459-1 | NI EN 459 |
| Traceability | Manufacture | 864/15/03/16 | | | 332/15/03/16 | 1556/v/23/11 | 1557/v/23/11 | 1556/v/23/11/15 | | | | 1339/15/12 | apr-16 |
| Aggregates for mortars | | AZ. 0-3 | AZ. 0-500 | C.P. 0-0,5 | C.P. 0-1,3 | TONAC. B. 0-1 | SIAL.M 0-50 | Granul. 500-800 | agg. | S.SILIC. | | CL-90-S | binder |
| CE conformity marking | | SI | | | SI | SI | SIAL.M 0-50 | SI | | SI | | SI | SI |
| Particle size curve | mm | 0-3 | 0-0,5 | 0-0,5 | 0-1,3 | 0-1 | 0-0,5 | 0,5-0,8 | | 0-3 | | C.HYDRATES | |
| Category | | GF85 | | | | 4 | 4 | 1 | | | | s. ventilated | |
| Thickness - Finesse | | Fine | | | | FP | MP | CP | | | | powder | |
| Reaction to fire | | Class A1 | Class A1 | Class A1 | Class A1 | Class A1 | Class A1 | Class A1 | | Class A1 | | | |
| Hardness | | 4,99 | | | 5,25 | 1,75 | 1,78 | 2,9 | 0 | | | 1,75 | 2,5 |
| Water absorption% | WA24 | 0,91 | | | 0,84 | 4,4 | 4,3 | 1,43 | 0 | | | 4,4 | 2,44 |
| APPARENT WEIGHT MASS | MVA | 1700 | 2000 | 1350 | 1270 | 1550 | 1570 | 1370 | 1160 | 1620 | Constituents | 530 | 810 |
| CHARACTERIZATION (p) | | | | | | $CaCO_3$ | $CaCO_3$ | $CaCO_3$ | aggregate | | $SiO_2$ | | 19% |
| $SiO_2$ | | 45,55% | | | SiO2/61,81% | 100% | 100% | 100% | | >90% | $Al_2O_3$ | | <3% |
| $Al_2O_3$ | | 14% | | | Al2O3/16,38% | | | | | | $Fe_2O_3$ | | <1% |
| $Fe_2O_3$ | | 5,14 | | | Fe2O3/5,49% | | | | | | $Ca(OH)_2$-CaO totale | >91% | >56% |
| CaO | | 10 | | | CaO/8,42% | | | | characterization | | Combined CaO | | >25% |
| Minor compounds | | | | | min./7.9 | | | | | | Free lime | | 17,10% |
| pH | | | | | pH 10,2 | | | | | | pH | ph 11,5 | |
| Specific surface | | | | | BET 2,80 $m^2/g$ | | | | | | Sup. spec. Blaine | | $cm^2/g$7710 |
| | | | | | | | | | | | Compr. strength 28 days | | Mpa 5,66 |
| Chromatic characteristics | | | | | rosso intenso | | | | territorial | | | whitish | Havana |
| Type of aggregate | | AZOLE | BLACK LAVA | C.PEST | C.PEST | T.WHITE | G.MORI | GRANULAR | x | Silica sand | type | | NHL 3,5 |
| Particle size curve | | 0-3 | 0-500 | 0-500 | 0-1,3 | 0-1 | 0-500 | 500-800 | 0-1 | 0-3 | | | |
| Grain size distribution | Sieves diameter | total | | | | | | | | | | | |
| | 3,15 mm | 100 | | | | | | | | 100 | | | |
| | 2,0 mm | 85,3 | | | | | | | | 99,6 | | | |
| | 1,6 mm | 69,9 | | 100 | | | | | | 85,17 | | | |
| | 1,0 mm | 51,9 | | 92,95 | 100 | | 100 | | | 65,26 | | | |
| | 500 mm | 33,7 | | 62,72 | 87,78 | 100 | 4,13 | | | 42,22 | | | |
| | 250 mm | 20 | | 29,64 | 72,05 | 42,81 | | | | 26,43 | | | |
| | 125 mm | 10 | | 6,87 | 46,46 | 9,88 | | | | 18,5 | | | |
| | 63 mm | 3 | | 4,06 | 24,88 | 5,49 | | | | 9,7 | | | |
| FREE LIME CHECK | H%-(N-O%-P%)=0 | | | | | | | | | | | | |

**Fig. 31.3** Formulary 3, containing the collection of raw material data used for the construction of natural mortars

| FINITURA | DESIGN | AZ. 0-3 | AZ. 0-500 | C.P. 0-0,5 | CP 0-1,3 | TONAC. B. | GIAL. M. | GRANUL. | X | SAND SILK | | BINDER |
|---|---|---|---|---|---|---|---|---|---|---|---|---|
| CURVA GRANULOMETRICA | mm | 0-3 | 0,05 | 0-05 | 0-1,3 | 0-1 | 0-0,5 | 0.5-0,8 | 0-X | 0-3 | | |
| | | A | B | C | D=X | E | F | G | CARATTERE | H | | L1 |
| INSERIRE DATI | VOLUME | 1 | 0 | 0 | 3 | 1 | 0,5 | 1 | 0 | 0,1 | AGGREGATES | 2,211 |
| 9,9 | %VOLUM | 14,24 | 0 | 0 | 42,86 | 14,24 | 7,14 | 14,24 | 0 | 1,42 | % VOLUME | 20,01 |

**Fig. 31.4** Formulary 4, containing the volumes of the compounds to be inserted and chosen for the specific designed natural mortar. The use of this formulary allows inserting the project data expressed in volume and the percentages of the aggregates and of the natural binder used

Following is an example of mortar obtained through the procedure illustrated and applied to the Church of Purity of Catania (Fig. 31.6).

Formulary 4 is applied with the subsequent project data:

D = X=42.86%—aggregate characterizing basic colors
A = 14.24%—aggregate inert basalt lava
G = 14.24%—white crystalline carbonate structural aggregate
F = 7.14%—yellow carbonate aggregate used for basic color variation X = D

| | | | | | | | | | | | | | |
|---|---|---|---|---|---|---|---|---|---|---|---|---|---|
| FINISHING 1 | MVA | AZ. 0-3 | AZ. 0-500 | C.P. 0-0,5 | C.P. 0-1,3 | TONAC.B. | GIAL.M. | GRANUL. | X | SAND SILK | | L1 |
| MVA AGGREGATES + BINDERS | 1298,11 | 252,82 | 0 | 0 | 423,3 | 210,18 | 107,82 | 164,2 | 0 | 22,96 | mva% | 54,33 |
| | | | | | | | | | | | | |
| FINISHING 1 | | AZ. 0-3 | AZ. 0-500 | C.P. 0-0,5 | C.P. 0-1,3 | TONAC.B. | GIAL.M. | GRANUL. | TER V. | SAND SILK | | L1 |
| MVA AGGREGATES + BINDERS | Kg | 308,2 | 0 | 0 | 516,02 | 256,21 | 131,43 | 200,16 | 0 | 27,99 | 1440,01 | 302,28 |
| | | | | | | | | | | | | |
| FINISHING 1 | HARDNESS | AZ. 0-3 | AZ. 0-500 | C.P. 0-0,5 | C.P. 0-1,3 | TONAC.B. | GIAL.M. | GRANUL. | X | SAND SILK | | L1 |
| MVA AGGREGATES + BINDERS | 3,57 | 0,74 | 0 | 0 | 1,75 | 0,24 | 0,12 | 0,35 | 0 | 0 | | 0,18 |
| | | | | | | | | | | | | |
| FINISHING 1 | ASS. WATER % | AZ. 0-3 | AZ. 0-500 | C.P. 0-0,5 | C.P. 0-1,3 | TONAC.B. | GIAL.M. | GRANUL. | X | SAND SILK | | L1 |
| MVA AGGREGATES + BINDERS | 2,12 | 0,14 | 0 | 0 | 0,28 | 0,6 | 0,3 | 0,17 | 0 | 0 | | 0,45 |
| | | | | | | | | | | | | |
| FINISHING 1 | Enter data | AZ. 0-3 | AZ. 0-500 | C.P. 0-0,5 | C.P. 0-1,3 | TONAC.B. | GIAL.M. | GRANUL. | X | SAND SILK | | L1 |
| MVA AGGREGATES + BINDERS | 1 | | | | | | | | | | | L1 |
| Incidence kg / sqm | 3 | kg | | kg | | kg | | kg | | kg | a/d/b | kg |
| kg total product | 3 | 0,45 | 0 | 0 | 1 | 0,41 | 0,21 | 0,36 | 0 | 0,04 | 3 | 0,31 |

**Fig. 31.5** Formulary 5, containing the immediate checks on the natural mortar designed by formulary 4

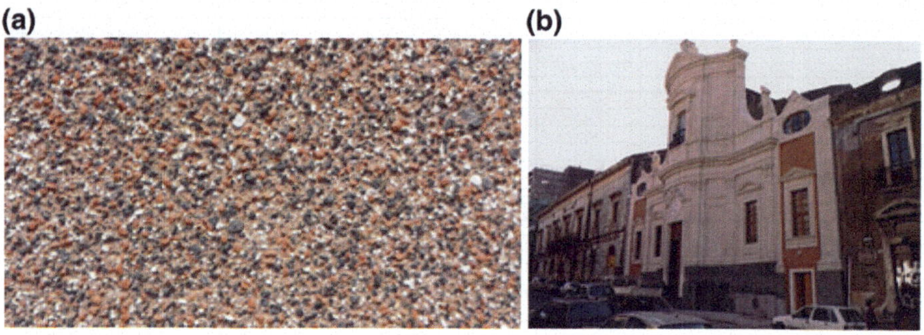

**Fig. 31.6** **a** close view; **b** perspective view

H = 1.42%—aggregate used for superficial chromatic characterization
Sum aggregates = 79.90%
New binder L1 = 20.01%.

## 31.5 Conclusions

The present process allows to produce lime-based natural mortars which are hydraulically active and obtainable without the addition of industrial substances, in compliance with the morphological characteristics of the historical buildings which require Conservation Restoration.

This process involves numerous advantages. In particular:

- the procedure guarantees the absence of non-combined residues, which will always be equal to zero in the formulated mortar. The process, therefore, allows to obtain a non-soluble dried mortar which guarantees high mechanical strength (reached in a short time and always with a class higher than CS2) and long durability;

**Fig. 31.7** Samples of hydraulically active natural mortars free from artificial pigments

- the colors will be obtained without the use of coloring additives, with a result that does not impact from the esthetic point of view;
- the procedures previously used to obtain natural hydraulic mortars based on hydraulic lime will be avoided, such as:
  - the slaughtering of the compounds;
  - the addition in the historical mortar of additives and/or industrial products that replaced and/or compensated for the absence of resistances typical of the only lime base compounds;
- the obtained mortar can also be used without preliminary laboratory tests. The proposed procedure can, in fact, be verified through post-work chemical and mechanical tests;
- the calculation and design process can also be used by nonexpert users in the field of Conservative Restoration, which will have the possibility to realize—even directly on-site and thanks to the use of formulary based on the illustrated procedure—mixes in which the reports between aggregates and binder are defined, to guarantee performances of high quality and not obtained with empirical approximation;
- the obtained compounds will always have a high hygroscopic capacity, which will guarantee, for the mortar applied in closed rooms, the ability to thermoregulate the relative humidity of the same environment;
- the obtained compounds will always be characterized by an excellent hydraulic capacity and high breathability, therefore suitable for application in historic walls that often suffer from the problem of rising damp.

Future implementations of the work are based on the possibility to obtain high albedo coefficient through the mortars object of this research (Fig. 31.7).

## References


1. The Circular Economy Concept—Regenerative Economy
2. Cradle to Cradle, The Product-Life Institute, Product-life.org, 14 November 2012
3. The Economics of the Coming Spaceship Earth, Eoearth.org
4. Zhijun, F., Nailing, Y.: Putting a circular economy into practice in China. Sustain. Sci. **2**, 95–101 (2007)
5. Nakamura, S., Kondo, Y., Nakajima, K., Ohno, H., Pauliuk, S.: Quantifying recycling and losses of Cr and Ni in steel throughout multiple life cycles using MaTrace-alloy. Environ. Sci. Technol. **51**, 9469–9476 (2017)
6. Circular Economy in Textiles and Apparel Processing, Manufacturing, and Design, The Textile Institute Book Series, pp. 1–11, Miguel AngelGardetti, 1—Introduction and the concept of circular economy (2019)
7. Iacovidou, E., Velis, C.A., Purnell, P., Zwirner, O., Brown, A., Hahladakis, J., Hopkins, J.M., Williams, P.T.: Metrics for optimizing the multi-dimensional value of resources recovered from waste in a circular economy: a critical review. J. Clean. Prod. **166**, 910–938 (2017)
8. Goodland, R.: The concept of environmental sustainability. Annu. Rev. Ecol. Syst. **26**, 1–24 (1995). https://doi.org/10.1146/annurev.es.26.110195.000245

# Chapter 32
# Assessment of the Efficiency and Reliability of the District Heating Systems Within Different Development Scenarios

**Aleksandrs Zajacs, Anatolijs Borodiņecs and Raimonds Bogdanovičs**

**Abstract** Renovation of multi-apartment buildings and district heating distribution system is aimed to reduce the final energy consumption. Directive 2012/27/EU of the EU on energy efficiency it is defined that as from 2014 3% of the public buildings owned and occupied by its central government should be renovated each year to meet at least the minimum energy performance requirements. Moreover, Directive 2010/31/EU and national regulation require the application of minimum requirements to the energy performance of existing and new buildings as well as increasing the number of nearly zero-energy buildings, which will have a significant impact on the consumption of thermal energy. In order to predict the influence of the future energy efficiency measures on the efficiency of the existing DH systems, District Heating Planning Tool was used for evaluation of the particular DH system. At the output, the tool provides twelve efficiency and balance indicators with the recommended permissible limits for the evaluation of DH system development scenarios. The tool validation is based on the results of comprehensive research of already renovated buildings. As a result of complex renovation decrease in thermal energy consumption can be achieved from 35 to 50% (average 42.3%). Consistent simulations of the proposed future development scenarios for a particular DH system with different building renovation rates of 3, 5 and 7% have shown a natural reduction in thermal energy consumption in a 15-year perspective by 13.5, 22.5 and 30%. Simulation results of the last three scenarios, which consider the use of renewable energy sources, have shown that $CO_2$ emissions have been considerably reduced by about 40% compared to the actual state and the share of biomass has reached 47% of the total fuel consumption for thermal energy production.

A. Zajacs (✉) · A. Borodiņecs · R. Bogdanovičs
Institute of Heat, Gas and Water Technology, Riga Technical University, Kipsalas Street 6A, Riga 1048, Latvia
e-mail: Aleksandrs.zajacs@rtu.lv



J. Littlewood et al. (eds.), *Sustainability in Energy and Buildings*, Smart Innovation, Systems and Technologies 163, https://doi.org/10.1007/978-981-32-9868-2_32

## 32.1 Introduction

Urban environment is a comprehensive resource that requires smart management and there is an increased need for the introduction of decarbonization policies, use of sustainable resources and development of sustainable cities with integrated smart disciplines—economics, energy sector, technologies, architecture and politics. Special attention also is paid to city resilience planning due to its importance for any city development as well as the national securitisation process by becoming better prepared and more responsible for national risk management, including independence of imported energy sources. By "city resilience", one typically understands sustainable energy management, securing stable energy supply, reducing energy consumption, and developing renewable energy sources. There are many worldwide and local documents that set the level of the planned reduction in energy consumption [1–3].

Documents that determine the importance of improving energy efficiency are approved in Europe and some of them are of global importance. Article 9 of Directive 2010/31/ES of the European Parliament, Member States shall ensure that by 31 December 2020, all new buildings are nearly zero-energy buildings. Under the Kyoto Protocol, by 2020 the European Union must reduce emissions by 20%, but since December 2015 the EU is committed to new ambitious goals of global climate change agreement—Paris Protocol. Ambitious national greenhouse gas reduction targets include 40% cut by 2030 compared to 1990 levels and specifying that the long-term goal should be to reduce global emissions by at least 60% below 2010 levels by 2050. The goals also include reaching 27% share of renewable energy consumption, 27% energy savings compared with the business as usual scenario and limiting a dangerous rise in global average temperature to below 2 °C compared with pre-industrial levels by 2030.

Achieving significant energy savings is only possible if there is a clear identification of priority sectors as well as mobilization of investment capital that can be easily accessed. The building sector is responsible for about 40% of energy consumption in the European Union; therefore, reducing the energy use in buildings and improvements in district heating (DH) and cooling sectors are top priorities to achieve a reduction in fossil fuel consumption and $CO_2$ emissions. Natural gas consumption could be cut by one-third if the renovation of buildings is speeded up. In the Northern Europe climate, it is found that human behaviour can lead to 50% higher heating demand and 60% higher heating power than those anticipated in the reference values in the standard calculations for energy demand patterns in energy efficient buildings.

In order to properly respond to future changes in energy demand and implement advantageous development of district heating sector, it is necessary to have a tool for evaluation of the planned DH system development scenarios based on a set of actual qualitative and quantitative studies on the district heating sector development trends and problems.

Since Planning tool is considered for assessment of large-scale systems for a long-term period (15 years), it is considered as a steady state model. Sameti and

Haghighat [4] endorses, that most models suffer from the very long computational time when large networks are taken into account. Dynamic modelling is necessary for prediction of daily or hourly variations, however, assessment of longer periods (years or decades) can be done using steady state models. In steady state models assuming the constant conditions for a given period, the corresponding heat load is estimated by well defined and standardized calculations. Fang and Lahdelma [5] notes, that static (steady-state) model works well if heat load doesn't change rapidly over time. An advantage of the static model is that it can be validated by computing the energy balance equation where the heat loss plus customer heat demands should equal the overall heat production. Duquette et al. [6] argue, that although the computational intensity is low for steady state DH grid models. The main strength of the steady state method is that it is possible to perform all necessary calculations using only data from heat supply utility and no prior knowledge from individual load components is necessary [7]. Gao et al. [8] foresee changes in production mode and management mode of district heating enterprise and Sayegh et al. [9] claims that it is not possible to create a universal solution for all EU members of renewable energy source integration into district heating. Each EU country should be analysed individually, taking into account all technical characteristics and emission considerations.

## 32.2 Methods

In order to estimate the possible reduction of heat energy consumption in buildings, which are connected to the district heating network, the research of the renovated buildings was performed within the framework of this research. Information about 55 buildings was collected: total area, total heated area, year of building, year of renovation, domestic hot water preparation (centralized or individual), hot water circulation (yes/no). Monthly heat energy consumption data for a 5-year period—2010–2014 were collected. The investigated buildings have different inner heating systems (one-pipe, two-pipe, with top distribution and with bottom distribution). Buildings are divided into 5 groups by the renovation year in order to evidently visualize the reduction in consumption after renovation.

Later on, data were corrected by degree-days in order to bring them to similar weather conditions in respect of outside air temperature and duration of the heating season. Additionally, to this, massive data array was evaluated statistically in order to approve it for further research. Standard deviations were calculated in order to understand how the measurements of one sample (one building) are spread around the mean of this sample. Since buildings are different by its construction, it is not correct assumption that heat energy consumption is a subject of direct compare, but since all buildings were constructed from the late '60s until late '80s we can assume that requirements of building codes of those days have complied and differences in thermal properties of building envelopes will not influence results dramatically. For all buildings, Standard Deviation (SD) was between 5.8 and 14.8 while the average of the mean values of each building was 28.8. Then Standard Errors (SE) for each building were calculated to define the possible difference of the mean of the sample (one building) from the mean of all population. For unrenovated buildings, SE was between 1.2 and 2.5 with average $\pm 1.8$ kWh/m$^2$ per month or $\pm 6.3\%$, and for all considered buildings SE was between 1.2 and 4.2 with average $\pm 2.1$ kWh/m$^2$ per month or $\pm 7.2\%$.

From all 55 measured buildings, only 26 were accepted for further research because of different issues. Only in 31 buildings renovation process was fully finished. 5 buildings were renovated in the year 2013/2014, so it was not possible to spot any changes in energy consumption at the moment of the research. Results of the other 26 buildings are presented in Fig. 32.1. A comprehensive renovation of buildings resulted in heat consumption reduction by approximately 40% from 130–140 to 75–80 kWh/m$^2$. Differences in the reduction of heat energy consumption are explained by the fact that different energy efficiency measures were implemented in different buildings.

The studied buildings were insulated according to the previous normative requirements, which were valid until April 2014. Since April 2014, Latvia has significantly reduced normative heat transfer values [10], as it is shown in Fig. 32.2.

The evaluation of installed capacity and the fuel used at existing heat sources was performed with the aim to follow the dynamics of reduction of thermal energy demand and the introduction of local renewable energy sources.

Analysing the statistical data of the Latvian Central Statistical Bureau, it was defined that the total installed heat capacity of all heating plants in Latvia reduced along with the total number of heating plants, as shown in Fig. 32.3.

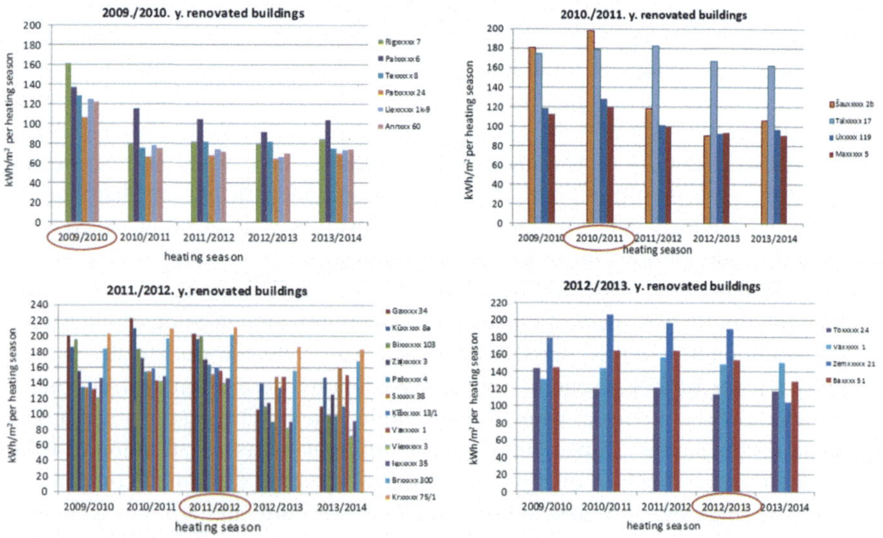


**Fig. 32.1** Heat energy consumption of multi-apartment buildings before and after renovation in the city of Riga (Legend indicatively shows the address of each building. Some letters are replaced with "x" to conform with regulations of personal data use)

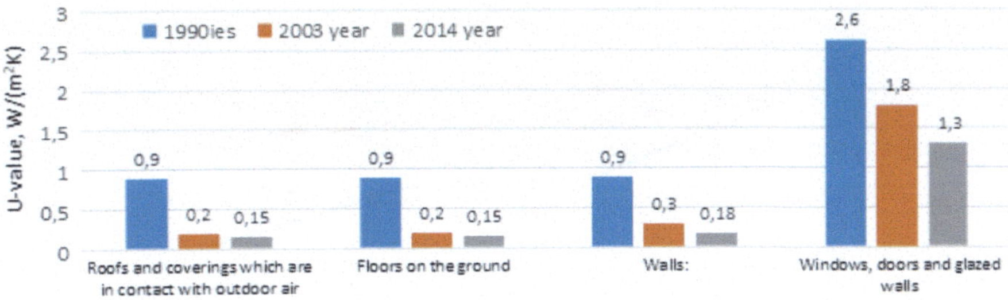


**Fig. 32.2** Comparison of normative U-values in Latvia since the 1990s

**Fig. 32.3** The number and the installed capacity of DH plants (left) and heating plants using wooden chips (right) in Latvia during the period from 2007 to 2014

During the last decade, there is a constant decrease in the total number and installed capacities of the heat plants. Despite the number and installed capacities of biomass heating plants were constantly growing, confirming the replacement of fossils by renewable fuels.

Another evidence of the reduction of heat energy demand is presented by the evaluation of the average heating loads of 8 DH plants in the city of Riga during the past decade. In order to compare data from different heating seasons, heat loads were recalculated to the same outside temperature of −20.7 °C. Results are presented in Fig. 32.4.

The calculated results showed that the biggest heat source HPs "IAT" had no significant heat load change during the last 10 years. It can be explained by the fact that the reduction of heat losses in the heating network and energy efficiency measures in the buildings are compensated with load growth due to connecting of new heat consumers. The weighted average heat load for all 8 heating sources reduced by 4% compared with 2006, and one of the reasons was a reduction of population in Riga, which has been declined by 3.3% over the last decade, according to the data of Riga City Council.

All previously mentioned research and information served as a base for the development and validation of the District Heating Planning Tool. Validation of the District Heating Planning Tool was performed for two DH system states. All system energy data was available for analysis from GIS-based software. Analysis of data, gathered

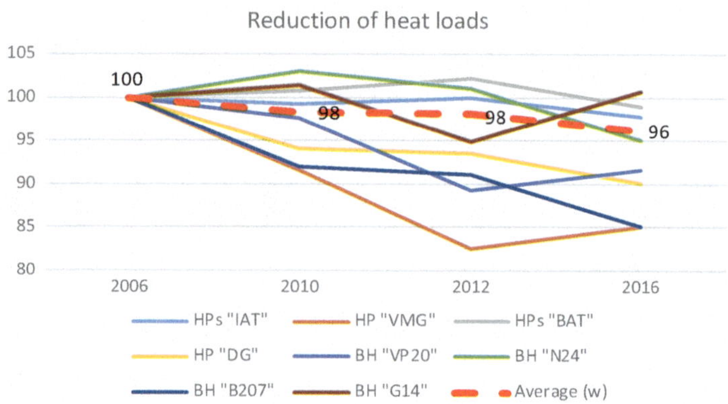


**Fig. 32.4** Heat load reduction for different heat sources

throughout the validation of the tool, confirms the reliability of the derived results with the mean absolute deviation between simulated and measured values less than 8%. Importance of the validation process is outlined by Åberg et al. [16]. District heating systems are not unified for all regions, states, cities and districts since the history and characteristics of buildings in each area are different [11]. One of the more complicated and responsible steps is modelling of the district heating system [12]. During modelling it should be taken into account, that every system will have its own features and characteristics since the behaviour of inhabitants will vary from place to place [13]. The planning tool mathematically describes heat energy production, distribution and consumption processes within one district heating system, as presented in Fig. 32.5.

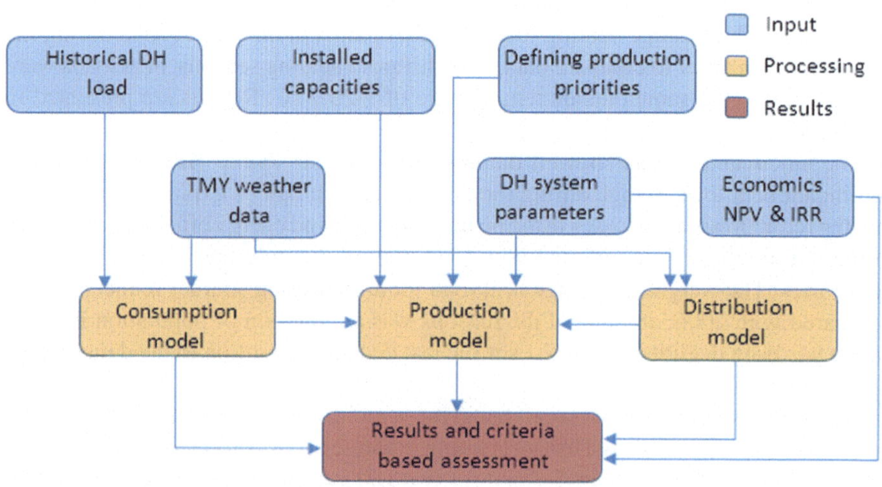


**Fig. 32.5** Workflow chart of the planning tool

Within the framework of this research were proposed, simulated and assessed five different district heating system development scenarios using developed District Heating Planning Tool. For the assessment was chosen existing district heating system with the following main characteristics:

Installed heat capacity—6 MW (2 × 3 MW natural gas-fired boilers);
Building stock—51069 m$^2$ of the heated area;
Distribution network—1962 m (outside 1138 m, basement—824 m, 58/42%).

Assessment of each development scenario was made for the 15-year time span. The main specifications of proposed DH development scenarios are described below:

a. Reference scenario—old distribution pipelines, production of heat only in the natural gas-fired the heating boiler, buildings are not renovated;
b. Realistic renovation—building renovation rate 3% of the buildings every year, the distribution network is renovated for 100%, the cogeneration unit is installed (15-year perspective);
c. Intense renovation—building renovation rate 5% of the buildings every year, the distribution network is renovated for 100%, cogeneration unit is installed, wooden biomass-fired water boiler 2 MW (15-year perspective);
d. Enhanced renovation—building renovation rate 7% of the buildings every year, the distribution network is renovated for 100%, cogeneration unit is installed, local solar collectors installed, temperature graph in distribution network 55/35 °C (15-year perspective);
e. Realistic renovation and biomass use—building renovation rate 3% of the buildings every year, the distribution network are renovated for 100%, cogeneration unit is installed, wooden biomass-fired water boiler 3 MW, flow/return temperatures in distribution network 55/35 °C (15-year perspective).

## 32.3 Results

Planning tool gives the possibility for a quantitative assessment of different energy efficiency criteria. There are twelve efficiency criteria, as shown in Fig. 32.6, there are: Electricity to Heat ratio (maximum possible el. energy/produced heat energy), Overcapacity (Installed heat capacity [MW]/max. possible demand), Fuel balance fossil versus renewable, Operational efficiency [%] heat and electricity (ratio of the total el.energy produced/heat energy produced by the plant during a period of time compared to the total potential electricity that could have been produced if the plant operated at 100 percent in the period), Load factor [%] heat and electricity (ratio between average load and peak load), Efficiency of fuel use [%], Emission factor [$tCO_2$/MWh], Total emissions $tCO_2$, Energy efficiency of the system (losses)[%] (Heat losses in the distribution network/Total produced energy), Heat density annual [MWh/m$^2$], useful heat energy/DH system area, Transmission effi-

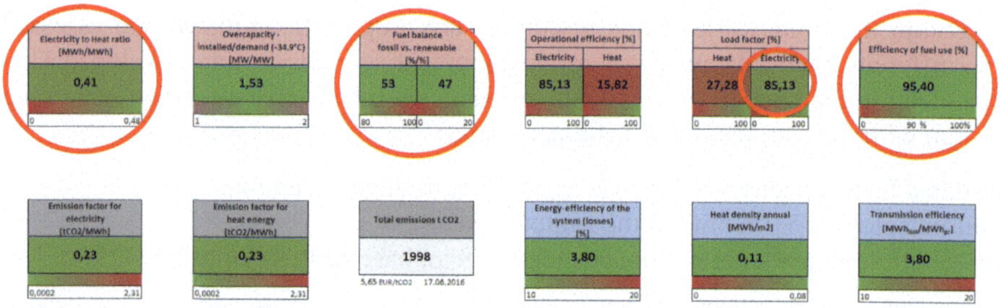


**Fig. 32.6** "Realistic renovation and biomass use" scenario criteria evaluation results (output results from district heating planning tool are given in SI units and could be recalculated manually)

ciency [MWhloss/MWhpr] (Heat losses in the distribution network/Produced energy on the heating plant.).

The output of the District Heating Planning Tool quantitatively visualizes results for each of efficiency criteria. Results for all five district heating system development scenarios were obtained during consistent simulations. Figure 32.6 represents results for "Realistic renovation and biomass use" scenario.

This development scenario is most attractive from a technological point of view since Load factor for cogeneration unit is high, compared with data from literature [14, 15], usually, these values are 50–60%. Load factor and operational efficiency for heat production are not so good since water boilers are operated only in periods when cogeneration unit cannot provide all required heat load, and this occurs rarely and only during the heating period. But the reduction of installed capacities is not an option if DH operator willing to comply with safety requirements for district heating supply systems.

Renovation of buildings reduce total heat demand and comparing with nonrenovated building stock with 10 231 MWh heat consumption per year, realistic renovation development scenario in 15 years' perspective has reduced total heat demand to 8850 MWh. The average heating season in Latvia accounts for 4824 h or 201 days.

Heat losses in the distribution network which are represented by transmission efficiency are relatively low only 3.8%. That shows that by reducing flow and return temperatures of the water in the distribution system it is possible to reduce heat losses more than by 35.6% or for 193 MWh (comparing with realistic renovation scenario). Main results for all five scenarios are presented in Fig. 32.7.

The share of renewable sources has reached 47% for the last development scenario, reducing total emissions, but natural gas-fired water boiler is used only about 260 h per year for coverage of peak load and this occurs only when outside air temperature slides below −22 °C (−7.6 °F). Recovering latent heat of the water vapour condensation in the combustion exhaust gas in wooden biomass-fired water boiler could increase the overall efficiency of fuel use up to 95.40% by lower calorific value.

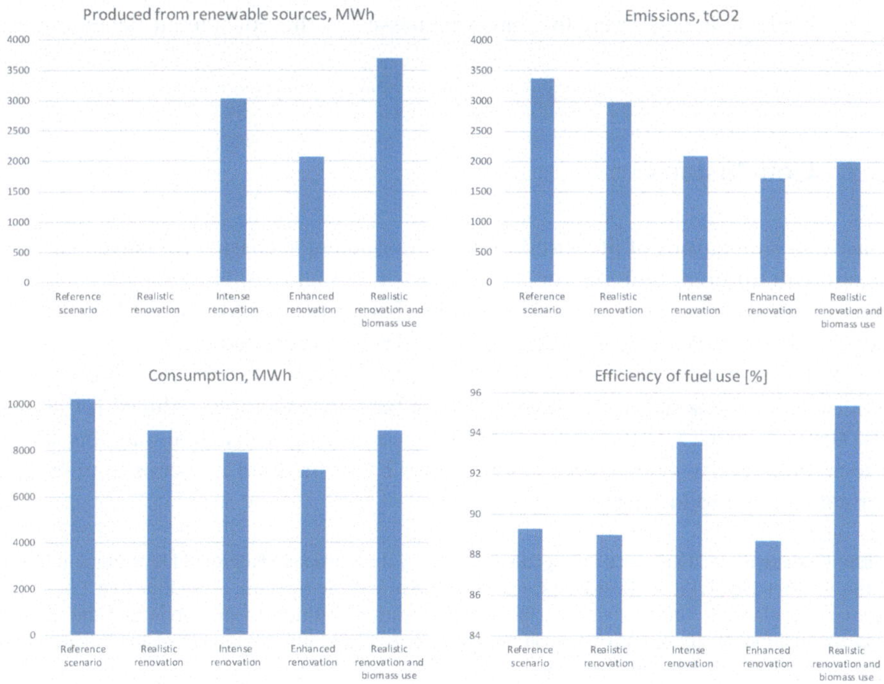


**Fig. 32.7** Comparison of six development scenarios of the DH system (output from District Heating Planning Tool)

## 32.4 Discussion

The developed planning tool addresses partly renovated post-Soviet district heating systems with poorly insulated DH distribution pipelines and partly renovated building stock. Robustness and reliability of the planning tool allow evaluation of the development scenarios considering the performance of the distribution systems, reduction of the thermal energy consumption and switch to renewable fuels.

In order to perform the development and adjustments of the District Heating Planning Tool, it was necessary to collect quantitative information across the production, distribution and consumption sectors of certain DH system. Previously performed qualitative and quantitative researches as well as comprehensive research of 55 renovated buildings in Riga served as a basis for the development of a methodology for DH Planning Tool in conditions of reduction of the thermal energy consumption. The planning tool is considered as a steady state model since it is foreseen assessment of large-scale systems for a long-term period. The main strength of the steady-state method is that it is possible to perform all necessary calculations using only data from heat supply utility and no prior knowledge from individual load components is necessary [7]. At the output, the tool provides twelve efficiency and balance indicators with recommended permissible limits which were proposed by the authors and evidently indicates strengths and weaknesses of the considered development pathways.

## 32.5 Conclusions

Consistent simulations of proposed future development scenarios for particular DH system with different building renovation rate of 3, 5 and 7% have shown a natural reduction of thermal energy consumption in 15-year perspective for 13.5%, 22.5% and 30% (1380 MWh, 2301 MWh and 3068 MWh) correspondingly.

Simulation results of the last three scenarios, where use of renewable energy sources was considered, $CO_2$ emissions have been considerably reduced for about 40% (from 3376 $tCO_2$ to the around 2000 $tCO_2$ comparing with actual state) but share of local biomass (wooden chips) has reached 47% of the total fuel consumption for thermal energy production.

**Acknowledgements** This work has been supported by the European Regional Development Fund within the Activity 1.1.1.2 "Postdoctoral Research Aid" of the Specific Aid Objective 1.1.1 "To increase the research and innovative capacity of scientific institutions of Latvia and the ability to attract external financing, investing in human resources and infrastructure" of the Operational Programme "Growth and Employment" (1.1.1.2/VIAA/2/18/344).

## References

1. Cheng, V., Steemers, K.: Modelling domestic energy consumption at district scale: a tool to support national and local energy policies. Environ. Model. Softw. (2011). https://doi.org/10.1016/j.envsoft.2011.04.005
2. Leichenko, R.: Climate change and urban resilience. Curr. Opin. Environ. Sustain. (2011). https://doi.org/10.1016/j.cosust.2010.12.014
3. Collier, M.J., et al.: Transitioning to resilience and sustainability in urban communities. Cities (2013). https://doi.org/10.1016/j.cities.2013.03.010
4. Sameti, M., Haghighat, F.: Optimization approaches in district heating and cooling thermal network. Energy Build. (2017). https://doi.org/10.1016/j.enbuild.2017.01.062
5. Fang, T., Lahdelma: Genetic optimization of multi-plant heat production in district heating networks. Appl. Energy (2015). https://doi.org/10.1016/j.apenergy.2015.09.027
6. Duquette, J., Rowe, A., Wild, P.: Thermal performance of a steady state physical pipe model for simulating district heating grids with variable flow. Appl. Energy (2016). https://doi.org/10.1016/j.apenergy.2016.06.092
7. Heller, A.J.: Heat-load modelling for large systems. Appl. Energy (2002). https://doi.org/10.1016/S0306-2619(02)00020-X
8. Gao, L., et al.: Technologies in smart district heating system. Energy Procedia (2017). https://doi.org/10.1016/j.egypro.2017.12.571
9. Sayegh, M.A., Jadwiszczak, P., Axcell, B.P., Niemierka, E., Bryś, K., Jouhara, H.: Heat pump placement, connection and operational modes in European district heating. Energy Build. (2018). https://doi.org/10.1016/j.enbuild.2018.02.006

10. Borodinecs, A., Zemitis, J., Sorokins, J., Baranova, D.V., Sovetnikov, D.O.: Renovation need for apartment buildings in Latvia. Mag. Civ. Eng. (2016). https://doi.org/10.5862/MCE.68.6
11. Aëšberg, M., Widén, J.: Development, validation and application of a fixed district heating model structure that requires small amounts of input data. Energy Convers. Manag. (2013). https://doi.org/10.1016/j.enconman.2013.05.032
12. Ziemele, J., Gravelsins, A., Blumberga, A., Vigants, G., Blumberga, D.: System dynamics model analysis of pathway to 4th generation district heating in Latvia. Energy (2016). https://doi.org/10.1016/j.energy.2015.11.073
13. Woods, P., Overgaard, J.: Historical development of district heating and characteristics of a modern district heating system. In: Advanced District Heating and Cooling (DHC) Systems (2015)
14. Kim, S., Kim, H.: A new metric of absolute percentage error for intermittent demand forecasts. Int. J. Forecast. (2016). https://doi.org/10.1016/j.ijforecast.2015.12.003
15. U.S. Energy Information Administration: The electric power monthly. Internet (2016). https://www.eia.gov/electricity/monthly/pdf/epm.pdf
16. Åberg, M., Widén, J., Henning, D.: Sensitivity of district heating system operation to heat demand reductions and electricity price variations: a Swedish example. Energy (2012). https://doi.org/10.1016/j.energy.2012.02.034

# Chapter 33
# Architects' Tactics to Embed as-Designed Performance in the Design Process of Low Energy Non-domestic Buildings

**Gabriela Zapata-Lancaster**

**Abstract** There is an urgent need to reduce the carbon emissions of the built environment and to foster practices that enable improved quality and efficiency in the building sector. However, research suggests that there are several barriers deterring the achievement of carbon reductions. Research shows that there are significant gaps between as-designed and in-use building performance. This work investigated the design process enacted to facilitate as-designed performance intentions. The study used ethnographic methods to explore the architect's actions and tactics observed during the real design process in 4 architecture practices. The analysis resulted in an alternative description of the building development timeline that illustrates the activities that designers undertook to facilitate as-designed performance. This work contributes to the understanding of what architects do to incorporate low energy considerations and how architects deliver as-designed energy performance in the design process.

## 33.1 Introduction

There is an urgent need to reduce the carbon emissions of the built environment and to foster practices that improve the quality and energy efficiency of buildings. The building sector is called to reduce its carbon emissions. However, research suggests that there are several barriers deterring the achievement of carbon reductions: lack of market demand for low energy buildings, the knowledge gaps and lack of skills to design and deliver low energy buildings [14, 16, 20]. Building performance evaluation studies in the UK have identified gaps between as-designed and in-use performance, suggesting that the as-designed performance predictions are rarely achieved during operation [3, 12]. The building industry tends to be slow in implementing performance-driven change [26]. The UK Zero Carbon [33] suggests that 'performance issues are much more concerned with the processes and cultures within

G. Zapata-Lancaster (✉)
Welsh School of Architecture, Cardiff University, Cardiff CF10 3NB, UK
e-mail: ZapataG@cardiff.ac.uk



J. Littlewood et al. (eds.), *Sustainability in Energy and Buildings*, Smart Innovation, Systems and Technologies 163, https://doi.org/10.1007/978-981-32-9868-2_33

the industry than with the model that is used to predict carbon emissions'. These challenges highlight the need to investigate the design process from the designers' perspective to support design teams to successfully deliver low energy buildings. This could potentially help to address systemic problems experienced by the building sector. Therefore, there is a growing interest in identifying how architects engage in building performance analysis and evaluation as a means to ensure better building performance and; more broadly, foster sustainability in buildings. Building performance evaluation is a set of methods to ensure that performance targets and building quality are delivered in different stages of the building life cycle, including design [21]. There are multiple benefits of building performance analysis in design; for example, to foster a better quality of building design [11]; to increase designers' knowledge about quality and value of design [32]; to promote usability and adaptive opportunities in buildings [4] and to inform design strategies [7]. Bachman [2] argues that the increased complexity of the built environment demands architects to consider 'evidence-based design, post-occupancy evaluation, commissioning and quality reassurance' in order to deliver sustainability. Empirical work suggests that architects undertake different investigations to increase the quality of buildings and to inform the design process using POE [17, 22]. Architects are encouraged to integrate design quality and building performance considerations [28]. Durosaiye et al. [13] identify some of the barriers to the application of POE by architects and highlight how POE could benefit architects. In studying how architects undertake building performance analysis, it is necessary to consider the nature of design problem solving. Design literature argues that the design process is ill-defined and exploratory in nature [23]. Designers explore problems following an open-ended semi-structured enquiry [8]. Architects engage in reflective conversations to gain a general understanding of design problems and explore acceptable solutions [25]. Architecture design solutions tend to be based on experiential knowledge [8, 9, 18, 25], precedents [19] and naive representations [1]. Another important aspect of the design process is that design has a social dimension [9, 10] that involves the negotiation of worldviews between design team members where design goals are unlikely to be fixed at the beginning of the process [6]. Given this context, this paper presents a study that investigated how architects facilitated as-designed building performance; focusing on tactics observed in the design process of low energy buildings.

## 33.2 Methodology

The work investigated the design process during early and detailed design phases; as per RIBA Plan of Work Stages 1–4 [24], referring as relevant to construction and operational phases (RIBA 5–7). The work aimed to explore how architects incorporated energy performance considerations in the design process. The research engaged with 4 architecture practices with experience in sustainable design, demonstrated by their portfolio and recognitions to their projects (RIBA, AJ100, CIBSE awards). The main research participants were the architects. The work also engaged with other design

**Table 33.1** Key energy performance indicators in the case studies (CS)

| | CS1 | CS2 | CS3 | CS4 | CS5 | CS6 |
|---|---|---|---|---|---|---|
| U-values | | | | | | |
| Roof | 0.15 | 0.18 | 0.18 | 0.15 | 0.15 | 0.10 |
| Walls | 0.15 | 0.26 | 0.26 | 0.18 | 0.16 | 0.16 |
| Glazing | 1.20 | 1.60 | 1.60 | 1.60 | 1.60 | 1.50 |
| Floors | | 0.32 | 0.19 | 0.21 | | 0.12 |
| Air tightness | 3 | 1 | 3 | 5 | 5 | 5 |
| EPC | 31 | 40 | 19 | 40 | 40 | 28 |
| % LZC tech | 10 | 20 | 30 | 15 | 15 | 10 |
| BREEAM rating | Excellent | Excellent | Outstanding | Very good | Outstanding | Excellent |

team members to identify their contribution to the design of low energy buildings: building services engineers, energy consultants and BREEAM assessors. The work investigated the design process of six non-domestic buildings [five schools (case studies 1–5)] and a multipurpose community/health centre (Case Study 6). Table 33.1 summarises the energy performance targets of the case studies.

The work applied an ethnographic research design. Ethnography was chosen because it allowed to study what happens in practice according to the perspective of the research participants [5, 27]. Ethnography focuses on action, meanings and processes enacted by practitioners [15]. The ethnographic research methods applied in this study included observational studies, semi-structured interviews and document analysis. The observational studies comprised shadowing of designers' work, observation of design team meetings to identify the key features of the design process in action. The interviews helped to elicit the perceptions of the research participants regarding the design development and the design activities. The document analysis investigated the evidence and methods used by designers to incorporate energy efficiency considerations in the building design proposals. Documents investigated included energy efficiency guidance, standards, trade literature, methods and models to estimate energy performance (i.e. energy simulation results, U-value calculations and energy reports), RIBA reports, drawings, plans, tender documents and design and access statements. The data discussion follows an 'ideal type' schema that describes the main characteristics and the underlying structure found in the case studies [15]. The ideal type of schema does not represent every detail of all observed case studies.

## 33.3 Findings

In order to support the delivery of as-designed performance, the design teams engaged in the following activities throughout the design process: (1) Defining the as-designed energy aspirations (RIBA 1, 4); (2) Increasing the qualitative and quantitative understanding of energy performance (RIBA 1–3); (3) Considering construction matters to inform the building design (RIBA 3–4); (4) Educating and 'enrolling[1]' the construction team (RIBA 4–5); (5) Materialisation of design intentions on-site (RIBA 5–6); (6) Learning from operation (feedback from RIBA 7). The activities included tactics to embed energy performance as discussed in the following subsections.

### 33.3.1 Defining the as-Designed Energy Aspirations

Designers suggest performance targets and aspirations as a means to create a collective vision of what the building should achieve and align the process accordingly. Statutory standards, regulations and guidance were found to be the main instruments to set the design aspirations (Table 33.2).

Architects invoked their prior experience on low energy buildings and precedents to propose the design aspirations. The design aspirations considered the relation between energy aspects and other factors such as cost implications of low energy design strategies, maintenance aspects, users' preferences and potential occupant behaviours during operation. In such a way, designers proposed design strategies that were feasible and had the potential to respond to the needs and preferences of various stakeholders; for example, clients and future building occupants.

**Table 33.2** Standards and guidance used in the case studies

| Guidance/standards | Case study |
|---|---|
| Approved documents; including Part L Conservation of fuel and power, Part E Resistance to sound, Part F Ventilation | All cases |
| Building rating systems BREEAM | 1, 2, 3, 4, 5, 6 |
| Building bulletins: 87: Guidelines for Environmental design of schools, 90: Lighting Design for Schools, 95: Schools for the Future design for learning communities, 98: Design Requirements for schools, 99: Areas, 100: Design for fire safety in schools, 101: Ventilation and indoor air quality in schools | 1, 2, 3, 4, 5 (relevant to schools only) |
| Design guidance by chartered and professional bodies i.e. RIBA, CIBSE | All case studies |
| Planning guidance and requirements by local authorities concerning low zero carbon technologies contribution, development plans and planning documents | All case studies |

[1]Enrolling refers to the alignment of the vision and goals of the construction team so they collectively deliver the as-designed performance intentions set by the design team.

*ARCH Case Study 2: 'Ultimately you put in all these systems but if the occupants don't know how to use them, then [the systems] might even be working less efficiently than the simple things that [the occupants] know how to use…'.*

The design aspirations were not static. They were revisited throughout the design process, particularly at points where regulations and planning requirements demanded evidence of the building performance (i.e. due to BREEAM certification for the planning application, Part L for building control).

### 33.3.2 Increasing the Qualitative and Quantitative Understanding of Energy Performance

Designers aimed to evaluate the implications of the design strategies in relation to the energy performance aspirations. This instance involved the qualitative and quantitative estimation of energy performance using a diversity of methods: experience-based advice between architects and building services engineers/energy consultants, rules of thumb, simplified in-house calculation tools. It should be noted that proprietary design simulation tools were not used by architects during the early design stage in the case studies. Simulation tools were used by the energy analyst or the building services engineers. Simulation tools were deployed more frequently in later design stages (RIBA 3–4) to produce energy performance estimations and to provide evidence for regulation compliance. An interesting aspect is that designers included considerations about users and operational performance while evaluating the potential energy performance of the buildings. In case of studies 2, 4 and 5, the architects developed workshops with the users to obtain data about occupancy, energy usage and preferences and identify opportunities for energy reduction. Designers thought that the workshops helped to encourage the ownership of the design aspirations by clients and users, enhanced the as-designed estimations in relation to likely occupancy patterns and informed the design strategies in the light of users' preferences and occupation practices. Designers also carried cost studies to determine the financial feasibility of different strategies and to seek funding opportunities for low zero carbon technologies (Case studies 2, 3 and 5). The cost estimations in early design tended to relate the energy aspirations to operational and maintenance costs. They helped to illustrate the potential benefits and savings of low energy buildings to the clients and future building users. For example, in cases 1, 3, 4 and 5 the design teams estimated the potential savings in energy bills resulting from low energy design strategies to show the benefits of low energy performance.

*BSE Case Study 3: 'They [the clients in a school building] wanted to have PVs on … They wanted to know how much it would save them potentially off their bills because they get just a set budget every year to run and maintain the school… the more we can reduce the energy consumption, the more they can spend on text books and things like that for the kids.'*

### 33.3.3 Considering Construction Matters to Inform the Building Design

The architects considered construction materials and methods, available low energy technologies, construction site tolerances and cost to inform the architectural details and tender specifications. Architects agreed that buildability was a key aspect to consider in the development of details in order to anticipate how details are delivered onsite and to establish safeguards to realise the as-designed performance during construction. The architects expressed that a poor understanding of detailing could lead to problems in achieving the expected as-designed performance in later phases. For example, the walls may be designed without considering the width needed to achieve the U-values (Case Study 4), certain types of glazing might not be feasible from the cost viewpoint (Case Study 6), louvres may be cost prohibitive (Case studies 1 and 2), the expected U-values of curtain walls may be difficult to deliver on-site (Case Study 1). Architects expressed that the tender documents served as an inscription of the requirements and a safeguard of the continuity of the design aspirations:

*ARCH Case Study 4: It's delivery team who don't understand everything that's in this final package... they are always under pressure to get the best price and I think it's about ensuring the [tender package] information is as rigorous as possible so that when you start choosing, for example the products that you're putting in the specifications, you know, and you're showing it on your drawings and that's going to the contractor, if these specifications here are as bullet proof and as watertight as possible then they don't have the chance to change it.*

Nonetheless, it was observed that the indicators in the tenders could be defined with different levels of precision. In general, there were explicit performance metrics such as U-values, G-values and airtightness. Nonetheless, other aspects that could ensure the execution of the details on-site might not be included, for example, the minimum standards of workmanship, tolerances on-site, site tests, remedial work. It was unlikely for tenders and detailed documentation to include metrics related to whole-building performance, i.e. energy consumption for the building. Only Case Study 4 referred to a specific target for energy consumption as a contractual requirement.

### 33.3.4 Educating and Enrolling the Construction Team

The case studies were procured by the Design and Build route; therefore, the transition between detailed design and construction involved the hand over of the work from the design team to the construction team led by the main contractor. Architects aimed to clarify the design rationale so the construction team was able to deliver the expected as-designed performance intentions. The architects perceived that the construction team had to understand the design rationale and commit to deliver the expected as-designed performance on site:

*ARCH Case Study 4: 'There's a reason why the building is like that, you know. And I think [achieving a low energy building] it's about educating people. I feel that contractors don't understand what [architects] we've been doing, to the depths that we've taken it. So you have someone, like well, we can do a frame in steel because it's cheaper. Hang on a sec, no, no, the frame isn't just there to support the building, the frame actually has other parts to play in the whole environmental model of what the building is. Yeah, and I think that's shown when we get the questions going, can we use this as an alternative? And you go, well it doesn't work for this reason, that reason and that reason. I can see you want to use it because it looks the same, but that's where the similarity ends. So it's about educating them [the construction team] to really understand … being more attentive to things like that.'*

### 33.3.5 Materialisation of the as-Designed Performance Intentions On-site

The architects recognised that they need to learn from the construction stage. Architects recognised that they learned about buildability and construction matters by comparing the as-designed estimations to the as-built performance achieved on site. The knowledge created during construction was recognised to inform design strategies, assumptions embedded in as-designed estimation models. It helped architects to consider construction matters in the building design. Nonetheless, it was observed that there was poor (or lack of) monitoring of the construction changes. Case Study 6 developed studies during construction to gather 'lessons learned' for architects. In Case Study 6, the architects investigated the sequence construction and challenges related to the thermal performance of the envelope. A construction diary was used to record detailed observations of the construction activities related to the walls, envelope and junctions. The diary recorded the construction of the wall, development of details on site, problems with the construction, feedback from the site team, unanticipated problems in delivering the original design proposals. The thermal performance of the envelope was to be measured after the completion of the construction. The architectural drawings would be updated and construction diaries would be uploaded on the architecture firm's database for future reference.[2] The architects reported that the lessons learned on a project basis were disseminated within the architecture firm to transfer knowledge from project to company level. The knowledge dissemination initiatives were primarily focused on exemplar projects excluding routine projects despite both types of projects offer lessons to inform future design processes.

[2]Due to the mismatch between project development timeline and research data collection period, the researchers did not have access to the as-built performance results nor the documents produced in this study to share lessons learned among architects.

### 33.3.6 *Learning from Operation*

Architects recognised that in-use performance could inform the design assumptions about occupancy, improve the as-designed estimation scenarios and evaluate the success of design strategies. However, little evidence was available to infer if and how the architects in the case studies obtained feedback from operational phases. A number of episodes suggest that design teams were informally learning from operational phases. For instance, case 4 used historical data of energy consumption to improve the as-designed performance estimation (i.e. input parameters in simulation models). Case Study 1 consulted the client about the energy patterns, preferences and maintenance practices of the future users to inform the energy simulation models. Information of in-use performance was available via design reviews and post-occupancy studies in Case Study 2. They helped to identify the challenges to achieve the expected as-designed performance during operation; for example, assumptions embedded in as-designed estimations did not reflect the actual occupancy patterns and preferences; misrepresentation of the wide range of technologies employed in classrooms. In Case Study 1, post project reviews were conducted to evaluate with the client the design process and the building after it was handed over.

## 33.4 Discussion

This empirical work has illustrated the design process, offering an alternative description of what architects do to ensure the continuity of as-designed performance intentions during the design process. It refers, as relevant, to construction and operational phases. The work identified five key activities enacted by architects to embed as-designed performance: (1) Defining the as-designed energy aspirations; (2) Increasing the qualitative and quantitative understanding of energy performance; (3) Considering construction matters to inform the building design; (4) Educating and enrolling the construction team; (5) Materialisation of design intentions on-site; (6) Learning from operation. The tactics helped designers to maintain the as-designed building performance aspirations. The observations in the case studies are aligned with propositions of design literature in terms of the social dimension of the design process and the negotiation of worldviews; highlighting that the design goals are not fixed in the process. During the design process, the design team members engaged in negotiations to find solutions that integrated energy requirements and other goals. This is aligned with the findings by [30] in terms of how design ambitions are negotiated in construction projects. The negotiations were not limited to the definition of the as-designed performance aspirations; architects engaged in negotiations with other design team members and with construction teams to ensure the continuity of the as-designed performance aspirations. In terms of estimation of as-designed performance, design teams used various tactics to improve the design estimations. They invoked experiential knowledge, rules of thumb and simplified calculation methods to have an indication of the potential energy performance of the buildings in early design. In later design stages, the estimation of performance was based on simulation tools. While this work focused on the design process, it identified that in order to ensure the achievement of as-designed energy performance aspirations, architects learned from construction and operational phases. In terms of the construction phase, the architects recognised the value of learning about buildability and construction matters. This brings attention to the need to foster learning between different phases of the building lifecycle. BIM is advocated as a tool that can support information management and transfer throughout the building lifecycle; however, there are many challenges in implementing BIM effectively [29]. Learning goes beyond the simple transfer of information. Construction and design management literature can enlighten efforts focused on the architect's knowledge and learning needs, considering the nature of architects' knowing and their preferences. Another important aspect identified in this work is that architects are willing to learn from the operation. They were using precedents, information about future occupants of the building to enhance the as-designed performance estimations and to inform the design strategies. Developments in monitoring, ubiquitous monitoring of existing buildings, big data analytics and other developments to measure, benchmark and understand the performance of existing buildings offer support in this direction.

The empirical data, albeit qualitative, show instances where knowledge about design, construction and operation was invoked by designers. Architects' tactics to embed as-designed performance relied on collective collaboration and aimed to foster collective knowledge about low energy design. The tactics illustrate what architects do during the design process in action to facilitate as-designed intentions, to ensure as-designed intentions in later building life cycle phases and to consider the wider project drivers and concerns: buildability, construction matters, operational performance, users' preferences, (capital and life cycle) cost. In situations where different expectations emerged between design team members and wider stakeholders; education and negotiation helped to align the design goals and supported the continuity of the as-designed performance aspirations. The tactics described help to identify what architects are doing to overcome performance gaps mapped by the UK Green Building Council [31]. The work shows that architects are interested in improving their understanding and knowledge of low energy design. They engage in building performance evaluation using different methods and types of evidence to inform their design in relation to performance goals. This work expands the examples of how architects use research and implement POE, identified by RIBA et al. [22].

## 33.5 Conclusions

This work has presented an alternative description of how architects embed as-designed energy performance in the design process. This paper highlights that architects committed to low energy design learn from the whole building life cycle to facilitate the achievement of as-designed performance intentions. Architects also adopt a champion role. They collaborate with design team members, construction team and other stakeholders to support the continuity of the as-designed performance aspirations and learn from the whole building life cycle process. The challenge for practice and research is to find effective ways to support architects to routinely include building performance analysis, evaluation and learning in the design process. Initiatives to foster the achievement of as-designed performance are advocated to consider designers' ways of knowing, preferred methods for analysis and evaluation; and, experiential knowledge. Future work will study how the intersection between architectural values and building performance analysis can inform mechanisms to enhance building quality and learning.

## 33.6 Limitations

This qualitative research included a small number of case studies; therefore, the findings are not generalizable. The paper does not aim to predict or prescribe the design process; it simply illustrates the circumstances found in the cases studies. Yet, the study is relevant as it identifies what architects do to facilitate as-designed performance intentions and expands the understanding of design in action in the context of low energy design and in relation to building performance.

## References

1. Akin, O.: Variants in design cognition. In: Eastman, C., McCraken, M., Newstetter, W. (Eds) Design Knowing and Learning Cognition in Design Education. Elsevier, Oxford (2001)
2. Bachman, L.R.: New professionalism: the post-industrial context. Build. Res. Inf. **41**(6), 752–760 (2013)
3. Bordass, B., Leaman, A.: A new professionalism: remedy or fantasy? Build. Res. Inf. **41**, 1–7 (2012)
4. Bordass, B., Leaman, A., Ruyssevelt, P.: Assessing building performance in use 5: conclusions and implications. Build. Res. Inf. **29**, 144–157 (2001)
5. Bryman, A.: Social Research Methods. Oxford University Press, Oxford (2008)
6. Bucciarelli, L.: Designing Engineers. MIT Press, Cambridge, MA (1994)
7. Carmona Andreu, I., Oreszczyn, T.: Architects need environmental feedback. Build. Res. Inf. **32**, 313–328 (2004)
8. Cross, N.: Design cognition: results from protocol and other empirical studies of design activity. In: Eastman, C., McCraken, M., Newstetter, W. (eds.) Design Knowing and Learning Cognition in Design Education, pp. 79–104. Elsevier, Oxford (2001)
9. Cross, N.: Designerly Ways of Knowing. Birkhäuser, Basel (2007)
10. Cross, N.: Natural intelligence in design. Des. Stud. **20**, 25–39 (1999)
11. De de Wilde, P.J.C.J.: Building Performance Analysis. Wiley Blackwell, NJ (2018)
12. De Wilde, P.: The gap between predicted and measured energy performance of buildings: a framework for investigation. Autom. Constr. **41**, 40–49 (2014)
13. Durosaiye, I.O., Hadjri, K., Liyanage, C.L.: A critique of post-occupancy evaluation in the UK. J. Hous. Built Environ. **34**(1), 345–352 (2019)

14. Fisher, J., Guy, S.: Re-interpreting regulations: architects as intermediaries for low-carbon buildings. Urban Stud. **46**(12), 2577–2594 (2009)
15. Hammersley, M., Atkinson, P.: Ethnography. Principles in Practice. Routledge (1995)
16. Hamza, N., Greenwood, D.: Energy conservation regulations: impacts on design and procurement or low energy buildings. Build. Environ. **1**(44), 929–936 (2009)
17. Hay, R., Samuel, F., Watson, K., Bradbury, S.: Post-occupancy evaluation in architecture: experiences and perspectives from UK practice. Build. Res. Inf. **46**(6), 698–710 (2018)
18. Lawson, B.: What Designers Know. Architectural Press, Oxford (2004)
19. Lawson, B.: Schemata, gambits and precedent: some factors in design expertise. Des. Stud. **25**(5), 443–457 (2004)
20. Osmani, M., O'Reilly, A.: Feasibility of zero carbon homes in England by 2016: a house builder's perspective. Build. Environ. **44**, 1917–1924 (2009)
21. Preiser, W., Hardy, A.E., Schramm. U. (eds.): Building Performance Evaluation From Delivery Process to Life Cycle Phases, 2nd edn. Springer, Cham (2018)
22. RIBA., Hay, R., Bradbury, S., Dixon, D., Martindale, K., Samuel, F., Tait, A.: Pathways to POE, Value of Architects. University of Reading, RIBA. Accessed in May 2019 (2016)
23. Rowe, P.: Design Thinking. MIT Press, MA (1987)
24. Royal Institute of British Architects. RIBA Plan of Work (2013)
25. Schön, D.: The Reflective Practitioner: How Professionals Think in Action. London (1983)
26. Sexton, M., Barrett, P.: Performance-based building and innovation: balancing client and industry needs. Build. Res. Inf. **33**, 142–148 (2005)
27. Silverman, D.: Doing Qualitative Research: A Practical Handbook. Sage, London (2005)
28. Stevenson, F.: Embedding building performance evaluation in UK architectural practice and beyond. Build. Res. Inf. **47**(3), 305–317 (2019)
29. Sturts Dossick, C., Neff, G.: Messy talk and clean technology: communication, problem-solving and collaboration using building information modelling. Eng. Proj. Organ. J. **1**(2), 83–93 (2011)
30. Tryggestad, K., Georg, S., Hernes, T.: Constructing building and design ambitions. Constr. Manag. Econ. **28**, 695–705 (2010)
31. UK Green Building Council: Delivering Building Performance Full Report (2016)
32. Whyte, J., Gann, D.M.: Closing the loop between design and use: post-occupancy evaluation. Build. Res. Inf. **29**, 460–462 (2001)
33. Zero, C.H.: Closing the gap between designed and built performance. Carbon Compliance for Tomorrow's New Homes. London. http://www.zerocarbonhub.org/full-lib (2010)

# Chapter 34
# How Much Does It Cost to Go Off-Grid with Renewables? A Case Study of a Polygeneration System for a Neighbourhood in Hermosillo, Mexico

**Moritz Wegener, Carlos Lopez Ordóñez, Antonio Isalgué, Anders Malmquist and Andrew Martin**

**Abstract** As governments and companies struggle to meet their own objectives for the energy transition, more innovative social and technological measures are needed to reduce Greenhouse Gas (GHG) emissions. For this purpose, an assessment of an off-grid polygeneration system, which can serve the electric and cooling demand of a neighbourhood in Hermosillo, Mexico, has been conducted. Energy computations have been done, the energy demand of one dwelling has been measured to ascertain the correctness of the computations, and a demand model for the entire neighbourhood has been created. Based on the model, an off-grid polygeneration system has been designed, which uses a biodiesel engine, PV panels, and an absorption chiller. The system has been optimized for its economic performance and is compared to the currently used system. The results show that the polygeneration system with higher energy efficiency could reduce GHG emissions down to 14%. However, electricity in Hermosillo is heavily subsidized making it harder for innovative systems to compete. Moreover, even without the state subsidies to the end user, the polygeneration system has still a nearly 30% higher Net Present Cost (NPC) than the conventional system over its project lifetime of 20 years. Nonetheless, with precise political incentives and further advances in the applied technologies, small-scale renewable polygeneration systems could become cost-efficient alternatives in the near future.

---

M. Wegener (✉) · C. L. Ordóñez · A. Isalgué
Architecture and Energy-School of Architecture, Polytechnic University of Catalonia (UPC), Barcelona, Spain
e-mail: mbgw@kth.se

M. Wegener · A. Malmquist · A. Martin
Heat and Power Technology, Department of Energy Technology, KTH Royal Institute of Technology, Stockholm, Sweden



J. Littlewood et al. (eds.), *Sustainability in Energy and Buildings*, Smart Innovation, Systems and Technologies 163, https://doi.org/10.1007/978-981-32-9868-2_34

## 34.1 Introduction

With the number of Greenhouse Gases (GHG) in the world's atmosphere still rising innovative and cost-effective measures to promote renewable energies are needed [1]. Motivated social communities initiating small-scale, local energy projects have been identified as crucial for developing technologies and processes for the renewable energy transition [2]. In this context, decentralized polygeneration systems, which serve several energy demands at once (e.g. electricity, heat and cooling), could be a more cost-effective alternative to the conventional energy provision based on fossil fuels. Combining this technology with renewable resources like biomass or solar can provide sustainable and autonomous energy systems with benefits but also challenges [3–5].

An example of a biomass-fired polygeneration system using absorption chilling (ACh) technology to serve a building complex has been given by Caliano et al. [6]. They conclude that these systems are highly affected by the electricity prices and the incentives set for cooling applications. Another bio-solar polygeneration system providing electricity, heat and cooling for a local community has been proposed by Chua et al. [7]. They found that with low renewable energy penetration, the system outperforms the reference model economically and ecologically.

While there have been studies using EnergyPlus and HOMER for electricity production in building clusters [8], no such study could be found where both engine tools have been used supplementing each other. Therefore, we present an innovative approach by combining the simulation tools EnergyPlus, which is employed for thorough demand analysis, and HOMER, which is subsequently used for economic optimization of an off-grid polygeneration system designed according to the demand analysis. The case study is performed for a block of 16 closely situated houses in the city of Hermosillo, Mexico, with high thermal and electrical demand fluctuations. Finally, the results of this polygeneration system are compared to the reference system, which conventionally relies on a connection to the electric grid and the transformation of electricity for thermal services. The findings are put into the political and technological context of the energy transition and local initiatives are proposed further adding to the novelty of the paper.

## 34.2 Methodology

The methodologic approach is visualized in Fig. 34.1. In the first step, energy consumption, energy demand and architectural data of one of the houses facing south has been measured and collected. Using DesignBuilder and EnergyPlus [9] a simulation model was constructed, which accurately represents the electric and thermal demand of the house on an hourly basis. Using this model, the energy demand for all 16 houses with their different exposure to the sun have been calculated and summed up.

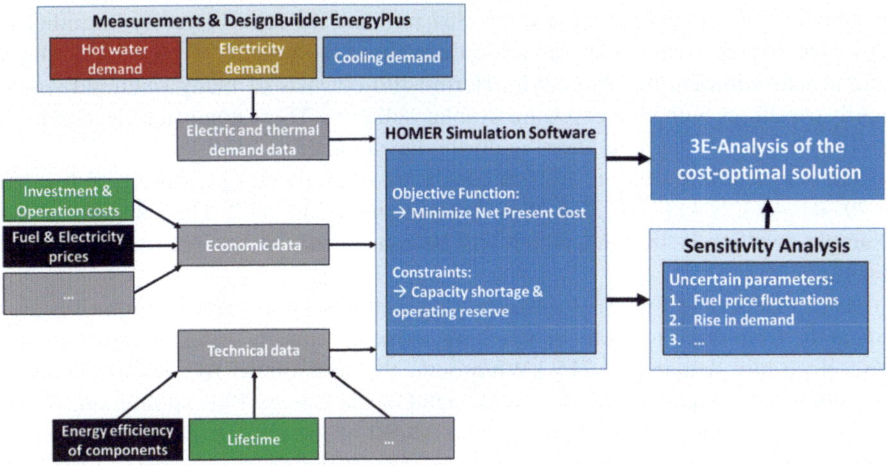


**Fig. 34.1** Methodology for optimal system and component sizing

In the second step, a renewable polygeneration system based on a biodiesel generator and PV panels has been designed, which satisfies the different energy demands without connection to the electric grid. Then the demand data together with technical as well as economic data from commercial, governmental and academic sources has been entered into the HOMER Energy software [10]. In HOMER, the system component sizes have been optimized for minimum Net Present Cost (NPC) under the restriction of having to satisfy the minimum capacity shortage and operation reserve. The NPC takes into account the current value of all foreseen present and future expenses for components and for the entire system over the project lifetime. To test the influence of different variables on the results, a sensitivity analysis has been conducted. The reference case has also been modelled to allow for comparison of the two systems.

Finally, the results are studied with respect to their environmental, economic and energetic repercussions for the dwellings inhabitants. The significance of the polygeneration system has been put into the context of the energy transition and local initiatives.

## 34.3 Site and System Description

### *34.3.1 Hermosillo: Electric and Thermal Demand of Dwellings*

Hermosillo features a hot desert climate with an average yearly temperature of 25 °C and a yearly oscillation of nearly 15 °C with temperatures reaching up to 45 °C in summer. The annual precipitation is 387 mm and the relative yearly humidity is 43%. The average daily solar radiation throughout the year is 5.71 kWh/($m^2$ day). Due to its traditional building style, Hermosillo consists of many scattered single family dwellings with high exposure to solar radiation. These conditions lead to high cooling demand for air conditioning during the summer months.

The scope of this study is a block consisting of 16 dwellings, which are located at 29° 04′ 40.1″N 111° 01′ 30.4″W and are shown in Fig. 34.2. The roof area of the example dwelling is 48.8 $m^2$, which will be considered the average roof area of all dwellings.

As can be seen in Fig. 34.3 the measured total electricity consumption rises significantly from May to October reaching its peak value in June. Subtracting the base electricity demand of 215 kWh/month, the so deduced AC electric demand resembles the calculated AC electric demand of the Energy Plus simulation. In the simulation, a Coefficient of Performance (COP) of the electric chillers (EChs) of 1.83 is assumed. Any differences can be mostly attributed to unexpected vacations or increased occupation of the dwelling. It should be noted that the Mexican government heavily subsidizes electricity costs during the hot season (May–October), which amplifies the rapid demand changes.

The reliability of the simulation can be estimated by calculating the Normalized Mean Bias Error (NMBE) and the Coefficient of Variation of the Root Mean Square Error (CVRMSE), as proposed by the ASHRAE Guideline 2002–14 [11]. The formulae for the NMBE and the CVRMSE for simulation and measured data are [12]:

$$\mathrm{NMBE} = 100 \times \frac{\sum_{i=1}^{n=12} (E_{si} - E_{mi})}{(n - p) \times \overline{E}_m} \tag{34.1}$$

**Fig. 34.2** Studied dwelling block in Hermosillo, Mexico

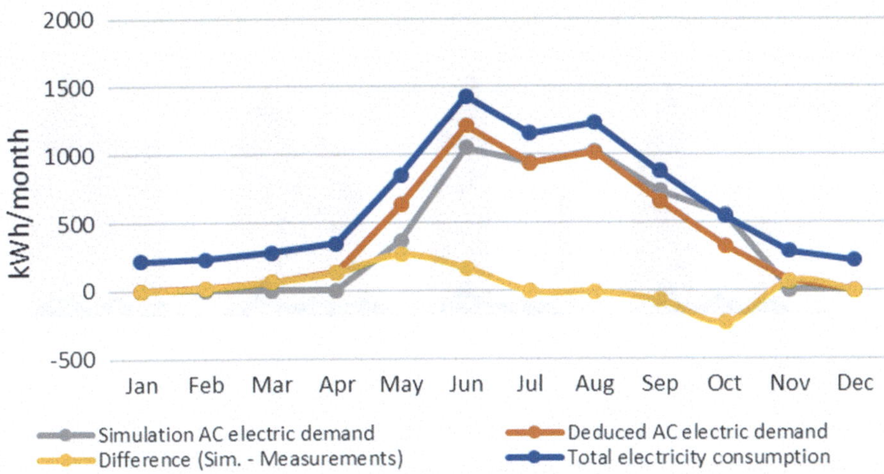


**Fig. 34.3** Electricity demand for electricity and air conditioning of one dwelling facing south

$$\text{CV RMSE} = 100 \times \left(\frac{\sum_{i=1}^{n=12}(E_{si} - E_{mi})^2}{(n-p)}\right)^{1/2} / \bar{E}_m. \tag{34.2}$$

where $E_{si}$. is the calculated value of the simulation for month i, $E_{mi}$ is the measured value for month i, $\bar{E}_m$ is the average value of measured monthly consumption, i is the particular month, n is the total amount of months and p = 1. With NMBE = 8.63% and CVRMSE = 30.57% the simulation values can be considered as accurate enough.

Using the simulation software and the measurement data, the electricity and the AC demands for all the dwellings of the building block have been calculated with the following dwelling orientations:

- Row houses: 6 houses facing south and 6 houses facing north
- Corner houses: 1 house facing South-East, 1 house facing South-West, 1 house facing North-East, 1 house facing North-West

Figure 34.4 shows the computed electricity demand for AC with the aforementioned COP of 1.83 and the remaining electricity demand for all 16 dwellings from Thursday to Sunday in June. For the remaining electricity demand, typical residential behaviour has been assumed. It can be seen that the AC electricity demand rises abruptly at 07.00 h and stays high until 09.00 h. The same behaviour is observed from 17.00 h until 24.00 h during weekdays and from 08.00 h until 24.00 h during weekends, which is when the dwelling is generally populated by the inhabitants.

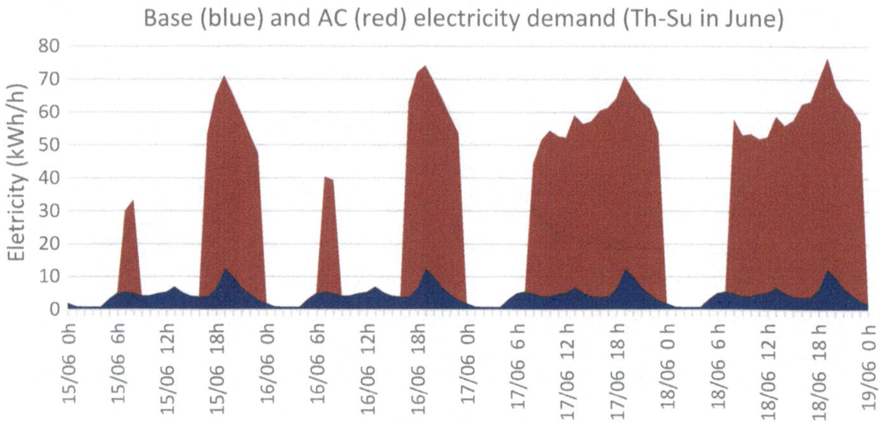


**Fig. 34.4** Computed total electricity and AC demand for all 16 dwellings in June

## *34.3.2 Polygeneration and Reference System*

In Fig. 34.5 the system layout of the proposed polygeneration system and the reference system are visualized. The proposed polygeneration system generates electricity using PV panels and a diesel engine fuelled with biodiesel. Batteries can store excess electricity and a converter is employed for electricity conversion. The demand for air conditioning can be met by a thermally driven absorption chiller (ACh) which during peak demand hours is supported by electricity-driven chillers (EChs). Apart from architectural design and measured data of the dwelling, scientific, governmental and commercial sources have been used to quantify key parameters. A list of these key parameters for the simulation and their references is shown in Table 34.1.

Whenever the ACh cannot satisfy the complete cooling demand during peak demand times, the amount of electricity needed for the EChs ($P_{el,EC}$) has to be

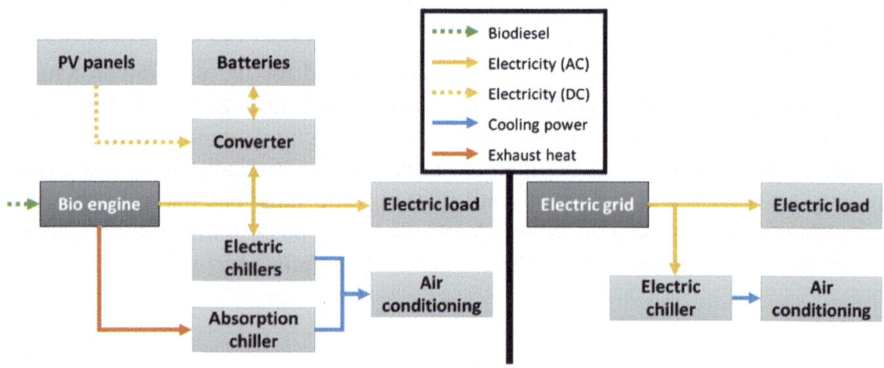


**Fig. 34.5** Bio-solar polygeneration system (left) and reference system (right)

**Table 34.1** Key technical and economic simulation parameters

| Component | Characteristics | Value |
|---|---|---|
| Diesel engine and generator | Capital costs | 500 USD/kW$_e$ [13] |
| | O&M costs | 0.025 USD/(op. hr · kW$_e$) [13] |
| | Mean electric efficiency | 35% |
| | Mean thermal efficiency | 45% |
| | Lifetime | 15,000 h |
| | Fuel price (biodiesel) | 0.8 USD/L [14] |
| Electric chillers (EChs) | COP | 1.83 |
| | Capital costs | Already installed |
| ACh equipment (incl. heat exchanger) | COP | 0.7 |
| | Capital costs (system) | 100 USD/kW [15] |
| | O&M costs (system) | 0.013 USD/kWh [15] |
| PV panels | Capital costs | 1500 USD/kW$_p$ [16] |
| | O&M costs | 0.15 USD/(kW year) [17] |
| | Max. efficiency | 15% |
| | Area per peak power | 8 m$^2$/kW$_p$ |
| Li-Ion batteries | Capital costs | 1000 USD/kWh [16] |
| | Minimum stage of charge | 20% [16] |
| Cooling system (tubing, pumps, control etc.) | Capital costs | 70,000 USD [18] |
| | O&M costs | 1000 USD/year |
| Electric grid: | Grid power price | 0.23 USD/kWh[a] |
| System parameters | Real discount rate | 3.85% |
| | Lifetime | 20 years |
| | Max. capacity shortage | 0 |

[a]The electric grid costs used are the production costs as stated on the electricity bill of the inhabitants. The inhabitants pay a lower price than the one stated here due to government subsidies after the production costs

calculated. Using energy balances for total electricity and total heat demand together with the energy efficiencies of the engine and the COPs of the different chillers, the following formula has been derived:

$$P_{el,EC} = \frac{Q_{th,t} - P_{el,R} \times COP_{AC} \times \frac{\eta_{th,eng}}{\eta_{el,eng}}}{COP_{EC} + COP_{AC} \times \frac{\eta_{th,eng}}{\eta_{el,eng}}} \quad (34.3)$$

where $P_{el,R}$ and $P_{el,EC}$ are the remaining electric demand needed without AC and the electric demand needed for the electric chiller; $Q_{th,t}$ is the total thermal demand; $\eta_{th,eng}$ and $\eta_{el,eng}$ are the thermal and the electric efficiency of the engine; $COP_{EC}$ and $COP_{AC}$ are the COPs of the EChs and of the Ach, respectively.

**Table 34.2** Component sizes and annual production/throughput of cost-optimal system

| Component | Size | Annual use/throughput |
|---|---|---|
| Generator | 61.0 $kW_e$ | 80.7 $MWh_e$/111.6 $MWh_{th}$ |
| PV | 16.5 $kW_e$/132 $m^2$ | 30.3 $MWh_e$ |
| Batteries | 51 kWh | 14.2 $MWh_e$ |
| System converter | 16.9 $kW_e$ | 40.3 $MWh_e$ (Inv. & Rect.) |
| ACh | 54 $kW_{th}$ | 53.5 $MWh_{th}$ |
| EChs | 104.3 $kW_{th}$ (57.1 $kW_e$) | 86.1 $MW_{th}$ |
| Electric grid | Unlimited | 118.1 $MWh_e$ |

## 34.4 Results

The component sizes and annual production of the cost-optimal polygeneration system are listed in Table 34.2. The minimum generator size has been determined to be 61 $kW_e$ to be able to satisfy the maximum thermal energy demanded by the ACh (54 $kW_{th}$) as well as the remaining electric demand. The engine would produce 80.7 $MWh_e$ or 72.7% of the total electricity produced per year as well as 111.6 $MWh_{th}$ of recoverable heat of which the ACh would use 76.4 $MWh_{th}$ to produce 53.5 $MWh_{th}$ for air conditioning. The biodiesel consumption would be 25,506 L operating 2784 h per year. The PV panels would supply 30.3 $MWh_e$ or 27.3% of the total electricity produced. It should be noted here that 18.7 $kWh_e$ of excess electricity could be produced, which could neither be stored in the 51 kWh batteries nor be used directly. This is mostly due to peak sunshine hours, where no occupants are in the dwellings (Fig. 34.4). An adequate control mechanism to manage these excess electricity loads should be installed. Additionally, a smarter control system could use this energy for increased efficiency. The roof area needed per dwelling would be 8.25 $m^2$, comfortably below the 48.8 $m^2$ available per dwelling.

The costs for each component are illustrated in Fig. 34.6. It should be remembered, that the electricity cost values used were before the government subsidies. The system costs over the project lifetime of twenty years amount to 593 kUSD for the entire neighbourhood block. The total capital costs amount to 191 kUSD, mostly due to the installation of the batteries and of the cooling system. However, the fuel costs outweigh the capital costs by far with 281 kUSD over the entire system lifetime.

Following the high impact of the biodiesel cost, the NPCs with varying biodiesel and electricity prices have been calculated for both systems and are shown in Fig. 34.7. In comparison, the polygeneration system is 219 kUSD more expensive than the reference system over the project lifetime of 20 years. Even when considering a 50% cheaper biodiesel price, the reference system would still be less expensive with the current costs of electricity production. However, the reference system is far more affected by rising fuel prices than the polygeneration system, which has a

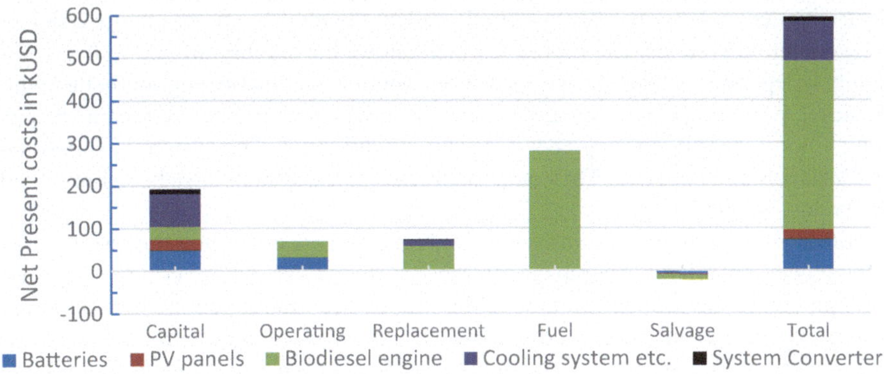


**Fig. 34.6** Cost of each component

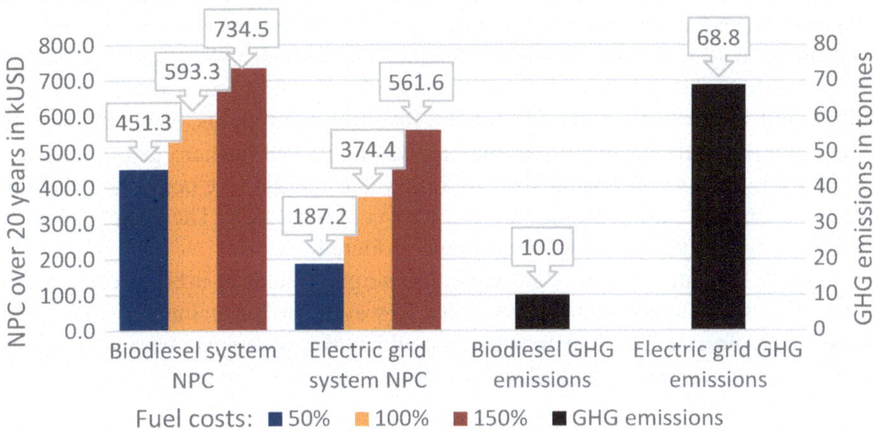


**Fig. 34.7** NPC costs with varying fuel prices and GHG emissions (GHG emissions for electricity: 0.582 kg/kWh [19]; GHG emissions for biodiesel: 0.12 kg/kWh [20])

higher level of autonomy due to the PV panels and the batteries. Figure 34.7 also illustrates the GHG emissions of each system. The proposed system would produce 58.8 tonnes of GHG emissions less per year or 14% of what the reference system produces annually.

## 34.5 Discussion and Conclusions

The energetic consumption of a one family dwelling in Hermosillo, Mexico, was studied and an energy demand model was created for an entire dwelling block using the data. During summer, the thermal demand increases significantly due to high electricity demands for the AC system. Based on the model, a more energy efficient polygeneration system was proposed, which allows the neighbourhood to live completely independent of the electric grid. The system would use an engine fuelled with biodiesel, providing electricity and exhaust heat, which would be used in an absorption chiller to supply cooling energy for air conditioning. The system is supported by PV panels and batteries. The component sizes of the system have been optimized for the lowest NPC.

Although the polygeneration system has higher energy efficiency, the results show that it is economically inferior to the reference system. Even over a period of 20 years, the high capital investment costs would not have been mitigated. Nonetheless, the system performs economically close to the reference system despite the low electricity production costs in Mexico. The polygeneration system also shows great ecologic potential, lowering the GHG emissions down to 14% compared to the reference system.

If designing a system for more dwellings, the relative capital and operation costs could be reduced further. Moreover, the AC system behaviour was not adjusted to the polygeneration system in the simulation leading to excess electricity and heat, which have to be dumped. Additionally, the demand for hot water was not considered, which could imply further savings for the dwelling inhabitants. With an entirely autonomous off-grid system, an extreme case was studied while many grid-connected designs with varying component sizes for renewable energies are possible and could combine economic and ecologic benefits to a certain degree. The effects of such systems on the electric grid should also be considered.

The study indicates that fully autonomous energy systems in urban areas are still a costly alternative to conventional systems. However, further technologic advances coupled with precise political incentives could make such systems a more cost-effective and much more sustainable alternative in the near future.

**Acknowledgements** This research has been conducted in collaboration between UPC (Universitat Politècnica de Catalunya) and KTH Royal Institute of Technology, funded through Erasmus Mundus Joint Doctoral Programme SELECT+, the support of which is gratefully acknowledged. A. I. acknowledges the Spanish project MOET_BIA2016-77675-R. Special thanks are given to Prof. Dr. Helena Coch for encouragement throughout the work.

## References

1. Le Quéré, C., Andrew, R.M., Friedlingstein, P., et al.: Global carbon budget 2018. Earth Syst. Sci. Data **10**, 2141–2194 (2018). https://doi.org/10.5194/essd-10-2141-2018
2. Hargreaves, T., Hielscher, S., Seyfang, G., Smith, A.: Grassroots innovations in community energy: the role of intermediaries in niche development. Glob. Environ. Chang. **23**, 868–880 (2013). https://doi.org/10.1016/J.GLOENVCHA.2013.02.008
3. Wegener, M., Malmquist, A., Isalgué, A., Martin, A.: Biomass-fired combined cooling, heating and power for small scale applications—a review. Renew. Sustain. Energy Rev. **96**, 392–410 (2018). https://doi.org/10.1016/j.rser.2018.07.044

4. Liu, H.: Biomass fuels for small and micro combined heat and power (CHP) systems: resources, conversion and applications. In: Small and Micro Combined Heat and Power (CHP) Systems, pp. 88–122 (2011)
5. Maraver, D., Sin, A., Sebastián, F., Royo, J.: Environmental assessment of CCHP (combined cooling heating and power) systems based on biomass combustion in comparison to conventional generation. Energy **57**, 17–23 (2013). https://doi.org/10.1016/j.energy.2013.02.014
6. Caliano, M., Bianco, N., Graditi, G., Mongibello, L.: Design optimization and sensitivity analysis of a biomass-fired combined cooling, heating and power system with thermal energy storage systems. Energy Convers. Manag. **149**, 631–645 (2017). https://doi.org/10.1016/J.ENCONMAN.2017.07.048
7. Chua, K.J., Yang, W.M., Er, S.S., Ho, C.A.: Sustainable energy systems for a remote island community. Appl. Energy **113**, 1752–1763 (2014). https://doi.org/10.1016/J.APENERGY.2013.09.030
8. Zhang, X., Lovati, M., Vigna, I., et al.: A review of urban energy systems at building cluster level incorporating renewable-energy-source (RES) envelope solutions. Appl. Energy **230**, 1034–1056 (2018). https://doi.org/10.1016/J.APENERGY.2018.09.041
9. DesignBuilder Software Ltd.: Home. https://designbuilder.co.uk/. Accessed 1 Feb 2019
10. Homer Energy: HOMER—Hybrid Renewable and Distributed Generation System Design Software. https://www.homerenergy.com/support/docs/3.10/glossary.html. Accessed 19 Feb 2019 (2019)
11. Reeves, G., Gillespie, K.L., Cowan, J.D., Frazell, C.: ASHRAE Guideline 14-2002: Measurement of Energy and Demand Savings (2002)
12. Wang, Z., Zhao, J.: Optimization of passive envelop energy efficient measures for office buildings in different climate regions of china based on modified sensitivity analysis. Sustainability **10** (2018). https://doi.org/10.3390/su10040907
13. Lau, K.Y., Tan, C.W., Yatim, A.H.M.: Photovoltaic systems for Malaysian islands: effects of interest rates, diesel prices and load sizes. Energy **83**, 204–216 (2015). https://doi.org/10.1016/J.ENERGY.2015.02.015
14. Riegelhaupt, E., Odenthal, J., Janeiro, L., Koper, M.: Diagnóstico de la situación actual del biodiésel en México y escenarios para su aprovechamiento (2016)
15. Coronado, C.R., Yoshioka, J.T., Silveira, J.L.: Electricity, hot water and cold water production from biomass. energetic and economical analysis of the compact system of cogeneration run with woodgas from a small downdraft gasifier. Renew. Energy **36**, 1861–1868 (2011). https://doi.org/10.1016/j.renene.2010.11.021
16. Schopfer, S., Tiefenbeck, V., Staake, T.: Economic assessment of photovoltaic battery systems based on household load profiles. Appl. Energy **223**, 229–248 (2018). https://doi.org/10.1016/J.APENERGY.2018.03.185
17. Fraunhofer ISE: Aktuelle Fakten zur Photovoltaik in Deutschland. www.pv-fakten.de. Accessed 1 Feb 2019 (2019)
18. Department of Energy & Climate Change: URN 15D/022—Assessment of the Costs, Performance, and Characteristics of Heat UK Networks. London (2015)
19. Factor de Emisión del Sector Eléctrico Nacional. https://www.gob.mx/cms/uploads/attachment/file/304573/Factor_de_Emisi_n_del_Sector_El_ctrico_Nacional_1.pdf. Accessed 5 Feb 2019 (2017)
20. Clark, D.H.: $CO_2$ emissions from biomass and biofuels. In: What Colour is Your Building? p. 10 (2013)

# Chapter 35
# Steps Towards an Optimal Building-Integrated Photovoltaics (BIPV) Value Chain in the Netherlands

**Ernst van der Poel, Wilfried van Sark, Yael Aartsma, Erik Teunissen, Ingrid van Straten and Arthur de Vries**

**Abstract** In the transition towards an energy neutral built environment, Building-Integrated Photovoltaics (BIPV) has been proposed as a way to combine energy generation with pleasing aesthetics and material savings. Despite these promising features, the adoption of BIPV has not yet shown the expected growth rate. Building on existing literature on the barriers for BIPV diffusion and based upon 27 qualitative interviews with relevant stakeholders, this research investigates current institutional, product-specific and value chain-specific conditions for a full-fledged and future-proof BIPV value chain in the Netherlands. Using the Theory of Change, practical steps to overcome these barriers were formulated. Per barrier, a prioritization value was given and this value was validated by three experts on the BIPV value chain. It can be concluded that to satisfy the institutional barriers, an interest group for the value chain is needed. Regarding product-specific conditions, the results show a strong need for research on the market demands for product characteristics. Finally, to satisfy the value chain-specific conditions, information and knowledge sharing across the whole value chain are required.

---

E. van der Poel (✉)
Wageningen University, 6708 PB Wageningen, The Netherlands
e-mail: e.vanderpoel@berenschot.nl

W. van Sark
Utrecht University, 3512 JE Utrecht, The Netherlands
e-mail: W.G.J.H.M.vanSark@uu.nl

E. van der Poel · Y. Aartsma · E. Teunissen · I. van Straten
Berenschot Groep B.V., 3526 KS Utrecht, The Netherlands
e-mail: y.aartsma@berenschot.nl

E. Teunissen
e-mail: e.teunissen@berenschot.nl

I. van Straten
e-mail: i.vanstraten@berenschot.nl

A. de Vries
Celstar B.V., 1066 NG Amsterdam, The Netherlands
e-mail: adevries@celstar.nl



J. Littlewood et al. (eds.), *Sustainability in Energy and Buildings*, Smart Innovation, Systems and Technologies 163, https://doi.org/10.1007/978-981-32-9868-2_35

## 35.1 Introduction

The solar energy market has experienced exponential growth over the past decade. By the end of 2017, 403 GW of PV installations was installed cumulatively [1] and in 2018, more than 100 GW was added to that total [2]. Substantial growth is also visible in the Netherlands, where 4.244 GWp of cumulative power was installed by the end of 2018, equal to an increase of 46% of total installed PV power compared to 2017 [3]. Within the solar industry, Building-Integrated Photovoltaics (BIPV) is considered a niche market. Several BIPV modules and products have been developed in recent years and are commercially available [4, 5]. BIPV entails functionally integrating a PV element into the building envelope. Advantages of BIPV include on-site electricity generation, material and resource savings through the integration of functions and enhanced aesthetic quality. Similar to the solar market as a whole, the BIPV market has witnessed growth in the past decade [1]. However, this growth has been very marginal compared to growth expectations and the potential ascribed to BIPV [6–9]. In the Netherlands, for instance, BIPV contributed to only 1% of the total installed PV power in 2014 [5]. A growth to a contribution of 3% of the total installed PV power in 2020 is expected [8].

It is clear that there are several barriers hindering the diffusion of BIPV. Previous research has focused on the barriers for BIPV diffusion in general [10–15], in Switzerland [9] and in the Netherlands [5, 16]. The identified barriers are dependent on the specific national context. A gap exists in the literature on what practical steps can be taken to overcome these barriers. This research provides a detailed description of practical steps that can be taken to overcome barriers to diffusion in de Dutch context. The Netherlands appears as an innovative hotspot for BIPV [1]. Therefore, it is interesting to investigate what conditions need to be satisfied for an optimal BIPV value chain in the Netherlands.

This study builds on the existing literature on barriers for BIPV diffusion within the Dutch context, by formulating conditions for an optimal BIPV value chain in the Netherlands and providing steps to satisfy these conditions. This paper will continue with a description of the current BIPV ecosystem and value chain in the Netherlands in Sect. 35.2. In Sect. 35.3, the method used in this study will be described. Section 35.4 includes the results of this study and discusses these findings. Lastly, Sect. 35.5 provides a conclusion and several recommendations.

## 35.2 BIPV in the Dutch Context

This section describes the current Dutch context for BIPV, by providing an overview of the BIPV ecosystem and value chain.

Figure 35.1 shows a stakeholder diagram for stakeholders within the BIPV ecosystem, as proposed by the Becquerel Institute [17]. The solid colored parts of the diagram represent established sectors, named in the small circles. The shaded colored

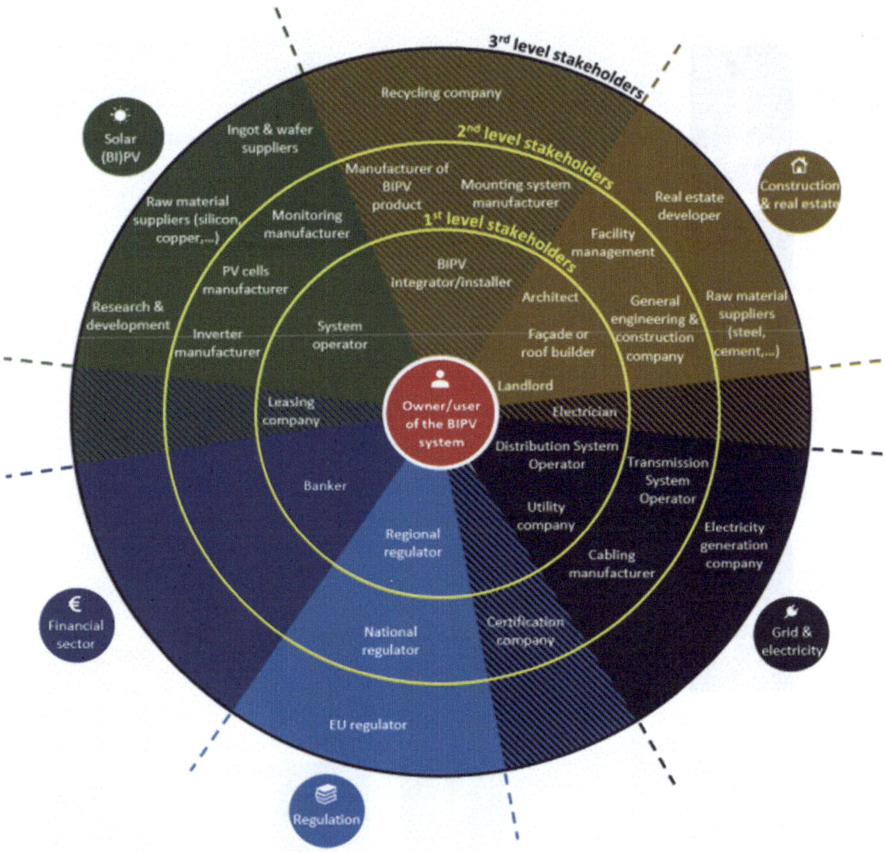


**Fig. 35.1** Stakeholders in the BIPV ecosystem [18]

parts of the diagram represent the overlap between adjacent sectors, thus creating a new field. Stakeholders that are part of the 1st level have the most direct contact with the owner/consumer, whereas stakeholders of the 3rd level have the least direct contact.

The Dutch ecosystem is characterized by well-developed research centers that are very active in the BIPV sector. Some even join new product initiatives. Furthermore, many small companies and start-ups can be found in the Dutch BIPV market, offering a wide range of different applications and products (for an overview of product categories see [4] for instance).

Figure 35.2 shows the value chain for BIPV in the Netherlands. A division is made between six categories: legal & regulations, research & development, production, financing, realization and exploitation & maintenance. The value chain is developed based on literature and data from interviews (see Sect. 35.3).

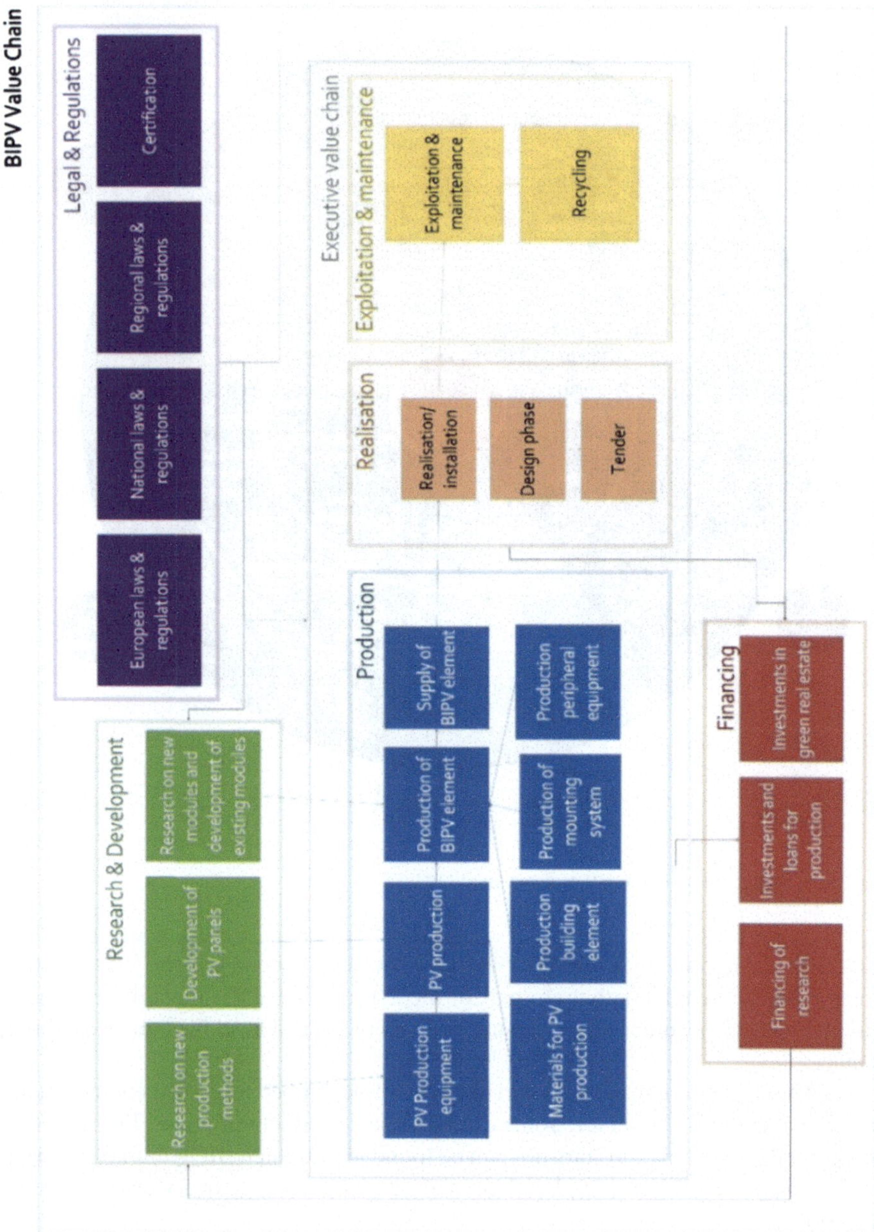


**Fig. 35.2** BIPV value chain (authors illustration)

## 35.3 Method

### *35.3.1 Data Collection*

For this study, three different data collection methods were used, which will be explained below briefly.

*Semi-structured interviews* For this study 27 semi-structured interviews were conducted with stakeholders that are active in the BIPV value chain in the Netherlands. These stakeholders were identified using the BIPV ecosystem diagram and value chain in Sect. 35.2. The final list includes 6 knowledge institutes, 5 BIPV module suppliers, 4 universities, 2 architects, 2 consultants, 2 housing corporations, 2 PV panel suppliers, 1 roofing company, 1 façade building company and 1 real estate developer.

*Literature research* This study made use of the existing literature on BIPV. Both Google Scholar and Scopus were used to search literature, based on the following search terms: 'BIPV', 'Barriers for BIPV diffusion', 'BIPV value chain' and 'BIPV ecosystem'. The literature was used to set-up a topic list for the semi-structured interviews and to complement findings from these interviews.

*Work sessions* Data gathered during two work sessions was used in this study. Both sessions were held at Berenschot, Utrecht. The first was on the 13$^{th}$ of March 2018 and was attended by 9 architects, 5 BIPV product suppliers and 3 knowledge institutions. The second session was on the 28th of November 2018 and was attended by 5 building companies, 7 BIPV suppliers and 4 knowledge institutions. The aim of these sessions was to identify and discuss gaps between the supply side and the demand side.

### *35.3.2 Data Analysis*

The gathered data from these methods were analyzed and clustered. The final conditions are issues mentioned by multiple sources. A prioritization value was given to each condition based on the number of times it was mentioned by the interviewees. This prioritization value was then validated with three BIPV experts. For the clustering of the conditions, the following categories, as proposed by Curtius [9], were used: institutional conditions, product-specific conditions and value chain-specific conditions. Using a simplified framework of the Theory of Change, practical steps to satisfy these conditions were then formulated. The simplified framework can be found in Fig. 35.3.

For this study, the Inputs and Outputs are not defined, because the identified steps and conditions are guidelines for the complete value chain. The executional details of these steps in terms of resources and monitoring is beyond the scope of this research.

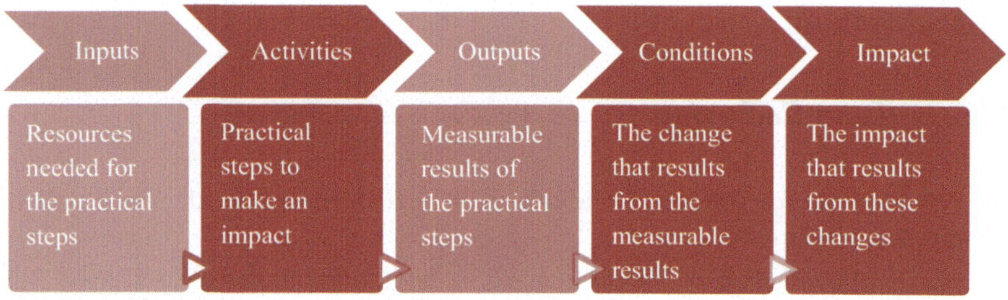


**Fig. 35.3** Simplified illustration of the theory of change (authors illustration)

## 35.4 Results and Discussion: Steps Towards an Optimal Value Chain

In this section, the findings from this study will be presented along the three different categories as mentioned in Sect. 35.3.

### *35.4.1 Institutional Conditions*

Regarding institutional conditions, the two main conditions that were found are 'high-performance requirements for buildings' and 'BIPV laws and regulations'. High-performance requirements for buildings are important because at this moment BIPV often has to compete with other features or measures in and around a building. High-performance requirements will make energy generation at the source, and thus BIPV, a necessity. Laws and regulations that are BIPV-specific are important because they can provide an incentive to apply BIPV. At this moment, several laws and regulations have the opposite effect, stimulating BAPV rather than BIPV. The main stakeholder group that should be involved in the process of satisfying these conditions is the interest group, which is understood to be an association of organizations that have similar concerns and with the main goal of influencing policy in their favor. However, currently, there seems to be no interest group of significant size for the BIPV value chain. Table 35.1 shows the identified institutional conditions, their prioritization value, practical steps to satisfy them and the main stakeholder groups involved per step. The range of the prioritization value is 1–5, with 1 being the lowest priority and 5 the highest.

### *35.4.2 Product-Specific Conditions*

'New applications', 'flexibility in the application' and 'simplicity & efficiency in the application' are the main product-specific conditions. In this study, it was found

**Table 35.1** Institutional conditions and practical steps to satisfy them

| Main condition | Sub-condition | Prioritization | Steps | Stakeholders |
|---|---|---|---|---|
| High-performance requirements for buildings | High requirements for new buildings | 5 | Cooperation BIPV suppliers and renovation companies | BIPV supplier, renovation company |
| | | | Stressing the importance of energy label A | BIPV interest groups, policymakers |
| | High requirements for existing buildings | 5 | Influence the process of the formulation of new BENG[a] requirements | Interest groups |
| BIPV laws and regulations | Possibility to include BIPV in service costs (housing corporation) | 3 | Lobbying activities with the Huurcommissie[b] | Interest groups |
| | Tax refund of BIPV system equal to BAPV system refund | 3 | Lobbying activities with the Ministry of Economic Affairs and Climate Policy | Interest groups |
| | Possibility to allocate BIPV energy in high-rise buildings | 3 | Formation of a 'think tank' to come up with innovative solutions | BIPV suppliers, installers, interest groups |
| | Clear product standards and regulations | 5 | Independent certification procedure to increase trust within the value chain | BIPV suppliers, certification companies, interest groups |

[a]The BENG (Bijna Energie Neutraal Gebouw, Near Zero Energy Building) requirements are currently being formulated and will replace the EPC values in the Netherlands [18]
[b]The Huurcommissie (Rents Committee) is an independent committee that provides owners and tenants with information, does research and may give a formal statement in case of conflict

that many features are already available when it comes to customizing a product. However, the exact demand for different features of potential customers remains largely unexplored. Research on these market demands helps to close the gap between supply and demand and may give suppliers more focus. Furthermore, in the design of a product, it is important to adapt the application to the existing practices within the value chain, so that the product will be more easily adopted by stakeholders along the chain. The most important stakeholder group that should be involved in the process of satisfying these conditions is the BIPV supplier. Table 35.2 shows the identified product-specific conditions, their prioritization value, practical steps to satisfy them and the main stakeholder groups involved per step.

**Table 35.2** Product-specific conditions and practical steps to satisfy them

| Main condition | Sub-condition | Prioritization | Steps | Stakeholders |
|---|---|---|---|---|
| New applications | New products for solar protection/blinds | 1 | Research on the demands and requirements for these products | Knowledge institutes, universities |
| | New products for balconies and galleries | 2 | Research on the demands and requirements for these products | Knowledge institutes, universities |
| Flexibility in the application | Flexible production process | 4 | Increase cooperation within the value chain to increase flexibility while keeping prices stable | Interest groups, knowledge institutes, BIPV suppliers |
| | Developments regarding color | 3 | Development of roof colors that match existing colors of conventional materials | BIPV suppliers, knowledge institutes |
| | | | Research on the demands of potential users of façade products | Knowledge institutes, universities |
| | | | Research on new methods that decrease coloring cost and efficiency loss | BIPV suppliers, knowledge institutes, universities |

(continued)

**Table 35.2** (continued)

| Main condition | Sub-condition | Prioritization | Steps | Stakeholders |
|---|---|---|---|---|
| | Developments regarding texture | 3 | New texture defining product combinations | BIPV suppliers |
| | | | Further development of thin film technology to increase texture possibilities | BIPV suppliers, knowledge institutes |
| | Developments regarding size | 4 | Research on the requirements of potential users regarding size | Knowledge institutes, universities |
| | Developments regarding transparency | 3 | Research on methods to increase transparency and decrease efficiency loss | Knowledge institutes, universities |
| Simplicity and efficiency in the application | Standardization | 4 | Increase cooperation within the value chain to stimulate standardization while maintaining flexibility | Interest groups, knowledge institutes, BIPV suppliers |
| | Correspondence of the product with existing practices | 4 | Streamlining BIPV installation with the isolation process | BIPV suppliers, building companies |
| | | | Development of prefab BIPV modules | BIPV suppliers, building companies |
| | | | Adapting BIPV installation to existing installation practices | BIPV suppliers, installers |
| | | | Adapting BIPV installation to existing roofing/façade building practices | BIPV suppliers, installers, roofers, façade builder |
| | | | Add BIPV products to BIM software | BIPV suppliers, interest groups, architects |

### 35.4.3 Value Chain-Specific Conditions

This study identified three main value chain-specific conditions, which are 'familiarity and trust in BIPV along the value chain', 'upscaling to larger projects' and 'sufficient knowledge and skills in the BIPV value chain'. The cooperation within the value chain can be improved by communication and knowledge sharing between the different stakeholders. By improving current communication methods, the trust in the product and the suppliers will grow, leading to a more frequent application of BIPV. Integration of BIPV in university courses and curricula can lead to a steady flow-through of (future) professionals who are interested in working in the sector. Courses and seminars of a more practical nature directed at different stakeholder groups can lead to increased knowledge about different products and systems and increased skill in applying them along the value chain. To make sure all stakeholders groups are involved and cooperation between them is facilitated, interest groups should take a leading role. Table 35.3 shows the identified value chain-specific conditions, their prioritization value, practical steps to satisfy them and the main stakeholder groups involved per step.

**Table 35.3** Value chain-specific conditions and practical steps to satisfy them

| Main condition | Sub-condition | Prioritization | Steps | Stakeholders |
|---|---|---|---|---|
| Familiarity with and trust in BIPV along the value chain | Clear and unambiguous provision of knowledge | 4 | Central database for BIPV products with comparable specifics | Interest groups, BIPV suppliers |
| | Familiarity and trust through reference projects | 4 | Research on how potential users want to be notified of reference projects | BIPV suppliers |
| | | | Free-to-visit open houses where information on the system is provided | BIPV suppliers |
| | Familiarity with BIPV possibilities of politicians | 4 | Inviting of politicians to project start of finish | BIPV suppliers, politicians |
| | | | Formation of a link between the value chain and the political arena | Interest groups, politicians |

(continued)

**Table 35.3** (continued)

| Main condition | Sub-condition | Prioritization | Steps | Stakeholders |
|---|---|---|---|---|
| | Familiarity with BIPV through education | 3 | Continuation and expansion of current University initiatives to introduce BIPV to curricula | Universities, knowledge institutes |
| Upscaling to larger projects | Sufficient scale size of suppliers | 4 | Introduction of BIPV to curricula, courses, classes and seminars to attract future personnel | Universities, BIPV suppliers |
| | | | Closer cooperation between suppliers and consultancies to create a decoupling point of responsibility | BIPV suppliers, solar consultancies |
| Sufficient now-ledge and skill within the BIPV value chain | Sufficient knowledge and skill with installers | 5 | Introduction of special courses for installers | BIPV suppliers, installers |
| | | | BIPV certification procedure for installers | Installers, certification companies |
| | Sufficient knowledge and skill with roofers | 4 | Integration of BIPV knowledge in roofing education | Roofers, BIPV suppliers |
| | Sufficient knowledge and skill with façade building companies | 4 | Integration of BIPV knowledge in façade building education | Façade building companies, BIPV suppliers |
| | Sufficient knowledge and skill with federations of spatial quality | 4 | Cooperation between suppliers and federations how BIPV can be applied within the boundaries set by the federation and the law | Federations of spatial quality, BIPV suppliers, interest groups |
| | Sufficient knowledge and skill with architects | 5 | Workshops, symposia and seminars aimed at the continuous training that is required of architects | Architects, BIPV suppliers, interest groups |

## 35.5 Conclusions

The diffusion of BIPV has so far remained below expectations of researchers and analysts. The body of literature on the barriers for this diffusion is growing. This study adds to this body of literature by determining conditions for a full-fledged and future-proof BIPV value chain in the Netherlands and identifying steps to satisfy these conditions. It was found that the main institutional conditions are 'high energy performance requirements for buildings' and 'BIPV-specific laws and regulations'. The main product-specific conditions are 'new applications', 'flexibility in the application' and 'simplicity & efficiency in the application'. Lastly, the main value chain-specific conditions are 'familiarity with and trust in BIPV along the value chain', 'upscaling to larger projects' and 'sufficient knowledge and skill within the value chain'. In order to satisfy these conditions, several practical steps need to be taken by different stakeholders. When these steps are undertaken, the value chain will come closer to an optimal form, in which the demand for BIPV is high, the BIPV supply matches this demand and there is cooperation in such a way that the BIPV supply can be delivered to the demand side fast and efficiently.

**Acknowledgements** This research is part of the 'Werkelijk Bouwen and BIPV' project', which is made possible by OPZuid, the province of Noord-Brabant and the EU. This project has led to the formation of BIPV Nederland. This study serves to support BIPV Nederland in its activities.

## References

1. International Energy Agency: Trends 2018 in Photovoltaic Applications: Survey Report of Selected IEA Countries Between 1992 and 2017, 23rd edn. International Energy Agency, Paris (2018)
2. Jäger-Waldau, A.: Snapshot of photovoltaics—February 2019. Energies **12**(5), 769–776 (2019)
3. CBS Statline: hernieuwbare elektriciteit; productie en vermogen. https://statline.cbs.nl/statweb/publication/?dm=slnl&pa=82610ned. Accessed 27 Mar 2019
4. SUPSI & SEAC: Building Integrated Photovoltaics: Product Overview for Solar Building Skins. SUPSI & SEAC, Eindhoven (2017)
5. Teunissen, E., Van den Hurk, L.: Roadmap Building Integrated Photovoltaics, 1st edn. Berenschot, Utrecht (2015)
6. International Energy Agency PVPS Task 15: Potential for Building Integrated Photovoltaics. International Energy Agency, Paris (2002)
7. Defaix, P., van Sark, W., Worrell, E., de Visser, E.: Technical potential for photovoltaics on buildings in the EU-27. Sol. Energy **86**(9), 2644–2653 (2012)
8. Transparency Market Research: Building Integrated Photovoltaics (BIPV) Market. Global Industry Analysis, Size, Share, Growth, Trends 2013–2019. Transparency Market Research, New York (2014)
9. Curtius, H.: The adoption of building-integrated photovoltaics: barriers and facilitators. Renew. Energy **126**, 783–790 (2018)
10. Yang, R., Zou, P.: Building integrated photovoltaics (BIPV): costs, benefits, risks, barriers and improvement strategy. Int. J. Constr. Manag. **16**(1), 39–53 (2016)
11. Heinstein, P., Ballif, C., Perret-Aebi, L.: Building integrated photovoltaics (BIPV): review, potentials, barriers and myths. Green **3**(2), 125–156 (2013)

12. Yang, R.: Overcoming technical barriers and risks in the application of building integrated photovoltaics (BIPV): hardware and software strategies. Autom. Constr. **51**, 92–102 (2015)
13. Koinegg, J., Brudermann, T., Posch, A., Mrotzek, M.: "It Would Be a Shame if We Did Not Take Advantage of the Spirit of the Times …": an analysis of prospects and barriers of building integrated photovoltaics. Gaia **22**(1), 39–45 (2013)
14. Temby, O., Kapsis, K., Berton, H., Rosenbloom, D., Gibson, G., Athienitis, A., Meadowcroft, J.: Building-integrated photovoltaics: distributed energy development for urban sustainability. Environ. Sci. Policy Sustain. Dev. **56**(6), 4–17 (2017)
15. Karakaya, E., Sriwannawit, P.: Barriers to the adoption of photovoltaic systems: the state of the art. Renew. Sustain. Energy Rev. **49**, 60–66 (2015)
16. van Horrik, M., Ritzen, M., Vroon, Z.: Belemmeringen voor BIPV: Opschaling en uitrol in de Nederlandse markt van gebouw geïntegreerde PV systemen. Rijksdienst voor Ondernemend Nederland, Heerlen (2016)
17. International Energy Agency PVPS Task 15: Transition Towards Sound BIPV Business Models. International Energy Agency, Paris (2018)
18. RVO: BENG indicatoren. https://www.rvo.nl/onderwerpen/duurzaam-ondernemen/gebouwen/wetten-en-regels-gebouwen/nieuwbouw/beng-indicatoren. Accessed 27 Mar 2019

# Chapter 36
# Citizen Engagement in Energy Efficiency Retrofit of Public Housing Buildings: A Lisbon Case Study

**Catarina Rolim and Ricardo Gomes**

**Abstract** In Portugal, there are about 120 thousand social housing and a large share of them are in need of some kind of rehabilitation. Alongside the technical challenge associated with the retrofit measures implementation, there is the challenge of involving the citizens in adopting more energy conscious behaviors. Within the Sharing Cities project and, specifically in the case of social housing retrofit, engagement activities with the tenants are being promoted, along with participation from city representatives, decision makers, stakeholders, and among others. This paper will present a methodology outlined to evaluate the impact of retrofit measures considering the citizen as a crucial retrofit stakeholder. The approach ranges from technical analysis and data monitoring but also conveys activities such as educational and training sessions, interviews, surveys, workshops, public events, and focus groups. These will be conducted during the different stages of project implementation; the definition process, during deployment and beyond deployment of solutions.

## 36.1 Introduction

Throughout the world, urban areas are growing both in size and population. This growth trend has led to challenges related to demographics, climate change, energy consumption, transport, among others, jeopardizing the resilience of urban systems and social wellbeing. Furthermore, increased energy needs pose some challenges to economic competitiveness, GHG emissions and other environmental problems, and security of supply. In this context, several solutions and city planning strategies have emerged to overturn such challenges and promote the sustainable development of cities, focusing on energy efficiency improvements and other factors. While these are crucial to tackle the impacts of urban growth, it is far from its full potential.

C. Rolim (✉) · R. Gomes
IN+, Center for Innovation, Technology and Policy Research, IST - Instituto Superior Técnico, Universidade de Lisboa, Avenida Rovisco Pais 1, 1049-001 Lisbon, Portugal
e-mail: catarina.rolim@tecnico.ulisboa.pt

R. Gomes
e-mail: ricardo.a.gomes@tecnico.ulisboa.pt



J. Littlewood et al. (eds.), *Sustainability in Energy and Buildings*, Smart Innovation, Systems and Technologies 163, https://doi.org/10.1007/978-981-32-9868-2_36

Of the sectors where energy efficiency can be improved, buildings present a high potential due to their long useful lifetime period and capacity for construction and technological updates.

### 36.1.1 Rehabilitation Needs

Portugal housing is found to be one of the most energy inefficient of all housing in the EU [1]. Between 29 and 33% of the Portuguese housing stock dating from the post-war period is in need of some kind of reparation [2–4]. According to ADENE around 29% of the building stock need some kind of retrofit measures.

In Lisbon, buildings represent around 62% of the primary energy consumption, and the residential buildings alone are responsible for 19% of the total primary energy in the city [5]. There is a large number of buildings requiring rehabilitation in Lisbon, from different typologies and constructions periods [3].

### 36.1.2 Retrofit Measures

In 2010, the European Commission has published an updated version of the Energy Performance of Buildings Directive, which emphasizes the need for "Energy Efficient Retrofitting" (EER) of existing buildings [6]. In general, EER focuses on the implementation of retrofitting measures in an existing building, aiming to reduce the total energy demand, while at least maintaining the users' comfort.

According to the Energy Road Map 2050 report [7] building retrofit plays an important role in reducing the environmental loads currently associated with the building stock. From the most commonly applied EER measures, it can be highlighted the installation of thermal insulation, glazing replacement, improvement of the building's airtightness, HVAC system replacement and the introduction of mechanical ventilation with heat recovery [8].

### 36.1.3 Social Housing in Portugal

In Portugal, there are about 120 thousand social housing, representing 3.3% of the total housing stock of permanent residences [9]. This is a public property owned and managed by municipalities, municipal housing companies, and a small share is owned by the Institute of Housing and Urban Renewal (IHRU), a government-run institute.

Regarding the aims, beneficiaries, and models of production, different periods may be identified in the history of social housing in Portugal [10]. The authoritarian regime (1933–74), during which it was promoted a residual percentage of social housing. The "Programa de Casas Económicas" (Low-cost houses program) (1933), "Programa de Renda Económica" (Low rent program) (1959–1969) and the "Fundo de Fomento da Habitação" (Housing Development Fund, 1969) did not answer both in number as in quality to the housing demand in Portugal. This lead to the widespread of illegal construction both in the suburbs and in the inner cities, where overcrowded dwellings with a lack of basic amenities, meant very poor housing conditions and the rapid degradation of buildings.

Although the investment on public housing after the 1974 revolution, in the 1980s high housing deficits persisted in Portugal. Pressured by the municipalities of Lisbon and Porto, in 1993 the central government launched a program of social housing construction specifically targeting people living in shanty towns. The PER (The Special Rehousing Program) enabled large volume construction within a short time and in limited areas [10].

Although the considerable number of rehabilitated houses in the last years (Table 36.1) there is still the need to retrofit a large number of buildings, since around 60 thousand dwellings were constructed before 2005 (Fig. 36.1).

The poor conditions of some social housing neighborhoods in Portugal are referred in different studies [12–14]. These conditions are intrinsically related to the concept of Fuel Poverty. Although its definition is not consensual, it can be referred as when "anyone who meets, in its housing, particular difficulties to have the necessary energy to meet its basic energy needs because of the inadequacy of its resources or of its housing conditions" [1, 15]. Additionally, it has been proven that fuel poverty can

**Table 36.1** Rehabilitated social dwellings in 2009, 2012, and 2015 [11]

| Year | 2009 | 2012 | 2015 |
|---|---|---|---|
| No. of rehabilitated dwellings | 8339 | 5247 | 9437 |

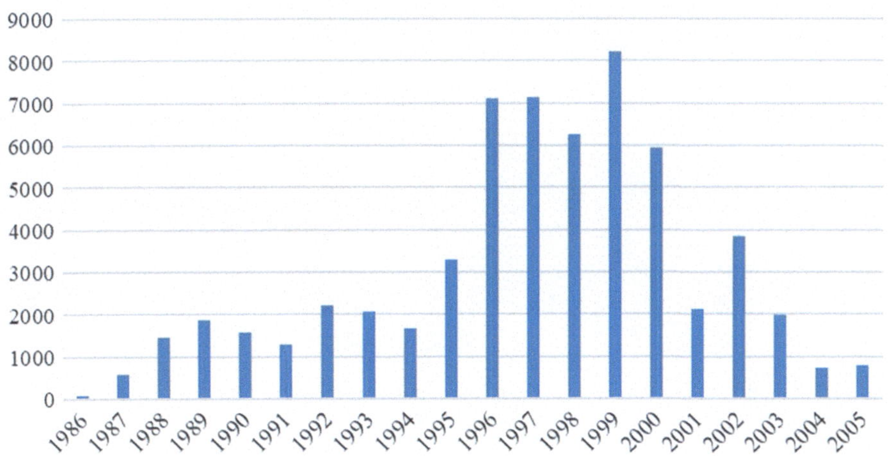


**Fig. 36.1** Number of social dwellings built between 1986 and 2005 (*source* [9])

affect mental wellbeing and social contact, as well as the development of children. It can be correlated with low household income, high energy cost, and energy inefficient homes. According to the alleviating fuel poverty in the EU report [16], Portugal is in the European group of countries with higher inability to keep homes adequately warm (Fig. 36.2).

Fuel poverty can be tackled by income increase, fuel prices regulation, and energy efficiency improvements in buildings. Energy costs are growing faster than household income, therefore, energy consumption subsidies and direct financial support for household heating cannot provide a long-term solution to the fuel poverty problem. EER can, however, give a long-term sustainable answer to it [16]. These measures address the root of the problem and result in reduced energy costs and/or improved thermal comfort in homes, create or maintain jobs, reduce illness, rehabilitate poor districts and therefore contribute to social inclusion [16]. Furthermore, the combination of retrofit measures with behavior change interventions leads to a higher reduction in total energy consumption than retrofit or behavior interventions applied separately [17].

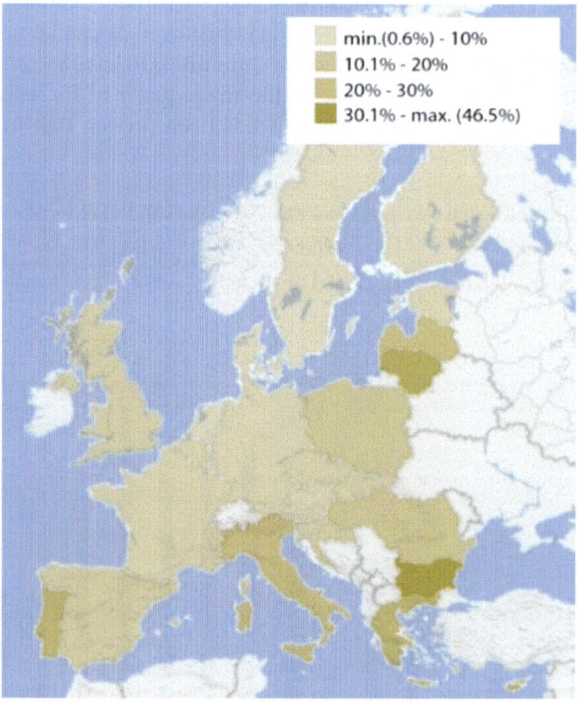


**Fig. 36.2** Inability to keep home adequately warm in Europe in 2012 [16]

## 36.2 Citizen Engagement in Behavior Changes

While there is an overall effort to improve building energy efficiency performance, through the implementation of best retrofit practices and deployment of local renewable energy solutions, it is important to consider that peoples' energy consumption patterns need to be adjusted and changed in order to accommodate them. Therefore, alongside the technical challenge to promote the transition to low carbon energy systems, there is also a social challenge that needs to be addressed. This transition will not be achieved without the engagement of citizens [18]. Therefore, it becomes essential to evaluate their behavior and culture but also to promote a bottom-up approach [19] by engaging citizens in the decision-making and designing tasks [20]. However, changing their behaviors can be a slow process and technology can play an important role in developing new urban services that generate high-impact behavior change. It is challenging to take full advantage of new solutions and people need custom-made feedback and information to embrace change [21]. Tailored information and interventions targeting different groups of people can help lead to successful adoption of energy-saving behaviors [22, 23].

When outlining solutions, it is essential to take into account citizens' participation, incentivizing an active role in the transformation of their living context contributing to the life of the city. Citizens can be engaged in this transformation not only by being part of the decision-making process (participating in co-design sessions of innovative solutions, shared learning initiatives, user research activities, etc.) but also during the implementation process of solutions and/or tools (participating in sessions or workshops, etc.) and after the deployment (participating in focus groups, interviews, surveys, etc.). Consequently, citizens become generators of data and information that can be applied for the development of new policies and new technologies. Including people in the design process, planning and management of these solutions have shown to improve the acceptability and increase their use [24]. Considering the challenges surrounding behavior change, engaging the citizen in the process of developing, implementing, and evaluating new solutions helps understanding people and behaviors and to identify challenges which, consequently, enables stakeholders to create exciting, meaningful, and prosperous systems and services creating the opportunities to change and establish new behaviors. Engaging citizens will also increase decision makers awareness and educate them on citizens experiences which, consequently might lead to the refinement, shaping, and creation of new policies and governance arrangements [25, 26].

## 36.3 Sharing Cities Lighthouse Project

Sharing Cities is an EU funded lighthouse program that is a proving ground for a better, common approach to making smart cities a reality, aiming to promote wide-scale deployment of smart city solutions, shifting the thinking of its citizens to decarbonized and local renewable energy sources. Additionally, the project has the ambition to promote the improvement of urban mobility, energy efficiency in buildings and carbon emissions by successfully engaging citizens in the transformation of their living context and fostering local level innovation, creation of new businesses and jobs. Nonetheless, while people play an important role in this process, changing their behaviors can be a slow process.

Within the Sharing Cities until 2019, the city of Lisbon will implement several retrofit actions in three different building types: public service buildings (historical municipality building and elementary school, both in the end stages of intervention), private residential buildings (one building completed and two with retrofit works taking place), and social housing buildings (retrofit intervention is taking place with expectations to be completed until the end of 2019). Taking this into account, the retrofit actions were selected for each building type considering their different needs and restrictions. The main actions aimed for a reduction of primary energy consumption, increase of comfort levels for the occupants and increase of renewable use. The improvement of the buildings performance is achieved through enhancements on the buildings envelope (e.g. façade insulation, roof insulation, windows replacement, and glazing replacement) and technical systems, with the improvement of lighting systems (installation of efficient LED light bulbs), replacement of HVAC systems and installation of heat pumps. The installation of PV panels will contribute to increase the use of renewable energy sources. Furthermore, the installation of the PV panels in this social housing block allows the development of one pilot study to analyze the viability and the potentialities of energy production at a community scale with these social characteristics. This pilot may allow evaluating if a PV installation could influence social tariffs, fight energy and fuel poverty implications, and influence future building legislation. Within the goals of this pilot are the evaluation of tenants' engagement and the attribution of benefits and responsibilities from PV management and production.

In order to assess and evaluate the impacts of these retrofit actions not only on the building performance but also in factors such as occupants' behaviors, business models, among others, a monitoring plan has been outlined for each of the buildings type. Figure 36.3 presents the generic approach considered in this study.

### *36.3.1 Lisbon Social Housing Case Study*

Within the Sharing Cities program, the Lisbon Municipality has selected a social housing unit to retrofit. The unit, covering a built area of 20,609 m$^2$, is composed of two housing blocks with a total of 10 buildings, comprising 248 dwellings inhabited approximately by 437 persons. A report presented by GEBALIS in 2012 [27] indicated that population living in this area has very low levels on income since 20.9% of the resident population was illiterate, 20.1% did not finish elementary school, and only 45.2% of the active population was employed. The main retrofit actions implemented in the unit focus on increasing and improving the energy efficiency of the

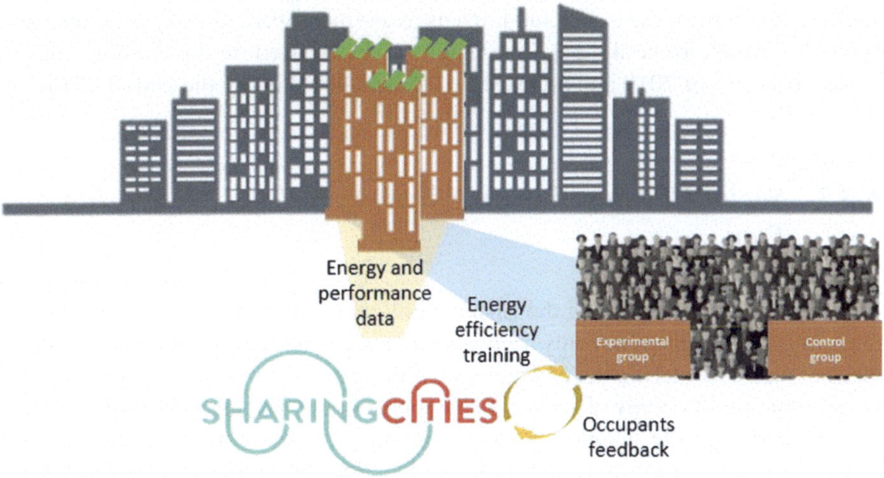


**Fig. 36.3** Approach overview considered in this study

buildings as well as the thermal comfort of the occupants. Additionally, PV panels will be installed on the roofs of the buildings increasing the buildings potential for renewable energy production. Table 36.2 presents the retrofit interventions for the Lisbon public housing buildings.

The strategy defined for improving the public housing buildings performance was developed focusing on improving tenants' quality of life. The assessment of potential improvements originated by the implementation of these retrofit measures was conducted based on a dynamic simulation of the building using EnergyPlus.

**Table 36.2** Summary of the type of retrofit interventions for the Lisbon public housing buildings

| Type of building | Address | Intervention area | Type of intervention | Description of intervention |
|---|---|---|---|---|
| Social housing | Quinta do Cabrinha | 17,212 $m^2$ | Insulation | Roof and façade exterior insulation with cork agglomerate 6 cm |
| | | | Windows | PVC frame with double glazing (4 mm + 16 mm air + 4 mm) |
| | | | Lighting | Replace lamps for LED in common areas |
| | | | Renewable energy | 869 PV panels |

Details on the retrofit measures and impacts assessment were already presented and explored in Monteiro et al. [28]. Intervention works started on the building sites in the first trimester of 2019 and are expected to be completed by the end of 2019.

### 36.3.2 Social Housing Energy Efficiency Retrofit Monitoring Plan

The monitoring plan regarding the Lisbon public housing consists of several implementation stages defined not only according to the retrofit intervention actions taking place but also considering the engagement of the tenants, by educating and promoting the adoption of energy efficiency behaviors at home. Engagement activities will be developed with the tenants (citizens) aiming to increase awareness on retrofit measures to improve comfort and promote the adoption of energy efficient behaviors. Even though the tenants were not involved in the definition of the design and of the retrofit measures to be deployed, this monitoring plan aims at engaging them during the intervention works and after the deployment. Their inputs will be essential when assessing the impacts of the intervention works not only in terms of energy consumption but also on comfort and quality of life. The methodology consists of five stages taking into account two specific groups:

- Experimental: a sample of tenants will be asked to participate as volunteers will receive education on energy efficiency and feedback on their energy consumption performance. Furthermore, a detailed questionnaire will also be conducted in order to make a socio-economic characterization of the volunteers, thermal and comfort levels, utilities and device usage characterization, attitudes, and perceptions towards energy efficiency and energy consumption.
- Control group: we will consider as a control group the tenants that will not volunteer to participate in the activities comprising the monitoring plan (educational sessions, feedback sessions, etc.). The control group will receive no information on energy efficiency and energy consumption patterns. This distinction will allow to assess the impacts of both energy efficiency education and retrofit interventions in energy consumption and comfort.

Figure 36.4 presents the monitoring plan that will be implemented. The number of activities with the tenants will be adjusted within the different monitoring and intervention stages. Nonetheless, an initial questionnaire will be conducted, as mentioned, in order socio-economic characterization survey will be performed to the participants belonging to the experimental group. An additional comfort survey and energy poverty assessment survey will be conducted to all habitants of the public housing unit retrofitted. Several energy efficiency and feedback sessions are planned. While the number of these sessions is not yet defined, at least 4–6 sessions will take place, coinciding with the several stages of the monitoring plan. Furthermore, focus groups or in-depth interviews will be conducted with the tenants in order to evaluate

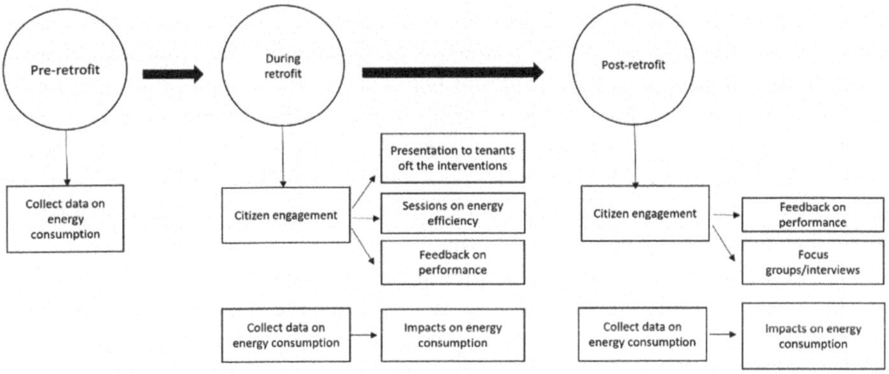


**Fig. 36.4** Social housing energy efficiency retrofit monitoring plan

the impacts of the activities promoted and their improvement, changes recommendations, satisfaction towards the retrofit, and educational sessions.

The energy consumption will be assessed through monitoring, and surveys will be conducted to the tenants to evaluate attitudes and perceptions towards behavior changes, comfort evaluation, improvement of quality of life, and satisfaction towards the interventions performed. The data collected, through energy meters installed at building and apartment level will also enable the development of tailor-made feedback reports to the participants. It is crucial to highlight that during the project, the privacy issues of the participants will always be considered. This monitoring and intervention plan will also convey the collaboration with local associations and the parish to promote community participation and facilitate the engagement process of the citizens. Furthermore, interviews with selected stakeholders and decision makers will also take place. These will allow assessing their willingness to replicate and/or scale-up such an approach to similar buildings in the city of Lisbon. This monitoring plan entails the collection of both quantitative data (e.g. energy consumption data) and qualitative data (e.g. tenants' inputs on behaviors adopted, on retrofit interventions, interviews with stakeholders, etc.), enabling performing with real world experimental data a statistical analysis based on multidisciplinary models [29].

## 36.4 Conclusions

The deployment of innovative solutions and the promotion of sustainable behaviors is essential to facilitate the move towards sustainable and smart cities. The participation of the city of Lisbon in the Sharing Cities program has enabled the city to invest and incentive the implementation of such measures. Among them, is the requalification of several types of buildings, including social housing units. Alongside the technical challenge associated with the implementation of such measures, there is the challenge of involving the citizens in the transformation process, namely adopting more energy efficient behaviors. The works developed within the scope of this project will be an opportunity to use real world experimental data, moving beyond simulation results, to assess the impacts of both retrofit and behavior change interventions on energy consumption. Furthermore, since the retrofit interventions and monitoring plan are being deployed in a social housing unit, this work will be an opportunity for the Lisbon Municipality to identify how to increase the impact of measures by engaging the tenants, and, consequently, have access to more accurate information and data to assess that can boost successful and lasting results. Finally, the results obtained can serve as a basis in the future for the development of new policies to promote building retrofit interventions and the adoption of more sustainable and efficient behaviors.

**Acknowledgements** The study was developed within the framework of the project SHAR-LLM, which has received funding from the European Union's Horizon 2020 research and innovation program under grant agreement No. 691895. Support from the IN+ strategic Project UID/EEA/50009/2013 is gratefully acknowledged.

## References

1. Healy, J.D.: Housing, Fuel Poverty and Health: A Pan-European Analysis (2004)
2. European Cooperation in the Field of Scientific and Technical, Improving the Quality of Existing Urban Building Envelopes: State of the Art (2007)
3. Instituto Nacional de Estatística, LNEC, Laboratório Nacional de Engenharia Civil: O Parque Habitacional e a sua Reabilitação. Análise e Evolução 2001–2011 (2013)
4. ADENE- Agência para a Energia: Energy Efficiency trends and policies in Portugal. ADENE – Agência para a Energia. Algés (2015)
5. Lisboa E-nova: Matriz Energética de Lisboa 2014 (2016)
6. Boermans, T., Grözinger, J.B., von Mante, N., Surmeli-Anac, A., Leutgöb, J.K., Bachner, D.: Assessment of cost optimal calculations in the context of the EPBD (ENER/C3/2013–414). Final report. BUIDE13705 (2015)
7. European Commission: Energy Roadmap 2050 (2011)
8. European Union: European Directive 2010/31/EU on the energy performance of buildings. Off. J. L **153**, 18 June 2010 (2010)
9. CET-ISCTE/IRIC-UPORTO/A. MATEUS ASSOCIADOS: Diagnóstico e proposta para uma estratégia de habitação 2008/2013 (2008)
10. Alves, S., Andersen, H.T.: Social housing in Portugal and Denmark: a comparative perspective. In: ENHR—European Network for Housing Research Conference, Lisbon (2015)
11. Instituto Nacional de Estatística: Available https://www.ine.pt/xportal/xmain?xpid=INE&xpgid=ine_indicadores&indOcorrCod=0009017&contexto=bd&selTab=tab2. Accessed May 2019 (Online)
12. Madeira da Silva, T., Marques Alves, P., Vasconcelos Magalhães, M., de Oliveira, M.J.: Edifícios de habitação social: diagnóstico e cenários de intervenção no edificado. Cidade, comunidades e territórios, 82–98 (2011)
13. Arnold, J., Farinha Rodrigues, C.: Reducing inequality and poverty in Portugal. Organisation for Economic Co-operation and Development. ECO/WKP(2015)76 (2015)
14. Gomes, R., Ferreira, A., Azevedo, L., Costa Neto, R., Laura, A., Silva, C.: Retrofit measures evaluation considering thermal comfort using building energy simulation: two Lisbon households. Adv. Build. Energy Res. (2018)

15. Magalhaes, S.M., Leal, V.M.: Characterization of thermal performance and nominal heating gap of the residential building stock using the EPBD-derived databases: the case of Portugal mainland. Energy Build. **70**, 167–179 (2014)
16. Bogdan Atanasiu, B., Kontonasiou, E., Mariottini, F.: Alleviating Fuel Poverty in the EU. BPIE—Buildings Performance Institute Europe (2014)
17. James, M., Ambrose, M.: Retrofit or behaviour change? Which has the greater impact on energy consumption in low income households? Procedia Eng. **180**, 1558–1567 (2017)
18. Chilvers, J., Pallett, H., Hargreaves, T.: Public Engagement with Energy: Broadening Evidence, Policy and Practice. UKERC/DM/2017/BN/01, London (2017)
19. Saunders, T., Baeck, P.: Rethinking smart cities from the ground up. Nesta Technical Report (2015)
20. Chourabi, H., Taewoo, N., Walker, S., Gil-Garcia, J.R., Mellouli, S., Nahon, K., Pardo, T.A., Scholl, H.J.: Understanding smart cities: an integrative framework system. In: 45th Hawaii International Conference on System Sciences (2012)
21. Rolim, C., Baptista, P.: Comparing drivers' self-perception on driving behaviour changes with real world driving performance data. Travel Behav. Soc. **11**, 86–92 (2018)
22. Seidl, R., Moser C., Blumer, Y.: Navigating behavioral energy sufficiency. Results from a survey in Swiss cities on potential behavior change. PLoS One, **12**(10) (2017)
23. Trombley, J., Halawa, E.: Can further energy reductions be achieved through behaviour changes in low income households? Energy Procedia **121**, 230–237 (2017)
24. Caird, S., Hudson, L., Kortuem, G.: A Tale of Evaluation and Reporting in UK Smart Cities. The Open University, UK (2016)
25. BUILD2LC Interreg Europe: New Energy Culture, Citizen Involvement and Energy Poverty—Topic Report (2017)
26. Bornemann, B., Sohre, A., Burger, P.: Future governance of individual energy consumption behavior change—a framework for reflexive designs. Energy Res. Soc. Sci. **35**, 140–151 (2018)
27. GEBALIS: Caracterização Sociodemográfica da Quinta do Cabrinha. Available http://www.gebalis.pt/Geral/Publicacoes/Relatorios_Outros/Lists/DocumentsList/estudocabrinha.pdf. Accessed 7 May 2019 (Online)
28. Sousa-Monteiro, C., Causone, F., Cunha, S., Pina, A., Erba, S.: Addressing the challenges of public housing retrofits. Energy Procedia **134**, 442–451 (2017)
29. Lopes, A.R.M., Antunes, C.H., Reis, A., Martins, N.: Estimating energy savings from behaviours using building performance simulations. Build. Res. Inf. **45**, 303–319 (2017)

# Chapter 37
# The Role of Thermal Insulation in the Architecture of Hot Desert Climates

**Carlos López-Ordóñez, Isabel Crespo Cabillo, Jaume Roset Calzada and Helena Coch Roura**

**Abstract** When we think about a dwelling's thermal comfort, one of the first things that come to mind is to improve the envelope of the building, usually through the use of thermal insulation. However, in hot desert climate cities, where there is a great temperature oscillation, both daily and annually, the behavior of thermal insulation is not that clear. The city of Hermosillo follows a scattered pattern of urban growth. It has a hot desert climate, an annual mean temperature of 25 °C with an annual mean oscillation of 15 °C; during the summer season it presents extreme temperatures of 40–45 °C, annual precipitation of 387 mm and relative humidity of 43%. The most accepted architectonic strategies to improve the thermal behavior of buildings in hot desert climates are solar protection by shading, especially glazed surfaces, and the use of thermal insulation in the building's envelope. This study aimed to assess the behavior of thermal insulation use in one storey dwellings. The results obtained show a reduction of energy loses through the envelope of the dwelling when more insulated, resulting in an interior thermal behavior not as favorable as was thought.

## 37.1 Introduction

Usually to enhance a building's thermal comfort, improving its envelope comes to mind, mainly through thermal insulation. But before going directly into the use of insulation, we need to take into consideration the environmental conditions we are located in.

Some doubts about the use of insulation in hot humid and hot dry climates appear, where the openness and lightness of the façade are needed for heat dissipation [1], or the use of selective glass, reflective painting and blinds is recommended [2], even different passive design strategies, such as evaporative cooling, solar chimneys, wind towers, and thermal mass [3].

C. López-Ordóñez (✉) · I. Crespo Cabillo · J. Roset Calzada · H. Coch Roura
AiEM, E.T.S. Arquitectura de Barcelona, Universitat Politècnica de Catalunya (UPC), Barcelona 08028, Spain
e-mail: carlos.fernando.lopez@upc.edu

 
J. Littlewood et al. (eds.), *Sustainability in Energy and Buildings*, Smart Innovation, Systems and Technologies 163, https://doi.org/10.1007/978-981-32-9868-2_37

In hot desert climate cities, where there is a great temperature oscillation, both daily and annually, the behavior of thermal insulation is also not that clear.

These cities are located in low medium latitudes, between 20° and 33° north and south latitude, where the greatest amount of the world population is concentrated [4]. A large number of these cities follow scattered urban growth: one-storey dwellings, great dependence on the use of the car, leading to very low urban densities.

In the American continent, cities such as Phoenix, Tucson, and Houston in the US and Hermosillo, Monterrey, and Mexicali or Ciudad Juarez in Mexico, follow this growth pattern [5]. It is in this framework, that the study of the behavior of thermal insulation in one-storey dwellings takes big importance due to its relation with the thermal comfort and energy consumption.

Energy consumption is one of the main challenges the world is facing right now, so addressing this issue is of big importance to every country subscribed to the international agreements addressing a climate change politics such as Horizon2020 [6, 7].

## 37.2 Objective

This study aims to assess the behavior of thermal insulation use in one-storey dwellings. While in the hot desert climate is practically mandatory to have air conditioning systems, this work is conducted taking into consideration a situation where a household cannot afford the air conditioning system. This is relevant since in this kind of climate there is a large number of developing countries.

The study is carried out in a city with an urban sprawl problem and continuous growth by building one-storey dwellings with almost no thermal insulation.

## 37.3 Methodology

This investigation focuses on simulations of a one-storey dwelling to see the thermal behavior throughout the year, but especially during the summer.

The study is carried out in three different scenarios: in the first one, a dwelling with thermal insulation only in the roof. In the second and third case, the characteristics of the envelope in the same dwelling are modified, thermal insulation is added to the different surfaces (roof + walls + windows) and (roof + walls + windows + floor).

The simulations are based on a dynamic calculate model using the DesignBuilder interface, which runs with EnergyPlus as a calculation engine. The weather data used from the EPW file (EnergyPlus Weather) was collected by the weather station of the International Airport of Hermosillo [8].

The geometry, materials, orientation, and the architectonic distribution of the model used in the simulations is the predominant building typology, the most common material used for walls construction is the concrete block ($U = 2.54\ W/m^2K$), the roof is made of prefabricated concrete joists and little EPS vaults, supporting a compression layer made of reinforced concrete ($U = 0.25\ W/m^2K$). The predominant colors are white for the roof and light colors for exterior walls. Usually, the interior height varies between 2.40 and 2.80 m, in this case, the height is 2.50 m [9].

For the windows, the most used material is a simple 4 mm clear glass ($U = 5.8\ W/m^2K$), becoming an important source of energy exchange between the interior of the dwelling and the exterior environment due to the incident solar radiation, it is common not to have shadowing devices on windows.

The model is located in a plot 8.0 m × 14.0 m meaning in 112 $m^2$, the distance between two houses is 0.90 m. Usually, all the dwellings have an outside parking space and a backyard. The lack of vegetation to provide shadow over the habitable zones, façade, and over the parking area is a constant (Fig. 37.1).

To validate the results of the simulations, a study carried out by I. Marincic, J. Ochoa and A. del Río in the city of Hermosillo was used. In this study, they carried out measurements campaigns of indoor temperature and relative humidity [10] obtaining similar data to the one obtained by the simulations (Fig. 37.2).

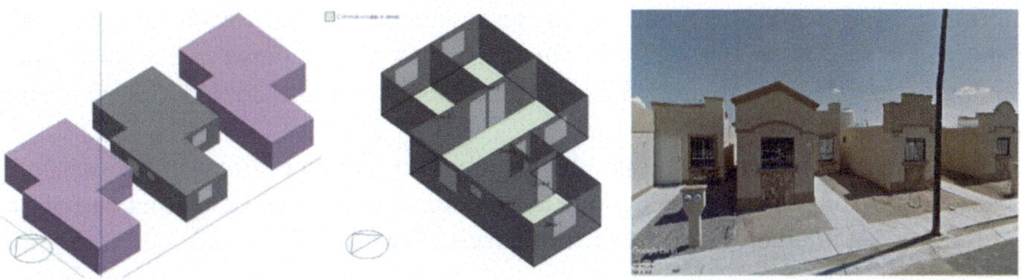

**Fig. 37.1** Left: simulation model. It has a north-south orientation; it is important to notice that there is no adjacency between the dwellings. Author's elaboration on DesignBuilder. Right: Photograph of the houses. Taken with GoogleEarth

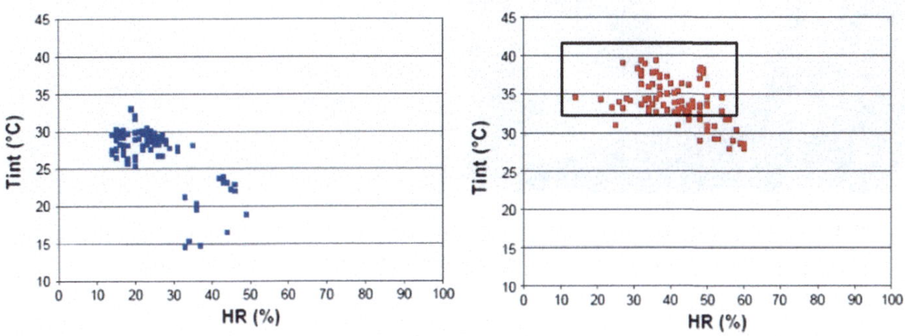


**Fig. 37.2** Indoor air temperature (Tint, °C) and relative humidity (HR, %) obtained during the measurement campaigns in winter (left) and summer (right) [10]

## 37.4 Case Study: The City of Hermosillo, México

The city of Hermosillo (lat. 29º05′N), located in the Mexican state of Sonora, follows a scattered pattern of urban growth. It has a hot desert climate (BWh, in the Köppen climate classification), an annual mean temperature of 25 °C with an annual mean oscillation of 15 °C, during the summer season it presents extreme temperatures of 40–45 °C, it has an annual precipitation of 387 mm and relative humidity of 43% [11].

As a desert city (is located towards the southern limit of the Sonora Desert), it has high levels of solar radiation (annual mean of 5.84 kWh/m$^2$ day), clear sky throughout the year [12]. Presenting a dust pollution problem, very characteristic of desert cities, posing a problem for natural ventilation: the windows are not open often [10] (Fig. 37.3 and Table 37.1).

It is important to note that in a hot desert climate, with a wide range of temperatures the comfort zone is wider than the one used in the international normative, such as ASHRAE, or in the classic bibliography (Olgyay and Givoni) [14, 15].

It has been conducted as a study and established a neutral temperature (nT) for interior comfort conditions in the city of Hermosillo: 32.2 $\pm$ 2.5 °C in summer while in the winter season is 26.9 $\pm$ 2.9 °C (in a restricted range), these values grow when moving to an expanded range: 32.2 $\pm$ 4.9 °C in summer and 26.9 $\pm$ 5.8 °C during winter.

The comfort range is wider than the usually seen in other climates; the reason is its strong relationship with the outside temperature oscillation, being this one of the most important characteristics of the hot desert climate [10].

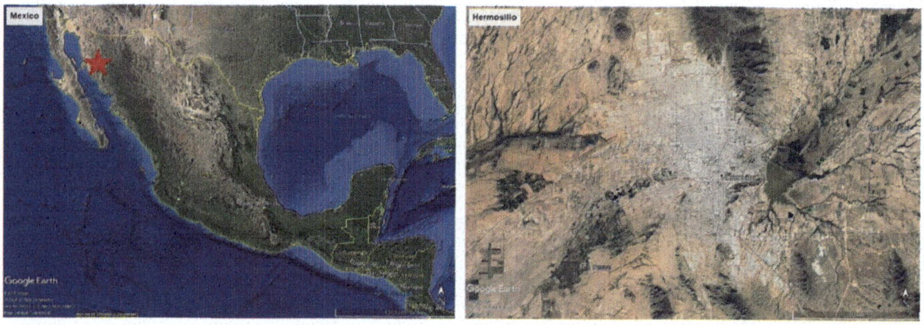


**Fig. 37.3** Location of Hermosillo in relation to Mexico (left), a satellite photograph of Hermosillo, spreading through the Sonoran Desert (right). Author's elaboration on GoogleEarth

**Table 37.1** Monthly data of average maximum temperature (AMT), mean temperature (MT), average minimum temperature (AmT) and relative humidity % (RH), and global horizontal radiation ($kWh/m^2$) [13]

| Month | Jan | Feb | Mar | Apr | May | Jun | Jul | Aug | Sep | Oct | Nov | Dec |
|---|---|---|---|---|---|---|---|---|---|---|---|---|
| AMT °C | 24.2 | 25.8 | 28.7 | 32.3 | 36.3 | 39.8 | 39.3 | 38.3 | 37.5 | 33.9 | 28.6 | 24.0 |
| MT °C | 17.2 | 18.5 | 20.9 | 24.1 | 27.9 | 31.8 | 32.5 | 31.9 | 31.0 | 26.9 | 21.3 | 17.1 |
| AmT °C | 10.2 | 11.3 | 13.1 | 15.9 | 19.4 | 23.8 | 25.8 | 25.6 | 24.6 | 19.8 | 14.0 | 10.2 |
| HR (%) | 48 | 44 | 40 | 34 | 31 | 34 | 48 | 53 | 48 | 42 | 43 | 49 |
| $kWh/m^2$ | 3.88 | 4.76 | 6.34 | 7.45 | 7.73 | 7.59 | 7.07 | 6.88 | 5.74 | 5.23 | 4.11 | 3.25 |

### 37.4.1 Social, Urban, and Energy Aspects

Since the 20th century, the city of Hermosillo has maintained an accelerated rate of urban growth, but it was in the 1990 decade that due to diverse economic aspects, there was a massive increment of the population and construction industry, hence, the housing development became a very appealing market.

The urban area in 1990 was 7421.8 ha, in 2015 it had increased to 17,430 ha, an increment of 134%, most of this area is occupied by housing. The city has 196,827 homes, of which 1337 don't have electricity and 6696 have a dirt floor [16].

In 2015 Hermosillo had a population of 884,273 inhabitants, but in 1990 the population was 448,966 inhabitants, the 41% of the population lives with less than 3 minimum wages per month, that is less than 300€, 23% receives a monthly income between 300 and 500€ [17].

It can be seen in Fig. 37.4 the growth that the city has had in the last 28 years, it has expanded to the north and south, principally low-income housing developments, dwellings ranging between 25 and 90 $m^2$.

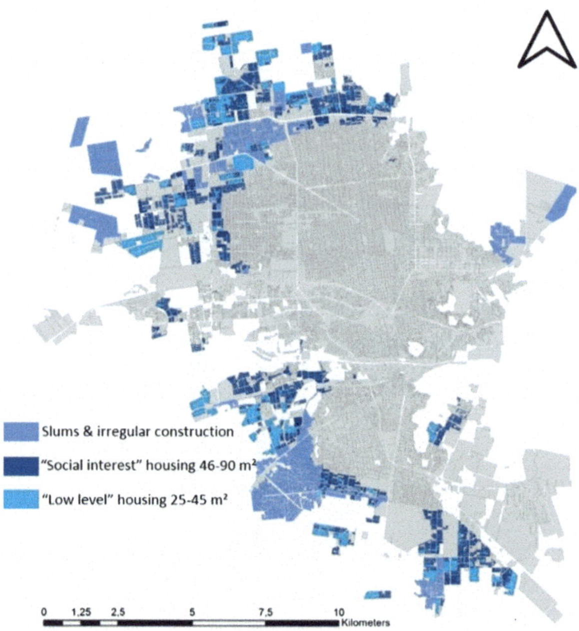


**Fig. 37.4** Location of the dwellings in which this study is focused (low-income households built after 1990 in Hermosillo). Author's elaboration with ArcGIS

## 37.5 Simulations and Results

The simulations were conducted in three different situations depending on the percentage of thermal insulation of the envelope of the dwelling, maintaining the same occupation (4 users, this is the only internal gains considered) a ventilation rate of 2 ac/h and an infiltration rate of 0.70 ac/h.

The cases described will be referred as case the "A" (insulation: roof), case "B", (insulation: roof + walls + windows) and case "C" (insulation: roof + walls + windows + floor). The roof and walls have an insulation thickness of 0.15 m and a U-value of 0.25 $W/m^2K$ and 0.22 $W/m^2K$ respectively. The insulation thickness of the ground floor is 0.05 m and a U-value of 0.55 $W/m^2K$, the improved windows are composed by two glass panels of 6 mm with an air gap of 6 mm, and a U-value of 3.10 $W/m^2K$.

To understand the behavior of the indoor temperature it is necessary to observe the behavior of the energetic balance (in this study: the gains and losses through the dwelling's envelope). For this purpose, the annual simulation of case "A" can give us the information needed.

In Fig. 37.5 it can be identified three elements that represent the majority of these exchanges: the walls, the windows (direct solar radiation gains), and the ground floors. The importance of these elements of the heat balance is that they represent 72% of the total exchanges of energy throughout the year, while is 78% during summer (Fig. 37.6).

It is important to notice that when the indoor temperature of the dwelling is higher than the outside temperature, the dwelling begins to release the interior heat through the envelope by long wave radiation.

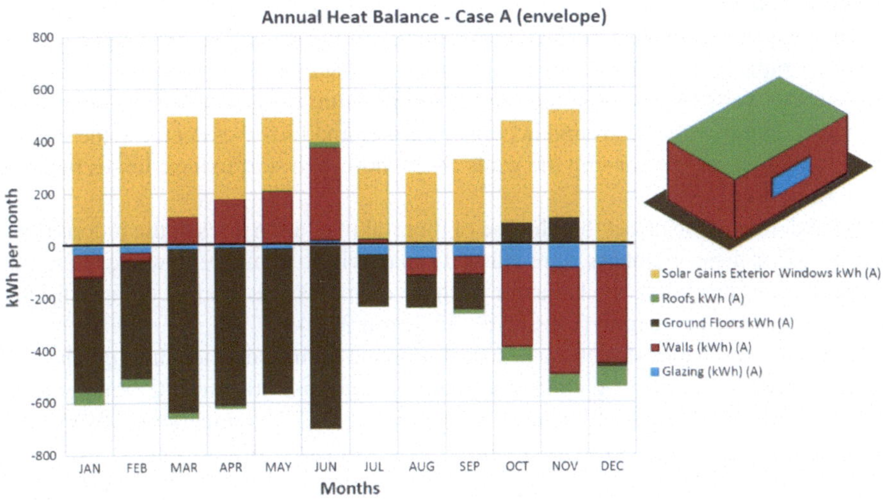


**Fig. 37.5** This chart shows the behavior of the energetic exchange that occurs through the envelope of the dwelling

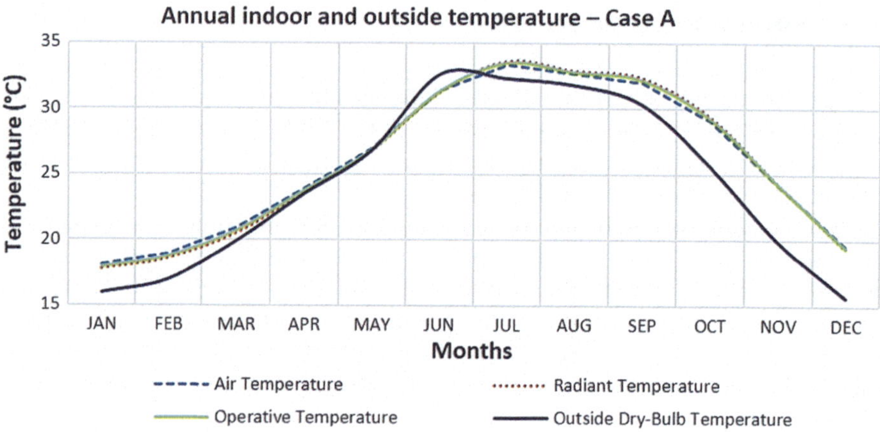


**Fig. 37.6** Chart with the temperature behavior of case "A". Indoor: air temperature, operative temperature, and radiant temperature. Outside: dry bulb temperature

While the gains from direct solar radiation (shortwave radiation) through windows do not seem affected by the influence of other elements of the envelope, the results obtained demonstrate that the energetic exchanges by long wave radiation through the walls and windows are closely related to the losses/gains through the ground, this behavior can be seen clearly during summer, e.g., in June the dwelling is losing through the ground the equivalent amount of energy to the one of the total gains of the dwelling's envelope, the wall's gains are the largest of the year and represent the 48% of June's total gains, this behavior changes in July when the losses through the ground start to decrease producing an opposite effect on the walls and windows energetic behavior.

This behavior is caused by the thermal inertia of the ground since its temperature is not as high as the outside air temperature, it makes easier the entry of energy through the walls by maintaining a lower temperature than outside. The next three months of summer as the ground becomes warmer and its losses decrease the indoor temperature increases, causing the walls and windows to start to have losses by long wave radiation.

To understand the effect that the use of thermal insulation does to the indoor thermal behavior these elements were studied separately to compare how the energetic exchanges are modified, the behavior of these elements was studied comparing the three cases at the same time (Figs. 37.7, 37.8, and 37.9).

The case "B" demonstrates the best performance during summer although the difference between the case "A" indoor temperature and case "B", and "C" temperature shown in Fig. 37.10 are small values. Whilst in case "A" the temperatures from June to September are: 31.7, 33.6, 32.9, and 32.2 °C, in case "B" the values, are: 29.5, 33, 32.6, and 32 °C, this results represent a reduction from "A" to "B" of 7, 2, <1, <1%. Case "C" presents the following values: 32.1, 33.6, 33.2, and 32.3 °C, unlike

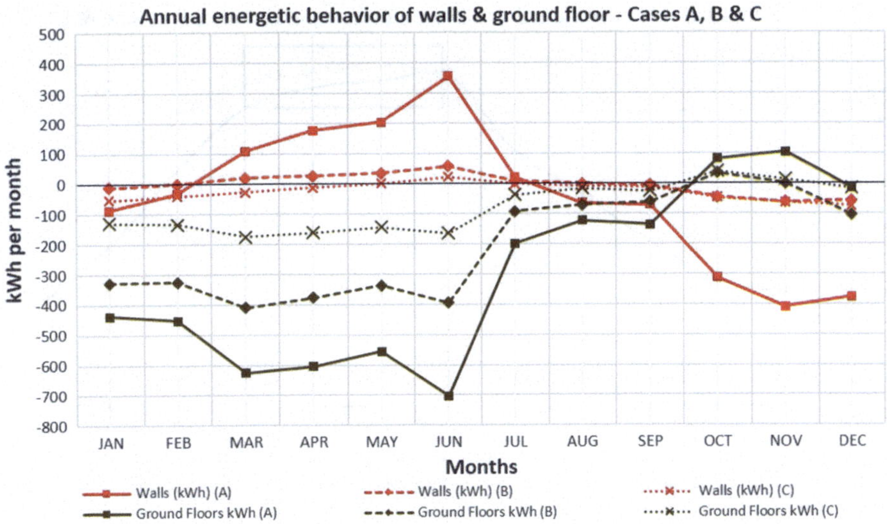


**Fig. 37.7** Comparison between the energetic behavior of the walls and the ground floor in all cases. The correlation among them can be seen through the year

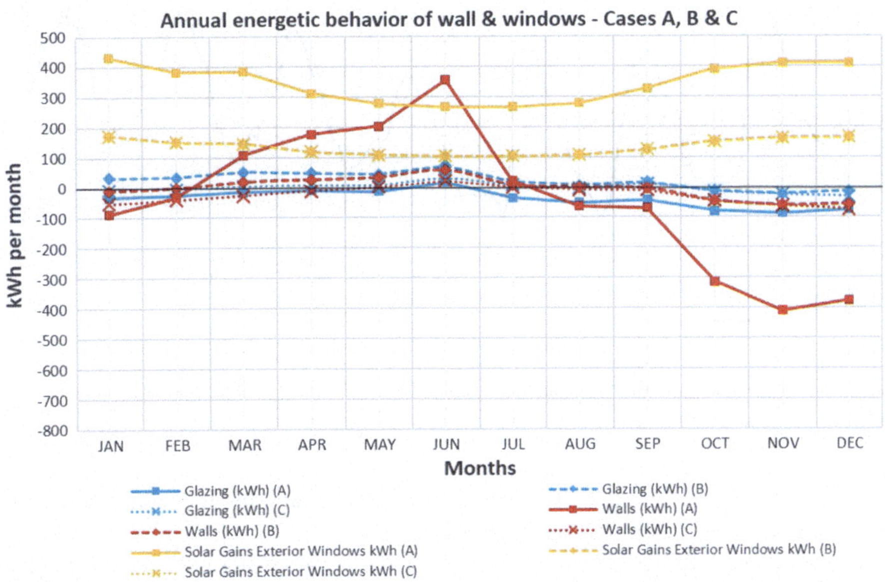


**Fig. 37.8** As explained previously, this chart shows how the direct solar gains through the windows act independently from the other elements of the dwellings envelope. Furthermore, it can be seen how the gains and losses reduced by increasing the level of thermal insulation

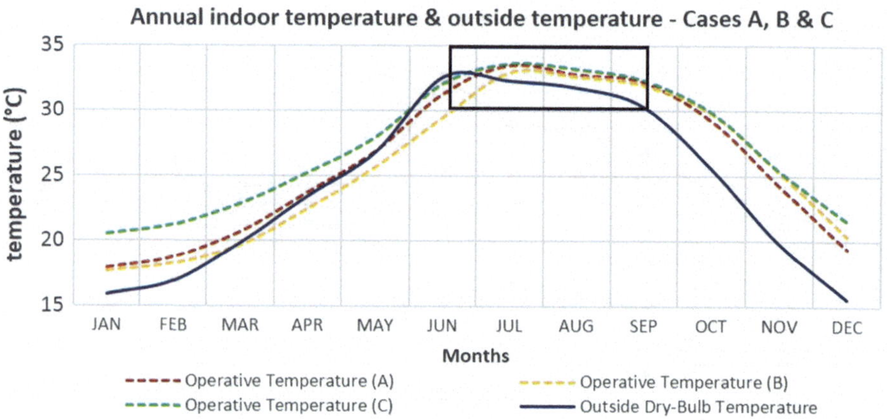


**Fig. 37.9** Comparison of the annual indoor operative temperature of the three cases with the outside temperature. During the summer the difference among the indoor temperatures reduce, getting very close to each other

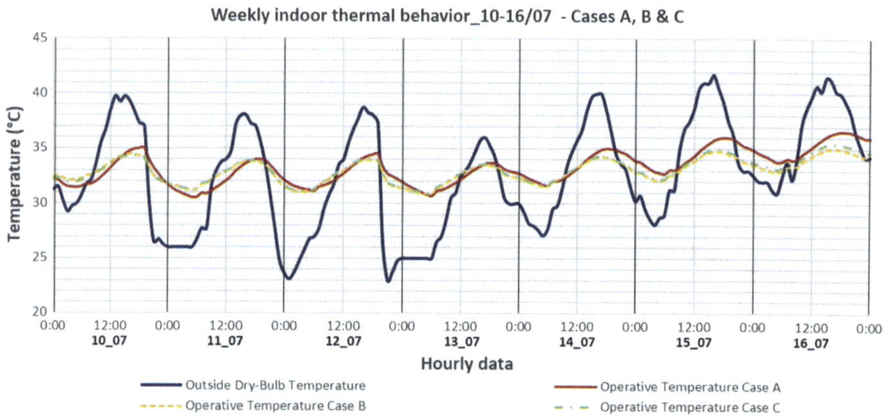


**Fig. 37.10** Weekly indoor thermal behavior in cases "A", "B" and "C", the data has been simulated hourly

case "B", this case offers a worse performance during summer, the values represent an increment from "A" to "C" of <1, 0, <1, <1%.

These values acquire more relevance when studying the thermal behavior hourly for a week of July when the higher temperatures take place:

As shown in Fig. 37.10 the behavior of the temperatures in the three cases is practically the same through the week, except when the temperature oscillation decreases (14-16/07) causing the dwelling to present difficulties to dissipate the heat at night.

It is interesting that case "A" presents a better delay in operative temperature during the last part of the day, but a higher operative maximum daily value in respect to cases "B" and "C". Another interesting result is given by analyzing the cases "B" and "C", although, they present the same behavior, when there is a lower outside temperature oscillation, case "C" has fewer losses in comparison to the case "B".

## 37.6 Discussion and Conclusions

The energetic behavior of the envelope of a "social interest" dwelling (60 $m^2$) in the city of Hermosillo has been studied in three different situations according to its level of thermal insulation. This is an attempt to identify and quantify how each situation affects the interior thermal behavior of the dwelling.

The results obtained indicate that the use of thermal insulation in this type of climate must be deeply studied before deciding to use it. This is especially important when the household cannot afford an air conditioning system.

It was found that case C (insulation: roof + walls + windows + floor) is counter-productive during the hot season, not allowing the dwelling to dissipate heat to the outside through its own envelope.

The behavior of cases A (insulation: roof) and B (insulation: roof + walls + windows) shows similitudes throughout the year. During the hot season, the performance of case B is slightly superior to case A: it warms less but at the same time its losses are lower, especially on days with a big thermal oscillation. These two cases have something in common: a great energetic exchange through the ground floor, resulting in something positive during the warm months since it works as a heat sink.

Is the use of thermal insulation in these climates necessary? According to the results obtained it seems that before using it for this type of dwellings, not only architectural strategies must be considered. As thermal results are not so conclusive, before deciding this investment, social considerations should be taken into account. The results are very different in conditioned or free-running dwellings. When conditioned, insulation will be positive [18], but it cannot be asserted the same in the case of free-running houses.

**Acknowledgements** This research has been supported by CONACYT – SENER (CVU 469347) and by the Spanish Ministry of Economy under the MOET project, code BIA2016-7765-R.

## References

1. Serra, R., Coch, H.: Arquitectura y Energía natural, 1st edn. Edicions UPC, Barcelona (1995)
2. Pellegrino, M., Simonetti, M., Chiesa, G.: Reducing thermal and energy consumption of Indian residential buildings: model validation by in-field measurements and simulation of low-cost interventions. Energy Build. **113**, 145–158 (2016)
3. Osman, M., Sevinc, H.: Adaptation of climate-responsive building design strategies and resilience to climate change in the hot/arid region of Khartoum, Sudan. Sustain. Cities Soc. **47**, 101429 (2019)
4. Datagraver. https://www.datagraver.com/. Accessed 15 March 2019

5. Smart Growth America. https://smartgrowthamerica.org/. Accessed 15 March 2019
6. Alalouch, C., Al-Saadi, S., AlWaer, H., Al-Khaled, K.: Energy saving potential for residential buildings in hot climates: the case of Oman. Sustain. Cities Soc. **46**, 101442 (2019)
7. Horizon 2020 The EU framework programme for research and innovation – energy. https://ec.europa.eu/programmes/horizon2020. Accessed 18 March 2019
8. DesignBuilder Software Ltd. https://designbuilder.co.uk/. Accessed 15 March 2019
9. Marincic, I., Ochoa de la Torre, J., Alpuche, M., Duarte, A., Vargas, L., González, I., Barrera, I.: La construcción actual de viviendas en Hermosillo y su adecuación al clima por medios pasivos. In: XXXV Congreso Nacional de Energía Solar. ANES, pp. 183–193. ANES, Chihuahua (2011)
10. Marincic, I., Ochoa, J., Del Río, J.: Confort térmico adaptativo dependiente de la temperatura y la humedad. ACE Archit. City Environ. **7**(20), 27–46 (2012)
11. López-Ordóñez, C., Crespo, I., Roset, J.: Urban space environmental conditions in hot-desert climate sprawl cities. In: II Congreso Internacional ISUF-H Zaragoza, pp. 117–128. Prensas de la Universidad de Zaragoza, Zaragoza (2018)
12. LEMA, Estación Meteorológica. https://lema-arq.unison.mx/. Accessed 15 March 2019
13. Servicio Meteorológico Nacional. https://smn.conagua.gob.mx/es/. Accessed 2019/03/15
14. Oglyay, V.: Design with Climate. Bioclimatic Approach to Architectural Regionalism, 1st edn. Princeton University Press, New Jersey (1963)
15. Givoni, B.: Man, Climate and Architecture. Elsevier Science Ltd, Wisconsin (1969)
16. IMPLAN, Programa de Desarollo Metropolitano de Hermosillo. http://www.implanhermosillo.gob.mx/programas-2/. Accessed 15 March 2019
17. INEGI. https://www.inegi.org.mx/. Accessed 15 March 2019
18. Idris, Y., Mae, M.: Anti-insulation mitigation by altering the envelope layers' configuration. Energy Build. **141**, 186–204 (2017)

# Chapter 38
# Performance of Different PV Array Configurations Under Different Partial Shading Conditions

**Haider Ibrahim and Nader Anani**

**Abstract** In this paper, variations of the adverse effects of different partial shading patterns on the performance of a photovoltaic (PV) array with different array configurations is investigated using MATLAB/Simulink. The configurations considered were: Series-parallel (SP), total-cross-tied (TCT), bridged-linked (BL), and honey-comb (HC). Each configuration was investigated using a 4 × 4 PV array of KYOCERA KC200GT modules. The results show strong correlation between these effects and the configuration scheme of a PV array. It was evident that the TCT configuration outperformed all others schemes under various partial shading patterns, while the SP scheme exhibited the least performance. Quantitative results of the maximum power yield in each configuration were used as the prime indicator of performance.

## Nomenclature

| | |
|---|---|
| $G$ | Solar irradiance (W/m$^2$) |
| $G_{STC}$ | Solar irradiance at standard test conditions (1000 W/m$^2$) |
| $I$ | Output current of a module (A) |
| $I_{sc,STC}$ | Short circuit current (A) at standard test conditions |
| $k$ | Boltzmann constant ($1.381 \times 10^{-23}$ J/K) |
| $K_v$ | Thermal coefficient of the open-circuit voltage (V/°C) |
| $K_i$ | Thermal coefficient of the short circuit current (A/°C) |
| $N_s$ | Number of the series-connected cells per module |
| $q$ | Electronic charge ($-1.602 \times 10^{-19}$ C) |
| $T$ | Temperature of the PV cell (K) |

H. Ibrahim
Department of Engineering and Design, University of Chichester, Bognor Regis Campus, Tech Park, Bognor Regis PO21 1HR, UK

N. Anani (✉)
Faculty of Science and Engineering, University of Wolverhampton, Telford Innovation Campus, Shifnal Road, Priorslee, Telford, TF2 9NT Shropshire, UK
e-mail: n.anani@wlv.ac.uk

 
J. Littlewood et al. (eds.), *Sustainability in Energy and Buildings*, Smart Innovation, Systems and Technologies 163, https://doi.org/10.1007/978-981-32-9868-2_38

$T_{STC}$ Temperature of a module at standard test conditions (25 °C or 298.15 K)
$V$ Output voltage of PV module (V)
$V_{oc,STC}$ Open-circuit voltage of the PV module at standard test conditions (V)

## 38.1 Introduction

At the heart of any PV power generation system, is the PV array. This consists of several interconnected PV modules/panels to furnish the required terminal voltage and current. Each PV module consists of several PV cells connected in series. The terminal IV characteristic of a PV array/module, under constant radiation and temperature, resembles that of a constant current source as shown in Fig. 38.1, which also includes the power-voltage (PV) curve. Clearly, to extract maximum power from a PV array, the array must be operated at its maximum power point (MPP). However, this point changes with temperature and incident radiation [1]. Consequently, a PV system uses a maximum power point tracker (MPPT) whose function is to track the MPP. The problem, of variations of the MPP, is further compounded since it is, in practice, difficult to ensure that all PV panels, making up a PV array, are subjected to the same level of incident radiation; an adverse condition known as partial shading (PS). This condition gives rise to a PV curve with multiple peaks, Fig. 38.2. Multiplicity of power peaks places onerous demands on the design of the MPPT due to the presence of local maximum points in addition to the global maximum power point [2]. Therefore, it is imperative that PS effects are fully understood prior to the design of any MPPT system. Further, the adverse effects of partial shading have been found to vary, not only with partial shading patterns, but also with the method individual modules within an array are interconnected. In this paper, the variation in the performance of a PV array with different configurations and different partial shading patters is investigated using an array of 16 KYOCERA KC200GT PV panels.

In general, partial shading reduces the total energy yield of a PV system and increases power loss [3]. Since in a PV module, all cells are connected in series, they all share the same current including those shaded cells. In a shaded cell, the module's

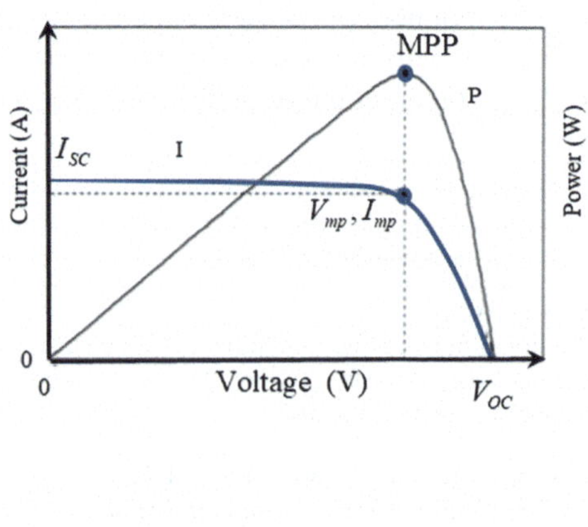


**Fig. 38.1** I-V and PV characteristics of a PV array

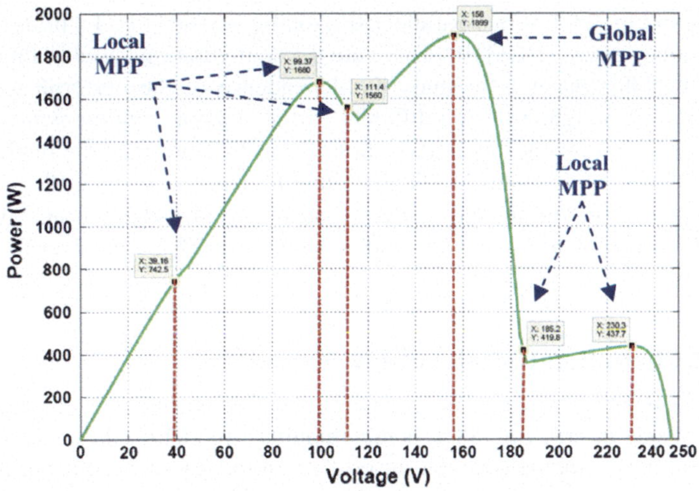


**Fig. 38.2** PV characteristics of a PV array under partial shading conditions

current will pass though the cell's parallel resistance causing a reverse bias voltage across its diode. This condition results in power loss and a hotspot that can damage the PV module [4]. This is often resolved using a bypass diode across, ideally, each cell but in practice, only certain number of cells or modules is bypassed. A bypass diodes provides an alternative current path for shaded part of array/module which leads to multiple maximum power points depending on the number of conducting and non-conducting bypass diodes [5, 6].

Another way to mitigate the effects of partial shading is to vary the configuration of the PV array depending on the existing shading pattern [7–9]. Figure 38.3 shows different PV array configurations schemes which have been reported in the literature

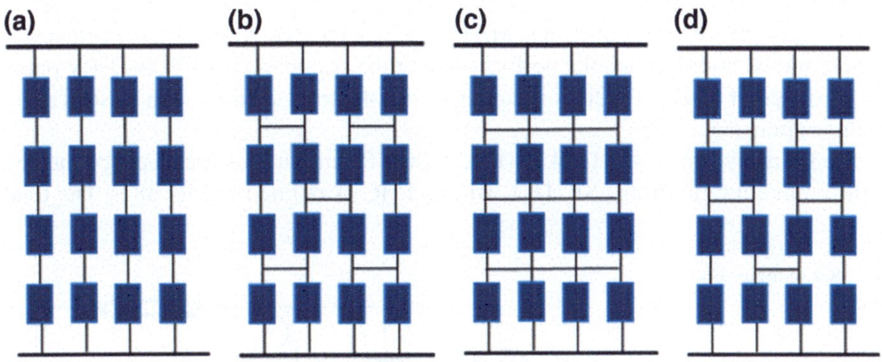


**Fig. 38.3** Different PV array configurations, **a** SP, **b** TCT, **c** BL, **d** HC

and defined in Fig. 38.3; series-parallel (SP), total-cross-tied (TCT), bridged-linked (BL), and honey-comb (HC) [10]. These are the configurations used in this work.

The aim of this paper is to study the dependence of the performance of different PV array configurations on different partial shading conditions using MATLAB/Simulink. Following this introduction, Sect. 38.2 presents the mathematical model of a PV module using MATLAB/Simulink and the various PV array configurations and different shading patters used in the investigation are explained. Section 38.3 presents the simulation results, while Sect. 38.4 concludes this work.

## 38.2 Mathematical Model of PV Module

The single-diode model, Fig. 38.4, is the ubiquitous model used to assess the performance of a PV module/array because of its simplicity and sufficient accuracy. The five parameters of the model are the value of the photocurrent $I_{ph}$, the saturation current and ideality factor of the diode $I_{sat}$ and $n$ respectively, and the series $R_s$ and shunt resistance $R_{sh}$ [11, 12]. The general implicit equation of the module current is

$$I = I_{ph} - I_{sat}\left[\exp\left(\frac{V + IR_s}{nN_s V_{th}}\right) - 1\right] - \frac{V + IR_s}{R_{sh}}, \quad V_{th} = \frac{kT}{q} \tag{38.1}$$

The photo and saturation currents depend on the radiation and temperature and may be expressed as [13]

$$I_{ph} = \frac{G}{G_{STC}}\left(I_{ph,STC} + K_i(T - T_{STC})\right) \tag{38.2}$$

$$I_{sat} = I_{ph,STC} + K_i(T - T_{STC})/\exp\left(\frac{V_{OC,STC} + k_v(T - T_{STC})}{nN_S V_{th}}\right) \tag{38.3}$$

Using the above three equations, MATLAB/SIMULINK software has been developed to model and simulate the performance of the commercial KC200GT PV panel [14] at different levels of radiations and patterns of partial shading with various array configurations.

The PV array consists of 16 (4 × 4) KC200GT PV modules, which were connected in different configurations; SP, TCT, BL, and HC as defined in Fig. 38.3. The total

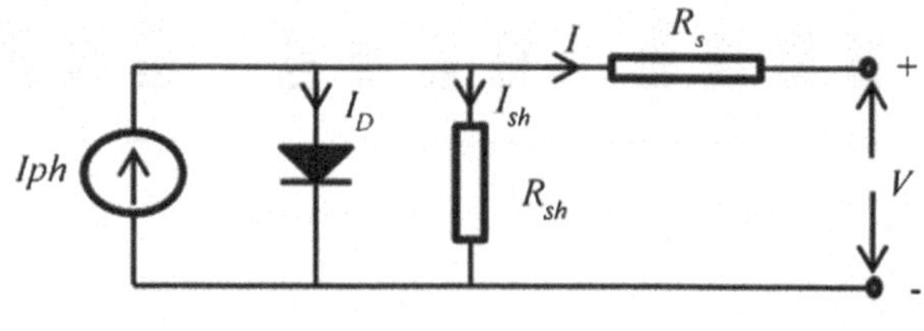


**Fig. 38.4** Single-diode model

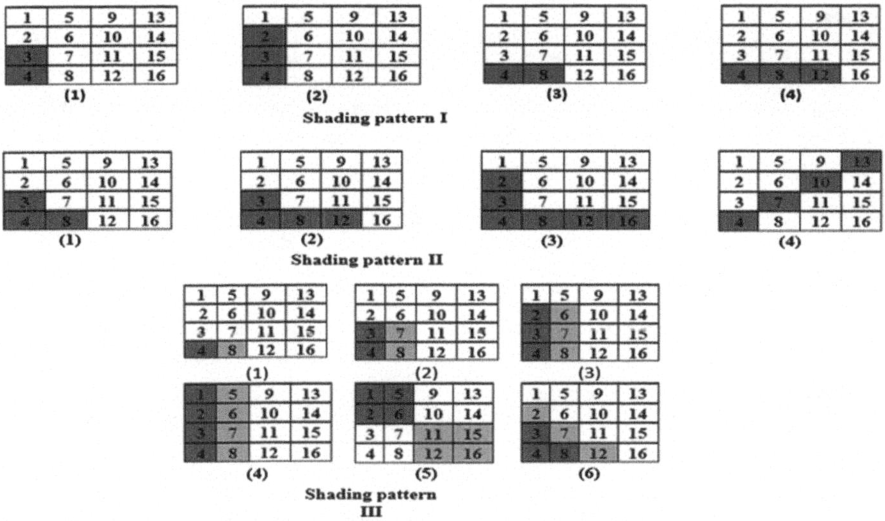


**Fig. 38.5** Different shading patterns: White (1000 W/m$^2$), Gray (600 W/m$^2$), Dark (200 W/m$^2$)

maximum power rating of the array is 3200 W (STC). The simulation was run to investigate the performance of different PV array configurations under three different shading patterns, referred to as Pattern I, II and III as defined in Fig. 38.5.

## 38.3 Simulation Results

Figure 38.6 shows the variations of the PV array characteristics with different configurations under partial shading pattern I using the four different cases (1–4) defined in Fig. 38.5. It is clear that for cases (1) and (2), the TCT has the best performance, while for case (3) and (4), the SP topology outperformed all others. Table 38.1 shows the results of maximum power, maximum voltage, and maximum current for partial shading pattern I, indicating that power losses can be reduced for case (1) and (2) with TCT topology while for case (3) and (4) the SP configuration provides the optimum performance.

Figure 38.7 shows the variations of the PV array characteristics with different PV configurations under partial shading pattern II using the four different cases defined in Fig. 38.5.

Table 38.2 shows that TCT topology produces maximum power in all four cases, therefore, power losses for this topology is the lowest. For case (1), (2), and (3), the HC topology exhibit better performance when compared with the SP topology, while in the case (4), the BL topology has similar performance to the TCT scheme. Figure 38.8 shows the simulation results for standard irradiance (1000 W/m$^2$) with two typical levels of partial shading (600 W/m$^2$) and (200 W/m$^2$). Table 38.3 shows

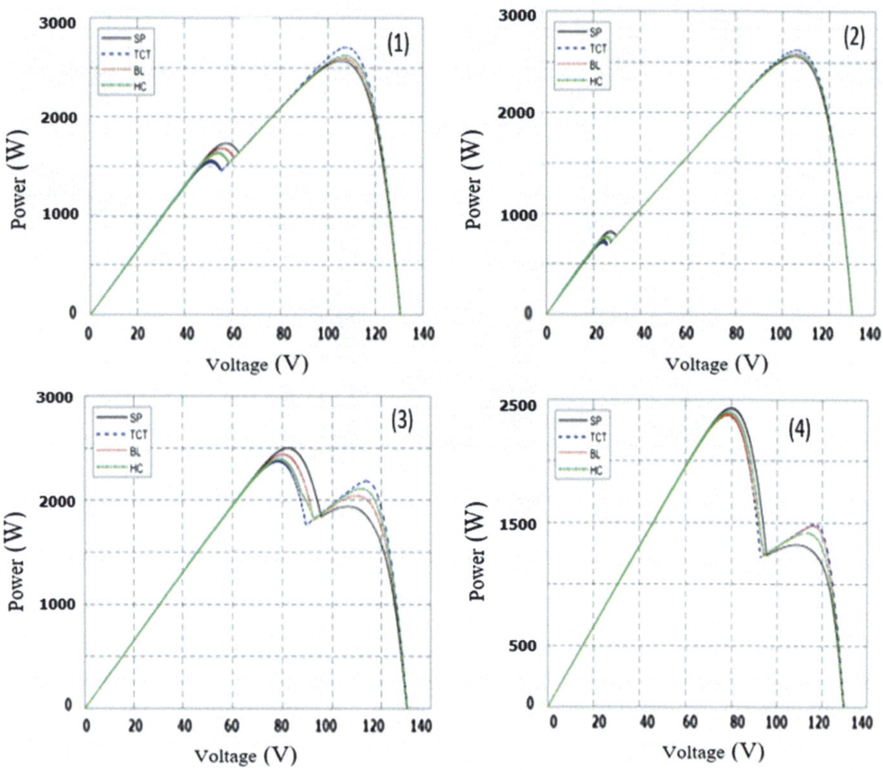


**Fig. 38.6** Variation of PV characteristics of the PV module under shading pattern I

**Table 38.1** Simulation results of shading pattern I

| Shading pattern I | Configuration | Pmax (kW) | Vmax (V) | Imax (A) |
|---|---|---|---|---|
| Case 1 | SP | 2.5684 | 105.5439 | 24.3349 |
| | TCT | **2.7039** | **107.7831** | **25.0865** |
| | BL | 2.5942 | 106.3367 | 24.3961 |
| | HC | 2.6198 | 107.2568 | 24.4255 |
| Case2 | SP | 2.5598 | 105.1516 | 24.3439 |
| | TCT | **2.6183** | **106.0719** | **24.6842** |
| | BL | 2.5844 | 105.8070 | 24.4256 |
| | HC | 2.5855 | 105.8074 | 24.4359 |
| Case 3 | SP | **2.5019** | **82.3202** | **30.3923** |
| | TCT | 2.3823 | 78.1074 | 30.5003 |
| | BL | 2.4435 | 80.3461 | 30.4122 |
| | HC | 2.4004 | 78.7641 | 30.4758 |
| Case 4 | SP | **2.4314** | **79.8201** | **30.4610** |
| | TCT | 2.3823 | 78.1087 | 30.4998 |
| | BL | 2.3821 | 78.1063 | 30.4982 |
| | HC | 2.4004 | 78.7659 | 30.4751 |

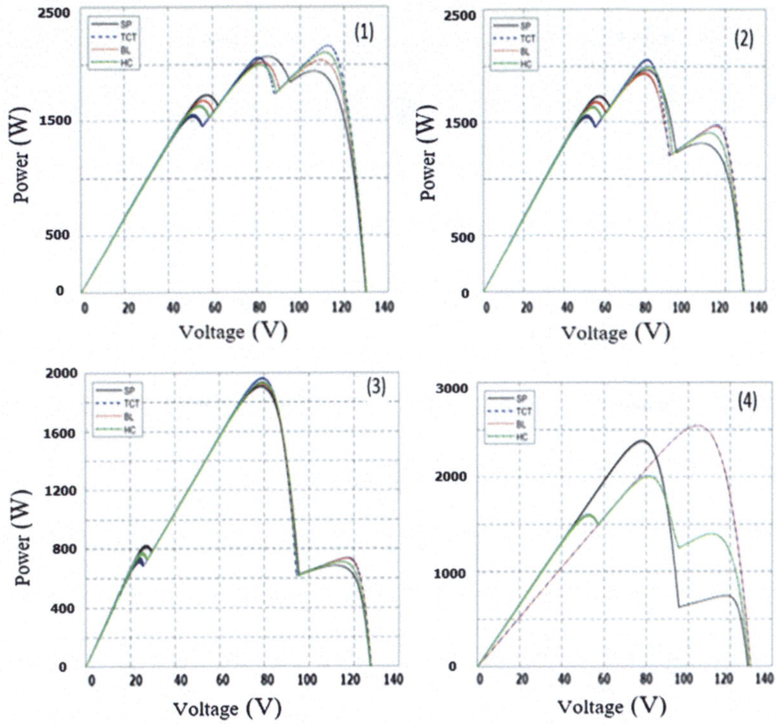


**Fig. 38.7** Variation of PV characteristics of the PV module under shading pattern II

**Table 38.2** Simulation results of shading pattern II

| Shading pattern II | Configuration | Pmax (kW) | Vmax (V) | Imax (A) |
|---|---|---|---|---|
| Case 1 | SP | 2.0688 | 85.3434 | 24.2409 |
| | TCT | **2.1620** | **112.7816** | **19.1698** |
| | BL | 2.0347 | 109.3601 | 18.6055 |
| | HC | 2.1009 | 111.2023 | 18.8926 |
| Case2 | SP | 1.9692 | 80.8720 | 24.3496 |
| | TCT | **2.0611** | **81.1339** | **25.4037** |
| | BL | 1.9437 | 79.4222 | 24.4730 |
| | HC | 1.9993 | 81.7924 | 24.4436 |
| Case 3 | SP | 1.9127 | 78.5023 | 24.3649 |
| | TCT | **1.9661** | **79.1585** | **24.8375** |
| | BL | 1.9332 | 78.8971 | 24.5028 |
| | HC | 1.9391 | 79.1595 | 24.4961 |
| Case 4 | SP | 2.3828 | 78.1085 | 30.5063 |
| | TCT | **2.5423** | **104.6257** | **24.2990** |
| | BL | **2.5423** | **104.6261** | **24.2989** |
| | HC | 2.0131 | 81.1358 | 24.8115 |

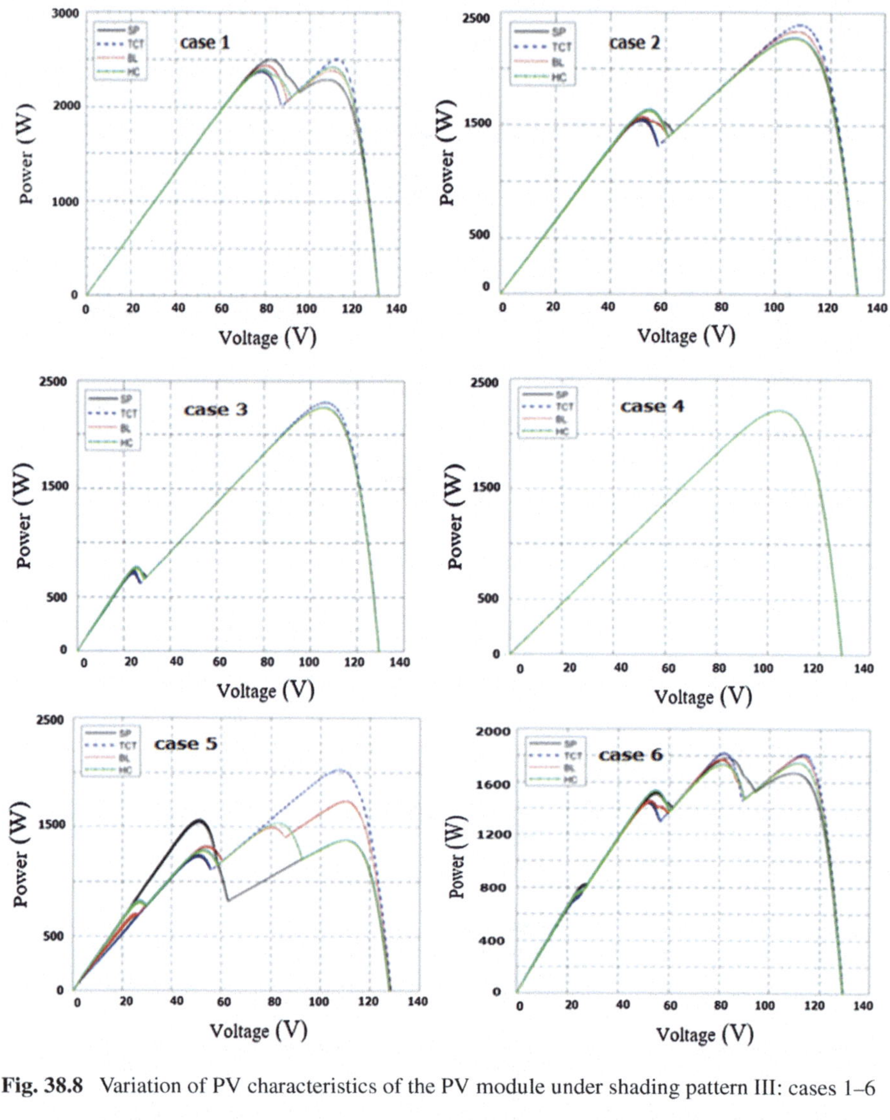


**Fig. 38.8** Variation of PV characteristics of the PV module under shading pattern III: cases 1–6

that the TCT configuration gives maximum power for all cases of partial shading patterns III. This indicates that it has the lowest losses among all other topologies. For case (1) the SP configuration exhibits good performance, however, it deteriorates in other cases expect case (3), in which all topologies have the same maximum power with this shading pattern. Cases (2), (3), (5), (6) show that BL configuration ranks second in its performance compared to the TCT.

**Table 38.3** Simulation results of shading pattern III

| Shading pattern III | Configuration | Pmax (kW) | Vmax (V) | Imax (A) |
|---|---|---|---|---|
| Case 1 | SP | 2.5019 | 82.3189 | 30.3928 |
| | TCT | **2.5112** | **112.2575** | **22.3700** |
| | BL | 2.4435 | 80.3445 | 30.4128 |
| | HC | 2.4278 | 110.1538 | 22.0401 |
| Case2 | SP | 2.2812 | 106.8603 | 21.3475 |
| | TCT | **2.3967** | **108.9657** | **21.9950** |
| | BL | 2.3420 | 107.4995 | 21.7557 |
| | HC | 2.2772 | 106.8596 | 21.3102 |
| Case 3 | SP | 2.2543 | 105.5443 | 21.3588 |
| | TCT | **2.3015** | **106.4666** | **21.6171** |
| | BL | 2.2523 | 105.6783 | 21.3128 |
| | HC | 2.2547 | 105.6778 | 21.3356 |
| Case 4 | SP | 2.2165 | 104.2269 | 21.2661 |
| | TCT | 2.2165 | 104.2269 | 21.2661 |
| | BL | 2.2165 | 104.2274 | 21.2660 |
| | HC | 2.2165 | 104.2274 | 21.2660 |
| Case 5 | SP | 1.5665 | 51.1285 | 30.6385 |
| | TCT | **2.0260** | **107.6532** | **18.8197** |
| | BL | 1.7424 | 110.2820 | 15.7995 |
| | HC | 1.5380 | 82.9732 | 18.5361 |
| Case 6 | SP | 1.7834 | 83.2373 | 21.4255 |
| | TCT | **1.8207** | **81.7910** | **22.2604** |
| | BL | 1.7975 | 112.6493 | 15.9566 |
| | HC | 1.7479 | 111.2066 | 15.7176 |

## 38.4 Conclusion

The paper presented an investigation of the performance of a PV array and its dependence not only on various partial shading patters, but also on the configuration of the modules within the array. Different array configurations were investigated; SP, TCT, BL, and HC using three different partial shading patterns. In general, the TCT configurations outperformed all others in most cases of varying shading patters. Although, the SP configuration is the most commonly used topology, its performance under most shading patterns, is inferior to other configurations. For shading pattern I, the TCT configuration has the highest maximum power when the shading progresses through one column of the PV array, while the SP array produce highest maximum power when the shading progresses through one row of the PV array. For the shading pattern II, the TCT has the best performance, while HC ranks second except for case (4) where BL array shows better performance than HC. For shading pattern III, the TCT array configuration has the best performance for all cases. Therefore, in practice it is necessary to vary the configuration scheme of a PV array as the pattern of shading varies.

## References

1. Burns, P., Anani, N.: Modelling and simulation of photovoltaic arrays under varying conditions. In: 2014 9th International Symposium on Communication Systems, Networks & Digital Sign (CSNDSP), Manchester (2014)
2. Balasubramanian, I.R., Gansen, S.I., Chilakapati, N.: Impact of partial shading on the output power of PV systems under partial shading condtions. IET Power Electronics **7**, 657–666 (2014)
3. Moballegh, S., Jiang, J.J.: Modeling, prediction, and experimental validation of power peaks of PV arrays under partial shading condtions. IEEE Trans. Sustain. Energy **5**(1), 293–300 (2014)
4. Silvestre, S., Boronat, A., Chouder, A.: Study of bypass diodes configuration on PV modules. Appl. Energy **86**(9), 1632–1640 (2009)
5. Yang, H., Wang, H., and Wang, M.: Investigation of the Relationship between Reverse Current of Crystalline Silicon Solar Cells and Conduction of Bypass Diode. Hindawi Int. J. Photoenergy **2012** (2012)
6. Wang, Y.J., Lin, S.S.: Analysis of a partially shaded PV array considering different module connection schemes and effects of bypass diodes. In: International Conference & Utility Exhibition on Power and Energy Systems, Pattaya City, Thailand (2011)
7. Picault, D., Raison, B., Bacha, S., De La Casa, J., Aguilera, J.: Forecasting photovoltaic array power production subject to mismatch losses. Solar Energy **84**(7), 1301–1309 (2010)
8. Gao, L., Dougal, R.A., Liu, S., Iotova, A.P.: Parallel—Connectd solar PV system to address partial and rapidly fluctuating shadow conditions. IEEE Trans. Indus. Elect. **56**(5), 148–1556 (2009)
9. Ramos-Paj, C., Bastida, J., Saavedra-Monte, A., Guinjoan-Gispert, F., Goez, M.: Mathematica model of total cross tied PV arrays in mismatching conditions. In: 4th Colombian Workshop on Circuits and Systems, Barranquilla (2012)
10. Humada, A.M., Herwan Bin Sulaiman, M., Hojabri, M., Hamada, H.M., Ahmed, M.N.: A Review on Photovoltaic Array Behavior, Configuration Strategies and Models Under Mismatch Conditions, vol. 11, pp. 4896–4903 (2016)
11. Ibrahim, H., Anani, N.: Evaluation of analytical methods for parameter extraction of PV modules. Energy Procedia **134**, 69–78 (2017)
12. Villalva, M.G., Gazoli, J.R., Filho, E.R.: omprehensive approach to modeling and simulation of photovoltaic arrays. IEEE Trans Power Electronics **24**(5), 1198–1208 (2009)
13. Ibrahim, H., Anani, N.: Variations of PV module parameters with irradiance and temperature. Energy Procedia **134**, 276–285 (2017)
14. KC200GT.: KYOCERA Corporation. Available: https://www.kyocerasolar.com/dealers/product-center/archives/spec-sheets/KC200GT.pdf. Accessed 10 Mar 2019

# Chapter 39
# Automatic Thresholding for Sensor Data Gap Detection Using Statistical Approach

**Houda Najeh, Mahendra Pratap Singh, Stéphane Ploix, Karim Chabir and Mohamed Naceur Abdelkrim**

**Abstract** Sensors are currently being used for applications in buildings. In sensor grids, a significant amount of sensor data may be lost. This paper tackles the issue of unreliable sensors in buildings. The common sensor faults known in the literature are bias and outliers. Occurrences of data gaps have not been given adequate attention in the research literature. A methodology based on statistical approach allows the automatic thresholding for data gap detection, i.e., abnormalities on the delay for a set of heterogeneous sensors in instrumented buildings. The efficiency of the method is evaluated on measurements obtained from a real building: an office at Grenoble Institute of technology with a large number of sensors.

## 39.1 Introduction

With the massive arrival of inexpensive communicating sensors, the building sector is undergoing an unprecedented revolution: the building is becoming smart, which means it offers new services to occupants related to safety, energy management, and comfort. Article 23 of the thermal regulation RT2012 in France requires the measurement of certain variables, which promotes the deployment of sensor grids in new buildings. In addition, research projects such as the ANR OMEGA show the interest of companies to guarantee overall performances, i.e., total real consumption, interior comfort, etc., after refurbishment. In addition to the different aspects of comfort and energy consumption, these sensors can also be used to estimate occupant practices essential for energy consumption by estimating the number of occupants per area and their metabolic contribution, their activities, and their routines. The concept of healthy sensors is explained in the literature [1–3]. Authors in [4] assumed that there

---

H. Najeh (✉) · S. Ploix
G-SCOP Lab, CNRS, UMR 5272, University of Grenoble Alpes, Grenoble, France
e-mail: houda.najeh@grenoble-inp.fr

H. Najeh · K. Chabir · M. N. Abdelkrim
MACS Lab, University of Gabes, LR16ES22 Gabes, Tunisia

M. P. Singh
Maersk Mc-Kinney Moller Institute, University of Southern, Odense, Denmark



J. Littlewood et al. (eds.), *Sustainability in Energy and Buildings*, Smart Innovation, Systems and Technologies 163, https://doi.org/10.1007/978-981-32-9868-2_39

are two groups of sensors: first that correctly measure structural responses (termed as «reference sensors») and second are failed (or uncertain) sensors. Affordable sensor grids in buildings are particularly sensitive to battery exhaustion, hardware, and other communications related failures [5]. As a result, data sets collected from these sensors often miss some records [6–8].

In the literature, many techniques are available for data analysis such as time series, statistical approaches, clustering methods, and many more [9–12]. However, all of them are interested only in bias [13] and outliers [14] fault types. The occurrence of data gaps has not been given interest from building researchers. Hence, there is a need for techniques to detect them, i.e., a need to know the periods of good operation of sensors. The possibility of time-delayed data becomes a reality. It means while after preprocessing the sensors report values regularly, reality shows that quite many values are missing [15–17]. Sensor measurements could have temporal and frequential representations [18]. In this paper, only the temporal representation is considered and sensor measurements are considered to be statistical time series. When the sensors are in ok states, the delay data have a distribution corresponding to the normal mode of operation and these distributions change when the sensor is faulty. In general, there is no regularly delayed data for a variable. Delays depend not only on the type of sensor but also on the measured values. The question that arises is from which delay can we say that the sensor becomes faulty? Hence, the automatic thresholding for data gap detection for heterogeneous sensors is a feasible paradigm for the instrumented residential environment. The objective of this paper is to solve the issue of unreliable instrumentation in buildings. It focuses on developing a method for automatic thresholding for data gap detection for heterogeneous sensors in instrumented buildings. An algorithm based on statistical approaches has been adapted to an office setting at Grenoble Institute of Technology which is equipped with 30 sensors.

This paper is organized as follow: Sect. 39.2 presents the problem statement and the proposed methodology. Section 39.3 presents the automatic thresholding with probability density function and Sect. 39.4 analyses the results for the office H358. Section 39.5 discusses the findings of the literature review, the proposed methodology, and its limits and the results. Finally, concluding remarks and future works are given in Sect. 39.6.

## 39.2 Problem Statement and Research Objective

Let us consider the evolution of raw sensor measurements for two different sensors: a door contact sensor and a $CO_2$ concentration sensor (see Fig. 39.1).

Many values are missing. Sensor values are not necessarily uniformly sampled. There are no regularly delayed data for a variable. Delays depend not only on the type of sensor but also on the measured values. The question that arises is from which delay can we say that the sensor becomes faulty? Hence, the necessity for auto-

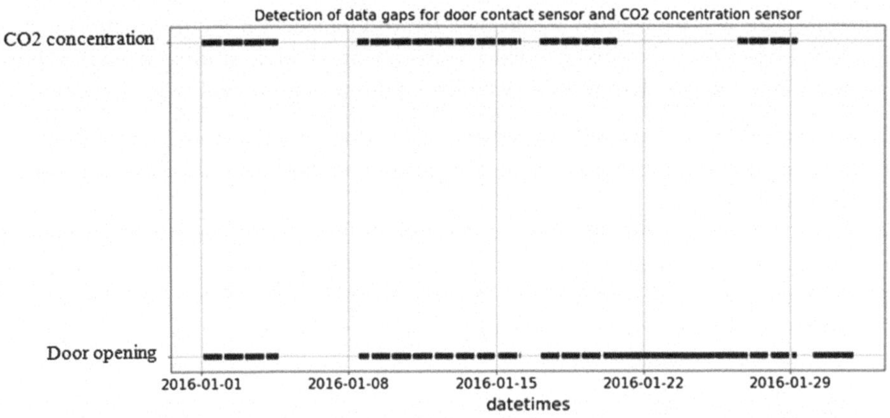


**Fig. 39.1** Time of recordings for door contact sensor and $CO_2$ concentrations sensor

matic thresholding for data gap detection for heterogeneous sensors in instrumented buildings.

**Definition 1: Data gap** Data gap means an abnormal change in the data delays sending by a sensor.

Let T be a sequence of times $T = t_0, t_1, \ldots, t_{n-1}$ with $\forall k, t_{k+1} > t_k$. The problem is to automatically detect abnormal delays defined by the following equation:

$$\Delta_k = t_{k+1} - t_k. \tag{39.1}$$

Let us define the sequence of delays:

$$\Delta = \{\forall k, \Delta_k = t_k - t_{k-1}, k \in (1, \ldots, n-1)\}. \tag{39.2}$$

Assuming that $\Delta$ is a stochastic variable satisfying a probability distribution function $\delta$ to be characterized by suitable algorithms.

A probability density function defined by (39.3) is most commonly associated with absolutely continuous univariate distributions. A random variable X has a density $f_X$, where $f_X$ is a nonnegative Lebesgue integrable function, if

$$P(a \le x \le b) = \int_a^b f_X \, dx. \tag{39.3}$$

A given time delay $D_k$ is abnormal if $D_k > d$ where d is a threshold value satisfying (39.4)

$$P(\Delta \ge \delta) \le \rho_{\%}, \tag{39.4}$$

where $\rho_{\%}$ is a low probability value like 1% of chance of a $\Delta_k$ to greater than δ. Equation (39.4) has to be solved to get $\delta = \delta(\rho_{\%})$, parameters coming from the characterization of the pdf. To solve this problem, a three-step methodology is proposed:

Step 1: The delay is considered as a continuous random process modeled by a stochastic variable. The first step is to identify the appropriate laws modeling the time series related to delays.
Step 2: Group by class the data in order to trace the actual observed distribution as a histogram. From the shape of the histogram, determine a distribution law that seems to describe the actual distribution. Then determine the parameters of this law.
Step 3: Solve the following equation to get the threshold th:

$$\int_0^{th} f_X dx = 0.99. \tag{39.5}$$

## 39.3 Automatic Thresholding with Probability Density Function

This section demonstrates the proposed methodology for automatic thresholding for data gap detection for a set of sensors.

This first step of the proposed methodology discussed what pdf to use with data analysis. The delay is a continuous variable. It can correspond to one of the continuous laws below (see Table 39.1)

According to the principles of each law of probability, it is concluded that the truncated normal law and the gamma law are adequate for the delay.

The second step of the proposed methodology is to compare the histogram representations of each of these laws. Figures 39.2 and 39.3 represent the histogram representation of the time delay and the pdf curve for the truncated normal and gamma distributions, respectively.

From these two figures, it is concluded that the truncated normal law is the most adequate for the time series of the delay. Indeed, the truncated normal law makes it possible to detect three abnormal delays (at epochtime = 139,880, epochtime = 288,171, epochtime = 596,927) while the gamma law makes it possible to detect only two abnormal delays (at epochtime = 139,880, epochtime = 288,171, epochtime = 596,927). So, it is concluded that the truncated normal law is the most adequate for the time series of the delay. In fact, there is an abnormal delay at epochtime = 596,927 which is detected by the truncated normal distribution and not detected by the distribution gamma.

**Table 39.1** Probability laws

| Law probability name | Adequacy for the delay | Support |
|---|---|---|
| Uniform, Normal, Student | No (because the domain of delay is $\mathbb{R}$) | $X \in [a, b]$, $-\infty < a < b < +\infty$ |
| Truncated normal, Gamma | Yes (because the domain of delay is $\mathbb{R}^+$) | $X \in [0, +\infty[$ |
| Beta | No (because the domain of delay [0, 1] and it's not meaningful to normalize time delays in case of failure) | $X \in [0, 1]$ |
| Exponential | No (because the order of magnitude of delay in ms is very high between 1,456,786,822,423 and 1,458,082,719,570 in the month of March 2016, the PDF is $e^{-\lambda x}$ which tends to 0) | $X \in [0, +\infty[$ |

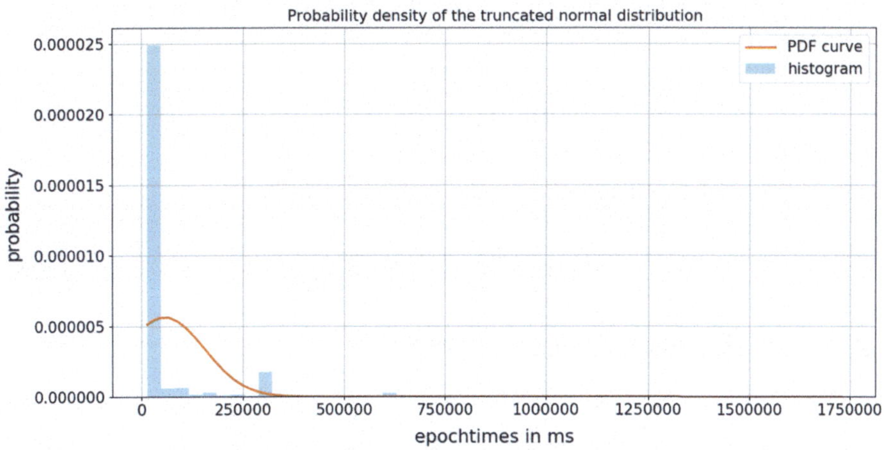


**Fig. 39.2** Fitted probability density of the truncated normal distribution

**The third step of the proposed methodology** is the calculation of thresholds for each time series related to a sensor. In probability and statistics, the truncated normal distribution is the probability distribution derived from the normal distribution.

Suppose $X \rightarrow N(\mu, \sigma^2)$ has a normal distribution and lies within the interval $X \in (a, b)$, $-\infty < a < b < +\infty$). Then X conditional on $a < X < b$ has a truncated normal distribution defined by (39.6).

$$f(x, a, b, \mu, \sigma) = \frac{\phi\left(\frac{x-\mu}{\sigma}\right)}{\sigma\left(\Phi\left(\frac{b-\mu}{\sigma}\right) - \Phi\left(\frac{a-\mu}{\sigma}\right)\right)}, \tag{39.6}$$

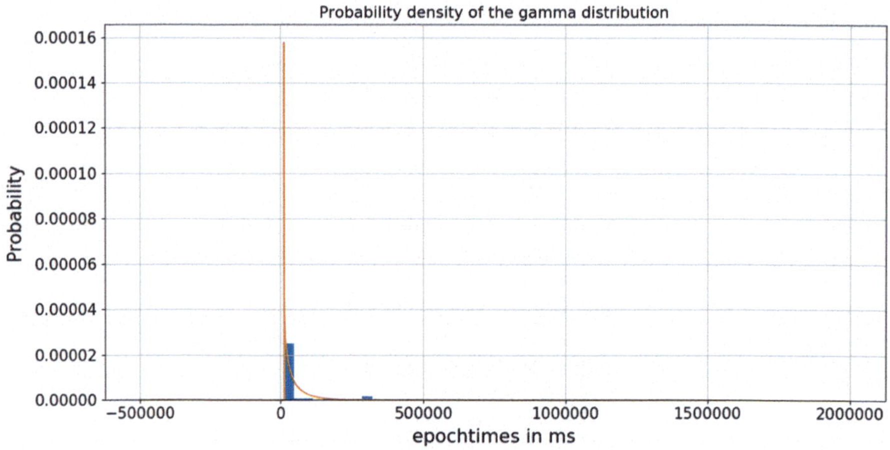


**Fig. 39.3** Fitted probability density of the gamma distribution

where $\phi$ is the probability density function of the standard normal distribution defined by (39.7)

$$\phi(\xi) = \frac{1}{\sqrt{2\pi}} e^{-\frac{1}{2}\xi^2} \tag{39.7}$$

and $\Phi$ is its cumulative distribution function defined by (39.8)

$$\Phi(x) = \frac{1}{2}\left(1 + \mathrm{erf}\left(\frac{x}{\sqrt{2}}\right)\right) \tag{39.8}$$

with erf is the error function defined by (39.9):

$$\mathrm{erf}(x) = \frac{1}{\sqrt{\pi}} \int_{-x}^{x} e^{-t^2} dt = \frac{2}{\sqrt{\pi}} \int_{0}^{x} e^{-t^2} dt \tag{39.9}$$

The delay is abnormal if there is a probability of, for instance, 1% of a delay greater than the threshold. The following equation has to be solved:

$$\int_{0}^{\mathrm{th}} f_X dx = 0.99. \tag{39.10}$$

In general, if the function f depends on a single parameter, so $f(x) = \int_0^x f'(t)dt$, where f′ is the derivative of x.

In our case, f depends on several parameters (x, $\mu$, $\sigma$, a and b). So, the derivative of $f$ with respect to x is $f_X$. Then, $f(x) = \int_0^{\mathrm{th}} f_X dx = 0.99$.

So, $f_X$ is the derivative of $f(x, a, b, \mu, \sigma)$ with respect to x and f is the primitive. The resolution of (39.10) yields

$$\mathrm{f(th)} - \mathrm{f(0)} = \frac{\phi\left(\frac{\mathrm{th}-\mu}{\sigma}\right)}{\sigma\left(\Phi\left(\frac{b-\mu}{\sigma}\right) - \Phi\left(\frac{a-\mu}{\sigma}\right)\right)} - \mathrm{f(0)} = 0.99; \quad \frac{\phi\left(\frac{th-\mu}{\sigma}\right)}{\sigma\left(\Phi\left(\frac{b-\mu}{\sigma}\right) - \Phi\left(\frac{a-\mu}{\sigma}\right)\right)} = f(0) + 0.99 \tag{39.11}$$

It becomes

$$\phi\left(\frac{\mathrm{th} - \mu}{\sigma}\right) = \sigma\left(\Phi\left(\frac{b - \mu}{\sigma}\right) - \Phi\left(\frac{a - \mu}{\sigma}\right)\right)(\mathrm{f(0)} + 0.99) \tag{39.12}$$

We apply the Ln

$$\frac{1}{\sqrt{2\pi}}e^{\left(\frac{-1}{2}\left(\frac{th-\mu}{\sigma}\right)^2\right)} = \ln(f(0) + 0.99) + \ln\left(\sigma\left(\Phi\left(\frac{b - \mu}{\sigma}\right) - \Phi\left(\frac{a - \mu}{\sigma}\right)\right)\right) \tag{39.13}$$

It yields

$$\mathrm{Ln}\left(\frac{1}{\sqrt{2\pi}}\right) - \frac{1}{2}\left(\frac{\mathrm{th} - \mu}{\sigma}\right)^2 = \ln(\mathrm{f(0)} + 0.99) + \ln\left(\sigma\left(\Phi\left(\frac{b - \mu}{\sigma}\right) - \Phi\left(\frac{a - \mu}{\sigma}\right)\right)\right) \tag{39.14}$$

It yields

$$\left(\frac{\mathrm{th} - \mu}{\sigma}\right)^2 = -2\mathrm{Ln}(\mathrm{f(0)} + 0.99) - 2\mathrm{Ln}\left(\frac{\Phi\left(\frac{b-\mu}{\sigma}\right) - \Phi\left(\frac{a-\mu}{\sigma}\right)}{\frac{1}{\sigma}}\right) + 2\mathrm{Ln}\left(\frac{1}{\sqrt{2\pi}}\right) \tag{39.15}$$

Then, the threshold is defined by

$$\mathrm{th} = \mu + \sigma \pm \left(\sqrt{-2\mathrm{Ln}(\mathrm{f(0)} + 0.99) - 2\mathrm{Ln}\left(\frac{\Phi\left(\frac{b-\mu}{\sigma}\right) - \Phi\left(\frac{a-\mu}{\sigma}\right)}{\frac{1}{\sigma}}\right) + 2\mathrm{Ln}\left(\frac{1}{\sqrt{2\pi}}\right)}\right) \tag{39.16}$$

where

$$\Phi\left(\frac{b - \mu}{\sigma}\right) = \frac{1}{2}\left(1 + \mathrm{erf}\left(\frac{\frac{b-\mu}{\sigma}}{\sqrt{2}}\right)\right) = \frac{1}{2}\left(1 + \mathrm{erf}\left(\frac{b - \mu}{\sqrt{2}\sigma}\right)\right); \Phi\left(\frac{a - \mu}{\sigma}\right) = \frac{1}{2}\left(1 + \mathrm{erf}\left(\frac{a - \mu}{\sqrt{2}\sigma}\right)\right);$$

$a = 0$ and $b = +\infty$

Equation (39.16) becomes

$$th = \mu + \sigma \pm \left( \sqrt{ -2Ln(f(0) + 0.99) - 2Ln\left( \frac{\frac{1}{2}\left(1 + erf\left(\frac{b-\mu}{\sqrt{2}\sigma}\right)\right) - \frac{1}{2}\left(1 + erf\left(\frac{a-\mu}{\sqrt{2}\sigma}\right)\right)}{\frac{1}{\sigma}} \right) + 2Ln\left(\frac{1}{\sqrt{2\pi}}\right) } \right) \tag{39.17}$$

with

$$f(0) = \frac{\phi\left(\frac{-\mu}{\sigma}\right)}{\sigma\left(\Phi\left(\frac{b-\mu}{\sigma}\right) - \Phi\left(\frac{a-\mu}{\sigma}\right)\right)} = \frac{\frac{1}{\sqrt{2\pi}} e^{\left(\frac{-1}{2}\left(\frac{\mu}{\sigma}\right)^2\right)}}{\sigma\left(\frac{1}{2}\left(1 + erf\left(\frac{b-\mu}{\sqrt{2}\sigma}\right)\right) - \frac{1}{2}\left(1 + erf\left(\frac{a-\mu}{\sqrt{2}\sigma}\right)\right)\right)} \tag{39.18}$$

In this section, a three-step methodology is used to find a threshold from which a delay is considered as abnormal. Delay is a continuous variable that follows the truncated normal law. The delay is calculated by solving Eq. (39.10).

In the following, we will test the calculation of this threshold on time series of delays for different variables

## 39.4 Results for H358 Office

### *39.4.1 Test Bed*

The example of study concerns an office H358 in Grenoble Institute of Technology. The setup for the sensor grids includes an ambiance sensing network, which measures temperature, $CO_2$ concentration, occupancy, power consumption, door and window positions, motions, and illuminance (see Fig. 39.4)

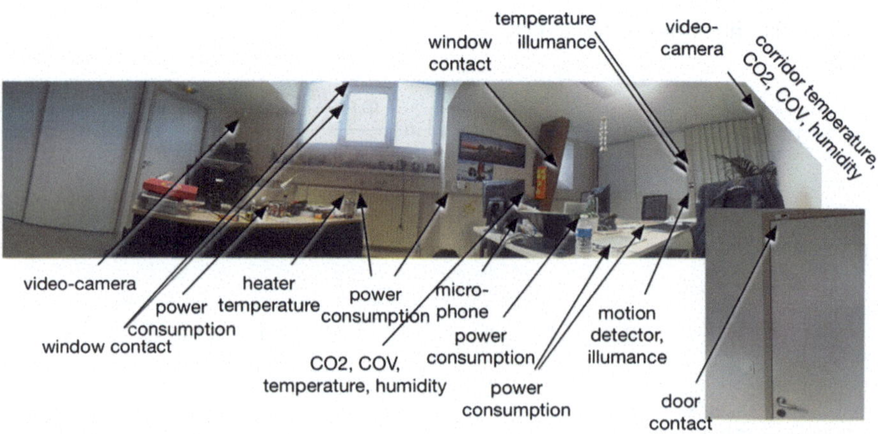


**Fig. 39.4** A Sensor test bed at Grenoble INP

### 39.4.2 Experimental Results

A data set covering 3 months from January 1, 2016 to March 31, 2016 has been used to detect the abnormalities on the delay and subsequently the data gaps from raw measurements of sensors. Figure 39.5 shows the detection of data gaps for the office $CO_2$ concentration sensor installed in the H358 office

The non-healthy periods for this sensor are the following:

[(2016, 1, 4, 14, 57, 5), (2016, 1, 8, 11, 42, 16)],
[(2016, 1, 16, 2, 21, 26), (2016, 1, 17, 2, 1, 28)],
[(2016, 1, 20, 16, 3, 19), (2016, 1, 26, 13, 13, 14)],
[(2016, 1, 29, 6, 39, 45), (2016, 2, 1, 15, 52, 1)],
[(2016, 2, 2, 11, 46, 55), (2016, 2, 23, 12, 22, 26)],
[(2016, 2, 24, 4, 22, 49), (2016, 2, 25, 11, 6, 24)],
[(2016, 3, 25, 18, 1, 6), (2016, 3, 29, 9, 33, 12)]

Figure 39.6 shows the detection of data gaps for six sensors (power sensor, illuminance sensor, door contact sensor, detected motions sensor, corridor $CO_2$ sensor, and temperature sensor) installed in the H358 office from January 1, 2016 to January 31, 2016.

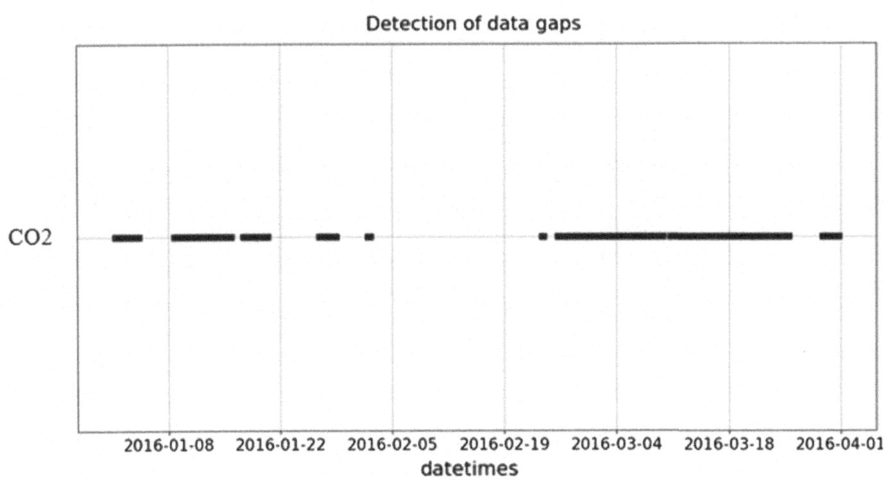


**Fig. 39.5** Detected data gaps for the $CO_2$ concentration sensor

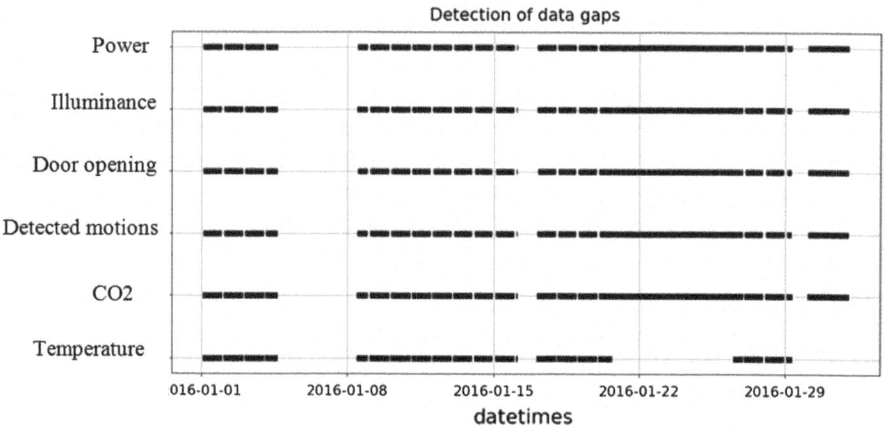


**Fig. 39.6** Detected data gaps for sensors in H358 office

The following table shows the threshold and the non-healthy periods, respectively, for each sensor (Table 39.2).

We conclude that the threshold depends on the measured variable.

**Table 39.2** Non-healthy periods for sensors in H358 office

| Measured variable | Intervals | Threshold (ms) |
|---|---|---|
| Door opening | [(2016, 1, 4, 14, 46, 33), (2016, 1, 8, 11, 59, 24)],<br>[(2016, 1, 16, 2, 10, 35), (2016, 1, 17, 2, 45, 28)] | 83,248,599.31 |
| Temperature | [(2016, 1, 4, 14, 57, 5), (2016, 1, 8, 11, 42, 16)],<br>[(2016, 1, 16, 2, 22, 56), (2016, 1, 17, 2, 1, 28)],<br>[(2016, 1, 20, 16, 3, 19), (2016, 1, 26, 13, 13, 14), | 38,101,174.48 |
| Corridor $CO_2$ | [(2016, 1, 4, 14, 57, 22), (2016, 1, 8, 11, 42, 33)],<br>[(2016, 1, 16, 2, 23), (2016, 1, 17, 2, 1, 12)],<br>[(2016, 1, 29, 6, 39, 51), (2016, 1, 30, 2, 1, 23)] | 19,695,541.07 |
| Detected motions | [(2016, 1, 4, 14, 57, 22), (2016, 1, 8, 11, 42, 33)],<br>[(2016, 1, 16, 2, 23), (2016, 1, 17, 2, 1, 12)],<br>[(2016, 1, 29, 6, 39, 51), (2016, 1, 30, 2, 1, 23)] | 61,472,466.89 |
| Power | [(2016, 1, 4, 14, 56, 43), (2016, 1, 8, 11, 48, 56)],<br>[(2016, 1, 16, 2, 11, 1), (2016, 1, 17, 2, 8, 58)],<br>[(2016, 1, 29, 6, 37, 8), (2016, 1, 30, 2, 4, 38)] | 35,154,742.37 |
| Illuminance | [(2016, 1, 4, 14, 52, 16), (2016, 1, 8, 11, 47, 11)],<br>[(2016, 1, 16, 2, 5, 30), (2016, 1, 17, 2, 21, 34)],<br>[(2016, 1, 29, 6, 32, 51), (2016, 1, 30, 2, 9, 7)] | 56,652,895.63 |

## 39.5 Discussion

In this paper, we have highlighted the issue of unreliable sensors in buildings. In fact, smart buildings are equipped with several sensors that collect large amounts of data that can be used for control, energy management, and diagnosis. Significant discrepancies between simulated and actual measured building performance also led to a move to uncover these discrepancies by analyzing the data collected through extended sensor grids. Since the diagnosis of building systems usually requires chronological data, ensuring the quality of sensor data is an important step in pretreatment.

Recent literature contributes methods for the detection and classification of sensor faults. Fault classification techniques vary: several existing fault taxonomies use different criteria for categorizing a fault. However, most researchers in the literature are interested only by the following known fault types: drift, outliers, and bias. The occurrence of data gap faults has also not been given an adequate span of attention in the research literature. This is the key objective of this paper. It focuses on developing a method for automatic thresholding for automatic data gap detection for heterogeneous sensors in instrumented buildings. An algorithm based on statistical approaches has been adapted to the H358 test bed.

The delay is a stochastic variable satisfying a truncated normal distribution and at a given time; the delay is abnormal if there is a law probability value like 1% of chance of a delay gather then the threshold. In this paper, only the percentage has to be specified. However, this value is chosen arbitrarily. Next works investigate the determination of this percentage by suitable algorithms.

Also, in this paper, only the data gaps due to a battery problem are studied. However, in a building system, the sensors inform the controller which, in function of these data, controls the actuators. Therefore, if the actuator is totally faulty, there will be no measurements taken by the sensor. So, the data gaps can have as origin a faulty actuator. Future works will be around diagnosing components themselves. The two steps, i.e., sensor diagnosis and actuator diagnosis, are decoupled.

## 39.6 Conclusion

This paper deals with the issue of unreliable sensors in buildings. In the beginning, a state of the art about sensor fault detection is presented. It is concluded that research literature focus only on classic fault types such as bias and outliers. This paper focuses on data gaps fault type and presents a method of automatic thresholding for data gap detection applying to heterogeneous sensors in instrumented buildings. The solution is based on the use of statistical approaches for detection of abnormalities on the delays.

In time series analysis, the goal is always to make the distribution of any continuous random variable compatible with a Gaussian distribution. In this work, we propose to check whether the delay is abnormal or not compared to a threshold. Before calculating the threshold for each variable, one must first identify the appropriate distribution law for the delay.

With the proposed method, data gaps in the measurements can be carried out whatever the number of sensors is. Only the threshold as percentage has to be specified. In this paper, only data gaps due to battery problem are studied. However, in buildings, if an actuator is faulty, there are no measurements taken by the sensor.

The next step is to diagnose components themselves. The two steps are decoupled.

**Acknowledgements** This work is supported by the French National Research Agency in the framework of the "Investissements d'avenir" Eco SESA program (ANR-15-IDEX-02) and by the ADEME in the framework of the COMEPOS project.

## References

1. Teixeira, A.M., Araújo, J., Sandberg, H., Johansson, K.H.: Distributed sensor and actuator reconfiguration for fault-tolerant networked control systems. IEEE Trans. Control Netw. Syst. **5**(4), 1517–1528 (2018)
2. Shi, L., Cheng, P., Chen, J.: Sensor data scheduling for optimal state estimation with communication energy constraint. Automatica **47**(8), 1693–1698 (2011)
3. Wang, X., Wang, Z., Xu, Z., Cheng, M., Wang, W., Hu, Y.: Comprehensive diagnosis and tolerance strategies for electrical faults and sensor faults in dual three-phase PMSM drives. IEEE Trans. Power Electron. (2018)
4. Li, X., Huang, D., Sun, Z.: A routing protocol for balancing energy consumption in heterogeneous wireless sensor networks. In: International Conference on Mobile Ad-Hoc and Sensor Networks (pp. 79–88). Springer, Berlin (2007)
5. Ni, K., Ramanathan, N., Chehade, M.N.H., Balzano, L., Nair, S., Zahedi, S., et al.: Sensor network data fault types. ACM Trans. Sens. Netw. (TOSN) **5**(3), 25 (2009)
6. Ramanathan, R.: Data envelopment analysis for weight derivation and aggregation in the analytic hierarchy process. Comput. Oper. Res. **33**(5), 1289–1307 (2006)
7. Werner-Allen, G., Lorincz, K., Ruiz, M., Marcillo, O., Johnson, J., Lees, J., Welsh, M.: Deploying a wireless sensor network on an active volcano. IEEE Internet Comput. **10**(2), 18–25 (2006)
8. Li, C.Y., Su, W.L., McKenzie, T.G., Hsu, F.C., Lin, S.D., Hsu, J.Y.J., Gibbons, P.B.: Recommending missing sensor values. In: 2015 IEEE International Conference on Big Data (Big Data) (pp. 381–390). IEEE (2015)
9. Le Gruenwald, M.H.: Estimating missing values in related sensor data streams. In: COMAD (2005)
10. Pan, L., Li, J.: K-nearest neighbor based missing data estimation algorithm in wireless sensor networks. Wireless Sens. Netw. **2**(02), 115 (2010)
11. Yuan, Y.C.: Multiple imputation for missing data: Concepts and new development (Version 9.0), SAS Institute Inc, Rockville, MD, vol. 49, pp. 1–11 (2010)
12. Gruenwald, L., Chok, H., Aboukhamis, M.: Using data mining to estimate missing sensor data. In: Seventh IEEE International Conference on Data Mining Workshops, pp. 207–212 (2007)
13. Yu, X., Fu, Y., Li, P., Zhang, Y.: Fault-tolerant aircraft control based on self-constructing fuzzy neural networks and multivariable SMC under actuator faults. IEEE Trans. Fuzzy Syst. **26**(4), 2324–2335 (2018)
14. Zhang, Y.: A novel outlier detection method for improving industrial process monitoring. In: 2018 Chinese Control And Decision Conference (CCDC) (pp. 1155–1159). IEEE (2018)
15. Kim, W., Katipamula, S.: A review of fault detection and diagnostics methods for building systems. Science and Technology for the Built Environment **24**(1), 3–21 (2018)

16. Zhou, J., Huang, Z.: Recover missing sensor data with iterative imputing network. In: Workshops at the Thirty-Second AAAI Conference on Artificial Intelligence (2018)
17. Mehmood, A., Zia, K., Muhammad, A., Kumar Saini, D.: Missing observation approximation for spatio-temporal profile reconstruction in participatory sensor networks. Int. J. Crowd Sci. **2**(2), 108–122 (2018)
18. Monte-Moreno, E., Chetouani, M., Faundez-Zanuy, M., Sole-Casals, J.: Maximum likelihood linear programming data fusion for speaker recognition. Speech Commun. **51**(9), 820–830 (2009)

# Chapter 40
# How the Position of a Root-Top One-Sided Wind Tower Affects Its Cross-Ventilation Effectiveness

**Mehrnoosh Ahmadi**

**Abstract** Despite their antiquity, wind towers have only recently gained popularity among the researchers and architects practicing natural ventilation, especially since their performance relies both on wind and stack effect. In studying natural ventilation using wind towers, different environmental parameters and their various geometrical elements have been taken into consideration previously, but the position of the wind tower with respect to the lower building has never been assessed. Therefore, this study employs CFD to understand how at various wind incident angles, three different positions of a one-sided wind tower with respect to the lower room's window can provide different cross-ventilation efficiencies in terms of mass flow rate, velocity magnitude, and mean age of air inside the room. Furthermore, this study shows how a case with a better airflow rate might not necessarily have the freshest air at the same time. Therefore, for a natural ventilation study to be reliable, it is necessary to evaluate several parameters simultaneously.

## 40.1 Introduction

Natural ventilation is a sustainable solution for providing good indoor air quality, improving thermal comfort, and saving energy at the same time [1]. It is, in fact, one of the passive alternatives for moderating the environmental impact of growing mechanical cooling systems [2]. One of the natural ventilation devices that has been the topic of several kinds of research in the recent few years is the wind tower a.k.a. windcatcher that was originally constructed in the Middle East over 2000 years ago [3].

Wind towers' performance similar to other ventilation devices depends on the pressure distribution around the building [4]. Therefore, the two driving forces for the wind tower are buoyancy and wind with the latter being the primary, providing almost three times more internal ventilation [5]. Assessing wind towers' performance as a combination of these two driving forces, wind towers are studied by various tools

M. Ahmadi (✉)
Polytechnic University of Turin, Viale Mattioli 39, 10125 Turin, Italy
e-mail: mehrnoosh.ahmadi@polito.it

 
J. Littlewood et al. (eds.), *Sustainability in Energy and Buildings*, Smart Innovation, Systems and Technologies 163, https://doi.org/10.1007/978-981-32-9868-2_40

such as wind tunnels, computational fluid dynamics, and smoke visualization tests to identify the effect of different environmental and geometrical elements including the number of openings, internal divisions, height, roof shape, etc. For instance, Montazeri, studying the cylindrical wind towers with different numbers of opening has concluded that in general, the wind towers' sensitivity against wind angle decreases by increasing the number of openings; however, the natural ventilation performance of those with less number of openings drops more considerably by increasing the wind angle [6]. Using smoke visualization, Kazemi has demonstrated that a curved-roof one-sided wind tower works better than the ones which are inclined with flat roofs [7]. Ghadiri has compared four-sided wind towers with different internal divisions and has realized that the towers with + shaped cross section perform relatively better than H-, X-, I- and K-shaped ones [8]. In another CFD study, Ghadiri has compared six four-sided wind towers with different heights ranging between 4.5 and 18.5 m and has concluded that ventilation performance of the taller towers is worse than the shorter ones even though the wind velocity on the windward side of the tower increases with height [9].

Furthermore, there have been studies focusing on the effect of a building's window size and position on cross ventilation using wind towers. For instance, Ghadiri has found that the best window position in a room ventilated by a four-sided wind tower is down the leeward wall opposite the wind tower [8]. Montazeri, studying one-sided wind-towers and zero wind incident angle, has compared the effect of window size. His CFD analysis has proven that as long as the window is smaller than the wind tower's opening, the increase in window size improves the cross ventilation. Whereas, when the window is larger than the wind tower's opening, the increase in window size does not affect the performance considerably [10].

Nevertheless, to date, the effect of a wind tower position with respect to a window on the cross-ventilation performance has not been assessed. On the other hand, in the majority of the previous studies on wind towers using CFD, the results were compared in terms of velocity, airflow rate, and temperature, while the indoor air quality and the contaminant dispersion have been ignored. Therefore, this study aims to evaluate how a wind tower's position affects its cross-ventilation effectiveness considering mass flow rate, indoor air velocity, and mean age of air as an indicator of the air freshness. The results of this study can help architects to properly apply wind towers to their designs based on their specific building orientation and the local wind direction.

## 40.2 Validation Study

### *40.2.1 Experimental Analysis*

The reference experimental study on a one-sided wind-tower model connected to a room with a window has been done in 2009 by Montazeri et al. [11]. The wind tunnel used in this experiment had a square cross section of 46 cm side and a length

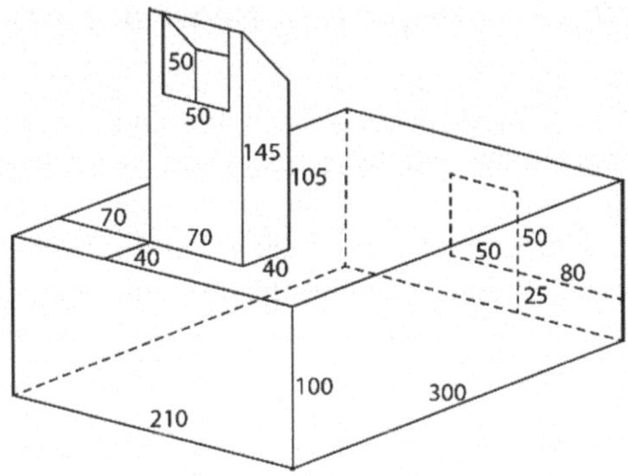


**Fig. 40.1** Dimensions (in cm) of the room and the wind tower model tested in the wind tunnel

of 3.6 m. The model in this analysis had been placed in the wind tunnel such that the room fell under the tunnel and the wind tower only was causing a maximum blockage of around 5%.

As part of the analysis, the static pressures on different internal surfaces of the wind tower had been measured using 23 pressure taps. In order to calculate the pressure coefficients at these 23 points, the reference static pressure was determined using a pitot tube placed at a 16.5 cm upstream of the wind tower at the height of 12 cm from the wind tunnel floor. The wind tower was then rotated by 15 degree intervals to see how different wind incident angles affect the pressure distribution inside the wind tower. The wind tower model's dimensions are reported in Fig. 40.1. The experiment had been done with a wind velocity of 20 m/s.

### 40.2.2 *CFD Analysis*

Gambit 2.4.6 was used to create two structured meshes of circa, four million hexahedral cells for two wind tower directions of 0° and 30°. The boundary conditions considered as part of this validation study include a uniform 20 m/s velocity with 3% turbulence intensity at the inlet of the tunnel and a zero gauge pressure of similar turbulence intensity at its outlet. The window of the room fallen out of the wind tunnel section is given a zero gauge pressure, too. The hydraulic diameter of the turbulent flow is 0.46 m considering the square cross section of the wind tunnel. All walls of the model and the wind tunnel have been considered as no-slip with zero roughness height.

Stead-state RANS equations were solved in ANSYS Fluent 18 with a standard wall function and realizable k-ε viscosity model that had been previously elaborated by Shih et al. in 1995 [12]. The SIMPLE scheme has been considered as the pressure–velocity coupling and the second-order discretization has been selected for pressure, momentum, turbulent kinetic energy, and turbulent dissipation rate.

### 40.2.3 Comparison Between CFD and Wind Tunnel Results

The comparison between CFD and experimental pressure coefficients have been done on the middle vertical line of each internal wall of the wind tower assuming a clockwise rotation of the tower with respect to the wind—i.e., green represents the middle vertical line on the left wall, black on the central wall, and red on the right one. In order to calculate the pressure coefficients using Eq. 40.1, the reference static pressure $P_0$ has been taken from the pitot tube, and the air density is considered $\rho = 1.225\ \text{kg/m}^3$.

$$C_p = \frac{P - P_0}{\frac{1}{2}\rho U^2} \tag{40.1}$$

As shown in Figs. 40.2 and 40.3, when the wind incident angle is 30° and 0°, the pressure coefficients produced inside the tower are close to those of the wind tunnel test.

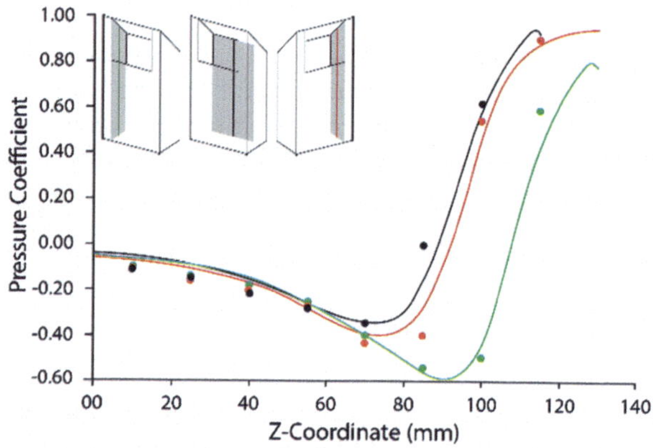


**Fig. 40.2** Comparison between the pressure coefficients derived from the CFD (lines) and the experiment (dots) at wind tower direction 30°

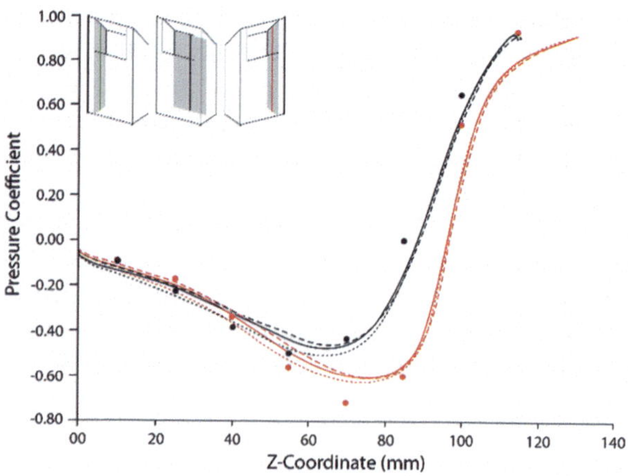


**Fig. 40.3** Comparison between the pressure coefficients on the mid (black) and right wall (red) resulted from the wind tunnel test (dots) and those of CFD derived from the coarse (dashed line), medium (full line), and fine (dotted line) meshes for wind incident angle 0

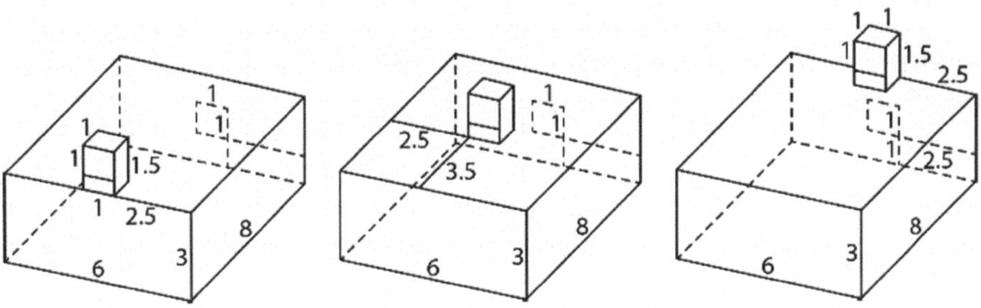


**Fig. 40.4** The room with three wind tower positions T4 (left), T5 (center), and T6 (right), and their relative dimensions in meters

### 40.2.4 *Grid Independence Analysis*

In order to make sure the solution is independent of the grid resolution, a grid sensitivity analysis has been done in the case with zero wind incident angle by coarsening and refining the mesh with a factor of $2^{1/2}$. The original grid contained 4,132,254 cells while the reproduced coarser and finer ones have, respectively, 1,391,827 and 11,134,616 cells. As demonstrated in Fig. 40.3, at this wind direction, the three grids reproduce the pressure coefficient values more or less with the same quality especially at the upper third height of the wind tower. In the mid-third of the tower, however, the finer grid generates closer values to the experimental results.

## 40.3 Description of the Case Study

In order to evaluate how the position of a roof-top wind tower affects its cross-ventilation performance, a one-sided wind tower with a square cross section of 1 $m^2$ and a height of 1.5 m having an aperture of 1 × 1 m was placed at three different positions on the roof of a room with the length, width, and height of 8, 6, and 3 m, respectively. As shown in Fig. 40.4, the tower was placed along the symmetry axis of the roof at the farthest (T4), central (T5), and closest (T6) distance from the window.

The three models were simulated in a domain with corresponding measurements of 98, 96, and 27 m, for wind incidents angles varying between 0 and 180 degrees at 30-degree intervals.

## 40.4 Computational Settings

According to the dimensions presented in the description three structured meshes of 3,447,444 (T4 and T6) and 4,098,360 (T5) cells were constructed in Gambit and simulated in ANSYS Fluent.

The models' dimensions and the solver settings are identical with those utilized in a 2018 paper by Montazeri [10] with the latter being similar to the validation study as well.

An atmospheric boundary layer (see Eq. 40.2) with a reference wind speed of $U = 3$ m/s at a height of 10 m is imposed using a user-defined function (UDF) applied on the domain's inlet and floor. The turbulent kinetic energy and turbulence dissipation rate are also calculated using the same UDF in accordance with Eqs. 40.3 and 40.4, respectively.

$$U(z) = \frac{u^*_{ABL}}{\kappa} \ln\left(\frac{z + z_0}{z_0}\right) \tag{40.2}$$

$$k(z) = 3.3\, u^{*2}_{ABL} \tag{40.3}$$

$$\varepsilon(z) = \frac{u^{*3}}{\kappa(z + z_0)} \tag{40.4}$$

Standard wall functions are used for the domain's floor which is assumed to be a terrain covered by large grass with the aerodynamic roughness length of $z_0 = 0.03$ m. As explained by Blocken et al. in 2007 [13], one requirement for a correct atmospheric boundary layer simulation is that the sand-grain roughness height $k_s$ has to be smaller than half height of the first cell adjacent to the domain floor. Therefore, in this study, $k_s = 0.006$ m is based on the relevant grid measurements. Equation 40.5 demonstrates the relationship between the three roughness parameters—roughness length, roughness height, and roughness constant in ANSYS Fluent. In the current case study, $C_s$ value of the domain floor is set to a maximum allowed number 7 using UDF.

$$k_s = \frac{9.793\, z_0}{C_s} \tag{40.5}$$

For all other wall surfaces, the standard wall functions, with $C_s$ value of 0.5 and $k_s$ of 0 have been utilized. A constant zero gauge pressure has been considered for the outlet of the domain and the lateral sides of the domain, according to the wind direction, have a symmetry boundary condition.

Furthermore, in computing the mean age of air, another user-defined function was applied to the room volume solely, solving the user-defined scalar (UDS) equation below (see Eq. 40.6), in which $\mu_{eff}$ is the effective air viscosity and $Sc_t$ is the Turbulent Schmidt number taken from Fluent's viscosity model settings as 1.2.

$$(2.88 * 10^{-5})\rho + \mu_{eff}/Sc_t \tag{40.6}$$

## 40.5 Results of the Simulations

### *40.5.1 Mass Flow Rate*

Comparing the mass flow rate values (kg/s) for different wind tower positions and different wind incident angles (see Fig. 40.5), it is evident that the tower position in which the highest mass flow occurs is T6. At this position, only at wind direction 0 and 90 the flow rate is slightly lower than the other two positions. Correspondingly, the least flow rate happens in tower T4 in which, only at 90 degrees the performance of the tower is better than the two other positions.

In general, the best wind directions for having a greater flow rate through the tower are 0°, 30°, 150°, and 180° and those for the worst are 60° and 90°.

### *40.5.2 Velocity Magnitude Distribution*

The contours of velocity magnitude were thoroughly assessed to understand the air movement pattern inside the room. In this paper, however, only those on the horizontal plane placed at 170 cm above the ground floor are demonstrated.

Comparing the three positions for wind incident angles 00 and 30, it is evident that the greater distance between the tower and the window results is a higher wind speed distribution inside the room. According to Fig. 40.6, T4 and T5 show an almost similar air movement at a standing person height while at T6, there is almost no air movement inside the room as the inletting air from the wind towers exhausts from the window immediately. At wind direction 60 and 90, a standing person inside the room feels almost no air movement. At 60°, there is a negligible air velocity underneath the tower while at 90 the room is completely still. As the wind passes from 90 and the window becomes active as an inlet, higher airspeed is felt inside the room. At 120 and 150 degrees, the case with the optimal velocity distribution is T5. In T5, in almost all parts of the room, the velocity is non-zero while at T6 the room core is majorly experiencing zero airspeed. Compared to other positions, T6 allows a speedier wind inside the room through the window and likewise the exhaust air

**Fig. 40.5** Mass flow rate (kg/s) values at tower aperture for different wind tower positions

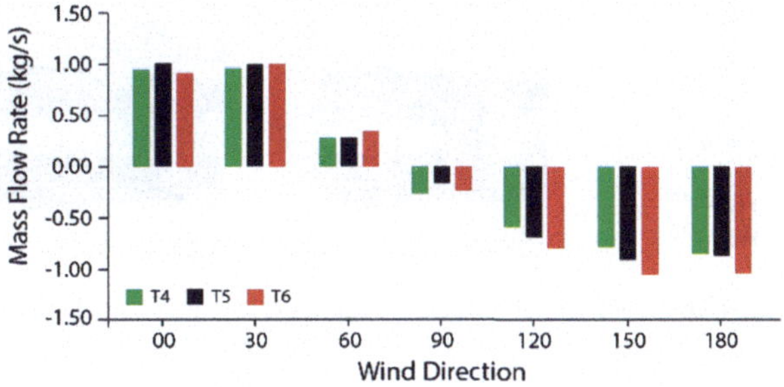

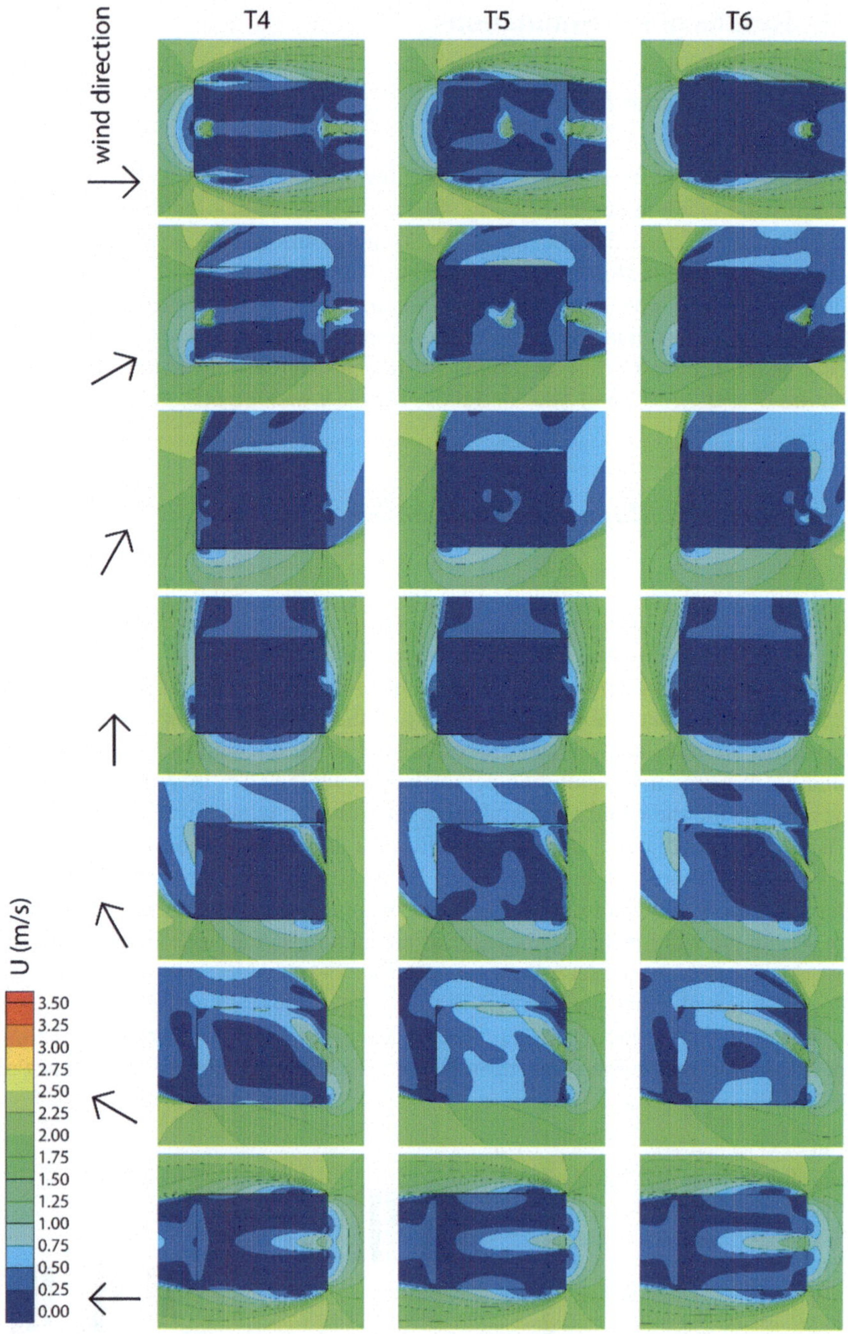


**Fig. 40.6** Contours of velocity magnitude on a horizontal surface placed 1.7 m above the ground

through the tower has a higher speed. At these incident angles, the worst velocity distribution is provided by T4 in which there is a minimal airspeed in front of the window. Finally, when the wind blows perpendicularly to the window and the wind towers are at their most leeward condition, the model that provides a better-spread air movement inside the room is T6 and case T4 is the one that has large parts of the room experiencing no air movement.

### *40.5.3 Mean Age of Air*

Demonstrated in Fig. 40.7, at 00° and 30° wind incident angle, the tower positions T4 and T5 provide much better air age in the room with respect to T6 position at a height of 170 cm from the ground. In T6, the gradient of the mean age of air has a wider range in the room and it arrives at about 480 s. As the wind deviates toward 90°, the freshness of the air inside the room for all three tower positions deteriorates. At 60° T6 and at 90, T5 take more time changing the air inside the room. In fact, these two conditions are those in which the air age arrives to 800 s and more. At 120, 150, and 180 wind incident angles, T4 is the worst position of all. At 120° and 180°, T6 provides the freshest air for a standing person while at wind direction 150, T5 is the best.

## 40.6 Conclusion

Comparing the three wind tower positions at each wind incident angle it can be observed that the velocity and the mean air age contours match as they both show that in wind directions where the window acts as the air inlet, there is more air movement with a superior distribution and quality inside the room. However, comparing the two contours for all wind incident angles, the above assumption is not valid. In fact, even though T5 provides the highest air velocity in the room at wind incident angle 50, the freshest air of all in generated by T6 at 180 degrees. In addition, while the wind velocity produced by the three tower positions at 60 degrees are similar, the mean age of air in the T6 position is much higher than the other two positions. Therefore, the contours of velocity magnitude and mean age of air should be both analyzed correspondingly.

Moreover, it should be stated that the contours of velocity and mean age of air do not necessarily match with the mass flow rate values measured at the tower's opening. In fact, at almost all wind direction, T6 provides the most, T4 the least, and T5 the medium mass flow rate values, while, according to the contours, at angles between 0 and 90, T5 provides the freshest and T6 the least fresh air. In addition, though at 0° and 30°, the three tower positions provide high mass flow rates similarly, according to the contours the T6 position is the one that has a considerably worse air quality and almost no air movement compared to T5 and T4.

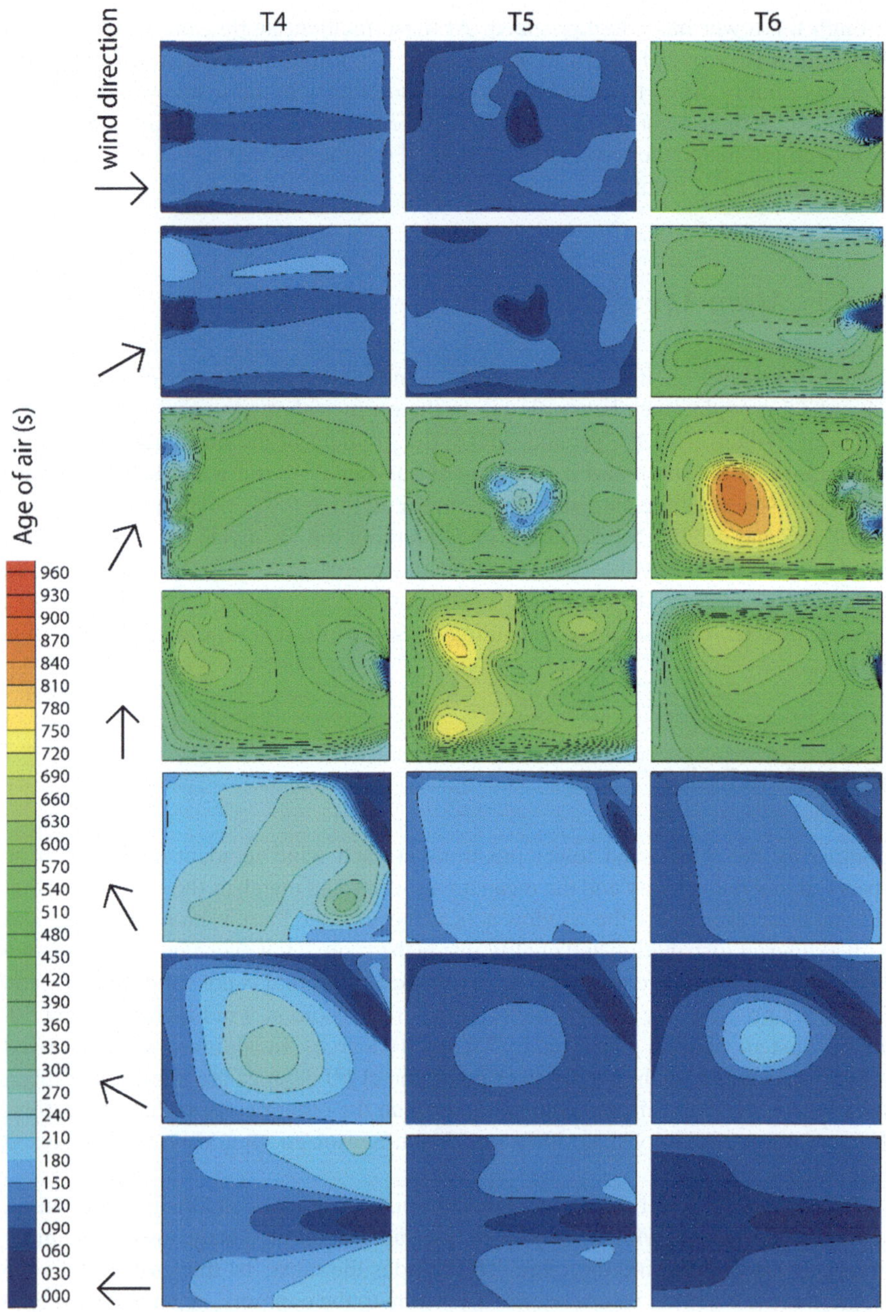


**Fig. 40.7** Contours of mean age of air on a horizontal surface placed 1.7 m above the ground

The mismatches between different results derived from the CFD analysis of a one-sided wind tower model working together with a window, attest that a more profound analysis on the natural ventilation behavior of wind towers is needed, especially in more complex cases where the number of inlets and outlets are more than one such as in multi-opening towers and of course in more complex building layouts.

## References


1. Heiselberg, P.: Natural ventilation design. Int. J. Vent. **2**(4), 295–312 (2004)
2. Chiesa, G., Grosso, M.: Breakthrough of natural and hybrid ventilative cooling technologies: models and simulations. Int. J. Vent. **15**(3–4), 183–185 (2016)
3. Vaziri, A.A.K., Bastani Parizi, M.E. (eds.): History of Kerman. Elm, Tehran (2006)
4. Battle McCarthy Consulting Engineers: Wind Towers, Detail in Building. Academy Press, West Sussex (1999)
5. Hughes, B.R., Cheuk-Ming, M.: A study of wind and buoyancy driven flows through commercial wind towers. Energy Build. **43**, 1784–1791 (2011)
6. Montazeri, H.: Experimental and numerical study on natural ventilation performance of various multi-opening wind catchers. Build. Environ. **46**(2), 370–378 (2010)
7. Kazemi, M.E., Dehghan, A.A., Dehghan, M.M., Mohagheghian, S.: Visualized flow structure around and inside of one-sided wind-catchers. Energy Build. **55**, 545–552 (2012)
8. Ghadiri, M.H., Ibrahim, N.L.N., Mohamed, M.F.: Performance evaluation of four-sided square wind catchers with different geometries by numerical method. Engineering Journal **17**(4), 9–17 (2013)
9. Ghadiri, M.H., Lukman, N., Ibrahim, N., Mohamed, M.F.: Applying computational fluid dynamic to evaluate the performance of four-sided rectangular wind catcher with different height. Res. J. Appl. Sci. Eng. Technol. **7**, 502–509 (2014)
10. Montazeri, H., Montazeri, F.: CFD simulation of cross-ventilation in buildings using rooftop windcatchers: Impact of outlet openings. Renew. Energy **118**, 502–520 (2018)
11. Montazeri, H., Azizian, R.: Experimental study on natural ventilation performance of one-sided wind catcher. Build. Environ. **43**(12), 234–244 (2008)
12. Shih, T.H., Liou, W.W., Shabbir, A., Yang, Z., Zhu, J.: A new k-ε eddy viscosity model for high Reynolds number turbulent flows. Comput. Fluids **24**, 227–238 (1995)
13. Blocken, B., Stathopoulos, T., Carmeliet, J.: CFD simulation of the atmospheric boundary layer: wall function problems. Atmos. Environ. **41**, 238–252 (2007)

# Chapter 41
# Cool Roofs with Variable Thermal Insulation: UHI Mitigation and Energy Savings for Several Italian Cities

**Maurizio Detommaso, Stefano Cascone, Antonio Gagliano, Francesco Nocera and Gaetano Sciuto**

**Abstract** The progressive modification of the land surface in the urban area during the last decades has involved an increase in urban air temperature compared to surrounding rural areas. The significant replacement of green areas with built surfaces characterized by high solar radiation favored the energy storage during the day and heat release in the nocturnal hours. The Urban Heat Island (UHI) is an emblematic effect of the increase of urban temperature which in turn affects the energy consumption in the building sector. Several passive building technologies are available to mitigate the UHI phenomenon and reduce the energy consumptions in buildings. In the Mediterranean climate, where the solar radiation control is the most important passive strategy in summer, cool roof technology is able to reduce the solar radiation absorbed by materials and to limit the outer superficial temperature in the presence of high solar loads. This paper presents a numerical analysis of the energy and environmental performance of an office building equipped with a cool roof for three different climate zones. Dynamic simulations highlight the different thermal behavior of the investigated cool roof in comparison to the standard roof, also as a function of the roof thermal resistance. Moreover, the Solar Reflectance Index (SRI) of the different roof configurations has been evaluated. It was found that slight or no insulated cool roofs has the lowest superficial temperatures and hence mitigates the UHI.

M. Detommaso · A. Gagliano
Department of Electrical, Electronics and Computer Engineering, University of Catania, Viale Andrea Doria 6, 95125 Catania, Italy

S. Cascone · F. Nocera (✉) · G. Sciuto
Department of Civil Engineering and Architecture, University of Catania, Viale Andrea Doria 6, 95125 Catania, Italy
e-mail: fnocera@unict.it

 
J. Littlewood et al. (eds.), *Sustainability in Energy and Buildings*, Smart Innovation, Systems and Technologies 163, https://doi.org/10.1007/978-981-32-9868-2_41

## 41.1 Introduction

The main effects of Urban Heat Island consist of human discomfort, environmental pollution, energy consumption for building cooling, peak of electricity demand, and consequently the rise of energy prices [1]. In order to counter the overheating of urban areas and reduce its impact on the communities are available several intervention policies in relation to space scale. It is possible to select three levels: urban, neighborhood, and microscale. In recent years, at microscale context, a wide set of countermeasures such as cool roofs [2], cool pavements [3], green roofs [4], and greenery of urban surfaces [5] are promoted because they are capable to reduce energy consumption and improve comfort conditions in buildings and surrounding spaces [6].

Cool materials are characterized by high values of solar reflectance ($r > 0.5$), which strongly reduce the amount of solar radiation absorbed by the roof outer layer, and by high infrared emissivity values ($\varepsilon > 0.8$), which contributes to release the heat accumulated during the day through an intensive radiant heat exchange at night [7].

Given the very important role consisting of mitigate solar gains, especially in the southern regions with a mild Mediterranean climate, analyses on cool coating were carried out to explore the possible effects of such a technology in the hot climate [8, 9].

Experimental measurements and numerical simulations of a school building located in Athens equipped with non-insulated and insulated roof surfaces have shown that a cool coating with a solar reflectance of 0.89 causes an important decrease in the annual cooling load [10]. Energy savings for space cooling by 40% for the uninsulated roof of the reference building and 35% for the insulated building roof were achieved [10]. A decrease in the temperature of the roof outer layer by about 25 °C was achieved after the application of the cool coating [10]. Dynamic simulations of a single floor building located in the south coast of Sicily, have shownd that the addition of the first step of thermal insulation in the standard roof and cool roof scenarios produced the highest energy saving for space cooling [11]. Numerical simulations of simple building models in temperate climate have proven that houses with little roof insulation and high solar reflectance provide the greatest energy savings. The authors have found that it is preferable to combine high emissivity and high reflectance with low thermal insulation [12].

This paper aims to investigate the effect of a cool roof on an office building by means an accurate dynamic simulation study. The impact of the cool roof on energy consumptions in relation to thermal insulation for three different weather conditions of the Mediterranean Climate was analyzed. In addition, the energy and environmental performance of a cool roof to mitigate the UHI phenomenon were evaluated and investigated comparing the Solar Reflectance Indexes (SRI) in the different roofs' configurations.

## 41.2 Methodology

An office building designed by Italian Thermo-Technical Committee (CTI) was chosen as Reference Building (RB). The energy needs for space heating ($PE_H$) and cooling ($PE_C$) as well as the Solar Reflectance Index (SRI) for the RB were calculated through DesignBuilder software [13] in three cities of Mediterranean climate characterized by different climate conditions: Milan, Rome, and Catania.

With the aim to test the veracity of the energy demand evaluated through dynamic simulations, such heating energy demand was compared with the one calculated in steady-state conditions by CTI in Milan. It is worth noticing that the average difference between steady-state calculations and dynamic simulations is below 10%.

Afterward, the RB was modeled adding a cool coating on the outer surface of the upper layer of the roof where the reflectance ($r_s$) was set equal to 0.8, while the standard roof has a solar reflectance of 0.2.

Moreover, three different thickness of the roof thermal insulation layer were investigated. Therefore, the RB was simulated under eight roof scenarios (see **Errore. L'origine riferimento non è stata trovata.**):

- Standard roof (SR), I($s_i = 0$), II($s_i = 0.5 \cdot s_{iref}$), III($s_i = s_{iref}$), IV($s_i = 2 \cdot s_{iref}$);
- Cool roof (CR), I($s_i = 0$), II($s_i = 0.5 \cdot s_{iref}$), III($s_i = s_{iref}$), IV($s_i = 2 \cdot s_{iref}$).

Here, $s_{iref}$ is the thickness of thermal insulation assumed equal to 8 cm in all the cities (Fig. 41.1).

The different roofs' thermal resistance can be assumed as reference for the following cases:

- $s_i = 0$, non-insulated roofs corresponding to buildings built before the 70s;
- $s_i = 0.5 \cdot s_{iref}$, averagely insulated roofs of buildings built between the 70s and 2005;
- $s_i = s_{iref}$, well-insulated roofs belonging to buildings built after 2005;
- $s_i = 2 \cdot s_{iref}$, high-insulated roofs.

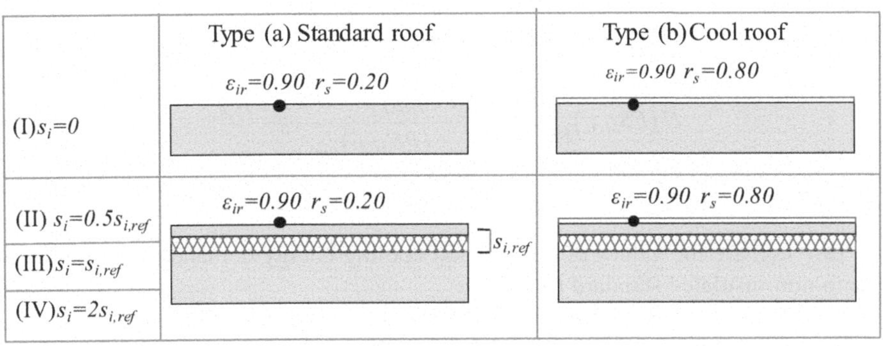


**Fig. 41.1** Types of investigated roof scenarios: **a** Roof without treatment; **b** cool roof

### 41.2.1 Thermal Calculation Model

The heat-balance model on the outer surface of standard and the cool roof was implemented in the DesignBuilder software which is the graphical interface for the numerical code Energy Plus [14]. The model evaluates the short-wave radiation incident on roof outer layer and reflects from its surface toward outdoor ambient, the long-wave radiation exchanged between roof outer surface and sky, and convective heat and heat transferred by conduction to the internal ambient. The heat and energy balance on the roof surface is based on Eq. [15] (41.1):

$$(1 - r_s) \cdot G(t) + h_{co}[T_{so}(t) - T_o(t)] + \sigma \cdot \varepsilon_{ir}\left[T_{so}^4(t) - T_{sky}^4(t)\right] = \frac{T_{so}(t) - T_i(t)}{R_{so\_i}} \tag{41.1}$$

where G(t) is the solar irradiance, $h_{co}$ is the convective coefficient [16], $T_o$ and $T_i$ are, respectively, the outdoor and the indoor temperatures, $T_{sky}$ is the sky temperature, ($R_{so\_i}$) is the thermal resistance from roof outer surface to the internal ambient and $T_{so}$ is the outdoor superficial roof temperature. The calculation method is based on the algorithm of Conduction Transfer Function (CTF).

### 41.2.2 Energy Demand

The energy savings for space heating and cooling $(ES_{H,C})_{I,k}$, for each roof configuration (I) and city (k), is calculated by the difference of the energy needs for the non-insulated standard roof SR0, k and the one of each roof scenario (I, k):

$$\left(ES_{H,C}\right)_{I,k} = \left(PE_{H,C}\right)_{SR0,k} - \left(PE_{H,C}\right)_{I,k} \tag{41.2}$$

Then, these energy savings $(ES_{H,C})_{I,k}$ are normalized with respect to $(PE_{H,C})$ of the standard roof scenario (SR0) in the same kth locality:

$$N\left(ES_{H,C}\right)_{I,k} = \frac{\left(PE_{H,C}\right)_{SR0,k} - \left(PE_{H,C}\right)_{I,k}}{\left(PE_{H,C}\right)_{SR0,k}} \tag{41.3}$$

where

- $(PE_{H,C})_{SR0}$ are the values of heating and cooling energy demand of the building with non-insulated standard roof;
- I = 1, 7 are the configurations for the standard roof (SR1–SR3) and cool roof (CR0–CR3), due to the different thickness of insulation, and k = 1, 3 are the three cities.

### 41.2.3 Solar Reflectance Index

Solar Reflectance Index (SRI) is a parameter introduced by the Green Building Council (GBC) [17] for measuring the combined effect of reflectivity and emissivity on the superficial temperature of a surface under specific solar and environmental conditions, defined as

$$SRI = 100 \cdot \frac{T_{sb} - T_{so}}{T_{sb} - T_{sw}} \quad (41.4)$$

where $T_{sb}$ and $T_{sw}$ are the outer temperatures of standard black and white surfaces, $T_{so}$ is the outer surface temperature of the current roof scenario. The standard black surface, SRI = 0, has $r_s = 0.05$, $\varepsilon_{ir} = 0.90$, while the standard white surface SRI = 100, has $r_s = 0.90$, $\varepsilon_{ir} = 0.90$ [18]. In this work, SRI is evaluated under actual weather conditions as

$$(SRI)_{I,k} = 100 \cdot \frac{(T_{sb})_{0,k} - (T_{so})_{I,k}}{(T_{sb} - T_{sw})_{0,k}} \quad (41.5)$$

where $(T_{sb})_{0,\,k}$ and $(T_{sw})_{0,\,k}$ are the outer surface temperatures of black and white surfaces of the standard roof with no thermal insulation, that is taken as a reference. $(T_{so})_{I,\,k}$ is the outer surface temperature of a I-ma roof configurations.

## 41.3 Case Study

The reference building is oriented toward cardinal points and it is bordered by the external environment through all its facades. The ground and first floors of the building are depicted in the layout of Fig. 41.2. The main geometric features of the building are reported in Table 41.1.

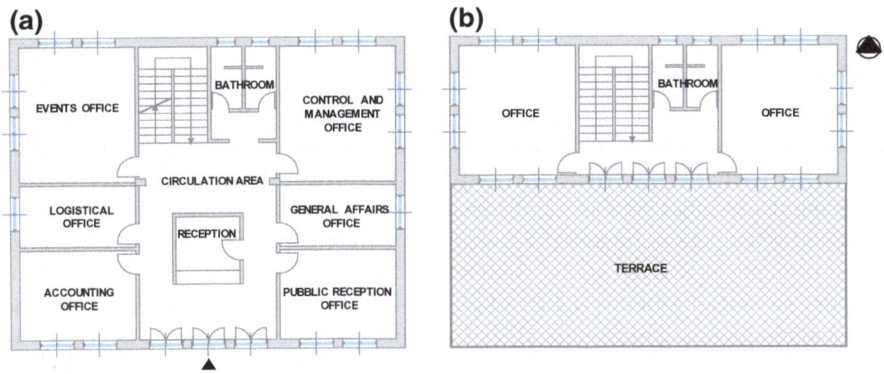


**Fig. 41.2** Layout of the two building levels: **a** Ground floor; **b** First floor

**Table 41.1** Geometric features of the reference building

| Geometric features | | | |
|---|---|---|---|
| Heated gross volume | V | 1309.44 | $m^3$ |
| Total external surface | S | 1122.65 | $m^2$ |
| Shape factor | S/V | 0.86 | $m^{-1}$ |
| Net floor area | $S_n$ | 289.21 | $m^2$ |
| Glazing surface | $S_w$ | 81.24 | $m^2$ |
| Roof surface | $S_r$ | 243.25 | $m^2$ |
| Roof/Total external surface ratio | $S/S_r$ | 0.22 | – |

**Table 41.2** Opaque envelope and glazing surface features

| Building components | SM ($kg/m^2$) | U ($W/m^2K$) | $U_{LIM}$-CT ($W/m^2K$) | $U_{LIM}$-RO ($W/m^2K$) | $U_{LIM}$-MI ($W/m^2K$) |
|---|---|---|---|---|---|
| Wall | 397 | 0.28 | ≤0.45 | ≤0.34 | ≤0.30 |
| Roof | 482 | 0.25 | ≤0.38 | ≤0.30 | ≤0.25 |
| Ground floor | 442 | 0.30 | ≤0.46 | ≤0.32 | ≤0.30 |
| Window | – | 1.80 | ≤3.20 | ≤2.00 | ≤1.80 |

The presence of occupants was assumed from 9.00 to 18.00 according to a typical workday of office space. A power density of 7.2 $W/m^2$ while 11 $W/m^2$ and 8 $W/m^2$ were considered for occupancy, lighting, and office devices, respectively. The air-conditioning (AC) system is represented by a condensing boiler ($\eta = 97\%$) for space heating and a chiller with Energy Efficiency Ratio (EER) of 2.50 for cooling purposes of the building. The heating and cooling systems have to keep an internal temperature of 20°C in winter and 26°C in summer, respectively. The limit U-values of the opaque and glazing components comply with the regulation limits for all three sites are reported in Table 41.2 while the thermophysical properties of the investigated roof scenarios are reported in Table 41.3.

### *41.3.1 Climate Data*

The dynamic simulations of RB under the eight proposed roof scenarios in the cities of Catania (Lat. 37.47), Rome (Lat. 41.80), and Milan (Lat. 45.44) were carried out using the weather database embedded in Energy Plus. All climate localities are classified as temperate climates (Cf) according to International Köppen classification. Milan is characterized by humid summers (Cfa) whereas Rome and Catania (Csa) have dry summers. Monthly values of global solar radiation on the horizontal plane ($H_h$) and outdoor air temperature ($T_o$) of weather database are depicted in Fig. 41.3.

**Table 41.3** Thermophysical properties of the investigated roof scenarios

| Thermophysical properties | | SR0 | SR1 | SR2 | SR3 | CR0 | CR1 | CR2 | CR3 |
|---|---|---|---|---|---|---|---|---|---|
| Roof thickness (s) | m | 0.27 | 0.31 | 0.35 | 0.43 | 0.27 | 0.31 | 0.35 | 0.43 |
| Insulation thickness | m | 0 | 0.04 | 0.08 | 0.16 | 0 | 0.04 | 0.08 | 0.16 |
| U-value | $W/m^2K$ | 1.94 | 0.55 | 0.33 | 0.18 | 1.94 | 0.55 | 0.33 | 0.18 |
| Surface Mass (SM) | $kg/m^2$ | 476 | 478 | 480 | 484 | 476 | 478 | 480 | 484 |
| Reflectance (r) | – | 0.20 | 0.20 | 0.20 | 0.20 | 0.80 | 0.80 | 0.80 | 0.80 |
| Emissivity (ε) | – | 0.90 | 0.90 | 0.90 | 0.90 | 0.90 | 0.90 | 0.90 | 0.90 |

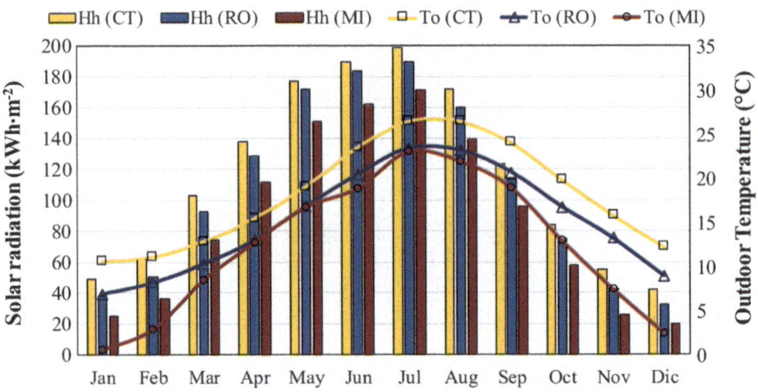


**Fig. 41.3** Main average weather data of Catania, Rome, and Milan

## 41.4 Results and Discussion

### *41.4.1 Energy Savings*

Energy demands for space heating are calculated considering that the heating system is switched on during the period from December 1 to March 31 in Catania, from November 1 to April 15in Rome, and from October 15 to April 15 in Milan whereas the cooling system operates from June 1 to September 30 in the cities considered. Figure 41.4a–c show the energy requirements and normalized energy savings to maximum energy demand ($ES_{H/C}/PE_{H/C,max}$) of the building, respectively, in the cities of Milan, Rome, and Catania, for both standard roof and cool roof, varying the thickness of the thermal insulation layer.

The results highlight that the addition of the first step of thermal insulation in the standard roof (SR) has produced the highest percentage of improvement in both heating and cooling energy performance. Indeed, a reduction of the energy needs for space heating of 9%, 11%, and 15% and cooling (7%, 2%) are achieved with respect to case SR0 in Milan, Rome, and Catania, respectively. Cool roof increases the energy demand for heating, not exceeding +5% for all configurations of cool roof and for all considered localities. This is certainly due to the high solar reflectance of the cool paint, which reduced the heat gains from the incident solar radiation on the building roof. On the contrary, during the cooling period, cool roof, without any thermal insulation, is significantly able to reduce the energy demand with decreases up to 50%. Reductions for space cooling of 52%, 42%, and 31% are achieved in CR0 scenario compared to standard roof (SR0) in Milan, Rome, and Catania, respectively. It has to be highlighted that the cool roof coupled with a moderate thickness of thermal insulation (II s = 4 cm) involved increases of cooling demand up to 32%. Further increases of the insulation thickness gave a very low incremental effect on cooling

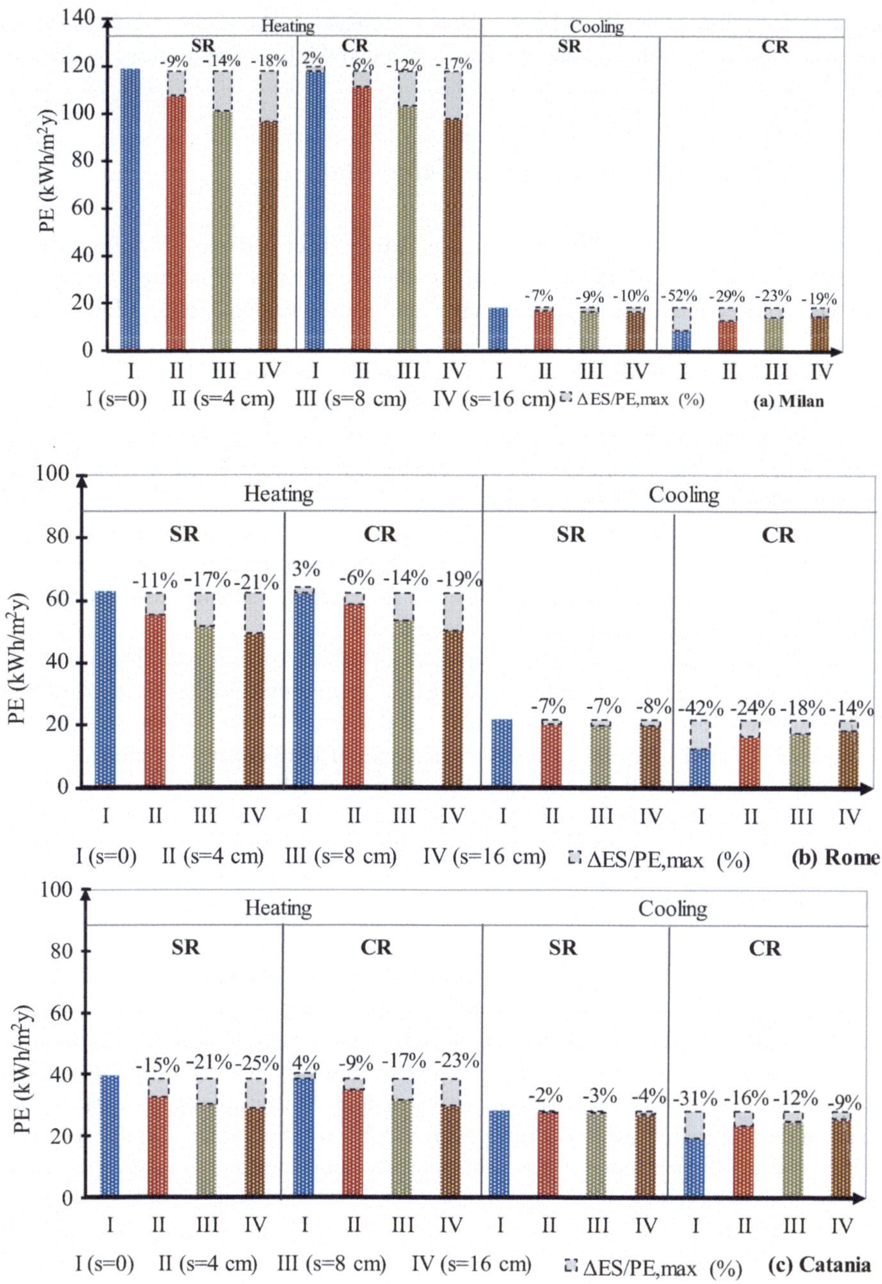


**Fig. 41.4** Energy demand for standard and cool roof: **a** Milan; **b** Rome; **c** Catania

demand. Therefore, for all cities, the higher increases/reductions of energy needs were recorded with roofs characterized by a reduced thickness of thermal insulation.

### *41.4.2 Solar Reflectance Index Calculation*

With the aim to evaluate the SRI indexes considering the effective roof configurations, SRI indexes were calculated under real weather conditions assuming as reference the data acquired when the maximum daily values of horizontal solar radiation ($H_h$) occur.

Table 41.4 shows the data of outer surface temperatures of black surfaces $(T_{sb})_{0,k}$ and white surfaces $(T_{sw})_{0,k}$ of the standard roof with no thermal insulation. $(T_{so})_{I,k}$ is the outer surface temperature of a I-ma roof configurations used to calculate SRI in all the investigated localities. As regard to the temperature $T_{sb}$, $T_{sw,}$ and $T_{so,}$ they were calculated in free-floating conditions.

From Table 41.4, it can be observed that the roof scenarios without any thermal insulation are characterized by higher values of SRI compared to the same insulated roof scenarios in all three cities. Indeed, the first step of thermal insulation (from CR0/SR0 to CR1/SR1) involves a decrease of SRI.

The negative values of SRI demonstrate that the scenarios of a thermal insulated standard roof are characterized by outer surface temperature higher than the one of a black surface posed over a not-insulated standard roof (SR0). This means that the increase of thermal insulation causes an increase of temperature on the roof outer layer and consequently the detrimental effect on the UHI.

## 41.5 Conclusion

This paper has analyzed the effects of a cool roof on an office building with several thermal insulation levels in comparison to a conventional roof for three different weather conditions. The results highlight that a cool roof without any thermal insulation offers the best performance during summer. Indeed, the cool roof (CR0) shows significant values of SRI and cooling energy demand lower compared to other scenarios and in all localities. Reductions even of 52% were obtained. On the contrary, an increase in heating demand was achieved in winter. Increase in heating needs of 4% can be considered negligible in relation to energy benefits achievable in summer. The presence of an insulation layer seems to counteract the cooling effect of cool paint as confirmed by the increase of cooling energy needs in the thermal insulated cool roof. Considering the different climatic conditions, the Payback Time of the economic investment will be lower for Catania, where the energy requirements for the cooling are higher and greater for Rome and Milan. Future research should assess the Payback Time of the economic investment, considering both the installation costs and the energy savings due to the cool roof.

**Table 41.4** Solar Reflectance Index (SRI) of standard and cool roof scenario

| Localities | $T_{sb,0}$ (°C) | $T_{sw,0}$ (°C) | SRI (–) | | | | | | | |
|---|---|---|---|---|---|---|---|---|---|---|
| | | | SR0 | SR1 | SR2 | SR3 | CR0 | CR1 | CR2 | CR3 |
| Milan | 63.18 | 16.07 | **12.76** | **−5.58** | **−7.75** | **−7.77** | **80.73** | **78.03** | **77.92** | **77.90** |
| | $(T_{so})_{I,k}$ (°C) | | 57.17 | 65.81 | 66.83 | 66.84 | 25.15 | 26.42 | 26.47 | 26.48 |
| Rome | 68.17 | 23.78 | **17.93** | **−0.97** | **−2.46** | **−2.95** | **88.33** | **82.27** | **82.02** | **81.93** |
| | $(T_{so})_{I,k}$ (°C) | | 60.21 | 68.60 | 69.26 | 69.48 | 28.96 | 31.65 | 31.76 | 31.80 |
| Catania | 72.81 | 21.57 | **17.33** | **0.18** | **−1.13** | **−1.78** | **86.24** | **80.82** | **80.58** | **80.27** |
| | $(T_{so})_{I,k}$ (°C) | | 63.93 | 72.72 | 73.39 | 73.72 | 28.62 | 31.40 | 31.52 | 31.68 |

**Acknowledgements** This work has been financed by the University of Catania within the project "Piano della Ricerca Dipartimentale 2016–2018" of the "Civil Engineering and Architecture Department" and Electrical Electronic and Computer Engineering Department"

## References

1. Hassid, A., Santamouris, M., Papanikolaou, M.F., Linardi, A., Klitsikas, N., Georgakis, C., Assimakopoulos, D.N.: The effect of the Athens heat island on air conditioning load. Energy Build. **32**, 131–141 (2000)
2. Gagliano, A., Detommaso, M., Nocera, F., Patania, F., Aneli, S.: The retrofit of existing buildings through the exploitation of the green roofs - a simulation study. Energy Proc. **62**, 52–61 (2014)
3. Akbari, H., Bretz, S., Hartford, H.: Peak power and cooling energy savings of high albedo roofs. Energy Build. **25**, 117–126 (1997)
4. Santamouris, M.: Cooling the cities – a review of reflective and green roof mitigation technologies to fight heat island and improve comfort in urban environments. Sol. Energy **103**, 682–703 (2014)
5. Cascone, S., Evola, G., Leone, C., Sciuto, G.: Vertical greenery systems for the energy retrofitting of buildings in Mediterranean Climate: a case study in Catania, Italy. IOP Conf. Ser. Mater. Sci. Eng. **415**(1) (2018)
6. Akbari, H., Konopacki, S.: Calculating energy-saving potentials of heat-island reduction strategies. Energy Policy **33**, 721–756 (2005)
7. Kültür, S., Türkeri, N.: Assessment of long term solar reflectance performance of roof coverings measured in laboratory and in field. Build. Environ. **48**, 164–172 (2012)
8. Synnefa, A., Santamouris, M., Akbari, H.: Estimating the effect of using cool coatings on energy loads and thermal comfort in residential buildings in various climatic conditions. Energy Build. **39**, 1167–1174 (2007)
9. Costanzo, V., Evola, G., Gagliano, A., Marletta, L., Nocera, F.: Study on the application of cool paintings for the passive cooling of existing buildings in Mediterranean climates. Adv. Mech. Eng. 1–10 (2013)
10. Synnefa, A., Saliari, M., Santamouris, M.: Experimental and numerical assessment of the impact of increased roof reflectance on a school building in Athens. Energy Build. **55**, 7–15 (2012)
11. Gagliano, A., Nocera, F., Detommaso, M., Evola, G.: Thermal behavior of an extensive green roof: numerical simulations and experimental investigation. Int. J. Heat Technol. **34**(special issue 2), S226–S234 (2016)
12. Gentle, A.R., Aguilar, J.L.C., Smith, G.B.: Optimized cool roofs: Integrating albedo and thermal emittance with R-value. Sol. Energy Mater. Sol. Cells **95**, 3207–3215 (2011)
13. DesignBuilder and energy simulation software, version 4, http://designbuilder.co.uk
14. EnergyPlus. The reference to energy plus calculation, Conduction Transfer Function solution algorithm. University of Illinois and California, Engineering reference (2011)
15. Muscio, A.: The solar reflectance index as tool to forecast the heat released to the urban environment: potentiality and assessment issues. Climate **6**(12), 1–21 (2018)
16. Costanzo, V., Evola, G., Marletta, L., Nocera, F.: The effectiveness of phase change materials in relation to summer thermal comfort in air-conditioned office buildings. Build. Simul. **11**(6), 1145–1161 (2018)
17. http://www.usgbc.org/DisplayPage.aspx?CMSPageID=1447
18. Akbari, H., Levinson, R., Berdahl, P.: ASTM standards for measuring solar reflectance and infrared emittance of construction materials and comparing their steady-state surface temperature. Lawrence Berkeley National Laboratory

# Chapter 42
# Green Space Factor Assessment of High-Rise Residential Areas in Harbin, China

**Ming Lu, Xuetong Wang and Jun Xing**

**Abstract** As an important part of urban green space, how to effectively guide the improvement of the ecological benefits of residential green space has become one of the important issues in current residential planning. Western countries such as Germany, Sweden, and the United States have successively used the "Green Space Factor" as a tool for ecological assessment, while in China's settlement planning, there is only one greening rate indicator as a guide. Based on the Green Factor of Seattle, USA, this study used Harbin high-rise residential area as the research object to establish a Green Space Factor assessment tool for Harbin. The evaluation indicators were screened, adjusted and assigned through expert questionnaires. Finally, an evaluation system with five categories and 25 evaluation indicators was established. Subsequent assessments of the ecological benefits of the selected nine high-rise residential areas were evaluated. From the evaluation results, it can be found that the advantages and disadvantages of the ecological benefits of the settlements are directly related to the allocation of the landscape elements in the settlements and have a certain relationship with the greening rate and the volume ratio of the settlements. The settlements with high volume ratio and low greening rate can adjust the landscape elements. The configuration can also achieve high ecological benefits. This study can provide strong support for guiding the construction of green infrastructure in high-rise residential areas in cold cities.

---

M. Lu · X. Wang (✉) · J. Xing
School of Architecture, Harbin Institute of Technology, 150006 Harbin, China
e-mail: 1228502580@qq.com

M. Lu
e-mail: hitlm@126.com

J. Xing
e-mail: junx@sina.com

Key Laboratory of Cold Region Urban and Rural Human Settlement Environment Science and Technology, Ministry of Industry and Information Technology, 150006 Harbin, China

 
J. Littlewood et al. (eds.), *Sustainability in Energy and Buildings*, Smart Innovation, Systems and Technologies 163, https://doi.org/10.1007/978-981-32-9868-2_42

## 42.1 Introduction

At present, in large- and medium-sized cities in China, due to the tightness of land resources and the government's policy of intensive land use, high-rise settlements have become the mainstay of newly built residential areas. Therefore, how to effectively improve the ecological benefits of green areas in high-rise residential areas has become one of the important issues that need to be solved in the planning of high-rise residential areas today [1–4].

The Green Space Factor is a series of mandatory assessment tools for urban green space ecological benefits implemented by some governments in foreign cities [5]. There is only one two-dimensional control index of "Greening Rate" on the evaluation index system of urban residential areas in China. It is only possible to achieve the planar greening effect as a whole and cannot judge the actual three-dimensional greening structure and the ecological benefits of the green space [6, 7]. Therefore, it is necessary to establish a three-dimensional evaluation index that can fully reflect the spatial structure of residential green space as a supplement to the current two-dimensional green space index [8–11]. In order to assess the ecological benefits of landscape green space in high-rise residential areas in Harbin. It provides a theoretical basis for the assessment of green space factors in high-rise residential areas in cold cities and guides the construction of landscaped green spaces in other areas of China.

## 42.2 Literature Review

Although the naming of Green Space Factor tools is different in different countries, the evaluation method is roughly the same as the purpose [12]. In 1997, Berlin implemented the "Biotope Area Factor" assessment tool to guide the construction of urban green infrastructure [13–15]. Subsequent Green Space Factors began to be implemented in several western cities, typically including the Malmö Green Space Factor in Sweden [16], Seattle Green Factor [17], US Fife Green factor, and so on [18].

The Seattle Green Factor is the Green Space Factor with the most comprehensive application and the widest impact. It is based on the Biotope Area Factor, and it has a direct impact on the construction of Green Space Factor indicators in other cities in the United States [19, 20]. Compared with other Green Space Factors, Seattle Green Factor has more detailed evaluation indicators and more scientific evaluation criteria. The goal of Seattle's green factor building is to optimize the appearance of the community; reduce stormwater runoff; regulate urban microclimates, reduce heat island effects; provide habitat for birds and beneficial insects; reduce urban crime rates [21].

At present, the localization and modification of the Green Space Factor in China only stay in the prospect of research in this field. This paper aims to draw on the western countries' restrictions on urban space green space construction and guidance methods as well as the reference to the "Seattle Green Factor" assessment method, combined with the actual situation of Harbin high-rise residential areas, to establish a tool for assessing the ecological benefits of urban settlements. The aim is to reduce the urban heat island effect, provide habitat for animals and plants, and improve the thermal factors in the settlement [22–24].

## 42.3 Methodology

### *42.3.1 Seattle Green Factor Assessment Method*

The elements of the Seattle Green Factor Assessment are divided into 7 categories, with a total of 21 assessment indicators. According to the level of ecological benefits provided by the corresponding weight assignment, the weight value is between 0.0 and 1.0. The Seattle Green Factor evaluation formula is SGF $=$ (landscape element 1 area $\times$ score coefficient 1 $+$ landscape element 2 area $\times$ score coefficient 2 $+$ $\cdots$ landscape element N area $\times$ score coefficient N)/total Land area. In the process of evaluation, if the landscape elements of the same plot meet the requirements of multiple evaluation elements, then the various landscape elements of the plot can be superimposed [25, 26].

### *42.3.2 Indicator Revision, Assignment, and Establishment*

Harbin has four distinct seasons and belongs to the severe cold regions of northern China. According to the status quo of high-rise residential areas in Harbin, two expert interviews were conducted, and 30 experts were interviewed, including 11 experts in landscape field, 10 experts in planning field, 2 experts in construction field, 2 real estate practitioners, 1 municipal expert and 4 designers.

The purpose of the first interview was to screen and adjust the Seattle Green Factor Assessment Indicators for the actual situation of high-rise residential areas in Harbin. Subsequently, the questionnaire interview was based on the characteristics of the Harbin cold city, and the scores were assigned to the evaluation indicators. Finally, the Green Space Factor assessment form for high-rise residential areas in Harbin is determined (see Table 42.1).

**Table 42.1** Green Space Factor assessment form for high-rise residential areas in Harbin

| Landscape type | Landscape elements | Score |
|---|---|---|
| Landscaped areas A | A1: Landscaped areas with a soil depth of less than 24 in. | 0.2 |
| | A2: Landscaped areas with a soil depth of 24 in. or greater | 0.6 |
| | A3: Bioretention facilities | 0.9 |
| Plantings B | B1: Mulch, ground covers, or other plants less than 2 in. tall at maturity | 0.2 |
| | B2: At maturity, shrubs or perennials are more than 2 ft. high, and each plant is 12 $ft^2$. (typically not more than 18 $ft^2$) | 0.4 |
| | B3: Small deciduous trees with a crown width of 8–15 ft., calculated as 75 $ft^2$ per tree | 0.4 |
| | B4: Small evergreen arbor crowns are 8–15 ft., calculated as 75 $ft^2$ per tree | 0.4 |
| | B5: Small or medium-sized deciduous trees with a crown width of 16–20 ft., calculated as 150 $ft^2$ per tree | 0.4 |
| | B6: Small or medium-sized evergreen trees with a crown width of 16–20 ft., calculated as 150 $ft^2$ per tree | 0.4 |
| | B7: Medium or large deciduous trees with a crown width of 21–25 ft., calculated as 250 $ft^2$ per tree | 0.5 |
| | B8: Medium or large evergreen trees with a crown width of 21–25 ft., calculated as 250 $ft^2$ per tree | 0.5 |
| | B9: Large deciduous tree with a crown of 26–30 ft., calculated at 350 $ft^2$ per tree | 0.5 |
| | B10: Large evergreen trees with a crown width of 26–30 ft., calculated as 350 $ft^2$ per tree | 0.5 |
| | B11: Large trees with a trunk diameter of 6 in. or more, calculated at 20 $ft^2$ per inch in diameter | 0.8 |
| Permeable pavement C | C1: Pervious pavement on 6–24 in. of soil or gravel | 0.2 |
| | C2: Permeable paving over at least 24 in. of soil or gravel | 0.4 |
| Water D | D1: Closed water landscape | 0.1 |
| | D2: Open water landscape | 0.7 |
| Bonuses E | E1: Planting medium height between 2 and 4 in. | 0.4 |
| | E2: Planting medium height is above 4 in. | 0.6 |
| | E3: Within 5 years into the area covered by vegetation | 0.6 |
| | E4: Drought-tolerant or native plant species | 0.3 |
| | E5: Meet at least 50% of the area that is irrigated annually by using collected rain or snow water | 0.3 |
| | E6: Landscaping visible to passersby from adjacent public right of way or public open spaces | 0.1 |
| | E7: Landscaping in food cultivation | 0.1 |

## 42.4 Result

### *42.4.1 Analysis Results of High-Rise Residential Areas in Harbin Based on Green Space Factor Assessment*

According to the size of the high-rise residential area, nine high-rise residential areas were selected as typical research cases. The typical residential areas selected have different volume ratios, different green rate and cover all evaluation factors as much as possible. The Green Space Factor assessment was carried out for the nine settlements. The higher the Green Space Factor value, the better the ecological benefit of the settlement. The evaluation results are shown in Table 42.2.

### *42.4.2 Analysis of the Impact of Landscape Elements on Ecological Benefits*

It can be seen from the evaluation process of the case settlement that the higher the score of the landscape element indicates the better the ecological benefit of the indicator is. As shown in Fig. 42.1, we can see that the use of any of the following 25 indicators can improve the overall ecological benefits of the settlement, but the ecological benefits provided by different landscape elements are different.

The bioretention facilities (A3) has the highest ecological benefits followed by large trees with a trunk diameter of 6 in. or more (B11) and an open water landscape (D2). The landscaped areas with a soil depth is 24 in. or greater (A2), and planting medium height is above 4 in. (E2) and wall greening (E3) provide high ecological benefits. For the effective improvement of the ecological benefits of settlements, the specific practices are as follows:

In the landscape area, we can see that the bioretention facilities (A3) has the highest weight value, so setting up bioretention facilities can effectively improve ecological benefits. The landscaped areas with a soil depth are of 24 in. or greater scores much higher than landscaped areas with a soil depth of less than 24 in. Therefore, under the condition that the landscape area is limited, increasing the thickness of the soil to 24 in. or more can maximize the ecological benefits of the area.

It can be seen from plant cultivation that different plant planting forms have a great impact on ecological benefits. If stratified planting is followed, that is, arbor–shrub—turf planting, the indicators can be superimposed to improve ecological benefits. Increasing the size of the tree to achieve a crown width of 21 ft. (B7) or more will also improve the ecological benefits. When the trunk diameter of the tree reaches 6 in. or more (B11), the ecological benefit is the highest at this time.

It can be seen from the paving that the ecological benefit of using the area in permeable paving over at least 24 in. of soil or gravel (C2) is the highest when the area of the pavement is limited.

**Table 42.2** Table of basic information on case settlements and composition of green space factors

| High-rise resident | Floor area ($m^2$) | Volume ratio | Greening rate (%) | Green Space Factor indicator composition | Score |
|---|---|---|---|---|---|
| Kaixuan Square | 32,000 | 2.1 | 20 | A1,A2,B1,B2,B3,B4,E4 | 0.20 |
| Walk around Paris | 185,840 | 2.25 | 36.42 | A1,A2,B1,B2,B3,B4,B5,B7,B9,C1,C2,D1,E4 | 0.38 |
| Xidian Home | 200,000 | 2.4 | 40 | A1,A2,B1,B2,B3,B5,B7,B9,C2,D1,E3,E4,E7 | 0.29 |
| Nanjun, Shannon Jarno Phase I | 82,166 | 2.45 | 43.36 | A1,A2,B1,B2,B3,B4,B5,B6,B7,C1,C2,D1,E4,E6 | 0.50 |
| Swan Bay | 156,107 | 2.53 | 40 | A1,A2,B1,B2,B3,B4,B5,B6,B7,C1,C2,D1,D2,E4 | 0.54 |
| Youth City Phase I | 61,843 | 3.33 | 38 | A1,A2,B1,B2,B3,D1,E4 | 0.27 |
| Zhonghai Huijing Xi'an | 97,049 | 3.7 | 45 | A1,A2,B1,B2,B3,B4,C2,D1,E4 | 0.53 |
| Yujintai | 25,780 | 4.8 | 30 | A1,A2,B1,B2,B3,B4,B5,B7,C2,D1,E4 | 0.42 |
| Zijincheng | 59,304 | 5.3 | 37 | A1,A2,B1,B2,B3,B4,B5,C1,C2,E4 | 0.36 |

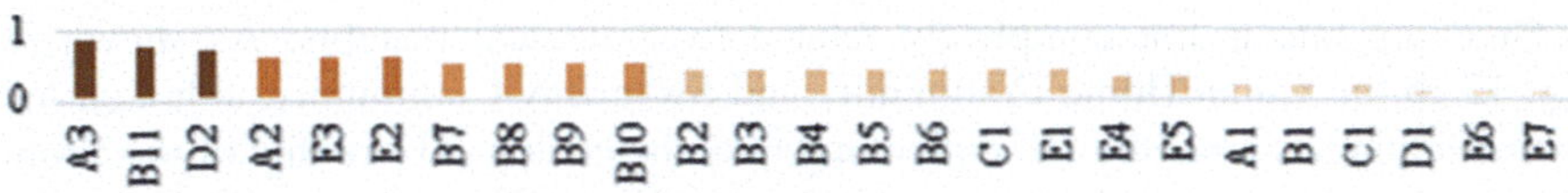


**Fig. 42.1** Indicator score distribution table

The open water landscape (D2) scores higher than the closed water landscape (D1), so using the open water landscape as the water landscape can maximize the ecological benefits of the region.

It can be seen from the bonus points that increasing the area of green roof (E2) and wall greening (E3) can effectively improve ecological benefits.

### *42.4.3 Relationship Between Greening Rate and Ecological Benefits*

After the case evaluation, it is found that the advantages and disadvantages of the ecological benefits of the high-rise residential areas are related to the greening rate. Residential areas with low greening rates also have poor ecological benefits. Residential areas with high greening rate have certain advantages in ecological benefits, but not all residential areas with high greening rate have good ecological benefits.

Greening rate is positively related to ecological benefits: As shown in Fig. 42.2a, the greening rate of the Kaixuan Square community is lower than that of other communities, so the ecological benefits of the Kaixuan Square community are also worse than those of other communities. As the greening rate of residential areas increases, the probability of increasing ecological benefits increases. For example, the greening rate of the Youth City Phase I community is lower than that of the Nanjun, Shannon Jarno Phase I community, and the ecological benefits of the Youth City Phase I community are lower than that of the Nanjun, Shannon Jarno Phase I community. Because of its high greening rate, Zhonghai Huijing Xi'an Community is also superior in terms of the advantages and disadvantages of ecological benefits.

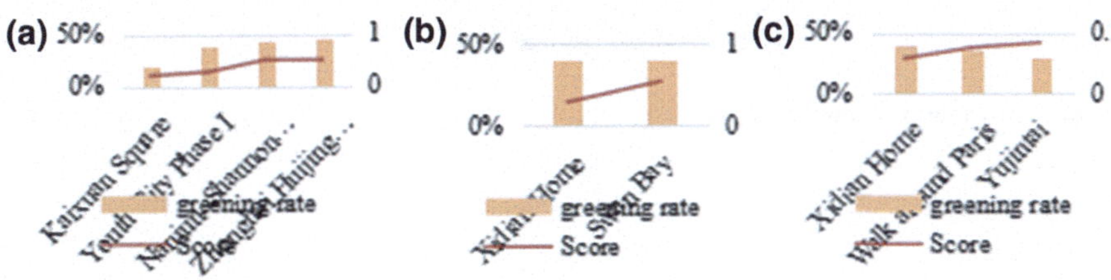


**Fig. 42.2** Relationship between greening rate and ecological benefits

Greening rate is not completely related to ecological benefits: As shown in Fig. 42.2b, in Xidian Home Community and Swan Bay Community, although the two communities have the same greening rate, the ecological benefits of the Swan Bay Community are significantly better than the Xidian Home Community.

Greening rate is negatively correlated with the ecological benefits: As shown in Fig. 42.2c, in the Yujintai Community, the Walk around the Paris Community, and the Xidian Home Community, although the greening rate increased, the ecological benefits decreased.

Subsequently, we analyzed the two landscape types with the largest proportion of Xidian Home Community, Walk around Paris Community, Yujintai Community, and Swan Bay Community, namely the landscape area and planting. As can be seen from Figs. 42.3 and 42.4, the greater the proportion of the landscape area occupied by landscaped areas with a soil depth of 24 in. or greater (A2), the higher the ecological benefits of the settlement. The greater the proportion of trees and shrubs planted, the higher the ecological benefits of settlements. Therefore, improving the greening rate of residential areas will help to improve the ecological benefits of settlements. However, improving the ecological benefits of settlements depend not only entirely on improving the greening rate of settlements but also on the allocation of landscape elements. For residential areas with low greening rate, the adjustment of landscape elements such as increasing the number of shrubs can effectively improve the ecological benefits.

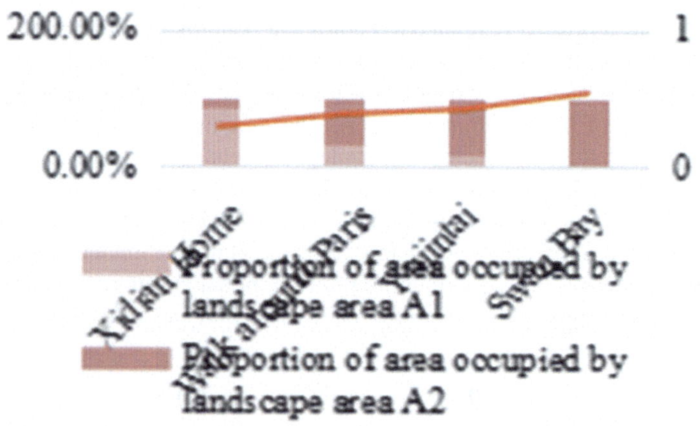


**Fig. 42.3** Area ratio of landscape area

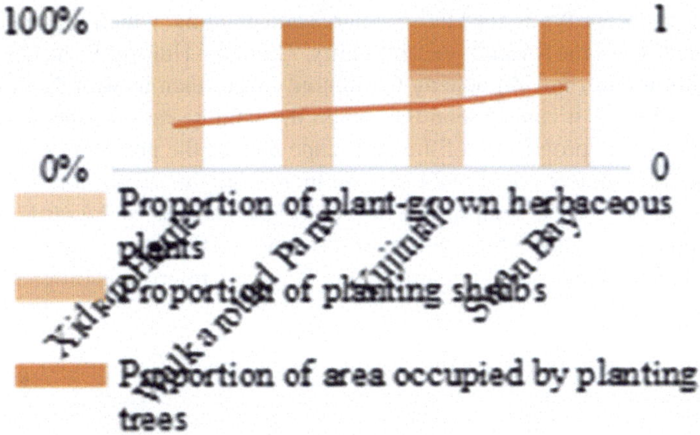


**Fig. 42.4** Proportion of plant area

### *42.4.4 Relationship Between Volume Ratio and Ecological Benefits*

It can be seen from the evaluation results that the advantages and disadvantages of ecological benefits are related to the level of volume ratio, as shown below.

The volume ratio is not completely related to the ecological benefits: As shown in Fig. 42.5a, when the volume ratio of the settlement is low, the volume ratio has little effect on the ecological benefits of the settlement. The ecological benefits at this time depend on the greening rate of the settlement and the configuration of the landscape elements.

The volume ratio is negatively correlated with ecological benefits: As shown in Fig. 42.5b, the residential area with a higher volume ratio has less green space to be used, so it is difficult to have a higher greening rate, resulting in a high-volume-ratio residential area that does not dominate the ecological benefits. The high volume ratio of Yujintai Community and Zijincheng Community is the main reason for the poor ecological benefits.

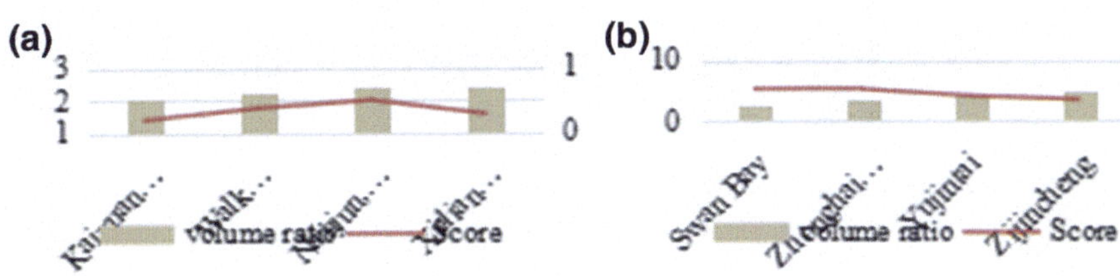


**Fig. 42.5** Relationship between volume ratio and ecological benefits

The two landscape types, the landscape area and planting, which accounted for the largest proportion of the Swan Bay Community, Zhonghai Huijing Xi'an Community, Yujintai Community, and Zijincheng Community were then analyzed. As shown in Figs. 42.6 and 42.7, it can be seen that when the volume ratio of the settlement is high, increasing the proportion of the landscape area of the green area with the soil thickness greater than or equal to 24 in., and increasing the proportion of large trees and shrubs planting, etc., can improve the ecological benefits of the settlement area. When the green area of the high volume ratio residential area is limited, increasing vertical greening, such as wall greening, can effectively improve ecological benefits.

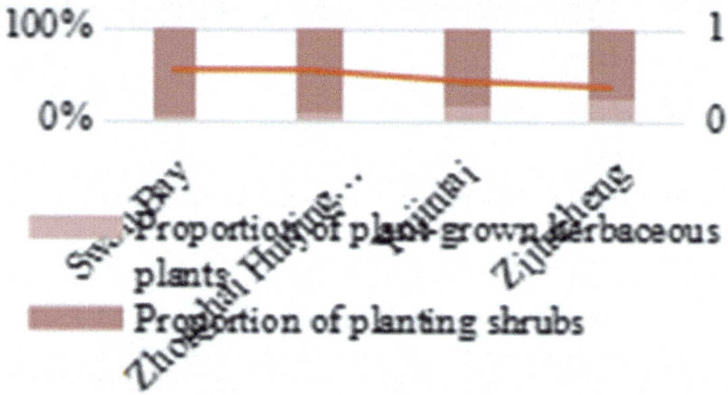


**Fig. 42.6** Landscape area ratio

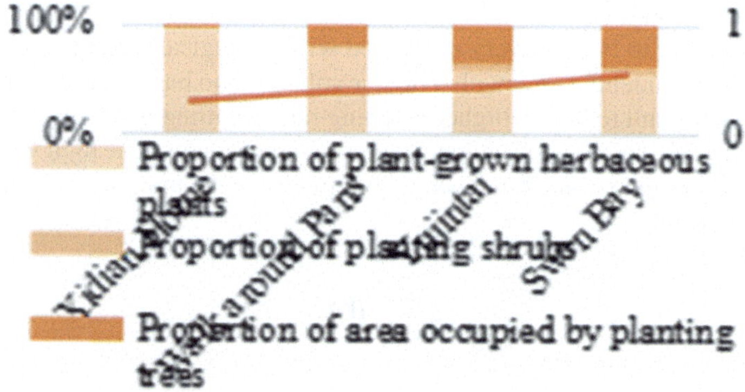


**Fig. 42.7** Proportion of the planted area

## 42.5 Discussion

### 42.5.1 *The Universality of the Green Space Factor Indicators and the Rationality of the Weight Values*

Due to the limitations of this study and the number of samples as well as the lack of support for the Harbin native plant database, there is no precise method. In addition, the weight adjustment and setting methods in this paper use the expert method and other methods, which have certain subjectivity. Therefore, some more quantitative and accurate related verification studies are needed to guide the scientific setting of index setting and weight value.

### 42.5.2 *Try to Develop Green Space Factor Indicators for Urban Land Use*

In this study, only the Green Space Factor tool is applied to the high-rise residential areas in Harbin. It is suggested that the application level of this type of tool can be extended to provide an urban development center.

### 42.5.3 *Limitations of Green Space Factor Methods*

Although the Green Space Factor Assessment Tool provides quantitative criteria for the development of urban green infrastructure, there is no restriction on its design tools, and there is no relevant assessment of the quality of the completed project. Therefore, project managers and project designers are required to conduct relevant processing to achieve the desired esthetic value and ecological service value.

## 42.6 Conclusions

This study is based on the Seattle Green Factor tool to construct a Green Space Factor tool for Harbin, a cold winter city, to assess the ecological benefits of high-rise residential areas in Harbin. Through the assessment of the ecological benefits of nine case settlements in Harbin, we draw the following conclusions:

The advantages and disadvantages of the ecological benefits of high-rise settlements are directly related to the allocation of landscape elements. The higher the scores of the landscape elements are, the better the ecological benefits are.

The ecological benefits of residential areas with low greening rates are not necessarily poor. For low green areas, the settlement can also achieve high ecological benefits by adjusting the configuration of landscape elements.

As the volume ratio of high-rise residential areas increases, the difficulty of providing better ecological benefits increases. For high-volume-dwelling settlements, high ecological benefits can also be achieved by adjusting the configuration of landscape elements.

**Acknowledgements** This study was supported by the National Natural Science Foundation of China (No. 51438005).

## References

1. De Lotto, R.: Assessment of development and regeneration urban projects: cultural and operational implications in metropolization context. Int. J. Energy Environ. **2**(1), 24–35 (2008)
2. Gómez-Baggethun, E., Gren, A.: Urbanization, biodiversity and ecosystem services: challenges and opportunities. Urban Ecosyst. Serves, 175–251 (2013)
3. McPhearson, T., Maddox, D. Gunther, B. & Bragdon, D.: Local assessment of New York City: biodiversity, green space, and ecosystem services. In: Urbanization, Biodiversity and Ecosystem Services: Challenges and Opportunities, pp. 355–383 (2013)
4. Kohsaka, R.: Developing biodiversity indicators for cities: Applying the DPSIR model to Nagoya and integrating social and ecological aspects. Ecol. Res. **25**(5), 925–936 (2010)
5. Ferguson, B.C., Frantzeskaki, N., Brown, R.R.: A strategic program for transitioning to a Water Sensitive City. Landscape & Urban Plan. **117**(9), 32–45 (2013)
6. Pan, J.: Analysis of human factors on urban heat island and simulation of urban thermal environment in Lanzhou city, China. J. Appl. Remote Sens. **9**(1), 095999 (2015)
7. An-Qing, M.A, Bing-Ran, M.A, Zhen, Z.: Driving Factors of the Thermal Environment of Coastal City: A Case Study in Qingdao. Periodical of Ocean University of China (2016)
8. Weng, Q.: Fractal analysis of satellite-detected urban heat Island effect. Photogram. Eng. Remote Sens. **69**(5), 555–566 (2003)
9. Gallo, K.P., Mcnab, A.L., Karl, T.R.: The use of a vegetation index for assessment of the urban heat island effect. Int. J. Remote Sens. **14**(11), 2223–2230 (1993)
10. Magee, N., Curtis, J., Wendler, G.: The urban heat island effect at Fairbanks, Alaska. Theoret. Appl. Climatol. **64**(1–2), 39–47 (1999)
11. Jack A.: The concept of ecosystem services in adaptive urban planning and design: a framework for supporting innovation. Landscape Urban Plan. 125 (2014)
12. Kong, F., Yin, H., Wang, C.: A satellite image-based analysis of factors contributing to the green-space cool island intensity on a city scale. Urban Forest. Urban Green. **13**(4), 846–853 (2014)
13. Huang, P.-S., Tsai, S.-M., Lin, H.-C., Tso I-M.: Do biotope area factor values reflect ecological effectiveness of urban landscapes? A case study on university campuses in central Taiwan. Landscape Urban Plan. **143** (2015)
14. Lakes, T., Kim, H.: The urban environmental indicator "Biotope Area Ratio". Ecol. Ind. **13**(1), 93–103 (2012)
15. Thierfelder, H., Kabisch, N.: Strategic urban development in Berlin-challenges for future urban green space development. Environ. Sci. Policy **62** (2015)
16. Gary, A.: Case study and sustainability assessment of Bo01, Malmö, Sweden. J. Green Build. 34–50 (2016)

17. Zhang W., Wang K.: Policy tools based on green infrastructure ecosystem service assessment, green space index research—taking the Berlin habitat area index and seattle green index as examples. Chin. Garden **33**(09):78–82 (2017)
18. Mckendry, C., Janos, N.: Greening the industrial city: equity, environment, and economic growth in Seattle and Chicago. Int. Environ. Agreements: Politics, Law Econ. **15**(1), 45–60 (2015)
19. Schaffler, A., Swilling, M.: Valuing green infrastructure in an urban environment under pressure—The Johannesburg case. Ecol. Econ. **86**, 246–257 (2013)
20. Lotto, R.D., Casella, V., Franzini, M.: Estimating the Biotope Area Factor (BAF) by means of existing digital maps and GIS Technology. In: International Conference on Computational Science & Its Applications. Springer, Cham (2015)
21. Mason, L.: Seattle's green factor. living architecture monitor magazine. Green Roofs Healthy Cities, 34–35 (2007)
22. Silberstein, R.P, Vertessy, R.A., Morris, J.: Modelling the effects of soil moisture and solute conditions on long-term tree growth and water use: a case study from the Shepparton irrigation area, Australia. Agricultural Water Manage. **39**(2–3), 0–315 (1999)
23. Bin W, Jingxu, W., Tao, Z.: Improvement of assessment methods for ecological effect of urban greenland. Urban Environ. Urban Ecol. (1997)
24. Han, B., Liu, H., Wang, R.: Urban ecological security assessment for cities in the Beijing–Tianjin–Hebei metropolitan region based on fuzzy and entropy methods. Ecol. Model. **S0304380014006280**, 318 (2015)
25. Hirst, J., Morley, J., Bang, K..: Functional landscapes: assessing elements of seattle green factor. International Report. The Berger Partnership PS (2008)
26. Seattle Department of Planning and Development: Seattle Green Factor. http://www.seattle.gov/dpd/Permits/GreenFactor. 22 Dec 2015

# Chapter 43
# Critical Review of Ageing Mechanisms and State of Health Estimation Methods for Battery Performance

**K. Saqli, H. Bouchareb, M. Oudghiri and N. K. M'Sirdi**

**Abstract** Battery Management System (BMS) is an essential component for lithium-ion battery-based devices. It provides a variety of functionalities that help to improve the overall lifespan of the battery, including states estimation algorithms. An accurate state of health is one of the essential features an advanced BMS provides, in order to track long-term performance and ensure reliable operation of the battery. This paper aims to provide a structured review of ageing mechanisms that affect the battery performance and also some of the most relevant algorithms that were used in the recent literature to predict the actual health of the battery. Finally, based on the different approaches that were studied, evaluation criteria are proposed to help judge the performance of State Of Health (SOH) estimation algorithm.

## Acronyms

| | |
|---|---|
| AEKF | Adaptive Extended Kalman Filter |
| BMS | Battery Management System |
| DEKF | Dual Extended Kalman Filter |
| EKF | Extended Kalman Filter |
| EV | Electric Vehicle |
| HEV | Hybrid Electric Vehicle |
| LS-EKF | Lebesgue Sampling Based Extended Kalman Filter |
| LTV | Linear Time-Varying |
| OCV | Open Circuit Voltage |

K. Saqli (✉) · H. Bouchareb · M. Oudghiri
Sidi Mohammed Ben Abdellah University - National School of Applied Sciences, Fez, Morocco
e-mail: saqli.khadijah@gmail.com

N. K. M'Sirdi
LIS - Informatics and Systems Laboratory (LIS CNRS 7020), Aix Marseille University CNRS, 13397 Marseille Cedex, France
e-mail: Kouider-Nacer.Msirdi@lis-lab.fr



J. Littlewood et al. (eds.), *Sustainability in Energy and Buildings*, Smart Innovation, Systems and Technologies 163, https://doi.org/10.1007/978-981-32-9868-2_43

| | |
|---|---|
| RUL | Remaining Useful Life |
| SEI | Solid Electrolyte Interface |
| SOC | State Of Charge |
| SOF | State Of Function |
| SOH | State Of Health |
| SPI | Solid Permeable Interface |
| SVR | Support Vector Regression |
| UKF | Unscented Kalman Filter |
| WTLS | Weighted Total Least Square |

## 43.1 Introduction

To cope with climate changing and global warming, communities have been promoting clean energy and they have been calling for a fossil-free society. According to the Environmental Protection Agency, 72% of carbon monoxide emissions come from automobiles only [1]. Thus, interest in Electric Vehicles (EVs) has grown fast.

Electric vehicles are much cleaner and cheaper to drive [2]. They usually use lithium-ion batteries as an energy source, this type of chemistries presents the advantages of high energy density and low self-discharge compared with other rechargeable cells [3]. However, the capability to perform correctly and to store energy tends to decrease with time due to the ageing phenomenon. Therefore, Battery Management Systems (BMSs) are needed to boost the battery performance and improve its safety.

In general, BMS is an electronic system that controls and monitors the battery for safe usage and long lifespan. It has three basic tasks: data acquisition, battery states estimation and electrical management of the measured input to take action when needed.

Besides cell balancing, an advanced battery management system needs to estimate the battery states, whether it is the state of charge, state of health or any other state indication parameter. The BMS uses the acquired data as an input of the prediction algorithms.

Battery State of Charge (SOC) is one of the essential parameters a BMS needs to determine. It represents an instant indication of the battery status that estimates the percentage of the available capacity. On the other hand, the State Of Health (SOH), as its name implies, is a parameter that reflects the battery health and its ability to convey a specific performance compared with a fresh battery [4]. Having control over ageing causes can help reduce the degradation effects, but once the SOH hits a certain percentage the battery can no longer be used for electric vehicles [5].

### *43.1.1 SOH Estimation in Literature*

Numerous surveys have been conducted in recent years to classify the SOH estimation methods [4, 6–9]. In order to have a clear vision of the different approaches that were developed or improved in the literature, Berecibar et al. [4] presented a detailed classification of battery SOH estimation methods, that arranges those approaches in two main categories depending on the way they were carried out. The strengths and weaknesses of each method were compared and analyzed, and the degradation mechanism was also presented. The authors of [6] review the estimation strategies and methodologies used in electric vehicles and classified them under three main categories that are: battery management related, vehicle energy management related and vehicle control related. Except for the battery State Of Charge (SOC), the rest of the estimation strategies are briefly discussed.

An extensive review of the methods for monitoring lithium-ion batteries in electric and hybrid vehicles is described in [7]. Based on an impressive 375 sources including technical and scientific literature, the estimation approaches for battery state of charge, state of health, capacity, impedance, available power and remaining useful life are provided.

Ungurean et al. [8] define a comprehensive, structured review of battery models and estimation methods used to determine the battery state of health. However, the ageing mechanisms of the battery are not presented. Same goes for Xiong et al. [9] who provide a detailed presentation of battery SOH estimation algorithms with emphasis on their limits and strengths.

Before starting to define the state of health estimation approaches it is important to first outline the ageing mechanisms since, without having a clear insight of the ageing factors, we can not appropriately determine the battery health.

## 43.2 Ageing Mechanisms

Degradation mechanisms affect battery health through reduction of total capacity and or expansion of equivalent series resistance, and observing those quantities that vary slightly and impact the battery performance as cell ages, is of great value to determine the battery state of health as accurately as possible.

Generally, unwanted side reactions and structural degradation that occur inside the battery cell lead to capacity and power loss, and the complexity of the Li-ion based batteries makes the processes of their ageing at anodes and cathodes even more complicated [10]. Researchers have proved that ageing mechanisms at anodes differ from those at cathodes [11]. Therefore, the knowledge of these ageing processes is mandatory to control battery health as well as other parameters.

### *43.2.1 Ageing in the Negative Electrode*

Depending on electrodes composition, safety properties and ageing ratio can differ from one cell to another. Therefore graphitic carbon materials are most widely used as anode material, and thus, many research papers were devoted in reviewing the ageing mechanisms of graphite-based anodes [12, 13].

Ageing effects can occur during storage just after manufacturing and during use, and can be observed at three different scales that are as follows:

- **Electrode/electrolyte interface**: The main ageing factor that considerably affects the graphite electrode properties happens once the electrolyte comes into contact with the graphite, especially during the first charge inside of a battery manufacturing facility [10]. The electrochemical stability window of the electrolyte does not include the voltages at which the graphite-based negative electrodes operate [14]. Hence a protective layer is developed as a result of electrolyte decomposition, known as Solid Electrolyte Interface (SEI) [15]. Once it is formed, it encapsulates the graphite and blocks electrons from transport, to prevent any further electrolyte degradation [15]. The porous structure of this passivating layer allows the transition of Li-ion from the electrolyte to the particle, adding resistance to the motion of Li-ion.
- **Within the electrode particles**: Ageing processes that occur inside the particles can have a minor negative impact on the electrode performance. The particles change volume when charging or discharging, due to the absorption and withdrawal of Li-ions [10], leading to structural changes that can cause particles to crack and be less ordered compared to the original structure, which enables further SEI growth and active material loss [16].
- **Composite electrode**: Volume changing affects the electrode porosity that facilitates the penetration into the bulk of the electrode. It is also considered as the main reason for the loss of contact between active cathode particles, that can be accelerated when overdischarging the battery cell, resulting in higher cell impedance and capacity loss [17].

### *43.2.2 Ageing in the Positive Electrode*

Ageing can also occur at the positive electrode and can affect battery performance, and as for the negative electrode the chosen material impact significantly the battery performance. In terms of performance cost and abundance, most Lithium-ion battery cells use Lithium manganese oxides with spinel structure and lithium nickel cobalt mixed oxides with layered structures as the cathode material [18]. An aged positive electrode material can have different effects that can be summarized as follows :

- **Film formation**: The electrolyte oxidization and the decomposition of the electrode particles form a film layer called Solid Permeable Interface (SPI) that has different composition and properties from the anode SEI [19]. High temperatures induce the dissolution of metals from the electrode into the electrolyte accompanied by $CO_2$ gas evolution breaking down the overall crystal structure of the electrode, leading to capacity loss [10].
- **Loss of active material**: The intercalation of lithium into the crystal structures of the cathode active materials causes mechanical stress that leads to strains of the metallic oxide particles and variation of their molar volume. These distortions are

known as a phase transition. They are often reversible, and they proceed as lithium ions are inserted/extracted in/from the positive electrode. The most used lithium-ion battery cells have a structured layer, therefore when the battery is overcharged the lithium particles that hold these layers apart are extracted and the layers can fall on each other reducing lithium storage sites and causing capacity loss [20].

- **Metal dissolution**: The dissolution phenomena of the metal from the electrode into the electrolyte results in capacity loss and can reprecipitate on the surface as a high-resistance film. The dissolved manganese particles move to the anode and merge into the SEI interface enhancing the self-discharge of the battery cell. In general manganese dissolution can occur at low potential, where trivalent manganese ions disproportionate into tetravalent and divalent ions, or at elevated potential known also as acid dissolution accelerated by hydrofluoric acid [10].

$$2\,\mathrm{Mn}(III) \longrightarrow \mathrm{Mn}(IV) + \mathrm{Mn}(II) \tag{43.1}$$

Battery cells age even when they are not used. Ageing can be separated into calendar ageing that represents degradation caused by battery storage, and cycle ageing that occurs when charging or discharging the battery.

## 43.3 Health Estimation

The estimation of the battery state of health is usually associated with the determination of two essential indicators that are battery capacity and internal resistance, which, respectively, reflect the energy and power potential of the battery cell [21, 22]. Different algorithms and processes can be applied to these parameters to track the degradation of the battery, and for better understanding of the performance of each approach, literature has proposed different classifications for these methods.

### *43.3.1 Internal Resistance*

Internal resistance stems from the fact that a battery is not an ideal voltage source. It represents a direct indicator of the battery health that serves to spot any defects tending to damage the battery, and that explains the interest of researchers to measure this parameter.

Ageing causes the battery to lose its capacity and increase its internal resistance. Yu et al. [23], define the battery internal resistance based on changes in the voltage curve while applying a discharge current, and by using Ohm's law, the internal resistance can be approximated as follows:

$$R = \frac{\Delta U}{\Delta I} \tag{43.2}$$

where $\Delta I$ represents the current step-change and $\Delta U$ is the corresponding voltage to the same step-change.

The internal resistance was proven to change with different factors such as battery size, chemistry, temperature and other parameters. Chen et al. [24] considered that the battery internal resistance depends on the state of charge, the current rate and temperature, they conducted a constant current test under different temperatures to provide continuous R-SOC results.

### 43.3.2 Coulomb Counting

Another way to estimate the battery SOH is by tracking the amount of charge that flows in or out during the charge–discharge process [8]. It first calculates the capacity at discharge by integrating the value of discharge current in time, so a high-precision current sensor is needed, then it uses this value to compute the SOH as follows:

$$SOH[\%] = \frac{Q_{max}}{Q_{nominal}} \times 100 \tag{43.3}$$

where $Q_{max}$ is the maximum available capacity and $Q_{nominal}$ is the measured capacity. The accuracy of this approach requires high monitoring and storing capacity and a high-precision current sensor. Being not affected by the temperature, DOD or C-rate gained her wide popularity among methods that estimate the battery SOH and also SOC [4].

### 43.3.3 Kalman Filtering

Kalman filtering is a sturdy algorithm that was used over history to estimate battery parameters and states. It is defined as an adaptive method that operates in two steps [4]. First, the algorithm predicts the current output variable, then, the estimation is updated to obtain more accurate results (Fig. 43.1).

Where $\hat{x}$ is the estimated state, F is the state transition matrix, B is the control matrix, u is the input vector, P and Q are respectively, the state and the process variance matrix, y is the output measurement, K is the Kalman gain, H is the measurement matrix and R is the measurement variance matrix.

Kalman filter can operate properly only when the system is linear. However, to deal with nonlinear systems extensions of this method have been developed such as Extended Kalman Filter (EKF), Unscented Kalman Filter (UKF), Adaptive EKF (AEKF) and Dual Extended Kalman Filter (DEKF) [8].

**Extended Kalman filter** Extended Kalman filter is used when the transition and the measurement equations are not linear. A linearization process is applied at every time step to approximate the nonlinear system with a Linear Time-Varying system (LTV)

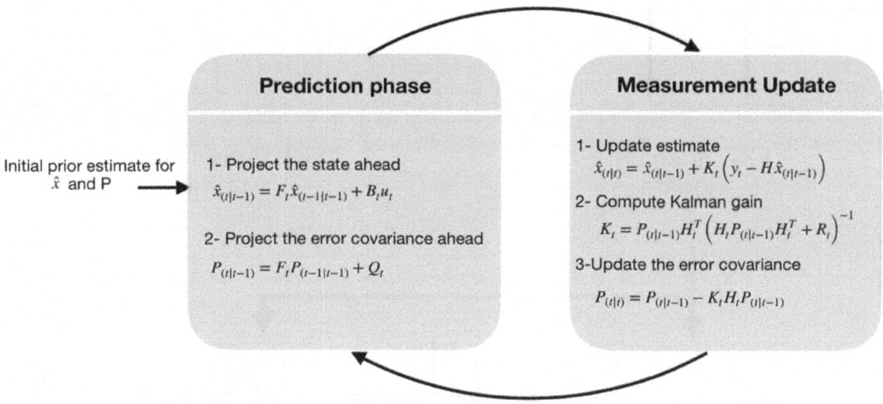


**Fig. 43.1** Process of kalman filter

[25]. The algorithm was widely used for battery modelling and states estimation [26, 27] thanks to its accuracy and speed.

In [28], the authors use a combination of extended Kalman filter and EIS algorithm to estimate the battery internal resistance and capacity to provide reliable information about its remaining stored energy.

Yan et al. [29] perform a co-estimation of battery SOC and SOH using Lebesgue Sampling based Extended Kalman filter (LS-EKF) to reduce the computation cost when the states change slowly. The SOH and RUL were first predicted using LS-EKF. Then the obtained results are used to estimate the battery SOC.

**Dual Extended Kalman filter** As the name suggests, Dual Extended KF (DEKF) is a combination of two extended Kalman filter that synchronously estimate the battery states and parameters [25] (see Fig. 43.2). N. Vassiliadis et al. propose a use case life cycle analysis of DEKF for battery SOC and SOH prediction. The paper investigates the performance of the DEKF algorithm including different SOH stages and load dynamics. A comparison of the performance of DEKF versus a simple EKF shows a partial improvement in SOC accuracy.

In Fig. 43.2 $\hat{x}$ represents the estimated state vector, $\hat{\theta}$ is the estimated parameter vector, $u$ is the input vector and $y$ is the measurement vector. The authors of [30] propose a combination of charging voltage curve and DEKF algorithm to estimate the battery SOC. The transformation of the charging voltage curve is used to determine the battery's internal resistance and capacity, then DEKF algorithm is used to estimate the battery SOC. The results of this combination show an improvement in accuracy and elimination of the measurement noise.

**Unscented Kalman filter** Unscented Kalman filter also known as the sigma point Kalman filter is an algorithm where the state distribution is specified using a minimal set of carefully chosen sample points [31]. Unlike the EKF that loses its accuracy when dealing with highly nonlinear systems, the UKF uses the true nonlinear models to estimate the results as accurately as possible [32].

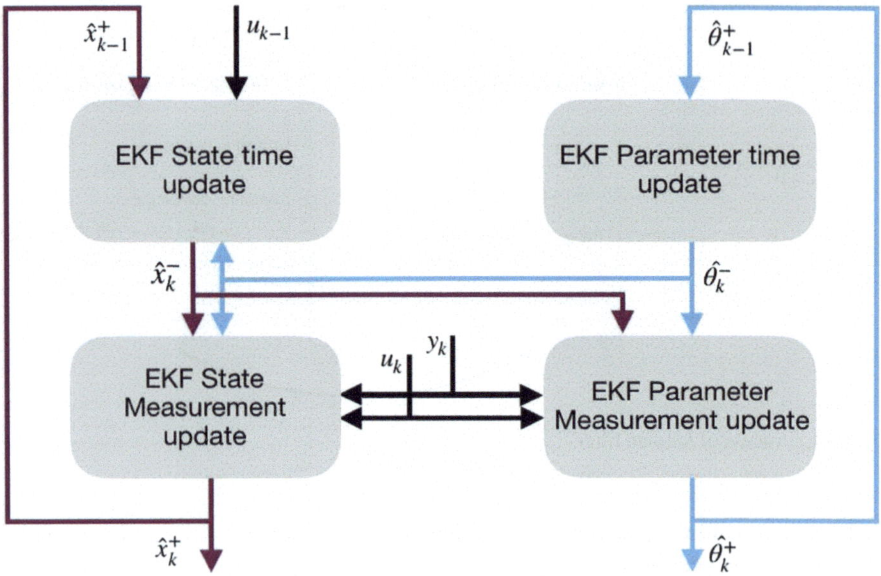


**Fig. 43.2** DEKF structure

Zhang et al. [33] used the UKF approach to predict the battery SOC, internal resistance and capacity, based on an improved Equivalent Circuit Model (ECM). The method was also used to evaluate different battery states (SOC, SOH, SOF), and was proved to be timesaving and efficient.

### *43.3.4 Fuzzy Logic*

Fuzzy logic is a mathematical concept that models complex nonlinear systems and solves uncertainties in a given problem. The method uses crisp sets or fuzzy sets that allow elements to be in a set with a degree of membership. A membership function is used to define the relationship between a member of a set and its degree of membership that range from 0 to 1 [34]. Taking the battery state of health as an example, we can define the membership functions outputs as *healthy*, *tolerable* and *not healthy*.

In [35], the author uses a fuzzy logic-controlled methodology to predict the battery SOH. The process is divided into two parts: first, the cell resistance and maximum capacity are determined based on voltage, current, temperature and time, second using the resistance and the maximum capacity values, SOH is predicted by applying fuzzy logic approach.

State of health estimation of Li-ion battery based on a fuzzy logic system was also presented in [36], using temperature and the absolute value of the current the fuzzy logic system is constructed, which enhances the SOH value on both the calendar and cycling life.

### 43.3.5 Support Vector Regression Algorithm

The support Vector Regression (SVR) algorithm is a nonlinear generalization that implements a structural risk minimization principle to get a good generalization on a limited number of learning patterns [37]. It adopts the original machine learning algorithm to predict the output of a given input.

In [38], the author used a novel support vector regression that takes the battery capacity as a state variable, and constant current and constant voltage as an input variable of the system to predict the battery SOH. Furthermore, the model was combined with a particle filter to reduce the errors caused by measurement noises and to provide more accurate and robust result.

Yang et al. [39] proposed a novel SOH estimation approach. They started by building an improved battery model, which combines OCV modelling and Thevenin equivalent circuit model. The battery SOC was estimated using the joint extended Kalman filter-recursive-least squares algorithm. Then a reliable SOH estimation result was obtained using a combination of the support vector regression approach and the particle swarm optimization algorithm.

### 43.3.6 Least Squares

The least square algorithm is a widely used procedure to estimate the parameters of a given system. The basic problem is to identify the best fit function that minimizes the sum of errors between measured output and system response.

To estimate the battery SOH, Pett [5] used an improved version of the RLS algorithm called the Weighted Total Least Squares estimation approach (WTLS). The author started by identifying the main drawback of using a standard least squares approach to estimate the battery health. Since this method doesn't take into consideration the uncertainty that we have on the input measurements (SOC) the author uses WTLS approach to determine the battery health with knowledge of both noises on the accumulate ampere-hour measurements and on the battery SOC. The method shows better results compared to a standard least squares algorithms and it is suitable for real-time implementation.

### 43.3.7 Evaluation Criteria

The previous section has presented some of the recent commonly used approaches in the literature to estimate the actual health of the battery. The accuracy of the predicted SOH is a significant parameter that authors try to prove. However, this parameter only is insufficient to evaluate the estimation algorithm results. Thus we propose the following criteria that can be used to judge the performance of a given algorithm:

- *Battery model*: When using an algorithm that predicts the battery states based on a specific battery model, the result precision can differ since some models lose their accuracy with the use, due to different ageing mechanisms.
- *Computation cost*: It is an indication of whether the proposed approach is suitable for a real-time application or not. In other words, if the method can be implemented in a BMS running an embedded system.
- *Time cost*: It defines the duration of the execution, the time needed to provide the result, which depends on the complexity of the algorithm and also the hardware specification.
- *Estimation precision*: Is an indication of the calculation error, defining how much the result diverges from the actual one.

An extensive comparison of health estimation algorithms is presented in [4]. The author concludes that there is no unique or perfect method to estimate the battery health. The choice of method should be selected based on the target application, what is needed to be estimated and how much error tolerance is acceptable.

## 43.4 Conclusion

Lithium-ion batteries have been widely used as the main energy storage solution for EVs. To make the most of this technology, BMS was used to manage the battery performance and ensure the safe operation of the cells.

State of health estimation is a crucial parameter that helps batteries operate in a safe condition. This review presented the ageing mechanisms that affects the lithium batteries during their life to help build a basic understanding of the different factors that accelerate the process of ageing and cause capacity fading. A description of the most relevant methods that were presented in the literature to estimate the battery internal resistance and capacity are provided. Finally, the author proposes some evaluation criteria that can help judge the performance of the health prediction algorithm.

## References

1. Victoria Environment Protection Agency (2018). https://www.epa.vic.gov.au/your-environment/air/vehicle-emissions-and-air-quality
2. Canepa, K., Hardman, S., Tal, G.: An early look at plug-in electric vehicle adoption in disadvantaged communities in California. Transport Policy (2019)
3. Zhang, C., Li, K., Mcloone, S., Yang, Z.: Battery modelling methods for electric vehicles—a review. In: 2014 European Control Conference (ECC) (2014)

4. Berecibar, M., Gandiaga, I., Villarreal, I., Omar, N., Van Mierlo, J., Van den Bossche, P.: Critical review of state of health estimation methods of Li-ion batteries for real applications. Renew. Sustain. Energy Rev. **56**, 572–587 (2016)
5. Plett, G.L.: Recursive approximate weighted total least squares estimation of battery cell total capacity. J. Power Sources **196**(4), 2319–2331 (2011)
6. Cuma, M.U., Koroglu, T.: A comprehensive review on estimation strategies used in hybrid and battery electric vehicles. Renew. Sustain. Energy Rev. **42**, 517–531 (2015)
7. Waag, W., Fleischer, C., Sauer, D.U.: Critical review of the methods for monitoring of lithium-ion batteries in electric and hybrid vehicles. J. Power Sources **258**, 321–339 (2014)
8. Ungurean, L., Cârstoiu, G., Micea, M.V., Groza, V.: Int. J. Energy Res. (2016)
9. Xiong, R., Li, L., Tian, J.: Towards a smarter battery management system: a critical review on battery state of health monitoring methods. J. Power Sources **405**, 18–29 (2018)
10. Vetter, J., Novak, P., Wagner, M.R., Veit, C., Moller, K.-C., Besenhard, J.O., Winter, M., Wohlfahrt-Mehrens, M., Vogler, C., Hammouche, A.: Ageing mechanisms in lithium-ion batteries. J. Power Sources **147**, 269–281 (2005)
11. Barré, A., Deguilhem, B., Grolleau, S., Gérard, M., Suard, F., Riu, D.: A review on lithium-ion battery ageing mechanisms and estimations for automotive applications. J. Power Sources **241**, 680–689 (2013)
12. Wu, Y.P., Rahm, E., Holze, R.: Carbon anode materials for lithium ion batteries. J. Power Sources **114**, 228–236 (2003)
13. Mekonnen, Y., Sundararajan, A., Sarwat, A.I.: A review of cathode and anode materials for lithium-ion batteries. In: 2016 IEEE. 978-1-5090-2246-5/16/
14. Zhang, S., Ding, M.S., Xu, K., Allen, J., Jow, T.R.: Understanding solid electrolyte interface film formation on graphite electrodes. Electrochem. Solid-State Lett. **4**(12), A206–A208 (2001)
15. Verma, P., Maire, P., Novák, P.: A review of the features and analyses of the solid electrolyte interphase in Li-ion batteries. Electrochimica Acta **55**, 6332–6341 (2010)
16. Vetter, J., Novák, P., Wagner, M., Veit, C., Möller, K.-C., Besenhard, J., Winter, M., Wohlfahrt-Mehrens, M., Vogler, C., Hammouche, A.: Ageing mechanisms in lithium-ion batteries. J. Power Sources **147**(1–2), 269–281 (2005)
17. Wang, Y., Guo, X., Greenbaum, S., Liu, J., Amine, K.: Solid electrolyte interphase formation on lithium-ion electrodes. Electrochem. Solid-State Lett. **4**(6), A68–A70 (2001)
18. Dubarry, M., Truchot, C., Liaw, B.Y., Gering, K., Sazhin, S., Jamison, D., Michelbacher, C.: Evaluation of commercial lithium-ion cells based on composite positive electrode for plug-in hybrid electric vehicle applications. Part II. Degradation mechanism under 2 C cycle aging. J. Power Sources **196**, 10336–10343 (2011)
19. Malmgren, S., Ciosek, K., Hahlin, M., Gustafsson, T., Gorgoi, M., Rensmo, H., Edström, K.: Comparing anode and cathode electrode/electrolyte interface composition and morphology using soft and hard X-ray photoelectron spectroscopy. Electrochimica Acta **97**, 23–32 (2013)
20. Wikner, E.: Lithium ion battery aging: battery lifetime testing and physics-based modeling for electric vehicle applications. Thesis for the degree of licentiate of engineering
21. Waag, W., Köbitz, S., Sauer, D.U.: Experimental investigation of the lithium-ion battery impedance characteristic at various conditions and aging states and its influence on the application. Appl. Energy **102**, 885–897 (2013)
22. Plett, G.L.: Dual and joint EKF for simultaneous SOC and SOH estimation (2004)
23. Yu, J., Mo, B., Tang, D., Yang, J., Wan, J., Liu, J.: Indirect state-of-health estimation for lithium-ion batteries under randomized use. Energies (2017)
24. Chen, Y., Liu, X., Yang, G., Geng, H.: An internal resistance estimation method of lithium-ion batteries with constant current tests considering thermal effect. In: IECON 2017-43rd Annual Conference of the IEEE Industrial Electronics Society (2017)
25. Plett, G.L.: Extended Kalman filtering for battery management systems of LiPB-based HEV battery packs Part 1. Background. J. Power Sources **134**, 252–261 (2004)
26. Afshar, S., Morris, K., Khajepour, A.: State of Charge estimation via extended Kalman filter designed for electrochemical equations. IFAC-PapersOnLine **50**(1), 2152–2157 (2017)

27. Jokić, I., Zečević, Ž, Krstajić, B.: State-of-charge estimation of lithium-ion batteries using extended Kalman filter and unscented Kalman filter. In: 23rd International Scientific-Professional Conference on Information Technology (2018)
28. Urbain, M., Raël, S., Davat, B., Desprez, P.: State estimation of a lithium-ion battery through Kalman filter. In: IEEE Power Electronics Specialists Conference (2007)
29. Yan, W., Zhang, B., Zhao, G., Tang, S., Niu, G., Wang, X.: Battery management system with Lebesgue sampling-based extended Kalman filter. IEEE Trans. Ind. Electron. **66**(4) (2019)
30. Wang, L., Dong, L., Liu, Q., Liu, L., Zhao, X.: State of charge estimation for $LiFePO_4$ battery via dual extended kalman filter and charging voltage curve. Electrochimica Acta **296**, 1009–1017 (2019)
31. Rampelli, M., Jena, D.: Advantages of unscented Kalman filter over extended Kalman filter in dynamic state estimation of power system network. In: Michael Faraday IET International Summit: MFIIS-2015, September 12–13, 2015, Kolkata, India
32. Sedighfar, A., Moniri, M.R.: Comparison of three well-known filters for the battery state of healt estimation application. In: 4th International Conference on Frontiers of Signal Processing (2018)
33. Zhang, F., Liu, G., Fang, L.: Battery state estimation using unscented Kalman filter. In: IEEE International Conference on Robotics and Automation Kobe International Conference Center (2009)
34. Patil, P.D., Patel, J.: A comprehensive review on fuzzy logic system. Int. J. Eng. Comput. Sci. **3**(11), 9160–9165 (2014). ISSN:2319-7242
35. Kim, J.: Fuzzy logic-controlled online state-of-health (SOH) prediction in large format $LiMn_2O_4$ cell for energy storage system (ESS) applications. In: 2014 IEEE International Conference on Industrial Technology (ICIT), February 26–March 1, 2014, Busan, Korea
36. Zenati, A., Desprez, P., Razik, H., Rael, S.: A methodology to assess the state of health of lithium-ion batteries based on the battery's parameters and a fuzzy logic system. In: 2012 IEEE International Electric Vehicle Conference
37. Basak, D., Pal, S., Patranabis, D.C.: Support vector regression. Neural Inf. Process.—Lett. Rev. **11**(10) 2007
38. Wei, J., Dong, G., Chen, Z.: Remaining useful life prediction and state of health diagnosis for lithium-ion batteries using particle filter and support vector regression. IEEE Trans. Ind. Electron. **65**(7), 5634–5643 (2018)
39. Yang, D., Wanga, Y., Pana, R., Chenb, R., Chen, Z.: State-of-health estimation for the lithium-ion battery based on support vector regression. Applied Energy (2017)

# Chapter 44
# Climate Adaptation of "Smart City" by Assessing Bioclimatic Comfort for UBEM

**Ilya V. Dunichkin and Irina N. Ilina**

**Abstract** The article considers the issues of forecasting the microclimate of cities and the wind energy potential of residential buildings applicable to Russia, Northern and Eastern Europe. The study analyzed the climatic structure of a large city, bioclimatic comfort, and presented a climatic analysis on the example of Moscow. The interrelation of a wind regime with climatic and town planning factors is specified. Possible approaches to the assessment of the wind energy potential of the building are presented. The foreign experience and classification of factors affecting the placement of wind power plants were analyzed. The possibility of detailing the microclimate data on the wind regime for the placement of wind power plants, taking into account urban improvement and landscaping, was indicated. The issue of primary binding of wind power plants in building on the basis of the wind power potential of buildings and territory was considered. The concept of the "Smart City" is considered to form a management system for wind and energy potential in urban construction and to assess the comfort of aeration for pedestrians with integration into Urban Building Energy Modeling (UBEM).

## 44.1 Introduction

The forecast of microclimate in urban development is necessary not only for the regulation of temperature and humidity conditions, gas contamination and dustiness of the air, substantiation of the density of the housing stock, increasing the efficiency of the use of the city territory, etc., but also for the multifactor evaluation of the wind energy potential of urban development and territory [1]. Boykova et al. [2] considered the dependence of city systems on external conditions and the need to introduce

I. V. Dunichkin (✉)
Moscow State University of Civil Engineering (National Research University), 26, Yaroslavskoye H, 129337 Moscow, Russia
e-mail: ecse@bk.ru

I. N. Ilina
National Research University Higher School of Economics, 2 Malaya Pionerskaya Ulitsa, 115054 Moscow, Russia



J. Littlewood et al. (eds.), *Sustainability in Energy and Buildings*, Smart Innovation, Systems and Technologies 163, https://doi.org/10.1007/978-981-32-9868-2_44

the "Smart Model" of a city, which can also be considered as the possibility for adapting the climate and generating energy from renewable sources: "Modern cities as complex systems will not be able to resist numerous challenges without having a management model capable of flexibly adapting to changing external conditions." [2]. Models and individual elements of the "Smart City" concept are being introduced in various climatic conditions, which makes it necessary to study the needs of the population in energy. Considering the description of the model of the "Smart City" in the book by Ilina et al., it is necessary to allocate one direction of development: "Smart environment" [3], since information technologies allow receiving data on bioclimatic comfort, wind speed, and wind directions. Monitoring of bioclimatic comfort and environmental safety for the "Smart City" allows analyzing wind energy potential and, based on this, making design decisions on the placement of low-power wind turbines in urban areas to improve energy efficiency and reduce the ecological footprint of a settlement. Thus, comfort in the "Smart City", based on wind speed, must be considered with energy potential.

Ishugah et al. wrote a review of the placement of wind turbines on the building. "This paper vividly captures the fact that wind resource has a great potential to be fully explored and developed in the urban environment. Although many new ideas and solutions that factor technical, economical and environmental sustainability in urban areas are coming up every day, challenges in design are gradually being solved to take advantage of urban low and turbulent wind speed characteristics, installation space challenges, vibration and noise reduction, among others" [4].

One of the main criteria for assessing the wind power potential in collective works under the leadership of Poddaeva et al. is described as "Information on the average annual wind speeds in the territory allows focusing on the future generation of electricity..." [5] and the probable real conditions in which the methods of assessing the microclimate are applicable to make a decision on the efficiency of wind power plants "Stagnant air zones indicate the inefficiency of the development of wind power plants, primarily horizontally axial type" [6]. Therefore, it is necessary to analyze the average annual wind speed along with other factors of the microclimate to make a decision on the placement of objects at the local level. In this regard, it is necessary to determine the relationship of the wind regime with the objects of the urban environment, physical, and technical factors that shape the climate at all levels. In turn, this entails the need to present a scheme of their interaction as a basis for assessing the wind energy potential. As a model of the air space of settlements, the development in the form of a two-tier structure based on the works of Serebrovsky was used [7].

All this determines the main goal of the research and gives an idea of working with many factors and types of design. The main goal of the research is to obtain, on the basis of a single set of data on the environment parameters of a residential microdistrict, two correlation estimates between each other: bioclimatic comfort and wind energy potential for joint design solutions of improvement and small wind turbines. Performing two analyses (bioclimatic and wind energy) based on the results of verified CFD modeling of airflow in the building is an important task of the research. The results of these analyses will make it possible to carry out correlation estimates of bioclimatic comfort and wind energy potential in the design process and to obtain information for making design solutions on the compensation of discomfort and the energy strategy for applying wind turbines to buildings. Subsequent research tasks include the classification and discussion of the factors of negative impacts and risks from the placement of wind turbines on buildings.

## 44.2 Three-Grade Urban Climate Method and Materials for Climate Forecasts

This research uses the method of structuring urban space according to the parameters of the air environment and two types of criteria for bioclimatic comfort and wind energy, due to the presence of analyzes in the main goal. Theory of method is based on the following:

(1) Gradation of the roughness of the underlying surface;
(2) Taking into account levels of urban climate and the climatic conditions in urban planning with the use of wind turbine technology.

Analysis of the airspace over the city shows that it can be represented as different layers—upper (up to 1000 m) and lower (up to 200 m). The open-air movement zone is represented by the upper layers. The zone of air movement with obstacles is represented by the lower layers. Kovalenko supplemented lower layers with a three-stage gradation of the roughness of the underlying surface (macro-, meso-, micro-) in his work "Urban climatology" [8], which in the lower surface layer of the air masses correspond to the main levels of the aeration regime as components of the microclimate and are the initial data for the assessment of the wind energy potential. The speed and direction of movement of air masses are formed under the influence of factors of the geographical environment only above the boundary of the lower (surface) tier.

They are divided into the macro-level, meso-level, and micro-level:

- Up to 200 m above the surface of the earth is influenced mainly by high-rise buildings (for example, houses of tower type)—a macro-level [9];
- Up to 25 m above the ground is influenced by the entire building (high-rise, medium-rise, and partially low-rise)—a meso-level [10];
- Up to 3 m above the surface of the earth is influenced by high-rise, medium-rise, and low-rise buildings, as well as hardscape elements and landscape gardening—a micro-level [11].

The axiom of the presented method is that each urban object affects the wind regime at the appropriate level while experiencing the impact of these levels, depending on the urban organization of space. Microclimate, as the term meteorology, here corresponds to the micro-level and meso-level of the aeration regime. The attached method of separating the expanses of a city allows observing changes in different climate levels and the effect of these changes on the wind turbine situation, bioclimatic comfort, and safety. This allows one to monitor and predict the risks of climate change and ensure the safety of citizens, such as in Moscow, the IT service sends SMS alerts to all phone numbers in the region about strong wind, snowdrifts, and ice on the streets.

In the work "City, architecture, people and climate", ed. by Myagkov, a generalized assessment of the wind climate of large cities in Central Russia "is given… the speed field above the city is almost always complex. Even in the case when the weather situation is as simple as possible (for example, cloudless sky and weak wind in the center of a large high-pressure zone), small local differences in the properties of the underlying surface, which include landforms, building parameters, cause disordered air currents. Therefore, broad generalizations about the influence of cities on the wind regime, found in the literature, can be, at best, starting points for understanding the mechanisms of this influence" [12]. This work once again confirms the need to structure the wind energy potential in accordance with the system of urban climate levels. Thus, the systematization of the elements of the air environment allows us to further perform a level assessment of the wind energy potential at macro-, meso-, and micro-levels.

The research presents materials on the climate and residential microdistrict of Moscow. This allows us to consider the three-grade urban climate method on the example of applying the concept of the "Smart City". In addition, this method distinguishes the micro-level of aeration regime for predicting bioclimatic comfort, and meso-, macro-levels for estimating wind energy potential. This distribution allows the method to be used to predict a microclimate in combination with the following studies:

- Climate analysis of the urban area;
- Physical tests of residential development models in a wind tunnel;
- Numerical simulation of airflow for residential buildings in the ANSYS CFX software.

## 44.3 Results of the Microclimatic Research and Estimate of the Wind Energy Potential of the Outer Space of "Smart City"

It should be noted that Moscow is a city of "two seasons" to address the issue of predicting the microclimate; therefore, with the architectural climatic analysis, two tasks are usually solved simultaneously—protection from wind and cold stress in winter and ensuring aeration and comfortable microclimatic conditions summer. Although analyzing the modern design practice, it should be noted that it does not fully comply with the classical recommendations of the Central Research Institute of Urban Planning and Foreign Scientists [13–15]. That raises the question of finding a tool to take into account the results of microclimate forecasting for the assessment of wind energy potential in the practice of designing. In this regard, another task arises and related to all seasons of the year is to determine the optimal locations for small wind turbines.

The stages of taking into account the climatic conditions in the planning of the new and the reconstruction of the existing and development are as follows:

- Identification of a complex of climatic and urban factors and the establishment of their interrelation in relation to a concrete stage of construction design and a town planning situation;
- Assessment of the complex of identified natural factors;
- Development of a system of requirements for the protection of the urban environment from the adverse effects of natural conditions;
- Development of proposals for the rational use of natural resources, including wind and solar energy;
- Development of urban planning and architectural and construction solutions that ensure the practical implementation of the identified conditions.

Urban development of the territory causes a number of consequences in the meso- and micro-climatic environment of the area. Their cause is the change in the conditions of meso- and micro-scale atmospheric air circulation as a result of the appearance of new construction volumes, the change in the thermal balance due to the emission of man-made heat and changes in the albedo (reflectivity) of the underlying surface, increase in the coefficient of surface runoff from the built-up area, etc. As a result of the studies of the Educational, Scientific and Production Laboratory for aerodynamic and aeroacoustic tests of the building structures of the "NRU MGSU," the most significant microclimatic processes were identified on a smaller scale, which are as follows:

- Changes in the conditions of insolation of the territory taken into account when placing photovoltaic panels in the building;
- Formation of zones of the steady decrease in wind speed, with polluted air;
- Formation of zones of thermal discomfort overheating in summer and hypothermia in winter;
- Formation of wind speed amplification zones in certain directions at the boundary of the building, optimal for the location of small wind power units.

All of the above data in Sect. 44.3 represent the disparate results of the study and recommendations without the possibility of controlling the application in the project documentation. Only IT technologies and the "Smart City" concept allow applying this disparate result as a unified system of the initial information for design [16], which will allow realizing and recording the real result on the comfort of settlements from the introduction of an integrated methodology for assessing the bioclimatic comfort of residential buildings area. Obtaining information on the location in the space of seasonal zones of wind amplification allows us to proceed to the assessment of the wind energy potential for the placement of small wind power units.

The presented results allow the analysis of parameters of urban space. This analysis is based on two criteria:

(1) Criteria of bioclimatic comfort for aeration of the territory (Flow rate in buildings from 0.5 m/s to 3 m/s in winter and from 1 to 5 m/s in summer);
(2) Criteria for the wind energy potential of the environment (The flow velocity is over 3–5 m/s depending on the type of wind turbine location and the quality standards of power generation).

## 44.4 Analysis of Correlation Between Bioclimatic Comfort of the Environment and Wind Energy Potential for UBEM

It is necessary to analyze in detail the results of studies conducted the example of the residential complex "Domashniy" in Moscow to determine the possible results of the application of the bioclimatic comfort assessment methodology and to identify the wind energy potential. The layout of the building was tested in a large wind tunnel gradient «NRU MGSU». The results of testing were used to verify numerical simulations that were performed in the Educational-Scientific-Production Laboratory for aerodynamic and aeroacoustic test building constructions.

As can be seen from the histogram in Fig. 44.1, most unfavorable wind directions for the comfort of pedestrians are rhumbas W, WNW, NW, SW, WSW. If we add information about the temperature during cold and warm seasons in Moscow to the pedestrian comfort data from wind speed, it is possible to determine the bioclimatic comfort in a residential complex. These data may be analyzed in the future with

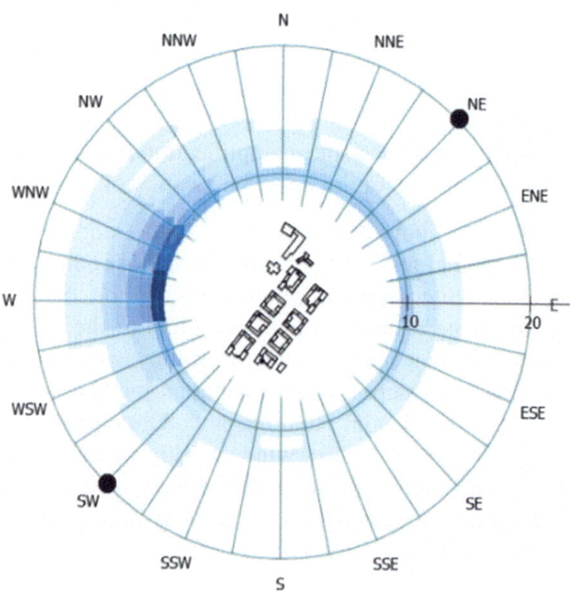


**Fig. 44.1** The histogram distributions of wind speed according to the orientation of the residential complex "Domashniy" (Moscow) to the cardinal

the full-scale operation of the sensor system in the development and the IT service feedback on the bioclimatic discomfort of the residents of the microdistrict. It can be concluded that the object is oriented unfavorably in terms of the impact of strong winds [17]. At the same time, the axis of ventilation from the northwest to the southeast is the most effective in terms of wind energy potential [18].

On the left side of Fig. 44.2, violet colors indicate zones of stable turbulence with minimal air exchange with other neighboring spaces. The same places on the right side of Fig. 44.2 have zones of reduced wind speeds, where harmful substances accumulate due to the indicated reduced air exchange. Thus, a system of closed turbulent jets and eddies in the courtyard spaces is formed by residential development. This identified process has a positive and negative effect on bioclimatic comfort:

(1) In seasons with a predominance of the northwest wind at high temperatures, overheating of the area and accumulation of harmful substances will occur, while ventilation of the yards will occur due to temperature convection when heated asphalt pavements buildings. For the same reasons, accumulation of snowdrifts is possible in the winter period which will require additional measures to clean up the area during operation;
(2) In the winter period with negative temperatures, with a decrease in the criteria for evaluating bioclimatic comfort to a range of 0.5–3 m/s, yards, we will have wind protection even with uncomfortable gusts of background airflow at a speed of over 6 m/s.

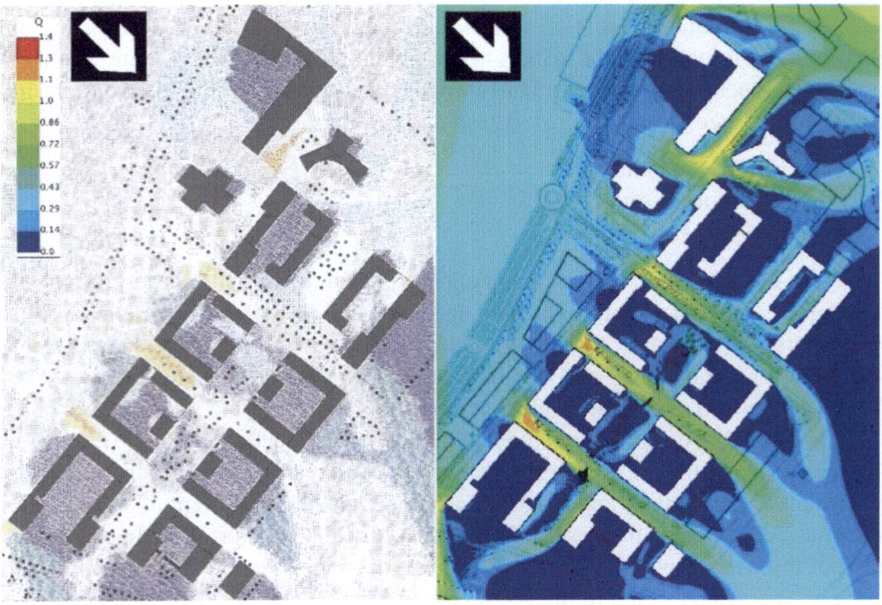


**Fig. 44.2** Distribution of the velocity amplitude and relative wind speed (V) in the vicinity of the object understudy in the northwest wind

The locations of discomfort zones on the territory have stable parameters in the northwest wind, are located between the adjacent facades of the residential complex, and are shown in yellow–green, Fig. 44.2 (Right side). The relative wind speed (V) in the places of discomfort reaches 1.1–1.2, which corresponds to about 4 m/s at ground level and more than 6 m/s on the facades of more than 10 m from the active surface at an average background speed of 3.6 m/s. Those buildings that are closer to 3 m from the zones of discomfort of pedestrians (increased stable wind speeds) have in the corresponding places of the facades a greater wind energy potential by 25–32%. Thus, the method of assessing bioclimatic comfort allows optimizing the design solution for the improvement of the yard, placing the entrance groups of buildings and obtaining information about the most effective surfaces of the garden for the integration of small wind power plants. The most favorable for the comfort of the residential complex are the directions of winds—N, S, SE, SSE, SSW—and the perpendicular facades will have indicators for the production of energy by small wind power plants by 12–46% lower.

## 44.5 Discussion of Wind Power Units Location on Buildings for UBEM

Typical examples of the location of wind power plants in the space of urban development can be considered in the work of Duffy [19]. The research work proves the possibility of the metric location of wind turbines. This data allows forecasting the microclimate to estimate the wind energy potential, taking into account the row-metric placement of equipment. Duffy divides the options for the location of wind turbines into three groups for installation on buildings:

- On the roof of the building;
- At the corner of the outer wall;
- On the middle part of the wall.

To this classification, you must add another group of wind turbines location. The fourth group of wind turbines is located inside the building structures, more precisely in the ventilation system of multi-storey buildings. The turbine operates on a stable airflow of a natural supply and exhausts ventilation system. This airflow rises from the lower rooms through the vertical ventilation duct. The wind turbine takes place at the level of the roof or attic, that is, at the top of this duct.

However, in Russian conditions, a number of other factors will also affect the placement of equipment, and hence the accuracy of the assessment of wind energy potential. According to the work of Prokhorova "the main factors affecting the placement of the wind power plant on buildings are the following:

- Climatic factor (wind speed in the lower tier of air masses, the use of maps of the wind situation);

- Wind mode (wind speed at the meso-level of the urban environment, data from weather stations);
- Planning situation of development (wind speed at the local level of the urban environment, numerical simulation data and physical experiments);
- Environmental limitations (the influence of the wind power plant on the aeration regime, the distribution of polluted air masses, noise impact);
- Architectural and structural limitations (design scheme and load-bearing capacity of structures);
- Architectural and aesthetic limitations (integrity of the composition of the building and wind power plant).

Such a complex orientation allows the further development of a typology on energy autonomy" [20]. These criteria, among which primarily the impact of noise, are the basis for assessing restrictions on the implementation of UBEM.

Noise impact is among environmental limitations due to the operation of wind turbines. Duffy has applied special design and design solutions to reduce noise and vibration from wind turbines: "optional swept tip blades for reduced noise and improved aesthetics, a set of optional rubber mounting shims to isolate vibrations from transferring into the building" [19]. Ishugah et al. confirm the possibilities of reducing vibration and noise in their studies, [4]. Manufacturers and many laboratories determine for small and micro wind turbines (power up to 20 kW) the absence of a significant effect of infrasound and a noise level of about 35–38 dB at a distance of more than 2 meters, and in the nearest closed rooms about 30 dB maximum. This allows the selection of small and micro wind turbines for installation on residential buildings while observing the maximum noise level of no more than 40 dB.

## 44.6 Conclusion

Thus, there is a typological relationship between the factors of microclimate comprehensively interacting with the wind power potential of the building and the territory of the city, in which it is necessary to add the constructive conditions of the building itself and the constructive limitations of small wind power plants, as well as safety requirements [21]. Summing up, it should be noted that according to a generalized assessment based on the prediction of the microclimate, for example, the city of Moscow, the primary objects of study are different storey buildings, first, the tower type buildings. In the microdistrict understudy, an increase in wind energy potential by 25–32% for facades along the flow was found. This is most often encountered when urban design of microdistrict, in which a low building that meets the airflow is located before the next high one. Since this is due to an increase in the turbulence of the airflow [22]. The decrease in the wind potential of facades is associated with the location of stagnant masses of air in the territory. This can reduce energy production by up to 46%. The adjustment of wind energy potential assessment tools, according to the analyzed materials, will allow not only assessing the potential of the development and the territory but also to detail these data for improvement and landscaping. This makes it possible to decide on the appropriateness of placing each wind turbine. The primary binding of wind power plants to buildings can be performed this way, which allows including these assessment methods and results in UBEM. Further development of a combined assessment of wind energy potential and bioclimatic comfort for climate adaptation is dictated by the integrated design of the ventilation system for buildings and wind turbines outside and inside the building in the ventilation system.

**Acknowledgements** The authors are grateful to Poddaeva and the team of the Educational-scientific-production laboratory for aerodynamic and aeroacoustic test building constructions (NRU MGSU) for advice. Dunichkin wrote Sect. 44.2, Sect. 44.3, Sect. 44.4, Sect. 44.5, Sect. 44.6. Ilina wrote Sect. 44.1.

## References

1. Baumüller, J., Hoffman, U., Reuter, U.: Climate booklet for urban development. References for zoning and planning. Retrieved 2005, from Baden-Württemberg Innenministerium (2005). http://www.staedtebauliche-klimafibel.de. Last accessed 03 Mar 2019
2. Boykova, M., Ilina, I., Salazkin, M.: The smart city approach as a response to emerging challenges for urban development. Foresight, Journal "Foresight and STI Governance" **10**(3 (eng)) (2016). NRU HSE, Moscow
3. Ilina, I.N., Plisetsky, E.E., Tkachenko, S.B.: "Smart City" and Global Trends in Urban Innovations. Publishing House "Uley", Moscow (2018)
4. Ishugah, T.F., Li, Y., Wang, R.Z., Kiplagat, J.K.: Advances in wind energy resource exploitation in urban environment: a review. Renew. Sustain. Energy Rev. **37**, 613–626 (2014)
5. Poddaeva, O.I., Dunichkin, I.V., Kochanov, O.A.: The main approaches to the study of renewable energy as an energy potential of the territories and the development. In: Herald MGSU, №10, pp. 221–228. NRU MGSU, Moscow (2012)
6. Poddaeva, O.I., Dunichkin, I.V., Prokhorov, T.V.: The spatial organization of the reconstructed residential development on the wind energy potential in the medium. In: Herald MGSU, №2, pp. 157–165. NRU MGSU, Moscow (2013)
7. Serebrovskiy, F.L.: Aeration Populated Areas. Publishing House "Stroyizdat", Moscow (1985)
8. Kovalenko, P.P., Orlova, L.N.: Urban Climatology. Publishing House "Stroyizdat", Moscow (1993)
9. Poddaeva, O.I., Buslaeva, J.S., Gribach, D.S.: Physical model testing of wind effect on the high-rise. In: Advanced Materials Research, vol. 1082, pp. 246–249. Trans Tech Publications (2015)
10. Poddaeva, O., Churin, P., Dunichkin, I.: Experimental study of wind loads on unique buildings and structures in Russia. In: MATEC Web of Conferences, vol. 86, p. 02012. EDP Sciences (2016)
11. Blocken, B., Stathopoulos, T., Van Beeck, J.P.A.J.: Pedestrian-level wind conditions around buildings: review of wind-tunnel and CFD techniques and their accuracy for wind comfort assessment. Build. Environ. **100**, 50–81 (2016)
12. Myagkov, M.S., Provincial, J.D., Konova, L.I., Litskevich, V.K.: City, Architecture, man and climate. In: Myagkov, M.S. (ed.). Publishing House "Architecture-S", Moscow (2006)
13. Guidelines for assessing and regulating the wind regime of residential development, Regulatory and Production Ed. «CR&DI (Central Research and Design Institute) for Urban Planning (Gosgrazhdanstroy)», Publishing House "Stroyizdat", Moscow (1986)

14. Lawson, T.V.: The wind content of the built environment. J. Ind. Aerodyn. **3**, 93–105 (1978)
15. Oke, T.R.: Street design, and urban canopy layer climate. J. Energy Build. **11**, 103–113 (1988)
16. Ilina, I.N.: Innovative approaches to the formation of models of integrated investment urban development of urban areas. Prop. Relat. Russ. Fed. № 1 (184) (2017)
17. Ricci, A., Kalkman, I.M., Blocken, B., Repetto, M.P., Burlando, M., Freda, A.: Local-scale forcing effects on wind flows in an urban environment. In: Proceedings, International Workshop on Physical Modelling of Flow and Dispersion Phenomena PHYSMOD2015 (2015)
18. Antoniou, N., Montazeri, H., Wigo, H., Neophytou, M.K.A., Blocken, B., Sandberg, M.: CFD and wind-tunnel analysis of outdoor ventilation in a real compact heterogeneous urban area: evaluation using "air delay". Build. Environ. **126**, 355–372 (2017)
19. Duffy, M.J.: Small wind turbines mounted to existing structures. Thesis for the Degree of Master of Science in Aerospace Engineering (2010). Georgia Institute of Technology, Atlanta, USA
20. Prokhorova, T.V.: Features and prospects of the development of wind in an urbanized environment. J. Volga Reg. Bulletin. **2**, 121–128 (2013)
21. Lazareva, I.V.: Urbi et orbi. The fifth dimension of the city. In: Proceedings of the Russian Academy of Architecture and Building Science. Series "Theoretical Foundations of Urban Planning". Publishing House "LENAND", Moscow (2006)
22. Ghiaus, C., Allard, F., Santamouris, M., Georgakis, C., Nicol, F.: Urban environment influence on natural ventilation potential. Build. Environ. **41**(4), 395–406 (2006)

# Chapter 45
# Adaptive Design to Mitigate the Effects of UHI: The Case Study of Piazza Togliatti in the Municipality of Scandicci

**Rosa Romano, Roberto Bologna, Giulio Hasanaj and Maria Vittoria Arnetoli**

**Abstract** The relationship between microclimate and built environment has been explored for almost 40 years, with reference not only to architecture but also to urban and to landscape design as instruments to increase the outdoor comfort. Recently, scientific studies on urban climate pay attention to Urban Heat Island (UHI) and related mitigation strategies as the main focus theme, also by developing new sustainable urban design assessment methods. Moving from these considerations, this paper aims to describe some results of the research 'Adaptive design and innovative technology for resilient regeneration of urban district depending on climate change', focused on the study of technological solutions that are able to improve the resilience of the built environment, also reducing the effects of the UHI. Furthermore, attention is given to the issue of urban space design, implemented as an instrument that is able to promote the regeneration of urban districts. In detail, this article presents two design proposals developed for Piazza Togliatti in Scandicci (Florence), analysing the results achieved (in terms of atmospheric temperature and predicted mean vote) by means of the ENVI-Met software.

## 45.1 Introduction

The 2030 Agenda of the United Nations recognises that climate change is one of the greatest challenges of our age [1]. Its negative impacts undermine the ability of the worldwide governments to implement sustainable development and achieve goals such as inclusion, security, eradicate poverty and hunger. Surface temperature is set to increase during the twenty-first century in all global warming scenarios assessed by the Intergovernmental Panel for Climate Change models [2]. This will lead to an increase in extreme events such as high temperatures, heavy rainfall and drought phenomena. A lot of research has shown how discomfort is linked to Urban Heat Island (UHI) that occurs increasingly in anthropic settlements all over the world [3, 4]. The UHI effect, in fact, is characterised by the development of noticeably higher

R. Romano (✉) · R. Bologna · G. Hasanaj · M. V. Arnetoli
Architecture Department DIDA, Florence University, Florence, Italy
e-mail: rosa.romano@unifi.it



J. Littlewood et al. (eds.), *Sustainability in Energy and Buildings*, Smart Innovation, Systems and Technologies 163, https://doi.org/10.1007/978-981-32-9868-2_45

temperatures in our cities compared with the countryside that directly surrounds them [5]. The intensity of UHI strongly depends on the local meteorological conditions, but also on the urban characteristics, such as the type of materials used for the construction of urban spaces and buildings and the presence (or absence) of green and blue infrastructures [3, 6]. Furthermore, it is important to remember how the UHI has several impacts on quality of life and health of citizens, increasing the risk of mortality, especially among the weakest sections of the population [7, 8]. Finally, it contributes to: grow the building energy consumptions for cooling [9]; worsen indoor and outdoor thermal comfort [7, 9]; raise the concentration of harmful pollutants in the air of our cities [9].

Starting from these considerations, the paper shows the first results of the National Research Project of Relevant Interest '*Adaptive design and innovation technology for resilient regeneration of urban district depending on climate change*' (PRIN AD), developed by the Universities of Reggio Calabria, Naples, Rome, Milan and Florence. The research aimed at investigating design methodologies and technological solutions to promote the resilient regeneration of three Italian urban districts (located in Florence, Milan and Naples), both in terms of reducing exposure to climate risks (flooding and water deficiencies, heatwaves and UHI) and social inclusion and economic sustainability. In detail, the article describes the analysis and evaluations developed to increase the resilience of the Piazza Togliatti, a degraded square located in the centre of Scandicci (a little city placed in the Southwest of Florence).

## 45.2 Research Methodology

With reference to the research methodology developed in the frame of the PRIN AD, the study was articulated as follows: (1) preprocessing the geometric and physical features of the Togliatti square, by ENVI-met; (2) simulating the urban climate at the current state in typical UHI conditions; (3) designing of two different suitable project scenarios; (4) preprocessing the geometric and physical features of two project scenarios through ENVI-met; (5) simulating the two project scenarios; (6) analysing the results and detecting the most effective strategies in terms of attenuation of the urban heat island, overall improvement of the urban microclimate and thermal comfort.

The holistic approach is the one that inspired the design of two scenarios. We tried to avoid and circumscribe the project proposal in each one, in fact, we developed the design solutions and analysed the benefits obtained within an increasing scale of project outputs. In other words, the technological solutions (shading devices, green infrastructures, etc.) proposed in the second scenario to improve the resilience of the Piazza Togliatti can be added, in a subsequent step, to the repaving strategies developed in the first one. The two scenarios, have been defined for enhancing the square's ability to react to the phenomena of climate change, included the UHI, but also with the aim of progressively transforming both its aesthetic image and its psychological perception by users, especially children and the elderly [10].

For these reasons, we carried out ENVI-Met simulations to measure and compare two indexes of the UHI's impacts on the urban environment: atmospheric temperature (°C) and environmental comfort (PMV index). In detail, we have used the ENVI-Met's application 'Biomet', that is able to carry out the value of the PMV of a specific user whose parameters related to sex, age, height and weight are inserted. Moreover, we have considered only the climatic conditions of UHI in our analysis. Therefore, the PMV is not always included between the range −3 and +3, but it sees higher values which identify worse environmental discomfort for the user that we have chosen: an elderly man.

## 45.3 The Case Study

Piazza Togliatti is a large, empty and degraded rectangular space, which showed many problems of liveability, accessibility and safety, as well as a negative level of climate resilience. Currently, in the square the following functional destinations are located (Fig. 45.1): a playground area, in the northern border; an outdoor area, in the southern border, used as a market in the morning and as a parking for the rest of the day. Citizens spend a lot of time in these two spaces during the whole year, although the environmental conditions—especially in the Summer months—do not guarantee sufficient comfort. For this reason, we decided to focus our requalification projects on the regeneration of this square, in order to increase not only its environmental performances but also its social potential as aggregative space. The square is actually characterised by the following physic characteristics: pavements made with 'hot' materials; green areas that appear in a fragmented composition; roads without trees and vegetation; absence of water elements; absence of green roofs or roofs made with cool materials.

**Fig. 45.1** Photos of Piazza Togliatti. On the left the playground area, on the right the parking and market area

## 45.4 Simulation Model

Compared to the methodological context described above, the analysis and simulation tools were developed with multiple objectives, involving different phases of the design process. In the case of Piazza Togliatti, the 'ex-ante' energy simulations allowed defining its behavioural profile in the existing configuration (showing its critical issues and vulnerabilities, but also the potential for its architectural renovation), and to analyse the benefits that each project proposal could achieve.

The urban microclimate, the atmospheric temperature (°C) and the environmental comfort (PMV index) were calculated by ENVI-met which is a prognostic model based on the fundamental laws of fluid dynamics and thermodynamics [11]. In particular, this software allows to perform a microclimatic analysis starting from input climatic data of speed and mean direction of the wind, and hourly course of temperature and humidity. It also provides the evaluation of the interrelation between buildings, vegetation and various surface coverings, and of their effects on the perceived microclimate [12]. The values of Fanger's indices calculated by ENVI-met are based on the modified Fanger heat balance equation developed by Jendritzky and Nubler for outdoor conditions [13].

In our research, the square model was created with the 'area input file editor' rotated 60° to the North, according to the main direction of the roads. The area was rendered with a $100 \times 100 \times 30$ (x-y-z) grids, keeping the following spacing: $dx = 3.00$ m; $dy = 3.00$ m; $dz = 2.00$ m. Ten buildings were included in the model with a height between 6 and 24 m, 15 m being the mean value. For the modelling of the vegetation, eight different trees types were employed (Table 45.1).

Furthermore, the model was kicked off especially regarding radiative properties of surface materials and of the two design scenarios (Table 45.2). Finally, the model included a part of the urban context near the square, since the software has some difficulties of execution of the calculations in the proximity of the edges.

Thanks to the collaboration with the Florence Institute of Biometeorology, the climatic data of the last 25 years were analysed [14], identifying the hottest day as 21 August 2011: Temperature minimum 20.7 °C at 6:00 a.m.; Temperature maximum 41 °C at 04:00 p.m.; Humidity minimum 14% at 01:00 p.m.; Humidity maximum 66% at 6:00 a.m.; 0.2 m/s wind speed from south direction.

The outputs related to atmospheric temperature and Predicted Mean Vote (PMV), were extrapolated for the worst microclimatic conditions (04:00 p.m.) and with

**Table 45.1** Principal features of 3D plants included in the model simulation

| Plants features | *Acer negundo* | *Pinus pinea* | Cypress | *Populus alba* | Tilia | *Populus nigra* | Privet |
|---|---|---|---|---|---|---|---|
| Height (m) | 11 | 15 | 7 | 7 | 25 | 18 | 20 |
| Width (m) | 9 | 11 | 3 | 5 | 15 | 13 | 15 |
| Albedo | 0.5 | 0.6 | 0.3 | 0.7 | 0.18 | 0.4 | 0.4 |

**Table 45.2** Materials in the current status and scenarios, with their albedo value

| Materials | Albedo | Current status | Scenario 1 | Scenario 2 |
|---|---|---|---|---|
| Asphalt | 0.1 | ✓ | ✓ | ✓ |
| Brick | 0.3 | ✓ | | |
| Dark concrete | 0.2 | ✓ | ✓ | |
| Red asphalt | 0.5 | ✓ | | |
| Loamy soil | 0.15 | ✓ | ✓ | ✓ |
| Grey concrete | 0.5 | | ✓ | ✓ |
| Beaten heart | 0.8 | | ✓ | ✓ |
| Grass | 0.2 | | ✓ | ✓ |

the parameters of the more disadvantaged users, represented by elderly men (age 70 years).

## 45.5 Investigated Scenarios

The impact of the main technological strategies, developed in the frame of the PRIN AD research in order to decrease the UHI effects, was evaluated, analysing the energy performances of three different model configurations (Fig. 45.2) listed below.

*Base Case*

The model was kicked off applying specific values of the urban environment of Piazza Togliatti, described in detail in the previous paragraph 4.

*Scenario 1*

The project developed in the first scenario aimed to improve environmental and social features of the square, without changing its actual uses and functions. For this reason, the retrofitting project concerned only the repaving action to create green spaces, particularly planting new trees in the parking-market zone. The materials for the repaving were chosen with a higher albedo than those existing. In order to build this simulation in ENVI-Met, the same input files of the *Base Case* were employed, modifying only the configuration file, which contains all the parameters set to the model.

*Scenario 2*

The project developed in the second scenario, starting from the repaving of the square (as done in the first scenario), was focused on: (1) the construction of solar shading devices in the parking-market area, where it will be possible to integrate technologies for producing renewable energy and to collect water; (2) the requalification of the bordering streets, converted to green corridors; (3) the use of water, integrating in the

**Fig. 45.2** Analysis of surface vegetation and soil in the current status and in the two scenarios. Legend of materials used in the models: 1 asphalt, 2 brick, 3 dark concrete, 4 red asphalt, 5 loamy soil, 6 grey concrete, 7 beaten heart and 8 grass

urban space of Piazza Togliatti a blue infrastructure in order to increase its outdoor comfort especially in Summer. Radiative properties were almost the same as *Base Case*. Also, in this case, a different configuration file (as for *Scenario 1*) was used.

## 45.6 Results

### *45.6.1 Atmospheric Temperature and PMV Index*

All the simulations were started on 21 August 2011 and the elapsed time was 24 h. Figure 45.3 presents thermal and PMV maps of the three scenarios analysed for the warmest moment of the day (04:00 p.m.). For horizontal views (x-y), all data are shown at pedestrian level height (z = 1.80 m, representing what people could perceive).

The results achieved with the simulations can be summarised as follows.

*Base Case*

The central area of the playground and of the parking-market zone are the hottest spaces into the square, due to the presence of asphalt and dark concrete, and due to the lack of vegetation, trees and shading devices. The maximum temperature is

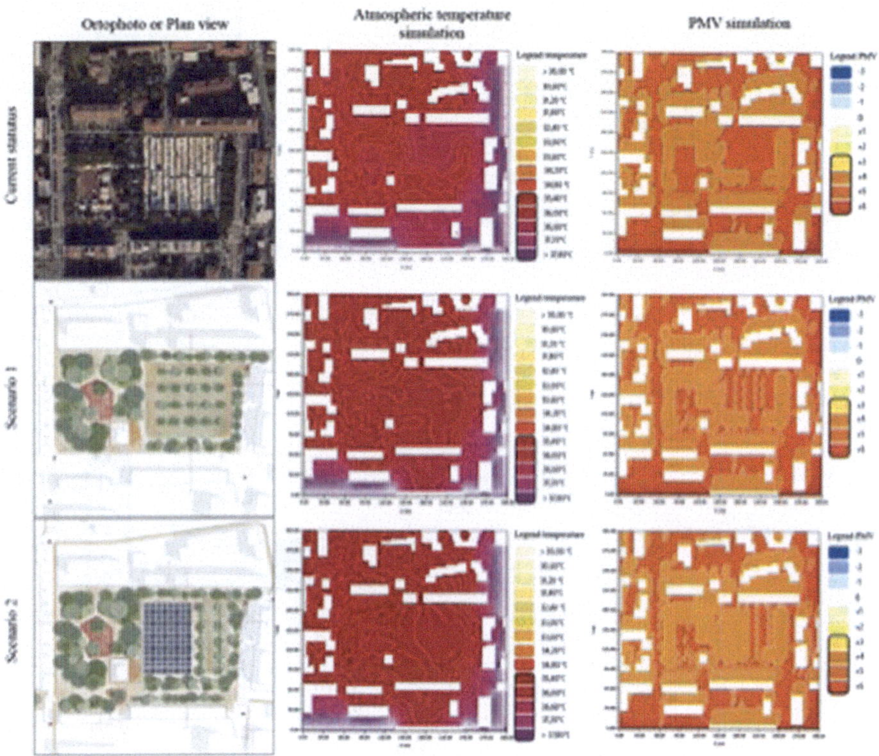


**Fig. 45.3** From left to right: square plan views; atmospheric temperature and PMV of elderly men (70 years old)

41.17 °C, while minimum, recorded in densely vegetated areas, is 35.83 °C. The PMV index shows a minimum value of 3.73 and a maximum of 6.94.

*Scenario 1*

In this scenario, the maximum temperature is 41.06 °C and the minimum temperature is 35.56 °C as usual recorded in the green areas. As shown in Fig. 45.4, a temperature reduction is achieved, due to the repaving strategies and the planting of new vegetation. The PMV index shows a minimum value of 3.67 and a maximum of 6.86, with a maximum reduction of −1.97 (Fig. 45.5) compared to the base case. The best results, in terms of temperature and PMV index, are recorded in correspondence of the central area of parking-market, where we have proposed to replace the existing dark pavement with cool materials in addition to planting 20 new trees in order to increase the shading areas.

*Scenario 2*

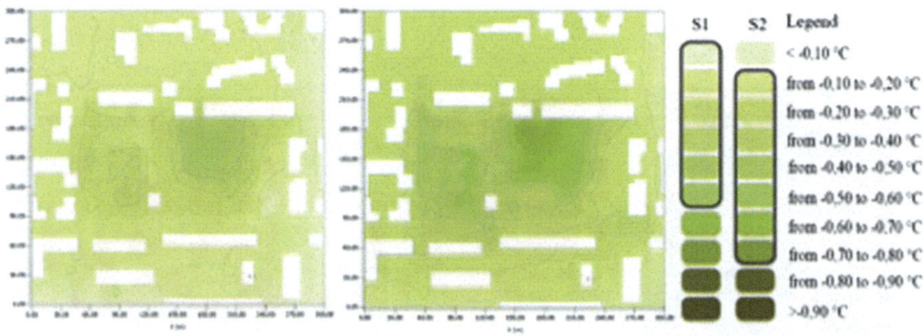


**Fig. 45.4** ENVI-Met® comparisons of the atmospheric temperature decreasing in scenario 1 and 2, x-y view at z = 1.80 m, at 04:00 p.m., 21 August 2011

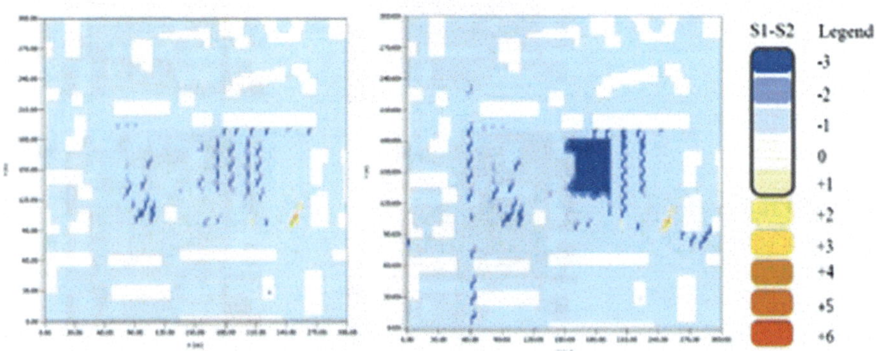


**Fig. 45.5** ENVI-Met® comparisons of PMV simulation in the two scenarios; x-y view at z = 1.80 m, at 04:00 p.m., 21 August 2011, parameters for PMV of elderly men (70 years old)

In this scenario, maximum temperature is 40.99 °C and minimum temperature is 35.39 °C. As shown in Fig. 45.4, a general decrease in the temperature is recorded, but it is deeper than in the first one. In detail, in the covered area of the parking-market, the atmospheric temperature is reduced by −0.74 °C, thanks to the presence of the shading adaptive devices. Further, the PMV index shows a minimum value of 3.60 and a maximum of 6.83, with a maximum reduction compared to the base case of −2.23 (Fig. 45.5).

### 45.6.2 Hourly Temperature Profile of Two Selected Spots

To better understand temperature oscillation simulated by ENVI-Met, and in order to compare the temperature in the different square areas, we selected two different points located in correspondence of the playground (Fig. 45.6) and of the parking-market (Fig. 45.7). Furthermore, we plotted an hourly temperature profile among the simulation period, using temperatures output by the software.

In this focused analysis, *Scenario 2* guarantees to achieve the best results, in terms of reduction of UHI phenomenon, in both spot points. Furthermore, it is interesting to note that the atmospheric temperature decreases in *Scenario 2* also in the playground area, although the design proposal for this zone is the same as *Scenario 1*. In other words, the simulation showed that increasing shading surfaces allowed reducing the atmospheric temperature throughout the project area and improving the urban microclimate also in the zone not directly involved in the regeneration actions.

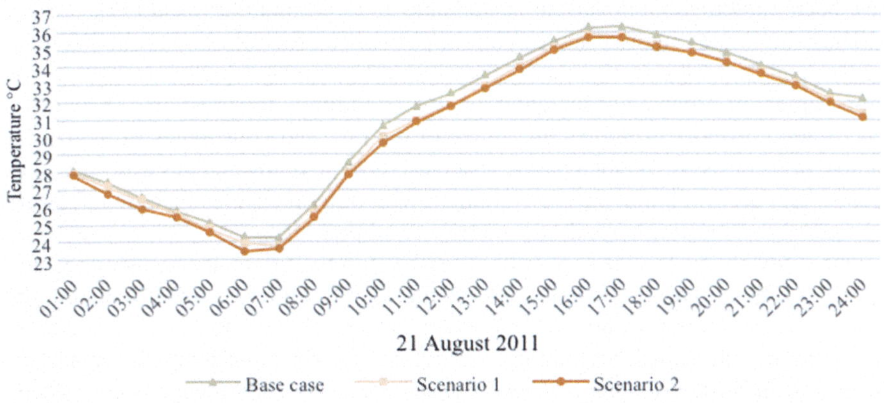


**Fig. 45.6** Hourly temperature profile of the selected spot situated in the playground area, at z = 1.80 m, at 04:00 p.m., 21 August 2011

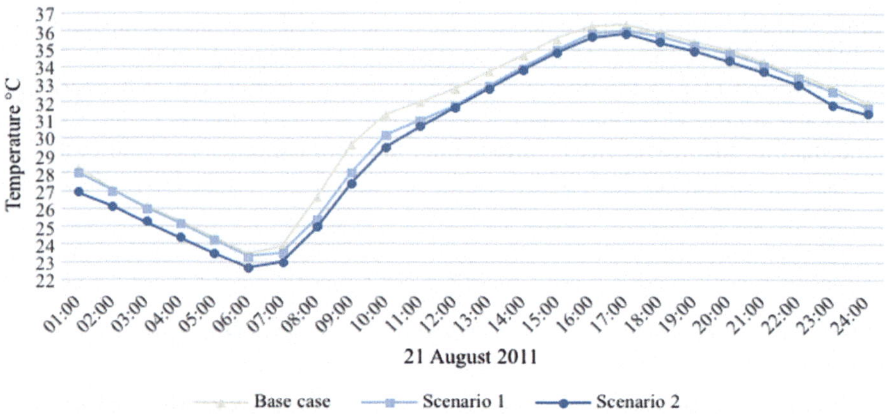


**Fig. 45.7** Hourly temperature profile of the selected spot situated in the market area, at z = 1.80 m, at 04:00 p.m., 21 August 2011

## 45.7 Conclusions

As mentioned above, in recent decades, urbanisation and regional climate changes have been occurring rapidly on a worldwide scale. Of the many aspects that are influencing the quality of urban life, one of the most essential is that of the microclimate of urban areas. In these circumstances, the value of the UHI has probably increased due to a variety of factors (building features and materials; more traffic and industrial activities; higher energy consumption, etc.). Accordingly, UHI mitigation strategies have been implemented in various sustainable cities around the world [15].

These strategies include identifying opportunities for urban greening, following the concept of the resilient and adaptive public spaces, which contribute to the improvement of urban environmental quality. In addition, the creation of new green spaces and corridors promotes the image of the city and contributes in improving the live of its citizens. When a public space needs to be renovated anyway, measures against heat accumulation can be combined with other measures that are necessary to improve the social, physical or economic condition [16].

Our study shows how it is possible to redevelop the urban space, using technological solutions that are actually able to improve the microclimate, reducing the atmospheric temperature and the PMV index, but also change its spatial and geometric configuration. Simple technological solutions (as repaving with cool materials, green and blue infrastructures and shading devices) can be implemented in a systemic way and stand a good chance of reducing the ill effects attendant upon the occurrence of UHI in the built environment. Finally, the possibility to use simulation tools allows the designers and the Public Administrations to know exactly the achievable results of these strategies.

## References


1. United Nations: Transforming our world: the 2030 Agenda for development (2015). https://sustainabledevelopment.un.org/content/documents/21252030%20Agenda%20for%20Sustainable%20Development%20web.pdf
2. IPCC: Climate Change 2014: Impacts, Adaptation, and Vulnerability. Part A: Global and Sectoral Aspects. Contribution of Working Group II to the Fifth Assessment Report of the Intergovernmental Panel on Climate Change. Cambridge University Press, Cambridge, United Kingdom and New York (2014)
3. Akbari, H., Cartalis, C., Kolokotsa, D., Muscio, A., Pisello, A.L., Rossi, F., Santamouris, M., Synnefa, A., Wong, N.H., Zinzi, M.: Local climate change and urban heat island mitigation techniques—the state of the art. J. Civ. Eng. Manag. **22**(1), 1–16 (2016)
4. Grimmond, C.S.B., Ward, H.C., Kotthaus, S.: How is urbanization altering local and regional climate? In: Seto, K.C., Solecki, W.D., Griffith, C.A. (eds.) The Routledge Handbook of Urbanization and Global Environmental Change. Routledge (2016)
5. Santamouris, M.: Energy and Climate in the Urban Built Environment. Routledge, Abingdon-on-Thames, UK (2013)
6. Dessì, V.: Urban materials for comfortable open spaces. World Renewable Energy Congress—Sweden; 8–13 May 2011; Linköping; Sweden, pp. 3300–3307 (2011)
7. Armstrong, B.G., Chalabi, Z., Fenn, B., Hajat, S., Kovats, S., Milojevic, A., Wilkinson, P.: Association of mortality with high temperatures in a temperate climate: England and Wales. J. Epidemiol. Commun. Health **65**, 340–345 (2011)
8. Hajat, S., O'Connor, M., Kosatsky, T.: Health effects of hot weather: from awareness of risk factors to effective health protection. Lancet **375**(9717), 856–863 (2010)
9. Santamouris, M.: Regulating the damaged thermostat of the cities—status, impacts and mitigation strategies. Energy Build. **91**, 43–56 (2015)
10. Arnetoli, M.V.: Climate resilience and urban regeneration. Innovative methods and tools for the project: the case study of public spaces in Scandicci. Degree Thesis, Department of Architecture, University of Florence (2018)
11. Ambrosini, D., Galli, G., Mancini, B., Nardi, I., Sfarra, S.: Evaluating mitigation effects of urban heat islands in a historical small center with the ENVI-Met® climate model. Sustainability **6**, 7013–7029 (2014)
12. Evola, G., Gagliano, A., Fichera, A., Marletta, L., Martinico, F., Nocera, F., Pagano, A.: UHI effects and strategies to improve outdoor thermal comfort in dense and old neighbourhoods. Energy Procedia **134**, 692–701 (2017)
13. Jendritzky, G., Nübler, W.: A model analysing the urban thermal environment in physiologically significant terms. Arch. Met. Geoph. Biokl. Ser. B. **29**(31), 313–326 (2018)
14. Crisci, A., Morabito, M., Nardino, M.: UNIFI IBIMET PRIN 2015 Climatic Database (Version 1.0) [Data set]. Zenodo (2018). http://doi.org/10.5281/zenodo.1292130
15. Rehan, R.M.: Cool city as a sustainable example of heat island management case study of the coolest city in the world. HBRC J. **12**, 191–207 (2016)
16. Kleerekoper, L., van Esch, M., Baldiri, Salcedo T.: How to make a city climate-proof, addressing the urban heat island effect. Resour. Conserv. Recycl. **64**, 30–38 (2012)

# Chapter 46
# The Correlation Between Urban Morphology Parameters and Incident Solar Radiation Performance to Enhance Pedestrian Comfort, Case Study Jeddah, Saudi Arabia

**Badia Masoud, Helena Coch and Benoit Beckers**

**Abstract** Many urban studies have evaluated the impact of urban morphology on the microclimate of cities, and it has been shown that the influence is substantial. One of the significant impacts on outdoor thermal comfort in hot climate cities depends on solar radiation. The city of Jeddah is expected to start constructing a subway in 2020 that is likely to drastically change the behaviors of its inhabitants. Therefore, this study aims to investigate the relationship between modern and contemporary morphology and the potential of incident solar radiation to enhance pedestrian comfort by avoiding transit under this harsh solar radiation. The authors have evaluated the performance of two urban layouts, with different building densities to investigate the extent to which morphological features of the built environment can affect solar penetration on the streets. Moreover, the study assessed the influence of various built parameters such as street geometry (H/W ratio, streets orientations) and the solar indicator (Sky View Factor). Results show that in the low rise building layouts, the orientation of streets and the H/W ratio are not reliable. Moreover, it is difficult to reduce solar radiation penetration on the streets by modifying the morphology. Therefore, other ways should be tested to shade from solar penetration. Furthermore, the authors have observed that for high rise building layouts, pedestrian comfort could be modified through the adjacent morphology.

B. Masoud (✉) · H. Coch
Architecture and Energy Research Group-School of Architecture, Polytechnic University of Catalonia (UPC), AV. Diagonal, 649-651, 08028 Barcelona, Spain
e-mail: Badia.ghassan.m@gmail.com; Badia_masoud@hotmail.com

B. Beckers
Department of Building and Public Works (ISABTP), Université de Pau et des Pays de l'Adour, Allée du Parc Montaury, 64600 Anglet, France



J. Littlewood et al. (eds.), *Sustainability in Energy and Buildings*, Smart Innovation, Systems and Technologies 163, https://doi.org/10.1007/978-981-32-9868-2_46

## 46.1 Introduction

The rapid and uncontrolled expansion of contemporary cities has led to environmental and social costs, that have become unsustainable, and it is now widely recognized that a large-scale approach to this problem and a new urban model are required [1]. It is commonly agreed that urban concentration, in contrast to dispersal, is a feature that makes the urban form more sustainable and resilient, as it reduces travel distances and, consequently, environmental impact [2]. The introduction of electricity as a new power transportation system during the Industrial Revolution brought a complete disengagement between energy sources and urban center locations. The subsequent expansion of transportation flexibility allowed for new urban structures, morphologies, and typologies [3]. Cities began a process of transformation related to horizontal and extensive urban growth, bringing with it the new concept of "urban sprawl" [4]. Car and road networks brought low-density structures with a lack of public space and social centers [5, 6]. Nonetheless, this urban morphology not only influenced the city structure from a social scope but also generates critical consequences for the environment [7]. The sun is the primary source of natural energy. In the 1970s; the solar radiation issue has caught the attention of architects and engineers. Among them, Paolo Soleri sharply criticized the waste of energies in megalopolises and introduced the concept of "Arcology", a specific exploration of solar effects concerning urban and architectural design [8]. 20 years later, Carl-Jochen Winter described the solar city as "an energetically self-contained settlement with solar irradiance as its main energy source" and considered what this big "energy converter" would look like and how it would have to be designed. In other words, he wondered what the esthetics and spatial organization of the urban fabric would be like in a solar city and which formal solutions would optimize the individual, technical, and social use of the sun [9].

The relationship between urban layout and solar radiation performance could provide crucial new input for the design process. Many previous researchers have investigated the relationship, and it has been proven that geometry could modify short and long radiation flux [10, 11]. Nevertheless, the majority of them focused on generic, archetypal urban forms without fully considering the morphological specificities that characterize the existing urban environments [12, 13]. Furthermore, as more than a quarter of urban areas are usually covered by streets, urban street design plays an important role in the urban climate. Urban streets vary in geometry as defined by the height/width ratio (H/W) and their orientation as defined by their long axis. These parameters directly influence the absorption, emission, and reflection of solar radiation, which has a significant impact on the temperature variations within the street as well as the surrounding environment [14, 15]. Moreover, the Sky View Factor is one of the geometric properties, that has major importance for urban climate applications since the long-wave radiation term is directly related to its value [16, 17].

This investigation sheds light on urban geometric layouts in the hot climate cities, of Jeddah/Saudi Arabia, as direct solar radiation leads to high temperatures. Jeddah's *modern* and *contemporary* urban layouts (Fig. 46.2) are characterized by wide streets and its urban form is classified as dispersed, where large parts of the streets are exposed to solar radiation. It is essential to study solar radiation performance in these types of layouts to enhance future pedestrian comfort. Masoud, Coch, and Beckers studied Jeddah's old compact layout (Alblad) and investigated the correlation between the Sky View Factor and the solar time interval of the horizontal surface (streets). They concluded that the Sky View Factor could be an indicator of solar access in the urban morphology, but in the streets with East–West orientation, it is not a reliable indicator of the solar radiation access, although it does give a sufficient approximation in streets with North–South orientation [18].

## 46.2 Methodology

The research establishes an evaluation and analysis by utilizing Heliodon2, a simulation software which provides possible data about the cumulative distribution of solar energy from the streets and the building envelopes (facades) [19]. Heliodon2 is a program that simulates the sun's path at a specific geographic site and provides graphical and numerical information about the evolution and the distribution of solar radiation, taking into account the shadows cast by neighboring obstructions. Two urban layout samples have been evaluated separately on 21 June (summer solstice) and 21 December (winter solstice). The information provided was the following: the Sky View Factor (%) and the solar flux (kWh/m$^2$/day) describing the quantitative distribution of incident radiation which enable us to identify specific aspects.

## 46.3 Case Study Description

### *46.3.1 Jeddah, Saudi Arabia Overview and Its Climate*

The city of Jeddah is located in the western part of Saudi Arabia, where it overlooks the east coast of the Red Sea. With 3.9 million inhabitants, Jeddah is considered as the second-largest city in Saudi Arabia in terms of population after Riyadh. Jeddah has the most important seaport of Saudi Arabia on the Red Sea. The city is at latitude 21° 32′ North and longitude 39° 10′ East. It has a hot climate with an average low/high temperatures typically varying between 18 and 39 °C, and rarely going below 16 °C or above 41 °C. Skies are almost always clear, with cloud cover varying between 2 and 18%. The relative humidity varies between 30 and 89%, so the air is mostly humid. Rainfall in Jeddah is generally sparse and usually occurs in small amounts in November and December. Jeddah is very close to the Tropic of Cancer, the sun almost reaches the zenith close to the summer solstice, reaching 45° at the winter solstice. The difference in day length between the two solstices is two hours and forty minutes. Solar height is up to 40° between 10:30 and 13:30 during the entire year, and the direct radiation is then up to 800 W/m$^2$ in sunny hours, reaching 1000 W/m$^2$ in the summertime [20].

### 46.3.2 Jeddah Map Layouts Description

The morphology orientation layouts in the city of Jeddah show the eligibility of an urban pattern as well as the consistency of the city's spatial organization, and it demonstrates how these patterns help us understand the amount of solar energy obtained in a particular morphology. Furthermore, on the map below in (Fig. 46.1) on the left, (diverse) each layout types are represented by a corresponding color, depending on its orientation. As a result of this observation, 60% of the urban layouts are oriented 90° towards North–South, 20% are 45° towards Northeast–southwest

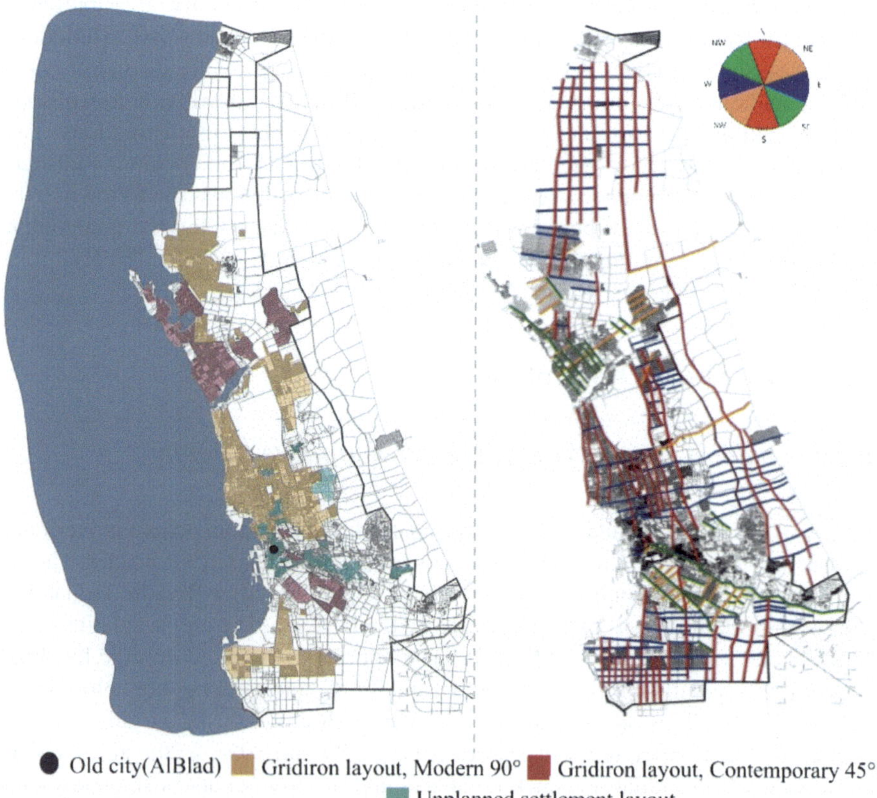


**Fig. 46.1** Jeddah map with different types of layouts and streets orientations. *Source* Author elaboration

and Northwest–Southeast, 15% are unplanned settlements, and 5% is the old area. The map on the right demonstrates some of the streets orientations in Jeddah city. The red lines represent the North–South streets orientation axis, the blue lines represent the East–West streets orientation axis, the green is the Northwest–Southeast, and the orange is the Northeast–Southwest orientation axis. The most dominant street orientations are the Red and blue lines (North–South and East–West).

### *46.3.3 Selected Layouts Description*

In order to evaluate the urban fabric, the analysis has to be limited, by reducing the number of variables to formal parameters—for example, choosing cases that have the same floor coverage ratio and different building intensities. This characteristic was evaluated using the floor area ratio and floor space index (GSI & FSI) [21] which quantifies the amount of floor space in relation to the gross land area [22]. The same approach was used in the (Compagnon) study of solar and daylight availability in an urban area in Fribourg (Switzerland). The floor area ratio constant was preserved and the performance of different hypothetical configurations was checked. Using the same strategy, the researchers identified two urban layouts: (AlSlamah) from the modern area in Jeddah that has a medium density and is characterized by high rise buildings, and another layout (Obhur AlShmalya) from the contemporary area of Jeddah that has a low density and is characterized by low rise buildings. The researchers created a 3D digital mock-up, using the cadastral information available, as shown in (Fig. 46.2). Both chosen urban layouts have a similar gridiron plan, but they have different characteristics, related to density, H/W, and plot orientation. The selected case in the modern area is a mixed-use area (residential and commercial), the building heights in the AlSlamah range from 4 to 7 stories and its street widths

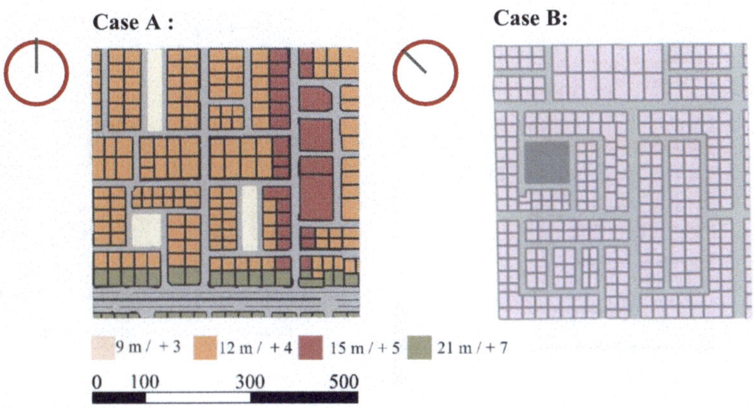


**Fig. 46.2** Modern and contemporary layout selection building heights Case (A and B) in Jeddah city, 500 × 500 m. *Source* Author elaboration

vary between 12, 15, 18, 20, and 50 m. The contemporary area (Obhur AlShamalya) is characterized by its 2–3 story detached villas, segregated from the commercial area, simple geometrically shaped patterns of streets designed for adequate vehicle movement, with width ranging from 10 to 20 m. Both layouts consist of lots with a typical size of 20 × 30 m and most blocks are 60 × 180 m. The district was laid out in a rectangular grid system, with rectangular and square lots. The architecture is distinctly American in style and built according to zoning regulations that stipulate distances from the street, and with detached rather than attached buildings. The simulation and analysis were on the streets and the envelope (façades) of the two layouts (Case A-The modern layout and the Case B-The contemporary layout).

## 46.4 Results and Discussion

In Figs. 46.3 and 46.4, both urban layouts on 21 June (summer solstice) show that incident solar radiation penetrates more on the horizontal layout (streets) and penetrates less on the vertical surfaces (facades). Due to the solar angle particularly with the absence of obstruction in low-density layouts Case B. In winter; however, the solar angle is lower than in summer; therefore, the incident solar radiation on the South–East and West facade in Case A is high. It is preferable to have more solar penetration in winter and less in summer. The results of the average incident solar flux on the horizontal surfaces (streets) for both case studies are similar in summer and in winter. The incident solar flux in general expresses the behavior of a portion of an urban layout. Case A receives 5.9 kWh/m$^2$/day and Case B receives 5.8 kWh/m$^2$/day. The results reveal that the amount received is the same, due to wide streets and plot

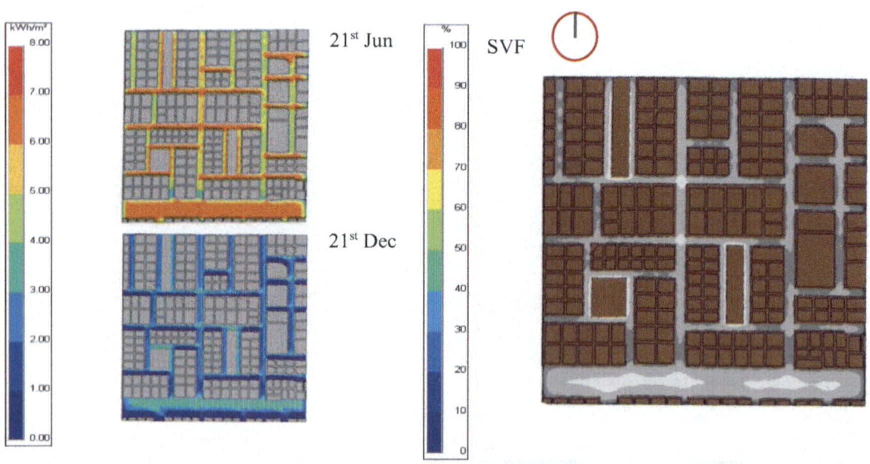


**Fig. 46.3** Case A on the left is the solar flux kWh/m$^2$/day in summer 21June and on the right is the Sky View Factor. *Source* Heliodon software

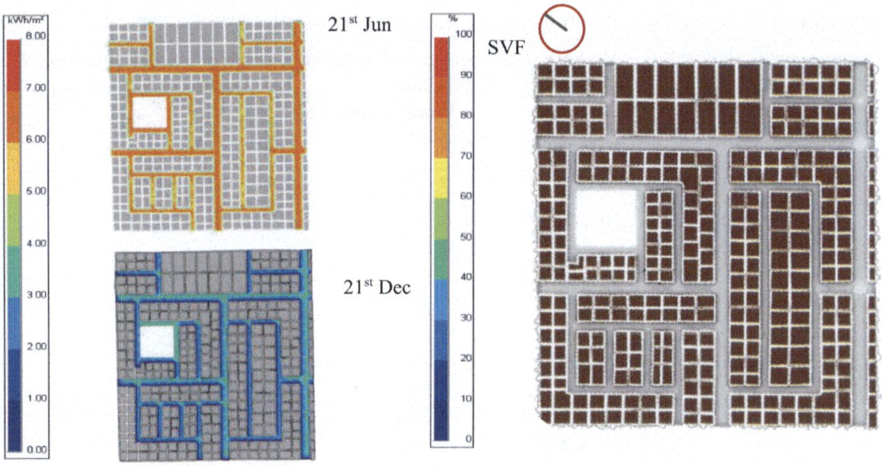


**Fig. 46.4** Case B on the left is the solar flux kWh/m$^2$/day and on the right is the Sky View Factor. *Source* Heliodon software

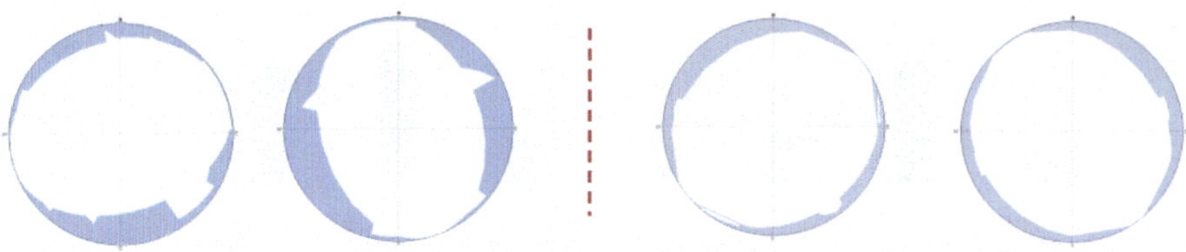

**Fig. 46.5** Sky View Factor of both case studies (Case A on the left and Case B on the right). *Source* Heliodon software

orientation. Nevertheless, as Case A is characterized by high rise building layout, the below analysis of each street orientation will reveal the different building heights impact on the horizontal surface (streets). Moreover, because the low rise building layout is characterized by a low H/W ratio and high Sky View Factor, incident solar radiation is unobstructed and penetrates from all directions (Fig. 46.5).

### *46.4.1 Case A Solar Flux Value and Sky View Factor*

In Case A, the solar energy value of each street orientation is depicted using SVF and sunshine duration. The spatial distribution of Sky View Factor describes the obstruction of sky exposure by each street. The simulation revealed in the below Table 46.1 and graph, Fig. 46.6, shows that the Sky View Factor increases from the center to the sides of the streets. The streets oriented North–South with an H/W ratio from 0.26 to 1 receive an average solar flux in the summer ranging from 7 to 2 kWh/m$^2$/day. And in winter from 3 to 1 kWh/m$^2$/day with an average Sky

**Table 46.1** Streets Average solar flux on each orientation, H/W ratio, SVF Case A

| Orientations | H/W ratio | AV. SVF (%) | Range AV. solar flux 21st June (kWh/m$^2$/day) | Range AV. solar flux 21 DEC (kWh/m$^2$/day) |
|---|---|---|---|---|
| N–S | 0.26 | 70 < 100 | 5–7 | 1–3 |
| | 0.60 | 60 < 90 | 5–7 | 1–3 |
| | 0.83 | 50 < 90 | 4–6 | 1–3 |
| | 1 | 40 < 60 | 2–5 | 1–3 |
| E–W | 0.46 | 70 < 90 | 6–7 | 0–2 |
| | 0.66 | 60 < 90 | 6–7 | 0–2 |
| | 1 | 50 < 60 | 6–7 | 0–2 |

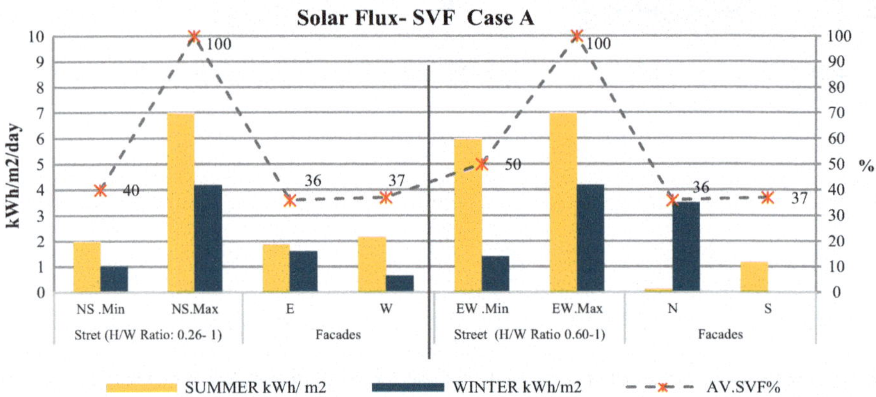


**Fig. 46.6** Case A. Average solar flux-SVF streets and facades. *Source* Author elaboration

View Factor between a minimum of 40% to a maximum of 100%. Additionally, on the East–West street orientation with an H/W ratio from 0.46 to 1, it receives an average solar flux ranging from 7 to 6 kWh/m$^2$/day. And in the winter ranging from 2 to 0 kWh/m$^2$/day with an average Sky View Factor between minimum 50% to a maximum 100%. Furthermore, in summer, the East–West orientation receives a higher solar flux value than the North–South orientation, and in winter the opposite is true.

## 46.4.2 *Case B Solar Flux Value and Sky View Factor*

In Case B, the layout is oriented at 45° and no obstructions are affecting the exposure of the layout surface to solar radiation. As a result of the low-density solar angle, and the plot orientation, Fig. 46.7 reveals that the average solar flux on the entire layout

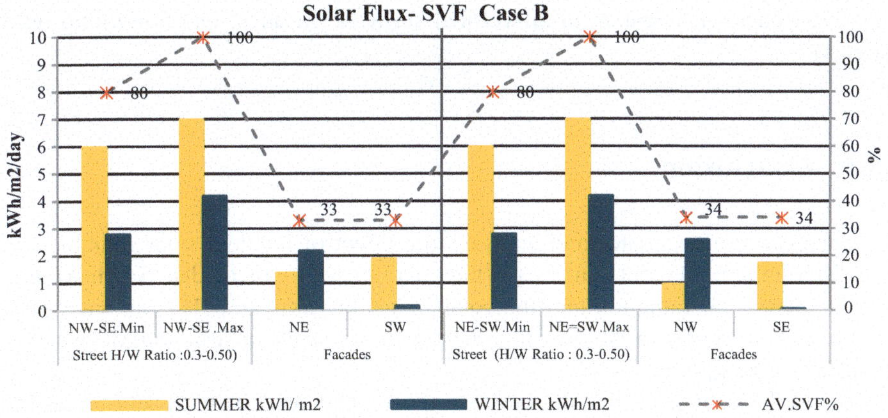


**Fig. 46.7** Case B. Average solar flux-SVF streets and facades. *Source* Author elaboration

is the same amount in general. It has a high solar flux value throughout the whole day on 21 June. Moreover, it shows that the average SVF is between a minimum of 60 and 80% on the streets oriented Northwest–Southeast. Northeast–Southwest orientation presents a higher average SVF, between 80 and 100%. Besides, solar radiation shows similar intensity due to the lack of obstruction. Consequently, the sun penetrates from all directions.

### *46.4.3 Facades Solar Flux and Sky View Factor*

Moreover, the overall quantitative analysis of the results in both Cases A and B in general shows that in the selected urban layouts, the vertical surfaces (facades) receive less solar radiation than the horizontal ones (streets). Additionally, vertical surfaces in both cases receive fluctuated solar radiation depending on the facade orientation and the solar angle. Accordingly, in (Case A), Fig. 46.3 on 21 June, the façade that obtains the most solar flux value is the West façade receiving 2.20 kWh/m$^2$/day with an average SVF of 37%. And the lowest in the North façade which receives 0.16 kWh/m$^2$/day with an average SVF of 36%. On 21 December, the North facade obtains the highest amount of solar flux value at 3.5 kWh/m$^2$/day. Furthermore, according to Case B, Fig. 46.4, on 21 June the Southwest façade receive the highest solar flux value it receives 1.90 kWh/m$^2$/day with an average SFV of 33%. And the Northwest façade receives the lower solar flux value of 0.97 kWh/m$^2$/day with an average SVF of 34%. Also, on 21 June, the Northwest receives the highest amount amongst the other facades of 2.6 kWh/m$^2$/day. And on 21 December, the Southeast facade receives the lowest incident solar energy receiving 0.067 kWh/m$^2$/day, because the solar angle on the summer solstice is higher than it is in the winter solstice. Additionally, due to the low density and dispersed urban layout, the average

Sky View Factor is constant in all vertical surfaces (facades) and horizontal ones (streets).

## 46.5 Conclusion

This investigation evaluates and analyzes the effect of incident solar radiation on two different urban layout densities in the hot climate city of Jeddah, Saudi Arabia. The incident solar radiation and the solar indicator (Sky View Factor) have been calculated considering the solar radiation intensity and solar obstructions determined by different urban intensities.

This analysis suggests that the most compact and dense urban layout such as the historic fabric of Jeddah (AlBalad) has less solar radiation penetration, and provides better pedestrian comfort than dispersed and less dense urban layouts. However, it might be difficult to copy the old layout and apply it on future urban developments considering other factors and human needs as privacy, security, transport, and health. In new urban layouts, it may be challenging to secure easy access to all places with public transportation to reduce the number of private cars.

Furthermore, evaluating and analyzing the low rise building of the contemporary urban layout, the researchers observed that the urban morphology parameters (H/W ratio, streets orientations) and solar radiation indicator SVF are not reliable. Incident solar radiation penetrates as no obstructions are available to protect from solar radiation. Therefore, as in these types of urban layouts, it is not possible to enhance pedestrian comfort by modifying the urban morphology and other solar radiation protection should be considered to protect from solar penetration especially on hot summer days.

Nevertheless, for future expansions and developments in Jeddah and other low latitudes, these results should be considered to enhance pedestrian comfort. And we suggest dispersed cities to consider compacting and designing solar protections in some areas of the city for pedestrians to travel under the shade, without needing to adjust all the gross morphology. The opportunity of performing rapid evaluations at the district scale and providing maps of the incident solar radiation energy within the city would undoubtedly ease the work of designers, planners, and decision-makers in identifying a priority of interventions on the built environment.

**Acknowledgements** The authors thank the Spanish Ministry of Education for supporting this work under project code: BIA2016-77675-R. And special thanks to the Ministry of Education of the Kingdom of Saudi Arabia for supporting and funding this work.

## References


1. Real, L: Density, city, residence. Densification techniques and anti-sprawl strategies. Editor, Gangemi, Roma (2008)
2. Banham, R.: The Architecture of the Well-Tempered Environment, 2nd Ed. University of Chicago Press: Chicago, IL, USA; The Architectural Press Ltd., London, UK (1984)
3. Forgaci, C., Van Timmeren, A.: Urban form and fitness: towards a space-morphological approach to general urban. In: Annual International Sustainable Development Research Conference. Norwegian University of Science and Technology (2014)
4. Ewing, R.: Characteristics, causes and effects of sprawl: a literature review. In: Marzluff, J., Shulenberger, E., Endlicher, W., Alberti, M., Bradley, G., Ryan, C., ZumBrunnen, C., Simon, U. (eds.) Urban Ecology An International Perspective on the Interaction Between Humans and Nature, 1st edn. Springer, USA, New York, NY, USA (2008)
5. Banham, R.: The Architecture of Four Ecologies. Allen Lane the Penguin Press, London, UK (1971)
6. Mumford, L.: New York et l'Urbanisme. Editions Seghers, Paris, France (1965)
7. Raydan, D., Steeners, K.: Environmental urban design. In: Santamouris, M. (ed.) Environmental Design of Urban Buildings, an Integrated Approach, pp. 1–35. Routledge. New York, USA (2006)
8. Lima, I.A.: Soleri: Architecture as a Human Ecology (2000)
9. Winter, C.: solar cities. Renew Energy **4**(1), 15–28 (1994)
10. Johansson, E.: Influence of urban geometry on outdoor thermal comfort in a hot, dry climate: a study in Fez Morocco. Build. Environ. **41**(10), 1326–1338 (2006). https://doi.org/10.1016/j.buildenv.2005.05.0222010.07.008
11. Lindberg, F., Grimmond, C.: Continuous Sky View Factor maps from high resolution urban digital elevation models. Clim. Res. **42**(3), 177–183 (2010)
12. Rasheed, A., Robinson, D., Narayanan, C., Lakehal, D.: On the effects of complex urban geometries on mesoscale modelling. In: Proceedings of the 7th International Conference on Urban Climate, Yokohama, Japan (2009)
13. Curreli, A., Serra-Coch, G., Isalgue, A., Crespo, I., Coch, H.: Solar energy as a form giver for future cities. Energies. https://doi.org/10.3390/en9070544 (2016)
14. Ali-Toudert, F., Mayer, H.: Effects of asymmetry, galleries, overhanging façades and vegetation on thermal comfort in urban street canyons. Sol Energy **81**(6) (2007)
15. Bourbia, F., Boucheriba, F.: Impact of street design on urban microclimate for semi arid climate (Constantine). Renew. Energy. https://doi.org/10.1016/j.renene.2009.07.017344 (2010)
16. Oke, T.R.: Canyon geometry and the nocturnal urban heat island: comparison of scale model and field observations. Int. J. Climatol, 237–254 (1981)
17. Middel, A., Lukasczyk, J., Maciejewski, R., Demuzere, M, Roth M.: Sky View Factor footprints for urban climate modeling. Urban Clim. 120–134. https://doi.org/10.1016/j.uclim (2018)
18. Masoud, B., Coch, H., Crespo, I., Beckers, B.: Effects of urban morphology on shading for pedestrians Sky View Factor (SVF) as an indicator of solar access. In: Passive and low energy architecture conference, Hong Kong (2018)
19. Beckers, B., Masset, L.: Heliodon2 software and user's guide. http://www.heliodon.net (2009)
20. Masoud, B., Coch, H., Beckers, B.: Sky access versus shading for pedestrian comfort in the hot tropical climate of Jeddah. In: First International Conference on Urban Physics, Quito-Galapagos (2016)
21. Pont, MB., Haupt, P.: Spacematrix, Space, Density and Urban Form. NAi Uitgevers ISBN 10:9056627422 (2012)
22. Compagnon, R.: Solar and daylight availability in the urban fabric. Energy Build. **36**, 321–328 (2004)

23. Middel, A., Häb, K., Brazel, A.J., Martin, C., Guhathakurta, S.: Impact of urban form and design on microclimate in Phoenix Urban Plan. 122, 16–28 (2014)
24. Shishegar, N.: Street design and urban microclimate: analyzing the effects of street geometry and orientation on airflow and solar access in urban canyons. J. Clean Energy Technol. https://doi.org/10.7763/jocet.2013.v1.13 (2013)

# Chapter 47
# Active Buildings in Practice

**Joanna Clarke, Paul Jones, John Littlewood and Dave Worsley**

**Abstract** This paper discusses the development and validation of a new Code of Practice for the design of Active Buildings, which will be the main output from a Professional Doctorate in the Sustainable Built Environment (D.SBE) research project being undertaken by the first author. It sets the context for the Code, related to UK, European and Global Climate Change targets, before describing the aspirations of the first author to develop the guidance document in conjunction with architectural practitioners, combining knowledge sharing and education with government targets for reducing energy consumption and associated carbon emissions from buildings. An explanation of the term 'Active Building' is included and the recently (2019) established Active Building Centre is also discussed to explain the need for the Code of Practice, which will be one of the main outputs from the centre at the end of its current 3-year programme ending in 2022. Progress on the development of the Code of Practice is discussed, including how the document is being tested with Architects and students, and how it could align with government policies, as well as enabling the roll-out of Active Buildings across the United Kingdom (UK).

## 47.1 Introduction

This paper discusses the development and validation of a new Code of Practice for the design of Active Buildings, which will be the main output from a D.SBE research project being undertaken by the first author [1]. The following paragraphs describe what an Active Building is and how it relates to the United Kingdom (UK) Government targets to tackle global climate change.

J. Clarke (✉) · P. Jones
Active Building Centre, Swansea University, SA1 8EN Swansea, UK
e-mail: Joanna.r.clarke@swansea.ac.uk

J. Littlewood
Cardiff Metropolitan University, Cardiff CF5 2YB, UK

D. Worsley
Swansea University, Swansea SA1 8EN, UK

 
J. Littlewood et al. (eds.), *Sustainability in Energy and Buildings*, Smart Innovation, Systems and Technologies 163, https://doi.org/10.1007/978-981-32-9868-2_47

In September 2018, the UK Government awarded £36 million of funding for clean energy innovation in the construction sector, through the 'Transforming Construction' challenge within their Industrial Strategy Challenge Fund (ISCF) [2]. This was awarded to an academic, industry and government consortium led by Swansea University, to prove scalable solutions to enable the construction industry to deploy the Active Building concept developed by SPECIFIC Innovation and Knowledge Centre, Swansea University, through a National Centre of Excellence, known as the Active Building Centre (ABC). An Active Building is described by 'Transforming Construction' as '*a building which integrates solar generation and storage technologies for both electricity and heat within its construction, and which is controlled by an intelligent system to optimise energy management and comfort for inhabitants. Active Buildings aim to be net energy generators and have the potential to utilise the surplus energy to trade*' [2]. The ABC will work to remove barriers to the large-scale adoption of Active Buildings on new developments throughout the UK, supporting the UK Government's mission to at least halve the energy use of new buildings by 2030, announced on May 2018 as one of their Grand Challenges [3]. Making buildings more efficient by embracing smart technologies will cut energy bills, reduce the demand for energy and boost the UK's economic growth whilst meeting targets for carbon reduction.

The aim of this paper is to demonstrate the need for design guidance to enable Active Buildings to be delivered in practice, which has been identified as necessary by the first author. This will be developed through this research project as a Code of Practice or Design Guide which aims to provide designers with guidelines to use when designing Active Buildings. This will be the ABC's first source of guidance on the design of Active Buildings and one of the main outputs of the centre. In developing the Code, the first author will use qualitative research methods to engage with practicing Architects and architectural students to help ensure the contents are relevant to Architects. Initial development and feedback are discussed in this paper, which will be of interest to Architects and others developing design guidance documents.

## 47.2 Context

Before embarking on the development of a Code of Practice for designing Active Buildings, the first author initially undertook a literature review to determine the political landscape, the current state of the construction industry and existing work on low energy buildings [1]. This is summarised in the following section.

The building and construction sector is responsible for approximately 30% of all global energy consumption and the associated greenhouse gases [4]. In the UK, heating and powering buildings accounts for 40% of energy use [2]. In addition to preventing rising global temperatures by reducing carbon emissions, Active Buildings could also help to significantly reduce fuel poverty [5]. In 2016, approximately 23% of households in Wales were living in fuel poverty, spending at least 10% of their income on energy costs [6]. To set the context for this project, the first author has undertaken a review of the energy-related targets the UK and Welsh Governments have committed to achieving, as well as some of the agreed Global targets, over a period of time from 2001 to 2018 [7–17]. The targets are set to be achieved within the time period 2020–2050.

In May 2019, the Committee on Climate Change produced a report entitled '*Net Zero: The UK's contribution to stopping global warming*' [18], which recognises that as a major contributor to global climate change, the UK now needs to take a leading role in tackling this. This aligns with the Transforming Construction mission, which enabling Active Buildings to be designed and delivered in practice will help achieve. In other related literature, Moseley [19] in his research into 'smart buildings' undertaken through the Horizon 2020 programme [20] highlights recent (2016) reports, as well as other directives, such as the EPBD [8] published 17 years ago in 2002, established to tackle climate change. The context for the 'smart building' projects described in his paper is very much aligned to the context for Active Buildings and there are many synergies between smart buildings and Active Buildings, which could be viewed as a type of smart building.

To further set the context for developing a Code of Practice for designing Active Buildings, the first author has also carried out research into the challenges in introducing innovation into the construction industry to enable low energy buildings, through a series of interviews and workshops with construction industry stakeholders. There are many factors for consideration in scaling the Active Building concept to enable it to be deployed on building projects across the UK. Some of the main challenges identified in this study were: time, procurement methods, cost, skills, early engagement of project delivery teams, aesthetics and planning restrictions, integration with Building Management Systems (BMS), maintenance worries, lack of Research and Development, lack of knowledge, and risk. The ABC was established to tackle these challenges and reduce the barriers to large-scale adoption of the Active Building concept. The first document to assist the ABC tackles these challenges and enable Active Buildings to be delivered in practice, by setting out clear design guidance, will be the Code of Practice for designing Active Buildings, developed through this research project and discussed in this paper.

## 47.3 Methods

Qualitative research methods were deployed in this project, with the main instruments used being semi-structured interviews and focus groups. In the first part of the pilot project, the first author developed a questionnaire to ask members of project delivery teams the main challenges in introducing innovation into the construction industry and what would help reduce these challenges or barriers. Rather than issuing the questionnaire via post or email, it was decided to use the questionnaire as a basis for semi-structured interviews and focus groups. In the first author's experience, interviews and focus groups provide more scope to probe issues, coupled with the fact that response rates to questionnaires can be low, so not completely reliable. Interviews and focus groups are often used in qualitative research as the main instrument for data collection [21].

Initially, the first author undertook several semi-structured interviews with Main Contractors and Project Managers. Each interview lasted 90–120 min and provided a good insight into the challenges from the point of view of those responsible for delivering building projects. However, the main focus of this project is on architectural practice and, with the first author's knowledge and experience as an Architect in practice, she is aware that Architects and other designers have limited time available to spend on non-fee-earning work. This led her to consider how sufficient meaningful data could be obtained, without encroaching significantly on Architects time, and in a way to provide some value to the architectural practice. Ultimately, the project aims to produce material that will help Architects and other designers to design Active Buildings. To aid this, one of the data collection methods identified was to develop a Continuous Professional Development (CPD) seminar or workshop, aligned with the Royal Institute of British Architects (RIBA) Core CPD Curriculum [22] which would provide a platform to discuss the issues explored in this research project. This would enable information to be gathered from groups of Architects via a session held within their own practice. To ensure the session was of value to the Architects, the seminar was split into two parts: the first describing Active Buildings, with case study examples; the second part taking the form of a workshop to discuss the challenges facing the construction industry in 2018/19 and the targets to reduce energy use of buildings that have been widely reported [15]. This seminar was also held with one building services consultancy. In total, 49 practitioners took part in these focus groups.

The workshops were recorded through sound recording, and transcribed following each session, enabling analysis of the results. As this research project has evolved, the questions asked in the seminars were adapted to reflect the project focus, i.e. to develop a Code of Practice for designing Active Buildings, which had not been identified as the main project output at the time of the first CPD. This aligns with the iterative process described by Groat and Wang [23] when discussing qualitative research strategies.

The next stage in the project was to develop the Code of Practice, using the feedback gleaned from the first part of the pilot project and testing the document as it develops, as described in the following sections.

## 47.4 The Code of Practice for Designing Active Buildings

According to the Collins English Dictionary, a Code of Practice is '*a set of written rules explaining how people working in a particular profession should behave*' [24]. Before embarking on the development of the Code of Practice for designing Active Buildings, the first author undertook background research, which formed the first parts of her Professional Doctorate studies, which included: determining and evaluating existing policies and legislation around energy and buildings—UK, Welsh Government and Global; Undertaking both a literature review to understand the context and the current state of the nation with regard to energy and buildings; Undertaking focus groups and interviews to determine some of the anticipated challenges in delivering Active Buildings and identify potential solutions.

The process of developing the guidance document is now underway and being tested with architectural practitioners and architecture students to ensure the guidance is suited to Architects, whilst also ensuring it aligns with the delivery programme for the ABC.

The main aim of the Code of Practice is to help Architects and other design professionals in the UK embarking on the design and construction of Active Buildings to understand how Active Buildings could help address some of the challenges facing the construction industry in reducing the energy consumption of buildings; and to understand key performance criteria for consideration when designing Active Buildings. The document aims to help Architects design buildings with lower operational energy use and lower operational energy costs. As well as meeting the needs of the building owners/occupiers, Active Buildings consider the current energy landscape and the need to significantly reduce the energy use of buildings as described earlier in the paper. The first iteration will focus mainly on new buildings in the UK but will subsequently be developed for retrofit of existing buildings and also buildings in other countries. The first author has established three different ways to test the Code of Practice during its development: with architectural practitioners; with architecture students; and with architects involved in Active Building projects. Testing the Code of Practice in this way has commenced and the results are discussed in the next section.

## 47.5 Results

### *47.5.1 Architectural Student Project*

The purpose of undertaking the project with architecture students was to work towards changing architectural practice by educating and influencing the next generation of practising Architects to understand the concept of Active Buildings and to ingrain the Active Building concept into all their designs. This is critical to ensure the mass uptake of Active Buildings by design professionals and the industry as a whole.

In December 2018, the first author approached a tutor of architectural students at an Architecture School in Swansea, UK, regarding trialling the Design Guide with students. This manifested as a project with a group of first-year architecture students, to design an Active House as part of their technology module, using the Design Guide developed. The students were issued with a project brief, the Active Building concept was explained to them through a presentation and tour of two Active Building demonstrators. The students were then issued with the Design Guide and supplementary information as it was deemed necessary and the first author attended weekly sessions with the students over a 3-month period (the duration of the project). Feedback was gleaned from the students at the end of the project through discussions and questionnaires to determine their experience of working with the guide. The project concluded with a presentation by each of the students and they also wrote an essay on the project as part of their assessment.

During the final presentation, it was clear that the students had gained a good understanding of Active Buildings. Their feedback was positive, and all felt that they understood Active Buildings and their purpose, i.e. to reduce the energy consumption of buildings. The students found that the guidance document was clearly structured, easy to follow and contained all the information they needed to design an Active House.

### 47.5.2 Architectural Practitioners

In December 2018, the author approached the Director of the Royal Society of Architects in Wales (RSAW)—the Welsh branch of the RIBA—to discuss opportunities to include a CPD session on Active Buildings to trial the Design Guide under development, as part of the RSAW CPD programme in 2019. Fortunately, the timing was good, and these sessions were included in the Spring programme to be held in March and April 2019. The session was also featured within the Wales Festival of Architecture 2019 [25]. Part of the Code of Practice was issued as a Design Guide and the Architects were asked to undertake a short design exercise to design an Active House using key design criteria set out in the document. They were then asked to provide feedback, both through a discussion and by completing questionnaires. The purpose of this exercise was to gain their feedback on the guide—the level of information provided, the structure and whether any additional information would be useful.

It was anticipated that, by linking to the RIBA Core CPD Curriculum, the RSAW CPD Programme and the Wales Festival of Architecture, it would be possible to engage with a larger number of Architects than that might be possible without these connections. However, the total number of Architects attending the sessions was fairly low, which could have been due to various factors, such as the timing of the session, the length of the session or perhaps the fact that it is not directly related to contracts or regulations. The feedback gained from the 10 participants was useful, nonetheless. The main feedback was that although the content was clear and informative, Architects would prefer it to be more visual—more diagrams, comparison tables, case studies and bullet points rather than paragraphs of text. Some of the Architects commented that although they were aware of some of the information included, the document acted as a useful reminder to them about key design principles such as the use of best practice environmental design. The main difference between practicing Architects and architectural students was that Architects are working on real projects with clients and have a duty to offer sound advice to clients on the best features to include in their buildings and crucially to get the best value for their clients. This led to discussions about anticipated additional capital costs and other challenges such as risk, aligning with the feedback from the earlier CPD sessions.

To summarise, Architects were seeking information that they could access quickly and use to advise their clients on the Active Building concept based on facts and figures that demonstrate the benefits of the approach. It was interesting that the students found the text useful, whereas practicing Architects wanted a less wordy document. The first author believes this could be due to the fact that Architects in practice already know a lot of the information presented from their education and experience, whereas the students were learning and did not have any design experience to draw on.

### *47.5.3 Architect of the Active Building Centre Headquarters*

In 2018, an architectural practice was engaged to design a new flagship building to house the ABC. Their brief was to design an 'Active Building', based on two Active Building demonstrator case studies—the Active Classroom [26] and the Active Office [27]. During the early design process of any building project, when often the only consultant engaged is the Architect, there are many issues that need to be considered and, as such, it is easy for elements of the design brief to be lost, unless these are written down in a document for use as a reference throughout the design process. At the time of designing the ABC headquarters, no Design Guide or Code of Practice had yet been developed and the Architects were working to a challenging budget for the size of building required, whilst also considering aesthetics, planning requirements, the need to meet building regulations [28] and office standards, such as the British Council for Offices (BCO) space standards [29]. Therefore, the 'Active' elements of the building could easily have been lost in the process, had the ABC team not been involved at every stage. On other projects, they would not necessarily be involved so, unless a clear brief or guidance is written down, there is no way of ensuring the delivered building will be an Active Building. A Design Guide or Code of Practice stipulating key design criteria and explaining why it is necessary to follow these is essential to the successful delivery of many Active Building projects across the UK. The first author has interviewed the Architect of the ABC headquarters to determine whether such a Design Guide would have been helpful in the design of the building. A draft of the Design Guide was issued to the Architect and then the Architect was questioned on the document to determine their views on whether it would be a useful document to aid the design process and on its content. The Architects comments are summarised as: It is clear what is meant by an Active Building; The content is good, but the document contains a lot of written information, organised in a sequential way. However, most Architects do not like reading lengthy documents, and representing the information more graphically via diagrams would be preferable to Architects; Use of case studies within the document would also be welcomed; The Architect agreed it would have been helpful to have such a document at the start of the project to design the ABC headquarters.

This concurred with feedback from the CPD sessions with groups of Architects. The results from all three sets of data collected will now be used to inform the further development of the guidance document, using an iterative process. For future projects, the Active Building Centre is involved with, it is envisaged that the Design Guide will be issued to the design team at the start of a project to provide clarity during the design phases. This will provide rigour to designs and enable reliable data to be collected on the performance of buildings, which can be measured against criteria set out in the Guide.

## 47.6 Discussion

The literature review and the launch of the Transforming Construction challenge (2018) [2] highlighted a need to change the way buildings are designed and delivered, to achieve the mission to 'at least halve the energy use of all new buildings by 2030' [2]. In line with this, the first author developed a methodology to identify a solution to enable the design of Active Buildings, which set out to achieve the mission. This involved interviewing and holding focus groups with Architects, Project Managers and Main Contractors to identify the challenges in introducing innovation into construction projects. As a response to the challenges identified, the first author made the decision to develop a Code of Practice for designing Active Buildings, which she has been testing with Architects. This exercise has proved that the design guidance is needed in order to deliver Active Buildings, aligned with the Transforming Construction mission and has provided feedback to aid its development. The next step is to develop the design guidance further using this feedback and to test it again once after it is further developed.

## 47.7 Conclusions

This paper has discussed the progression of a D.SBE project focused on developing a Code of Practice for designing Active Buildings. It has described what an Active Building is, set out the context for Active Buildings and explained the process of developing the guidance document together with the iterative process being used to test it with architectural practitioners as it develops.

In order to achieve the aims of the ABC within the timescale of 3 years, it is critical to develop some guidance to aid the process of rolling out the Active Building concept. In this respect, the first author's research project and the development of the Code of Practice for designing Active Buildings will be a valuable contribution for helping to achieve the centre's aims. This is an exciting opportunity to shape the way that all new buildings in the UK, and subsequently other countries globally, will be designed and is the first guidance of this kind to be developed. By testing the guidance documents with architectural practitioners, the first author has the opportunity to ensure the final document, which will be developed over the next few years, will act as an essential document for all designers of new buildings in the UK in the first instance contributing to achieving the goals of the Transforming Construction challenge [2].

**Acknowledgements** The first author wishes to acknowledge support of UKRI who fund the ABC, as part of the 'Transforming Construction' challenge within the Industrial Strategy Challenge Fund (ISCF).

## References

1. Morgan, J., Littlewood, J., Jones, P., Wilgeroth, P.: Testing and validation of 'building as power station technologies' in practice, to maximise energy efficiency and user comfort and minimise carbon emissions. In: International Conference on Sustainability in Energy and Buildings, Chania, Crete (2017)
2. Department for Business, Energy and Industrial Strategy (BEIS). 2018[a]. News story: Swansea receives £36 million UK government funding for its clean energy tech breakthrough. https://www.gov.uk/government/news/swansea-receives-36-million-uk-government-funding-for-its-clean-energy-tech-breakthrough, Accessed 12th Dec 2018
3. BEIS. 2018[b]. The Grand Challenge Missions. https://www.gov.uk/government/publications/industrial-strategy-the-grand-challenges/missions, Accessed 12th Dec
4. Laski, J., Burrows, V.: From Thousands to Billions: Co-ordinated action towards 100% Net Zero Carbon Buildings by 2050 (2017). http://www.worldgbc.org/news-media/thousands-billions-coordinated-action-towards-100-net-zero-carbon-buildings-2050, Accessed 22nd June 2017
5. Jones, M.: Energy-positive design 'could cut home fuel bills by 60%' (2017). http://www.bbc.co.uk/news/uk-wales-south-west-wales-40865141, Accessed 22nd Octo 2017
6. Spragg, J., Evans, S.: Research Briefing: Low Carbon Energy in Wales. National Assembly for Wales, Cardiff, UK (2017). https://seneddresearch.blog/2017/05/22/new-publication-low-carbon-energy-in-wales/, Accessed June 2017
7. HM Government: Environmental taxes, reliefs and schemes for businesses (2017). https://www.gov.uk/green-taxes-and-reliefs/climate-change-levy, Accessed 18th Dec 2017
8. European Parliament and Council of the EU: Directive 2010/31/EU of the European Parliament and of the Council of 19 May 2010 on the energy performance of buildings (2010). http://eur-lex.europa.eu/legal-content/EN/TXT/?qid=1513631349697&uri=CELEX:32010L0031, Accessed 18th Dec 2017
9. European Commission. 2017. 2020 climate and energy package. https://ec.europa.eu/clima/policies/strategies/2020_en, Accessed 18th Dec 2017
10. HM Government: Climate Change Act 2008 (2008). https://www.legislation.gov.uk/ukpga/2008/27/pdfs/ukpga_20080027_en.pdf, Accessed 23rd Aug 2017
11. HM Government: The UK Renewable Energy Strategy (2009). https://www.gov.uk/government/uploads/system/uploads/attachment_data/file/228866/7686.pdf, Accessed 2nd Sept 2015
12. Welsh Assembly Government: Climate Change Strategy for Wales (2010). http://gov.wales/docs/desh/publications/101006ccstratfinalen.pdf, Accessed 18th Dec 2017
13. European Parliament and Council of the EU: Directive 2012/27/EU of the European Parliament and of the Council of 25 October 2012 on energy efficiency (2012). http://eur-lex.europa.eu/legal-content/EN/TXT/?qid=1399375464230&uri=CELEX:32012L0027, Accessed 18th Dec 2017

14. DEFRA. 2013. Environmental Reporting Guidelines: Including mandatory greenhouse gas emissions reporting guidance. https://www.gov.uk/government/uploads/system/uploads/attachment_data/file/206392/pb13944-env-reporting-guidance.pdf, Accessed 18th Dec 2017
15. HM Government: Construction 2025: Industrial Strategy: Government and Industry in Partnership. Department for Business, Innovation and Skills (2013). https://www.gov.uk/government/publications/construction-2025-strategy, Accessed 17th Feb 2015
16. National Assembly for Wales: Well-being of Future Generations (Wales) Act 2015 (2015). https://futuregenerations.wales/wp-content/uploads/2017/01/WFGAct-English.pdf, Accessed 24th Nov 2015
17. Department for Transport: The Road to Zero (2018). https://www.gov.uk/government/publications/reducing-emissions-from-road-transport-road-to-zero-strategy, Accessed 19th Dec 2018
18. Committee on Climate Change: Net Zero—The UK's contribution to stopping global warming (2019). https://www.theccc.org.uk/publication/net-zero-the-uks-contribution-to-stopping-global-warming/, Accessed 3rd May 2019
19. Moseley, P.: EU support for innovation and market uptake in smart buildings under the horizon 2020 framework programme. Buildings **4**(7). https://www.mdpi.com/2075-5309/7/4/105, Accessed 2nd May 2019
20. European Commission. No date. What is Horizon 2020?. https://ec.europa.eu/programmes/horizon2020/en/what-horizon-2020, Accessed 2nd May 2019
21. Saunders, M., Lewis, P., Thornhill, A.: Research Methods for Business Students. Pearson Education Limited, Harlow, UK (2012)
22. RIBA: RIBA CPD Core Curriculum. https://www.architecture.com/-/media/files/cpd/cpd-core-curriculum/riba-cpd-core-curriculum-2018.pdf, Accessed 14th Nov 2017
23. Groat, L., Wang, D.: Architectural Research Methods. Wiley, Hoboken, New Jersey, USA (2013)
24. Collins: Definition of 'code of practice' (2019). https://www.collinsdictionary.com/dictionary/english/code-of-practice, Accessed 24th Mar 2019
25. RSAW: Wales Festival of Architecture (2019). https://www.architecture.com/-/media/files/regional/rsaw-festival-of-architecture-flier-2019-edit.pdf?la=en. Accessed 12th May 2019
26. SPECIFIC: The Active Classroom (2018). http://www.specific.eu.com/assets/downloads/casestudy/Active_Classroom_Web_Case_Study.pdf, Accessed 22nd Feb 2019
27. SPECIFIC: The Active Office (2018). http://www.specific.eu.com/assets/downloads/casestudy/Active_Office_Case_Study.pdf, Accessed 22nd Feb 2019
28. Planning Portal: Building Control (2019). https://www.planningportal.co.uk/info/200128/building_control, Accessed 24th Mar 2019
29. British Council for Offices: About the BCO (2019). http://www.bco.org.uk/Research/BestPracticeGuides/Guide-to-Specification.aspx, Accessed 24th Mar 2019

# Chapter 48
# Urban Climate and Health: Two Strictly Connected Topics in the History of Meteorology

**Chiara Bertolin and Dario Camuffo**

**Abstract** In the history of meteorology, urban climate and health were strictly related to each other. After the discovery of the first liquid-in-glass thermometer (1641) able to provide accurate and repeatable readings, the foundation of observational Networks favored the standardization of instruments, thermometric scales, and observational practices, making possible the confront of readings. In the earliest instrumental period (1654–1769) beginning with the *Medici Network*, only a few instruments were weatherproof, and most of the instruments were suitable for indoor use, mostly in urban environments. This fact favored health and occupational medicine studies. The mature instrumental period (1770–1859) began with the *Societas Meteorologica Palatina*, Mannheim, when outdoor exposition was possible. This period gave a strong impulse to studies concerning agriculture and epidemics. The modern Meteorology (1860–today) began with permanent international cooperation. Weather observations were strictly standardized all over the world, for climate and forecasting purposes. Urban climate, healthy environments, monument conservation, and other applied fields were developed as new disciplines.

## 48.1 Introduction

This paper is an introductory note to weather observations, urban climate, and health that in the history of Meteorology were strictly related to each other. The material on which this study is based is huge stuff of the scientific literature in the early instrumental period (i.e., seventeenth–nineteenth century) and the original instrumental records. This material will be discussed and commented under a holistic view, fol-

C. Bertolin (✉)
Faculty of Engineering, Department of Mechanical and Industrial Engineering, Norwegian University of Science and Technology (NTNU), Trondheim, Norway
e-mail: chiara.bertolin@ntnu.no

D. Camuffo
Institute of Atmospheric Sciences and Climate, National Research Council, Padua, Italy
e-mail: d.camuffo@isac.cnr.it



J. Littlewood et al. (eds.), *Sustainability in Energy and Buildings*, Smart Innovation, Systems and Technologies 163, https://doi.org/10.1007/978-981-32-9868-2_48

lowing the methodologies of science and history and inserted in the context of the actual research in the field.

The earliest observations started in urban environments, often in the observers' house. The technological limits of instruments, i.e., vulnerable to adverse weather or weatherproof, were crucial for outdoor exposures. When outdoor exposures were impossible, the attention was focused on indoor applications, first of all with medical purposes on health; when outdoor exposure was possible, the key application was agriculture and farming.

To this aim, we should consider three main periods, i.e., the earliest instrumental period (1654–1769), the mature instrumental period (1770–1859), and the modern instrumental period (1860–today), as follows.

## 48.2 Historical Overview Concerning the Development of Weather Observations and Related Applications

### *48.2.1 Observations in the Earliest Instrumental Period: 1654–1769*

In Florence, in the first half of the seventeenth century, the Prince Leopold and his brother the Grand Duke of Tuscany Ferdinand II de'Medici, the scientist Galileo Galilei, and his pupils gave a tremendous impulse to the development of science with innovative ideas and new instruments including the liquid-in-glass thermometer in 1641, the most famous type being the Little Florentine Thermometer (LFT, Fig. 48.1a), and the barometer in 1643 [1–3]. The Grand Duke founded the first scientific academy, the Accademia del Cimento (i.e., the Experiment Academy, active from 1657 to 1667), and the first international network of coordinated observations, called the "Medici Network", with identical LFTs, exposure, schedule, and sampling protocols. This was the first meteorological network, composed of 11 stations in Europe, some of them in urban and others in rural or mountain sites. The two primary stations, Florence and Vallombrosa made three-hourly observations from 1654 to 1670 with two LFT exposed outdoors, hung on the same wall but one LFT facing north and one facing south. The LFT was fully made of glass and the scale in enamel beads directly sealed on the tube. The liquid was a spirit, hermetically sealed inside [4]. For this reason, it was weatherproof, i.e., resistant to sunshine, humidity, rainfall, frost and was soon nicknamed "Weather Glass".

In each station, the observations were recorded on daily logs, in standard tabular form, with the indication of the site and the date in the heading, and then a matrix with columns ordered per sampling time, Northern and Southern exposure thermometers and general weather notes; rows were composed of the related readings and notes. The logs were then regularly sent, i.e., daily or weekly, to the Grand Duke. Aim of the Network was to gather observations to give an answer to some fundamental questions of Science, especially Physics, to understand weather variables and for future climate

**Fig. 48.1** **a** (Left) Three Little Florentine Thermometers (LFT) with scale in 50° Galileo (°G) made of enamel beads directly sealed to the tube. The LFT were used in the Medici Network (1654–1670). By courtesy of Museo Galileo—Institute and Museum of the History of Science, Florence. **b** (Center) A thermometer with shape similar to the Florentine thermometers, but with the scale DD separate from the tube and fixed to a wooden framed tablet (CCCC). The glasswork, i.e., bulb (A) and tube (BB) is fixed to the same tablet. This thermometer is usually temporarily exposed on a loggia [9]. **c** (Right) Some decorative thermometers at the turn between the seventeenth and the eighteenth century, with gilt frame and tube shaped like a spiral, a triangle, and a star, all for indoor use only [9]

purposes [4–7]. In 1641, the year in which Galileo died, Robert Boyle (1627–1691) joined the academy in Florence and remained in Italy for 4 years. Twenty years later, he made his experiments on the perfect gases [8] with an accurate replica of the LFT. However, in general, it was extremely difficult and expensive to build weatherproof thermometers all with the same scale. Popular thermometers were based on a poor technology, being composed of two separate parts: the glass tube with the bulb filled with the liquid (e.g., oil, spirit, quicksilver) and the wooden tablet with the scale glued or drawn on it (Fig. 48.1b).

The tube was fixed to the tablet with an iron wire. However, the wooden tablet, shrinking and swelling, was not resistant to external exposures, except for short-term sampling or good weather. Practically, for the limited technological features of the early thermometers only indoor measurements were possible [6]. Considered the internal use, the manufacturers did their best to make attractive these instruments, caring for the decoration and shape not only of frames but also of the glasswork (Fig. 48.1c).

In this earliest instrumental period, indoor use of meteorological instruments was considered from a positive point of view for some advantages as we will explain. Useful parameters developed by the authors of this paper to recognize if readings were originally taken inside, outside, or in a room with a limited ventilation, are (1) the daily span between readings or, if only one reading is available, (2) the day-by-day variability [10] or (3) the normalized temperature difference (NTD), i.e., the ratio between the observed diurnal temperature range (or a given portion of it) and the same but measured outdoors in the 1961–1990 reference period [11]. In the case of historic buildings made of stone or brick, when historically a room was unheated and facing north, the indoor temperature range was limited compared to the external range, and in a first approximation, a single indoor reading was considered representative of the daily average of the outdoor temperature. This is the case of the example reported in Fig. 48.2, with the reconstruction of the temperature observations (Fig. 48.2c) performed in Bologna by Galeazzi in 1716–1726 in room F within the Poggi Palace, headquarter of the Restless-spirit Academy, later transformed into Institute and Academy of Sciences and Arts (IASA) [12]. In Fig. 48.2a, the façade of the Poggi Palace facing NNW (330°) and the window of the F room is encircled. In Fig. 48.2b, the Rooms B, C, D, E, F, (cyan) that hosted thermometers and barometers, on the second floor of the Poggi Palace, are highlighted. Figure 48.2c reports the indoor–outdoor relationships for the Rooms C and D (red line—exposed SE) and for Room F (blue line—exposed NNW) in Poggi Palace. These transfer functions are given by the best-fit equations of the plot made with the indoor readings versus the outdoor ones over their common period. The fit for Room F highlights a small difference of +1,76 °C between the indoor/outdoor average temperatures.

In this early period, most scientists defined the daily mean temperature as the average between the maximum and the minimum levels, i.e., the median of the range that was close to the average over 24 h (T24 h) [13] and in the case of unheated buildings, the difference between the temperature readings at 14:00 and 22:00 can be considered as a room fingerprint parameter, strictly linked to the room ventilation,

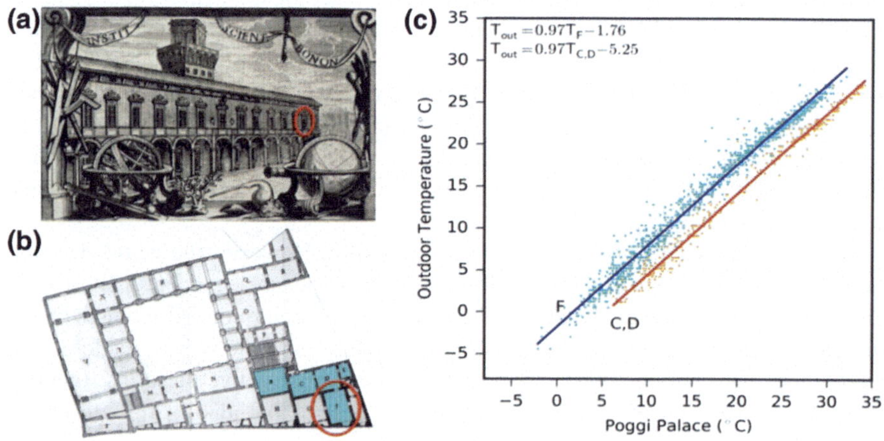


**Fig. 48.2** **a** Poggi Palace, headquarter of the Restless-spirit Academy in Bologna, Italy in the eighteenth century. The facade is facing NNW (330°) and the window of the observation room (F) is encircled. **b** Rooms B, C, D, E, F, (cyan) at the second floor of Poggi Palace. **c** Indoor–outdoor relationships for the Rooms C and D (red line, SE exposure), and for Room F (blue line, NNW exposure) in Poggi Palace

solar exposure, and use. Its value is approximately half of the daily span [11, 12]. Theories about connections between locations, weather, and health arose since the antiquities but it was since the first half of the seventeenth century that most physicians considered particularly critical temperature and smoke measurements as they thought lungs were responsible for moving the blood around—throughout the body. In fact, people generally lived in bad conditions in unheated and often unhealthy houses that caused the common cold, flu, pneumonia, and breathing diseases for a combination of cold, smoke from the fireplace filling houses, dampness, and molds. All of the above reasons constituted a strong argument in favor of indoor readings in manned, i.e., urban, areas.

The Italian physician Bernardino Ramazzini, who was interested in the relationship between meteorological factors and epidemics, gave another relevant contribution in this direction. Every year he carefully documented all unusual climatic events, such as rainy periods and dryness, cold or hot seasons, acid fog and acid dew caused by aerosols of volcanic eruptions, and so on. The climate and the extreme events of each year were summarized and discussed in yearly publications, the "Ephemerides Barometricae" (i.e., "Weather diaries", [14, 15]) and "Constitutiones Epidemicas" (i.e., "List of, and rationale for, the diseases of the year", [16]). In these publications, he included comments, and supposed links between weather, yield and health. He also took regular daily observations of air pressure, wind direction (in quarters), state of the sky, and precipitation (only frequency) for the calendar year 1694 [15–17]. In addition, he made a careful treatise about the diseases of all kinds of workers and attempted to establish cause–effect relationships between environmental factors, air quality, lifestyle, and illness typical of the various activities. His study "De Morbis Artificum Diatriba" (i.e., "Diseases of Workers", [18]) was a milestone in this field and Ramazzini was considered the father of Occupational Medicine.

The second successful international meteorological network was established in 1723, when James Jurin, secretary of the Royal Society, London, invited some scientists to join his new meteorological network. The network was active from 1724 to 1735 and observations were published in the Philosophical Transactions, London. Jurin [19] established precise norms for instruments (thermometer, barometer, and rain gauge), operational methodology, exposure, and reading times. Unfortunately, he suggested measuring temperature indoors, in a north-facing room where fire was never, or hardly ever lit, following the practice introduced in 1660 by John Locke himself. We can imagine that the key reason to determine this choice was that the early thermometers were not resistant to outdoor conditions; in addition, indoor sampling was easier and feasible for a large number of observers.

### *48.2.2 The Mature Instrumental Period: 1770–1859*

In the second half of the eighteenth century, the technology was improved and the vulnerable wooden tablets with the iron wire were substituted with metal or ceramic items or enclosed inside an additional glass tube, making them resistant to outdoor environments and weather injuries. This made possible to take outdoor records [20] and some meteorological networks were founded, as follows. In Germany, Wolfgang Goethe, when he was minister of the Grand Duke of Weimar (1775–1832), organized an efficient regional network. In 1778, in Baden, Johann Lorenz Bockman established another regional weather network composed of 16 stations, which also had a short life. In France, under the patronage of Louis XIV, the court physician Felix Vicq d' Azyr set up a network on behalf of the newly founded Societé Royale de Médecine (Royal Society of Medicine), Paris. The aim of the network, composed of 22 stations, was to establish a link between weather and health, by collecting meteorological records and medical reports concerning man and animal epidemics that happened in France and other countries. For the period 1777–1786, Father Louis Cotte collected and published summaries of the meteorological records in the official journal of the Society, named "Histoire" (i.e., "History") and made the same in his book "Traité de Météorologie" (i.e., "Treatise on Meteorology") [21]. In Germany, the Prince Elector Karl Theodor von Pfalz and his secretary John Jacob Hemmer, founded the "Societas Meteorologica Palatina", Mannheim [22]. This international network, active from 1781 to 1792, was composed of 39 sites over Europe except for England and the Iberian Peninsula; they included three stations in Russia, one in Greenland and two in Massachusetts. The Societas established the operational methodology and the observations schedule. It also distributed some instruments following the style of the Accademia del Cimento, or specified their characteristics, following the style of the Royal Society, London. The observations were published in the Ephemerides Societatis Meteorologicae Palatinae from 1783 to 1795 (records from 1781 to 1792, [23]). In Padua, Italy, Giuseppe Toaldo professor of Astronomy and Meteorology was continuing the series of instrumental observations that started in 1725 by Giovanni Poleni following the invitation by Jurin [19]. Toaldo studied if it was possible to establish statistical relationships between meteorological variables, weather extremes, solar and lunar forcing, agriculture, yield and farming activity, diseases, number of dead or born people. To this aim, he wrote a number of books, e.g., "Della vera influenza degli Astri, delle Stagioni e mutazioni di Tempo" (i.e., "The true influence of heavenly bodies on the seasons and weather changes") [24] and "La Meteorologia applicata all'Agricoltura" (i.e., "Meteorology applied to agriculture") [25]. Toaldo joined the networks of the Societé Royale de Médicine, Paris and the Societas Meteorologica Palatina, Mannheim. In 1778, he founded the "Giornale Astro-Meteorologico" (i.e., "Journal on Astronomy and Meteorology issues") with the aim of stimulating colleagues and amateurs to perform meteorological observations and to correspond with him; in this way he formed a national network composed of 32 observers and the records were summarized in the journal. This network disappeared after Toaldo died, i.e., in 1797.

### *48.2.3 The Modern Instrumental Period: 1860–Today*

In the 1860s, national meteorological services were established in several countries. In 1860, George Biddel Airy (Greenwich Observatory) and Urbain Jean-Joseph Le Verrier (Paris Observatory) signed an important international agreement for cooperation to collect British and French observations and forecast storms close to these two countries. This was the first step to establish a permanent transnational cooperation that was prepared with international conferences in Leipzig (1872), Vienna (1873), and Rome (1879). The International Meteorological Committee was founded in 1873 in Vienna, and Christophorus Buys Ballot was the first president. The International Meteorological Committee was the ancestor of the World Meteorological Organization, i.e., the specialized agency of the United Nations established in 1950 with 160 Members. After that date, Meteorology was a mature discipline supporting general climate studies. Specific topics on applied fields were considered separately although related by strong links. New studies started with these disciplines, e.g., boundary layer climates [26], urban climate [27, 28], indoor climate [29–31] healthy buildings [32–34], microclimate and conservation of cultural heritage [35, 36], and microclimate and microbiological habitat dangerous for mold development [37–40].

## 48.3 Discussion on Further Use of Historical–Meteorological Data in UHI Studies Nowadays

Since the beginning, meteorological observations have been repeatedly organized in networks that were fundamental to homogenize instruments, calibrations and scales, observational protocols, and exposure to obtain comparable data. In addition, networks established and strengthened contacts between a number of leading scientists and stimulated them to collect high quality, well-documented, and homogeneous records. Most of these early observers lived in urban areas, some others in rural or remote sites. For the limited instruments technology available to the observers from the 1660s to 1770s, they took most of the observations indoors. Fortunately, some of them took parallel outdoor readings, with a temporary external exposition of the thermometer, or with good ventilation of the room [4, 6, 11, 12, 44]. In many cases, the observers continued the records after the closure of the network, thus generating long-term instrumental series, some of arrive till date[12, 21, 41, 42].

Therefore, in the origins, the history of meteorology, the urban climate, and indoor environments were strictly linked together. Nowadays, these research domains are disjoint as demonstrated in Table 48.1 by the overview of the main context of these disciplines organized using specific keywords.

Within the generic framework of urban climate research, nowadays the interest inherently to monitor and measure indoors is a recent topic (ranging 2012–2018 period) with a few contributions. Works in the literature are mainly focused on the

**Table 48.1** Overview of data, methods and Urban Climate and UHI domain that have been analyzed to highlight the state-of-the-art and the potential of using historic climate data for an advancement of the research in UHI domain

| Specification | Context overview |
|---|---|
| *Urban climate; indoor; measures/monitoring* | |
| Data | Indoor/outdoor climate [43]<br>Climate change (CC) [44]<br>UHI data and collection on mitigation actions [45]<br>Vegetation and streets geometry data [47]<br>Urban development data and CC adaptation actions [46] |
| Methods | Hygrothermal Building Simulation [43]<br>CC scenarios [44]<br>Methods for UHI analysis [45]<br>Empirical analysis [47]<br>Numerical analysis [46] |
| Urban climate and UHI domain | Indoor climate scenario [43]<br>Influence on summer cooling costs and heat stress [44]<br>UHI and related mitigation actions [45]<br>Interaction between vegetation and street geometry microclimate [47]<br>Interaction between urban development and CC adaptation capabilities [46] |
| *Urban climate; indoor; measures/monitoring; health* | |
| Data | Hazardous airborne particles [48, 49]<br>Climate data at micro scale [53]<br>IAQ [50–52] |
| Methods | Risk assessment (RA) analysis [48, 49]<br>Case studies [51–53]<br>New tool for biomonitoring and bioaccumulation [50] |
| Urban climate and UHI domain | RA of hazardous airborne particles in roadside dwellings [48] or due to short-term exposure and asthma diseases in urban climate [49]<br>Microclimate observation network for urban unit [53]<br>Assessing categories of IAQ in housing [52] or in school environment with new tools [50]<br>Challenges on IAQ field monitoring [51] |
| *UHI; meteorological; measures/monitoring; health* | |
| Data | UHI intensity and T data [54]<br>Satellite data Landsat-8 TIRS and health data [55]<br>HUMIDEX [57]<br>Urban water bodies and T data [56]<br>Spatio-temporal data of urban development and T [58]<br>Land surface and air T data [59] |

(continued)

**Table 48.1** (continued)

| Specification | Context overview |
|---|---|
| Methods | Data analysis [54, 60] and their application to a case study [57–59]<br>RA for health using space technology [55]<br>Experimental [56]<br>Geospatial techniques [58] |
| Urban climate and UHI domain | Reduced UHI intensity under warmer conditions [54]<br>Assessing UHI effect and its impact on health using space technology [55]<br>Microclimate evaluation of a case-study city using HUMIDEX [57]<br>Influence of urban water bodies on outdoor thermal environment [56]<br>Spatio-temporal analysis of UHI [58]<br>Relationship of land surface and air T as UHI indicators [59] |

estimation of indoor climate scenarios. The data processed are both indoor and outdoor measurements with eventually the estimation of Urban Heat Island (UHI) and heat stress. These data are compared with the results of hygrothermal building simulations to demonstrate the benefit of simulation tools in identifying and discussing details of the expected levels of indoor comfort and the influence of the outdoor heat stress on the increase of summer cooling costs [43, 44]. In addition, Table 48.1 clearly highlights how methods used for UHI analysis [45] are applied to improve both the understanding of the phenomenon and of the influence of climate change effects [44–46] at the urban scale. The final aim being the analysis and evaluation of possible future adaptation measures to climate change [46], e.g., the control of urban vegetation and streets geometry [47] with repercussions in modifying internal/external climate in buildings.

If the keyword "health" (as in the earliest instrumental period in historical climatology) is added to the previous literature review research, the interest of the scientific community completely changes direction, moving from urban building physics, comfort, climate change, and adaptation measures toward indoor air quality (IAQ), risk assessment, and diseases at urban scale (see Table 48.1). Even these research topics, however, are quite recent (ranging 2010–2018 period) and not common in the literature. What emerges is that authors exploit concentration and spatial spread of hazardous airborne particles at urban scale both indoors and outdoors [48, 49]. Outdoors are very much focused on pollution from traffic, indoors more on a summative assessment of IAQ [50–52] to connect short-term exposure effects of airborne particles with asthma or other types of diseases. These topics require the use of a risk assessment approach when elaborating the data. In addition, the works found in the literature are very much focused on urban scale case studies [48, 49, 51, 53] and very often in populous cities in China or Asia. Challenges in extensive spatial coverage, regular temporal sampling, IAQ or risk categories assessment are very much discussed.

Keeping the interest on health and on meteorological measures and monitoring but focusing the research challenges from urban climate to the UHI, the literature review (Table 48.1) reports as the interests of the research community becomes more devoted to exploiting new types of data and methods to reconstruct and analyze—through the use of specific indicators—the UHI intensity [54], to assess the UHI impact on health [55], and the possible mitigation measures to reduce it [56]. Among the new proposed data, one can find satellite data, health data as UHI proxy [55], interconnected climate data as HUMIDEX [57], geophysical data as urban water bodies [56], spatio-temporal data of urban development [58], and land surface data [59]. Finally, the proposed data analysis methods ask for expertise in risk assessment, experimental approach, and geospatial techniques. Within the context just traced, historical data are nowadays collected and analyzed to

- better detect and eliminate the UHI effects in long-term temperature records [60];
- assess the climate change effect at urban scale linking it with urbanization, land use, and cover changes [60–62];
- improve the understanding in urbanization impacts analysis increasing the temporal scale of climate data back in time;
- precisely understand the local warming in historical cities centers taking advantage of climatic records that cover more than one century [4, 5, 11, 12, 17, 20, 42, 61];
- improve calibration when integrating diverse sources of data as in the case of satellite and air temperature data from continental meteo stations [63];
- analyze—thanks to the multiple daily readings in the past decades—the present-day nocturnal temperature cooling rate response with respect to the historical ones taking into consideration urban land use and land cover changes [62, 64];
- investigate the interaction between the outdoor environment and the indoor microclimate;
- reconstruct the spatio-temporal estimation of historical particulate matter concentration and meteorological variables;
- reconstruct important achievements for the history of science.

As an example of the potentiality in applying multi-century historical climate data to UHI studies, Fig. 48.3 reports the map of Padua in the eighteenth century when Padua had 38,000 inhabitants, and Poleni, Morgagni, and Toaldo were taking their observations [42]. Today the town has ca. 210,000 inhabitants and it has expanded more than five times if one includes the industrial area. The evolution of the UHI may constitute a problem for the undesired bias when one is interested to know the air temperature of the undisturbed air circulation. To have homogenized and corrected temperature series as that published in [41] going back to the beginning of the eighteenth century can provide a mine of useful information and valuable data in the field of urban climate and UHI research. On the other hand, as demonstrated in Table 48.1, the urban environment (e.g., UHI, urban climate, and atmospheric pollution), the human health, the thermal comfort of people, the indoor climate in healthy or unhealthy buildings (i.e., the Indoor Air Quality (IAQ)), the bio-adaptation of various systems, the impact of atmospheric agents on man and monuments included climate change (CC), and many others specific studies fall outside of this umbrella

**Fig. 48.3** **a** Map of Padua by Giorgio Valle (1784). Dots represent the location of the weather observations carried out by Giovanni Poleni, GianBattista Morgagni, Giuseppe Toaldo and Vincenzo Chiminello. Cultivated fields extend all around the walls. **b** Padua today from the sky. The historic city (HC) is the red triangle in the middle, contoured by the yellow road running on the side of the city walls. Pink area is the recent town development. The white area (IA) on the right side is the Industrial Area. Green: cultivated fields

and need specific instruments and observational methodologies. It is clear that a teleconnection exists between the global, the regional climate, and the urban or the indoor climate, and that the link depends not only on researcher interests and data collection, and methodological application (e.g., mathematical functions to calculate air exchange, heat, and moisture transfer across the building envelope) but also on additional variables introduced by man, e.g., building use and HVAC systems.

## 48.4 Conclusions

Further studies are needed to improve the knowledge of how they are related to each other: for example, how the general climate will influence the above specific items, and vice versa; how the individual behavior will affect the sustainability, the general circulation, and the future scenarios. The link between external weather, urban and indoor climate is of primary relevance, especially if one considers that about half of the human population lives in built environments. Finally—as presented in this paper with the historical instrumental series in their epochs (e.g., from the seventeenth, eighteenth to nineteenth centuries), observational sites originally were in the open countryside; nowadays they can be classified as belonging to historic city centers. Therefore, these data become very valuable to be analyzed to achieve a better understanding of the UHI in connection with the urban development. Nevertheless, other records or new data processing methods of existing historical meteorological series have not yet been systematically explored although great efforts have been done recently through EU-funded research projects, for example, IMPROVE (Improved understanding of past climatic variability from early daily European instrumental) and MILLENNIUM (European climate of the last millennium), and as publications [4, 5, 11, 12, 17, 20, 41, 42]. In our opinion, they are good sources to improve the achievements in urban climate, UHI, and interdisciplinary-related research fields, for example, for the assessment of indoor condition in relation to sick buildings syndrome, estimating buildings air exchange rates differences before (i.e., in the past) and after (i.e., nowadays) energy efficiency retrofitting interventions, and changes in IAQ and comfort levels over the centuries or temporal differences (i.e., past versus present) in original historical building materials performances.

**Acknowledgements** This study is part of the EU-funded Climate for Culture project (Grant 226973).

## References

1. Segni, A., Magalotti, L.: Diario delle sperienze dal 19 Giugno 1657, Handwritten Manuscript Gal 262, Biblioteca Nazionale Centrale, Florence, pp. 1657–1667 (1657)
2. Magalotti, L.: Saggi di naturali esperienze fatte nell'Accademia del Cimento. Giuseppe Cocchini, Florence (1667)
3. Targioni Tozzetti, G.: Notizie degli aggrandimenti delle Scienze Fisiche accaduti in Toscana nel corso di anni LX del secolo XVII, Tomo I. Bouchard, Florence (1780)
4. Camuffo, D., Bertolin, C.: The earliest temperature observations in the World: the Medici Network (1654-1670). Clim. Chang. **111**, 335–363 (2012). https://doi.org/10.1007/s10584-011-0142-5
5. Middleton, W.E.K.: The experimenters. A study of the Accademia del Cimento. J. Hopkins Press, Baltimore (1971)
6. Camuffo, D., Bertolin, C.: The earliest spirit-in-glass thermometer and a comparison between the earliest period of the Central England Temperature series and the instrumental observations of two Italian stations of the Medici Network, active 1654–1670. Weather **67**(8), 206–209 (2012). https://doi.org/10.1002/wea.1913
7. Camuffo, D., Bertolin, C.: The world's earliest instrumental temperature records, from 1632 to 1648, claimed by G. Libri, are reality or myth?. Clim. Chang. **119**(3), 647–657 (2013). https://doi.org/10.1007/s10584-013-0742-3
8. Boyle, R.: New Experiments Physico-Mechanical, Touching the Spring of the Air and Its Effects, 2nd edn. Hall, Oxford (1662)
9. D'Alancé, J.: Traittéz des Barométres, Thermométres et Notiométres ou Hygrométres. Maurret, Amsterdam (1707)
10. Cocheo, C., Camuffo, D.: Corrections of systematic errors and data homogenisation in the Padova series (1725–today). Clim. Chang. **53**(1–3), 77–100 (2002)
11. Camuffo, D., della Valle, A., Bertolin, C., Santorelli, E.: The Stancari air thermometer and the 1715–1737 record in Bologna, Italy. Clim. Chang. **139**, 623–636 (2016). https://doi.org/10.1007/s10584-016-1797-8
12. Camuffo, D., della Valle, A., Bertolin, C., Santorelli, E.: Temperature observations in Bologna, Italy, from 1715 to 1815: a comparison with other contemporary series and an overview of three centuries of changing climate. Clim. Chang. **142**(1–2), 7–22 (2017). https://doi.org/10.1007/s10584-017-1931-2
13. Camuffo, D.: Errors in early temperature series arising from changes in style of measuring time, sampling schedule and number of observations. Clim. Chang. **53**(1–3), 331–354 (2002)

14. Ramazzini, B.: Ephemerides barometricae Mutinenses anni MDCXCIV una cum disquisitione causae ascensus ac descensus Mercurii in Torricelliana fistula, juxta diversum aeris statum. Capponi and Pontiroli, Modena (1695)
15. Ramazzini, B.: Ephemeris Barometricae, in Beranardini Ramazzini Operum Omnium, vol. 3, pp. 273–623, Conzatti, Padua (1718)
16. Ramazzini, B.: Constitutiones Epidemicas Anni 1690, 1691, 1692, 1693, 1694, in: Beranardini Ramazzini Operum Omnium, vol. 2, pp. 1–195. Conzatti, Padua (1718)
17. Camuffo, D., Bertolin, C., Jones, P.D., Cornes, R., Garnier, E.: The earliest daily barometric pressure readings in Italy: Pisa, 1657–8 and Modena, 1694 and the weather over Europe, 1694. Holocene **20**(3), 1–13 (2009)
18. Ramazzini, B.: De Morbis Artificum Diatriba. Van de Water, Padua (1703)
19. Jurin, J.: Invitatio ad observationes Meteorologicas communi consilio instituendas a Jacobo Jurin M. D. Soc. Reg. Secr. et Colleg. Med. Lond. Socio, Philosophical Transactions 379, pp. 422–427, London (1723)
20. Camuffo, D.: History of the long series of the air temperature in Padua (1725-today). Clim. Change **53**(1–3), 7–76 (2002)
21. Cotte, L.: Traité de Météorologie contenant 1. l'histoire des observations météorologiques, 2. un traité des météores, 3. l'histoire & la description du baromètre, du thermomètre & des autres instruments météorologiques, 4. les tables des observations météorologiques & botani-cométéorologiques, 5. les résultats des tables & des observations, 6. la méthode pour faire les observations météorologiques, pp. 635. Imprimerie Royale, Paris (1774)
22. Hemmer, J.J.: Descriptio instrumentorum meteorologicorum, tam eorum, quam Societas distribuit, quam quibus praeter haec Manheimii utitur, Ephemerides Societatis Meteorologicae Palatinae, Tomus 1, pp. 57–90. Mannheim (1783)
23. Colacino, M.: Toaldo and the Societas Meteorologica Palatina, in: L. Pigatto, (Ed.), Giuseppe Toaldo e il suo tempo, Centro per la Storia dell'Università di Padova e Osservatorio Astronomico di Padova, pp. 979–996. Padua (2000)
24. Toaldo, G.: Della vera influenza degli Astri, delle Stagioni e mutazioni di Tempo, Saggio Meteorologico, 1st edition, Manfrè. Stamperia del Seminario, Padua (1770)
25. Toaldo, G.: La Meteorologia applicata all'Agricoltura. Storti, Venice (1775)
26. Oke, T.R.: Boundary Layer Climates. Methuen, London (1978)
27. Landsberg, H.E.: The Urban Climate. Academic Press, New York (1981)
28. Rosenzweig, C., Solecki, W.D., Hammer, S.A.: Climate change and cities. First Assessment Report of the Urban Climate Change Research Network. Cambridge University Press, Cambridge (2011)
29. Mc Intrye, D.A.: Indoor Climate. Applied Science Publishers, London (1980)
30. Givoni, B.: Climate Considerations in Building and Urban Design. Wiley, New York (1998)
31. Rohinton, E.: An Urban Approach To Climate Sensitive Design. Strategies for the Tropics. Spon Press, New York (2012)
32. Holdsworth, B., Holdsworth, W.J., Sealey, A.: Healthy Buildings: a Design Primer for a Living Environment. Longman, London (1992)
33. Bluyssen, P.M.: The Indoor Environment Handbook: How to Make Buildings Healthy and Comfortable. Earthscan Publications Ltd., Taylor & Francis Group, London (2009)
34. Bokalders, V., Block, M.: The Whole Building Handbook: how to Design Healthy, Efficient and Sustainable Buildings. Earthscan Publications Ltd., Taylor & Francis Group, London (2010)
35. Camuffo, D.: Microclimate for Cultural Heritage—Conservation, Restoration and Maintenance of Indoor and Outdoor Monuments, 2nd edn. Elsevier, Amsterdam (2013)
36. Camuffo, D., Bertolin, C.: Unfavourable microclimate conditions in exhibition rooms: early detection, risk identification, and preventive conservation measures. J. Paleontol. Tech. **15**, 144–161 (2016)
37. Warscheid, T., Braams, J.: Biodeterioration of stone; a review. Int. Biodeterior. Biodegrad. **46**, 343–368 (2000)
38. Sedlbauer, K.: Prediction of mould growth by hygrothermal calculation. J. Build. Phys. **25**, 321–336 (2002)

39. Caneva, G., Nugari, M.P., Salvadori, O.: Plant Biology for Cultural Heritage—Biodeterioration and Conservation. The Getty Conservation Institute, Los Angeles (2009)
40. Sterflinger, K., Pinzari, F.: The revenge of time: fungal deterioration of cultural heritage with particular reference to books, paper and parchment. Environ. Microbiol. **14**(3), 559–566 (2012). https://doi.org/10.1111/j.1462-2920.2011.02584.x
41. Camuffo, D., Jones, P (eds.): Improved Understanding of Past Climatic Variability from Early Daily European Instrumental Sources, pp. 392. Kluwer Academic Publishers, Dordrecht, Boston, London (2002)
42. Camuffo, D., Bertolin, C.: Recovery of the early period of long instrumental time series of air temperature in Padua, Italy (1716–2007). J. Phys. Chem. Earth **40–41**, 23–31 (2012). https://doi.org/10.1016/j.pce.2010.02.002
43. Tudiwer, D., Höckner, V., Korjenic, A.: Greening Aspang-Hygrothermal building simulation for analysis and evaluation of various scenarios related to indoor climate. Bauphysik **40**(3), 120–130 (2018). https://doi.org/10.1002/bapi.201810015
44. Hooyberghs, H., Verbeke, S., Lauwaet, D., Costa, H., Floater, G., De Ridder, K.: Influence of climate change on summer cooling costs and heat stress in urban office buildings. Clim. Chang. **144**(4), 721–735 (2017). https://doi.org/10.1007/s10584-017-2058-1
45. Mahdavi, A., Kiesel, K., Vuckovic, M.: Methodologies for UHI analysis. In: Musco, F. (eds.) Counteracting Urban Heat Island Effects in a Global Climate Change Scenario, pp 71–91. Springer, Cham (2016). https://doi.org/10.1007/978-3-319-10425-6_3
46. Fahmy, M.: Climate change adaptation for mid-latitude urban developments. In: 28th International PLEA Conference on Sustainable Architecture + Urban Design: Opportunities, Limits and Needs—Towards an Environmentally Responsible Architecture, PLEA 2012 (2012)
47. Kannamma, D., Sundaram, A.M.: Empirical study on the interaction between vegetation and street geometry elements towards ambient urban microclimate. Indian Concr. J. **89**(7), 20–27 (2015)
48. Chan, C.S., Ranasinghe, R.S.A., Ho, S.S.H., Ho, K.F., Yim, S.H.L., Sugathapala, A.G.T., Lee, S.C., Hung, W.T., Huang, Y., Zhang, H.: Evaluation of hazardous airborne carbonyls in five urban roadside dwellings: a comprehensive indoor air assessment in Sri Lanka. Atmos. Pollut. Res. **9**(2), 270–277 (2018). https://doi.org/10.1016/j.apr.2017.10.002
49. Jung, K.H., Perzanowski, M., Rundle, A., Moors, K., Yan, B., Chillrud, S.N., Whyatt, R., Camann, D., Perera, F.P., Miller, R.L.: Polycyclic aromatic hydrocarbon exposure, obesity and childhood asthma in an urban cohort. Environ. Res. **128**, 35–41 (2014). https://doi.org/10.1016/j.envres.2013.12.002
50. Protano, C., Owczarek, M., Antonucci, A., Guidotti, M., Vitali, M.: Assessing indoor air quality of school environments: transplanted lichen Pseudovernia furfuracea as a new tool for biomonitoring and bioaccumulation. Environ. Monit. Assess. **189**(7), 358 (2017). https://doi.org/10.1007/s10661-017-6076-2
51. Ho, S.S.H., Huang, Y., Lee, S.-C., Cao, J.: Challenges on field monitoring of indoor air quality in China. Indoor and Built Environ. **26**(4), 576–584 (2017). https://doi.org/10.1177/1420326X15626991
52. Clark, M.L., Reynolds, S.J., Burch, J.B., Conway, S., Bachand, A.M., Peel, J.L.: Indoor air pollution, cookstove quality, and housing characteristics in two Honduran communities. Environ. Res. **110**(1), 12–18 (2010). https://doi.org/10.1016/j.envres.2009.10.008
53. Lai, X., Tang, Y., Li, L., Chan, P., Zeng, Q.: Study on microclimate observation network for urban unit: a case study in a campus of Shenzhen, China. In: Physics and Chemistry of the Earth (2018). https://doi.org/10.1016/j.pce.2018.08.003
54. Scott, A.A., Waugh, D.W., Zaitchik, B.F.: Reduced urban Heat Island intensity under warmer conditions. Environ. Res. Lett. **13**(6), 64003 (2018). https://doi.org/10.1088/1748-9326/aabd6c
55. Sagris, V., Sepp, M.: Landsat-8 TIRS data for assessing urban heat island effect and its impact on human health. IEEE Geosci. Remote Sens. Lett. **14**(12), 8107666 (2017). https://doi.org/10.1109/LGRS.2017.2765703
56. Syafii, N.I., Ichinose, M., Wong, N.H., Kumakura, E., Jusuf, S.K., Chigusa, K.: Experimental study on the influence of urban water body on thermal environment at outdoor scale model. Procedia Eng. **169**, 191–198 (2016). https://doi.org/10.1016/j.proeng.2016.10.023

57. Rožnovský, J., Litschmann, T., Středová, H., Středa, T., Salaš, P., Horká, M.: Microclimate evaluation of the Hradec Králové City using HUMIDEX. Contrib. Geophys. Geod. **47**(3), 231–246 (2017). https://doi.org/10.1515/congeo-2017-0012
58. Vyas, A., Shastri, B., Joshi, Y.: Spatio-temporal analysis of UHI using geo-spatial techniques: a case study of Ahmedabad city. Int. Arch. Photogramm. Remote Sens. Spat. Inf. Sci. ISPRS Arch. **40**(8), 997–1002 (2014). https://doi.org/10.5194/isprsarchives-XL-8-997-2014
59. Schwarz, N., Schlink, U., Franck, U., Großmann, K.: Relationship of land surface and air temperatures and its implications for quantifying urban heat island indicators—An application for the city of Leipzig (Germany). Ecol. Ind. **18**, 693–704 (2012). https://doi.org/10.1016/j.ecolind.2012.01.001
60. Dienst, M., Lindén, J., Saladié, Ò., Esper, J.: Detection and elimination of UHI effects in long temperature records from villages—A case study from Tivissa. Spain. Urban Clim. **27**, 372–383 (2019). https://doi.org/10.1016/j.uclim.2018.12.012
61. Fortelli, A., Scafetta, N., Mazzarella, A.: Local warming in the historical center of Naples. Int. J. Heat Technol. **34**(2), S569–S572 (2016). https://doi.org/10.18280/ijht.34S252
62. Noro, M., Busato, F., Lazzarin, R.M.: Urban heat island in Padua, Italy: experimental and theoretical analysis. Indoor Built Environ. **24**(4), 514–533 (2015). https://doi.org/10.1177/1420326X13517404
63. Zhang, P., Bounoua, L., Imhoff, M.L., Wolfe, R.E., Thome, K.: Comparison of MODIS land surface temperature and air temperature over the continental USA meteorological stations. Can. J. Remote. Sens. **40**(2), 110–122 (2014). https://doi.org/10.1080/07038992.2014.935934
64. Chow, W.T.L., Svoma, B.M.: Analyses of nocturnal temperature cooling-rate response to historical local-scale urban land-use/land cover change. J. Appl. Meteorol. Climatol. **50**(9), 1872–1883 (2011). https://doi.org/10.1175/JAMC-D-10-05014.1

# Chapter 49
# The Impact of Stakeholder Preferences in Multicriteria Evaluation for the Retrofitting of Office Buildings in Italy

**Giuseppe Pinto, Alfonso Capozzoli, Marco Savino Piscitelli and Laura Savoldi**

**Abstract** Buildings are responsible for about 26% of the total final energy consumption in Italy. Therefore, building retrofitting represents an opportunity to achieve economic and environmental benefits. However, a challenging task is the application of robust methodologies for evaluating cost-optimal retrofit measures. The paper evaluates, in terms of multiple criteria-based approach, several retrofitting alternatives selected for a typical office building in Italy. The alternatives are evaluated considering economic, environmental, and technical requirements and are compared by means of a Stochastic Multicriteria Acceptability Analysis (SMAA) method, able to consider uncertainties in the criteria evaluation. Three different stakeholder preferences are analyzed and compared with the aim to point out the importance of preference information in multicriteria analysis. The results highlight that, when the preference is the investment cost, for the case study analyzed the most suitable solution is represented by a gas boiler and electricity withdrawn from the market. On the other hand, when the operational cost has the same or more importance than the investment cost, the best solution is represented by a micro-CHP coupled with PV plant. Lastly, the analysis highlights that the main driver of a building retrofit is of economic nature and that, depending on the actors involved, a precise study of preference information could influence the outcome of the analysis.

## 49.1 Introduction

The ambitious target set by European Union to cut carbon emissions increased the attention toward the energy efficiency in buildings, which in Italy represents 26% of total final energy consumption [1]. Improving building energy performance in the urban environment could help meeting sustainability targets set by the European Union for 2020 [2]. Nowadays, the replacement rate of existing buildings is limited to 1–3% per year. Considering that a large part of European building stock was built

---

G. Pinto · A. Capozzoli (✉) · M. S. Piscitelli · L. Savoldi
Politecnico di Torino, Turin, Italy
e-mail: alfonso.capozzoli@polito.it



J. Littlewood et al. (eds.), *Sustainability in Energy and Buildings*, Smart Innovation, Systems and Technologies 163, https://doi.org/10.1007/978-981-32-9868-2_49

before 1960s, the potential economic and environmental savings using retrofitting measures are very high [3].

During the last years, the research focused on the development of clean technologies for microscale application, including micro-CHP [4] and heat pumps [5]. In particular, micro-generation proved to be effective for increasing the efficiency of production by matching thermal and electrical demand.

The complexity of the sustainability problem increases the need for adopting Multicriteria Decision Analysis (MCDA) methods in the energy field [6]. MCDA has been applied to large-scale energy systems [7, 8] and more recently to buildings [9, 10]. The aim of the study is to identify the most suitable retrofit alternatives for an office building, considering economic, environmental, and technical criteria. Since the problem may be influenced by several parameters and assumptions, the Stochastic Multicriteria Acceptability Analysis (SMAA) method, introduced by Lahdelma et al. [11] was employed. This method is specifically designed to consider uncertainty related to criteria measurements.

The SMAA has been applied to several decision-making problems, including energy policy assessment [12], sustainable solution for buildings [13], heating choices for residential area [14], and integration of carbon-neutral technologies into district heating system [15]. A recent survey on the application of SMAA can be found in [16].

In detail, the paper is aimed at assessing the importance of stakeholder preferences when a retrofitting scenario analysis is performed for a typical office building. Three different preferences were analyzed and compared.

The paper is organized as follows: Sect. 49.2 introduces the methods used for performing the retrofitting analysis. Then, Sect. 49.3 describes the case study and highlights criteria and preference information used, while in Sect. 49.4 the alternatives proposed are presented. Section 49.5 presents the results, while discussion and conclusion are given in Sect. 49.6.

## 49.2 Methods

### *49.2.1 Simulation of Energy Needs and Alternatives*

The energy demand of an archetype of office building was evaluated using EnergyPlus 8.0.0 [17], an open-source software able to estimate thermal and electrical energy needs of the buildings. The assumptions adopted to model the system were made according to the Italian legislation and previous literature [18, 19]. A more detailed description of the model can be found in [20, 21].

The retrofit measures, described below, were simulated using Matlab [22], in order to evaluate the main input necessary for the SMAA analysis. The model was formulated to satisfy the energy demand over a full year of operation while optimizing the energy efficiency according to different criteria.

### 49.2.2 Stochastic Multicriteria Acceptability Analysis

To evaluate and compare the alternatives proposed, the JSMAA [23], an open-source version of the SMAA method [24], was used. SMAA has been developed for situations where neither criterion nor weights are precisely known.

The problem is formulated as a set of m alternatives $\{x_1, x_2, \ldots, x_m\}$ to be evaluated in terms of $n$ criteria. The decision-makers' preference structure is represented by a value function $u(x_i, w)$. Uncertain criteria are represented by a matrix of stochastic variables $\xi_{ij}$ with a probability density function $f(\xi)$. The value of each alternative is obtained through an additive value function:

$$u(x_i, w) = w_1 * u_{i1} + w_2 * u_{i2} + \cdots + w_n * u_{in} \tag{49.1}$$

where the partial values $u_{ij}$ are evaluated through linear scaling with respect to the best and worst outcome into the range [0,1].

Preference information provided by decision-makers about weights are represented by suitable weight distribution $f(w)$ in the feasible weight space W, where weights are nonnegative and normalized to 1:

$$W = \left\{ W \in R^n | w \geq 0\, and \sum\nolimits_{j=1}^{n} w_j = 1 \right\} \tag{49.2}$$

Since both criteria and weights are represented by distributions, Monte Carlo simulation was used to analyze the problem. A value of K = 10,000 simulations was considered, obtaining a variance of around 1%. During the K simulations, several statistics figures were computed, including

- $B_{ir}$: the number of times alternative $x_i$ obtained rank $r$;
- $W_i$: sum of the weight vectors that made alternative $x_i$ most preferred.

Moreover, two descriptive measures, the rank acceptability index ($b_i^r$) and the central weight vector, were evaluated.

The rank acceptability index $b_i^r$ measures the probability that the alternative $x_i$ obtains a certain rank $r$. The rank acceptability indices are estimated as follows:

$$b_i^r \approx \frac{B_{ir}}{K} \tag{49.3}$$

The rank acceptability is useful to identify the most acceptable alternative. In fact, alternatives with high acceptability for the best ranks are identified as acceptable solution. On the other hand, alternatives with low or zero acceptability index for the best ranks are excluded from further analyses.

The central weight vector $w_i^c$ is the expected center of gravity (centroid) of the favorable first rank weights of an alternative. It represents the preferences of a DM that supports that alternative and may be useful for the decision-makers to understand how different weights correspond to different choices. The central weight vector is estimated as follows:

$$w_i^c \approx \frac{W_i}{B_{i1}} \tag{49.4}$$

## 49.3 Case Study

The paper analyzes a retrofit process of an archetype of office building considering various preference information. The office building is located in Turin (45.07 °N, 7.67 °E) that is characterized by a humid subtropical climate and around 2500 heating degree day. In particular, the analysis considers an archetype of an office building constructed in Italy during the 1946–1970 period. According to Rollino [25], the typical office building in Italy until 1970 has a cellular plan, with no more than six floors and located in an urban environment, surrounded by a lower building. For the case study here, a four-floor office building was analyzed.

As shown in Fig. 49.1, the electrical load is characterized by a constant baseload and a small seasonal variation, while the heat profile has no baseload and a strong seasonal variation, due to the temperature variation that influences the heat demand. Moreover, the office is used only in the weekdays; therefore, during the weekend there is no heat or electrical demand.

### *49.3.1 Criteria and Preference Information*

The multicriteria evaluation considers economic, environmental, and technical criteria to analyze and compare various retrofit solutions. The criteria selected here are summarized and briefly described in Table 49.1.

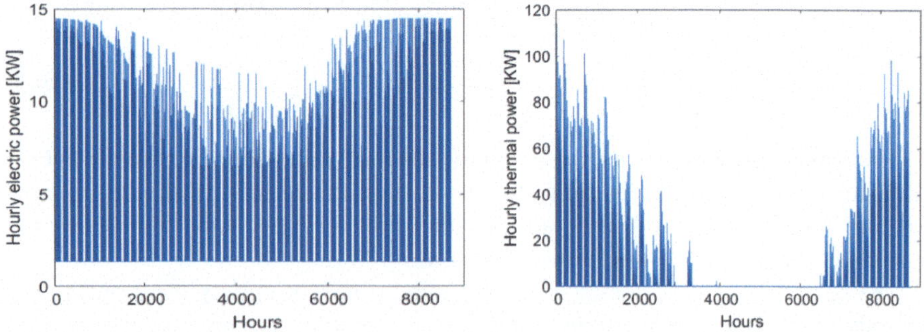


**Fig. 49.1** Simulated electric and thermal load of the building during the year

**Table 49.1** Criteria chosen for the analysis

| Criteria | Description |
|---|---|
| Investment cost (INV) | Cost of initial investment for the alternative |
| Operational cost (OP) | Cost of the system including fuel, electricity, maintenance |
| $CO_2$ emissions ($CO_2$) | The $CO_2$ emissions during the year |
| Flexibility (FLEX) | Fast response adjustment of production volume |

**Table 49.2** Decision-makers' preference information

| Scenarios | 1 | | 2 | | 3 | | 4 |
|---|---|---|---|---|---|---|---|
| Scenario 1 | INV | > | OP | > | $CO_2$ | ~ | FLEX |
| Scenario 2 | INV | ~ | OP | > | $CO_2$ | > | FLEX |
| Scenario 3 | OP | > | INV | > | $CO_2$ | > | FLEX |

Lifetime cost, which includes both investment and operational cost, is very often used in retrofit projects; however, merging two criteria could introduce dependent uncertainties and lead to a loss of information, reducing the possibility to express stakeholder preferences. Therefore, since the main goal of the analysis is the impact of stakeholder preferences, the two criteria were analyzed individually.

After the criteria selection, the methodology focuses on preference information provided by decision-makers. Three scenarios were considered:

- Scenario 1: It gives priority to the investment cost. This may represent the preference of a small–medium company that analyzes the possible retrofit of the building.
- Scenario 2: The priority is given both to operational and investment costs, considered of equal importance. This may represent the preferences of a medium–large company.
- Scenario 3: It gives priority to the operational cost.

The preference information is summarized and ordered in Table 49.2.

## 49.4 Design of the Retrofit Alternatives

The base case includes a gas boiler, used to satisfy the thermal demand, while the electricity is withdrawn from the market to cover the electrical demand (A1).

The retrofit solutions are based on the adoption of three technologies: solar-based systems, biomass boiler, and internal combustion engine (ICE) cogeneration units. The alternatives are described in the following:

- *Photovoltaic plant*: A solar power plant was integrated on the roof. Based on solar irradiance data [26], two alternatives were analyzed: A2 includes a 10 kW PV plant, sized to cover part of the electrical demand, while A3 considers a 20 kW PV plant, dimensioned to cover the demand and sell electricity surplus. Other three alternatives were studied, coupling the 10 kW plant with a cogeneration unit (A6) and two different PV systems of different sizes (10 and 20 kW) with biomass boiler (A8, A9).

- *CHP unit*: Two sizes of CHP unit, respectively, 10 and 20 kW [27], were integrated with the gas boiler (A4, A5). The smallest size aims to partially cover the electrical demand, while the 20 kW CHP was designed to cover the electrical demand and sell the surplus.
- *Biomass boiler*: The alternative A7 replaces gas boiler with biomass (*pellet*) boiler, assessing economic and environmental benefits.

### *49.4.1 Data and Assumptions*

In the following, technical and economic data, such as emission factors and costs of the different technologies, are introduced to evaluate each criterion. Then, the assumptions on the uncertainties of the model are briefly discussed.

As far as the technical data and assumptions are concerned, the investment costs considered for each alternative were assessed according to [28] and reduced by the incentives, taken from [29]. To evaluate the operational cost, for each alternative, the model takes into account fuel costs, electricity costs (withdrawn and sold to the grid), and other variable costs (O&M) and subsidies, taken from [30–33]. Looking at environmental and technical criteria, $CO_2$ emissions were evaluated using emission factors according to [34]. Flexibility of the plant was ranked according to the type of technology and its size, where 1 represents the best alternative and 9 the worst.

Concerning the uncertainties assumptions, the buildings energy models are affected by several types of uncertainties, including technical and demand uncertainties. To simulate the variabilities of investment cost, fuel and electricity prices, and heat and electricity demand, an uncertainty of $\pm 10\%$ was considered for each criterion according to [14]. Uncertainties affect costs and emissions and are represented through a uniform distribution. The flexibility was supposed to be not affected by uncertainty, since it depends only on the type of technology.

## 49.5 Results

### *49.5.1 Simulation Output for Different Alternatives*

In Table 49.3, the investment and operational costs, the emissions, and the flexibility for each alternative are shown. In particular, these values are associated with the uncertainties previously discussed and used as input for SMAA simulations, coupled

**Table 49.3** Output of the simulation for the alternatives

| Alt | INV [k€] | OP [k€] | $CO_2$ [t/y] | FLEX |
|---|---|---|---|---|
| A1 | 0 | 10.5 | 27 | 8 |
| A2 | 9 | 9 | 22.5 | 6 |
| A3 | 18 | 8.5 | 20 | 4 |
| A4 | 6.3 | 9.5 | 21 | 3 |
| A5 | 12.6 | 8.5 | 21 | 2 |
| A6 | 21.6 | 5 | 11 | 1 |
| A7 | 5 | 10 | 15 | 9 |
| A8 | 14 | 8.7 | 10 | 7 |
| A9 | 19 | 8.2 | 8 | 5 |

with the preference information for each scenario. The table highlights that each alternative leads to a decrease in operational costs and emissions with respect to the base case. Alternative A6 halves operational costs, while alternative A9 reduces significantly the emissions.

### 49.5.2 SMAA Results

The rank acceptability index $b_i^1$ is the probability that the alternative $i$ is the most preferred considering the uncertainties information; therefore, it was used as an indicator to identify the best solution.

As shown in Fig. 49.2, the most acceptable alternatives are A1 and A4, with acceptability values of 57% and 25%, respectively. These results underline how, even if there is high space for improvement, the choice of many companies is influenced by the investment cost, even when incentives are provided. The most acceptable alternative excluding the base case is represented by a small CHP, able to reduce both electrical and heat costs with a limited investment cost. The central weights shown in Fig. 49.2 represent typical preferences that make each alternative the most preferred. In particular, A1 is preferred if the investment cost obtains ~60% of weight, while A4 is preferred when the weights are more homogeneous.

Figure 49.3 shows how, changing preference information, as in Scenario 2, the results of the analysis are different. In fact, in this scenario, the most acceptable alternatives are A6 and A1, with 40 and 35% acceptability value, respectively. Moreover, as can be seen from the weights presented in Fig. 49.3, A6 is more preferred with respect to the previous scenario if operational costs obtain more importance from stakeholder, while A1 even with the same central weights obtains less acceptability value.

From Fig. 49.4 it can be seen that the most preferred alternative for Scenario 3 is clearly A6, with an acceptability value of 74%, followed by alternative A1 and A7

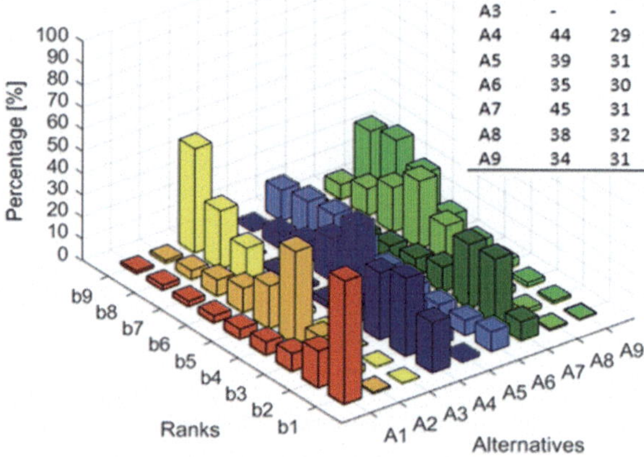


Central weights

| Alt | Inv | Op | CO2 | Flex |
|---|---|---|---|---|
| A1 | 59 | 25 | 8 | 7 |
| A2 | 42 | 36 | 12 | 10 |
| A3 | - | - | - | - |
| A4 | 44 | 29 | 11 | 16 |
| A5 | 39 | 31 | 10 | 20 |
| A6 | 35 | 30 | 17 | 18 |
| A7 | 45 | 31 | 19 | 5 |
| A8 | 38 | 32 | 25 | 5 |
| A9 | 34 | 31 | 25 | 10 |

**Fig. 49.2** Rank acceptability indices and central weights for Scenario 1 (%)

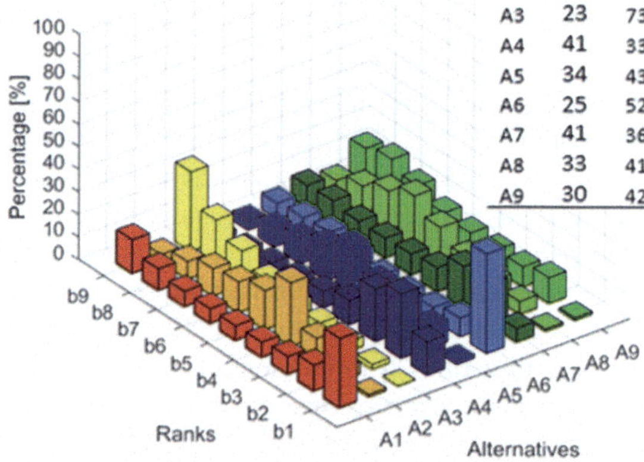


Central weights

| Alt | Inv | Op | CO2 | Flex |
|---|---|---|---|---|
| A1 | 56 | 28 | 11 | 4 |
| A2 | 37 | 45 | 13 | 5 |
| A3 | 23 | 73 | 2 | 2 |
| A4 | 41 | 33 | 17 | 9 |
| A5 | 34 | 43 | 15 | 8 |
| A6 | 25 | 52 | 15 | 7 |
| A7 | 41 | 36 | 19 | 4 |
| A8 | 33 | 41 | 21 | 5 |
| A9 | 30 | 42 | 21 | 7 |

**Fig. 49.3** Rank acceptability indices and central weights for Scenario 2 (%)

with 8%. As highlighted by the central weights, A7 is the most acceptable solution if more importance is given to environmental criteria. On the other hand, if more importance is given to investment cost, then A1 is identified as most acceptable solution.

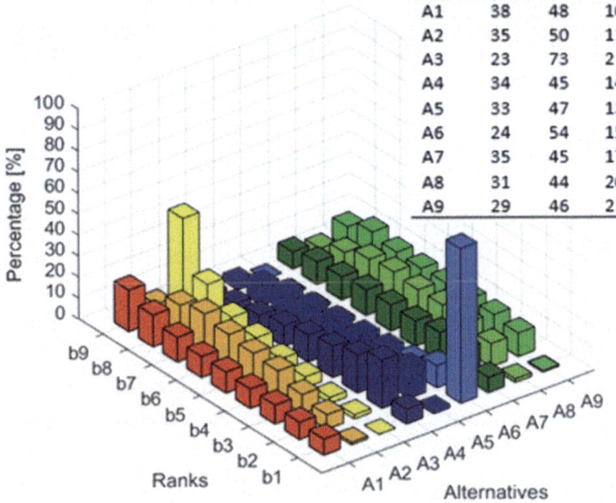


**Central weights**

| Alt | Inv | Op | CO2 | Flex |
|---|---|---|---|---|
| A1 | 38 | 48 | 10 | 3 |
| A2 | 35 | 50 | 11 | 4 |
| A3 | 23 | 73 | 2 | 2 |
| A4 | 34 | 45 | 14 | 6 |
| A5 | 33 | 47 | 13 | 7 |
| A6 | 24 | 54 | 15 | 7 |
| A7 | 35 | 45 | 17 | 3 |
| A8 | 31 | 44 | 20 | 4 |
| A9 | 29 | 46 | 21 | 5 |

**Fig. 49.4** Rank acceptability indices and central weights for Scenario 3 (%)

## 49.6 Discussion

The presented paper aimed at identifying the best retrofit alternatives in the presence of uncertainties and considering three scenarios with different preference information. SMAA method was used for its capacity to consider uncertainties, differently from other MCDA methods. The analysis is focused on the impact of stakeholder preferences, considering three scenarios. In the first scenario, the possible preferences of a small company were considered, highlighting how even if there are profit margins for the investment, according to the high importance given to the investment cost, the use of the current heat technology and withdraw electricity from the grid is preferred. On the other hand, scenarios that emphasize the importance of reducing operational cost suggest as best solution the coupling of a CHP and PV plant. Moreover, further analyses showed that assigning a high weight on $CO_2$ criteria, the best solution remains the coupling of a CHP and PV plant, followed by the coupling of biomass and PV.

The previous scenario could be useful for policy-makers, interested more in environmental aspects. Note that the method is able to overcome the difficulty of assigning precise weights in MCDA, using ordinal preference information; the uncertainties assessment could be improved using, e.g., the methodology presented in [15]. The use of such method could help decision-makers in understanding the impact of their opinion on the analysis.

Future works may consider the application of the method to residential buildings, try to include other technologies to the system, as electric and thermal storage, or shifting the analysis toward environmental criteria helping policy-makers in their decisions. As far as the MCDA methodology presented here is concerned, further improvements could move in the direction of handling partially conflicting preference information.

## 49.7 Conclusion

In this study, the stochastic Multicriteria Acceptability Analysis (SMAA) was selected for its ability to identify the best alternative in the presence of uncertainties. SMAA method was used to identify sustainable alternatives to produce both electricity and heat in office buildings. Then, the study compared three scenarios consisting of three different stakeholder preferences. The analysis highlighted how, even if there is an increasing interest for the environmental aspects, the main drivers of feasibility study are of economic nature, and that the investment cost is still the hardest aspect to face.

The paper highlighted the impact of stakeholder preferences for the retrofit of an office building, pointing out how different preferences can lead to different results and, consequently, how a precise study of stakeholder preferences could influence the outcome of the analysis.

## References

1. International Energy Agency IEA: Energy efficiency indicators highlights (2016)
2. European Commission: Europe 2020. A European strategy for smart, sustainable and inclusive growth (2010)
3. Europe's buildings under the microscope. A country-by-country review of the energy performance of buildings
4. Murugan, S., Horàk, B.: A review of micro combined heat and power systems for residential applications. Renew. Sustain. Energy Rev. **64**, 144–162 (2016)
5. Bagdanavicius, A., Jenkins, N.: Power requirements of ground source heat pumps in a residential area. Appl. Energy **102**, 591–600 (2013)
6. Wang, J.-J., Jing, Y.-Y., Zhang, C.-F., Zhao, J.-H.: Review on multi-criteria decision analysis aid in sustainable energy decision-making. Renew. Sustain. Energy Rev. **13**(9), 2263–2278 (2009)
7. Maxim, A.: Sustainability assessment of electricity generation technologies using weighted multi-criteria decision analysis. Energy Policy **65**, 284–297 (2014)
8. Matteson, S.: Methods for multi-criteria sustainability and reliability assessments of power systems. Energy **71**, 130–136 (2014)
9. Chinese, D., Nardin, G., Saro, O.: Multi-criteria analysis for the selection of space heating systems in an industrial building. Energy **36**(1), 556–565 (2011)
10. Wang, E.: Benchmarking whole-building energy performance with multi-criteria technique for order preference by similarity to ideal solution using a selective objective-weighting approach. Appl. Energy **146**, 92–103 (2015)
11. Lahdelma, R., Hokkanen, J., Salminen, P.: SMAA—Stochastic multiobjective acceptability analysis. Eur. J. Oper. Res. **106**(1), 137–143 (1998)

12. Rahman, M.M., Paatero, J.V., Lahdelma, R.: Evaluation of choices for sustainable rural electrification in developing countries: a multicriteria approach. Energy Policy **59**, 589–599 (2013)
13. Loikkanen, O., Lahdelma, R., Salminen, P.: Multicriteria evaluation of sustainable energy solutions for Colosseum. Sustain. Cities Soc. **35**, 289–297 (2017)
14. Kontu, K., Rinne, S., Olkkonen, V., Lahdelma, R., Salminen, P.: Multicriteria evaluation of heating choices for a new sustainable residential area. Energy Build. **93**, 169–179 (2015)
15. Pinto, G., Abdollahi, E., Capozzoli, A., Savoldi, L., Lahdelma, R.: Optimization and multicriteria evaluation of carbon-neutral technologies for district heating. Energies **12**(9) (2019)
16. Pelissari, R., Oliveira, M., Ben Amor, S., Kandakoglu, A., Helleno, A.: SMAA methods and their applications: a literature review and future research directions. Ann. Oper. Res (2019)
17. Crawley, D.B., et al.: EnergyPlus: creating a new-generation building energy simulation program. Energy Build. **33**(4), 319–331 (2001)
18. EN 15232:2012 Energy performance of buildings—impact of building automation, controls and building management
19. Ballarini, I., Corrado, V.: Analysis of the building energy balance to investigate the effect of thermal insulation in summer conditions. Energy Build. **52**, 168–180 (2012)
20. Cascone, Y., Capozzoli, A., Perino, M.: Optimisation analysis of PCM-enhanced opaque building envelope components for the energy retrofitting of office buildings in Mediterranean climates. Appl. Energy **211**, 929–953 (2018)
21. Cascone, Y.: Optimisation of opaque building envelope components with phase change materials. Politecnico di Torino (2017)
22. The MathWorks: MATLAB 2016b
23. Tervonen, T.: JSMAA: open source software for SMAA computations. Int. J. Syst. Sci. **45**(1), 69–81 (2014)
24. Lahdelma, R., Miettinen, K., Salminen, P.: Ordinal criteria in stochastic multicriteria acceptability analysis (SMAA). Eur. J. Oper. Res. **147**(1), 117–127 (2003)
25. Luca, R.: Fabbisogni energetici per edifici caratterizzanti il terziario in Italia: aspetti termici e illuminotecnici. Politecnico di Torino (2012)
26. PVGIS: Photovoltaic geographical information system. http://re.jrc.ec.europa.eu/pvgis/. Accessed 29 Oct 2018
27. Asja Group: TOTEM energy. http://www.totem.energy/wp-content/uploads/TOTEM_ST.pdf. Accessed 02 Aug 2019
28. Badami, M., Chicco, G., Portoraro, A., Romaniello, M.: Micro-multigeneration prospects for residential applications in Italy. Energy Convers. Manag. **166**, 23–36 (2018)
29. ENEA: Ecobonus2018 (2018). https://finanziaria2018.enea.it/index.asp. Accessed 02 Nov 2019
30. AIEL: Costo del pellet: scenario sul mercato nel 2019. http://www.aielenergia.it/public/rassegna_stampa/378_1febbraio2019biomassapp.it.pdf. Accessed 12 Mar 2019
31. GSE: Ritiro dedicato. https://www.gse.it/servizi-per-te/fotovoltaico/ritiro-dedicato. Accessed 03 Jan 2019
32. EIA: Updated buildings sector appliance and equipment costs and efficiencies
33. Badami, M., Camillieri, F., Portoraro, A., Vigliani, E.: Energetic and economic assessment of cogeneration plants: a comparative design and experimental condition study. Energy **71**, 255–262 (2014)
34. Covenant of Mayors: The emission factors. https://www.eumayors.eu/IMG/pdf/technical_annex_en.pdf. Accessed 02 Apr 2019

# Chapter 50
# Study of the Effect of Different Configurations of Bypass Diodes on the Performance of a PV String

**Haider Ibrahim and Nader Anani**

**Abstract** In this paper, the impact of different configurations; overlapped and non-overlapped, of bypass diodes on the performance of a photovoltaic (PV) module under varying levels of shading, are investigated using a simulation approach alleviating the need for complex modelling using numerical techniques. MATLAB/Simulink models have been developed for 18 series-connected solar cells in the commercially available KC200GT PV module. Simulation results show that the non-overlapped configuration of bypass diodes outperforms the overlapped scheme in terms of the energy yield and the number of local maximum power points in the power-voltage characteristic of a PV module. The overlapped scheme can lead to an increased number of local maximum power points placing onerous demands on the design of the maximum power point tracker required to track the global maximum power point.

## 50.1 Introduction

Concerns over the harmful effects of fossil fuels and the sustainability and security of their supplies and prices, led to accelerated interest and huge investment in renewable energy resources, such as wind, solar and tidal. Solar energy has been popular due to its reduced complexity compared to, say, tidal energy, and lends itself to localized generation of electrical power. In an electric solar energy generation plant, the energy in the sunlight is converted to electricity utilizing the photovoltaic (PV) effect. This is done using a PV cell manufactured from a semiconductor material such as Silicon. However, a PV cell provides a limited terminal voltage, typically 0.5 V and, therefore, a number of cells are normally manufactured in an integrated module, also

H. Ibrahim
Department of Engineering and Design, University of Chichester, Bognor Regis Campus, Tech Park, Bognor Regis PO21 1HR, UK

N. Anani (✉)
Faculty of Science and Engineering, University of Wolverhampton, Telford Innovation Campus, Shifnal Road, Priorslee, Telford, TF2 9NT Shropshire, UK
e-mail: n.anani@wlv.ac.uk

 
J. Littlewood et al. (eds.), *Sustainability in Energy and Buildings*, Smart Innovation, Systems and Technologies 163, https://doi.org/10.1007/978-981-32-9868-2_50

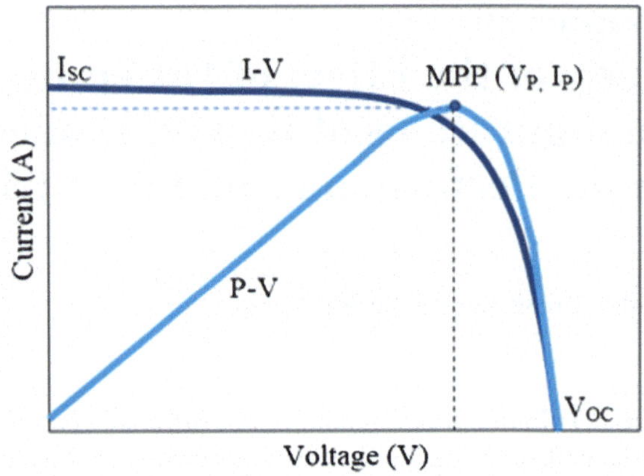


**Fig. 50.1** I-V and P-V characteristics of a PV cell/panel

known as a PV panel, in which all cells are connected in series to provide higher terminal voltage. Further, to provide larger amount of power, panels are connected in series/parallel combinations making a PV array. The terminal I-V characteristic of a PV cell/panel/array resembles that of a current source as depicted in Fig. 50.1, which also includes the power-voltage (P-V) curve. The latter is characterized by its maximum power point (MPP). It is evident from the P-V curve that, for maximum efficiency, a PV panel/array must be operated at the MPP. However, this MPP varies dynamically with the amount of radiation and the ambient temperature [1]. For this reason, a maximum power point tracker (MPPT) is deployed in any PV system whose function is to continuously match the operating point with the current MPP [2]. Further, in practice, it is difficult to ensure that all PV cells in a module receive same level of irradiance, a phenomenon referred to as partial shading (PS), which can arise due to obstacles such as clouds, dirt, buildings, etc. Partial shading has an adverse effect on the performance of a PV panel, as explained in the next section, and its mitigation requires the use of bypass diodes [3]. However, the use of these diodes leads to a P-V curve with multiple power peaks, Fig. 50.2, which further complicates the design of a PV system.

### *50.1.1 Partial Shading and Bypass Diodes*

A simple method of visualizing the effect of partial shading is to consider a PV module, which consists of n cells connected in series as shown in Fig. 50.3a with the *n*th cell represented by its single-diode equivalent circuit [4]. Under normal conditions of uniform irradiance, the output voltage is V (V) and the output current is I (A). If the *n*th cell is shaded its short circuit current will be reduced to zero and the current from the rest (n − 1) cells will go through its parallel resistance impressing a reverse voltage across its diode giving rise to joules heating and a hot spot which can eventually damage the cell and hence the whole panel [5]. In less extreme cases, the

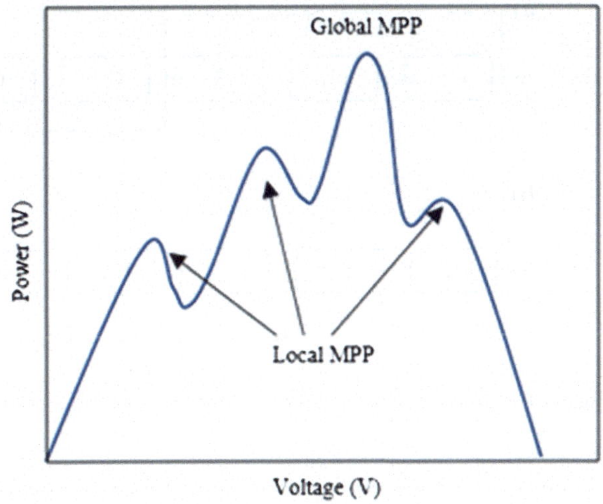


**Fig. 50.2** P-V characteristic with multiple peaks

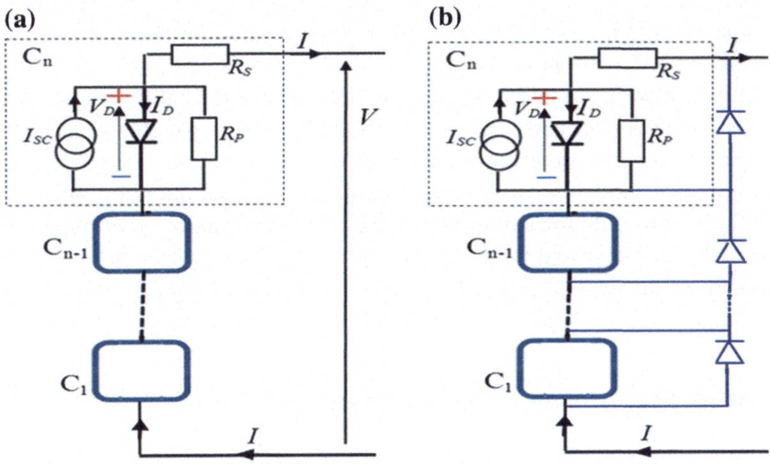


**Fig. 50.3** A PV module with n-cells in series

shaded cell instead of adding energy to the module, it is reducing the energy yield produced by the module [6]. To mitigate the effects of shading, anti-parallel bypass diodes can be deployed, during the manufacturing process, as shown in Fig. 50.3b, ideally, one diode for each cell. Under normal conditions, i.e. with no shading, the voltage rise across the cell will reverse bias the bypass diode and hence it acts like an open-circuit. However, when the cell is shaded, the bypass diode is forward biased with the negative effect of producing a voltage drop across it, however, this is only about 0.6 V. In practice, manufacturers don't manufacture modules with bypass diodes across each individual cells. Instead, bypass diodes are usually placed across groups of series connected cells making up a string. That is, a PV module consists

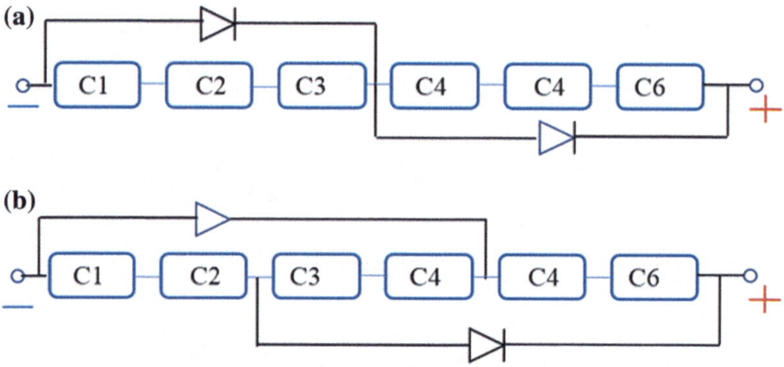


**Fig. 50.4** **a** Non-overlapped and **b** over-lapped configurations of bypass diodes

of number of strings equipped with one bypass diode for each string. Since a bypass diode has some negative impact on the maximum power of the photovoltaic system, it is essential to study their effects on the performance of a PV system. There are two common configurations of bypass diode implemented on photovoltaic module; overlapped and non-overlapped schemes as depicted in Fig. 50.4 [7].

The objective of this paper is to study the effect of different bypass diodes configurations on the performance of the PV string (18 cells of KC200GT panel) under varying partial shading conditions using MATLAB/Simulink. This mitigates the need for complex modeling using numerical methods [8]. Following this introduction, Sect. 50.2 presents the two common different configurations of bypass diodes reported in the literature and their simulation results. Finally, Sect. 50.3 summarizes the outcomes of this work.

## 50.2 Modeling and Simulation Results

Two bypass diodes are connected across the PV string which consists of 18 solar cells of KC200GT module [9] in two different configurations, non-overlapped and overlapped, to study the effect of bypass diodes configurations under different levels of partial shading. MATLAB/Simulink model was constructed for each cell using the single diode model [10]. The simulation is divided into two parts covering each configuration of bypass diodes.

### *50.2.1 Non-overlapped Bypass Diodes*

In order to assess the effect of the non-overlapped configuration of bypass diodes on the performance and shape of the I-V and P-V characteristics of a PV module under

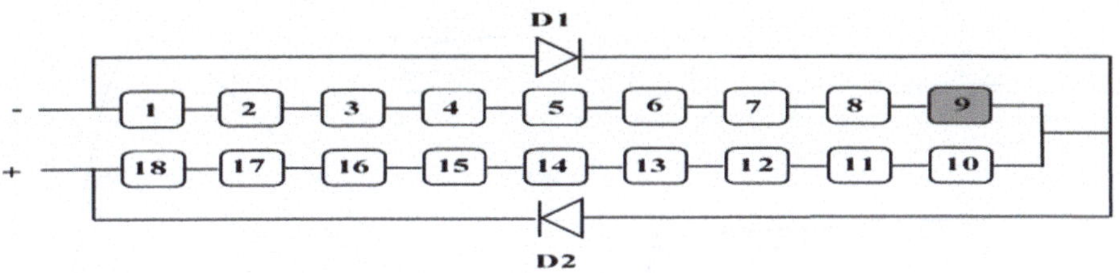


**Fig. 50.5** PV String with two non-overlapped bypass diodes

different levels of shading, the non-overlapped configuration shown in Fig. 50.5 which includes two bypass diodes; D1 and D2, was used for the simulation. The simulation was performed with 100% irradiance for all cells, then it was reduced to 75, 50 and 25% for cell number 9 only, which is shown shaded in Fig. 50.5. The simulation results are presented in Fig. 50.6 which includes the I-V and PV characteristics of the PV string and the current through bypass diode D1.

As can be seen, a new local maximum power point appears at low voltage values; lower than the voltage that corresponds to the maximum power point under conditions

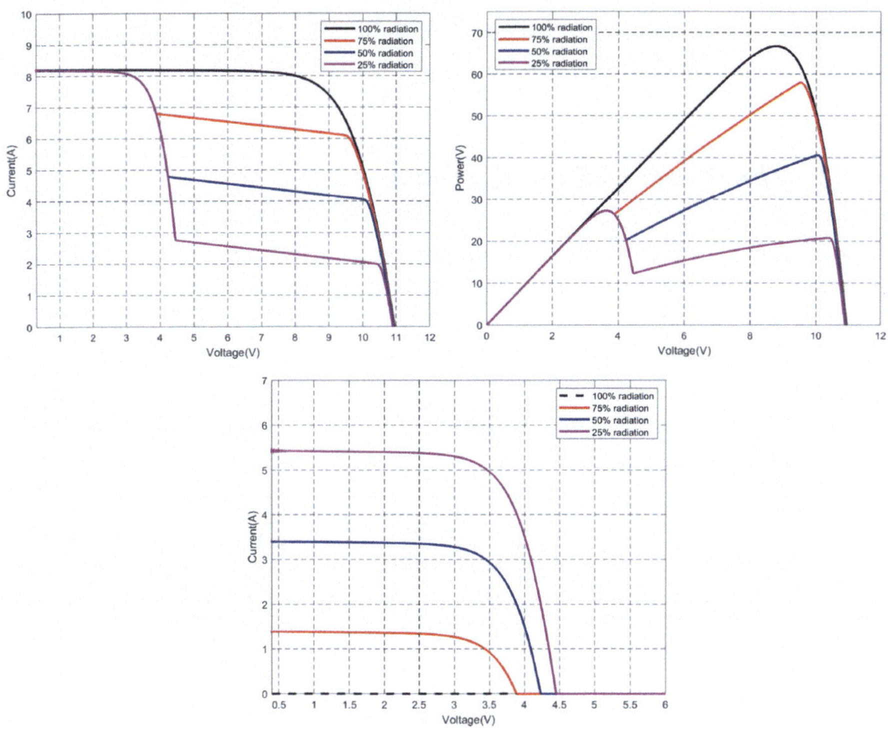


**Fig. 50.6** The characteristics of the PV string and current through D1under different levels of partial shading

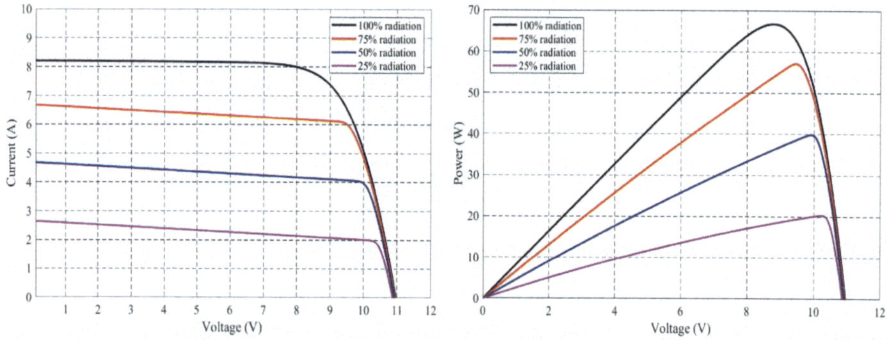


**Fig. 50.7** Characteristics of PV string under different illumination over one solar cell of each branch covered by bypass diodes

of uniform irradiance. The actual value of this low voltage increases with increasing the level of irradiance and the corresponding maximum power point moves to the right of the P-V curve. In another investigation, cells 9 and 18 were subjected to the same level of shading whilst the rest of the cells were exposed to 100% irradiance. The results are presented in Fig. 50.7. It is evident in this case that no local maximum power points appear indicating that both bypass diodes are in non-conduction state. In this case, the panel behaves as if there were no bypass diodes.

### *50.2.2 The Overlapped Configuration*

To investigate the effect of partial shading using the overlapped bypass diode configuration, the scheme shown in Fig. 50.8 was used in the simulation. In this scheme cells 5-10 are overlapped using the two bypass diodes D1 and D2. Cells 1-10 are bypassed by D1 while cells 5-18 are by bypassed by D2. This is a non-symmetrical

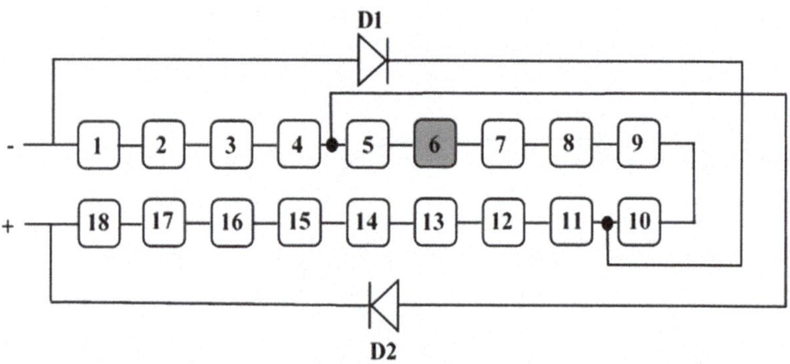


**Fig. 50.8** PV String with two overlapped bypass diodes

overlapping configuration. It was found that the effect of partial shading on any one of the non-overlapped cells is similar to the performance of the PV string covered by non-overlapped bypass diodes as in the in the previous subsection. The diode bypassing the shaded solar cells enters in conduction state and the I-V and P-V characteristics of the PV string show the same trends observed in Fig. 50.7. However, when the overlapped cell 6 was shaded, the PV panel demonstrated different I-V and P-V characteristics as shown in Fig. 50.9. As can be deduced from these characteristics, for low voltage values both bypass diodes are in conduction, and the current flowing through both diodes D1 and D2 is the same, leading to an increase in the short circuit current of the PV string. This is because the PV panel current is the sum of the currents in each diode. As the voltage increases, bypass D2 is reverse biased reducing the output current of the PV string. Further, it is evident that the overlapped scheme increases the number of local maximum power points which compounds the complexity of the maximum power point tracking.

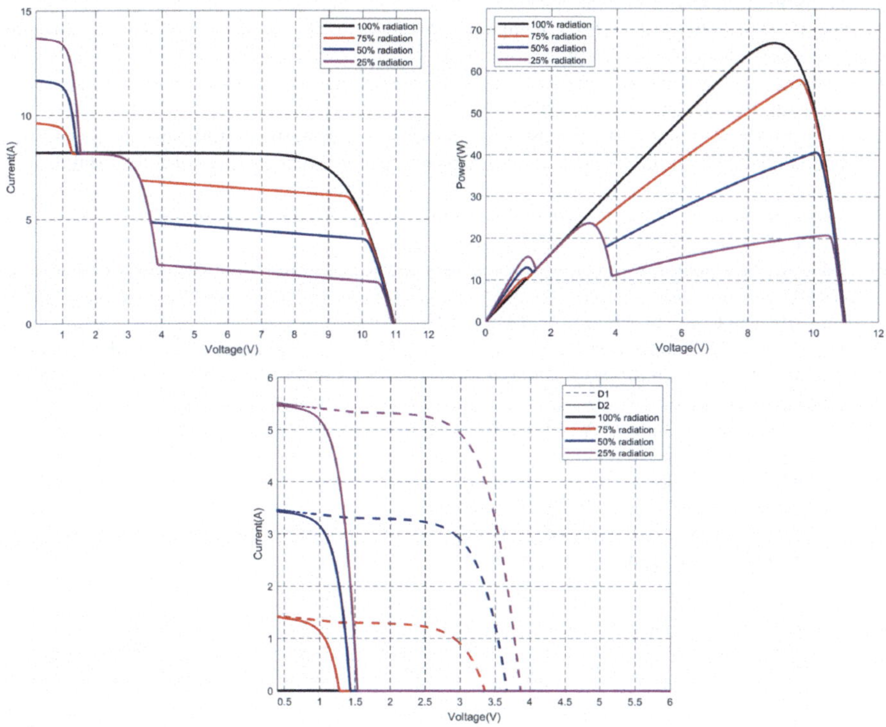


**Fig. 50.9** The characteristics of the PV string and current through D1 and D2 under different levels of partial shading

## 50.3 Conclusions

A simulation study was performed to assess the performance of a PV panel varying levels of partial shading using the overlapped and non-overlapped configurations of bypass diodes. This alleviates the complexities associated with numerical methods. The investigation showed that, in general, the performance of a PV module depends on the level of partial shading, the number of shaded cells and the configuration of the bypass diodes. The overlapped configuration can increase the number of local maximum power points when the shaded cell is overlapped.

## References

1. Ibrahim, H., Anani, N.: Variations of PV module parameters with irradiance and temperature. Energy Proced. **134**, 276–285 (2017)
2. Eltawil, M.A., Zhao, Z.: MPPT techniques for photovoltaic applications. Renew. Sustain. Energy Rev. **25**, 793–813 (2013)
3. Silverstre, S., Chouder, A.: Effects of shadowing on photovoltaic module performance. Prog. Photovolt. Res. **16**(2), 141–149 (2007)
4. Ma, T., Yang, H., Lu, L.: Development of a model to simulate the performance characteristics of crystalline silicon photovoltaic modules/strings/arrays. Solar Energy (Elsevier) **100**, 31–41 (2014)
5. Sathyanarayana, P., Ballal, R., Sagar, P.L., Kumar, G.P.S.: Effect of shading on the performance of solar PV panel. Energy Power **5**(1A), 1–4 (2015)
6. Mäki, A., Valkealahti, S.: Power losses in long string and parallel-connected short strings of series-connected silicon-based photovoltaic modules due to partial shading conditions. IEEE Trans. Energy Convers. **27**(1), 173–183 (2012)
7. Díaz-Dorado, E., Suarez-Garcia, A., Carrilloo, C., Cidras, J.: Influence of the shadows in photovoltaic systems with different configurations of bypass diodes, in *International Symposium on Power Electronics, Electrical drives, Automation and Motion*, Pisa (2010)
8. Petrone, G., Ramos-Paja, C.A.: Modeling of photovoltaic fields in mismatched conditions for energy yield evaluations. Electr. Power Syst. Res. **81**(4), 1003–1013 (2011)
9. "KYOCERA KC200GT panel," [Online]. https://www.kyocerasolar.com/dealers/product-center/archives/spec-sheets/KC200GT.pdf
10. Ibrahim, H., Anani, N.: Evaluation of analytical methods for parameter extraction of PV modules. Energy Proced. **134**, 69–78 (2017)

# Chapter 51
# Developing Management Guidance for Government Funded Dwelling Retrofit Schemes to Improve Occupant Quality of Life

**D. Jahic, J. R. Littlewood, G. Karani, A. Thomas, J. Atkinson and J. Kirrane**

**Abstract** This paper discusses a methodology for gathering empirical data on occupant views that live in dwellings retrofitted under the Arbed two scheme in wales, UK, and also the views of stakeholder's that delivered and managed the retrofit scheme between 2012 and 2014. The analysis of this data is being used to develop guidance for managing government funded dwelling retrofit schemes, contributing to occupant quality of life. The methodology used consists of Causal Loop Diagrams, Theory of Change and IDEFØ mapping to systematically map out relationships between activities extracted from the findings of two case studies, demonstrating key issues for investigation. This multi-tool process has been critical within the process of developing key clusters. The key clusters are fed into the IDEFØ mapping system, from analysing the complex tacit knowledge data in order to produce guidance documents to be used in government funded residential retrofit schemes. This paper will be useful for academics, landlords of housing stock undergoing retrofit measures, scheme managers undertaking large scale residential energy and retrofit upgrades, and government agencies funding retrofit upgrades to dwellings.

## 51.1 Introduction

Wales has seen two completed strategic energy performance funding programs known as Arbed 2010–2011 (Arbed one) and 2012–2014 (Arbed two) targeting exist-

D. Jahic (✉) · J. R. Littlewood · J. Atkinson
Sustainable and Resilient Built Environment (SuRBe) Group, Cardiff Metropolitan University, Cardiff CF5 2YB, UK
e-mail: dejahic@cardiffmet.ac.uk

G. Karani
Environmental Public Health Group, Cardiff Metropolitan University, Cardiff CF5 2YB, UK

A. Thomas
School of Management, Cardiff Metropolitan University, Cardiff CF5 2YB, UK

J. Kirrane
Melin Homes, Pontypool NP4 0XJ, UK

 
J. Littlewood et al. (eds.), *Sustainability in Energy and Buildings*, Smart Innovation, Systems and Technologies 163, https://doi.org/10.1007/978-981-32-9868-2_51

ing dwellings for retrofit energy efficiency upgrades [1]. Arbed three commenced in 2018 and will be completed in 2021 [2]. The UK, and in particular Wales has some of the worst and inefficient dwellings from an energy perspective within Europe, resulting in significant fuel poverty and potential impacts upon occupant quality of life [3–5].

The operations, stakeholder involvement and management of the three Arbed scheme varies, and so do the processes and procedures of managing such projects, potentially applying to other UK wide or regional government funded retrofit projects. In this instance, there is clear evidence from literature and primary data that there is a need to develop best practice guidance for operating large scale Government funded projects, contributing to occupant quality of life. There have been no studies in the UK that have been published which investigate the impacts of dwelling upgrades and decision making of stakeholders in Government funded retrofit schemes, upon occupant quality of life [6, 7]. Neither have there been any studies published that uses IDEFØ mapping of such retrofit schemes to analyse and illustrate the interview date, potential contributions to knowledge and practice. Hence, this project is attempting to fill this gap by interviewing key Arbed two stakeholders to assess specific aspects of the project management and delivery. This paper discusses the first element of this work which is stakeholder interviews, Causal Loop Diagram (CLD) [8], Theory of Change (ToC) [9] analysis and how this is fed into the IDEFØ for mapping. This methodology will help to deliver process and guidance maps for the management of Government funded dwelling retrofit programs, contributing to occupant quality of life.

## 51.2 Background and Context

The Institution of Engineering and Technology (IET) stated in 2018 that "The most efficient way for the UK to achieve 2050 standards in social housing is for the providers to invest money that would have been spent over 30 years on repairs and maintenance, on long-term refurbishment." [10]. The report also supports the idea that "Deep retrofit of housing should not be considered purely from an energy efficiency and carbon emissions standpoint but should be fully integrated into the plans for a thriving and resilient local economy that meets the current and future needs of citizens" [ibid]. The awareness of worldwide greenhouse gas usage and energy consumption is a common topic both in the news and regulation changes [10, 11]. With Brexit a topical subject in 2018/19 in the UK news, energy consumption, the UK's growing housing stock, fuel prices and fuel poverty are all factors in debate and discussion [12]. Considering the uncertainty of future weather conditions, fuel levels and economic stability, there has never been a more urgent time to retrofit existing housing stock, as 80% of the homes UK habitants will be living in by 2050 have already been built [10].

This KESS2 Ph.D. research project is undertaken in collaboration with Melin Homes, a registered social landlord. Melin Homes whose Being Greener team managed the Arbed two scheme in south Wales [6, 13]. This KESS2 project takes into consideration the sensitive social and economic matters in 2018/19, and investigates country wide government funded retrofit projects, to ensure best practice of improving occupant quality of life, including health and wellbeing, thermal comfort and fuel poverty in a holistic manner. With rising fuel prices in 2019, there are questions about whether or not current (2019) retrofit schemes of dwellings are taking into consideration variables which may impact the alleviation of occupant fuel poverty in the future. This Ph.D. will address this need through stakeholder interviews engaged in retrofit schemes and occupants in retrofitted dwellings in Wales. The findings are being used to develop a methodology for combined occupant and stakeholder engagement, and to produce guidance informing the management of future retrofit schemes for quality of life. In addition, the study will help with current and continuous development of retrofit solutions, indicators, guidance, and procedure mapping documents to allow for best practice. Furthermore, it will provide data for ongoing and continuous learning and research for retrofit projects in line with occupant health and wellbeing, thermal comfort and fuel poverty.

### *51.2.1 Gap in Knowledge and Practice for Occupant Quality of Life*

The UK has set targets to meet the Paris Agreement, starting in 2020, aiming to limit the increase in global average temperature to 1.5 °C [14]. In addition to this, publicly available specification (PAS) 2050 standards [10, 15] backs this up and aims at reducing global emissions by 80–95% [ibid]. PAS 2030 is a standard setting out requirements all Green Deal and Energy Company Obligation (ECO) installers must comply with [16]. PAS 2030 standards are being updated, and a release of the newly refined PAS 2035 standards prepared by the British Standards Institution (BSI) Retrofit Standards Task Group are set to be published mid 2019 [17]. In addition to this, Wales has seen the Well-Being of Future Generations Act legislated (WbFGA) and effective since 2015, focusing on seven pillars to enhance wellbeing goals [18]. What is currently missing in the above literature and guidance is information regarding occupant quality of life living in dwellings that are retrofitted from government funded dwelling retrofit schemes. In particular, managing such projects contributing to occupant quality of life while adhering to the literature discussed above. Therefore, the gap in knowledge is a methodology to collect views on such matters from occupants and stakeholder engaged in managing and delivering the retrofit scheme, in turn will deliver guidance and process models with regards to delivering occupant quality of life within government funded retrofit dwellings. The project seeks to make contribution to knowledge and practice by developing and testing a methodology for obtaining empirical data about the health and wellbeing of occupants, and also the decision making by key stakeholders following the management of Government funded dwelling retrofit schemes. In addition, to developing and testing a methodology using a tacit knowledge analysis tool known as IDEFØ mapping, to analyse the complex decision making of stakeholder groups. Using these findings could help to develop guidance for the decision making of Government funded dwelling retrofit schemes that aim to contribute to occupant quality of life.

## 51.3 Methodology

### *51.3.1 Case Study One: Hybrid Short Form 36 Health Survey*

Following the findings from the Hybrid Short Form 36 Health Survey (HSF36) on Arbed two occupants and results interpreted using the RAND scoring system [19], and labelled case study one, it was demonstrated the Arbed two occupant responses following retrofit measures, occupants had a more positive result to '*Physical functioning, Role limitations due to physical health, Energy/fatigue, Emotional wellbeing, Social functioning, and General health*'. Only their '*Role limitations due to emotional problems*' and '*Pain*' scored the same when comparing before retrofit measures and after retrofit measures of the Arbed two project. The successful outcome of case study one method using the HSF36 demonstrated that the Arbed two scheme overall had a positive impact on occupant quality of life. What was also clear and demonstrated in the findings [19] was the need to explore stakeholder involvement of the Arbed two project to gather further evidence of the management and delivery processes and whether there were impacts on occupant quality of life. This second stage of empirical data collection is referred to as case study two hereafter with an approach focusing on stakeholders engaged in Government funded dwelling retrofit schemes in the UK.

### *51.3.2 Case Study Two: Arbed Two Stakeholder Engagement*

Case study two concentrates on undergoing Arbed two stakeholder interviews investigating aspects of project operations and management, occupant wellbeing, fuel poverty, project aims, and further suggestions for the Arbed two project. The questionnaire was approved by ethics at Cardiff Metropolitan University in August 2018. Case study two interview questionnaire is based on the stakeholders' skills, opinions and experience around the Arbed two project. 42 stakeholders were identified by the industrial partner's database of the Arbed two scheme; and 14 of the 42 (33% of sample group) stakeholders were interviewed by the first author between September 2018 and November 2018. The interviews have collected a range of valuable data which is outlined in the results section below. Case study two uses the findings and learning from case study one, where a short, open ended interview questionnaire was developed by the first author. This allowed for discussion between the lead researcher and the Arbed two stakeholders, ensuring relevant information was provided for the potential of understanding gaps in Arbed two scheme delivery. The qualitative findings have been analysed and the complex tacit knowledge data had to be extracted and analysed in stages to ensure the data was used appropriately and accurately. Due to the complexity of the tacit knowledge, it was decided to build on earlier work by Littlewood [20] who managed the interviews of stakeholders who were engaged in the government funded sustainable urban regeneration of the Swansea High Street project by using the analysis tool IDEFØ mapping [7].

### *51.3.3 IDEFØ Mapping Tool*

To map and compare the complex tacit knowledge generated by the interviews in case study one and two IDEFØ mapping has been used. IDEFØ mapping is an advanced level of sophistication for knowledge encoding [21]. IDEFØ offers a language able to model complex functions and thus the use of IDEFØ is justified by the fact that one of its most important features is that it gradually introduces greater levels of definitions (or detail) within a complex process. In fact, whether as a communication tool, IDEFØ enhances involvement and consensus decision-making through simplified graphical devices. At the same time IDEFØ is an analysis tool, it assists the modeller in identifying what functions are performed, and what is needed to perform those functions, and (consequentially) it highlights what the current system does, and what the current system does not [22] (Fig. 51.1).

The mapping of decision making in Government funded dwelling retrofit schemes is a matter as complex as the activity of dwelling retrofit itself. Therefore, the author has used Toledo and Littlewood's (2013) approach to IDEFØ mapping [20] with a twist of incorporating CLD and ToC to the process. Tacit knowledge is complex in its nature, but to try and compare between stakeholders' personal and professional beliefs is even more complex [23]. As such IDEFØ mapping provides one method to explore stakeholder knowledge and the authors suggest the vision of the Arbed two retrofit scheme impacting upon occupant quality of life. Open coding will be used to

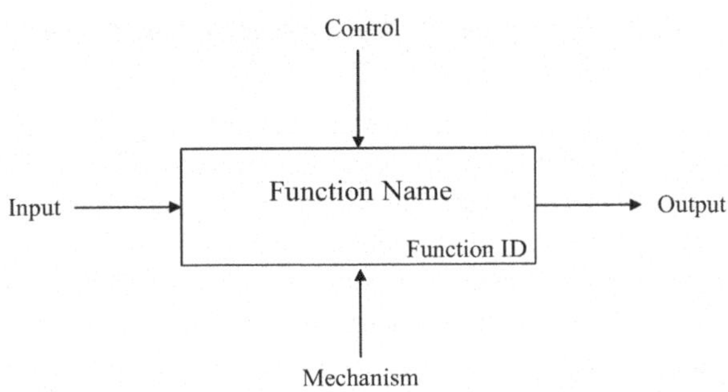


**Fig. 51.1** Elementary information of a basic IDEFØ diagram [22]

break down and categorise the data from the case studies with the assistance of the CLD [8] and ToC diagrams [9] mentioned in Sect. 51.4 [21].

## 51.4 Interpretation and Results

Case study one and two has brought together interesting data including tacit knowledge data, which has led the use of a multi-tool analysis approach using CLD [8], ToC [9] and IDEFØ [21]. This has led to the development of decision-making maps. This section discussed the process and methods used to analyse case study two interview data and once the interviews are completed will lead to the development of guidance documents.

### *51.4.1 Causal Loop Diagram (CLD) and Theory of Change (ToC)*

CLD demonstrates in a visual format how key activities or variables interconnect within a system [8]. CLD's are an ideal way of demonstrating looped connections such as the relationship between activities, for example within the Arbed two project. Data collected demonstrated that the key nodes of the CLD in the context of case study one and two were: occupants, quality of life, quality of work, management and ill health. The nodes and edge relationships of the CLD are marked with positive or negative links to demonstrate the type of relationship they have with each-other. These relationship links are beneficial for developing a ToC to logically map out causal linkages between detailed outcome boxes derived from CLD nodes and edges. This approach was used to exercise how the Arbed two project is contributing to occupant quality of life, and if there were further operational steps or requirements which needed to be undertaken.

### *51.4.2 Case Study Two Results Overview*

From the 14 stakeholder interviews, the key project findings and issues related to; technical knowledge, project management skills and tools, communication, quality of processes, and workmanship. In addition, the organisational structure of the project teams proved unsuitable during enabling works and at the beginning of the Arbed two scheme, demonstrating that there was a lack of ownership and communication relationships. With this data, and from the development of the ToC including case study one data, a requirement to developing appropriate project team structures for adequate communication, roles and responsibilities, and quality of work were key outcomes. In addition, a requirement for staff training and adequate resourcing to ensure adequate management, knowledge and experience is in place were also critical factors of ensuring effective retrofit projects with contribution to occupant quality of life.

Quality of work [4] and the need for technical knowledge is something that requires further investigation due to the recurring feedback provided during case study two stakeholder interviews. The risk in this instance is that there is a case of large mitigation costs following completion of such projects, resulting in a reduction of occupant quality of life, but also the result of poor project management [ibid]. The adaptation and implementation of guidance, specifically a process map using IDEFØ mapping may help to assist in managing critical issues of initial remedial works, quality management and mitigation risks and costs.

### *51.4.3 Interpretation of Results Using IDEFØ*

One of the central aims of using IDEFØ mapping allows the author to determine the key triggers and issues in Wales for the dwelling retrofit schemes. A first iteration of an IDEFØ map has been created (Fig. 51.2).

The first approach looks at the control and functions of the IDEFØ map from a researcher's perspective, reviewing the process where certain functions, inputs

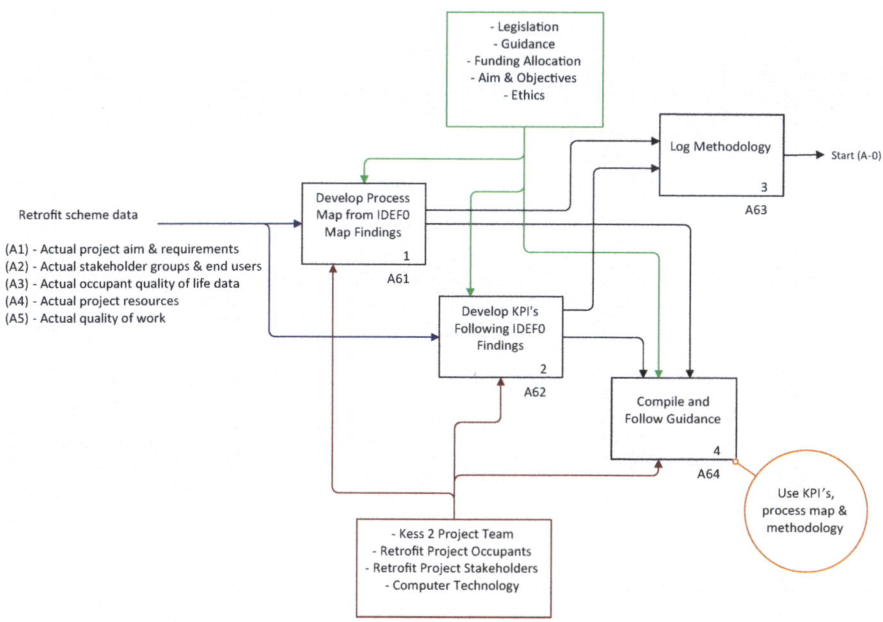


**Fig. 51.2** IDEFØ layer A6 (first iteration)

or outputs may have been missed out or showing lack of detail, ensuring a fully functioning and robust mapping system is developed. Figure 51.2 demonstrates the overall outcome of the map from a wider project perspective, signifying clear process of achieving a methodology, KPI's and the process map using actual outcome data from layers A1 to A5 illustrated in Fig. 51.3.

The key detail in this first iteration is the process of logging the methodology journey and re-routing the outcomes back to layer A-0 to undertake a second iteration for validation using controls and functions from a project viewpoint, managing a government funded retrofit project contributing to occupant quality of life. Five key clusters which are explored and will be validated by the stakeholders from case study two (in June 2019) and modified as required are seen in Fig. 51.3, layer A0. The five titles have been developed using case study one and two data, critically detailing steps to undertake the project in an appropriate and systematic manner, minimising issues and gaps recorded from project data evidence. Greater layers of definition have been mapped out under each of the key clusters using this data. The map benefits from a visually and easy to use language where project teams can follow the process from project start to completion. More specifically, it could help with pre-planning tasks as well as process aids for specific team groups. Layers A2 and A3 for example can assist a project team with information on how to contribute in improving occupant quality of life in the overall scheme, how this may be linked to financial implications or project resources, and who the team groups for communication may be. This is achieved through the connections between title boxes, inputs, outputs, controls and mechanisms.

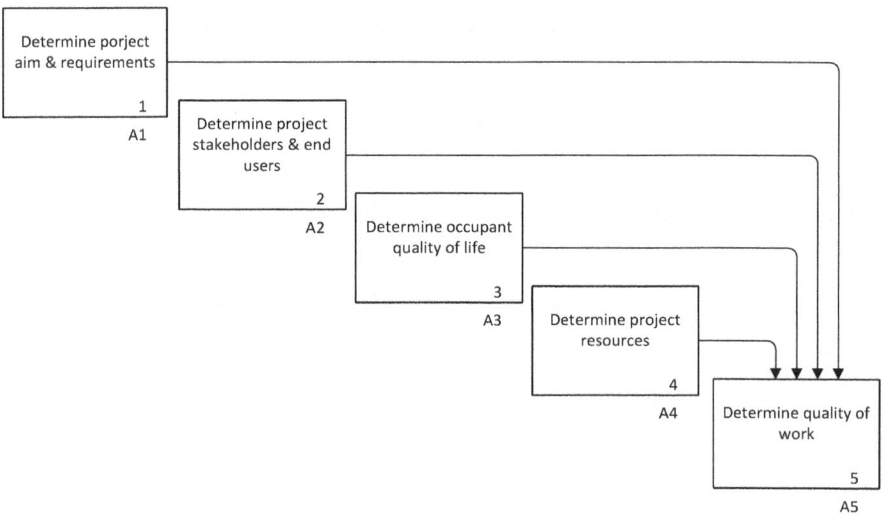


**Fig. 51.3** IDEFØ layer A0 (first iteration)

## 51.5 Discussion

This paper criticised the importance of developing guidance documents, specifically a process map for use in government funded dwelling retrofit schemes, to improve occupant quality of life. It is shown in literature that current standards and specifications do not illustrate guidance or procedures contributing to occupant quality of life when retrofitting dwellings. Learnings from Arbed one and two demonstrated critical issues relating to key management and project operations of dwelling retrofit schemes highlighted by the CLD and ToC process. The WbFGA legislated 2015 demonstrates seven pillars to enhance wellbeing goals [18], one of which is 'a healthier Wales', applicable to this project. It is clear that the WG fundamental direction of ensuring the WbFGA sustainable direction is developed through '*five ways of working*' [24] which looks at long term needs, integrated approaches for wellbeing objectives, involving diversity in decision making processes, collaborative working for sustainable solutions, and prevention of root causes and issues. This paper and the guidance being developed through IDEFØ mapping is playing a fundamental role in ensuring '*a healthier Wales*' pillar is contributed to through the five ways of working.

With a variety of stakeholder groups involved in the project process throughout case study one and two stages, covering North and South Wales, this project has a wealth of tacit knowledge data being fed into a multi-tool process which has never been used in the UK to map processes for government funded retrofit schemes, to improve occupant quality of life using IDEFØ. Using the HSF36 health survey as well as Arbed two stakeholder interview has provided the project with data for multi-disciplinary use adding value to academic practice and industry through applied research. In addition, the development and industry use of the IDEFØ mapping tool has the potential to have an astounding contribution to the way government funded retrofit schemes are run, contributing to the WbFGA goals, gaps in literature and contribution to occupant quality of life.

## 51.6 Conclusion

This paper has discussed the process undertaken to gather empirical data on occupant quality of life, as well as stakeholder engagement for the Arbed two scheme. The use of a multi-tool process has exemplified to be a key process in reducing bias outcomes within the IDEFØ map. This is achieved by developing key clusters from case study 1 and 2 data, forming the CLD, which in turn was used to structure relationships and change theories between the functions to form a ToC diagram, derived from the key clusters. In addition, the multi-tool process has helped in predetermining relationships and routes within the project process and in turn, easily fed into the mapping of the IDEFØ. The IDEFØ process outcome within Fig. 51.2 has demonstrated the need of such visual decision process tool for government funded retrofit schemes, contributing to occupant quality of life. The appropriate use of this

IDEFØ map within such retrofit schemes can play a vital role as a project management tool for stakeholders to ensure key functions are undertaken appropriately with the correct tools in place; controls and mechanisms following the input of the mapping system.

**Acknowledgements** The KESS2 Ph.D. Scholarship has been supported by the Low Carbon, Energy and Environment Grand Challenge Economic Area, administered by the Welsh European Funding Office in conjunction with Melin Homes.

## References

1. Welsh Government.: Arbed—strategic energy performance investment programme. http://gov.wales/topics/environmentcountryside/energy/efficiency/Arbed/?lang=en (2013). Accessed 22 February 2018
2. Arbed am Byth.: Helping to warm homes across Wales. http://arbedamby-th.wales (2019). Accessed 25 January 2019
3. Welsh Assembly Government.: Arbed strategic energy performance investment programme. http://www.communities-first.org/eng/news/Arbed_strategic_energy_performance_investment_programme/ (2010). Accessed 27 March 2017 (Not available)
4. Atkinson, J., Littlewood, J.R., Geens, A.J., Karani, G.: Did ARBED I save energy in Wales' deprived dwellings. Energy Procedia **63** (2015)
5. Atkinson, J., Littlewood, J., Geens, A., Karani, G.: Relieving fuel poverty in Wales with external wall insulation. Eng. Sustain. **170**(2), 93–101 (2017)
6. Littlewood, J.R., Karani, G., Atkinson, J., Bolton, D., Geens, A.J., Jahic, D.: Introduction to a Wales project for evaluating residential retrofit measures and impacts on energy performance, occupant fuel poverty, health and thermal comfort. Energy Procedia **134**, 835–844 (2017)
7. Littewood, J.R., Davies, G.: The sustainable regeneration of the Swansea high street—a cohesive community. Sustain. Energy Build. Res. Adv. J. **6**(1), 35–43 (2017)
8. Svirakova, E.: Economic development of company in creative cluster. In: Semmelrock-Picej, M.T., Novak, A. (eds.) Proceedings of the 9th European Conference on Management Leadership and Governance, pp. 274–282. Klagenfurt, Austria (2013)
9. Taplin, H.D, Clark, H., Collins, E., Colby, C.D.: Theory of Change, Technical Papers: A Series of Papers to Support Development of Theories of Change Based on Practice in the Field, pp. 1–23. Act Knowledge, New York, NY (2013)
10. IET.: Scaling Up Retrofit 2050: Why a Nationwide Programme to Upgrade the Existing Housing Stock is the Only Way for the UK to Achieve its Carbon Saving Goals. UK (2018)
11. GOV UK.: 2010 to 2015 government policy: energy efficiency in buildings. https://www.gov.uk/government/publications/2010-to-2015-government-policy-energy-efficiency-in-buildings (2015). Accessed 16 April 2019
12. British Broadcasting Corporation.: Climate change: where we are in seven charts and what you can do to help. https://www.bbc.co.uk/news/science-environment-46384067 (2018). Accessed 3 December 2018
13. Bolton, D.: Sustainability homes, people, business: maximising the impact of retrofit. https://www.all-energy.co.uk/__novadocments/54282?v=635376502894700000 (2015). Accessed 11 April 2019
14. United Nations Framework Convention on Climate Change.: The Paris agreement: essential elements. https://unfccc.int/process-and-meetings/the-paris-agreement/the-paris-agreement (2018). Accessed 17 January 2019
15. PAS 2050.: Specification for the assessment of the life cycle greenhouse gas emissions of goods and services. BSI 2011: ISBN 978 0 580 71382 8, ICS 13.310; 91.190 (2011)

16. PAS 2030.: Improving the energy efficiency of existing buildings. Specification for installation process, process management and service provision. BSI 2017: ISBN 978 0 580 82569 9, ICS 91.120.10/91.140.10 (2017)
17. Each Home Counts.: Each home counts implementation plan—standards. http://www.eachhomecounts.com/wp-content/uploads/2018/04/Implementation-Plan-Standards.pdf (2018). Accessed 2 April 2019
18. Welsh Government.: Well-being of future generations (Wales) Act 2015. https://gov.wales/docs/dsjlg/publications/150623-guide-to-the-fg-act-en.pdf (2015). Accessed 16 April 2019
19. Jahic, D., Littlewood, J.R., Karani, G., Atkinson, J., Bolton, D.: Evaluating occupant wellbeing in retrofitted dwellings with the short form 36 questionnaire. In: Smart Innovation, Systems and Technologies Series, vol 131. SEB18 International Conference Australia, Springer, UK. Book Journal (2018)
20. Toledo, L., Littlewood, J.R.: An (un)attainable map of sustainable urban regeneration, chapter 59. In: Håkansson, A., Howlett, R.J. (eds.) Sustainability in Energy and Buildings-Proceedings of the 4th International Conference on Sustainability in Energy and Buildings (SEB'12), vol. 13, pp. 637–648. Springer, Heidelberg, Germany (2012)
21. National Institute of Standards and Technology Integration Definition for Function Modelling (IDEFØ). Federal Information Processing Standards Publications, vol. 183. http://www.idef.com/pdf/idef0.pdf (1993). Accessed 16 April 2019
22. Knowledge Based Systems Inc. IDEFØ. http://www.idef.com/IDEF0 (2010). Accessed 16 April 2019
23. Tsoukas, H.: Complex knowledge: studies in organizational epistemology. Oxford University Press Inc., New York, NY (2005)
24. SPSF 1: Core guidance: Shared purpose: shared future statutory guidance on the well-being of future generations (Wales) act 2015. https://gov.wales/sites/default/files/publications/2019-02/spsf-1-core-guidance.PDF (2016). Accessed 16 April 2019

# Chapter 52
# Innovative User Experience Design and Customer Engagement Approaches for Residential Demand Response Programs

**Matteo Barsanti, Letizia Garbolino, Muhammad Mansoor, Giulia Realmonte, Rita Zeinoun, Francesco Causone and Valentina Fabi**

**Abstract** The increasing share of intermittent sources is making it more difficult to guarantee a real-time balance between demand and supply on the electricity grid. To decrease the dependency from fossil fuel generation, a change in paradigm is required: from supply following demand whenever it occurs to demand following generation when it is available. Demand response (DR) programs enclose all practices that allow demand to take part in actively managing the grid. According to this perspective, the residential sector hides a huge still unexploited flexibility resource. Therefore, utilities and aggregators need to address weak customer engagement and a lack of regulation in order to employ innovative business models for harnessing residential DR programs potential. Within this paper, some of these challenges are investigated, with the view to improve the design of an appropriate engagement strategy and an incentive scheme to involve residential customers. The innovation consists in the development of a questionnaire as a tool to understand customers' behavior and preferences, so as to consequently design customized solutions. Finally, a first-order approximation techno-economic analysis is conducted to contextualize the actual incentives for the single customer.

---

M. Barsanti (✉) · M. Mansoor · G. Realmonte · F. Causone
Department of Energy, Politecnico di Milano, via Lambruschini 4, 20156 Milan, Italy
e-mail: matteo.barsanti@mail.polimi.it

L. Garbolino
Department of Architecture and Design, Politecnico di Torino, v.le Pier Andrea Mattioli 39, 10125 Turin, Italy

R. Zeinoun
Department of Architecture and Urban Studies, Politecnico di Milano, via Bonardi 3, 20133 Milan, Italy

V. Fabi
Department of Energy, Politecnico di Torino, c.so Duca degli Abruzzi 24, 10129 Turin, Italy

 
J. Littlewood et al. (eds.), *Sustainability in Energy and Buildings*, Smart Innovation, Systems and Technologies 163, https://doi.org/10.1007/978-981-32-9868-2_52

## Nomenclature

| | |
|---|---|
| DCE | Discrete choice experiment |
| DESWH | Domestic electric storage water heater |
| DR | Demand response |
| DUoS | Distribution use of system |
| FFR | Firm frequency response |
| STOR | Short-term operating reserve |
| TNUoS | Transmission network use of system |
| WTP | Willingness to pay |

## 52.1 Introduction

According to the urgent need to decarbonize our economy, new solutions have to be found in order to provide the required flexibility [1] for integrating intermittent renewable power sources, such as solar and wind. Among the new sources of flexibility, demand response (DR) is becoming a cost-effective solution, pushed by electrification of transport and heating/cooling sectors and new supporting technologies.

By definition, DR refers to any change in electrical consumption done by consumers with respect to their usual patterns during peak demand times and in response to electricity price variations over time [2]. These changes in consumption can be caused either by manual operation of customers (i.e. manual DR) or by a remote control of the loads (i.e. automated DR). DR should include a form of incentive payments to the final customers, so as to trigger their active participation [3].

Considering the status of DR in European electricity market [4], it can be noticed that, while at industrial- and commercial-level DR practices encounter an appropriate legal framework (e.g. UK, Switzerland, Belgium and France), at residential level they are still in test phase. Nevertheless, the majority of theoretical demand response potential lies with residential consumers, hindered by low consumer engagement with energy-related activities, and a lack of regulation specifically designed for this customer segment. In recent years, several reports have studied evidences of consumer engagement for residential DR in UK through surveys, trials and pilot schemes [5–7]. Three critical elements to unlock the potential of DR are the maximization of value proposition, consumer awareness about their energy choices and the risk reduction of negative influences [8], but scientific evidences about customer engagement and economic potential of DR programs are still limited.

The focus of this paper is to investigate some of the challenges utilities and aggregators need to address in order to harness the potential coming from manual and automated load shifting at household level. In particular, a general procedure has been depicted to facilitate the design of an appropriate engagement strategy and incentive scheme to involve residential customers and offer this flexibility as a service on the electricity market. The innovation consists in the development of a questionnaire as a tool to understand customers' behavior and preferences, so as to consequently design customized solutions for DR programs. The questionnaire was rolled out in Italy to perform a first validation, by considering respondents' feedbacks and consistency of the results. At the same time, a step-based framework has been defined to develop DR programs in the residential market, so as to involve progressively more and more people.

Section 52.2 describes the methodology adopted to prepare the questionnaire and the engagement strategies, together with the procedure applied for the techno-economic assessment of residential DR. Section 52.3 describes the questionnaire development and the results obtained after the first roll-out. Finally, Sect. 52.4 contains the detailed techno-economic analysis followed by the conclusion in Sect. 52.5.

## 52.2 Methodology

One of the main problems of energy engagement campaigns directed toward residential customers is the high drop-out rates, as people do not perceive benefits or a clear incentive to change their daily habits. In order to avoid this, a step-based approach is expected to have a twofold benefit. On the one hand, each step is simplified, and customers can better embrace the change, feeling it less demanding or disruptive; on the other hand, it allows updating the following steps according to new technologies, trends and local requirements.

Customers are involved step by step, starting with an extensive customer engagement phase, followed by the actual DR program. Despite being interlocked and sharing the main actors, each of the progressive steps will have a specific business model and value proposition. A brief description of the steps is provided as follows:

1. *User experience design*. The environment where the program will be implemented is analyzed. Customer characteristics and preferences are outlined through a questionnaire, in order to tailor the services and the communication strategies.
2. *Customer engagement and awareness*. Customer activity and community involvement are emphasized. Services that are offered are not limited to energy-related ones, to further stimulate interest of customers through additional drivers.
3. *DR programs*. Once residential customers have understood the value of being "active", they can contribute to the grid management through load shifting. DR can be implemented both manually and automatically.

### 52.2.1 Questionnaire and Engagement Strategies

A questionnaire has been developed to evaluate the preferences of end-users and their willingness to pay for a number of energy-related services. The questionnaire survey is considered as a tool for identifying customer segments and tuning the service offering accordingly. Within the questionnaire, a discrete choice experiment (DCE) has been selected as a quantitative technique to investigate customers preferences over different service attributes based on a marginal valuation method. DCEs allow assessing the willingness to pay (WTP) for having the preferred attribute included in a product or service [9]. With respect to the methodologies where attributes are ranked or rated by individuals (i.e. conjoint analysis), stated preference methods (i.e. choice experiments) are closer to the real-life situation of any shopping experience, where the customer is asked to make a choice among products with marginal substitutions. By presenting attributes as embedded in a service, results are less affected by strategic answering (i.e. moral arguments), indicating which attributes the respondents are more sensitive to [10].

Customer engagement strategies are implemented via communication channels (i.e. mobile application, web portal) where end-users interact with energy providers and manage their energy-related services. "Idea flow" and gamification strategies play a major role in engaging users on the long run. Idea flow [10] refers to the propagation of behaviors and beliefs through a social network by means of social learning and social pressure: people are inclined to take action according to their peers' behavior. In addition, gamification and individual rewards effectively stimulate users, and prevent the program from being considered as a boring duty. Competition among peers triggers these strategies, hence reward schemes are based both on individual achievement and comparative data, and saving challenges are set up.

Despite the important role of social pressure in customer engagement, the economic dimension remains the dominant one. In this regard, a techno-economic pre-analysis was developed. The aim is to outline critical aspects that may affect manual and automated DR economic potential at residential level now and in the future, rather than an accurate estimation of their present value.

The approach applied to both types of DR program can be summarized as follows: first, determine household appliances that are able to provide load flexibility and their capacity; second, understand which application each type of demand-side management may be more suited to; once the DR use is selected, identify the economic unitary value obtained by selling the flexibility as a market product or by increasing operational saving; then, based on these data, estimate the overall potential revenues, that must be eventually compared with the expected costs of the programs.

## 52.3 Questionnaire and Results

The choice experiment included in the questionnaire has been designed following a case study where interviewees select their preferred scenarios for landscape development [11]. The attributes included in each scenario represent the main features of a product or service and its price: each of them can have different levels representing the various states of that feature. On analyzing the results, it can be identified which attributes levels the users are more inclined to pay for. The software Gretl has been used to run a multinomial logit model and provide coefficients for the WTP calculation. These coefficients are then used in the Eq. (52.1) defining the WTP as "minus the ratio between the estimate of the coefficient for the attribute of interest and the tax coefficient" [11, 12]. The resulting WTP values represent the additional (positive or negative) amount of money the customers are willing to pay for having one specific feature in the service, with respect to a base level.

$$WTA_a = -\left(\frac{\beta_a}{\beta_{interest\ rate}}\right) \tag{52.1}$$

### 52.3.1 Questionnaire Development and Description

Our questionnaire "Evaluation of preferences for Monitoring Services and Management of Electricity Consumption in residential buildings" has three main objectives: identify customer segments, investigate user awareness and learn the most-valued features. Consequently, the survey has been divided into multiple sections:

- Characteristics of the user, including questions about age, gender, educational level, number of household members, income, dwelling type, size and construction year, energy consumption and use of the main electric appliances
- Energy awareness, including open and multi-choice questions about climate change, pollution and energy waste, while investigating the actions performed to save energy in their houses
- Dual-choice tasks, where the interviewee has to choose between packages with different services and prices. A total of 22 packages—coupled into 11 choice tasks—are built upon combinations of different levels of the five selected attributes, according to Table 52.1. The combination process, an "orthogonal design" by IBM SPSS software, allows getting meaningful results through a limited number of combinations [11].

Each package includes one level from each attribute. The base monthly fee relates to similar services offered by European companies. On top of that, the fee includes the lease of devices, where present within the offer.

The questionnaire was spread out as an online form and sponsored by word-of-mouth. A first group of 30 respondents provided direct feedback about the intelligi-

**Table 52.1** Dual-choice tasks: attributes levels

| Attribute | Level A | Level B | Level C |
|---|---|---|---|
| User interface | Notifications via SMS; services management via account on website | Notifications and services management via in-home display | Notifications and services management via mobile application |
| Base service | Reports on historical consumption data and real-time feedback | Level A + short-term advice about behavioral change | Level B + automated control of some electric appliances |
| Additional rewards | Gamification and data comparison within a virtual community | Inefficiency detection of household electric appliances | Long-term advice (e.g. retrofit measures, appliances replacement) |
| Rewards | Bonus for public transport and shared sustainable mobility | Discounts on products purchased via approved platforms or sellers | Discounts on the energy bill |
| Monthly fee | 5€ | 11€ | 16€ |

bility of the survey. In parallel, the overall structure was analyzed by an expert, highlighting specific issues. The questionnaire was adjusted according to these feedbacks and then spread out to a larger group of people living in Piemonte and Lombardia (Northern Italy). The achieved number of respondents was 83.

### *52.3.2 Questionnaire Results*

Because of the small and to some extent uneven group of respondents (i.e. age distribution), the survey cannot be considered an accurate market analysis. However, some considerations about the market segment and the questionnaire potential can be discussed. *Respondent's age*, *Household income*, and *Household members* are set as the clustering criteria, and three values (respectively 29 years old, 25,000€, 2 people) are chosen for splitting the respondents' group into categories. These clusters are employed when representing the electric appliances use patterns (Fig. 52.1) and the WTP (Fig. 52.2). From these charts, market strategies insights can be derived. For instance, the clusters showing a peak of appliances use in the evening (younger respondents, higher-income households, smaller households) are those with a stronger preference for appliances' automated control. Smaller households, whose appliances use pattern suggests the house is mostly empty at daytime, are those showing the lowest preference toward a fixed in-home-device.

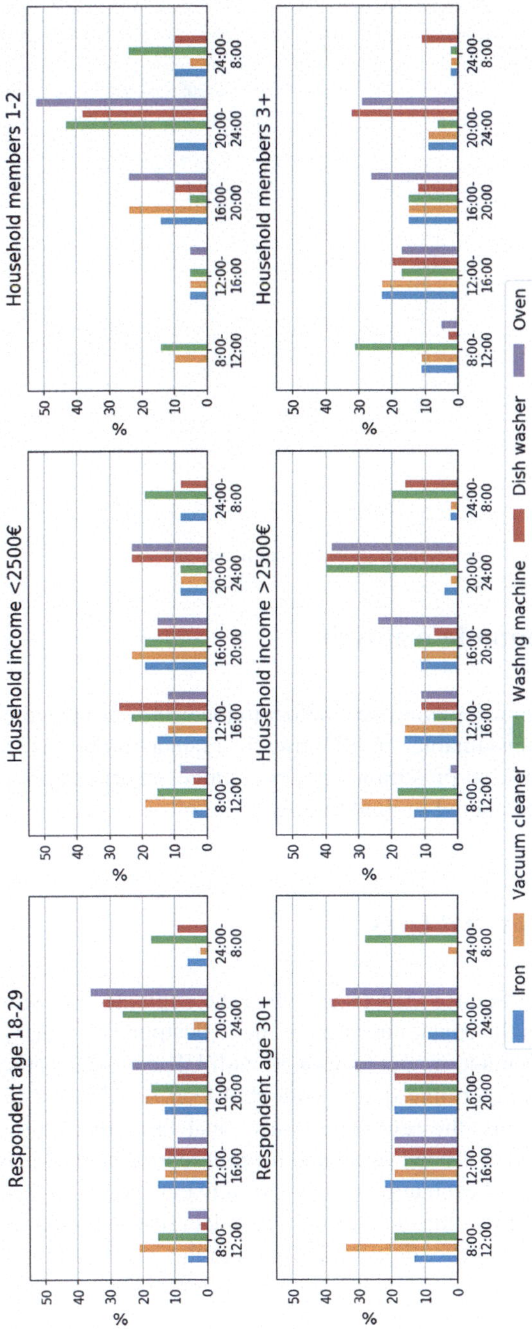


**Fig. 52.1** Respondents' electric appliances use pattern by cluster

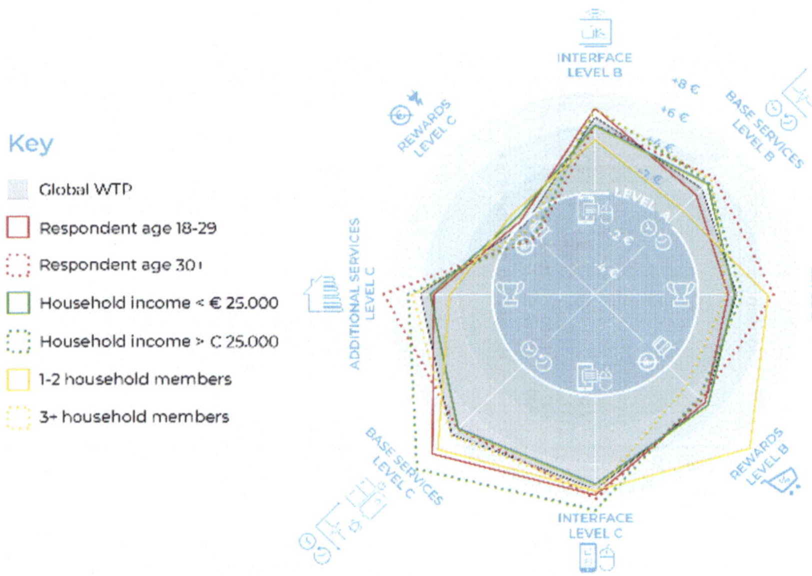


**Fig. 52.2** Respondents' WTP represented globally and by cluster

## 52.4 Techno-economic Analysis

This section aims to generally evaluate whether manual and automated DR programs are economically self-sustainable, highlighting the main assumptions and uncertainties behind the analysis. No presumptions about value accuracy are meant, so that the results are to be considered limited to their order of magnitude.

### *52.4.1 DR Sources and Uses*

Demand flexibility can be provided either when the appliance has some degree of energy storage or when the consumers are willing to postpone its usage [13]. A high-resolution stochastic demand model calibrated with UK data and statistics, CREST [14], has been used to determine daily residential load profiles and to assess load capacity. For our analysis, the reference is a typical three-residents UK terraced household. Load demand has been simulated for four representative days of the year.

The flexibility extent of each appliance is strongly dependent on dwellers' habits and load patterns; hence it is important to use real data from a similar context to the one under analysis. In our case, field evidence from the past trials has been used to determine the response that can be achieved with different DR programs [5, 15, 16], see Table 52.2.

**Table 52.2** DR uses and resources parameters. *s: summer; n: not summer; y: yearly*

| Use | Price | DR use time window | Total load | Fridge | Dish washer | Dryer | Washing machine | Water heater | Heat pump |
|---|---|---|---|---|---|---|---|---|---|
| Wholesale price spread | 50 £/MWh | Daily | 6–15% | – | – | – | – | – | – |
| Cash-out imbalance | 40 £/MWh | Daily | 6–15% | – | – | – | – | – | – |
| Short-term operating reserve (STOR) | 15 £/kW/y | 7:00–14:00 s<br>10:00–14:00 n<br>17:00–23:00 y | – | 0% | 78% | 78% | 78% | 50% | 50% |
| Firm frequency response (FFR) | 50 £/kW/y | Daily | – | 100% | 78% | 78% | 78% | 100% | 100% |
| Transmission network use of system (TNUoS) | 46 £/kW/y | 17:00–18:00 w | – | 0% | 78% | 78% | 78% | 100% | 100% |
| Distribution use of system (DUoS) | 26 £/kW/y | 11:00–14:00 s<br>16:00–19:00 w | – | 0% | 78% | 78% | 78% | 33% | 33% |

In order to couple flexibility sources with their appropriate use, the framework developed in [13] has been readapted and improved for our purposes (Fig. 52.3). This framework selects among all the possible DR uses (represented by different colored field in the chart) only the ones whose technical requirements (i.e. notice period, duration and frequency, represented by the different axis) can be guaranteed by the source (different class of appliances). Events duration and notice period are the main requirements for direct load control practices (i.e. automated DR), because end-users are not directly involved. Differently, in manual DR programs events regularity acquires importance, affecting the response rate in the short and long run [17].

A reasonable assumption is to consider that the conditions (prices and requirements) imposed by the current UK regulation for industrial and commercial DR programs will be kept unchanged also for residential applications, at least in the short-term scenario. It is possible that over the coming years, new regulatory arrangements (e.g. network charges) could lead to different requirements and prices for DR.

### 52.4.2 Revenue Evaluation

Considering customer concerns for short notice period and frequent DR requests, and observing the DR use selection framework (Fig. 52.3), manual DR programs seem to provide operational saving from the supplier's perspective. The flexibility capacity can be indicated as a percentage of the daily consumption [18] (Table 52.2), and it is remunerated in terms of energy shifted. Differently, in case of automated DR, flexibility is valued in terms of power. The temporal dimension (i.e. the synchrony between flexibility demand and its availability) was included in the methodology developed by Wegner et al. [18]. The flexible power $P_{k,j}$ is set as the average consumption of chosen appliances over the DR time window defined in [19, 20] (Table 52.2), while the load flexibility share $c_{k,j}$ is defined as the fraction of power that can be actually shifted in time depending on DR use (minimum duration) and source (shiftability), see Eq. (52.2). For each DR use potential, revenues $R_{k,i}$ are evaluated as the product of the average power $P_{k,j}$, the temporal factor $c_{k,j}$ and the DR price $p_{DR,k}$, see Eq. (52.3)

$$c_{k,j} = \frac{\Delta t_{flex}}{\Delta t_{DRuse}} \tag{52.2}$$

$$R_{k,i} = P_{k,j} \cdot c_{k,j} \cdot p_{DR,k} \tag{52.3}$$

A remarkable aspect is the possibility to aggregate the flexibility sources in a portfolio for providing diverse services, according to economic and technical optimization strategies and to the needs of the energy system. As there is not a considerable literature evaluating the interaction of different DR sources and uses into a flexibility portfolio, the two extreme cases of perfect and absent integration have been analyzed.

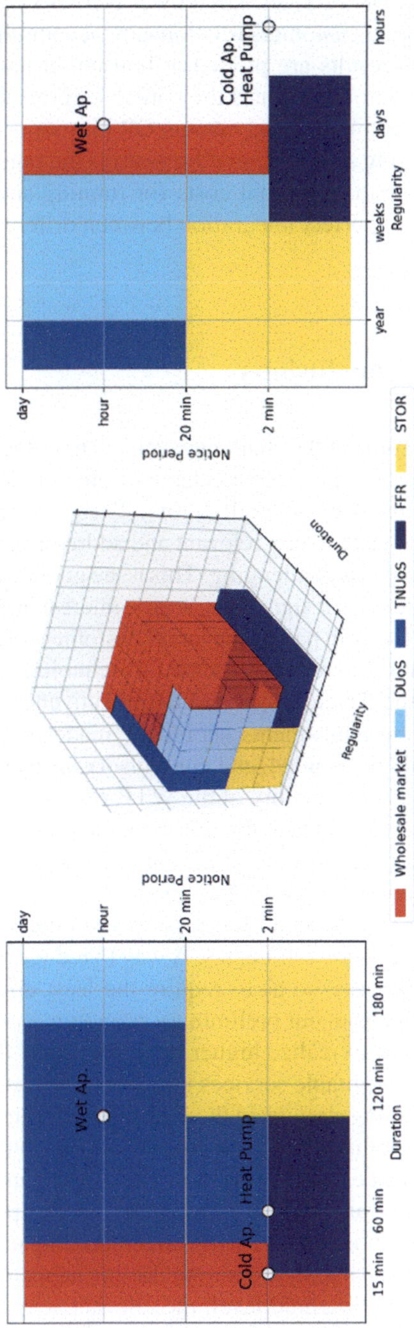


Fig. 52.3 DR framework

The costs involved will vary significantly depending on the assets and the specific arrangement used to provide DR services. UK Power Networks [21] states that the main cost term is due managing customers relationship, accounting for 105 *[£/y/p]*.

In Fig. 52.4, the analysis results are plotted at household level for the manual DR programs, while for the automated ones they are disaggregated by DR use and DR source. Finally, both the automated and manual DR programs revenues are compared with a reference operating cost [21]. In all scenarios the potential revenues are not sufficient to cover the high operational costs for running such programs. This occurs even in the best case of perfect integration between different uses of demand flexibility.

## 52.5 Discussion and Conclusions

The present work addresses some of the challenges related to DR programs in the residential segment by investigating possible mechanisms and engagement strategies. However, the analysis is limited by several factors. The number of questionnaire respondents is too small to be statistically relevant and addressing a part of the population—namely elderly people—is challenging. This issue goes beyond the reach-out potential of a questionnaire tool, as it signals the scarce interest in such programs and the related technical barrier for a big part of the target customers. Spreading the questionnaire only through digital channels represented a technological barrier in itself, and this aspect may affect the development and the effectiveness of DR programs. For future applications, some adjustments should be made on the questionnaire. Open-ended questions should be avoided, and compulsory answers reduced in order to shorten it. To achieve a higher reliability of the WTP, respondents should state for each choice task whether they would actually purchase the preferred solution or not.

Moreover, each age cluster demonstrates different needs and drivers to participate in these initiatives. As the economic reward that can be returned to end-users results limited, it cannot be adopted as the main leverage to keep them involved and it is key to engage customers with various strategies. A way to reduce high drop-out rates is automating most of the process, so as to require the least effort and time from customers. In addition, results from our preliminary economic assessment show that automated DR programs allow to realize higher revenues as the flexibility capacity can be used to provide more profitable services to the energy system.

This is likely to be the winning strategy in the long term to achieve a firm response that can be leveraged as a flexibility asset on electricity markets, but it requires a complete renovation of all appliances as they need to be "smart" and connected. This process started few years ago (and it is still going on) with respect to new energy efficiency labeling; therefore, it is not likely to happen once again in the short term only relying on market mechanisms. In order to make this happen and to fully leverage the flexibility potential at household level, the relevant actors and stakeholders (suppliers, aggregators and system operators) must incentivize it properly so as to create the conditions for DR programs to work. At this point, it should be considered

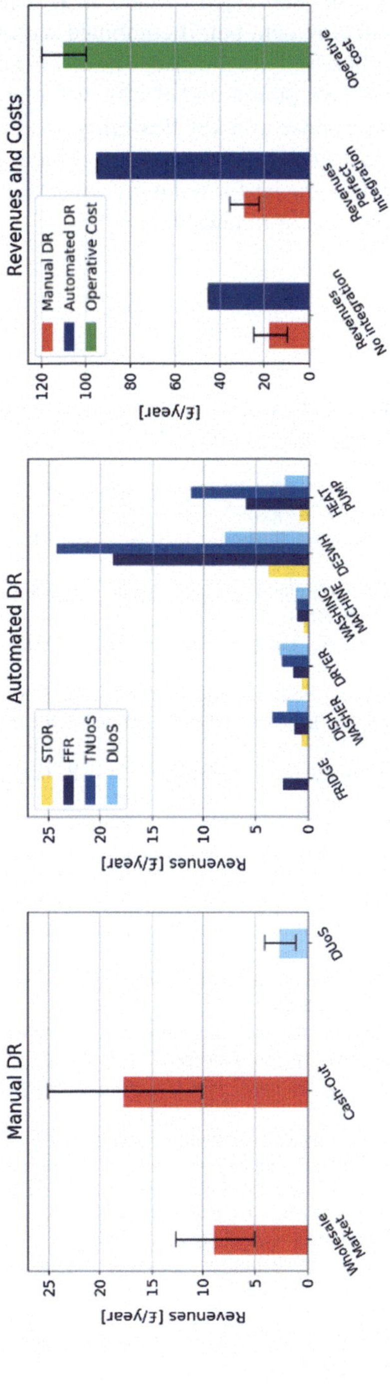


**Fig. 52.4** DR programs revenues and cost estimations

whether the flexibility impact of household load shifting will be cannibalized by the deployment of storage systems and electric vehicles at scale. DR is not likely to substitute storage systems deployment but, if combined with them, it can reduce the overall costs for providing flexibility to the grid. The electrification of residential demands due to the diffusion of heat pumps and electric vehicles will open new scenarios for the role of demand response as a key flexibility asset of the power sector. Nevertheless, this solution must not be the reason to promote high electrification rates in the residential sector. It is meant for reducing stress and strains on the grid and its framework is the penetration of renewables.

## References

1. Crosbie, T., Broderick, J., Short, M., Charlesworth, R., Dawood, M.: Demand response technology readiness levels for energy management in blocks of buildings. Buildings **8**(2), 13 (2018)
2. Albadi, M.H., El-Saadany, E.F.: A summary of demand response in electricity markets. Electr. Power Syst. Res. **78**(11), 1989–1996 (2008)
3. IEA: The Power to Choose: Demand Response in Liberalised Electricity Markets. OECD Publishing (2003)
4. Eid, C.: Demand response in Europe's electricity sector: market barriers and outstanding issues. INIS **48**(8), INIS-FR—170-0182 (2015)
5. Parrish, B., Heptonstall, P., Gross, R.: The potential for UK residential demand side participation. HubNet, Tech. Rep. (2016)
6. Hledik, R., Gorman, W., Irwin, N., Fell, M., Nicolson, M., Huebner, G.: The value of TOU tariffs in Great Britain: insights for decision makers. Citizen Advice, Final Report 1 (2017)
7. Chase, A., Gross, R., Heptonstall, P., Jansen, M., Kenefick, M., Parrish, B., Robson, P.: Realising the potential of demand-side response to 2025: A focus on small energy users. Department for Business, Energy & Industrial Strategy (UK), Sum. Report (2017)
8. Carmichael, R., Gross, R., Rhodes, A.: Unlocking the potential of residential electricity consumer engagement with demand response. Energy Futures Lab (Imperial College London), Briefing Paper (2018)
9. Mangham-Jefferies, L., Hanson, K., Mcpake, B.: How to Do (or Not to Do) … Designing a discrete choice experiment for application in a low-income country. Health Policy Plan. **24**(2), 151–158 (2009)
10. Pentland, A.: Social Physics: How Social Networks Can Make Us Smarter. Penguin Publishing Group (Kindle version), New York (2014)
11. Bottero, M., Cozza, G., Fontana, R., Monaco, R.: Choice experiments: an application for the corona verde landscape in Turin (Italy). In: Gervasi, O., et al. (eds.) Computational Science and Its Applications—ICCSA 2017. Lecture Notes in Computer Science, 10406, pp. 532–546. Springer, Cham (2017)
12. Haab, T.C., McConnell, K.E.: Valuing Environmental and Natural Resources: The Econometrics of Non-Market Valuation. Edward Elgar Publishing, Northampton (2002)
13. Frontier Economic, LCP, Sustainability First: Future potential for SDR in GB (2015)
14. McKenna, E., Thomson, M.: High-resolution stochastic integrated thermal–electrical domestic demand model. Appl. Energy **165**, 445–461 (2016)
15. Carmichael, R., Schofield, J., Woolf, M., Bilton, M., Ozaki, R., Strbac, G.: Residential consumer attitudes to time-varying pricing. "Low Carbon London" LCNF project (Imperial College London), Tech. Rep. A2 (2014)
16. KEMA: 2005 Smart Thermostat Program Impact Evaluation prepared for San Diego Gas and Electric Company, California, US (2006)

17. Parrish, B., Heptonstall, P., Gross, R.: How Much Can We Really Expect from Smart Consumers?. HubNet, Position Paper (2015)
18. Wegner, M.S., Hall, S., Hardy, J., Workman, M.: Valuing energy futures; a comparative analysis of value pools across UK energy system scenarios. Appl. Energy **206**, 815–828 (2017)
19. Hong, J., Kelly, N.J., Richardson, I., Thomson, M.: Assessing heat pumps as flexible load. Proc. Inst. Mech. Eng. Part A J. Power Energy **227**(1), 30–42 (2013)
20. Pagliuca, S., Lampropoulos, I., Bonicolini, M., Rawn, B., Gibescu, M., Kling, W.L.: Capacity assessment of residential demand response mechanisms. In: 46th International Universities' Power Engineering Conference (UPEC), pp. 1–6, Soest, Germany (2011)
21. UK Power Networks: Industrial and Commercial Demand Response for Outage Management and as an Alternative to Network Reinforcement. "Low Carbon London" LCNF project, Tech. Rep. A4 (2014)
22. Adamowicz, W., Boxall, P., Williams, M., Louviere, J.: Stated preferences approaches for measuring passive use values: choice experiments and contingent valuation. Am. J. Agr. Econ. **80**(1), 64–75 (1998)

# Chapter 53
# Sustainability Issues in Context of Indian Hill Towns

**Pushplata Garg and Harsimran Kaur**

**Abstract** Ensuring sustainability of urban centres, particularly in environmentally sensitive hill areas, is important for protecting the quality of their natural and built environment, as well as wellbeing of its people. Although the basic principles of sustainable urban development remain the same in all contexts, the type and extent of problems and issues of sustainability vary significantly due to the generic differences between plains and hills. This requires a deeper understanding of the intrinsic characteristics of their natural environment as well as built environment; and the impacts of one on the other through actual cases. This paper attempts to understand and identify the critical sustainability issues of Indian hill towns based on literature review and observational studies of selected hill towns of North India. Some of the issues of sustainability in hill towns in India include those of hill instability, fragile ecology, proneness to natural hazards, inaccessibility, presence of natural resources, and visual incompatibility of built forms. Proper understanding of these issues is essential for ensuring sustainable development of hill towns and their surrounding regions.

## 53.1 Introduction

Ensuring sustainable development has become a primary objective for protecting the natural environment as well as the wellbeing of people [1, 2], and the basic concepts, dimensions, and issues of sustainability have been reasonably researched and understood over the last few decades. However, due to greater adverse impact of unsustainable development on larger cities, the focus of research in the domain has been on these, most of which are located in plain areas, and not in all environmentally sensitive areas, such as Himalayas which has been recognized amongst the important eco-sensitive zones with fragile ecology, lower carrying capacities

P. Garg (✉) · H. Kaur
Department of Architecture and Planning, Indian Institute of Technology Roorkee, Roorkee, India
e-mail: pushpfap@iitr.ac.in; lncs@springer.com

H. Kaur
e-mail: hkaur@ar.iitr.ac.in



J. Littlewood et al. (eds.), *Sustainability in Energy and Buildings*, Smart Innovation, Systems and Technologies 163, https://doi.org/10.1007/978-981-32-9868-2_53

and threshold limits [3], facing several threats and challenges due to natural and anthropogenic activities. Mountain ecosystems play an important role in shaping the sustainable development of Indian Himalayan region by providing the rich environmental heritage/ecosystems in terms of goods and services to half of the global population [4]. Considering its uniqueness, the research and development efforts in this mountain system require special attention in terms of site planning, design, and development to ensure long-term sustainability of the region and "the fragility of mountain ecosystems makes the impacts of unsustainable development more severe and more difficult to correct than in other areas of the world" [3]. Promoting high quality research with adequate field data support has been identified as the need of the hour to find solutions for the prevailing environmental problems that the Indian Himalayan region (IHR) is facing. due to its remoteness and inaccessibility [4–7]. Understanding the issues of sustainability in Himalayan hill context, particularly in the urban areas that are specific to the context and different from those in larger cities in plains, has become crucial for ensuring their sustainable development.

Therefore, this paper attempts to understand and identify various problems and issues related to sustainability in Indian hill towns that lie in the lesser Himalayan region in context of their natural environment, resources and socio-cultural character to ensure the sustainable development. This paper presents a brief review of literature on sustainable development in general and in context of hill areas, understanding of the problems and issues in context of Indian hill towns based on specific case studies and identification of sustainability issues.

## 53.2 Literature Review

Sustainable development, defined as "meeting the needs of the present without compromising the ability of future generations to meet theirs", is presented as an integrated approach for addressing concerns about a range of environmental and socio-economic issues [8]. Environment, equity, and economics [9, 10] are the three main dimensions of sustainability, also known as the three main pillars of sustainability. Environmental pillar is the most recognized aspect of sustainability that aims at promoting and enhancing environmental quality (reducing air, water, and waste pollution), preventing the physical and visual deterioration/degradation of the environment and preserving ecology, whereas equity is the social aspect of sustainability that promotes community's health and welfare, education (accessibility to facilities and services) and empower communities to participate in the process to take action for health improvement and environment around them and focuses on "empowerment, accessibility, participation, sharing, cultural identity, and institutional stability". The economic pillar relates to growth, productivity, and development [11]. On considering these together, the three Es of sustainability represent a balanced approach, and if any pillar is weak, then the system as a whole is unsustainable.

Implications and influence of each of these dimensions of sustainability vary at different levels, that is, macro-level (national, regional, and global/international), meso-level (settlement), and micro-level (individual, household, building, site, area, and community), each having different, as shown in Table 53.1.

Further, measurable indicators are needed that can assess the impact of (urban) development on sustainability, and hence various urban sustainability assessment tools such as LEED-ND, BREEAM Communities, CASBEE-UD, GBI Township, IGBC Green Townships, GRIHA-LD have been developed [12]. However, none of these are specific to hill context, sustainability issues of which are significantly different than those of plains.

However, rapid urban population growth, industrialization, urban sprawl, poverty, housing shortage, water shortage, inadequate sanitation resulting in heaps of garbage and unhygienic conditions, pollution (air, water, soil or land, solid waste, noise), increase in greenhouse gases, urban heat island due to energy guzzlers, traffic congestion, and management (waste, energy, water supply) are major sustainability concerns of urban settlements located in plains; sensitive environmental context, precious natural resources, fragile ecology, lower threshold limits/carrying capacities, constraints on urban development, tourism and infrastructure, ensuring water security, deforestation, management of waste (solid, hazardous, water, pollution of air, land), forest management, creating attractive built environment, excessive tourism, infrastructure for increasing permanent and floating, population (housing and tourism), transportation are the important concerns in hill areas, making the sustainability issues in the two contexts different [13].

Prevalent site planning and development practices in hill settlements/areas are not sustainable and have the potential to adversely affect the mountain environment and development [4]. In spite of the generic differences between the settlements in hills and in plains, an approach towards the planning has largely been the same [13]. Also,

**Table 53.1** Implications and categories of dimensions of sustainability at different levels

| Dimensions of sustainability | LEVELS | | |
|---|---|---|---|
| | Macro (global, national) | Meso (community, sector/settlement) | Micro (Area, Site & Building) |
| Environmental | MDG's, UN-SDG's, emission targets, Global climate policy, Protocols & Agreements INDCs | Land use and Ecology, Climate & Energy, Water&Waste Management, Pollution Reduction | Energy, Water &Waste, Recycling & Reuse, Site & Landscaping, Materials and technologies |
| Equity/Social | | Transportation & Connectivity, Community Planning & Development, Housing, Infrastructure, Open Spaces | Occupant Comfort, Inclusive Environments, Access to Facilities, Participation & Control, Education, Health & Safety |
| Economic | | Growth potential, Efficiency / Rationality, Innovation | Local Economy, Efficiency of Use, Adaptability & Flexibility, Ongoing &Capital costs Innovation |

Note: environmental social economic applies to all dimensions planning and design of hill settlements has been largely neglected in India. Today, majority of the hill towns and cities are confronted with a number of environmental issues relating to rapid development on steeply sloping areas, inappropriate use of natural resources, insufficient water sources, natural drainage disturbances, pollution of streams and lakes, cutting of trees, overcoming hazards, land degradation, landslides, climate change impacts in highlands, lack of access to basic infrastructure and insufficient road system, which are crucially relevant to consider for sustainable development of hill towns [14–16].

Modern buildings constructed with non-local materials, construction techniques and design criteria, different from local and traditional practices are not only unsympathetic to the local landscape and architectural character, but are also resulting in massive levelling of hill slopes and cutting of trees. Increased instances of hill stability, soil erosion, disturbances in surface and sub-surface hydrology and changes in micro-climate manifested in frequent occurrences of landslides, building failures, pollution and siltation of water bodies, water insecurity due to drying up of natural springs, loss of vegetation cover, increased temperature and lesser rainfall are a frequently occurring phenomenon along with increased construction activity in most of the hill towns adversely affecting the environmental quality [17].

Considering the importance of Indian Himalayan region (IHR), need for "mountain-specific research", to build an integrated approach keeping all the disciplines in centre stage [7], focusing on a context-specific area "horizontal (sectors and sections of society) and vertical (spatial scale of village, block and district) integration in microplanning process", and addressing issues of theoretical and practical significance [7, 15] have been highlighted by number of researchers. The Prime Minister of India also expressed the "need for a fresh analysis of the problems of the hill states and hill areas of the country in a manner that these areas do not suffer in any way on account of their peculiarities" [5] and the National Mission for Sustaining the Himalayan Ecosystem (NMSHE) was launched by Government of India. Though a considerable research has been done separately on specific aspects of hill areas/settlements in different disciplines, very little work has been done towards understanding specific issues of sustainability in hill towns, particularly those based on study of specific hill towns in India [17], and towards an integrated mountain environment and development [5, 6].

## 53.3 Methodology

This paper is an exploratory research using primary and secondary sources of information and their analysis. Secondary sources included research papers/articles, reports, books and websites on urban sustainability in general and in the context of hill areas. Primary studies included on-site observations and surveys, interviews, and discussions with domain experts and residents on problems in selected hill towns, namely New Tehri, Nainital, and Almora—three small towns located in the state of

Uttarakhand in India. Based on the above the sustainability issues of hill towns have been identified.

## 53.4 Results and Discussions

Based on review of literature and studies on problems and issues of environmental degradation in selected hill towns, it is observed that problems and issues of sustainability in hill towns vary considerably because of their peculiar setting and conditions (natural environment) and anthropogenic activities. Accordingly, sustainability issues of Indian hill towns, identified as gap in knowledge, have been identified. These include issues related to slope instability; disturbance to natural drainage patterns; conservation of natural resources; accessibility; housing, infrastructure and community facilities; visual incompatibility of built forms; tourism; and heritage.

### *53.4.1 Slope Instability*

Hill instability is the most important problem affecting sustainability of Himalayan hill towns like Nainital, where the presence of unplanned civic structures, different types of discontinuities (e.g. joints and faults), and neotectonic activity in the area are the main reasons which affect the stability of the town [18]. Stability of hill slopes is affected by three interrelated natural factors—the intrinsic natural characteristics geologic structure and lithology (soil/rock properties), hydrology (water), and the area gradient [19]; external natural factors such as cloudbursts, heavy rainfall, and seismic activities; and anthropogenic activities such as unplanned/inappropriately planned development, location and siting of buildings and roads, inefficient drainage, removal of protective natural vegetative cover, and loss of wildlife habitat that contribute to surface run-off and accelerated soil erosion, slope failure, landslides, unsuitable site development practices requiring excessive cutting and filling of slopes for the construction of roads that lead to steepening of slopes, which in turn adversely affect the stability of the hills.

### *53.4.2 Disturbance in Natural Drainage Pattern*

Drainage plays a crucial role in stability of hill slopes and environmental degradation. Natural drainage patterns in many hill towns have changed due to unplanned/unregulated development through the indiscriminate cutting of hills and buildings built across drainage courses, leading to multiple problems in the form of flash floods (due to cloud-bursts), water logging, landslides/subsidence of slopes/creep movements/building failures, soil erosion (due to increased run-off), and poor water quality. Prevalent unsuitable planning and building practices, as observed in most of the hill towns studied, involve excessive slope cutting generating large amounts of debris, adversely affecting stability of slopes; the increased land surface exposure aggravates the process of surface erosion by many folds; and the eroded sediments deposited in the downstream drainage system cause choking of drains when dumped in/near drains, in turn leading to landslides and pollution and siltation of water bodies downstream. Further, the retaining walls constructed (as remedial measure) without sufficient weeping holes also adversely affect natural drainage in the area. Inappropriate site development and construction practices also result in increased area of impervious surfaces that decrease infiltration of water and increase run-off due to reduction of soft/open areas [20]. Uncontrolled/unplanned construction activities during rainy season also obstruct the pattern of natural surface and sub-surface drainage by blocking the natural watercourses that affect the stability of the slopes.

### *53.4.3 Conservation of Natural Resources*

Conservation of the rich natural resources of the Himalayan hill region, including that of its settlements such as water sources (lakes, rivers, springs) [21], dense and quality vegetation, salubrious climate, and scenic beauty—all of which result in the ecological aesthetic and socio-economic significance of the area/s is important. However, the prevalent planning approach and building practices are resulting in depletion of/deterioration in quality of such resources, as manifested in siltation and drying up of lakes, rivers, and natural springs; loss of vegetation cover; increased temperature; lesser rainfall; and loss of unique landscape and built environment character [17].

### *53.4.4 Accessibility*

Inaccessibility is the most common issue of mountain areas due to steep slope, altitude, overall terrain conditions, and periodical seasonal hazards (e.g. landslides, flooding, snow, storms, etc.) [22], resulting in less economic development of the regions as well as causing inconvenience within the towns. The natural hill terrain leads to linear pattern of development (road, land use) and new residential areas being developed away from the central area, resulting in longer distances, more efforts and time taken required to reach work centres, town centres, and community facilities. These areas also lack in availability of public transportation and its facilities. Adjacent to vehicular roads in hills are considered to be most suitable locations for constructing hotels, offices, shops, restaurants, and even houses. Commercial activities being carried out along the roads invariably spill over it, causing problems in the smooth flow of traffic/creating traffic bottlenecks, putting enormous pressure on road network. Owing to inadequacy of space and with areas around subjected to intense development, future widening of roads is not feasible.

### 53.4.5 Housing, Infrastructure, and Community Facilities

Increased demand for housing, infrastructure, and community facilities required to cater to the needs of growing resident population, and influx of tourists is exerting pressure on the existing housing and infrastructure facilities (water supply, sewage, solid waste collection, and disposal) [23]. Many hill towns in India are facing problems of shortage of water supply, poor water quality, and water pollution (especially during peak tourist season); inadequate sewage system, inefficient solid waste collection and disposal system due to absence of adequate litter bins; lack of public awareness; poor administration; and garbage can be seen littered all over in public areas, on roads and even on hill slopes. Further, no efforts are made to recycle it which causes unhealthy and unsightly conditions, visual pollution, and blocking of drains. Moreover, provision of infrastructure in hills is also costly as well as complicated due to scattered development, physiographic, and geologic constraints. Laying of sewer network requires excavations and the alternate construction of sewage disposal systems such septic tanks, soak pits (due to adverse effects of water seepage on hill stability) may cause saturation of the soil, resulting in geological problems such as landslides, ground slippage, or sinking hazard [24]. In spite of the extensive generation of electricity in the region, many towns face power shortage. Most of the towns lack recreational facilities (insufficient and small parks are there, that too without proper maintenance). Health facilities are adequate in some towns but emergency care is unsatisfactory. There is lack of accessibility of basic emergency vehicles—police, fire, and ambulance due to absence for roads for vehicular movement. Other community facilities and shopping areas, though adequate, are not of high quality in some cases like Rishikesh, Nainital, Almora. However, facilities for garbage collection, gas supply as well as vendors/hawkers in most of the residential areas are severely constrained. A large number of houses have insufficient exposure to sun, inadequate open spaces, and dampness (except in New Tehri), whereas most of the traditional/old houses have large front terraces that get enough sun and compensates for insufficient sunlight in inner rooms. Dampness is the main problem in buildings and is a cause of damage.

### 53.4.6 Visual Incompatibility of Built Forms

Visual aesthetic experiences and the image of hill towns are affected by its landform, topography, road pattern, development pattern, architectural character, and vegetation. Buildings built without proper visual aesthetic considerations have resulted in loss of beauty and character of the towns such as Shimla, Darjeeling, Nainital,

Mussoorie, New Tehri [25–29], where modern buildings constructed with non-local materials, construction techniques and design criteria, different from local and traditional practices are unsympathetic to the local landscape and architectural character, and the built environment is dominating the natural landscape character of hill towns. Buildings such as houses, hotels, offices, built over past few years, standout against the slopes due to their heavier building bulk and insensitive design, especially roofs and facades. Many hotel buildings and shops in New Tehri are built opening directly onto the roads and have no open space/setbacks in front, unlike buildings built by the Tehri Hydro Development Corporation (THDC). These buildings have incompatible form and character of building, which not only degrades the visual image of the town but also adversely affect the scenic views/vistas by obstructing views of scenic resources such as lake, surrounding valleys, and hill slopes from roads.

### *53.4.7 Tourism*

Himalayan region is known for various activities such as trekking, mountain climbing, sightseeing, winter sports, and most importantly, for the pilgrimage tourism. It includes centres of pilgrimage like Badrinath, Kedarnath, Yamunotri, and Gangotri and popular tourist destinations like Mussoorie, Rishikesh, Nainital, Almora, Ranikhet, and New Tehri. Owing to government's initiative to promote tourism and socio-economic development, hills are now witnessing enormous influx of tourists. However, most of the places lack adequate facilities of transport, accommodation, waste disposal, and other amenities for the ever-growing number of tourists and pilgrims that visit them every year. Also, there is a lack of regulatory mechanism for infrastructure creation, management for controlling the tourist inflow in such sites. As a result, not only their sensitive ecosystems and carrying capacities are being compromised upon, aesthetic appeal and cultural values of these places are also getting lost [30, 31].

### *53.4.8 Heritage*

Hill areas have large reservoir of valuable built and natural heritage in the form of forts/palaces, temple precincts/temple, sites of scenic beauty, bio-sphere reserves, lakes, meadows, dense forests, and wild life sanctuaries. Owing to extensive tourism, uncontrolled development of hill towns, and fires, number of heritage buildings have been tempered with and/or have been lost. Uncontrolled tourism has also resulted in causing irreparable damage to the heritage areas due to rampant construction around heritage buildings. Absence of any regulatory mechanism, such as appropriate model building bye-laws and development control regulations, has not only made the implementation of existing law tardy but also led to choking of the areas around heritage sites [32]. Unauthorized encroachments have destroyed important public spaces, available in front of or around these heritage sites and precincts. Unplanned and substandard development has contributed its share in destroying the valuable treasures of the areas/towns.

Most of the problems and issues discussed above are due to insufficient understanding of the issues of the hill contexts, specifically those affecting their sustainability. This requires a deeper understanding of the intrinsic characteristics of the natural environment, as well as the demands of the aspiring citizens for development and necessary built environment; the interdependence of natural and built environment; and the impacts of one on the other. The findings of the study are based on the observational studies of only few selected hill towns of North India, which could be further substantiated through detailed studies of more hill towns. Further, sustainability assessment tool/s can also be developed to measure the sustainability performance of hillside development.

## 53.5 Conclusion

Sustainability of hill towns is crucial not only for ensuring the quality of natural and built environmental and wellbeing of the community but also for ensuring the sustainability of the ecologically sensitive hill regions surrounding the settlements. The extent and significance of sustainability issues vary considerably between those in plains and hills. The main issues concerning sustainability of Indian hill towns relate to slope instability; disturbance to natural drainage patterns; conservation of natural resources; accessibility; housing, infrastructure and community facilities; visual incompatibility of built forms; tourism; and heritage. Proper understanding of these is essential for ensuring sustainable development in towns in environmentally sensitive hill areas.

**Acknowledgements** The authors would like to thank Indian Institute of Technology Roorkee (IITR), Uttarakhand, India and Ministry of Human Resource and Development (MHRD) for providing platform and grant in completing the research.

## References

1. UNCED: Agenda 21—World Summit on Sustainable Development (WSSD). Rio de Janerio, Brazil (1992) (ebook)
2. Bai, X., Nath, I., Capon, A., Hasan, N., Jaron, D.: Health and wellbeing in the changing urban environment: complex challenges, scientific responses, and the way forward. Curr. Opin. Environ. Sustain. **4**(4), 465–472 (2012)
3. Nilsson, C., Grelsson, G.: The fragility of ecosystems: a review. J. Appl. Ecol. **32**(4), 677–692 (1995)
4. Kumar, K.: Brief about Mountain Division (5th Unit of the Institute) (2019). http://gbpihed.gov.in/beta/Mountain_Division_detail.php. Last accessed 29 Feb 2019 (Blog)

5. Prime Minister's Task Force on Micro, Small and Medium Enterprises, Government of India. Released by Deputy Chairman, Planning Commission, India (2010). http://dcmsme.gov.in/Final_Report.pdf. Last accessed 29 May 2019 (Online)
6. National Mission for Sustaining the Himalayan Eco-System: Government of India Department of Science & Technology. Ministry of Science & Technology, New Delhi (2010)
7. Negi, G.C.S., Rawal, R.S., Sharma, S., Kumar, K., Dhyani, P.P.: Need for strengthening mountain specific research. Curr. Sci. **106**(5), 659–661 (2014)
8. World Commission on Environment and Development—Our common future. Oxford University Press, vol. 383, Oxford (1987) (Report)
9. Kirkby, J., O'Keefe, P., Timberlake, L.: The Earthscan Reader in Sustainable Development. Earthscan, London (1995)
10. Basiago, A.D.: Economic, social and environmental sustainability in development theory and urban planning practice. Environmentalist **19**, 145–161 (1999)
11. Khan, M.: Sustainable development: the key concepts, issues and implications. In: International Proceedings on Sustainable Development Research Conference, Sustainable Development, vol. 3, no. 2, pp. 63–69. Manchester, UK (1995)
12. Kaur, H., Garg, P.: Urban sustainability assessment tools: a review. J. Clean. Prod. **210**, 146–158 (2019)
13. Pushplata, Kumar A.: Building regulations: a means of ensuring sustainable development in hill towns. Asian J. Environ. Res. Dev. **7A**(1A), 553–560 (2012)
14. Mahat, T.J.: Issues in Sustainable Mountain Development: The Himalayan Experience. International Centre for Integrated Mountain Development (ICIMOD), Kathmandu, Nepal (2004)
15. Price, M.F., Jansky, L. F., Iatsenia: Key Issues for Mountain Areas. United Nations University Press, Tokyo, Japan (2006)
16. Dung-Gwoma, J., Jugub, A.: Characteristics and planning challenges of hilltop settlements in Jos Metropolis, Nigeria. Urban Plan. Landsc. Environ. Des. **2**(2), 129–149 (2017)
17. Pushplata: Urban design matrix for hill towns. Ph. D. thesis, Department of Architecture and Planning, Indian Institute of Technology, Roorkee, India (2000) (Unpublished)
18. Sah, N., Kumar, M., Upadhyay, R., Dutt, S.: Hill slope instability of Nainital City, Kumaun Lesser Himalaya, Uttarakhand, India. J. Rock Mech. Geotech. Eng. **10**(2), 280–289 (2018)
19. Sidle, R.C., Pearce, A.J., O'Loughlin, C.L.: Hillside Stability and Land Use, 2nd edn. America Geophysical Union, Water Resources Monograph, vol 11 (1986)
20. Hosseinzadeh, S.R.: The effects of urbanization on the natural drainage patterns and the increase of urban floods: case study Metropolis of Mashhad-Iran. Int. J. Sustain. Dev. Plan. **1**, 423–432 (2005)
21. Pushplata: Urban development and its impact on natural water springs in a hill town—a case study Almora. Spatio Econ. Dev. Rec. **13**(3), 25–34 (2006)
22. Pushplata and Anbalagan R.: Geo-environmental assessment for determining suitability of buildable land in context of a hill town. J. Eng. Geol. **14**, 391–399 (2008)
23. Pushplata: Planning and Design of housing in hill towns. In: 36th IAHS World Congress on Housing Sciences, Kolkata, India (2008)
24. American Society of Planning Officials: Hillside Development. Planning Advisory Service, Illinois, Chicago (1968). https://www.planning.org/pas/reports/report126.htm. Last accessed 21 Apr 2018 (Online)
25. Pushplata and Vishwamitter: Sustainable development of hill towns through integrated environmental conscious planning. In: International Conference on Civil Engineering for Sustainable Development, Roorkee (1997)
26. Kumar, A., Pushplata: City profile: Shimla. Cities **49**, 149–158 (2015)
27. Pushplata: Management of pollution and siltation of lakes in hill towns—case study Nainital. J. Ind. Build. Congr. **12**(1) (2005)
28. Dasgupta, S., Pushplata: Transformation of architectural character of Darjeeling. J. Ind. Inst. Arch. **81**(11), 39–44 (2015)
29. Monalisa: A methodology for character appraisal of hill towns. Ph. D. thesis, Department of Architecture and Planning, Indian Institute of Technology, Roorkee, India (2015) (Unpublished)

30. Pushplata and Vishwamitter: Promotion of tourism and conservation of hill towns. In: Conference on Methodology for Conservation of Heritage, C.B.R.I., Roorkee (2002)
31. Pushplata and Vishwamitter: Eco-synergism and planning for tourism in hill towns. In: International Conference on Creating Environments for Tourism, Las Vegas, USA (1994)
32. Kumar, A.: Area specific building regulations for hill towns. Ph. D. thesis, Department of Architecture and Planning, Indian Institute of Technology, Roorkee, India (2014) (Unpublished)

# Chapter 54
# Studies on Thermal Performance of Advanced Aerogel-Based Materials

**Jürgen Frick, Marina Stipetić, Oliver Mielich and Harald Garrecht**

**Abstract** The present work describes the development and assessment of aerogel-based brick fillings, as well as aerogel blow-in insulation with optimized thermal performance. The brick filling was developed within the EU Horizon 2020 project Wall-ACE. The focus of this work was the assessment of the brick filling developed by the industrial partners in the project to present a new class of bricks with outstanding thermal performance. Within the EU-project EFFESUS, an aerogel-based blow-in insulation was developed by an industrial partner and tested by several research partners. A feasibility study at a heritage building in Glasgow and large-scale laboratory tests were performed to assess the material with respect to thermal performance, applicability and removability.

## 54.1 Introduction

Traditional insulation materials reached their optimization limit of thermal performance. Therefore, new generations of materials are nowadays developed, for example, vacuum insulation panels (VIP), different composite materials. Silica-based aerogel is as well a very good thermal insulator. It is produced as granulate or blanket with thermal conductivity usually between 0.015 and 0.025 W/(m K).

Within the EU-project Wall-ACE,[1] granular aerogel was used to develop a mineral brick filling with low thermal conductivity. Thermal conductivity of developed aerogel-based brick filling must be lower than for standard brick filling materials, for example, mineral wood, perlite, polystyrene, wood wool (whose thermal conductivity is higher than 0.030 W/(m K)). Hollow bricks with standard brick filling can reach thermal conductivity always higher than 0.070 W/(m K). Further optimization

[1] EU-H2020 project Wall-ACE—Wall Insulation Novel Nanomaterials Efficient systems, Grant Agreement No. 723574, https://www.wall-ace.eu/.

J. Frick (✉) · M. Stipetić · O. Mielich · H. Garrecht
Materials Testing Institute, University of Stuttgart, Pfaffenwaldring 32, 70569 Stuttgart, Germany
e-mail: juergen.frick@mpa.uni-stuttgart.de

 
J. Littlewood et al. (eds.), *Sustainability in Energy and Buildings*, Smart Innovation, Systems and Technologies 163, https://doi.org/10.1007/978-981-32-9868-2_54

**Fig. 54.1** Left: section of hollow brick. Right: aerogel-based filling material

of thermal performance of hollow brick can be reached with developed aerogel-based brick filling.

The blow-in insulation developed in the EU-project EFFESUS[2] used aerogel blankets cut in small cubes to generate a blowable insulation like mineral wool blow-in insulation. This is the first time that this type of aerogel material was used as blow-in insulation. The advantage compared to other applications which used aerogel granulate as blow-in material [1] is the better resistance against setting.

The results of both projects were recently published [2, 3]; therefore the present publication gives an extended summary of the main results with focus on the assessment of thermal performance of the two materials.

## 54.2 Materials and Methods

### *54.2.1 Hollow Brick Filled with Aerogel-Based Filling Material*

Hollow brick filled with aerogel-based filling material was developed within project Wall-ACE from companies Leipfinger-Bader KG and quick-mix Gruppe GmbH & Co. KG. The development was mainly focused to improve thermal resistance of filled hollow bricks. This was basically reached by aerogel-based brick filling with lower thermal conductivity than filling materials nowadays used. Here the developed aerogel-based brick filling (see Fig. 54.1, right) is based on studies of insulation plasters (e.g. [4–6]) because of their similar material performance. The company quick-mix Gruppe GmbH & Co. KG developed four mixtures of filling with different percentage of aerogel (X, X + 5%, X + 13% and X + 23% of aerogel in mixture). Used aerogel particles were produced from Enersens [7]. Furthermore, geometry and

[2]EU-FP7 project EFFESUS—Energy Efficiency for EU Historic Districts' Sustainability, Grant Agreement No. 314678, http://www.effesus.eu/.

clay mixture of brick were improved from Leipfinger-Bader KG. The developed brick geometry (see Fig. 54.1, left) of hollow brick has large-sized holes and medium thick web in order to gain optimal combination of thermal conductivity and load-bearing capacity. Moreover, it is easier to fill bricks with pasty aerogel-based substance when they have larger holes.

Studies on the thermal performance of bricks were performed at Materials Testing Institute (MPA), University of Stuttgart. Thermal conductivity of clay and developed mixtures of aerogel-based filling was measured in laboratory. Measurements of thermal conductivity were performed on dry samples in two-plates-apparatus with guarded hot plate (see Fig. 54.2) according to EN 12664 [8] for clay and according to EN 12667 [9] for aerogel-based filling. Mean temperature of measurements was 10 °C and temperature difference ($\Delta T$) was 11 K. For each performed measurement in two-plates-apparatus with guarded hot plate, two samples of the same material with comparable thickness were used. Measurement uncertainty of used device is 2%. Longitudinal ribs of brick were cut and polished to 200 × 200 × 5–7 mm square specimens before measurement. Required sample thickness of measurement device is between 10 and 40 mm. For this reason, two layers of clay specimens were used for measurement in order to get a specimen with thickness higher than 10 mm.

**Fig. 54.2** Two-plates-apparatus with guarded hot plate for measurement of thermal conductivity

Thermal conductivity of aerogel-based filling was measured for samples shown in Fig. 54.1 (right). Aerogel-based filling has very low density and that specimen would be compressed by heavy equipment in measurement device. Therefore, specimens of aerogel-based filling were placed in a purpose-built frame for keeping constant thickness along sample and to avoid change in density of measured specimen.

Parametric studies on the thermal conductivity of the developed hollow brick were performed with 2D numerical simulation using the software BISCO from Physibel. The numerical simulation is based on EN ISO 6946 [10]. Here, the developed brick geometry was used. Emissivity of brick for complete parametric study was 0.9. Mean temperature for determination of thermal conductivity was set to 10 °C and temperature difference $\Delta T$ between outer layers was 20 K. Thermal conductivity of clay and filling of brick varied in parametric study. Higher values of thermal conductivity in parametric study are comparable with their commercially available materials. Thermal conductivity lower than 0.25 W/(m K) for clay and lower than 0.03 W/(m K) for filling of brick is beyond state-of-the-art. For comparison of filled and unfilled hollow bricks, a parametric study was additionally performed for bricks without filling. Here, air in the brick web was taken as still with thermal conductivity of 0.025 W/(m K). The main reason for parametric study was to show influence of both varied parameters to thermal conductivity of developed hollow brick. It can also be seen which combinations of clay and brick filling result in thermal conductivity beyond state-of-the-art for hollow bricks (lower than 0.070 W/(m K)).

### *54.2.2 Aerogel Blow-in Insulation*

The industrial partner A. Proctor Group Ltd. within EFFESUS has developed Spacefill®, a highly efficient insulation for use in historic buildings with solid wall construction and existing wall cavities. Spacefill uses this empty space in the masonry wall, so does not encroach on valuable room space. As it uses the existing installation methods, this is also relatively simple, with minimal redecoration required. This would be a major advantage to the homeowner. Spacefill is derived from aerogel insulation which has a class leading thermal conductivity, but is also breathable and water resistant—all properties which are ideal for cavity insulation. Ways in which to install this insulation using existing blowing in techniques and equipment have been researched.

The initial product are cubes cut from a blanket material (Spacetherm®) with a specified thermal conductivity of 0.015 W/(m K) [11]. The idea is to recycle cut rest material from other blanket applications for a new product.

The product was tested at the Materials Testing institute in Stuttgart (MPA) to assess the optimal density with the best thermal performance. A procedure according to EN 14064-1 [12] for loose-fill mineral wool products was followed for the tests because of the similarity of the products and use. The material was blown with a professional blowing machine (Neuhaus Dämmtechnik: Model ISO-Standard) to produce realistic material for the thermal performance testing according to EN 14064-

**Fig. 54.3** Blowing of the material at MPA

1, Annex C2.2 (see Fig. 54.3). Owing to the blowing process the cubes are fluffed (see Fig. 54.4).

The blown cubes are put in a measurement box (80 × 80 × 10 $cm^3$) with an inner volume of dimensions 50 × 50 × 10 $cm^3$. The density was defined by the known volume (25 l) and the weight of the material. The bottom and the top of the measurement box were covered with a thin foil to keep the material in place (see Fig. 54.4). As reference, the non-blown cubes as delivered were prepared in the same way. The samples were stored in climate 23 °C and 50% relative humidity until the measurement.

Thermal resistance (R) and thermal conductivity (λ-value) were measured according to EN 12667 [9] using two-plates-apparatus with guarded hot plate (Quade measurement). Several densities of the blown material and the non-blown reference were tested. For the measurement two samples with identical density were used. The conditions were as follows: mean temperature $T_{mean} = 10$ °C and temperature difference $\Delta T = 11$ K. The measurements were performed in indoor climate at around 23 °C and 40–60% relative humidity.

The uncertainty of the measurement according to EN 12677 [9] is 2%. Owing to the low numbers of samples, no uncertainty according to material fluctuations at the same density could be given. The blow process for the test material was performed once; therefore the tested material was as homogenous as possible.

**Fig. 54.4** Top: fluffed cubes after the blowing process. Left: detail. Right: card box with blown material. Down: preparation of the measurement samples. Left: weighing. Right: finish with foil

## 54.3 Results and Discussion

### *54.3.1 Hollow Brick Filled with Aerogel-Based Filling Material*

Measured thermal conductivity of clay is lower than 0.20 W/(m K). For four developed mixtures of aerogel-based brick filling (different percentage of aerogel), thermal conductivity varies between 0.023 and 0.031 W/(m K). Here, thermal conductivity was lower with higher percentage of aerogel in brick filling.

Results of the parametric studies on the thermal conductivity of the developed hollow bricks (numerical simulation) are shown in Table 54.1. The results of parametric study were grouped in thermal conductivity of hollow brick which are state-of-the-art ($\lambda \geq 0.070$ W/(m K)) and beyond state-of-the-art ($\lambda < 0.070$ W/(m K)). It can be seen that filling of hollow bricks has positive influence to thermal performance of bricks. The thermal conductivity of the developed hollow bricks filled with brick filling of thermal conductivity from 0.04 W/(m K) is 50% lower compared to the unfilled state.

**Table 54.1** Results of parametric study on thermal conductivity of hollow brick for varied thermal conductivity of clay and brick filling

| Thermal conductivity (filling of brick) [W/(m K)] | 0.04 | 0.035 | 0.03 | 0.025 | 0.022 | 0.02 | Air |
|---|---|---|---|---|---|---|---|
| Thermal conductivity (clay of brick) [W/(m K)] 0.35 | 0.073–0.084 (state-of-the-art) | | | 0.063–0.068 | | | 0.177 |
| 0.3 | 0.072–0.077 (state-of-the-art) | | 0.056–0.067 | | | | 0.167 |
| 0.25 | 0.070 (state-of-the-art) | 0.050–0.065 | | | | | 0.150 |

Reaching of thermal conductivity lower than 0.070 W/(m K) is only possible with fillings which have thermal conductivity lower than 0.035 W/(m K). The parametric study shows that the developed hollow bricks are beyond state-of-the-art based on experimental results of thermal conductivity for its developed components ($\lambda_{clay}$ = 0.20 W/(m K); $\lambda_{filling}$ = between 0.023 and 0.031 W/(m K)).

### 54.3.2 *Aerogel Blow-in Insulation*

The results of the thermal conductivity measurements are shown in Fig. 54.5. The blown test material showed a better thermal performance compared to the non-blown

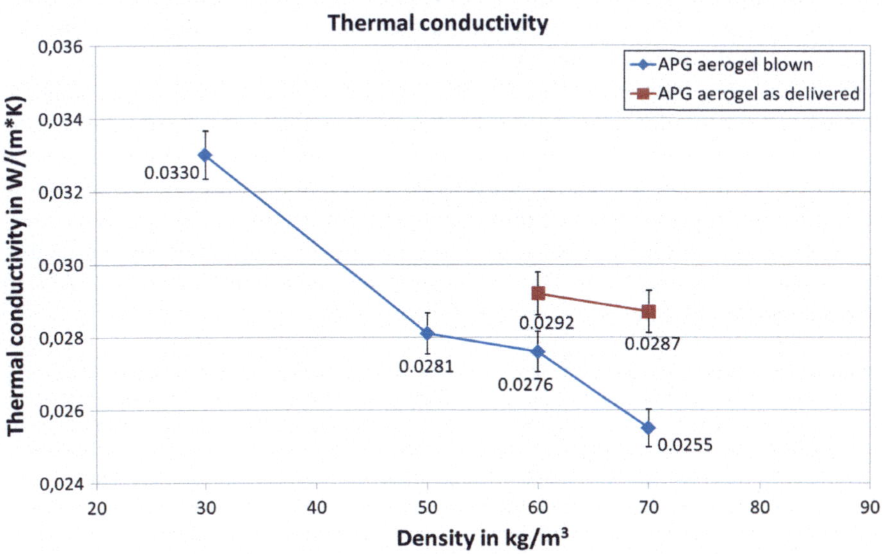


**Fig. 54.5** Thermal conductivity of blown and non-blown samples at different densities

material. The optimum performance was achieved at a density of 70 kg/m$^3$ with a thermal conductivity of $\lambda = 0.0255$ W/(m K). The non-blown reference material as delivered reached only $\lambda = 0.0287$ W/(m K). The thermal conductivity decreases with higher density for the blown test material than the non-blown reference material. In addition, a difference between the two materials occurred. The blown test material has a lower thermal conductivity compared to the non-blown reference material at similar densities. On the other hand, the specified value of the blanket material ($\lambda =$ 0.015 W/(m K) which was the base for the cubes is still lower.

The decrease of thermal conductivity with higher density could be explained by the better performance of the material compared to the surrounding air. This explains as well the difference between the blown test material with a better package due to the fluffing and the non-blown reference material. The optimal density which would probably correspond to the blanket material was not yet reached. It is questionable whether the blowing machine could generate such a high density in a cavity. Large-scale laboratory tests gave comparable results to the measured value [3]. Further assessment of water absorption and fire behavior was performed (see [3]).

## 54.4 Conclusion

Studies on the thermal conductivity of developed hollow bricks with various aerogel-based fillings were experimentally and numerically performed. Numerical simulation shows that bricks filled with developed aerogel-based filling achieve a thermal conductivity between 0.050 and 0.065 W/(m K) depending on the percentage of aerogel in the brick filling. The developed bricks with aerogel-based filling have a thermal performance beyond the state-of-the-art. Further studies on other material properties of the developed bricks are ongoing.

On product level, Spacefill®, as a blow-in insulation, is developed well enough to undergo certification for market entry. Owing to a missing harmonized standard, the way for certification and CE marking will be a European (ETA) or national technical assessment. Still some challenges must be tackled, like the dust production during the application process, which can be covered by a health and safety assessment within the certification. A set of test procedures for in-house and external quality control will be prepared in the form of a European assessment document (EAD) to guarantee the quality and performance of the product.

**Acknowledgements** The research leading to these results has been performed within the Wall-ACE project (https://www.wall-ace.eu/) and received funding from the European Community's Horizon 2020 Programme (EEB-01-2016: Highly efficient insulation materials with improved properties) under grant agreement no. 723574.
The EFFESUS (Energy Efficiency for EU Historic Districts' Sustainability) project (https://www.effesus.eu/) has received funding from the European Union's Seventh Programme for research, technological development and demonstration under grant agreement no. 314678.

## References

1. Energieeffizienz mit Kerndämmung, baublatt.ch. (2011) https://www.baublatt.ch/marktnotizen/energieeffizienz-mit-kerndaemmung. Last accessed 29 May 2019
2. Stipetic, M., Heizinger, V., Krupski, S., Zech, O., Özkan, H., Frick, J.: Experimental and numerical study on the performance of various filled hollow bricks. In: Proceedings of 13th Conference on Advanced Building Skins, Bern, Switzerland, pp. 513–518, 1–2 Oct 2018
3. Lucchi, E., Becherini, F., Concetta Di Tuccio, M., Troi, A., Frick, J., Roberti, F., Hermann, C., Fairnington, I., Mezzasalma, G., Pockelé, L., Bernardi, A.: Thermal performance evaluation and comfort assessment of advanced aerogel as blown-in insulation for historic buildings. Build. Environ. **122**, 258–268 (2017)
4. Buratti, C., Moretti, E., Belloni, E., Agosti, F.: Development of innovative aerogel based plasters: preliminary thermal and acoustic performance evaluation. Sustainability **6**, 5839–5852 (2014)
5. Ibrahim, M., Biwole, P.H., Achard, P., Wurtz, E., Ansart, G.: Building envelope with a new aerogel-based insulating rendering: experimental and numerical study, cost analysis, and thickness optimization. Appl. Energy **159**, 490–501 (2015)
6. Júlio, M., Soares, A., Ilharco, L.M., Flores-Colen, I., de Brito, J.: Aerogel-based renders with lightweight aggregates: correlation between molecular/pore structure and performance. Constr. Build. Mater. **124**, 485–495 (2016)
7. Enersens aerogel particles. http://enersens.fr/en/home/silica-aerogel-particles/. Last accessed 29 June 2018
8. EN 12664—thermal performance of building materials and products—determination of thermal resistance by means of guarded hot plate and heat flow meter methods—dry and moist products of medium and low thermal resistance (2001)
9. EN 12667—thermal performance of building materials and products—determination of thermal resistance by means of guarded hot plate and heat flow meter methods—products of high and medium thermal resistance (2001)
10. EN ISO 6946—building components and building elements—thermal resistance and thermal transmittance—calculation method (ISO 6946:2007) (2017)
11. SPACETHERM® Performance specification. https://www.proctorgroup.com/assets/PerformanceSpecs/SpacethermPerformanceSpecification.pdf. Last accessed 29 May 2019
12. EN 14064-1, Thermal insulation products for buildings. In-situ formed loose-fill mineral wool (MW) products. Part 1: Specification for the Loose-fill Products before Installation (2010)

# Chapter 55
# Design of an Adsorption Refrigeration Machine with an Auxiliary Heater for $CO_2$-Neutral Air-Conditioning of E-Vehicles

**Sebastian Haas, Stefan Weiherer and Michael S. J. Walter**

**Abstract** The automotive industry has to face major changes over the next few years. Alternative driving systems, like battery-powered vehicles, have to be developed to reach the $CO_2$ targets that the German government has specified. When the battery of an electric car is charged with regenerative energy, the $CO_2$ emissions while driving are zero. In addition to potentially zero emissions, the efficiency of electric cars is impressively high. However, there are still some major problems with electric cars. The charging times are relatively high, the driving range is too short and, moreover, the range also decreases when the interior climate control system is used. The automotive industry works to overcome the range of problems by developing better batteries and more efficient drivetrains. The focus of our research lies more on the energy-intensive loads, such as the heating and cooling unit of electric cars. In the past an auxiliary heater, powered with bioethanol, was mounted in the trunk of a Renault ZOE R240. Bioethanol is used to keep the emissions of the car as low as possible as it is nearly or even complete $CO_2$ neutral. The mounted system was evaluated in test drives during cold seasons in Germany. The results showed clearly that the substitution of the on-board heating system leads to a significant range extension. This paper deals with the installation and application of this auxiliary heating system to the research car. Furthermore, the possible combination with an adsorption chiller (used for cabinet-cooling) is described.

## 55.1 Introduction

Electric-powered cars offer great advantages over conventional combustion-driven cars. Especially, the potentially zero emissions with a well-to-wheel view [1] and the high efficiency of the electric drivetrain [2] has to be mentioned. Despite the advantages, electric cars have some disadvantages, such as limited range and charging possibilities and long charging times. Various studies confirmed that the range of electric cars is highly affected by the usage of heating and cooling unit by the

S. Haas (✉) · S. Weiherer · M. S. J. Walter
University of Applied Sciences Ansbach, Residenzstraße 8, 91522 Ansbach, Germany
e-mail: Sebastianhaas@posteo.de



J. Littlewood et al. (eds.), *Sustainability in Energy and Buildings*, Smart Innovation, Systems and Technologies 163, https://doi.org/10.1007/978-981-32-9868-2_55

passengers. For example, Kim et al. showed that the usage of an electric heater in battery electric cars increases the energy consumption up to 30% [3]. Kiss et al. pointed out that the climate control of electric cars can reduce the vehicle efficiency and range by more than 50% [4]. Meyer et al. noted that not just the range is affected by energy-intensive heating and cooling systems but also safety [5]. Furthermore, Lohse et al. carried out that the energy consumption of a BEV in city-type driving at a temperature around −7 °C is twice as high as at temperature of around 22 °C [6]. The range is also affected by the battery temperature control [7]. Therefore, we focus on the substitution of the heating and cooling unit of the Renault ZOE R240 by using an auxiliary heating system and an adsorption refrigeration machine (to provide cooling). The auxiliary heating system is used to substitute the on-board heating unit of the car, while the adsorption refrigeration machine (powered by the hot air from the auxiliary heating system) is used to substitute the on-board cooling unit. A study by SEMPRINI et al. showed that the usage of an adsorption machine can be used to cool the interior air, using hot air at a temperature of around 300 °C [8].

The following paragraph shows how dramatically the range drops in different temperature areas. Under optimal conditions (outdoor temperature around 20 °C, adjusted ambient temperature 20 °C, ECO mode on [max. velocity 94 km/h, reduced heating and cooling of the passenger compartment]) the Renault ZOE R240 has a range of 170 km. Owing to the experience we made with the car, a realistic range between 120 km in summer (hot day, operating climate control) and 80 km in winter (cold day, operation on-board heating system) is possible (see Fig. 55.1). A clear need for action can be derived from this.

The high decrease of the range can easily be explained by the energy-intensive heating and cooling of the passenger compartment. The range loss in winter is even higher than it is in summer due to the fact that the battery capacity decreases at lower temperatures [7]. UMEZU et al. pointed out that the heating of the passenger

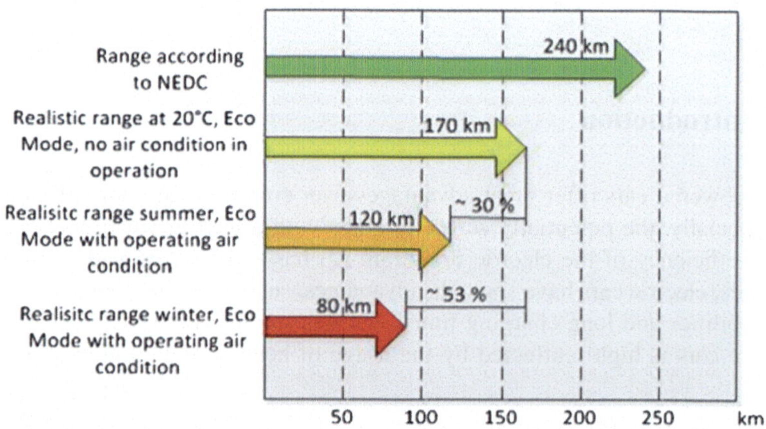


**Fig. 55.1** Driving range of Renault ZOE R240 depending on the weather conditions [18]

compartment reduces the range of a MITSUBISHI i-MiEV dramatically [9]. This study in combination with our own experiences indicates that there is a lot of potential for a range extension by substituting the air-conditioning system of an electric car.

Owing to these arguments, we installed an auxiliary heating system in our research car. The test of the system quantified the range extension through battery relief [10]. In this paper we will provide relevant insights on the installation and test of the heating system as well as discuss the challenges arising when adding an adsorption machine to provide cooling.

The first section of this paper covers the conceptual design and implementation of the auxiliary heating system, their details and selected problems which occurred during the installation. Section 55.2 gives an overview of the evaluated results of test drives made by the research team and also describes experiences made at a long-distance trip from Ansbach (Germany) to Seinäjoki (Finland). Section 55.3 shows the potential of a combination of the auxiliary heating system and the adsorption refrigeration machine.

## 55.2 State-of-the-Art

### *55.2.1 Current State-of-the-Art Air-Conditioning in Electric Vehicles*

Electric cars currently available on the market are sold mainly with commercial air-conditioning systems as they are used in combustion vehicles. These air-conditioning systems need electric energy to operate. Owing to insufficient heat losses of BEVs compared to internal combustion engines, a lot of energy is needed to provide hot air in winter [10]. This energy is taken from the battery and is therefore no longer available for driving operation. Currently, four heating and two cooling concepts are used for BEVs:

- Heating
  - Electrical high-voltage air heaters
  - Electrical high-voltage water heaters
  - Electrical low-voltage air heaters
  - Heat pump heaters.
- Cooling
  - Heat pump
  - Compression air-conditioner.

The advantage of high-voltage heaters is that they can be used to heat up air or water depending on the requirements. Normally, these heaters operate at a voltage of 400 V. High-voltage heaters are very energy-efficient even though they reduce the driving range of the vehicle up to 49% [11]. Low-voltage heaters do not need as much energy as high-voltage heaters but also have a lower heat output. Also heat pump heaters need electrical energy to operate, but they can also be used as cooling unit [12].

### *55.2.2 Alternative Air-Conditioning Technologies in Electric Vehicles*

In this section a short overview about regenerative technologies for heat and electricity generation in BEVs is given.

Fuel cells: Fuel cells do not emit $CO_2$ while operating and if the hydrogen is produced with regenerative energy the overall $CO_2$ emissions are zero. Furthermore, about 40% of the energy stored in the hydrogen is converted in waste heat, which can be used for heating the passenger compartment [13]. The electrical energy released during the conversion of hydrogen and oxygen into water can be used to run a cooling unit in summer.

Solar panels: To create auxiliary electrical energy, solar panels or foils can be used. The energy produced by the solar modules can be stored in a secondary battery and is usable for auxiliary loads like fans, light or area heaters. Especially, the use in summer time for a preconditioning is useful [14, 15].

Fuel-operated heaters (FOHs): Auxiliary heaters can be used to heat up air and water by using either diesel, gas or biofuels. With an auxiliary heater the heating section of the air-conditioning unit can be substituted. The low fuel and electric energy demand for the operation of these auxiliary heaters makes them suitable for electric cars. There are heaters available on the market which can be easily implemented into nearly any air-conditioning system. MIMURO et al. [16] and RIESS et al. [10] showed that FOHs are a possible solution for the heating of the passenger compartment without the usage of energy out of the battery. Also, manufacturers are working on the implementation of FOHs in electric vehicles. EBERSPÄCHER implemented a FOH-prototype, also used in our project, in a MITSUBISHI i-MiEV to substitute the on-board heating unit [16].

## 55.3 Research Question

As mentioned in Sect. 55.2 alternative air-conditioning systems seem to be a good possibility to increase the range of BEVs. Thus, two questions arise:

1. Is it possible to design and construct a combination of auxiliary heater and adsorption refrigeration machine which fits in commercial cars?
2. Will the cooling power of the system be high enough to create a comfortable temperature in the passenger compartment even on hot summer days?

Fig. 55.2 Renault ZOE (R240)

To answer these questions the work of the last months will be classified to determine future improvement suggestions for the auxiliary heating system. Furthermore, the potential of a combination of auxiliary heater and adsorption refrigeration machine is analyzed.

## 55.4 Research Vehicle: Renault ZOE (R240)

For practical testing and research, the University of Applied Sciences Ansbach owns a Renault ZOE (R240) which is a battery-powered vehicle (BEV) (see Fig. 55.2). In Table 55.1 all relevant technical details of the Renault ZOE are shown.

## 55.5 Installation and Occurring Problems

### *55.5.1 Installation and Application*

The centerpiece of the application is a diesel-operated auxiliary heater from EBERSPÄCHER GmbH which is based in Esslingen, Germany. The used auxiliary heater is named Airtronic D2 and has a maximum power output of 2.2 kW. Inside the heater

**Table 55.1** Technical details of the Renault ZOE (R240)

| Technical data | Renault ZOE R240 |
|---|---|
| Electric drivetrain | Three-phase synchronous generator |
| Cooling system | Air cooled |
| Battery | 22 kWh |
| NEDC range | 240 km |
| Power output | 65 kW (88 HP) peak |
| Max. torque | 220 Nm |
| Max. speed | 135 km/h |
| Energy consumption | 13.3 kWh/100 km |
| Air conditioning system | Heat pump |

diesel is burned to heat up the air taken in by an impeller. The hole system is a parking heater with only little modifications at its periphery. The diesel variant is used due to legal regulations in relation to fuel-tank systems in Germany. To run the system with bioethanol, only the software of the heater has to be overwritten and the fuel-tank has to be exchanged.

The whole auxiliary heating system, except the control unit and the hot air hoses, is mounted in/under the trunk of the Renault ZOE. The combustion air is taken in from under the car through a hole in the bottom plate. The exhaust gas is also led through the bottom plate to the outside and cleaned by an oxidizing catalytic converter and silencer. The air intake for the hot air is placed behind the back bumper so that no exhaust gas is sucked into the passenger compartment. In addition, fresh air from outside the vehicle is required to ensure that the air inside the vehicle does not become too dry. The hot air is led to the ventilation system of the Renault ZOE in insulated hoses with a diameter of 60 mm. The standard fuel-tank from Eberspächer is covered by a self-developed stainless-steel protection. This protection is needed to ensure that no fuel is phased out to the inside of the car. It is also used to ventilate the fuel-tank to get out gases which diffuse through the plastic (see Fig. 55.3).

The control unit is the so-called "EasyStart Timer" from EBERSPÄCHER. It is wired to the auxiliary heater and a temperature sensor which is mounted near the steering wheel. The control unit allows the driver of the car to operate the heater while driving.

### *55.5.2 Problems While Installation and Operation*

During the installation of the auxiliary heater some problems occurred and had to be overcome. One of the biggest problems was the available space in the Renault ZOE. Normally those kinds of heaters are mounted in the engine compartment of the car. This was not possible because there was no room for the heater and its peripheral

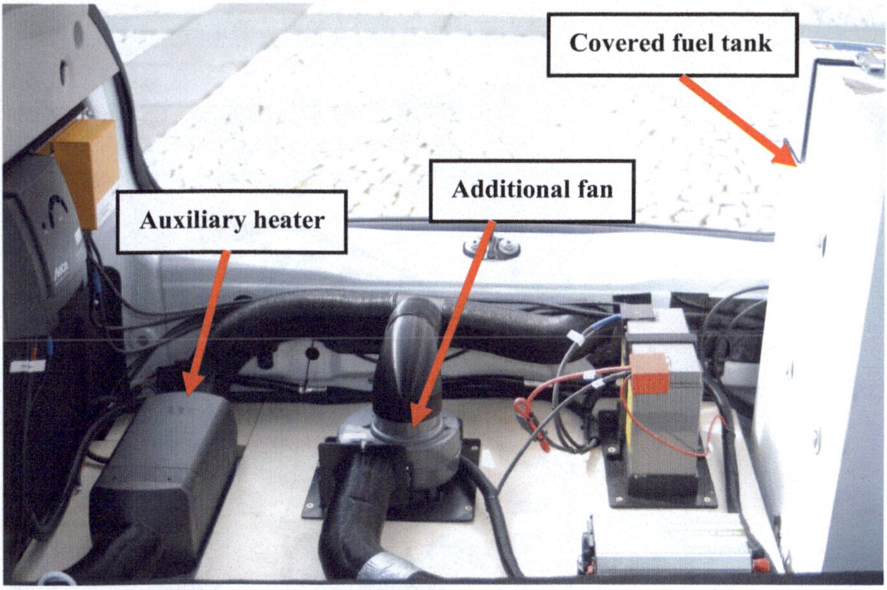


**Fig. 55.3** Auxiliary heater, covered fuel-tank and air hoses

components. Furthermore, the fuel-tank had to be covered due to legal regulations. This made an installation in the engine compartment impossible (see Fig. 55.4).

After the heater was installed the hot air hoses had to be laid to the ventilation system of the ZOE. Here another space problem made it nearly impossible to fit the hoses under the seats (see Fig. 55.5).

During the operation of the system, a lack of ventilation power was detected. This led to too little heating of the passenger compartment. To solve this problem an additional, solar-powered fan was installed (see Figs. 55.6 and 55.7). This fan rises the pressure from the back and ensures that enough hot air is getting to the front of the car.

After a long test run, several of the 3D-printed air ducts were molten (see Fig. 55.8). The problem occurred due to a backlog (backflow of hot air into the auxiliary heater) of hot air. This resulted in an emergency shutdown of the whole heating system and destruction of the air ducts. To solve this problem, the hot air outlets in the passenger compartment were relocated to make sure there is no backlog of hot air by blocking objects.

**Fig. 55.4** View into the engine compartment

## 55.6 Data Analysis and Personal Experiences

In this section the data analysis and also the personal experiences of the research team are described. The design of the experiments which led to a series of test drives in cold winter months is not part of this paper.

### *55.6.1 Data Analysis*

The recorded driving data are stored in two different .txt files on an Android tablet. These files are converted with an application for Microsoft Excel. The application is specially programmed for this task. The driving data are then sorted by categories (such as battery temperature, state of charge, current flow rate or velocity). In combination with the time stamps of the measurements, the data are converted into graphs [17].

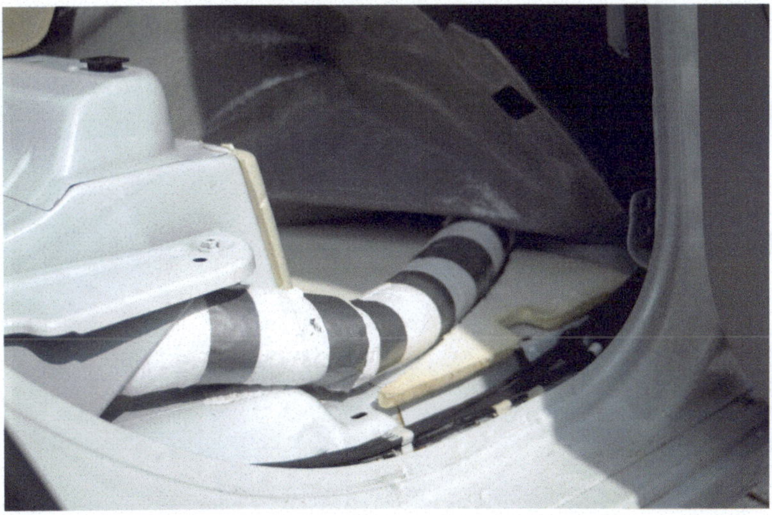

Fig. 55.5 Hot-air hoses from the trunk to the front co-passenger seat

Fig. 55.6 Auxiliary fan with air hoses and solar battery

**Fig. 55.7** Solar panel in the windshield (operated only when car is parked)

**Fig. 55.8** Molten air duct

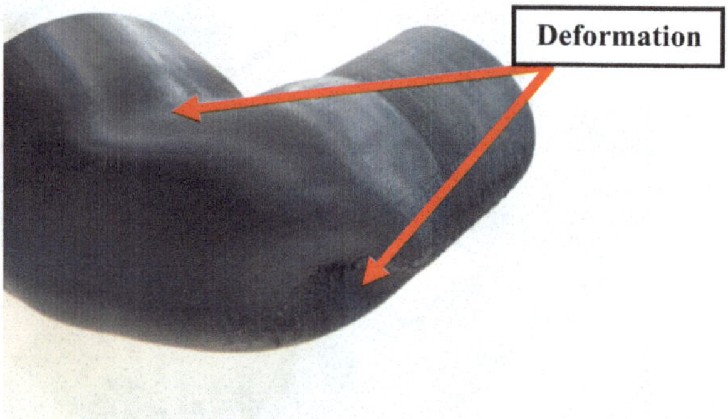


Before the test drives, a high energy drive (ECO-mode off, on-board heater on, interior temperature 27 °C) was made to create a reference (see Table 55.2). All following the test drives were compared with this high energy drive. The results of a test drive under harsh winter conditions (outside temperature –13 °C, adjusted interior temperature 22 °C) are shown in Fig. 55.9. The green/dotted graph shows a test drive in ECO-mode with the on-board heater substituted by the auxiliary heater. The red/dashed graph represents test drive in Eco-mode and the on-board heating unit set to a temperature of 22 °C with air taken from outside. It can be seen that at the end of the test drive with the auxiliary heating system (green/dotted graph) the state of charge (SOC) is around 10% higher than after the test drive with the on-board heating system running at the same outdoor starting temperature. This shows that

**Table 55.2** Comparison of test-driving setup

| Technical data | Economical mode | High energy test drive |
|---|---|---|
| Power train | ECO-driving mode | NORMAL-mode |
| On-board heating | OFF | ON |
| On-board fan | Level 2 | Level 2 |
| Auxiliary heater | ON | OFF |
| Driving speed | Max. 94. km/h | Max. 94. km/h |
| Air intake | Interior air | Outside air |
| Lights | ON | ON |

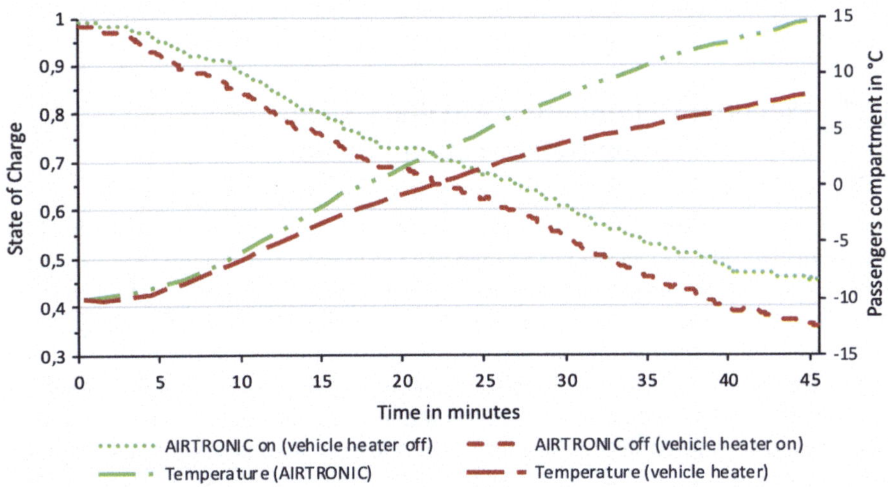


**Fig. 55.9** Comparison of two test drives (outside temperature –13 °C, SOC = 1 (battery fully charged), SOC = 0 (battery fully discharged))

the substitution of the on-board heating system leads to a higher range and a better comfort feeling. When the green/dotted graph is compared with the high energy drive mentioned before, the SOC increase rises to 25%. Also, the interior temperature is significantly higher after the test drive with the Eberspächer "Airtronic" on. Besides, this led to a better comfort feeling of the passengers (Fig. 55.9).

## 55.6.2 *Personal Experiences*

We made around 30 test drives (50 km each) in winter months to generate data about our installed auxiliary heating system and in addition to that a long-distance drive to Seinäjoki, Finland (round 3500 km). During this time, we were able to gain a lot of personal experiences concerning the usage of the Renault ZOE which will be described in the following:

- Comfort rating: In Fig. 55.9 it can be clearly seen that the ambient temperature after a driving time of around 45 min is 8 °C when the on-board heating system is used. This very low temperature leads to a bad comfort of the passengers.
- Range: Especially in the cold winter months the range drops dramatically when the on-board heater is used. This creates a bad feeling when it comes to long-distance drives and pressure for the passengers to not use the heating system.

Our personal experience was that many advantages of electric vehicles are particularly clouded by the poor comfort and the fear of long distances especially in the winter months.

## 55.7 Potential Combination Auxiliary Heater and Adsorption Refrigeration Machine

The results of the test drives showed that the application of an auxiliary heating system in electric cars can lead to a range extension in winter months. The effect on the range would be even greater when the car is used in a Nordic country like Norway or Finland where the average temperature is lower than in southern Germany.

To extend the functionality of the system, we are currently developing a climate system for battery electric vehicles (BEV) that is able to heat and cool the passenger compartment and the battery-pack. This would increase the range in winter and summer times, too. A possible solution for the cooling problem is a combination of the auxiliary heating system and an adsorption refrigeration machine. An experimental and numerical investigation of SEMPRINI et al. [8] showed that an adsorption refrigeration machine filled with zeolite can be used to produce cool air in a temperature range (12–20 °C) which is needed in the passenger compartment. Also, the hot air temperature for the process could be reached with the Eberspächer "Airtronic" if it is optimized for that purpose.

To create a lightweight, cheap and efficient cooling, respective heating system parts of the "Airtronic D2" have to be optimized. The built-in fan has to be changed to create more pressure and the housing of the heater has to be redesigned to fulfill the future tasks. Further, the complete adsorption refrigeration machine has to be scaled and designed on the basis a simulation of the system which is done by another employee at the University of Applied Sciences Ansbach. One of the major challenges will be to design a system that fits into a commercial electric vehicle without major conversions on the car itself. Furthermore, the sizing of the adsorber surface and the whole periphery needs to be done before the construction work.

Also, the fuel-tank system of the auxiliary heater needs to be optimized because it is not approvable for the use with (bio)ethanol. It is also not legal to mount a fuel-tank in the passenger compartment/trunk of car, like it was done in our project, so the whole tank has to be redesigned to fit in the engine compartment. Otherwise only a special approval for research cars is reachable.

This will be the next step toward a substitution of the heating/cooling unit of electric cars.

## References


1. Wang, M., Han, J., Dunn, J.B., Cai, H., Elgowainy, A.: Well-to-wheels energy use and greenhouse gas emissions of ethanol from corn, sugarcane and cellulosic biomass for US use. Environ. Res. Lett. **7**(4) (2012)
2. Manescu, V. (Paltanea), Paltanea, G., Horia, G., Scutaru, G., Peter, I.: High efficiency electrical motors state of the art and challenges. Rev. Roum. Sci. Techn. Électrotechn. et Énerg. **62**, 14–18 (2017), Bucarest
3. Kim, K., Lee, W.-S., Kim, Y.-Y.: Investigation of electric vehicle performance affected by cabin heating. J. Korea Acad. Ind. Coop. Soc. **14**(10), 4679–4684 (2013)
4. Kiss, T., Lustbader, J., Leighton, D.: Modeling of an electric vehicle thermal management system in MATLAB/Simulink. In: SAE 2015 World Congress & Exhibition (2015)
5. Meyer, N., Whittal, I., Christenson, M., Loiselle-Lapointe, A.: The Impact of Driving Cycle and Climate on Electrical Consumption & Range of Fully Electric Passenger Vehicles. p. 11
6. Lohse-Busch, M.D., Rask, E., Stutenberg, K., Gowri, V., Slezak, L., Anderson, D., et al.: Ambient temperature (20 °F, 72 °F and 95 °F) impact on fuel and energy consumption for several conventional vehicles, hybrid and plug-in hybrid electric vehicles and battery electric vehicle. In: SAE 2013 World Congress & Exhibition (2013)
7. Rugh, J.P., Pesaran, A., Smith, K.: Electric Vehicle Battery Thermal Issues and Thermal Management Techniques, p. 40
8. Semprini, S., Asenbeck, S., Kerskes, H., Drück, H.: Experimental and numerical investigations of an adsorption water-zeolite heat storage for refrigeration applications. Energy Procedia **135**, 513–521 (2017)
9. Umezu, K., Noyama, H.: Air-conditioning system for electric vehicles (i-MiEV). In: SAE Automotive Refrigerant & System Efficiency Symposium (2010)
10. Riess, C., Walter, M.S.J., Weiherer, S., Haas, T., Haas, S., Salceanu, A.: Heating an electric car with a biofuel operated heater during cold seasons—design, application and test. ACTA IMEKO **7**(4), 48–54 (2019)
11. Beetz, K., Kohle, U., Eberspach, G.: Beheizungskonzepte für Fahrzeuge mit Alternativen Antrieben. ATZ - Automob. Z. **112**(4), 246–249 (2010)
12. Peng, Q., Du, Q.: Progress in heat pump air conditioning systems for electric vehicles—a review. Energies **9**(4), 240 (2016)
13. Klell, M., Eichlseder, H., Trattner, A.: Wasserstoff in der Fahrzeugtechnik. Springer Fachmedien Wiesbaden, Wiesbaden (2018)
14. Großmann, H.: Pkw-Klimatisierung, 2nd edn. Springer, Berlin, Heidelberg (2013)
15. Haas, S.: Planung, Konstruktion und Einbau eines mit Solarenergie betriebenen Lüftungssystems zur Vorkonditionierung des Innenraums eines Elektrofahrzeugs. Bachelors thesis, University of Applied Sciences Ansbach (2018)
16. Mimuro, T., Takanashi, H.: Fuel operated heaters applied to electric vehicles. Int. J. Autom. Technol. **8** (2014)
17. Riess, C., Walter, M.S.J., Weiherer, S., Groper, M.: Evaluation and quantification of the range extension of battery powered electric vehicles in winter by using a separate powered heating unit. In: 2018 International Conference and Exposition on Electrical And Power Engineering (EPE), pp. 0075–0080, Iasi (2018)

18. Haas, T., Walter, M.S.J., Weiherer, S., Haas, S.: Increasing the driving range of electric vehicles using secondary energies—a review. In: Proceedings of the 22nd IMEKO TC4 International Symposium & 20th International Workshop on ADC Modelling and Testing, Iasi, pp. 325–330 (2017)

# Chapter 56
# Research into the Possibility of Achieving the NZEB Standard in Poland by 2021—Architect's Perspective

**Anna Bac**

**Abstract** The document presents studies on the level of awareness of Polish architects regarding the possibilities of improving the energy efficiency of buildings, the use of energy performance, and the achievement of the NZEB standard by 2021. The survey was conducted among 118 architects involved in the preparation of project documentation required by Polish law for newly built facilities. The article presents the background and methodology of the conducted research, including the design process and selected architectural and construction procedures conditioning the real improvement of the energy efficiency of the Polish construction industry. The results of surveys regarding architects' opinions on the practical application of energy performance as the basic tool for energy evaluation and their approach to the issues of efficient energy management in buildings were presented. These results were confronted with the current architectural and construction procedures in Poland. The conclusions were drawn up and the weaknesses of the latest regulations and routines were presented. The recommendations for their improvement were formulated, showing the urgent necessity of preparing and educating architects in order to effectively implement the EU climate and energy policy. The results of the research were presented during expert meetings at the Ministry of Investment and Economic Development (MIED) in November and December 2018, aimed at improving the legal regulations related to the implementation of the new Directive 2018/844/EU into the national law.

## 56.1 Introduction

The issue of improving energy efficiency is one of the key problems in the twenty-first century architecture. Legislative activities, as well as project practices and the development of investments in many countries, are focused more on sustainable energy management, both on an architectural and urban scale. The European Union's

A. Bac (✉)
Faculty of Architecture, Wroclaw University of Science and Technology, B. Prusa Str. 53/55, Wrocław, Poland
e-mail: anna.bac@pwr.edu.pl

 
J. Littlewood et al. (eds.), *Sustainability in Energy and Buildings*, Smart Innovation, Systems and Technologies 163, https://doi.org/10.1007/978-981-32-9868-2_56

policy clearly aims to achieve the increasingly ambitious goals described in the 2030 climate and energy framework. Currently, it assumes the reduction of greenhouse gas emissions by at least 40% compared to 1990 levels, achieving at least a 40% share of energy from renewable sources in total energy consumption in the EU, and improving energy efficiency by at least 32.5% [1]. One of the milestones of the European Parliament and European Council towards that objective is the new Energy Performance of Buildings Directive (EPBD) of 9 June 2018 [2] which will replace the binding Directive 2010/31/EU on the energy performance of buildings [3] and Directive 2012/27/EU on energy efficiency [4]. Member countries are obliged to implement it in the national law until March 2020, in the meantime carrying out preparatory works, like National Climate and Energy Plans (NECP). New Directive 2018/844/EU also encourages member countries to improve management and control systems for climate and energy strategies, as well as social consultations of energy objectives and their implementation [5].

As part of these activities, in Poland at the end of 2018, the Ministry of Investment and Economic Development (MIED), represented by the Department of Architecture, Construction, and Geodesy (DACG), organised a series of expert meetings to discuss the initial concept of changes in Polish law prepared by the government to implement the new, above-mentioned Directive 2018/844/EU. In the meetings involved were the representatives of professional organisations preparing energy performance certificates, and consultative institutions; representatives of higher education institutions dealing with environmental engineering, civil engineering and architecture, research centres dedicated to energy; and representatives of institutions directly and indirectly related to legislation in Poland, as well as representatives of industry and organisations involved in sustainable development. Almost 40 experts from all over Poland participated in four sessions.

During these meetings, several sets of issues were discussed, mainly regarding the current methodology for calculating energy performance, the appearance of energy performance certificates in terms of their social acceptance, the availability of certificate inventory, and issues of investment development included in the research presented in this article.

## 56.2 Background

The research described in the article was conducted for the above-mentioned expert meetings in MIED. The research was planned to express the point of view of Polish architects on the impact of energy conservation policies and legislation on current building design procedures. At the same time, the research was to show the approach and preparation of the Polish architects' community for the perspective of achieving the nearly-zero energy-building standard (NZEB) for all newly designed construction from 2021 onwards.

The author of the article is a practising architect specialised in sustainable and energy-saving housing and public buildings, as well as an architecture academic. For nearly 10 years, she has been observing design works and the process of investment development in Poland in terms of the ideas and principles of sustainability and the pursuit of low-energy life [6–8]. From her own observations and design experience, it appears that it is extremely easy not to adhere to Polish law on issues related to energy efficiency, and that current energy conservation policies and legislation procedures do not provide proper support for designers.

In Poland, architects are usually responsible for preparing the complete design documentation, coordination, and obtaining a building permit. The project is preceded by the investment planning and conceptual phase. In typical conditions, the starting point for design is the functional programme and an architectural concept design.

In Poland, the most important documents regulating the effective energy management in buildings are: 'Technical Requirements' (TR) [9] and 'Regulation of detailed scope and form of a building permit design' [10]. According to the methodology for determining the energy performance of a building and energy performance certificates [11], the energy performance of a building consists of two documents: the designed energy performance and energy performance certificate. The first is the only one tool required by Polish law, according to energy performance evaluation in buildings. The certificate is to be carried out after the completion of construction and submitted to the building supervision authorities. There are two more important acts: 'Construction Law' [12] and 'National plan aimed at increasing the number of buildings with low energy consumption' [13].

## 56.3 International Context

Various studies based on construction stakeholders' survey demonstrate very similar drivers and barriers faced in implementing building energy efficiency in different cultural and legislative contexts. The subject of the analysis is the impact of current energy conservation policies and legislation on building design practice and the role of an architect in energy conservation strategies in projects.

British research shows that there are several factors that can influence the implementation of energy legislation and policies in building the design process. These are categorised as: financial drivers and non-financial incentives, government and legislative drivers, education, communication, and training [14]. Research conducted in central and southern European countries shows that the key barriers in promoting sustainable communities are related to achieving the energy efficiency and renewable energy supply. These are categorised as: social barriers, economic barriers, market barriers, and training barriers [15]. In developing countries, behavioural and production barriers are being mentioned [16]. In general, high importance is attributed to changes in users' behaviour that result in increased energy savings in buildings [17].

Research describes how contemporary design strategies change in order to reach a high performance building by using various types of simulation tools. According to the current developments of building information modelling (BIM), the most effective decisions influencing future building performance are made at early design stages [18]. The latest European research confirms the well-known findings of Olgyay and Lechner from the 1960s and 1990s—the reduction of building energy needs proper architectural design. Research found that the energy-saving potential of architectural design decisions varies from 63 to 76%, depending on the climate. The maximum impact could be achieved by strategising the building elements, such as orientation, form, opening, sun shading devices, and materials [19].

The architect's role in energy efficiency is very important. However, there are differences in the methods used to achieve it, from BIM approach, through optimising simulation tools to the 'creative instinct' of architects [20]. For example, in developing countries, an architect's awareness of building envelope technologies for energy-efficient building applies to using LED lightning, window attachments, external overhangs, and photovoltaic roofs [21]. The architects are aware of other aspects of sustainable design, such as the environmental impact of building materials. Yet, they do not adopt green design approaches [22].

The U.S. Department of Energy's (USDOE) Building Energy Codes Program (BECP) and the American Institute of Architects (AIA) prepared 'Building Energy Codes Resource Guide: Commercial Buildings for Architects'. The aim of the guide is to promote codes that will provide for energy and environmental benefits and help foster adoption, implementation of, and compliance with those codes [23].

## 56.4 Research Methods and Data Collection

Based on the knowledge and practice in the preparation of project documentation and the knowledge of architectural and construction procedures, an initial set of issues for analysis was prepared. A series of conversations with other architects was carried out, which made it possible to identify key problems related to the implementation of the NZEB in Polish formal and legal realities. A survey of questions and answers was chosen as well as an open-ended question. Pilot studies were carried out on a group of 15 architects, and based on that, corrections were made to the survey. It was assumed that the approach of 'standard' and 'green' architects may be different; therefore, two groups of respondents were selected. This type of research is unique in Poland.

The survey was prepared as a Google Form [24] and was available online for 21 days. It consisted of a short introduction, 15 closed-ended questions (eight of which were multiple-choice and seven single-choice answers), and a respondent's background data. An 'open-ended answer' (marked in the table by *) was provided. The questionnaire covered two issues: current Polish energy conservation policies and legislation in the practice of building design and an architect's awareness and the willingness of applying energy conservation strategies in design practice.

Two organisations that associate practising architects were selected as target groups for conducting the survey. The first one is the Lower Silesian Regional Chamber of Polish Architects (LSRCPA) which associates nearly 1400 members.

Membership in the chamber is mandatory for every licensed architect in Poland. An invitation to complete the survey was sent to all members of the chamber, 99 people replied, which is 7% of members, including 95 licensed architects and 4 unlicensed architects. People who completed the questionnaire are mostly (63.6%) owners or co-owners of design offices performing the functions of leading architects or project managers, working in offices of up to 10 people.

The other institution is the Polish Green Building Council (PLGBC), associated with the World Green Building Council (WGBC), an international organisation working for green buildings, conducting trainings in green construction, and supporting the implementation of sustainable development goals. PLGBC has sent an invitation to complete the survey to about 550 members who are architects or have connections with architecture. A total of 23 completed surveys were obtained, which is less than 4% of the associates, including 14 licensed and 9 unlicensed architects. Most of them are leading architects (47.6%), project managers (28.6%), assistants (19.5%), and a person who is involved in energy modelling; generally—people working in design offices of up to 10 people (61.4%).

## 56.5 Survey Results

The level of response was poor—only ca. 5.5% of people belong to two professional architects' organisations. More questionnaires were filled in by members of the LSRCPA, although according to the research, their commitment to improving energy efficiency in Poland is smaller than that of the architects belonging to PLGBC. A smaller percentage of respondents from this more conscious group of architects can be connected to a fairly large number of emails, newsletters, or surveys, and other types of correspondence generated as part of the association's activities. Nevertheless, the presented research gives a picture of the situation in Poland and is very important in shaping future methods of improving energy efficiency.

Key survey questions with responses divided into two groups of respondents are shown in Table 56.1.

### *56.5.1 General Interpretation of the Results*

First of all, the results of the survey confirmed that there is a difference in the approach to design between architects affiliated and non-affiliated in the so-called 'green' organisation. Obviously, the former are better acquainted with the goals of sustainable development and familiar with the issues of energy efficiency. They have more awareness and stronger motivation to use formal and legal tools for energy optimisation of buildings. The most disturbing fact is that almost 60% of the respondents, both 'standard' and 'green', believe that Poland will not be able to achieve the NZEB quality for newly built objects by 2021.

**Table 56.1** Opinions expressed in surveys, *open-ended answers marked

| Subject and answer | Answers of LSRCPA members (%) | Answers of PLGBC members (%) |
|---|---|---|
| *Effectiveness of the designed energy performance in the process of improving the energy efficiency of buildings in Poland* | | |
| Ineffective | 39.3 | 33.3 |
| Partly effective | 47.5 | 61.9 |
| Effective | 5.1 | 0 |
| Redundant | 8.1 | 0 |
| I don't know | 0 | 4.8 |
| *The use of designed energy performance to improve the energy efficiency of a building in the design process* | | |
| Yes | 57.1 | 70 |
| No | 36.3 | 20 |
| I don't know | 1 | 5 |
| Sometimes* | 2 | 0 |
| Only as a necessary annex to the project* | 2 | 0 |
| No, because the characteristics doesn't help in that* | 1 | 5 |
| *The use of designed energy performance to improve the energy efficiency of a building at the stage of* | | |
| Functional programme | 19.5 | 27.8 |
| Preliminary architectural concept design | 35.1 | 50 |
| Interdisciplinary concept design | 40.3 | 61.1 |
| Building permit design | 71.4 | 66.7 |
| Executive design | 19.5 | 22.2 |
| Construction details | 19.5 | 11.1 |
| I don't know | 0 | 0 |
| *Obtaining recommendations for changes in the project to improve the results of designed energy performance* | | |
| Yes | 53 | 61.1 |
| No | 42.9 | 28.6 |
| I don't know | 1 | 4.7 |
| Independent preparation and revision of the project* | 2 | 5.6 |
| *The use of recommendations for changes in the project to improve the results of designed energy performance* | | |
| Yes | 48.6 | 61.1 |

(continued)

**Table 56.1** (continued)

| Subject and answer | Answers of LSRCPA members (%) | Answers of PLGBC members (%) |
|---|---|---|
| Partly | 31.1 | 22.2 |
| Only in building services | 8.1 | 5.6 |
| No | 16.2 | 11.1 |
| I don't know | 0 | 0 |
| *Designing according the principles of energy conservation is* | | |
| Important and I apply them in my design practice | 59.2 | 61.9 |
| It is not important and I apply them to a small extent in my design practice | 9.2 | 4.7 |
| Redundant | 4.3 | 4.7 |
| Unnecessary because I use the TR | 14.3 | 4.7 |
| I don't know | 0 | 0 |
| Important but depends on external factors* | 13 | 24 |
| *Ensuring efficient energy management in designed objects is provided through* | | |
| Devices | 63.2 | 65.2 |
| Object management | 12.2 | 26.1 |
| Avoiding heat bridges and other solutions in the field of building physics | 88.8 | 78.3 |
| Thermal insulation coefficient values of external partitions smaller than those indicated in TR | 73.5 | 73.9 |
| Architectural solutions, incl. zoning, orientation, size and parameters of glazing | 66.3 | 69.6 |
| Others | 0 | 8.6 |
| I don't know | 0 | 0 |
| *Achieving the NZEB standard for newly built facilities by 2021 in Poland* | | |
| Is likely | 25.2 | 26.1 |
| Is impossible to obtain | 59.6 | 56.6 |
| Is redundant | 8.1 | 4.3 |
| Is certain | 2 | 0 |
| I don't know | 3 | 0 |
| The construction market is not prepared for it* | 0 | 8.6 |

### 56.5.2 *Results*

The results of the conducted research show that only 60% of architects apply the rules of energy conservation in design practice. At the same time, almost the same number of professionals believes that achieving the NZEB standard by 2021 is impossible. The most commonly used solutions to reduce energy consumption in the range from 88.8 to 63.2% on average are: avoiding heat bridges and other solutions in the field of building physics, thermal insulation coefficients of external partitions less than those indicated in TR, and architectural solutions, among others zoning, orientation, size, and parameters of glazing; and the selection of energy-saving devices.

Almost 70% of the respondents calculate energy performance at the permit design stage, when all important design decisions were already made and energy efficiency is limited to improving the thermal insulation of building partitions, which is not related to designed architecture. It is worth noting that green architects use the designed energy performance as a tool for initial evaluation of the project at the stage of its early functional programme and at the stage of architectural and interdisciplinary concept design. As many as 36% of the surveyed architects state that the designed energy performance is not effective in improving the energy efficiency of buildings, but about 63% of designers use it to improve the energy balance of buildings. As mentioned above, this is usually done only at the stage of building permit design which is limited to juggling with U-value of partitions and glazing parameters, or the parameters of computable air flow stream for room ventilation. Over a half of the designers obtain recommendations for changes to the project from specialists elaborating the designed energy performance, of which 50% is used, but as many as 14% of the designers omit them.

**Issues of open-ended questions**. Many responses, from 13 to 24%, indicate that designing according to the principles of energy saving is important but it depends on external factors. The costs and the lack of investor's interest are mentioned. According to 8.6% of architects, ensuring efficient energy management in designed buildings is provided through renewable energy sources (RES). Also, 8.6% of architects believe that the construction market in Poland is not prepared for achieving the standard for newly built facilities by 2021. The following reasons were mentioned: the lack of effective tools for the verification of energy-saving design solutions, the lack of specialist training for designers and contractors, and the fact that contractors often do not implement energy-saving solutions provided by the project.

## 56.6 Discussion Based on Practice Experience

According to the author's practice experience, the basic problems of the design process, which are disclosed in Table 56.1, can be divided into key inter-related areas described as follows:

*Legally binding regulations in Poland.* The clauses in TR defining the project requirements are quite general. The formulas 'building should be designed and constructed in a way limiting the risk of overheating', 'should be designed', 'recommended airtightness', and so on [9] are used. Similarly, the regulation regarding the detailed scope and form of a building permit design contains general requirements that define 'rational use of energy', 'rational use of highly effective alternative systems', and so on [10], for which there is no design template and criteria for selecting optimal solutions, for example, $CO_2$ emission. There are also no requirements for the form of elaborating the designed energy performance that can range from half a page to four pages.

*Methods of control.* In Poland, the building authorities verify the compliance of a project and its realisation with the requirements of the law. However, they do not have the right to verify the design in terms of its accordance with regulations, for example, in the field of energy saving. They control only the site development plan.

Since March 2015, building control inspectors are no longer obliged to verify energy performance certificates, which allows for purely theoretical fulfilment of the requirements specified in TR. During the construction phase its participants, designers, investors, or contractors can therefore change material and even installation solutions without examining their consequences for the designed energy performance. Controls of certificates reported in the central register carried out by MIED are limited to inspecting project materials only, while the actual building parameters are not verified [25]. Energy performance certificates that, by law, belong to sales or rental documents for residential properties are in fact overlooked and occur only in 5% of the secondary market transactions [26]. The primary market for residential real estate looks better, as developers want to ensure the proper status for their investments. However, the certificates diverge from reality through fictitious RES installations that guarantee achieving the required primary energy ratio (EP).

Additionally, from January 2017 only architectural, construction, gas installation, and ventilation projects are required to obtain building permit, while projects of heating installations, air-conditioning, and electrics—directly related to energy efficiency—are not required by building authorities.

## 56.7 Conclusions and Implications

According to the presented research, the level of awareness of Polish architects regarding the possibilities of improving the energy efficiency of buildings, the use of designed energy performance, and the achievement of the NZEB standard by 2021 for newly built facilities, is quite limited. It probably results from the lack of any activities preparing the architects for the effective implementation of the EU climate and energy policy.

The comparison of the opinions of 'standard' and 'green' architects showed differences in the responsible approach to energy management in designed objects. The latter group shows a higher level of awareness and the need to use an energy performance tool to optimise buildings. Designed energy performance is an underestimated tool for initial energy optimisation of buildings due to the calculation methodology that arouses many doubts. Energy performance certificates do not fulfil their informative function; they do not trigger the desired market mechanisms and social preferences for energy-efficient buildings.

Educational activities should be undertaken immediately to raise the awareness among architects. A number of legislative measures should also be taken, such as changes in the methodology of calculating energy performance and more detailed formal and legal provisions, as well as changes in the occupancy permit issued by building control inspectors.

Future research may concern an action plan for design with sustainable energy management approach.

**Acknowledgements** The author would like to thank the LSRCPA and the PLGBC for the recommendation and invitation to complete surveys that made it possible to carry out the research described in the article. Also the author would like to express her gratitude to the architect Kajetan Sadowski for his valuable comments to the survey.

## References

1. https://ec.europa.eu/clima/policies/strategies/2030en. Last accessed 21 Apr 2019
2. https://eur-lex.europa.eu/legal-content/EN/TXT/PDF/?Uri=CELEX:32018L0844&from=EN. Last accessed 21 Apr 2019
3. https://eurlex.europa.eu/legalcontent/EN/TXT/PDF/?uri=CELEX:32010L0031&from=EN. Last accessed 21 Apr 2019
4. https://eurlex.europa.eu/legalcontent/EN/TXT/PDF/?uri=CELEX:32012L0027&from=EN. Last accessed 25 Apr 2019
5. https://ec.europa.eu/energy/en/topics/energy-efficiency/energy-performance-of-buildings. Last accessed 20 Apr 2019
6. Edwards, T.: Rough Guide to Sustainability, 3rd edn. RIBA Publishing, London (2010)
7. A Green Vitruvius—Principles and Practice of Sustainable Architectural Design, Eartscan (2008)
8. Bonenberg, W., Kaplinski, O.: The architect and the paradigms of sustainable development: a review of dilemmas. Sustainability **10**(1). https://www.mdpi.com/2071-1050/10/1/100 (2018). Last accessed 25 Mar 2019
9. http://prawo.sejm.gov.pl/isap.nsf/download.xsp/WDU20170002285/O/D20172285.pdf. Last accessed 14 Feb 2019
10. http://prawo.sejm.gov.pl/isap.nsf/download.xsp/WDU20180001935/O/D20181935.pdf. Last accessed 18 Feb 2019
11. http://prawo.sejm.gov.pl/isap.nsf/download.xsp/WDU20150000376/O/D20150376.pdf. Last accessed 18 Feb 2019
12. http://prawo.sejm.gov.pl/isap.nsf/download.xsp/WDU20170001332/U/D20171332Lj.pdf. Last accessed 14 Feb 2019
13. http://prawo.sejm.gov.pl/isap.nsf/download.xsp/WMP20150000614/O/M20150614.pdf. Last accessed 14 Feb 2019
14. Adeyeye, K., Osmani, M., Brown, C.: Energy conservation and building design: the environmental legislation push and pull factors. Struct. Surv. **25**(5), 375–390. https://doi.org/10.1108/02630800710838428 (2007). Last accessed 25 Mar 2019

15. González, M.J., Montero, E., Aguilar, F., Implementation of international energy engineering e-courses for promotion of sustainability buildings and communities. In: Central Europe towards Sustainable Building, CESB10, Education and Information, Prague (2010)
16. Addy, M.N., Adinyira, E., Koranteng, C., Architect's perception on the challenges of building energy efficiency in Ghana. Struct. Surv. **32**(5), 365–376, https://doi.org/10.1108/SS-03-2014-0014 (2014). Last accessed 20 Mar 2019
17. Fotopoulou, E., Zafeiropoulos, A., Terroso-Saenz F., et al.: Providing personalized energy management and awareness services for energy efficiency in smart buildings. Sensors **17**(9), 2054, Basel (2017)
18. Hamedani, M.N., Hamedani, M.N.: Smith, R, Evaluation of performance modelling: optimizing simulation tools to stages of architectural design. Procedia Eng. **118**, 774–780 (2015)
19. Naboni, E., Malcangi, A., Zhang, Y., Barzon, F.: Defining the energy saving potential of architectural design. Energy Procedia **83**, 140–146 (2015)
20. Ganiyu, S.A., Olufemi, A.: Energy conservation in the built environment: the roles of architects, In: Proceeding of the 1st Annual Conference on Mineral Exploration, Exploitation and Sustainable Environmental Development, School of Earth and Mineral Sciences, Federal University of Technology, Akure, pp. 358–365 (2015)
21. Akinola, A., Adeboye, A., Oluwatayo, A., Alagbe O.: Survey dataset on architect's awareness and adoption of building envelope technologies for energy efficient housing in Lagos State, Data in Brief. https://doi.org/10.1016/j.dib.2018.06.093i (2018). Last accessed 20 Feb 2019
22. Ofori, G., Kien, H.L.: Translating Singapore architects' environmental awareness into decision making. Build. Res. Inf. **32**(2004/1), 27–37 (2010)
23. Building Energy Codes Program, American Institute of Architects, Building Energy Codes Resource Guide: Commercial Buildings for Architects (2011)
24. https://docs.google.com/forms/u/0/. Last accessed 20 Jan 2019
25. https://www.data.gov.pl/dataset/591. Last accessed 11 Apr 2019
26. Sadowski, K.: Private Archive Containing Statistics of the Calculated Energy Performance, Wroclaw (2019)

# Chapter 57
# Experimental Analysis of the Hygrothermal Performance of New Aerogel-Based Insulating Building Materials in Real Weather Conditions: Full-Scale Application Study

**Timea Béjat and Didier Therme**

**Abstract** Aerogel-based building materials represent a promising new direction to achieve remarkable energy performance by applying new layers on existing or newly built walls. Expectations of thermal performance for new aerogel materials are necessarily high. Nevertheless, it is hard to match very promising laboratory thermal performance with full-scale, real weather exposed wall performance. In this paper, a detailed experimental analysis shows the application of ISO 9869 standard method to a newly built low U-value wall and discusses its possible extension in order to reduce the measurement time necessary to obtain stabilized values. Moisture content measurements complete the heat flux data on both sides of the wall. As expected, the moisture content significantly affects the thermal performance, even several months after the construction of the new wall. Therefore, the stabilized U-value had not been reached after 6 months of measurement. At the end of the paper, some recommendations highlight the importance of the drying process.

## 57.1 Introduction

By the end of 2020, Europe aims to reach the *near-zero energy* consumption level for new buildings as fixed by the Directive on the energy performance of buildings [1]. This energy performance level suggests developing more efficient building materials to reach this challenging objective. One way to do so is by developing aerogel-based building materials. Within the European H2020 project, WALL-ACE [2], several building products have been developed for new and retrofit applications.

Aerogel-based renders and plasters reach very high thermal performance according to laboratory measurements [3–5]. However, their in situ performance remains lower than those values obtained under controlled conditions [6]. This 'energy-performance gap' between designed and built structures is an issue the building

T. Béjat (✉) · D. Therme
Univ Grenoble Alpes, CEA, LITEN, DTS, LIPV, INES, 38000 Grenoble, France
e-mail: timea.bejat@cea.fr

 
J. Littlewood et al. (eds.), *Sustainability in Energy and Buildings*, Smart Innovation, Systems and Technologies 163, https://doi.org/10.1007/978-981-32-9868-2_57

sector must solve, in order to estimate a building's operating costs efficiently. It depends not only on the behavior of building's occupants but also on its envelope characteristics [7].

The assessment of this gap, which can reach 26–30% for new thermal insulating plasters [8], starts by applying standardized methods to determine the in situ performance of built walls. ISO 6946 [9] supplies a simple method to calculate the R value of known wall structures. ISO 9869 [10] supplies the average method to obtain a wall's U-value, its thermal transmittance, by measuring the temperature difference on both sides of the structure and the thermal heat flux on the inside, as the indoor air temperature variation is expected to be lower than the outside one. This experimental set-up is easy to install and gives reliable performance data; however, a gap between designed and in situ measured data remains present as it was reported by several studies [11–13]. We propose an extension to this method as suggested by [14] by using outside heat flux measurements to determine the final U-value of studied wall. This added sensor helps reducing the convergence time between outside and inside thermal transmittance values. Nevertheless, the used outside heat flux sensor should be protected from direct solar radiation. At the location of presented measurements, the greatest quantities of rain fall on north-oriented facades. For these two reasons, we chose to study north-oriented walls only.

We present the moisture performance of the same wall structure, which has a coupled effect on the heat transfer within the wall, thus directly affecting its thermal performance.

## 57.2 Experimental Set-Up

Within the WALL-ACE project, several full-scale walls were built at CEA LITEN's INCAS experimental platform in France (Le Bourget du Lac). In FACT (façade testing facility) two north facing walls were built on the ground floor (Fig. 57.1). In test cell N°1 the outside render was added to a classic brick wall with plasterboard inside rendering. Within the WALL-ACE project, other aerogel-based building products were installed too at INCAS platform, but this paper presents only the performance of the outside render, thus the wall is installed in test cell N°1. Figure 57.2 presents the wall's structure and the position of sensors.

Figure 57.3 presents the sensors installed between the brick wall and the outside aerogel-based render before the render was applied at point B.

Heat flux and temperature sensors were installed on the wall:

- Hukseflux HFP01 plate heat flux sensors [15] were used at three different heights of the wall. Their thickness is 5.0 mm and their diameter is 80.0 mm, as presented in Fig. 57.3. These sensors have a range of ±2000 W/m$^2$ with an accuracy of ±5% and their sensitivity lies within 60–61 μV/(W/m$^2$).
- Inside and outside temperatures were recorded by first class T type thermocouples, with accuracy of ±0.5 °C, measuring within the interval of −25 to 100 °C [16].

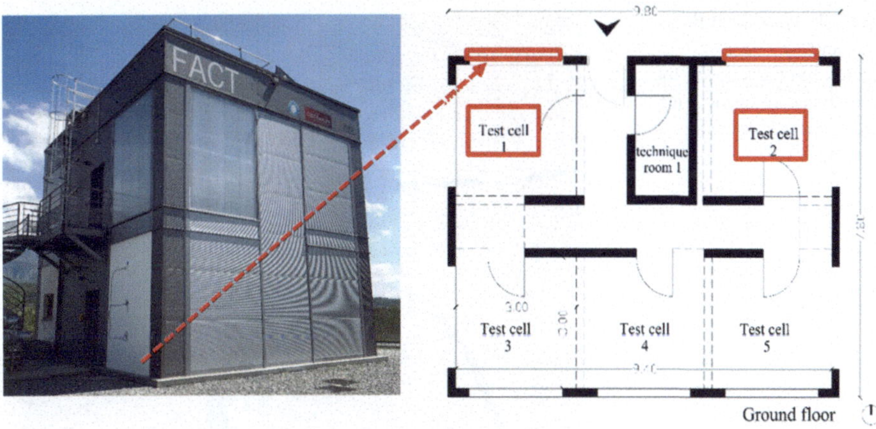


**Fig. 57.1** FACT with the two test cells on the north façade, CEA LITEN's INCAS experimental platform, France

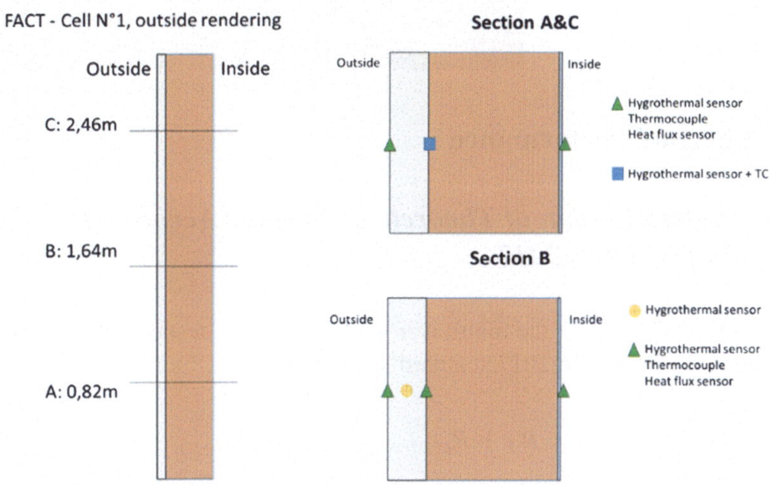


**Fig. 57.2** Sensor position in test cell N°1 wall (outside render)

- Data recording for both types of sensors was carried out by an Agilent data logger (34980A) [17].

The presented data cover the measurement period from October 2018 to mid-May 2019 in a pre-Alpes climate. Weather data is recorded as well as diffuse solar radiation on the north facing wall, but not presented here because of lack of space. The brick wall was built during spring 2018 and the first layer of the outside render was applied in July 2018. We considered the drying process of the total wall structure in the fourth chapter of this paper.

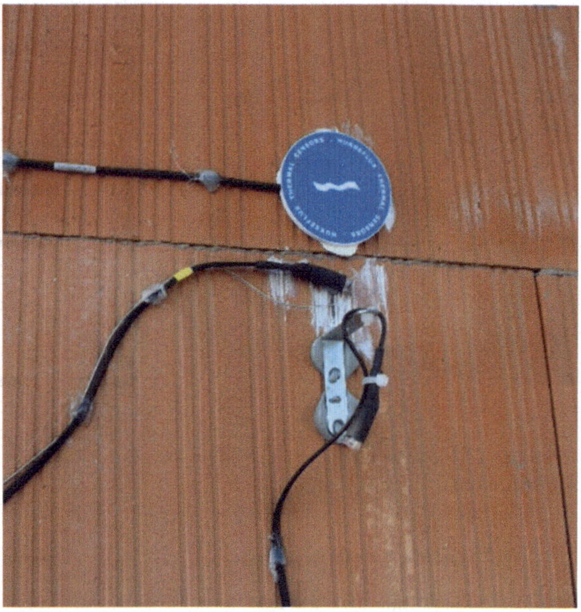

**Fig. 57.3** Used heat flux meter and thermos-hygrometer on the outside surface of the wall (before outside render's installation)

## 57.3 Thermal Performance

### *57.3.1 Determination of Theoretical U-Value According to ISO 6946:2017*

The theoretical U-value of the multilayer wall structure of test cell N°1 is obtained according to the ISO 6946:2017 standard's expression:

$$R_{tot} = \frac{1}{U} = R_{si} + R_{plasterb} + R_{brick} + R_{render} + R_{se} \tag{57.1}$$

where $R_i$ is the thermal resistance of each layer as designed and $R_{si}$ and $R_{se}$ are the interior and exterior superficial resistances, respectively $R_{si} = 0.13$ and $R_{se} = 0.04$ [(m$^2$K)/W] for horizontal heat flux which is the case here [9]. Table 57.1 presents the wall's composition with theoretical thermal performance values.

### *57.3.2 Determination of the In Situ U-Value of the Wall*

As a first step, we use ISO 9869-1:2014 to assess the in situ U-value of the studied wall. The standard method is called the heat flow meter (HFM) method, which proposes two possibilities for the data analysis: the average or the dynamic method.

**Table 57.1** Wall characteristics at test cell N°1 (brick wall with outside render)

| Layer | Thickness (cm) | Thermal conductivity (W/m K) | Density (kg/m$^3$) | Specific heat (kJ/kg K) | R (m$^2$ K/W) | Theoretical U-value (W/m$^2$ K) |
|---|---|---|---|---|---|---|
| Outside aerogel-based render | 10 | 0.04 | na | na | 2.5 | 0.365 |
| Brick | 20 | 1 | 1380 | 0.8 | 0.2 | |
| Plasterboard | 1.3 | 0.325 | 11.3 | 1.54 | 0.04 | |

Here we limit our application to the average method, with the plan of comparing these results with the results obtained using the dynamic method, before the end of the WALL-ACE project.

The installation followed the recommendation of the standard: we avoided the boundaries of the wall and ensured good surface contact by adding plaster between the surface and the sensor to fix it on. It is recommended that sensors be covered by a material that has the same thermal emissivity as the rest of the wall, but this was not done in this application, as on the IR thermography image the surface temperature difference remained within an acceptable range.

We obtained the in situ R-values as follows by measuring surface temperatures of the wall:

$$R_{wall} = \frac{\sum_{i=1}^{j} (T_{iint} - T_{iext})}{\sum_{i=1}^{j} q_{int}} \quad (\mathrm{m^2\,K/W}) \tag{57.2}$$

The U-value is as follows, which takes into consideration the surface convective heat exchange:

$$U = \frac{1}{R_{tot}} \quad (\mathrm{W/m^2\,K}) \tag{57.3}$$

To compare the convergence of our results with those of [14], we decided to obtain U-values of the wall by using an outside heat flux meter instead of that installed on the inside. For this reason, a second data analysis according to Eq. (57.3) was carried out.

$$R'_{wall} = \frac{\sum_{i=1}^{j} (T_{iint} - T_{iext})}{\sum_{i=1}^{j} q_{ext}} \quad (\mathrm{m^2\,K/W}) \tag{4}$$

Measurements started on October 2018 and planned to finish in October 2019.

The standard summarizes the main criteria and conditions to stop taking measurements as follows:

- The minimum test duration is 72 h if the temperature is stable around the HFM (without interruption);
- Rc values variation obtained over the last two days do not exceed 5%.

As the last criteria have not been satisfied, measurements continue until the temperature difference between the outside and inside stay over 10°C as recommended by [19]. It can be induced by fixing a very high indoor temperature limit in the test room to create a significant temperature gradient within the wall.

Hereafter we present a one-week measurement period in February 2019 with rather stable temperature variations outside.

As Fig. 57.4 showed, in situ U-values, determined by using indoor or outdoor heat flux data, are not identical. The curves tend to converge but the stabilized values have not yet been reached. Stahl et al. [19] recommend several years of long measurement campaign which is not possible here within a European project. A clear difference between the bottom and the top part of the wall is visible, with section C giving U-values closer to the theoretical one with indoor heat flux measurements. This is due to the temperature difference observed in the test room. As Fig. 57.5 shows, a temperature difference of about 3° is present between point A and C inside the room. To assess its impact on the drying process, we present moisture performance as follows.

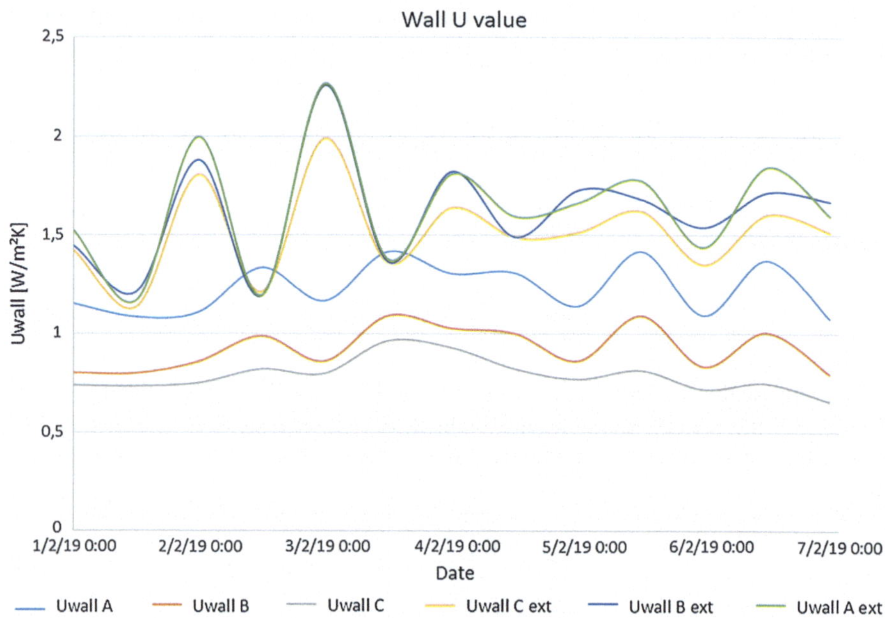


**Fig. 57.4** U-value at A, B and C position

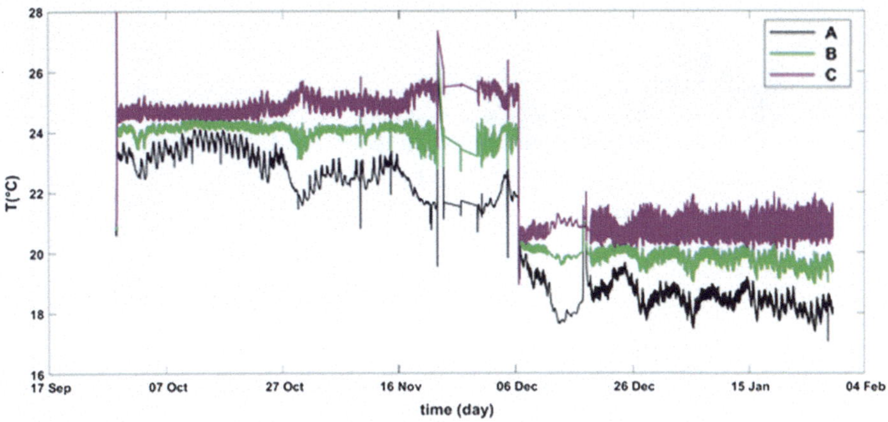


**Fig. 57.5** Indoor temperature in test cell

## 57.4 Moisture Performance

During the first few weeks of measurement, a temperature set point of 24 °C was fixed in order to create a significant difference between outside and inside temperatures. Then, as the outside weather became colder, a lower set point (18 °C) was fixed, and there was still no moisture generation inside. We consider the ventilation rate as 0.5 ACH, a standard value for the residential sector in France. To record the humidity content of the studied materials, we used Sensirion SHT75 sensors [18]. These sensors supply data of relative humidity and temperature simultaneously.

Temperature, relative humidity and absolute humidity values recorded in the center of the wall are presented in the following figures. As expected, in point C (upper part of the wall), the wall is dryer than in the rest of the structure. This emphasizes the fact that drying should be taken into consideration for newly built structures, as U-values depend on the moisture content of the final structure. Therefore, a time shift between the construction phase and the thermal performance measurement should be allowed before determining its in situ U-value in full-scale applications or in real buildings.

As Figs. 57.6, 57.7 and 57.8 present, point C is the driest measurement point of the wall. We consider the measured U-values at this point to compare in situ behavior to the theoretical performance. Table 57.2 presents an average value of the aforementioned 7 days measurement campaign of February 2019 of both measured U-values and the theoretical value.

As one can see, the measured U-values overestimate the theoretical one. It emphasizes once again the fact that the wall is still moist, and more drying is necessary before the determination of the final value. If we consider Figs. 57.4 and 57.8 together, we can see that more the wall section contains water, more its measured U-value fluctuates according to the higher heat flux transmitted at the same temperature difference. It explains the difference between values recorded at A, B and C points.

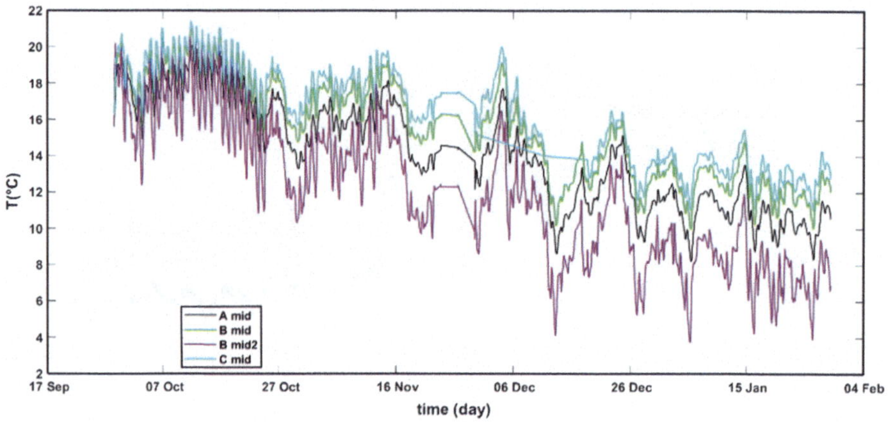


**Fig. 57.6** Temperature in the wall (between brick layer and the render)

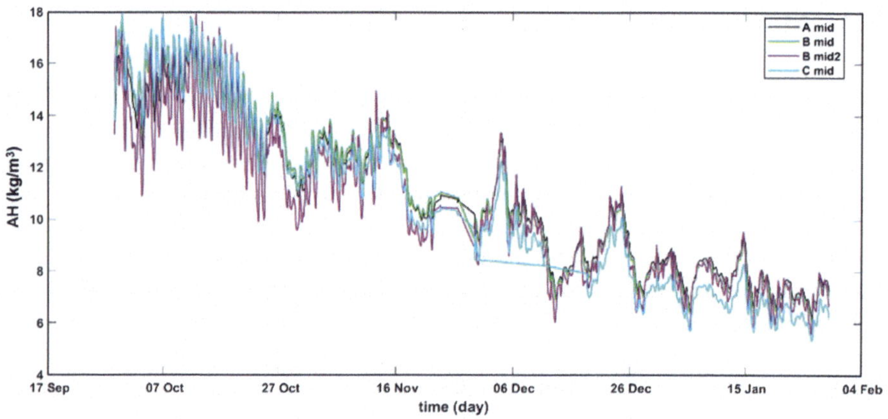


**Fig. 57.7** Absolute humidity in the structure (moisture content of the material surrounding the RH sensors)

## 57.5 Conclusions and Perspectives

A brick wall covered by an aerogel-based render was studied experimentally. In situ U-values were determined according to current available standards. We compared theoretical values with measured ones. We used both outside and inside heat flux data to determine wall's U-value. At this stage of the measurements, there is still too much variation in measured data, thus ISO 9689 criteria of less than 5% variation between two consecutive days is not fulfilled, and final U-value had not been reached yet. A clear influence of the moisture content of the wall was highlighted. At point C of the wall, where the moisture content is the lowest, the measured U-value is the closest to the theoretical one.

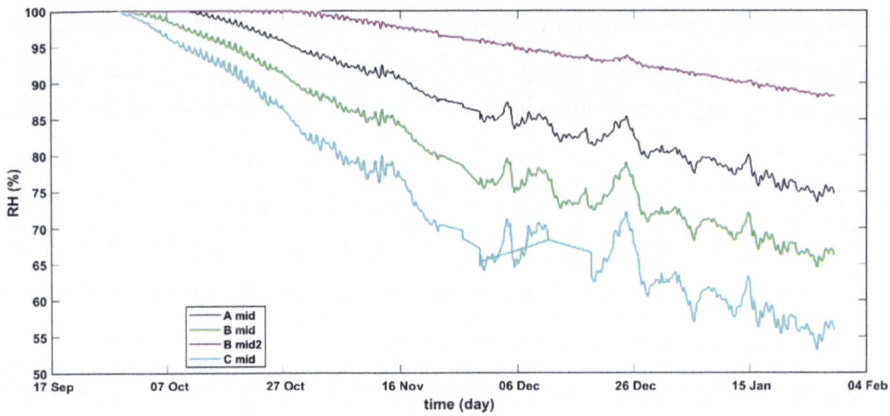


**Fig. 57.8** Relative humidity in the wall (between brick layer and the render)

**Table 57.2** Wall U-value

| In situ U-value ($q_{in}$) (W/m$^2$ K) | In situ U-value ($q_{ext}$) (W/m$^2$ K) | Theoretical U-value (W/m$^2$ K) |
|---|---|---|
| 0.78 | 1.6 | 0.365 |

As a perspective of this study, a complete analysis over a longer period is planned. The dynamic method of the ISO 9869 standard as well as the ASTM C1046-95(2013) standard methodology will be compared to the presented methodology. In addition, a comparative study will follow with other aerogel-based products, installed in the frame of the WALL-ACE project at INCAS experimental platform.

**Acknowledgements** The research leading to these results has been performed within the WALL-ACE project (www.wall-ace.eu) and received funding from the European Community's Horizon 2020 Work Program (H2020/2014–2020) within the Energy-efficient buildings 2016–2017 call, under the grant agreement N° 723574.

## References

1. European Union Directive 2010/31/EU of the European Parliament and of the Council of 19 May 2010 on the energy performance of buildings (recast) L153/13-35 (2010). https://eur-lex.europa.eu/LexUriServ.do?uri=OJ:L:2010:153:0013:0035:en:PDF. Accessed 30 Apr 2019
2. WALL-ACE EU H2020 Project. https://www.wall-ace.eu/
3. Ibrahim, M., Nocentini, K., Stipetic, M., Dantz, S., Caiazzo, F.G., Sayegh, H., Bianco, L.: Multi-field and multi-scale characterization of novel super insulating panels/systems based on silica aerogels: thermal, hydric, mechanical, acoustic, and fire performance. Build. Environ. **151**, 30–42 (2019)
4. Fantucci, S., Fenoglio, E., Grosso, G., Serra, V., Perino, M., Dutto, M., Marino, V.: Retrofit of the existing buildings using a novel developed aerogel-based coating: results from an in-field

monitoring. In: 7th International Building Physics Conference, IBPC2018, Syracuse, NY, USA (2018)
5. Stahl, T., Zimmermann, B.M., Wakili, K.G.: Thermo-hygric properties of a newly developed aerogel based insulation rendering for both exterior and interior applications. Energy Build. **44**, 114–117 (2012)
6. Wakili, K.G., Dworatzyk, C., Sanner, M., Sengespeick, A., Paronen, M., Stahl, T.: Energy efficient retrofit of a prefabricated concrete panel building (Plattenbau) in Berlin by applying an aerogel based rendering to its façades. Energy Build. **165**, 293–300 (2018)
7. Van den Brom, P., Meijer, A., Visscher, H.: Performance gaps in energy consumption: household groups and building characteristics. Build. Res. Inf. **46**(1), 54–70 (2018)
8. Fenoglio, E., Fantucci, S., Serra, V., Carbonaro, C., Pollo, R.: Hygrothermal and environmental performance of a perlite-based insulating plaster for the energy retrofit of buildings. Energy Build. **179**, 26–38 (2018)
9. ISO 6946:2017, International Organization for Standardization: Building components and building elements—thermal resistance and thermal transmittance—calculation methods (2017)
10. ISO 9869-1, International Organization for Standardization: Thermal insulation—building elements—in-situ measurement of thermal resistance and thermal transmittance—Part 1: Heat flow meter method (2014)
11. Gaspar, K., Casals, M., Gangolells, M.: In situ measurements of façades with a low U-value: avoiding deviations. Energy Build. **170**, 61–73 (2018)
12. Gaspar, K., Casals, M., Gangolells, M.: A comparison of standardized calculation methods for in situ measurements of façades U-value. Energy Build. **130**, 592–599 (2016)
13. Soares, N., Martins, C., Gonçalves, M., Santos, P., Simoes da Silva, L., Costa, J.J.: Laboratory and in-situ non-destructive methods to evaluate the thermal transmittance and behavior of walls, windows, and construction elements with innovative materials: a review. Energy Build. **182**, 88–110 (2019)
14. Rasooli, A., Itard, L.: In-situ characterization of walls' thermal resistance: an extension to the ISO 9869 standard method. Energy Build. **179**, 374–383 (2018)
15. Hukseflux Thermal Sensors, User Manual HFP01. https://www.hukseflux.com/uploads/product-documents/HFP01_HFP03_manual_v1721.pdf. Accessed 30 Apr 2019
16. TC Direct. https://www.tcdirect.co.uk/
17. https://www.keysight.com/en/pd-429828-pn-34980A/multifunction-switch-measure-unit?nid=-33260.536894538.00&cc=FR&lc=fre
18. https://www.sensirion.com/en/
19. Stahl, T., Wakili, K.G., Hartmeier, S., Franov, E., Niederberger, W., Zimmermann, M.: Temperature and moisture evolution beneath an aerogel based rendering applied to a historic building. J. Build. Eng. **12**, 140–146 (2017)

# Chapter 58
# A Working Methodology for Deep Energy Retrofit of Residential Multi-property Buildings

**Cecilia Hugony, Maria Elena Hugony, Francesco Causone and Eugenio Morello**

**Abstract** In many EU countries, the existing private condominium buildings represent an important energy problem to the attainment of the EU energy objectives for 2020 and beyond. 41.7% of EU population lives in flats that are responsible for 68% of total energy use in buildings. Deep renovation of residential private buildings is, therefore, a paramount and critical issue for the achievement of the EU targets for building stocks. Where property is fragmented among many small owners, like in condominiums, the decision process is the major obstacle to the uptake of an adequate deep renovation. In "Sharing Cities", a project granted by H2020 program under smart cities and communities initiatives, this barrier has been overcome through an innovative methodology based on citizens' participation and the collaboration between public and private sector. This paper describes the methodology and reports the last results to-date for the lighthouse city of Milan.

## Nomenclature

| | |
|---|---|
| DR | Deep renovation |
| EE | Energy efficiency |
| MPB | Multi-property building |
| RES | Renewable energy sources |

C. Hugony (✉) · M. E. Hugony
TEICOS UE S.r.l., Via Caviglia 3/a, 20139 Milan, Italy
e-mail: c.hugony@teicosgroup.com

F. Causone
Department of Energy, Politecnico di Milano, via Lambruschini 4, 20156 Milan, Italy

E. Morello
Department of Architecture and Urban Studies, Politecnico di Milano, via Bonardi 3, 20133 Milan, Italy

 
J. Littlewood et al. (eds.), *Sustainability in Energy and Buildings*, Smart Innovation, Systems and Technologies 163, https://doi.org/10.1007/978-981-32-9868-2_58

## 58.1 Introduction

Italy has 12 million residential buildings; over 60% of them have been built more than 40 years ago, before the issue of any regulation on energy efficiency in buildings. According to Istat [1], 49% of the occupied dwellings are located in multi-property (or condominium) buildings.

To foster energy renovation of the building stock in accordance with European directives, Italy adopted specific instruments:

- New legislations with mandatory energy efficiency targets for new buildings and, recently, new regulations for energy performance certifications and energy targets for existing and new buildings;
- VAT reduction to 10% for building retrofit;
- Tax credit for 65/70/75% of the investment in energy efficiency renovation.

Despite of this encouraging incentive framework, deep renovation (DR) represents just 0.4% of the total incentive allocated and most of the funding (68% in 2013) is spent for windows replacement only [2]; 34% of the incentives are given to the three or more story high buildings [3], and according to ENEA, just 1.4% of incentives have been requested for energy efficiency improvement of the opaque building envelope [4]. Moreover, an analysis of CRESME reports that in the last 15 years just 16% of the total surface of opaque building envelopes has been renovated or maintained [5].

This scenario shows that "the incentives (alone) do not contribute to a real reduction of energy needs in existing buildings" [6], but other actions are required. The principal barriers to energy efficiency (EE) in residential multi-property buildings (MPB) have been analyzed in the literature [7]; in particular, the major ones are:

- The lack of information to final clients about energy efficiency;
- The lack of specific financing instruments for energy efficiency in buildings (the capital investment is still not affordable for the majority of flat owners considering the payback time with current incentives);
- The difficult decision-making process in multi-property buildings;
- The uncertainty of consistent economic savings after renovation.

Despite these difficulties, the ambitious goal of energy renovation of private residential buildings is implicitly confirmed in all planning tools, both at a municipal and regional level. As an example, the action plan for sustainable energy of the municipality of Milan (SEAP [8]), and the Energy and Environmental Regional plan (PEAR [9]) establish goals for a reduction of emissions of global warming gases up to 55% in the period 2015–2020. Without the contribution of private buildings these and further objectives are not achievable.

This paper reports a successful attempt made in order to overcome the major barrier to energy renovation in residential multi-property buildings, that is, the decision-making process. To do so, we developed an innovative methodology within the EU-granted project "Sharing Cities" that led, in about 3 years, to the deep energy retrofit of five private buildings in Milan, for a total area of 24,000 m$^2$. The following sessions describe the methodology and the major results to-date.

## 58.2 Methodology

### *58.2.1 Overall Strategy*

In "Sharing Cities" [10] a specific task of the city of Milan was to realize deep energy renovations of MPB for about 21,000 $m^2$ of conditioned area. Among the several objectives of the project, this was considered one of the most challenging, since the flat owners had to pay by themselves for the majority of retrofit measures, with only 15% incentives on the total amount of the works, via EU funding.

Recognizing the pivotal role of people in this big challenge, an engagement and design strategy has therefore been developed, based on the direct participation of flat owners. It had the specific aim to increase people awareness on economic and environmental impacts of energy retrofit actions. The strategy was based on four pillars:

- Virtuous competition among owner's communities;
- Co-design process of the energy measures with the flat owners;
- Environmental and energy monitoring of the retrofitted building and training on behavioral adaptation to post-retrofit scenario.
- Development of a specific financial instrument to sustain the investment with a private bank.

In particular, the bank credit was accorded to the condominium (not to the single flat owner), with first rates due at the end of the renovations works (together with State's incentives), considering a cover factor of 100% of the costs and low interest rates (3.5%).

The entire process has been applied to 20 condominium buildings in the south-east area of Milan, with the ambitious goal of concluding real retrofit works on five of them.

### *58.2.2 Engagement Methodology*

The methodology adopted for the engagement of flat owners in the 20 buildings followed a precise sequence of steps:

- Building nominations through a public call;
- Feasibility study on respondents (52 buildings) and selection of pilot buildings (20 buildings);
- Co-design process on selected owners' communities (20 buildings);
- Technical design development;
- Sharing and discussion of the energy audits with flat owners;
- Financial instruments development and proposal;
- Monitoring system design and deployment;
- Post-retrofit guide sharing.

The municipality of Milan issued a public call in order to select the most appropriate buildings. A total of 52 buildings presented their nomination to the public call, and preliminary energy audits were undertaken to identify the buildings meeting the requirements of the project. Of the 52 candidates, 20 buildings resulted eligible from a technical point of view (see Fig. 58.1).

In order to overcome the barriers described in Session 1, a co-design process was set up by the principal project's partners: Politecnico di Milano, Poliedra, Legambiente, Comune di Milano and TEICOS, supported by FCC London, the work package leader. The co-design process accounted for three main meetings and an extra-meeting with the so-called "energy champions" of MPBs. The energy champions were groups of owners, identified in each building and present at the meetings, particularly pro-active and prone to collaborate with the project's partner to develop a proper energy renovation of their building.

The technical design process, developed by engineers and architects, had to go in parallel with the co-design process, in order to provide material to be shared and discussed with the owners' communities. The scheme reported in Fig. 58.2 tries to represent the complexity of this "double process" in terms of number of meetings and parallel co-design activities.

Specific communication tools (see Fig. 58.3) were developed in order to give people comprehensible information on the physical status of their buildings and on the opportunities of improvement. The reciprocal comprehension, made possible by the

**Fig. 58.1** Twenty buildings that approved their participation to Sharing Cities

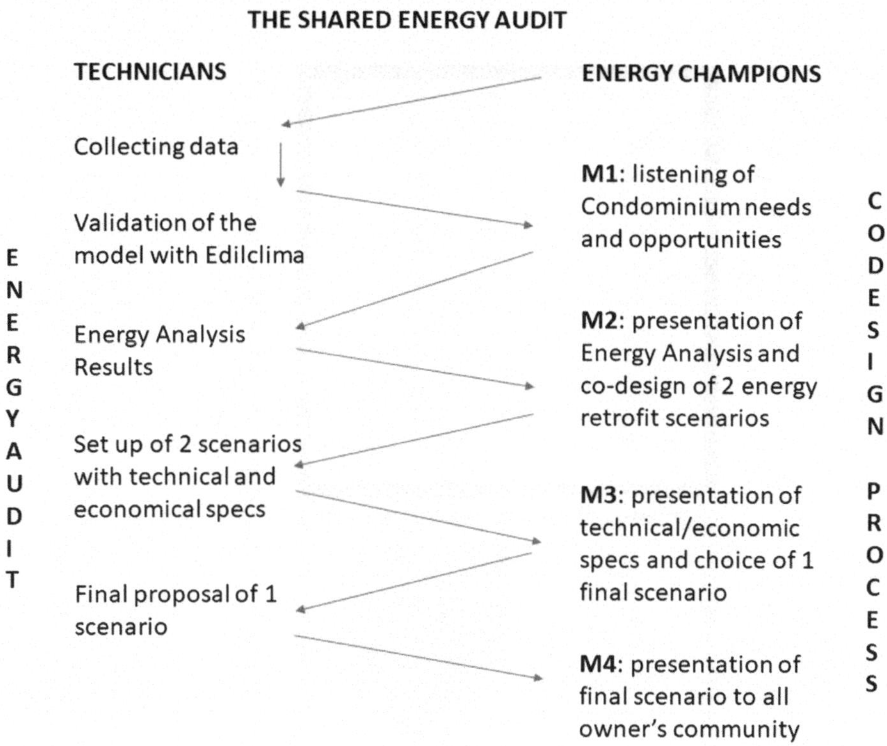


**Fig. 58.2** The double-process scheme of co-design and technical design

developed communication tools, allowed for an effective sharing of the information contained in the energy audit that would have resulted, otherwise, as a technical and not comprehensible document for the majority of the people.

At the end of the co-design process, among 20 buildings, five approved the defined measures. During the process, a financial scheme was presented to the building owners, including all the incentives and integrated with a specifically developed bank loan proposal: with long term (7 years), low interest rates (3.5%) and a cover factor of 100% of the costs.

A monitoring system was installed in 33 apartments of the five buildings, in order to assess the effect of the energy retrofit measures, and/or the change of behavior of flat owners after the implementation of the overall methodology in terms of raising awareness on the sustainability issues related to energy consumption in buildings. Besides energy consumption, the internal environmental quality is monitored with sensors of temperature, relative humidity, pressure, illuminance and total volatile organic compounds (TVOCs). Eventually, a user guide providing information on how operating and properly behave in a retrofitted building, in order to obtain energy saving together with high indoor comfort conditions, was developed together with

**Fig. 58.3** Example of one of the communication tools provided to flat owners: “EE measures postcards”: measures possible for building envelope (red), roof (green) and energy system (blue)

Legambiente, an Italian ONG working on environmental and sustainability topics. This was shared with all flat owners participating in the co-design process, recognizing the pivotal role of the occupant on the actual building performance.

## 58.3 Application and Results

The goal of the co-design activity was to create a shared project of retrofit measures to be implemented in each building. During the workshop, the 20 buildings have been divided into three main working groups. Table 58.1 shows the summary of groups.

A total of nine workshops (three meetings per group) with 160 residents (energy champions) from the selected 20 buildings were performed (see Fig. 58.4).

As a result of the co-design process and of the interaction between flat owners, architects and engineers, 40 different scenarios for buildings intervention have been developed, starting from the information gathered in the energy audit (each building first choose two scenarios: one more realistic and the other one more ambitious in terms of EE).

All the final mix of interventions chosen by the flat owners (see Table 58.2) included the thermal insulation of the opaque envelope, usually both vertical and horizontal, combined with PV system installation on the roof and/or new heating systems (gas heat pump or condensing boiler).

**Table 58.1** Summary of group of condominiums

| Group A | Group B | Group C |
|---|---|---|
| Beatrice d'Este | Fiamma | Altaguardia |
| Insubria | Mercalli | Archimede |
| Martini | Oglio | Barzoni |
| Passeroni | Ortles | Cirene |
| Quadronno | Pampuri | Muratori |
| Soave | Ripamonti | Scheiwiller |
| | Verro | Benaco |

**Fig. 58.4** Some pictures from workshop #2

**Table 58.2** EE measures chosen by building owner's during the co-design

| Buildings | Roof insulation | PV | Attic insulation | Green roof | External walls insulation | External walls in-cavity insulation | Basement insulation | LED lighting | Remote control of heating system | Heating system replacement | Other |
|---|---|---|---|---|---|---|---|---|---|---|---|
| Beatrice d'Este | x | x | | | x | x | x | x | x | x | |
| Insubria | x | x | | | x | | x | | | | |
| Martini | | x | | | x | | x | x | | | |
| Passeroni | x | x | | | x | | x | | x | x | |
| Quadronno | x | x | | | x | x | | | | | |
| Soave | x | x | | | x | | x | x | x | | |
| Fiamma | x | x | | | | x | x | x | | | |
| Mercalli | x | x | | | x | | | | x | x | |
| Oglio | | x | x | x | | x | x | | | | |
| Ortles | x | x | | | x | | x | x | | | |
| Pampuri | x | x | | | x | | | | x | x | |
| Ripamonti | | | | | x | x | | x | | | |
| Verro | x | x | | | x | | | x | x | | Windows replacement |
| Altaguardia | x | x | | | x | | x | x | | | |
| Archimede | | x | x | | x | | x | | x | | |
| Barzoni | | x | x | | x | x | x | x | x | | |
| Cirene | | x | | | x | x | x | x | | | |

(continued)

**Table 58.2** (continued)

| Buildings | Roof insulation | PV | Attic insulation | Green roof | External walls insulation | External walls in-cavity insulation | Basement insulation | LED lighting | Remote control of heating system | Heating system replace-ment | Other |
|---|---|---|---|---|---|---|---|---|---|---|---|
| Muratori | | | | | x | | | | | | |
| Scheiwiller | x | x | | | x | | x | x | | | Solar thermal panels |
| Benaco | | x | x | | x | x | x | x | | | |

After the co-design process, the retrofit works have been effectively approved by the condominium board for five buildings: Via Tito Livio 7, Via Fiamma 15/1, Via Verro 78 B/C, Via Passeroni 6, Via Benaco 26 (see Table 58.3). The pathway was short and followed the principle of "first come, first served". Indeed, the incentive provided by EU funding was covering only up to 21,000 $m^2$ of conditioned area; this has been a powerful driver to speed up the decision-making process.

All the site works have been concluded by the end of 2018, excluding Via Benaco, which is going to be concluded by August 2019. Since the entire process started in mid-2016, with the municipality's public call for nomination, we reached and exceeded the ambitious goal, and will complete the process with 24,310 $m^2$ of retrofitted surfaces in residential MPBs in about 3 years.

The retrofit works aimed at saving nearly 50% of energy (including all final energy uses), according to Sharing Cities' targets.

Considering that for the majority of the buildings the actual energy savings are expected in one-year time, after a whole heating season monitoring, some early observations can be done for space heating in the condominium of Via Verro 78 B/C, whose renovation works were concluded in May 2017. Table 58.4 reports the building energy use for space heating (delivered energy) before the retrofit and after the retrofit; it also includes the output of energy simulations, performed for energy certification, thus using a quasi-steady-state approach according to standard UNI EN 13790 (in force at the time of the design) and the national standard series UNI TS 11300.

The actual energy use for the heating season 2018–2019 (15 October to 15 April) is higher than expected (i.e. higher than energy simulation results). The reason for

**Table 58.3** Building specifications

| Name | Year of construction | Number of floors | Number of apartments | Total conditioned area ($m^2$) |
|---|---|---|---|---|
| Via Passeroni 6 | 1963 | 4/6[a] | 50 | 6.260 |
| Via Tito Livio 7 | 1960 | 7 | 25 | 2.049 |
| Via Verro 78 B/C | 1979 | 5 | 36 | 3.857 |
| Via Fiamma 15/1 | 1967 | 7 | 15 | 3.314 |
| Via Benaco 26 | 1960 | 6 | 141 | 8.830 |

[a]Via Passeroni is a complex condominium composed by 3 buildings of 4 floors, 5 floors and 6 floors, respectively

**Table 58.4** Heating energy use in Via Verro 78 B/C

| Pre-retrofit (delivered) | Energy simulation | Post-retrofit (delivered) |
|---|---|---|
| 65.11 kWh/$m^2$ y | 37.08 kWh/$m^2$ y | 48.44 kWh/$m^2$ y |

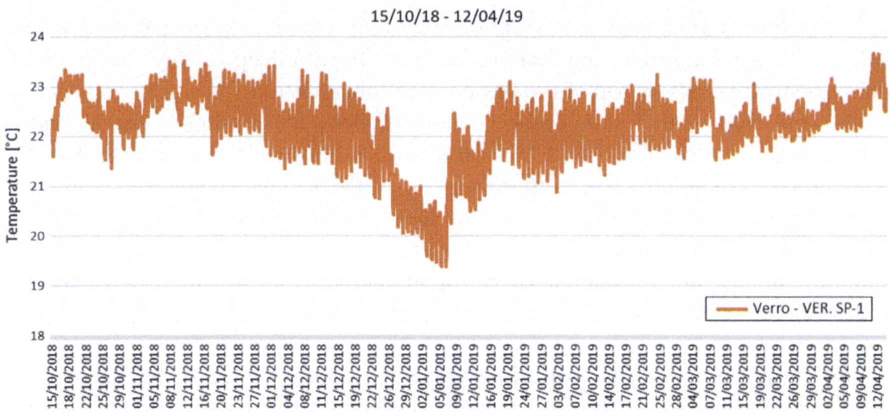


**Fig. 58.5** Indoor air temperature in a flat of Via Verro (post-retrofit)

this discrepancy depends on many factors: reference and actual weather, simulation input errors and occupant behavior. The trend of the indoor temperature in a reference apartment (see Fig. 58.5) is useful to identify the large role played by occupant behavior in particular. Figure 58.5 shows that the indoor temperature in winter is always beyond the target comfort temperature of 20 °C, which means that the thermoregulation (i.e. thermostatic valves) was not properly used after the renovation works. If, on the one hand, this demonstrates that the real energy consumption is almost in line with the energy simulation output, excluding the effect of thermostatic valves, on the other hand, it shows that the post-retrofit campaign to inform the users about a proper building operation was not (yet) properly effective.

## 58.4 Discussion and Conclusion

The experience reported in this paper shows that non-technical barriers, which often represent the major limit to energy retrofit in residential multi-property buildings, may be overcome through an approach based on co-design and people engagement. The success of the methodology depends on some key factors:

- Information sharing and co-design;
- Timely and easy proposal of financial aspects;
- Human-centric approach.

Moreover, all the five renovated buildings presented in the paper reported trigger points for renovation that helped accelerating the process: asbestos removal on the roof (via Fiamma and Via Passeroni), water infiltration from the roof (Via Verro) and façade decay (Via Benaco, Via Tito Livio, Via Passeroni, Via Verro). Identifying urgencies further than energy renovation may help convincing owners to proceed with works, showing them the benefits deriving from a deep energy retrofit, with little investment costs on top of the basic urgent intervention.

The very heart of the methodology is, nevertheless, the sharing of energy audits and the direct confrontation on technical solutions between flat owners and architects/engineers. The effective comprehension of the technical measures and their added value is fundamental in order to achieve owners' commitment. People felt involved in the ambitious challenge of deep renovation also due to the institutional commitment and direct participation of the municipality, of the university and of a NGO (Legambiente) in the activities. Having these partners on board helped creating trust in the process.

In order not to hinder the procedure, financial aspects should be presented only at the end of the technical process, once the owners agreed on the technical benefits of the retrofit measures and they are ready to properly weigh their economic value.

In general, the flat owners should become the pivotal point of the retrofit. The experimented methodology showed us that four meetings are enough to help the designer or promoter understanding the peculiar building community, owners' expectations and initial ideas, paving the way to a more effective dialog and cooperation.

The only part of the methodology that, at the moment, did not show totally effective was the post-retrofit campaign to inform the users about a proper building operation. This informational and educational process should be better structured considering the wide variety of users and habits that might be encountered: retired people, students, working population, singles, families, locals, immigrants, and so on.

In conclusion, the proposed methodology showed to be effective and could be further repeated, perhaps creating a more structured approach based on a larger and continuous commitment of the municipality or other professional actors such as a "process facilitator".

**Acknowledgments** The study was developed within the framework of the project SHAR-LLM, which has received funding from the European Union's Horizon 2020 research and innovation program under grant agreement no. 691895.
The design and co-design activities (including energy audit and simulations, retrofitting measures and financial budget and tools) have been promoted and led by TEICOS as part of the Sharing Cities project. Energy audits, simulations and design activities have been implemented by TEICOS. The co-design methodology was developed by the Department of Architecture and Urban Studies (Dastu) of the Politecnico di Milano with TEICOS and the contributions of Legambiente and Poliedra. The workshops and engagement activities have been organized and carried out by TEICOS with Legambiente, Dastu, Poliedra and supported and hosted by the Municipality of Milan and the local Municipality 4. The monitoring system was designed and installed by Future Energy and TEICOS with the support of the Department of Energy of the Politecnico di Milano that also carried out the data analysis.

## References

1. ISTAT Homepage. https://www.istat.it/it/files/2015/12/C18.pdf. Accessed 15 May 2016
2. Nocera, M.: Le detrazioni fiscali del 55–65% per la riqualificazione energetica del patrimonio edilizio esistente nel 2013, p. 8. Roma, ENEA (2015)
3. Nocera, M.: Le detrazioni fiscali del 55–65% per la riqualificazione energetica del patrimonio edilizio esistente nel 2013, pp. 21–22. ENEA, Roma (2015)
4. Nocera, M.: Le detrazioni fiscali del 55–65% per la riqualificazione energetica del patrimonio edilizio esistente nel 2013, p. 26. Roma, ENEA (2015)
5. D'Assandris, P.: Valutazione della convenienza e dell'impatto economico dell'isolamento termo acustico degli edifici. In: Riqualificare gli edifici, una necessità per il rilancio del paese. Le opportunità del risparmio energetico per l'economia ed il benessere dei cittadini. Conference 2014, p. 35. FIVRA, Roma (2014)
6. Renovate Italy Homepage. https://renovateitaly.files.wordpress.com/2015/05/position-paper_completo_26ott15.pdf. Accessed 31 Oct 2015
7. RSE: Edifici energeticamente efficienti: un'opportunità, pp. 91–95. ALKES, Milano (2015)
8. AMAT: Piano di Azione per l'Energia Sostenibile (PAES) del Comune di Milano. Documento di indirizzo per lo sviluppo del piano. Comune di Milano, Milano (2014)
9. PEAR Homepage. http://www.energialombardia.eu/. Accessed 18 May 2017
10. Sharing Cities Homepage. http://sharingcities.eu/. Accessed 24 Apr 2019

# Chapter 59
# Considering Institutional Logics in Building Performance Evaluation Research

**Sonja Oliveira and Magdalena Baborska-Narożny**

**Abstract** The purpose of this paper is to reflect upon potential analytical benefits of institutional theory in the study of building performance evaluation (BPE) practice. BPE studies have mostly been based on descriptive empirical insights with little conceptual underpinning on theoretical observation. Where theory has been drawn upon, it has tended to be largely based on socio-technical approaches, with focus mostly placed on the user or the client. Designers' experience of BPE and a multilevel approach are mostly overlooked. The analysis draws upon mixed data, including documentary evidence, and focus on group session with experts in BPE, as well as use of theory in empirical settings. The findings from the study would enable initial development of potential institutional theory approach, that is, the study of BPE practice. There are also methodological implications into ways new conceptual approaches could be considered in their application across a range of fields.

## 59.1 Introduction

### *59.1.1 Evaluation Practice and Building Performance*

Developing evaluative understandings associated with building performance has been a growing concern in the built environment domain. Improving building performance evaluation is viewed by the UK government agencies and policy-makers as a key way to reduce carbon emissions in the built environment [1]. Research has focused the attention primarily on developing useful evaluative tools and criteria. Methods drawn upon in BPE have also traditionally tended to be mostly empirical

S. Oliveira (✉)
Department of Architecture and the Built Environment, University of the West of England, Bristol BS16 1QY, UK
e-mail: sonja.dragojlovic-oliveira@uwe.ac.uk

M. Baborska-Narożny
Wroclaw University of Science and Technology, ul. B.Prusa 53/55, 50-317 Wroclaw, Poland



J. Littlewood et al. (eds.), *Sustainability in Energy and Buildings*, Smart Innovation, Systems and Technologies 163, https://doi.org/10.1007/978-981-32-9868-2_59

and descriptive, with concerns primarily focused on quantifying aspects of occupant behaviour that are attributed to energy consumption or ambient environment quality [2].

Whilst evaluation of building performance based on empirical measurements and tests is relatively well developed and widely applied, critical analysis of potential theoretical approaches to the study of feedback is less well developed [3]. Exceptions include use of social practice theory focusing on users' experience and engagement with feedback concerning mostly issues of comfort and energy use [4–7]. Lowe et al. [8] suggest that though there is wide recognition that buildings can be viewed as complex socio-technical systems, few studies in BPE have applied that view. Their study illustrates a case study approach drawing on a socio-technical conceptual framework, where the application of theory provided a necessary lens through which social components of the case were analysed. In addition to Lowe et al. [8], recent work by Tweed and Zapata-Lancaster [9] considered the use of architectural theory, primarily phenomenology, calling for richer investigations into the meanings associated with users' and designers' experience of building performance and feedback.

The broader sociological domain has been interested in evaluation, evaluative understandings and practices beyond the use of tools. Evaluative understandings and practices are viewed as an important aspect of the knowledge-making process in 'gatekeeping, filtering and legitimating knowledge' [10]. Evaluative practices are defined as a way of 'assessing how an entity attains a certain type of the worth' [11]. Evaluative tools are viewed as some of the constraints that shape evaluative practices [12]. Scholars have considered how evaluation unfolds in different settings and between different disciplines [11]. In most studies, quantification is viewed as a formalizing of evaluative practices viewing quantifiable measures as an inevitability of evaluation [13]. In creative fields where quantifiable evaluation is rarely used, the focus is on peer review, the role of critics and those carrying out the evaluation [11].

A number of scholars have indicated that examining how understandings on evaluation develop is important in terms of gaining insights into how evaluative practices evolve [11, 13]. One theoretical domain that has been concerned with how understandings develop, become taken-for-granted and infused with value has been institutional theory. The primary concern of institutional theory, and particularly institutional logics, has been in examining how understandings on an issue develop, become infused with value and ultimately institutionalized [14]. Scholars have initially examined how these institutionalized understanding influence actors by invoking stability.

The purpose of this paper is to reflect upon the potential analytical benefits of institutional theory as a potential theoretical lens to the study of BPE practice. The analysis draws upon a workshop format that explored potential analytical benefits of a range of relevant theoretical frameworks, with this paper focusing on discussions that examined institutional theory only. The paper concludes by discussing implications for research in the field of BPE, as well as illustrating benefits to practice, education and policy in the built environment.

### 59.1.2 Evaluation—Practices and Logics

Evaluation is viewed as the process through which knowledge is made, shared and transformed [15]. Lamont [11] draws attention to the rise in benchmarks, assessment models and performance standards, often viewed as representative of evaluative practice, to the detriment of other alternative views. Her review of evaluation across different domains calls for studies that examine evaluative practice in a range of diverse settings. Evaluative practices including the use of tools and constraints such as codes, categories and standards are viewed across social sciences as the underlying foundation to social and intellectual activity [10]. The use of evaluative tools and constraints is seen as one aspect of evaluation, whereas evaluative practices are viewed as a way of assessing the worth of an entity [11].

Evaluative practices are seen as a complex array of rules, distinctions and social conventions that underpin social behaviour [15]. For some scholars evaluation is about the negotiation of values where value is seen as the 'merit of a product in terms of its overall estimation in which it is held' by those who evaluate [16]. Lamont [11] indicates evaluative practices can undergo several estimation processes such as categorization and legitimation. At minimum, evaluative practices are concerned with categorization by determining the group into which an entity belongs [12]. In most domains categorization is determined through quantitative means by classifying, ranking and ordering [13].

Scholarship on evaluation views opinions and decisions to be guided by mainly formal influences such as contextual factors, disciplinary views or the setting. Literature on institutional logics views evaluative opinions and decisions to be guided by logics seen as guiding principles that provide the content for particular understandings. Although the institutional logics scholarship does not explicitly discuss evaluation, it offers some valuable insight into how evaluation is shaped through societal, industry or professional understandings on a particular issue. Lamont [11], though not referring to logics, highlights the importance of analysing broader evaluative influences. The following section discusses how influences on evaluation are viewed through scholarship in institutional logics and the research in the wider sociological domain.

### 59.1.3 Institutional Logics Guiding Evaluation

Institutional logics viewed as 'material-symbolic languages' [17] are argued to provide content to actors on defining new or redefining existing understandings on evaluation. Logics are defined as 'organizing principles that govern the selection of technologies, define what kinds of actors are authorized to make claims, shape and constrain the behavioral possibilities of actors, and specify criteria of effectiveness and efficiency'. [18]. However, an overarching construct that ties logics together is the inclusion of both material and symbolic elements [19]. Material elements are primarily seen as structures and practices, whilst symbolic elements are identified as institutional myths through which the meaning of material practices travel [19].

A review of recent scholarship on the ways the material and symbolic elements of logics have been studied is summarized in Table 59.1. The table shows how the elements have been defined by key authors, as well as how the elements are seen to be linked. Recent research in institutional logics and evaluation has highlighted the importance of empirically exploring phenomena related to the symbolic and material elements in logics [20]. The seminal work of Friedland and Alford [17] was one of the first to illustrate the value of examining the symbolic and material elements that shape logics. Jones et al. [20] also highlight the ways the material and symbolic elements underpinning the multiple institutional logics shape the content for a new category in architecture, enabling the emergence of 'modern architecture'. An institutional logics perspective recognizes that the interplay between the material and symbolic is key to the development of evaluative understandings [18].

The evaluative processes such as categorization and legitimation that shape these understandings are a key mechanism by which institutional logics shape cognition. Evaluative constraints such as categories and criteria are suggested to be shaped by institutional logics [17]. The concept of institutional logics was first introduced by Friedland and Alford [17] as a way of addressing the lack of societal context

**Table 59.1** The symbolic and material elements in logics

| Symbolic elements (definition) | Relationship to material | Scholarly work on these topics |
|---|---|---|
| As meanings embodied in identities | At the field level | Rao et al. (2003) |
| As institutional myths | Through which meaning of material practices travel | Townley (2002) |
| As cultural building blocks | Embedded in structures and practices | Zilber (2002) |
| As institutional resources | Furnishing guidelines for practical action | Rao et al. (2000) |
| Cognitive maps | As symbolic immunity | Lepoutre and Valente (2012) |
| Cultural frames | Shaping network structures | Lizardo (2006) |
| Material elements (definition) | Relationship to symbolic | Scholarly work on these topics |
| As the 'world of action' | The field level | Mohr and Duquenne (1997:309) |
| Structures and practices | | Thornton et al. (2002) |
| Routines, relationship systems and artefacts | Material immunity | Lepoutre and Valente (2012) |
| As objectified cultural constructions | | Delemestri (2009) |

in most institutional research at the time. Friedland and Alford [17] argued that societal context and more significantly society moderated the decisions, actions and behaviours of actors at multiple levels.

From that initial conception of logics as societal orders at family, religion or market levels, recent research has developed views of logics examined at professional and industry levels [14]. Whilst the concept of institutional logics has not been applied in the domain of BPE studies, research by Oliveira and Sexton [21] drawn upon the concept in the study of sustainability evaluation in architecture awards. There has also been an increasing use and application of institutional theory in the construction sector as a whole. Institutional logics are particularly helpful as a conceptual analytical tool in better understanding how evaluative practice is shaped by different actors (for instance, by different members of the design team in the context of BPE or across different design firms). It would also be beneficial to understand how feedback from BPE is experienced at different institutional levels and what material and symbolic elements of logics actors draw on.

The following section discusses methods drawn upon in the exploration of theoretical tools for the study of BPE.

## 59.2 Methods

Drawing upon Alvesson and Karreman's [22] suggested approaches to exploring analytical benefits of theory, in particular empirical settings, the study used a workshop format with key experts to examine the key questions. The expert participants were grouped in two discussion forums, initially focusing on (1) building users' perspectives and (2) designers' approaches to BPE. Participants were asked before attending the workshop to consider possible analytical benefits of theoretical approaches that could be developed to study evaluative practices in the context of building performance research, including social practice [23]; institutional theory [14], Gabriel De Tarde's social theory [24] on imitation and Actor Network theory [25], amongst others. See Table 59.2 for an outline of the participants.

**Table 59.2** Overview of workshop participants

| Disciplinary domain | Participant | Type of forum participation |
|---|---|---|
| Architecture | C | Designers |
| Building engineering | M | Users |
| Environmental science | T | Users |
| Energy behaviour | R | Designers |
| Sustainability consultancy | B | Designers |
| Sociology | L | Users |
| Institutional theory/Architecture/Energy behaviour | S | Designers |

At the end of the two parallel forum discussions, conclusions and observations were recorded as described in detail in Sect. 59.3. The recorded discussions and photos of diagrams used in the discussions were analysed thematically in order to determine the most appropriate theoretical approach. The themes that evolved from the preliminary analysis are discussed in the following section. Particular attention is placed on the potential usefulness of institutional logics.

## 59.3 Building Users and Designers—Empirical Issues and Theoretical Tools

The two forums explored a range of theoretical tools, with discussions in both forums largely highlighting the potential usefulness of institutional logics. The following sections illustrate how institutional logics could be drawn upon in the study of two empirical settings explored in the forums.

### *59.3.1 Building Users—The Institutional Micro Perspective?*

The discussion forum focused on the institutional at micro level, exploring the users' engagement in building performance feedback in the domestic setting. Feedback practice in the context of domestic BPE was viewed primarily through devices such as smart meters and energy monitoring devices. 'Feedback' was discussed as potentially needing to be viewed as 'the independent variable' and energy use is the 'dependent one'—how does it change if one 'applies feedback'.

Discussions highlighted the potential analytical benefits of institutional logics to better understand the dynamics and drivers that shape engagement with feedback devices. Key questions that emerged included:

- How is engagement with BPE shaped in the home—between different household members, across different household types, between different countries?
- What motivates (or not) engagement in BPE?

In the first question, it may be helpful to consider drawing on material and symbolic elements the logics that users draw on to justify or maintain engagement with BPE. Institutional logics would enable socio-cultural insights into meanings that users draw on to maintain or disrupt engagement with BPE. The second question draws attention to the logics that support initial interest and motivation to engage with feedback (rather than maintain or disrupt it). Institutional logics are a particularly helpful analytical tool to better understand meanings that actors draw on to sustain or disrupt particular actions. Jones et al. [20] highlight the ways the material and symbolic elements underpinning the multiple institutional logics shape the content for a new category in architecture, enabling the emergence of 'modern architecture'. An institutional logics perspective recognizes that the interplay between the material and symbolic is key to the development of evaluative understandings [17].

Overall, users' involvement and integration both in developing tools and the use of tools offer important insights, in terms of how evaluative influences are discussed. The design and development of tools is seen to be significantly influenced by users' conceptions and expectations regarding both BPE and the performance outcomes. The focus on users, however, may exclude other actors' views such as those who evaluate, including designers, clients and researchers.

### *59.3.2 Building Designers—The Institutional Organizational Levels*

The discussion forum focused on analytical benefits of institutional logics in examining how designers engage or not in building performance evaluation. The following questions were emphasized:

- How do designers' approach BPE (logics) during design?
- How do meanings on BPE develop in building projects (organizational cultures and processes)?
- What are the effects of no or too much BPE?

In the first question, it may be helpful to start to consider logics that shape designers' response to 'effects of design decisions' during design as well as certification, assessment models or evaluation during and after building handover. Logics can be seen to be shaped by societal and/or professional contexts and are enacted through evaluative practices presented through categories, constraints and criteria. For instance, evaluative constraints such as categories and criteria are suggested to be shaped by institutional logics at societal, professional or industry levels. Initially, logics were mainly seen to occur at societal levels. Friedland [17] argued that societal context and more significantly society moderated the decisions, actions and behaviours of actors at multiple levels. From that initial conception of logics as societal orders at family, religion or market levels, recent research has developed views of logics examined at professional and industry levels.

Most studies examining how new evaluative constraints such as categories emerge suggest changes in logics enable transformation in evaluative understandings, and thereby contribute to new categories being negotiated. For instance, Glynn and Lounsbury [26] examine institutional logics played out in critics' reviews of orchestral performances over time. Glynn and Lounsbury's [26] study specifically focuses on how shifts in logics shape the discourses of critics and consequently their evaluation. The traditionally embedded aesthetic logic was analysed over time to find that it transformed to commercially oriented market logic. The move from an aesthetic to a commercial focus was explained as occurring in response to a musicians' strike with post-strike reviews appealing to more commercial aspects of the performances. This change from an aesthetic to market logic was found to contribute to new evaluative practices being developed which accounted for the commercial turn.

The second and third questions could thereby further explore the logics identified in question 1, over time as well as between different actors in a design team or across diverse design firms.

## 59.4 Conclusion

The implications of this paper are twofold. First, the paper illustrates a potential method through which initial ideas on helpful analytical tools in a particular empirical setting can be developed. The workshop format shows a beneficial approach to the examination of theoretical tools. It is acknowledged that the sample in this workshop discussed in this paper is limited and that future work may extend the sample and enable more focused discussion on one theoretical approach. Second, the paper discusses the analytical benefits of institutional theory and logics to the study of BPE—a greatly overlooked issue.

Overall, discussions in the workshop focused on the dynamics of BPE, how it occurs, what makes it, who the actors are and the effects they have through their engagement (or lack of). Presentation of BPE was discussed in both forums as a potential barrier to engagement to both designers and users. The scale and context of analysis enabled a broader discussion of suitability of theoretical tools such as institutional theory in the study of the societal and professional underpinnings to BPE.

Tools and constraints are emphasized as the primary lens into evaluation. Even when alternative conceptualizations of BPE are proposed, there is still a sense that this would involve the need for new tools that would support new design frameworks. An emphasized focus on tools and constraints can provide a restricted one-dimensional view of evaluation [14]. The paucity of analysis in different empirical settings may be limiting the understandings on building performance evaluation in the architectural domain. The wider sociological domain and institutional literature on logics has engaged with evaluation and evaluative practices in diverse settings offer helpful conceptual framing.

Policy has tended to see that answers lie in proposing new tools, rules and regulations. Most of the research has similarly viewed that answers lie in enhancing current tool designs or creating entirely new evaluation tool types or methods. Tools are also seen as technical, scientifically-driven conceptions. Answers may lie in getting away from tools as objectified conceptualizations and allowing for multiple evaluative understandings to surface. Both research and policy have tended to focus exclusively on users, occupants' views and user participation. A deeper engagement with those who evaluate may also give some extended insights. Understanding institutional logics that guide evaluative decisions for instance would help better understand how criteria on specific areas of BPE are enacted and justified.

In addition, BPE issues are rarely discussed in the context of other evaluative concerns and yet the design process counters for multiple considerations from aesthetics, innovation and value. Discussing BPE in the context of other evaluative issues would offer extended perspectives on evaluation and performance of buildings. Considering evaluation as a process rather than primarily as the enactment or application of various tools could also extend current approaches to studying evaluation.

## References

1. Palmer, J., Terry, N., Armitage, P.: Building Performance Evaluation Programme: Findings from Non-Domestic Projects. Innovate UK (2016)
2. Göçer, O., Hua, Y., Göçer, K.: Completing the missing link in building design process: enhancing post-occupancy evaluation method for effective feedback for building performance. Build. Environ. **89** (2015)
3. Brown, Z., Cole, R., Robinson, J., Dowlatabadi, H.: Evaluating user experience in green buildings in relation to workplace culture and context. Facilities **28**(3/4), 225–238 (2010)
4. Chiu, L.F., Lowe, R., Raslan, R., Altamirano-Medina, H., Wingfield, J.: A socio-technical approach to post-occupancy evaluation: interactive adaptability in domestic retrofit. Build. Res. Inf. **42**(5), 574–590 (2014)
5. Vischer, J.C.: Applying knowledge on building performance: from evidence to intelligence. Intell. Build. Int. **1**(4), 239–248 (2009)
6. Tweed, C.: Socio-technical issues in dwelling retrofit. Build. Res. Inf. **41**(5), 551–562 (2013
7. Coleman, S., Robinson, J.B.: Introducing the qualitative performance gap: stories about a sustainable building. Build. Res. Inf. **46**(5), 485–500 (2018)
8. Lowe, R., Chiu, L.F., Oreszczyn, T.: Socio-technical case study method in building performance evaluation. Build. Res. Inf. **46**(5), 469–484 (2018)
9. Tweed, Chris, Zapata-Lancaster, Gabriela: Interdisciplinary perspectives on building thermal performance. Buil. Res. Inf. **46**(5), 552–565 (2018)
10. Camic, C., Gross, N., Lamont, M.: Social Knowledge in the Making. The University of Chicago Press, Chicago (2011)
11. Lamont, M.: Toward a comparative sociology of valuation and evaluation. Ann. Rev. Sociol. **38**(21), 1–21 (2012)
12. Zuckerman, E.: The categorical imperative: securities analysts and the illegitimacy discount. Am. J. Sociol. **104**(5), 1398–1438 (1999)
13. Wijnberg, N.: Classification systems and selection systems: the risks of radical innovation and category spanning. Scand. J. Manag. **27**, 297–306 (2011)
14. Powell, W.W., DiMaggio, P.J.: The New Institutionalism in Organizational Analysis, p. 1991. University of Chicago Press, Chicago (1991)
15. Douglas, M.: How Institutions Think. Syracuse University Press, New York (1986)
16. Moeran, B., Garsten, C.: Business anthropology: towards an anthropology of worth. J. Bus. Anthropol. **2**(1), 1–8 (2013)
17. Friedland, R., Book review: Thornton, Ocasio and Lounsbury: The institutional logics perspective: a new approach to culture, structure and process. Management **15**(5) (2012)
18. Lounsbury, M., Geraci, H., Waismel-Manor, R.: Policy discourse, logics and practice standards: centralising the solid-waste management field. In: Hoffman, A., Ventresca, M.J. (eds.) Organizations, Policy and the Natural Environment, pp. 327–345. Stanford University Press (2002)
19. Thornton, P., Ocasio, W.: Institutional logics and the historical contingency of power in organizations: executive succession in the higher education publishing industry, 1958–1990. Am. J. Sociol. **105**(3) (2008)

20. Jones, C., Maoret, M., Massa, F.G., Svejenova, S.: Rebels with a cause: formation, contestation and expansion of the de novo category 'modern architecture' 1870–1975. Organ. Sci. **23** (2011)
21. Oliveira, S., Sexton, M.: Conflict, contradiction, and concern: judges' evaluation of sustainability in architectural awards. ARQ: Arch. Res. Q. **20**(4), 325–332 (2016)
22. Alvesson, M., Karreman, D.: Qualitative Research and Theory Development: Mystery as Method. Sage Publications (2011)
23. Shove, E., Pantzar, M., Watson, M.: The Dynamics of Social Practice: Everyday Life and How it Changes. Sage (2012)
24. De Tarde, G., Wolf, J.: Die Gesetze der Nachahmung. Suhrkamp (2009)
25. Akrich, M., Callon, M., Latour, B., Monaghan, A.: The key to success in innovation Part I: the art of interessement. Int. J. Innov. Manag. **6**(02), 187–206 (2002)
26. Glynn, M.A., Lounsbury, M.: From the critics' corner: logic blending, discursive change and authenticity in a cultural production system. J. Manag. Stud. **42**(5) (2005)

# Chapter 60
# Ecology of Heat Pump Performance: A Socio-technical Analysis

**Lai Fong Chiu and Robert Lowe**

**Abstract** The UK government's heat strategy is to reduce emissions from buildings "to virtually zero by 2050" through a combination of technologies including heat pumps (HPs). As part of this strategy, it introduced the Renewable Heat Premium Payment (RHPP) scheme to incentivise the installation of HPs in the residential sector. Using a socio-technical approach and case study method developed by the authors in the field of energy research and building, this paper explores the reasons for variation in performance of HPs supported by this scheme. Twenty-one sites/households were selected for investigation. Owing to limited space, this paper does not seek to present all cases, but instead focuses on key insights from five cases that were originally thought to perform poorly. The findings highlight how the complex ecology of a socio-technical system in determines performance. We will show that system performance emerges from the dynamic interaction of monitoring system, heat pump system configuration and occupants' heating practices, and heating load factor. Limitations, practical implications, and scope for future research are briefly discussed.

## 60.1 Introduction

The UK domestic sector accounts for approximately 29% of final energy consumption, with space and hot water heat accounting for approximately 80% of this. Enshrined in the sectoral plan of the government's Carbon Plan [1] and Heat Strategy [2] is the goal of reducing emissions from this sector to close to zero through a combination of improved energy efficiency and deployment of low-carbon heating. Heat pumps are seen as a key technology for achieving this, and the UK government set up the Domestic Renewable Heat Incentive (RHI) scheme and the Renewable Heat Premium Payment (RHPP) to promote the installation of HPs in this sector [3]. In parallel, it set up a field trial to understand the performance of HPs installed under the RHPP scheme. This trial collected data for 700 of the c. 14,000 HPs installed

L. F. Chiu (✉) · R. Lowe
University College London, London WC1H 0NN, UK
e-mail: laifong.chiu@ucl.ac.uk



J. Littlewood et al. (eds.), *Sustainability in Energy and Buildings*, Smart Innovation, Systems and Technologies 163, https://doi.org/10.1007/978-981-32-9868-2_60

under the scheme, forming the largest HP field trial undertaken in the UK to-date. Performance in the field is critical in terms of the economic competitiveness of the technology, its acceptability to dwelling owners and occupants, and to enable the UK government to demonstrate compliance with the UK's target for renewable heat under the EU Renewable Heat Directive [4]. Currently, the performance of HPs is defined by two indices, the coefficient of performance (COP) and the seasonal performance factor (SPF). The Renewable Heat Incentive scheme regulations [5] require that both air source heat pumps (ASHPs) and ground source heat pumps (GSHPs) should attain COP = 2.9 and SPF = 2.5 at boundary H4, to qualify for support under the scheme. A team led by the second author of this paper was commissioned to analyse remote monitoring data from the field trial and to undertake detailed case studies of a sample of 21 installations using a socio-technical approach with the aim of improving understanding of remote monitoring systems, quality of metadata, reasons for variations in performance, and occupant satisfaction with their HP systems. Owing to limited space, this paper does not seek to present all cases, but instead focuses on key insights from five cases that were originally thought to perform poorly.

## 60.2 Research Design and Methods

The present authors' previous research in building performance has shown the necessity of taking a socio-technical system perspective to empirical investigation of the performance of energy technologies in the field [6]. This has resulted in the development of a socio-technical case study method to collect and integrate data from people and technical systems [7] that are interconnected, mutually adaptive, and co-constituted. Below is a brief outline of the design and methods used for data collection, analysis, and interpretation. A full report on the RHPP field trial case studies is available online [8].

### *60.2.1 Sampling Strategy and Collection of Data*

Based on statistical analysis of remote monitoring data, 117 householders and 31 registered social landlords (RSLs), representing a range of HP performances and geographic locations across the UK, were identified for initial contact. Positive responses were received from approximately one-third of the householders, and almost half of the tenants contacted by RSLs. Following a further round of selection, occupants of 21 dwellings, 7 under the ownership of RSLs and 14 owner-occupied, were recruited to take part in case studies. Site visits of 2–3 h duration were made by a team that included one technical researcher and at least one social researcher, who together recorded detailed characteristics of each dwelling, configuration, and arrangements of HP systems and interviewed occupants about their life-style, perceived thermal comfort in relation to the installation and operational experience of and satisfaction with the HP during a room-by-room "walk-through" of the dwelling [8: 42].

### 60.2.2 Analysis

In the first instance, both quantitative and qualitative data were entered into a master matrix. Analytic matrices were then constructed [8: App. 1] based on an investigative logic of HP performance. This analytic framework gave primacy to thermodynamic constraints, expressed qualitatively by the equation:

$$\text{COP} \approx 0.5 \times 330/\Delta\text{T}$$

(where ΔT is the difference between the HP's source and output temperatures), to the configuration of system components and their interactions with built-form and heat loss taken from Energy Performance Certificates (EPCs), to impacts of human behaviours such as commissioning of systems by installers or occupants themselves, and to occupants' operating strategies in the context of lifestyles and costs.

Table 60.1 summarises the 16 cases in terms of types of HPs and their performances. Analysis was initially performed on a sub-sample of 10 cases selected from the 21 cases—these are arranged horizontally in Table 60.1. Because this sub-sample included only a single RSL dwelling, the sub-sample was then expanded to include all 7 RSL dwellings in the 21, bringing the total number of cases subjected to detailed comparative analysis to 16. The additional six cases are arranged vertically in Table 60.1. Estimates of SPFs for all systems were refined during the analysis.

**Table 60.1** Cases classified by heat pump type, SPF band, and tenure

"well performing" heat pumps (SPF > 2.5)
"poorly performing" heat pumps (SPF < 2.5)
Heat pumps reclassified as "well performing" following analysis
RSL-owned dwellings shown in *italics*.
"AS" denotes air source and "GS" denotes ground source heat pump.

| | | | | | | | | | |
|---|---|---|---|---|---|---|---|---|---|
| | | | | | | | | | *CS02 AS* |
| | | | | | | | | | *CS03 GS* |
| | | | | | | | | | *CS04 GS* |
| | | | | | | | | | *CS05 GS* |
| | | | | | | | | | *CS06 AS* |
| CS18 GS | CS13 GS | CS19 AS | CS14 GS | CS20 AS | CS16 GS | CS12 GS | CS15 GS | CS09 AS | *CS07 AS* |
| | | | | | | | | | *CS08 AS* |

## 60.3 Factors Influencing Performance

### *60.3.1 Monitoring Systems and Data Quality*

Seven "poorly performing cases", CS02, CS03, CS07, CS09, CS12, CS15, CS16, were initially suspected by the quantitative team of having monitoring system issues. A combination of revised algorithms for automatic selection of monitoring data and analysis of case study data led to SPFs for CS07, CS09, and CS15 being revised upward to above SPF = 2.5, leaving only four poorly performing cases.

Metering issues were one of a number of problems that affected CS07, including a report from the occupants that their HP had suffered a "blow out" which was corroborated by inspection of the monitoring data. These problems appeared to be resolved following replacement of the faulty external unit. Monitoring issues were also suspected in CS09; monitoring data indicated that the HP exhibited a persistent but unexplained reduction in mass flow in the primary heating circuit as recorded by the heat meter flow sensor. However, in the interview, the occupants reported that there had been a flat battery in the monitoring system in the initial period following installation which, once detected, had been replaced. The SPF for this system was re-estimated based on data that was deemed not to have been affected, resulting in reclassification as well as performance. CS15 had a monitoring issue due to faulty installation of equipment, but this was corrected after its discovery.

> When sensors were placed, [the] fitting was wrong – it was running backwards for a long time – then they came back and altered it; they'd put the sensors on the wrong pipes; this happened [maybe] around 18 months ago.

CS12 was suspected to have potentially serious and unresolved problems with the monitoring system, which were confirmed during the site visit (see Sect. 3.2). Metering issues were suspected in CS16 due to a low SPF, but subsequent analysis suggested that this was better explained by the low load factor of the heating system (see Sect. 3.3), which in turn arose from the occupants' heating practices. Finally, metering issues were also suspected in CS13 and CS18, due to data showing heat output with no electricity input for short periods.

### *60.3.2 System Configuration and Heating Practices*

The five well performing cases, CS13, CS14, CS18, CS19, and CS20, with SPF well above 3 seemed to have well-insulated systems. Well-insulated internal pipework was also found in two of the reclassified cases, CS15 and CS09, suggesting an association with performance.

Metering data for CS09 indicated the presence of resistance heating for domestic hot water. This appears to correspond to CS09's everyday heating practices. As a couple of retirees, the occupants lived mostly in the kitchen during the day-time, retreating in the evening into a small study to watch television; hence the living room was rarely used. The temperature in the kitchen tended to be set at 18 °C, as "movement [of] the sun & heat from cooking [would] make it comfortable [enough]". The kitchen also housed a large electric oven (an Aga). It is possible that the occupants had been using the electric Aga to provide hot water but were not aware that this might be costly.

Believing that they could keep themselves comfortable with lower costs, occupants of CS09 controlled the temperature in the dwelling by switching their HP on and off and by altering its flow temperature. They said that they did not find the temperature so controllable using the thermostat. The researchers observed that the thermostat was sited in an unheated lobby with a door often closed between the lobby and the living space. This might have caused a higher than set-point temperature in the rest of the dwelling, prompting the occupants to adopt the on–off control strategy to keep the temperature down. From the documentation, less-than-adequate airtightness of the dwelling was discovered by pressurisation test at the time of installation of the HP. This might have added to the poor performance recorded at the outset.

CS12 was another poorly performing case with SPF initially estimated as 0.7. The heating system was characterised by un-insulated pipework (see Fig. 60.1).

CS12 was a GSHP. Its hot water cylinder was housed in a cupboard within a heated utility room. There was underfloor heating installed on the ground floor but radiators upstairs. Initially, it was thought that internal uninsulated pipework should not affect

**Fig. 60.1** CS12—uninsulated pipework associated with heat pump

performance. However, CS12's room thermostat was in the kitchen immediately adjacent to the heated utility room. The room thermostat in the kitchen could have been affected both by heat gain from cooking and uninsulated pipes in the adjacent utility space. Either factor might in turn have increased the tendency of this system to cycle.

Monitoring data showed that heat demand for space and water heating was below electricity input in every month for two consecutive years. An inquiry was made into whether the data could be subject to metering error. Two features of the monthly mean electricity and heat demand plot suggest it might have been. First, the extreme seasonality of hot water demand and the persistence of space heating demand through the summer suggested that there may have been a misallocation of total HP heat output between space and water heating—this appeared likely in the light of a photograph taken on site of the monitoring installation (Fig. 60.2), showing an auxiliary temperature sensor on one of the primary circulation pipes, which had been installed with insulation between it and the pipe. This could explain why the level of both space and water heating recorded was implausibly low, even in a well-insulated dwelling with a single occupant. The total heat output of the HP in January was sufficient to raise the internal temperature of the dwelling by about 4 °C. Without other sources

**Fig. 60.2** CS12—pipework showing insulation between temperature sensor and pipe. The auxiliary temperature sensor for the heat meter is just right-of-centre in a pocket. *Source* BRE

of heat, this would have brought the mean internal temperature in January to around 8 °C. Solar gain in January is negligible, and internal free heat gains would have been unlikely to add more than a few degrees centigrade to the internal temperature.

The site visit took place in December. Researchers observed that set-point temperatures of most room thermostats in this dwelling were in the range of 17–18 °C. Internal temperatures were not recorded, and it was therefore difficult to determine whether set-point temperatures were reached or not. But the occupant stated that she was "comfortable". The manufacturer's documentation showed the system installed in this dwelling was fitted with either a 6 or a 9 kW "additional electric heater". Given doubts about the monitoring system, it was not possible to determine from the data collected what proportion of HP output was produced by electric resistance heating. However, plotting heat output from the HP against electricity input from available physical monitoring data showed that a proportion of the heat output of this system was derived from resistance heating, and that a proportion of the electricity input resulted in no measured heat output. This is consistent with a statement made by the occupant during interview that the booster had, probably accidentally, been turned on by a plumber, but it was not until a "huge bill" (around £1100/year) arrived that she realised there was a problem. The occupant reported that she tried and managed to switch the immersion heater off on the control panel with the help of her father and the manual. This illustrates the complex issues involved in measuring HP performance. The presence of multiple problems—a low thermostat set-point and the accidental triggering of resistance heating—might have played a part in lowering the performance.

### *60.3.3 Low Load Factor and Performance*

CS16 was a newly built farm house of 314 m$^2$ gross floor area. The EPC showed high ratings for all aspects of the dwelling and its systems, and an overall EPC rating of A. But the monitoring data showed an SPF of only 1.7.

In an attempt to understand the reasons why this HP performed poorly, the researchers first checked whether the design capacity of the HP (12 kW) was adequate to meet the heat demand of the dwelling. The calculation suggested that there was sufficient capacity in the HP to heat the dwelling down to external temperatures of around 0 °C. Attention then turned to the monitoring data. It was noted that the envelope of HP electricity consumption dropped abruptly by roughly 3 kW in late March 2014, coupled with a drop in the system output temperature. A possible explanation is that the booster heater was turned off at this point.

Inspection of the mean monthly energy flows in CS16 prompted further questions. The split between DHW and space heat changed by a large amount between the two years. DHW went up. Space heat went down by about the same amount. Total heat output stayed roughly the same.

From the analysis of the interview data, there appeared to be a range of issues affecting performance, driven by occupant changes to their control strategy. Despite having photovoltaic panels installed under a generous feed-in-tariff (FIT), the occupants did not feel that these payments were sufficient to offset the cost of their energy bill. They reported that since the HP had been installed, their electricity bill "goes up hugely during the winter quarter", to around £720 in the 2015 winter quarter and approximately one-third of that in the 2015 summer quarter. They estimated their annual bill was approximately £1900 that year. In order to save money, they decided to use the HP only to provide background heating and tried out different ways to heat the house. These included: turning down most of the room thermostats to 16 °C (exceptions were the hallways and *en suite* bathroom, which were set at 21 °C); having their electric oven (an Aga) set to its "snooze" setting (110 °C, with a corresponding continuous electricity consumption of approximately 600 W) to provide background resistance heating in the living room–kitchen; and using two wood-burners (one in the snug and one in the lounge) to supplement the heating every night in winter, and during the day if it was cold. Moreover, the dwelling was a farmhouse and the occupants' life as farmers might also help to explain how lower temperatures were tolerated as a result of wearing warm clothing inside the house. It was observed that they and their two dogs spent most of their days in-and-out of the house. Despite sophisticated zoning, their external doors were opened and shut quite often, making precise control of temperature in the house difficult. This, coupled with a rather high electricity consumption resulting from frequent use of appliances such as a tumble drier, a dishwasher, and a washing machine as well as electric power showers, makes it unsurprising that their electricity bill was high. But, in the light of a growing awareness of the consequences of their lifestyle, the occupants had given up trying to control their internal temperature using room thermostats and were instead "experimenting" with using secondary heating devices such as the two wood-burners and the oven mentioned above. It is likely that the use of the Aga, the power showers, tumble dryer, and the wood-burning stoves, none of which were controlled by a room thermostat, would have restricted the operation of the HP during the monitoring period. All of this begins to explain why the HP delivered less than 10,000 kWh/year of heat, less than 40% of the more than 26,000 kWh/year estimated by the EPC assessor for this house.

Further light was shed on the poor performance of CS16 by a comparison with two other cases with similar physical characteristics. All three dwellings were over 290 $m^2$ floor area (roughly 3.5 times UK mean floor area per dwelling, and the largest among the 16 case studies). The estimated heat annual demands recorded on the EPCs for these houses were: CS14, 16,800 kWh/a; CS16, 26,300 kWh/a; CS18, 20,400 kWh/a.

The GSHP installed in CS14 was rated at 12 kW; CS16 and CS18 were fitted with GSHPs from the same manufacturer. All systems had an integrated auxiliary heat unit, to back up and supplement the output from the HP. All three dwellings had underfloor heating. Based on the monitoring data, the mean heat demands of CS14 and CS18 were shown to range between 3.3 and 7 kW in the period from November to February, yet CS16's mean heat demand was less than half this, between 1.3 and 2.7 kW. Setting aside the possibility of issues with data logging for the moment, the question is whether the low heat demand of CS16 might have contributed to the low SPF. This working hypothesis was put to the test by plotting the monthly mean COP versus load factor for these three cases (Fig. 60.3). The graph suggests that:

- The low SPF of 1.7 for CS16 is in fact likely to be correct and can be explained by the low heat load factor in this dwelling: the HP performance profile plotted in this way is more or less indistinguishable from the installation in CS14, which has an SPF of 3.5;
- Monthly COP rises monotonically with heat load factor up to 0.3 for CS14 and 0.4 for CS18—at higher load factors, monthly COP falls, perhaps suggesting the onset of electric resistance heating.

Note that the CS16 GSHP had a fixed speed compressor, but it is not known whether ground loop and primary circuit circulation pumps were fixed or variable speed. If fixed speed, this would also tend to reduce part load performance.

Thus, it appears that efficiency is a function of load factor, and that it may therefore be affected by behavioural and social factors, such as a requirement to save energy or reduce high fuel bills, or the use of secondary heating not metered by the HP monitoring system.

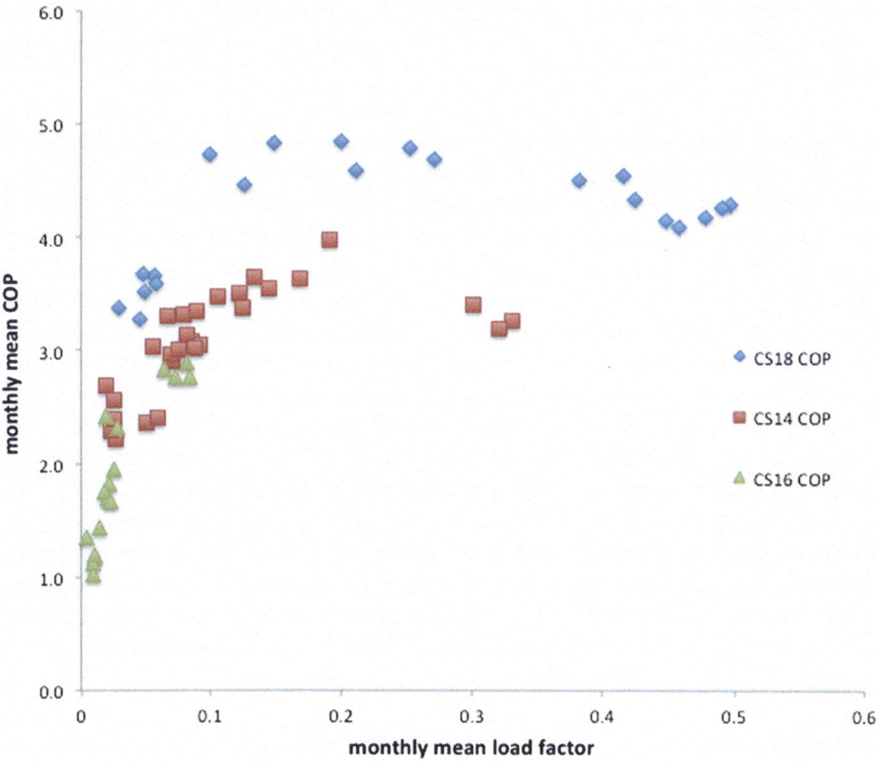


**Fig. 60.3** COP versus heat load factor (monthly means) CS14, CS16, and CS18

## 60.4 Discussion

Using a socio-technical approach that aims to go beyond simply understanding either users' behaviours or the technical system, this paper attempts to illustrate the complex ecology of interactions and relationships within and between both systems in determining performance. The metric of HP performance is itself the product of a socio-economic-technical system (the Renewable Energy Directive) that sets the boundary between good and poor performance. Each HP's operational performance is determined from data collected by a monitoring system, the performance of which depends on quality of installation. The detailed configuration/installation (i.e. arrangements of physical components) of the HP in the dwelling prefigured occupants' heating/ventilation practices. These determine the heat load, which impacts on the HP performance. This in turn appeared to reinforce the original behaviour of the occupants. The plot of performance against heat load factor provided a way to encapsulate the dynamic relationships between a constellation of contextual factors, such as thermal efficiency, lifestyle, perceived comfort, occupants' practical understanding and skills in controlling the HP through manipulating thermostat, and flow rate or the use of non-thermostatically controlled secondary heating, as well as unintentional human decisions that resulted in an increase of electric resistance heating. It appears that unexpectedly high electricity bills could be a cue for remedial measures, triggering actions that led to reduced heat load, and beginning a positive feedback cycle.

If we take seriously what this study has revealed, we will have to approach policy implementation, monitoring, and evaluation of low carbon heat technologies differently. While SPF might be a useful metric for determining the overall success of policies to support HPs, the complex paths of learning by which performance is achieved and sustained cannot be ignored. It is possible that an overly simple conceptualisation of uncertainties around HP performance, reflected in energy models, might have been one factor contributing to the recent scaling back of UK government support for the RHI programme. A richer understanding of performance in the real world that captured the process of inter-adaptation between technology and people [6] might help to avoid such outcomes in the future, and lead to more innovative policy and more effective interventions to improve performance. For example, many of the issues raised in this paper could be addressed by reconceptualising heat as a service and supporting the development of customer care packages.

It has not been possible in this short paper to expand on the theoretical and methodological developments that have underpinned the analysis, but references have been given. However, the socio-technical analysis presented here challenges the conventional separation of physical and social disciplines in applied energy research. A research programme that opened opportunities for multi-disciplinary collaboration would be one way to improve the environment for learning.

**Acknowledgements** The authors acknowledge support from UK Research and Innovation through the Centre for Research into Energy Demand Solutions, grant reference EP/R035288/1. The analysis of data from the RHPP Field Trial was undertaken by RAPID-HPC under contract to BEIS, and with support from the Centre for Energy Epidemiology, grant reference EP/K011839/1. Support for the Case Studies Report was provided by Eleni Oikonomou, Colin Gleeson, Jenny Love, Jez Wingfield and Phil Biddulph.

## References

1. HM Government: The Carbon Plan: delivering our low carbon future. Presented to Parliament Pursuant to Section 12 and 14 of the Climate Change Act 2008 (2011)
2. DECC: The Future of Heating: Meeting the Challenge (2013). https://www.gov.uk/government/publications/the-future-of-heating-meeting-the-challenge. Accessed 14 May 2019
3. DIRECTIVE 2009/28/EC of the European Parliament and of the council of 23 April 2009 on the promotion of the use of energy from renewable sources and amending and subsequently repealing Directives 2001/77/EC and 2003/30/EC. https://eur-lex.europa.eu/legal-content/EN/TXT/PDF/?uri=CELEX:32009L0028&from=EN. Accessed 14 May 2019
4. DECC: Domestic renewable heat incentive: the first step towards transforming the way we heat our homes, July 2013. UK Department of Energy and Climate Change. https://assets.publishing.service.gov.uk/government/uploads/system/uploads/attachment_data/file/212089/Domestic_RHI_policy_statement.pdf (2013). Accessed 19 June 2018
5. Renewable Heat Incentive Scheme Regulations 2018. https://www.legislation.gov.uk/ukdsi/2018/9780111166734/pdfs/ukdsi_9780111166734_en.pdf. Accessed 14 May 19
6. Chiu, L.F., Lowe, R., Raslan, R., Altamirano-Medina, H., Wingfield, J.: A socio-technical approach to post-occupancy evaluation: interactive adaptability in domestic retrofit. Build. Res. Inf. **42**(5), 574–590 (2014)
7. Lowe, R., Chiu, L.F., Oreszczyn, T.: Socio-technical case study method in building performance evaluation. Build. Res. Inf. **46**(5), 469–484 (2018)
8. Lowe, R., Chiu, L.F., Oikonomou, E., Gleeson, C., et al.: Analysis of data from heat pumps installed via the renewable heat premium payment (RHPP) scheme. Case Studies Report from the RHPP Heat Pump Monitoring Campaign, Mar 2017. https://www.gov.uk/government/publications/detailed-analysis-of-data-from-heat-pumps-installed-via-the-renewable-heat-premium-payment-scheme-rhpp

# Chapter 61
# An Evaluation of Offsite Timber Frame Manufacturers in Wales, UK

**F. Zaccaro, J. R. Littlewood, R. Lancashire, G. Newman and D. Hedges**

**Abstract** The Knowledge Economy Skills Scholarship two (KESS2) doctoral project undertaken by the first author has contributed research towards work package 4 (WP4) of the Home-Grown Homes project (HGHP), focused on disrupting the challenges that can lead to a supply chain of home-grown timber (UK) for high performance and healthy homes. This paper discusses the element of research within WP4 that has evaluated the manufacturers of timber frame (TF) construction systems manufactured offsite. The Welsh Government's Innovative Housing Programme introduced in 2017 is helping to immensely incentivize the shift to offsite manufacturing (OSM) in Wales where increased systemization is requiring a profound re-think of how Wales conceives and delivers and shifts towards nearly zero-energy dwellings, from 2020. Context to the UK need to drive efficiencies in housing supply and performance is given, highlighting the current and future drivers for the development of OSM. This paper will be of interest to researchers engaged in projects to disrupt conventional thinking and investigate the challenges of using softwood timber, grown and processed offsite, for their own country and for the export markets within Europe.

## 61.1 Introduction

The European Commission (EC) has always aimed to investigate some of the environmental challenges, such as depletion of natural sources, reducing dependency on non-renewable resources and mitigating climate change [1], particularly in a year which has seen evidence of considerable climate change [2]. Within this context,

F. Zaccaro (✉) · J. R. Littlewood
Sustainable and Resilient Built Environment (SuRBe) Group, Cardiff Metropolitan University, Cardiff CF5 2YB, UK
e-mail: fzaccaro@cardiffmet.ac.uk

R. Lancashire
TRADA, Chiltern House Site, High Wycombe HP14 4ND, UK

G. Newman · D. Hedges
Woodknowledge Wales, Ffarm Moelyci, Tregarth, Bangor LL57 4BB, UK

 
J. Littlewood et al. (eds.), *Sustainability in Energy and Buildings*, Smart Innovation, Systems and Technologies 163, https://doi.org/10.1007/978-981-32-9868-2_61

the construction sector contributes 42% of final energy consumption, 35% of greenhouse (GHG) emissions, 50% of extracted materials and, in some regions, 30% of water consumption [3]. According to the European provisions, the Welsh Government (WG) in 2015 sets out the Well Being of Future Generations Act (2015) to share a common vision of a nation for a sustainable future, with seven pillars of excellence to achieve. Notwithstanding the WG intentions, in 2019, Wales is facing several challenges such as climate change, poverty, health inequalities, job security and growth requirements [4] that will affect the actual and the future generations. In this context, woodlands and trees have a role to play in enabling the achievement of various international and domestic climate targets such as the 2015 Paris Agreement on climate change [5]. The construction and the offsite manufacturing (OSM) sectors have the possibility of improving the social, economic, environmental and cultural well-being of Wales [4] with a collaborative and collective approach based on agreed goals and strong general commitments. Given the shortage of housing and the environmental challenges of carbon reduction, cost and speed of build, and life-time affordability in energy cost performances, the timber frame (TF) systems meet more than all these pressing needs [6]. The development of the KESS2 doctoral project, co-funded by the European Social Fund (Low Carbon and Materials stream) in collaboration with Woodknowledge Wales (WKW), is presented in this paper, discussing the adoption of innovative engineered timber and OSM techniques in Wales, and the use of TF and their current practice.

## 61.2 Context to Wales' Timber Supply Chain

Wales, with 14.3% (306,285 ha) of woodland cover in 2011 [7] and despite the ambitious desire to increase forest area by 100,000 hectares (ha) by 2030 [8], has created just 3500 ha of new woodland between 2010 and 2016 [9]. The Welsh forest industry must be incentivized by a larger market for timber products, thus the construction industry could help to link the demand for new woodlands while showing the capacity of wood to be utilized within high performance building. Through the mechanism of demand and supply, more trees will be planted, acting as carbon sink, which helps to reduce the climate change disastrous effects. In addition, there is an increased need to drive efficiencies in housing supply and performance with the current and future skills shortages in the UK construction industry (as articulated in the Farmer Review—Modernise or Die) and help to encourage a shift to OSM [10]. In 2016, the TF construction method accounted for 30.7% of all new builds recorded in Wales [11], with a total capacity of 2990 units/year, generating £28 million, and a willingness to expand to over 4300 units [6]. The WG's Innovative Housing Programme (IHP) introduced in 2017 with £100 million is helping greatly incentivize the shift to OSM in Wales, by funding 100% of the cost of innovation [12]. The advantages of OSM over conventional construction of housing include quicker completion, greater quality of finish, less defects and minimal onsite duration [13]. The performance tends to be much nearer to design aspirations than conventional construction techniques [14]. The move to OSM and increased systemization is requiring a profound re-think of how Wales conceives and delivers industrialized housing [15]. In 2018, the open-panel TF systems accounted for around 71% of the total market and are the largest proportion of UK timber production, followed by closed-panel TF systems (11%) [16]. Therefore, it is necessary to recognize the current structure of the manufacture while assessing the Welsh OSM timber construction systems in use, with a focus on engineered timber solutions for the building fabric, that is, exterior walls, floors, ceiling and roofs in innovative homes. One pan-Wales's wide research project that is attempting to disrupt the thinking around using timber grown and processed in the UK for OSM is the Home-Grown Homes project.

## 61.3 Home-Grown Homes Project

WKW was appointed in 2018, in partnership with the SuRBe group at Cardiff Metropolitan University (CMU), Timber Research and Development Association (TRADA) and Coed Cymru, by Powys County Council (PCC) to lead the delivery of an ambitious exemplar construction programme to provide demand-led stimulus for forest sector development in Wales: the HGHP. The HGHP represents the next development of the entire Welsh timber supply chain with the objective of creating jobs in growing, harvesting, processing and manufacturing of homes from natural timber resources. HGHP is funded by Welsh Government through the rural development programme [17] and aims to build an important collaboration with building clients, developers, contractors and the timber supply chain, providing a compelling business case for expansion of timber construction while driving the growth of local OSM and the use of home-grown timber [18]. The HGHP has seven work packages (WP) with WP4 aimed at 'more and better local manufacturing' that is led by TRADA in conjunction with the SuRBe group at Cardiff Metropolitan University; see [19–21] for more information.

## 61.4 Home-Grown Homes Project Work Package 4 and KESS2

WP4 of the HGHP is supporting the business case for increasing the volume and quality of construction systems supplied from Welsh manufacturers using home-grown timber [19–21]. The central feature of WP4 is to explore opportunities to increase systemization and standardization of Welsh manufactured solutions for improved cost and performance outcomes. The position of the TF OSMs in Wales is discussed in this paper following an evaluation between June 2018 and March 2019, by the authors of this paper. The KESS2 Ph.D. study being undertaken by the first author aims to learn from the literature published on OSM using timber, and also the TF sector in Wales in order to develop a pattern book (PB) for developers of affordable dwellings for rent and sale that achieve nearly zero-energy to zero-energy dwellings. The PB to be produced by the KESS2 project [19, 21] is in two parts. Part one of the PB (PB_1) is to model/calculate and validate thermal bridge details for two types of TF OSM exterior wall systems: (a) a closed panel with synthetic non-breathable insulation manufactured by Sevenoaks Modular [22] and (b) a closed panel with natural breathable insulation, which the OSM is yet to confirm. The development of PB_1 is being undertaken between April 2019 and December 2019. Part two of the PB (PB_2) is to conduct dynamic thermal modelling using the validated thermal bridge details from the two TF OSMs, for three dwelling types, which are forecasted to be in demand across Wales and to be developed by the housing association partner (Wales and the West Housing Association) in order to achieve nearly zero-energy and zero-energy standards from 2020. The development of PB_2 will be undertaken between January 2020 and May 2021.

## 61.5 Evaluation Methodology

In order to evaluate the TF OSMs operating in Wales, the researchers used a multi-methodological approach after Dainty [23] to collect both qualitative and quantitative data. The research has established the key characteristics of the sector, identifying what is manufactured from timber in Wales and, at the same time, obtaining the manufacturers' views on the future of the industry. Three different methodologies have been used, after ethics approval was granted by CMU in 2018. These included: first, a desktop study which developed an assessment matrix to evaluate the main features and innovation of Welsh TF OSMs through their websites, to identify key features and to collate the sample population to investigate. Secondly, five organizations were selected for interviews to cover a diverse group in terms of company age, size, location and customer base, and part of a pilot study for the OSM operating in Wales. Availability of Directors at each sample has been an important criterion for selection, since most of the population identified in the desk-based interview were not interested or too busy to be interviewed. In order to collect what is often tacit knowledge, the interview process was expected to take between 2 and 3 hours, to include also a guided tour of each OSM's facilities and production line/s. The questionnaire (with 13 open questions) used in the interviews sought to identify data about the organizations and their work, as well as opinions about the challenges and opportunities ahead. The interview questionnaire was designed to capture both quantitative and qualitative information, allowing comparative analysis and enabling a specified structure of answers within the freedom of speech left to the participant. Thirdly, a questionnaire was developed, delivered by email and completed without the researchers present. This second questionnaire included refined/additional questions following the analysis of the results of the pilot stage (two) and was developed due to the budget and time constraints of the KESS2 project, as typically half to one day was taken in the second stage data collection with up to four researchers present.

In particular, this second questionnaire allowed specific queries about lean practices, collaboration with the other supply chain actors, barrier to further expansion, quality checks, availability of specific competences and machines, sustainability measures in use and future development plans. The questionnaire was articulated in 18 structured closed questions, with possibility to insert comments. The final investigation phase aimed to cover a larger example of the population, enabling the researcher to draw general conclusions on the state-of-the-art of the Welsh TF OSMs.

A range of screening techniques have been used and the selected 36 from a possible 56 TF OSMs in Wales were identified. In summary, the screening included identifying members of trade bodies, such as the Structural Timber Association (STA) and TRADA, then analysis of websites and testing phone numbers, plus purging organizations that traded under several brands. The total number of independent and active TF manufacturer in Wales selected with the filtering operations specified was 36.

## 61.6 Results

The results of the three phases have shown that TF OSM organizations in Wales have operated up to 25 years and are classifiable as small to medium size enterprises (SMEs). These SMEs tend to work with low profit margins (around 3/4%), thus without any significant investment in innovation and research. They are linked to the traditional TF offsite techniques, such as TF panelling with/without insulation and without any finished linings applied [24], and completely detached to the modern industrialized automated process seen in the more advanced Scottish manufactures [25]. It appears that the pace of change in the sector has been slow and marginal, with small tweaks to design and/or manufacturing methods rather than significant innovation in product or process. As a consequence, it can be argued that significant changes are often driven by cost savings or a need to comply with new regulations. The Welsh TF manufacturers can offer different services: from manufacture of planar elements (consisting of roofs, walls and floors and deemed the main construction typology used in Wales) to delivery and erection of the production with specific transport/lifting equipment. The production is mainly focused on open and pre-insulated panels but there are some relative product innovations: Closed panel technology is becoming established and produced, but still by very few manufacturers; modular/volumetric house production starts to be used on specific project where high performance and continuous repetitiveness are required. The manufacture of such products is linked to the building typology to be constructed and to the speed of construction required. Too often, the approach required by customers prevents any kind of realistic systemization, standardization or repetition in design and manufacture, so each project is effectively bespoke, requiring a tailored manufacturing process each time. A bespoke solution is inevitably more time-consuming to set up and carry out, making it difficult to achieve any economy of scale.

It is apparent that there is a serious lack of knowledge about timber and TF manufacturing outside of the sector—clients, designers and funders seem to fail to see its added value, and as a result, do not value its potential. This is the reason why the manufacturer is involved at the end of the design chain, rather than near the start. So, for example, customers who want a TF delivered 'as soon as possible' or have already laid a foundation slab before approaching a manufacturer appears to occur in many projects, confirming that the TF OSM is one of the final design choices in the construction of buildings using timber in Wales. This behaviour prevents any real influence over design and specification and often sees the discussion focus more on cost than on quality. Consequently, the manufacturers are only building what they have been asked to. Early engagement with manufacturers would help improve the quality of the end product.

The manifested fear for new investment seems to be due to the Welsh TF construction market share, accounting for 30.7% according to the STA statistics during 2016 [11]. Further uncertainty is brought by Brexit and the consequent repercussion on the timber supply market after the closure of the EU border [26]. The risk of higher importing cost and the effect on sterling is believed of great impact on the actual raw timber market. The British timber market is heavily dependent on import and scarcely supplied with native trees. A lot of improvements are needed on the forestry side, increasing the woodland cover. The importance of creating and sustaining a consistent flow of activity for manufacturers and their employees is important and crucial for any kind of effective production process to succeed. As a result, the cashflow needs to be maintained and business investment decisions are to be made.

Finding people with the right skills needed is a challenge; principally not only in TF design but also in erection, when sub-contracted. Manufacturers report difficulties in accessing good quality groundwork services and often must spend a great deal of time packing and adapting frames before or during erection on site. Consequently, there are various discussions taking place with organizations which might be able to help meet the skills gap through apprenticeships, training and employment support [27], but these are yet to mature. Some manufacturers are large enough to employ their own high-skilled professionals: designers and structural engineers are growing their own talent with practice, although there is a perception that TF manufacturing suffers from the same negative perception of boom/bust construction industry cycles, struggling to attract young people. Manufacturers said that a long-term commitment from government to fund, for example, social housing and/or TF skill development programme was key to confidence.

The design, when delivered by the in-house team or by sub-contracted specialized engineers, uses computer-aided design (CAD) specific software; the building information modelling (BIM) method seems to be far from a daily use, but a lot of initiatives have been taken to satisfy regulation requests. The survey found little evidence of the use of BIM computer software tools with few manufacturers being asked to use it. However, a range of bespoke and generic software and electronic solutions are being employed to design and then manufacture frame components. There is little sign of widespread use of tablets on site for inter-communication and updates with the OSM.

The manufacturing processes happen without the assistance of computer numerical controlled (CNC) machines, nor any degree of automation; exception made for one manufacturer which showed on the shop floor an automated CNC (computer numerical control) cutter; the production hardly relies on manual and semi-automatic machines controlled by skilled workers. All materials used are PEFC (Programme for the Endorsement of Forest Certification) or FSC (Forest Stewardship Council) certified, which represent the protocols adopted for wood and supply chain certification [28]. These certificates ensure that the timber is purchased from well-managed forests and/or recycled materials, ensuring the sustainable harvesting of the trees. The manufacturers have shown little to no use of home-grown timber. Whist there is interest in sourcing home-grown timber, it appears that there is only occasional demand from clients. This is as a consequence of supply problems, with most using timber from Scandinavia and Eastern Europe. The quality of home-grown timber in the past had been a concern for some manufacturers. It is also clear that the use of timber cladding, fascia's or windows is limited, with other materials perceived to be cheaper and of lower maintenance. The importance of protecting TF immediately after manufacture, during delivery and erection against the consequences of poor weather (typical in Wales, with driving and prolonged spells of rain), would appear to vary. Storing of raw materials and manufactured TFs occurs to be generally external without coverage, exposing the production to the weather and potentially damage, even though the panels are deemed safe because protected by the membrane applied, acting as perceived water barriers. In exposed conditions, timber can reach a moisture content higher than the equilibrium moisture content. As opposite, when installed, moisture will be desorbed which can cause wood to shrink, resulting in dimensional changes that can harm the performance [29]. There are examples where some of the manufacturers store the raw timber and manufactured TFs under cover. As a result, for some manufacturers there are limitations on the space available for storage or growth, where delays in construction programmes mean that even with just-in-time delivery TF stock is often stored outside under the weather, awaiting delivery to site for up to 6 months. At the same time, moving a business to larger premises comes at a cost and brings risks if the market declines.

The OSM TF industry could gain substantial improvements in efficiency including lean concept within the production [15]. In 2019, Welsh manufacturers use a basic lean approach. The customization seems to be matched by all the manufacturer specificities, which offer the client an 'ad hoc' product, considering the performance required (e.g. thermal, structural, acoustic, fire, etc.). The shop floor is flexible and adaptable to case by case solutions; the semi-skilled to multi-skilled workforce can customize and respond to unplanned changes and continuously improve the process through practice. The reduction of timber waste is not always managed; only one of TF manufacturers used a combined heat and power boiler, which burns factory wooden by-products to produce heat and to dry fresh green chipped wood, while electricity power is used to feed a small extent of the factory machines. Just-in-time production concept are apparently adopted by few manufacturers, who produce, deliver and erect subsequently; this process seems not to be perfectly implemented, as the manufacturers still require large stocking space or anyway keep a lot of the production and supplied product in the yard stocking space because of last minute site delays. Transfer concept of lean production from manufacturing to construction is still difficult because of the great variety of products, but not impossible: the TF industry need to be more efficient and better organized to supply the future housing requests. Standardization, quality control and reduction of uncertainties [30] seem to be a pre-requisite to build affordable and efficient houses.

To improve the performances of the industry in Wales, it is necessary to develop a close collaboration with research, supply chains and academic institutions. Some manufacturers already collaborate proactively with other manufacturers and with constructors, building solid and long commercial relations, based on workloads and niche jobs, where the adaptation of the shop floor is not convenient. This reflects the outcome of a Finnish study [31], where manufacturers have shown particular reluctance for change and risk-taking operations. Only one of the samples was collaborating with a University, in a 3-year Knowledge Transfer Partnerships (KTP) project aiming to improve competitiveness and productivity through the better use of knowledge and technology [32]. These kinds of experiences are necessary in a slow-changing sector like construction, and are useful to build up the knowledge to deal with problems and move forward the industry.

## 61.7 Discussion

The TF OSM industry Wales appears to be stuck in the middle of a highly competitive procurement route, driven by low cost and late engagement from clients, but demanding quality and speed. There is often friction between TF OSMs and their clients, leading to many challenges and creating buildings with compromises. Many projects are not designed with TF in mind and are converted in final design decisions. At the same time, late changes obstruct the workflow and create waste [30], reducing some of the TF benefits and making structural design of the building unnecessarily more difficult and expensive. Contract terms stacked against TF OSMs by their clients and payment delays expose them to risk, particularly should the client get into financial difficulties. It is important that the Welsh TF OSMs build bridges with buildings contractors and developers, so that they understand the benefits they can offer, how it fits alongside other trades and how efficiencies and performance can be realized. TF OSMs have so much more to offer the client if early engagement and collaboration are possible. End users of buildings often do not know or care what construction systems are adopted or understand the benefits that TF can offer. If, through education, clients could be led towards specifying TF OSM, further benefits for all involved could be realized. Recent examples of some UK-based TF manufacturers partnering with housing associations and developers demonstrate that this is possible [33].

The drive towards modern methods of construction (MMC) means that parts of buildings made in a factory will be arriving on site to a standard of finish and tolerance that the construction industry is not used to, in the UK. Site alterations are necessary to fit and avoid exposure to weather or slow construction adding cost. All sectors of the wider construction industry will need to improve procedures to meet these new challenges presented by the TF needs, to be more proactive for the long-term benefit of the product. With an ageing workforce and the demand for more homes to be built every year, the construction industry is slowly walking into an impossible situation. Carrying out repetitive tasks in a factory environment where quality can be managed and site skills reduced may offer the answer. Many items made in factories are mass produced; however, although some developers use standard house types, many buildings are unique. This leads to less efficient factory production and a higher risk of mistakes as each building becomes its own prototype. There is great potential for lean production methods used more generally in manufacturing to be applied to TF factories [30]. TF has been built in factories for many years and is well placed to raise its market share, particularly if it can move to the next level of factory production.

There is currently little excitement about using Welsh timber in the TF industry as long as it is timber of adequate strength and quality. If the food industry can create a demand and premium for a home grown offering [34], there is a potential through marketing of local timber to do the same. Some self-builders who have access to local timber go to great lengths to use it and feel better for the experience. Further work on the forestry side is needed to ensure that supply can sustain future demand. With short-term costs and regulations driving innovation, changes are often slow in the construction industry. New government legislation, such as the recent announcement that gas boilers and hobs will be banned from new properties from 2025 [35], will see rapid changes in building fabric performance and TF is well placed to capitalize on these. If we can educate clients and contractors, demonstrate performance and create a demand for home-grown timber, the buildings of tomorrow could be entirely TF. Findings of the engagement with the TF OSMs are informing the development of the PB in collaboration with Wales' largest developer of affordable homes [36] and one TF OSM [32].

## 61.8 Conclusions

This paper has given context and explicated WP4 of the HGHP and the KESS2 doctorate study, both of which are investigating the challenges in using home-grown timber to deliver high quality, high performance and healthy homes for Wales. The findings from the survey will be used in the following phases of the project where the author will review different OSM systems in terms of cost, building performance and the applicability to the Welsh context in terms of housing requirement and potential for the use of home-grown timber now and into the future. Next challenges and phases of the KESS2 project as well as the potential outputs have been discussed.

**Acknowledgements** The KESS2 Ph.D. Scholarship has been supported by the Low Carbon, Energy and Environment/Advanced Engineering Grand Challenge Economic Areas Sectors, administered by the Welsh European Funding Office in conjunction with WKW Ltd.

## References

1. EC: Innovating for sustainable growth: a bioeconomy for Europe. Ind. Biotechnol. (2012)
2. UNFCCC: United Nations Climate Change Summit. United Nations (2019). https://www.un.org/en/climatechange/un-climate-summit-2019.shtml
3. EC: Roadmap to a resource efficient Europe (2011). http://ec.europa.eu/
4. WG: The well-being of future generations (Wales) Act 2015 (2015)
5. WG: Woodlands for Wales. The Welsh Government's strategy for Woodlands (2018)
6. EGAN consulting: Timber frame housing manufacture in Wales and its capacity and capability (2017)
7. Forestry Commission: NFI 2011 woodland map Wales (2011)
8. Osmond, J., Upton, S.: Growing our woodlands in Wales (2012)
9. National Assembly for Wales: The Welsh Government's progress on climate change mitigation. Annual report of the climate change, environment and rural affairs committee (2018). www.assembly.wales
10. Farmer, M.: The farmer review of the UK construction labour model (2016). https://www.gov.uk/government/publications/constructionlabour-%0Amarket-in-the-uk-farmer-review
11. Structural Timber Association: Annual survey of UK structural timber markets. The UK house-building market 2016 (2017)
12. WG: Innovative housing programme (IHP) 3, 0–20 (2019)
13. Pan, W., Gibb, A., Dainty, A.: Offsite modern methods of construction in housebuilding perspectives and practices of leading UK housebuilders. Strategies (2005)
14. Hetherington, D.: Delivering new homes—a future off-site?. Northern House Consort 2016. https://www.northern-consortium.org.uk/2016/04/29/delivering-new-homes-a-future-off-site/
15. Höök, M.: Lean culture in industrialized housing—a study of Timber volume element prefabrication. Environ Eng. (2008)
16. STA: Timber frame construction market UK (2019)
17. Arwain: Home Grown Homes (2018). http://www.arwain.wales/en/news/archive/article/home-grown-homes.html
18. Newman, G., Binding, T.: Feasibility study for the development of the Home-Grown Homes supply chain development service. 2017
19. WoodKnowledge Wales: The Home Grown Homes project. http://woodknowledge.wales/prosiect-cartrefi-o-bren-lleol-home-grown-homes
20. Littlewood, J.R., Zaccaro, F., Wilgeroth, P., Whyman, T., Karani, G., Evans, N., et al.: Systemised offsite manufactured timber dwelling typologies from UK forestry supply chains, a transition to nearly zero carbon homes in Wales. Smart Innov. Syst. Technol. **131**, 425–434 (2019)
21. Zaccaro, F., Littlewood, J., Wilgeroth, P., Whyman, A., Newman, G., Lancashire, R., et al.: An introduction to systemised offsite manufactured and engineered timber dwelling typologies from Welsh and UK forestry supply chains, enabling transition to nearly zero carbon homes in Wales. Sustain. Ecol. Eng. Des. Soc. Conf. Dublin 734–742 (2018)
22. BBA: Prefabricated timber frame panels—Triso-warm external wall panels (2016)
23. Dainty, A.: A call for methodological pluralism in built environment research. In: CIB World Building Congress. Build a Better World. Postgraduate Plenary Lecture (2010)
24. Hairstans, R., Sanna, F.: A Scottish perspective on timber offsite construction. Offsite Archit. Constr. Futur. Taylor & Francis, Routledge (2017). http://researchrepository.napier.ac.uk/Output/931004
25. CCG Construction and Manufacturing Group CCG (2019). http://c-c-g.co.uk/
26. Michael, P.: What will be the impact of BREXIT on the UK construction industry ? 5–6
27. Welsh Government: Woodlands for Wales action plan (2016). http://gov.wales/docs/drah/publications/160223-woodlands-for-wales-action-plan-en.pdf
28. Coulson, J.: Sustainable use of wood in construction. Int. Wood Prod. J. (2014)
29. Shmulsky, R., Jones, P.D.: Wood and water. For. Prod. Wood Sci. 141–173 (2019)

30. Johnsson, H., Sardén, Y.: Industrialised timber housing: from trial to production. In: Proceedings of 24th Annual Conference on Association of Researchers in Construction Management, ARCOM 2008, pp. 155–164
31. Hurmekoski, E., Pykäläinen, J., Hetemäki, L.: Long-term targets for green building: explorative Delphi backcasting study on wood-frame multi-story construction in Finland. J. Clean. Prod. **172**, 3644–3654 (2018)
32. Littlewood, J.R.: Innovative construction & offsite manufacture—Trisowarm (2019). https://surbe.org/innovative-construction-offsite-manufacture/
33. McColgan, C.: Work starts on £10 million Cardiff social housing development (2018). https://businessnewswales.com/work-starts-on-10-million-cardiff-social-housing-development/
34. More People: MorePeople | 7 in 10 UK consumers want homegrown produce (2018). https://www.morepeople.co.uk/knowledge-centre/article/7-in-10-uk-consumers-want-homegrown-produce/801845826
35. Harrabin, R.: Gas heating ban for new homes from 2025. BBC News (2019). https://www.bbc.co.uk/news/science-environment-47559920
36. Welcome to Wales & West Housing (2019). https://www.wwha.co.uk/

# Chapter 62
# Building Performance Assessment Protocol for Timber Dwellings—Conducting Thermography Tests on Live Construction Sites

**J. R. Littlewood, D. Waldron, G. Newman, D. Hedges and F. Zaccaro**

**Abstract** This paper introduces the pan-Wales (UK) Home-Grown Homes (HGH) project (2018 to 2020) which focusses on three areas of improvement for delivering high performance, affordable and healthy homes. The HGH project is funded by Powys County Council, through the European Regional Development Fund's Agricultural Stream. The HGH project is being delivered by Woodknowledge Wales in a consortium with Cardiff Metropolitan University (CMU), TRADA and Coed Cymru, with seven work packages. 'More and Better Homes from Wood' (work package (WP) WP3) focusses on the assessment of building performance for dwellings using timber, and is being delivered by a multi-disciplinary team at CMU through the Sustainable and Resilient Built Environment (SuRBe) group. This paper discusses the context and need for the HGH project as Wales launched its low carbon agenda in March 2019. The focus of this paper on introducing the building performance assessment (BPA) protocol to be implemented by SuRBe across several housing case studies in Wales, through the design, in-construction and occupancy phases, to address thermal and fire (TaF) performance issues, and impacts on occupants' quality of life, comfort and safety. Preliminary results of in-construction testing on a live construction site are presented, with the challenges of conducting thermography tests whilst construction is in progress and weather conditions in spring in the UK (April 2019). This paper will be useful for academics, architects, building contractors, housing developers and professionals undertaking building performance assessment and evaluation on live construction sites.

---

J. R. Littlewood (✉) · D. Waldron · F. Zaccaro
The Sustainable & Resilient Built Environment Group, Cardiff Metropolitan University, Cardiff CF5 2YB, UK
e-mail: Jlittlewood@cardiffmet.ac.uk

G. Newman · D. Hedges
Woodknowledge Wales, Ffarm Moelyci, Tregarth, Bangor LL57 4BB, UK



J. Littlewood et al. (eds.), *Sustainability in Energy and Buildings*, Smart Innovation, Systems and Technologies 163, https://doi.org/10.1007/978-981-32-9868-2_62

## 62.1 Introduction

This paper introduces the HGH project, WP3, and the building performance protocol being adopted as part of work on housing case studies across Wales. The main aim of this protocol is to gather evidence with a methodology that can be replicated and validated as it is undertaken on live construction sites. The suitability and reliability of in-construction tests are evaluated. This article describes the development of the protocol, the main concepts behind it and provides a case study to exemplify the challenges, advantages and limitations of investigations of this nature. The objective is to contribute towards the improvement of buildings, in order to tackle the issues related to the performance gap and the safety gap, and consequently improving the efficiency of buildings' and occupants' safety and well-being.

## 62.2 The Need for Building Performance Assessment of Timber Dwellings

### *62.2.1 Context to Wales and the Need for Low to Nearly Zero Carbon Dwellings*

Since 2008, the Climate Change Act has been a relevant part of the UK's commitment to develop a number of strategies towards reducing greenhouse gas (GHG) emissions [1]. The established targets (in 2008) were to achieve at least 34% GHG reduction by 2020 and 80% by 2050, against 1990 baseline [1, 2]. The energy use in the residential stock account for more than 25% of energy use and carbon dioxide ($CO_2$) emissions in the UK [1, 2]. As it was eloquently described in the 'Farmer Review – Modernise or Die' [3] and the UK Government plays an active role towards the development of better new homes, it has the potential to influence the way buildings are delivered [3]. This can be better achieved by having clear and transparent mechanisms in place, to ensure the quality of the end-product (buildings). This process will create a long-term sustainable construction industry and will have quantifiable benefits in terms of productivity, quality and efficiency [3]. More recently, in February 2019, a publication by the Committee on Climate Change (CCC) and its Adaptation Committee reported that 'UK's legally-binding climate change targets will not be met without the near-complete elimination of greenhouse gas emissions from UK buildings' [4].

Therefore, it is evident that there is a significant need to increase the quality and efficiency of dwellings in the UK. Research evidence supports the need for clearer methods towards truly achieving better buildings in a more transparent manner [5]. The construction industry should follow specific practices that will help not just to 'aim' to achieve better quality of buildings, but to 'ensure' high standards and that truly efficient and safe buildings are being delivered. This could be achieved by having further mandatory compliance tests to ensure that buildings are safe and efficient [5]; that is, in terms of fire safety as well as thermal efficiency, this will also help to reduce fuel poverty and increase quality of life of residents.

In 2017, the Welsh Government (WG) (Wales is one of four sovereign states with some unitary authority in the UK) introduced the Innovative Housing Programme (IHP) [6]. The IHP funding is aimed at helping to incentivise further research towards increasing locally sourced timber construction, by investing in local resources, skills and industries [6]. The HGH project is an example of this investment. Furthermore, by 2020, the WG's target for new affordable dwellings is 20,000 and the traditional forms of construction with brick/block exterior walls are both unsustainable from a materials and climate change perspective, but are also slow to construct compared with timber construction [7].

### *62.2.2 Context to the Performance Gap and the Safety Gap*

One of the greatest problems of new housing is the performance gap, where defects in the construction process can impact upon thermal performance and operational energy use for heating once occupied—a major finding of the UK Government's Building Performance Evaluation programme in 2016 [8, 9]. There is significant and compelling evidence on the existence of the performance gap in buildings and its impact on the energy efficiency of dwellings and thermal comfort [10–13]. The performance gap can be described as the 'discrepancy between the measured and the theoretical energy performance' [13], also described as 'a "Knowledge Gap" between off-site designers and on-site constructors and within both the policy makers and administrators, including local government building control officers' [5]. There are many factors inducing the presence of the performance gap in the building industry; some of these factors are the inconsistencies in test guidelines, financial pressures, time constrains that lead to short-term fixes [5], as well as the lack of compulsory monitoring standards, particularly at the construction stage, where many issues can be concealed.

Littlewood has gone a step further and identified and entitled a 'safety gap' [5, 13–15], and through his work, co-funded by Sustainable Construction Monitoring and Research Ltd, and Cardiff Metropolitan University (CMU), he developed and implemented a non-destructive building fire safety assessment, measurement and reporting BFS&R protocol that illustrates where the construction process and products can impact on inadequate fire safety performance, particularly with a lack of or inappropriate fire stopping and compartmentation. Problems with passive fire protection measures were demonstrated catastrophically in 2017 by the Grenfell Tower tragedy [16]. In 2019, The UK Fire Industry Association (FIA) highlighted that close to 40% of all new homes built in the UK are unsafe with zero fire stopping installed to prevent the spread of fire, smoke and toxic gas [17]. These issues are creating significant concerns regarding the safety of new build homes. Therefore, the imperative importance is to find solutions to these two significant issues: the safety gap and the performance gap, or thermal and fire performance (TaF) [18], which are in many cases linked by unwanted air leakage pathways inherent following inadequate workmanship on site. TaF performance is being investigated by the CMU's team as part of their contribution to the HGH project, which commenced in April 2018 [18].

## 62.3 Methodology

### 62.3.1 Work Package (WP) 3: More and Better Homes

The BPA protocol for the HGH project is discussed below, and covers the design, in-construction [1] and operation phases of timber dwellings in Wales, which has been developed and is to be implemented as part of WP3 'More and Better Homes from Timber'. WP3 aims to develop, enhance and prove the business case for timber construction, including offsite manufacturing (OSM) with particular attention on architectural technology and building fabric performance [19]. Its focus is on supporting a number of active housing projects with local authorities, housing associations and housing developers by proving guidance on how to assess building performance and impacts on occupants' safety and well-being, through design, construction and occupation. This is to help clients understand the role of building performance in the procurement of future housing projects and ultimately in developing better, more energy efficient, and healthier and safer homes. In doing so, the HGH project will be able to establish the value of this kind of assessment and the challenges in carrying it out, and as such, complements the WG's Innovative Housing Programme (IHP) since it undertakes building performance assessment during the design stage and also the construction stage on site, with a focus on thermal and fire (TaF) performance; the habitation stage focussed on occupants' well-being and comfort. The focus of the IHP project's [20] building performance monitoring will be primarily energy focussed (but in May 2019 this has not as yet been tendered for delivery). So both projects are mutually beneficial to clients, their design teams and dwelling occupants.

A brief description of the different research practices and processes described in the BPA protocol developed by the CMU team based on their previous experience that is being used in the HGH project is summarised here: at the design stage—dynamic thermal modelling (3D) and thermal bridge analysis (2D); at the construction stage—in-construction testing: Littlewood's smart fire performance test to prevent the safety gap, thermography for assessing the fabric thermal performance and air permeability tests; at the operation stage—occupants' interviews with short form (SF)-36 questionnaire and diaries to measure well-being and quality of life. This BPA protocol aims to assist parties involved at any stage of the construction process, in order to increase the efficiency and quality of buildings. This will be achieved by following a series of exploratory processes that contribute towards a better understanding of the individual buildings and their elements. As a result, it is aimed to further identify and collate evidence on the issues at the design and construction stages that impact on the performance and safety gaps, and as such, occupants' quality of life.

Assessing thermal performance during the construction process was first developed by Littlewood in 2009, as part of a suite of tools known as in-construction testing (iCT) to assess workmanship defects on construction sites, which if present could be rectified before completion [1]. In 2019, iCT is being used to assess both thermal performance and also fire performance, and the next section of the paper discusses a case study in the use of thermography known as an iCT_Th test.

### *62.3.2 Building Performance Assessment (BPA) Protocol—In-construction Testing: Thermography ICT_Th*

Thermography is 'the science of acquisition and analysis of thermal information from non-contact thermal imaging devices' [21]. Thermal 'information' is captured by an infrared (IR) camera, by measuring the intensity of IR radiation emitted from the surface of the object under observation. An IR camera is adjusted by the operator to convert the measured intensity of infrared radiation emitted by an object into a surface temperature (these adjustments compensate for the material and surface properties of the object and the environmental conditions in which it is observed). Hence, the resulting thermal image provides a two-dimensional map of temperature difference over the surface of the object. Thermography is an established technique within the construction industry for checking the continuity of insulation, identifying sources of air leakage, thermal bridging, workmanship, detecting and mapping moisture in a building [22]. The output when using an IR thermal camera is a thermogram, which illustrates hot and cold areas as different colours [23]. In most instances qualitative thermographic building surveys provide sufficient information, for example to identify: sufficient, correctly installed and thereby continuity of insulation; occurrences of thermal bridges; sources of air leakage, particularly at critical construction junctions; moisture and damp within an element; hidden components, such as pipes and wall ties; and electrical faults [24, 25]. As a minimum before an iCT_Th test, the developer and building contractor are asked information, as follows. Client: drawings, building plans, sections and elevations; construction details, including 2D/3D models (BIM) if available; details of location; altitude of the building; specification of materials; heating and ventilation systems; availability for meetings.

Building contractor: Details of the type of insulation, external cladding; facilitating access on site in a safe and efficient manner; suitable access to electricity supply, that is transformers, extension leads; occasional observation of the heaters to make sure they remain activated for 8 h before the test; collaborating with the research team to some extent, to ensure that the test can be carried out, that is helping to ensure that the equipment can be safely installed and researchers can gather the necessary data and undertake the necessary test under specific conditions required for the test. The iCT_Th test process includes three stages, a pre-test, on-test and post-test; each process is discussed in more detail here [1] but is summarised next. The pre-test process is undertaken a couple of weeks before the actual test and includes a walk-around site to determine the test dwellings, meeting and discussing the test with the construction staff and if they have identified any particular issues that they would like explored during the test. In addition, the pre-test phase is to identify any issues with supply of power and locations of transformers and cable lengths, or gaining access to the test dwellings after sunset in terms of trip hazards that could impact upon health and safety and so in essence this is part of a risk assessment.

The case study presented in this paper is a pilot study for the implementation of the construction stage assessment using the BPA protocol and thermography, iCT_Th. The aim was three-fold: first, to assist the developer and building contractor understand whether there were inadequate workmanship issues in the test dwelling and secondly for them to understand what is required for a thermography test and the limitations in April in south Wales (related to weather and hour of day for the test). Thirdly, to gain an appreciation by the researchers of working together and also the challenges of conducting a thermography test on a live construction site, and with varying weather conditions. The case study building uses an OSM timber frame system for exterior walls and includes five blocks of four floors, of either one or two bedroom apartments. The construction site is located in Cardiff, Wales and has been in progress for approximately 6 months and is due for completion at the end of 2019. The initiation of the iCT_Th test process was commenced in March 2019 and took until the 16 April 2019 to be completed. As part of the pre-test process, the researchers conducted several visits to the construction site to identify the test dwellings, location and availability of power and the equipment needed to pre-heat the one bedroom flats—in order to record a delta-T of 10 °C (as recommended by Pearson [26]). The equipment used for the test was a thermal camera, heaters and extension cables. The day of the test took several days and weeks to arrange due to unexpected instances on site; see Discussion section. In addition, the weather patterns during spring can be unpredictable in the UK and with daylight saving coming to an end on the 31 March 2019, this meant that the minimum time to commence an external iCT_Th was delayed until 21:00 (1 h after sunset).

## 62.4 Results—ICT_Th

On the day of the iCT_Th test, unfortunately during the day the weather changed and rain started, which precluded an external test (to some relief of the site manager who was due to remain at work until 22:00). An internal iCT_Th test was undertaken in one flat as the building in question restricted the use of two (connected) 110 V/32 A extension cables connected to a transformer on the ground floor (a consideration when the flat on the first floor). Whilst the team has two 110 V/32 A electric heaters, without a splitter (an electrical socket with one male 110 V/32 A connection and two female 110 V/32 A connections), it was only possible to connect one 110 V/32 A 3 kW heater). Ideally, following Pearson [26] the heater would have been activated 24 h prior to the iCT_Th test. However, since the building is timber frame, and due to fire safety regulations, and as there is no site personnel on site between 18:00 and 07:00 the next day, it is not permitted to leave any electrical heating devices unattended between these hours. So the heater was activated by one of the researchers at 07:30 on the test day, and verified as being working by the site manager at various times during the day. The environmental conditions measured during the test are listed here. External and internal temperatures and relative humidity are 15.4 °C/52.7% and 23.6 °C/55.5%. The Delta T was not ideal at 8 °C, however, as Hopper et al. [27] have reported results that can still be evident with a Delta T limited to 5 °C. This is one of the challenges of conducting iCT_Th tests when balanced against changing weather patterns, the end of daylight saving, the availability of case studies for testing, and also the goodwill of site staff; who often have to remain on site late into the evening whilst the test is undertaken due to health and safety procedures. It should be noted that the construction site staff and their housing developer client from the case study could not have been more accommodating and helpful to the research team. The analysis of thermograms considered this issue in the margin of error.

During the iCT_Th test both thermograms and digital images were captured simultaneously with the Flir E60 camera and also images with a digital camera, in order to analyse the performance of the test dwelling currently under construction [28]. One difficulty during the test was the materials of the insulation exposed to the room was finished with a highly reflective material; and the plasterboard finish was not fixed. Thus, this affected the reliability of the thermal camera in terms of the thermogram captured. It is noted that materials with high emissivity values have a low reflectivity, and materials with low emissivity have a high reflectivity [21]. When conducting thermography tests, only materials with high emissivity provide reliable readings, since materials with low emissivity tend to reflect the temperature of surrounding objects and materials [25]. Thus, materials with low emissivity can produce misleading results [27].

Figure 62.1 illustrates some of the thermally tuned thermograms and digital images from the test dwelling.

However, since the iCT_Th test was to be repeated once the insulation was covered as part of the construction process, and one of the aims of the test was to understand the challenges which can be encountered for both the researchers, developer and building contractor the test proceeded. Despite the circumstances with the insulation material, the findings are being documented as learning one of the outputs of the HGH project so that other developers, timber frame manufacturers and researchers can learn from the process. Indeed, the results are useful to illustrate when not to rely on the results of a thermography test to identify continuity of insulation and unwanted air leakage as they can be misleading. Another TaF test as part of the BPA protocol that the researchers are planning is the BFS&R protocol, which is not affected by material characteristics. The process undertaken to analyse the thermograms involved thermally tuning each image before interpreting the results. For internal surveys, any colours indicating extremities of cold entering the building are potential issues (dark blues, purples and black) [28]. But, it is acknowledged that the accuracy of the tuning of the thermograms relies on the environmental conditions being recorded accurately and with calibrated equipment; the camera was in focus;

**1st floor apartment – Internal view of external wall, North Façade (Bedroom)**

**Digital Image 1** illustrates the corner of one of the external walls which is North facing.

**Thermogram 1** illustrates the overshadowing effect of the high reflective coating of the insulation

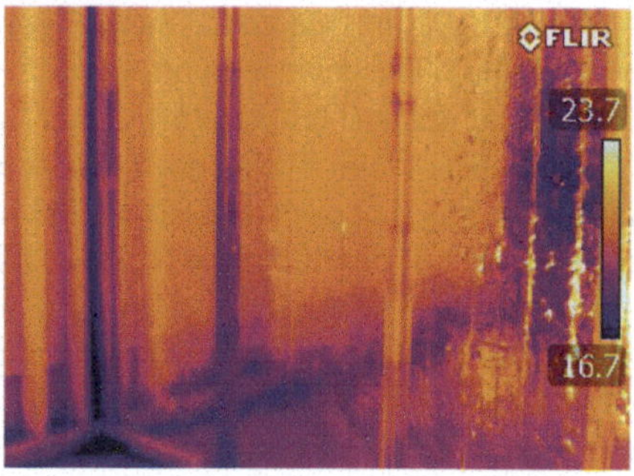


**1st floor apartment – Internal view of external wall, West Façade. Ceiling to wall junction and window detail. (Bedroom)**

**Digital Image 2** Junction detail

**Thermogram 2** Junction detail

**Fig. 62.1** Digital images and thermograms

and the reflective and atmospheric temperatures are recorded for each surface under investigation by the thermal camera. Temperature bars to the right-hand side of each thermogram provide a reference to the scale used (For ease of comparison, all internal images have been given the same temperature range of 10–22 °C and the external images a range of 1.5–3.5 °C.) [28].

## 62.5 Discussion

The development of the test has shown important results to the thermography practitioners. Some of the main learning outcomes from the case study outlined in this paper are described here. It is imperative when conduction a iCT_Th test to always plan extra time that might be expected when conducting a thermography test on a completed building. This is due to the fact that there are many variables affecting the successful undertaking of the test, allowing time for setting up of the equipment (fan heaters, extensions, and temperature and humidity probes) and site checking (feasibility of test, reachability and safety of the specimen, electric power availability and minimal interfering with site operation). Weather conditions must be monitored to avoid non foretasted rain, a frequent occurrence in Wales phenomena (increasing the heat flow exchange internal to external) and direct sun exposition (overcast sky is the ideal condition). Previous assessment of the surface materials of walls should be thoroughly undertaken. If high reflective materials are exposed, it will be better to undertake the thermography test after the material is suitably covered, that is after plasterboard or other finishing coverings with non reflective surfaces is installed. In conclusion, the researchers have learnt that the variables entering the thermography test are high and can be difficult to control for the success of the survey. It is extremely difficult to prevent the unexpected; however, good management and a protocol to follow are very useful. The protocol must contain information about the issues that can be encountered, addressing them and suggesting ways to solve the most likely problems to be encountered. Continuing to validate these methods will be of extreme value to this field of research. The HGH research team is aiming to repeat the test in the autumn of 2019 once the plasterboard has covered the reflective insulation.

## 62.6 Conclusion

This paper has given context to the HGH project, the BPA protocol and challenges and benefits of implementing thermography tests on live construction sites. Performance problems (particularly fire safety) are more important considerations in timber buildings, which are design, construction and manufacturing challenges. This study is adding further evidence of the suitability and advantages of implementing in-construction testing in the development of efficient and more sustainable buildings, by analysing the challenges of implementing these practices in real live projects. This is a step forward towards the evaluation of the overall benefits of undertaking these tests, in terms of achieving greater energy efficiency of buildings, increasing thermal comfort and occupants' safety and well-being. This can be achieved by tackling important issues such as the safety gap and the performance gap and finding viable solutions.

## References

1. Littlewood, J.R.: Assessing and monitoring the thermal performance of dwellings (Chapter Four). In: Emmitt, S. (ed.) Architectural Technology: Research & Practice Editor. Wiley Blackwell, Oxford, UK (2013)

2. Palmer, J., Cooper, I.: United Kingdom housing energy fact file. Department of Energy & Climate Change. Online access to official document developed by Cambridge Architectural Research, with data provided by BRE (2012). https://assets.publishing.service.gov.uk/government/uploads/system/uploads/attachment_data/file/201167/uk_housing_fact_file_2012.pdf. Accessed on 7 May 2019
3. Farmer, M.: The Farmer review of the UK construction labour model. Online access to official document developed by Cast Consultancy, published by the Construction Leadership Council (CLC) (2016). http://www.constructionleadershipcouncil.co.uk/wp-content/uploads/2016/10/Farmer-Review.pdf. Accessed 7 May 2019
4. Committee on Climate Change: UK housing: fit for the future? (2019). https://www.thecccorg.uk/wp-content/uploads/2019/02/UK-housing-Fit-for-the-future-CCC-2019.pdf. Accessed 9 May 2019
5. Littlewood, J.R., Smallwood, I.: Testing Building fabric performance and the impacts upon occupant safety, energy use and carbon inefficiencies in dwelling. Energy Procedia **83**, 454–463 (2015)
6. WG: Welsh Government Consultation on transition to low carbon homes (2017). http://www.senedd.assembly.wales/documents/s71460/Consultation. Accessed 9 May 2019
7. Anon: Offsite manufacturing—a real solution—buy better build more (2017). https://chcymru.org.uk/uploads/events_attachments/Offsite_Maufacturing_-_A_Real_Solution.pdf. Accessed 9 May 2019
8. InnovateUK: building performance evaluation programme: findings from domestic projects making reality match design (2016). https://www.gov.uk/government/uploads/system/uploads/attachment_data/file/497758/Domestic_Building_Performance_full_report_2016.pdf. Accessed 9 May 2019
9. Johnston, D., Farmer, D., Brooke-Peat, M., Miles-Shenton, D.: Bridging the domestic building fabric performance gap. Build. Res. Inf. **44**(2), 147–159 (2014). https://www.tandfonline.com/doi/full/10.1080/09613218.2014.979093
10. CIBSE Energy Performance Group: Carbon bites (2012). https://www.cibse.org/getmedia/55cf31bd-d9eb-4ffa-b2e2-e567327ee45f/cb11.pdf.aspx. Accessed 9 May 2019
11. Meneses, et al.: Predicted vs. actual energy performance of non-domestic buildings: using post-occupancy evaluation data to reduce the performance gap. Appl. Energy **97**, 355–364 (2012). https://doi.org/10.1016/j.apenergy.2011.11.075. Accessed 9 May 2019
12. Dollar, T., Edwardsm, P.: Builders' book—an illustrated guide to building energy efficient homes. Zero Carbon Hub (2016). http://www.zerocarbonhub.org/sites/default/files/resources/reports/Zero%20Carbon%20Hub%202015%20FINAL%20REV%202910_WEB.pdf. Accessed 9 May 2019
13. Littlewood, J.R., Alam, M., Goodhew, S.: A new methodology for the selective measurement of building performance and safety. Energy Procedia **111**, 338–346 (2017)
14. Littlewood, J.R.: Smart fire performance—assessment of occupant safety in specialised dwellings (Chapter 42). In: Sustainability in Energy and Buildings 2018 SEB'18. Smart Innovation, Systems and Technologies. Springer, vol. 131 (2018). https://doi.org/10.1007/978-3-030-04293-6
15. Littlewood, J.R., Smallwood, I.: In-construction tests show rapid smoke spread across dwellings. J. Eng. Sustain. **170**(2), 102–112 (2017). Themed issue on sustainability in energy and buildings – part 2
16. BBC: London fire: a visual guide to what happened at Grenfell Tower (2018). http://www.bbc.co.uk/news/uk-40301289. Accessed 10 May 2019
17. FIA: New build homes are not fire safe. https://www.fia.uk.com/news/new-build-homes-are-not-fire-safe.html. Accessed 5 May 2019
18. Littlewood, J.R.: Assessing thermal, acoustic and fire performance in reality. Keynote presentation at WoodBuild 2018, Cardiff Metropolitan University, 14th June 2018. https://woodknowledge.wales/wkw-resource/event/presentation/dr-john-littlewood. Accessed 15 May 2019

19. Woodknowledge Wales: Home Grown Homes (2019). http://woodknowledge.wales/prosiect-cartrefi-o-bren-lleol-home-grown-homes. Accessed 6 September 2019
20. WG: Innovative Housing Programme (IHP) (2019). https://gov.wales/innovative-housing-programme. Accessed 15 May 2019
21. Infrared Training Center: Thermography Level 1 Course Manual. Infrared Training Center, Sweden (2010)
22. Atkinson, J., Littlewood, J.R., Geens, A., Karani, J.: Did ARBED I save energy in Wales' deprived dwellings? Energy Procedia J. **83**, 444–453 (2015)
23. Lechner, N.: Heating, Cooling, Lighting-Design Methods for Architects, 2nd edn. Wiley, Chichester, UK (2001)
24. Thomsen, K.E., Rose, J.: Analysis of Execution Quality Related to Thermal Bridges. Danish Building Research Institute, Denmark (2009)
25. Atkinson, J.: Evaluating exterior wall insulation. Unpublished Ph.D. Thesis, Cardiff Metropolitan University, Cardiff, UK (2015)
26. Pearson, C.: A BSRIA guide to thermal imaging of the building fabric. BSRIA, UK, BG/9/2011 (2011)
27. Hopper, J., Littlewood, J.R., Geens, A., Counsell, J.A.M., Taylor, T., Thomas, A., Evans, N.: Assessing retrofitted external wall insulation using infrared thermography. Struct. Surv. J. **30**(3), 245–266 (2012)
28. Philip, B. Jones, P., Littlewood, J.: Caerau heat network. Work package 9. Internal Report (2018)

# Chapter 63
# Understanding Residential Fuel Combustion Challenge—Real World Study in Wroclaw, Poland

**M. Baborska-Narożny, M. Szulgowska-Zgrzywa, A. Chmielewska, E. Stefanowicz, N. Fidorów-Kaprawy, K. Piechurski and M. Laska**

**Abstract** High emission linked with residential combustion for heating purposes is a pressing challenge for most cities in Poland. There are multiple uncertainties linked with the development of economically feasible and socially acceptable strategies for transition towards environmentally friendly domestic heating systems. Local municipalities lack basic inventories of heating systems in use or evaluation of the performance of retrofit buildings. Attitudes towards the planned changes are unknown. A field research that attempts to close this gap is presented. A novel approach based on building performance evaluation experience used multiple data sources and mixed research methods to learn about the heating in 422 occupied dwellings in Wroclaw, Poland. The research covered an urban quarter of historic tenements. The share of different heating systems, that is, solid fuel-based, gas, electric and district heating was established. An insight into the factors underpinning inhabitants' attitudes towards heating their homes and potential change resulting from solid fuel ban implementation was gained. New themes emerged from the study revealing key areas of concern for solid fuel burning households, for example, thermal comfort and mould issues, fuel transport issues and high operational cost of water heating. None of these themes is present in current narrative advocating change. The majority of solid fuel burning households (86%) interviewed proved to be willing to change their heating system. This suggests the key barrier to change in this building typology is not down to negative social attitudes but other constraints discussed in the paper. Conclusions include recommendations for future studies aimed at estimating the scope of challenge linked with solid fuel bans in domestic sector and change in narrative aimed at convincing households to swap solid fuel combustion for other options.

M. Baborska-Narożny (✉) · M. Szulgowska-Zgrzywa · A. Chmielewska · E. Stefanowicz · N. Fidorów-Kaprawy · K. Piechurski · M. Laska
Wroclaw University of Science and Technology, Wyb. S. Wyspiańskiego 27, Wroclaw, Poland
e-mail: magdalena.baborska-narozny@pwr.edu.pl

 
J. Littlewood et al. (eds.), *Sustainability in Energy and Buildings*, Smart Innovation, Systems and Technologies 163, https://doi.org/10.1007/978-981-32-9868-2_63

## 63.1 Introduction

Poor air quality, in particular in cities, is a global challenge. Scientific evidence indicates that high concentrations of particulate matter (PM10, PM2.5), nitrogen dioxide ($NO_2$), benzo[a]pyrene (BaP) and other chemical species cause damage to human health [1, 2]. Excessive emission of carbon dioxide ($CO_2$) causes negative climate changes. According to 2016 World Health Organisation (WHO) data, 33 out of 50 European most polluted cities were in Poland [3], revealing acute accumulation of the challenge. Annual mean concentration of PM2.5 in Poland in 2016 was 21.6 $\mu g/m^3$, which is second highest of all EU countries and over twice the WHO ambient annual mean guideline of 10 $\mu g/m^3$ [4]. Poland needs to comply with the national exposure reduction target (NERT) in order to reach exposure concentration obligation of 20 $\mu g/m^3$ based on the Ambient Air Quality Directive [5]. In 2018, the EU Court of Justice handed down judgment regarding most severe particulate matter exceedance in Poland. The EU Commission explains that that judgement confirms '*the European Commission's view that persistent exceedances require more effective measures to be taken by the Member States concerned to limit the exceedances to the shortest possible period*' [6, p. 9]. The EU delegates development of efficient strategies to tackle the problem to Member States.

The key source of fine particulate matter (PM10, PM2.5) high emissions is linked with residential fuel combustion for heating purposes. In Poland the fuel used in inefficient domestic boilers is mainly coal or wood [7], however, concerns are raised that burning waste, including plastic waste also occurs. It is estimated that about 29% of Polish households have individual coal burning installations [8]. After decades of marginalising the problem by public authorities at all levels, low air quality has gained visibility in public discussion and has been gradually taken up by the policy makers, notably at a local level. In cities urban activists brought pollution to public attention and voiced criticism for lack of action. Similar scenario was repeated in all major Polish cities. In Wrocław, a city of nearly 700 thousands inhabitants, at the end of 2017 Lower Silesian Regional Assembly passed Local Regulation indicating a legally binding roadmap to solid fuel ban by 2028 for households within the municipality boundaries [9]. Given that there are over 300 thousand dwellings, and ca. 30 thousands solid fuel furnaces in social housing alone, Wroclaw faces an urgent and complex challenge. There are multiple uncertainties linked with the development of economically feasible and socially acceptable strategies for transition towards environmentally friendly domestic heating. This is due to lack of knowledge on the exact scale of the problem, both in terms of investment cost or change in operational cost of heating, for example, number of heating installations requiring intervention is unknown. Most local municipalities across Poland lack basic inventories of heating systems in use or evaluation of the performance of the already retrofit buildings. Attitudes towards the planned changes are also unknown.

Selected results of a field research that attempted to close these gaps in knowledge are presented. The research was part of model revitalisation project aimed at developing methods and strategies to improve quality of life in deprived urban areas of Wrocław [10]. Poor air quality was one of the challenges tackled. Specifically, the aim was to test a method of gaining knowledge about domestic heating systems in use in a quarter of tenements. Attitudes towards implementation of the solid fuel ban were also explored.

The following sections introduce the scope of the study and building performance evaluation derived field research methods applied. Key findings are followed by discussion on the themes emerging from the study that challenge prevailing narratives present in public discourse. Conclusions include recommendations for upscaling studies aimed at estimating the scope of challenge linked with solid fuel bans in domestic sector. In particular, mixed methods approach was recommended, as well as broadening the narratives intended to win support for solid fuel ban among the affected households.

## 63.2 Scope and Methods

Wrocław and other municipalities in Poland need to estimate the scale of heating-related intervention needed to improve air quality. The research presented here was planned as the first step, allowing for testing the methods and gain insight into quantitative and qualitative aspects of the challenge. Exploring water heating was also included as it is the second highest after heating contributing factor in domestic energy consumption. The sample was preselected by Wrocławska Rewitalizacja Sp. z o.o. based on earlier multi-level analysis of the urban context [11].

### *63.2.1 Analysed Area Characteristics*

The study aimed to learn about the space and water heating systems used in 422 occupied apartments in Wroclaw, Poland. All apartments are located within an urban quarter comprising 29 buildings, mostly historic tenements built between 1880 and 1912 (Fig. 63.1). Among the historic tenements, only two went through deep retrofit. There were also two in-fill buildings completed in the twentieth century (1990).

The quarter is in deprived urban area of Przedmieście Oławskie; however, it is very well located with public transport available and proximity to historic city centre. The presence of many grade II listed buildings poses specific challenge for building renovations and energy efficiency measures. Collecting feedback on the performance of the few retrofit buildings against the prevailing building stock waiting for investment was of particular interest. Backyard elevations are particularly in poor shape; many missing plaster, with signs of damp and leaking gutters. Most roofs require repairs. Few were damp-proofed but without adding thermal insulation. The total floor area of the buildings is over 24,000 $m^2$, including almost 23,000 $m^2$ for residential and 1500 $m^2$ devoted to commercial use. About 6% of apartments were

**Fig. 63.1** Analysed quarter. Plan and photos (photo Ewelina Stefanowicz)

vacant. In terms of ownership type, 300 apartments were social housing and 122 were privately owned.

### *63.2.2 Methods*

Multiple data sources and both qualitative and quantitative research methods were applied. Key data sources were databases regarding technical features of the building stock in possession of the municipality and facility managers (FM) and semi-structured interviews with the inhabitants conducted in conjunction with an expert walk-through in the apartments. The databases covered, for example, addresses or ownership type of dwellings and relevant FM contact details. FM had technical data, for example, building audit documents, dwellings' floor areas, renovation works and inventories. Utility companies were also approached to share data about energy and water consumption. The selection of methods was based on in-depth building performance evaluation experience [12]. Development of surveys was based on extensive energy audit experience in tenements [13]. Below interview scope is presented and data collection process introduced. This is followed by the scope of information asked from FM and the effectiveness of this data source.

**Semi-structured interviews**: All 422 apartments were approached to participate. General information about the study was provided through posters in communal areas, whereas details regarding the aims and scope, funding, data processing, and so on were included in information letters delivered to each household. Whenever the inhabitants agreed, short interview was accompanied by a walk-through inside the apartments.

The interviews with walk-through covered fabric and systems-related issues, as well as inhabitant feedback. The questions and observations focused on the type of space and water heating systems, the type of fuel used, building envelope properties or visible mould. The inhabitants were asked about perceived pros and cons of

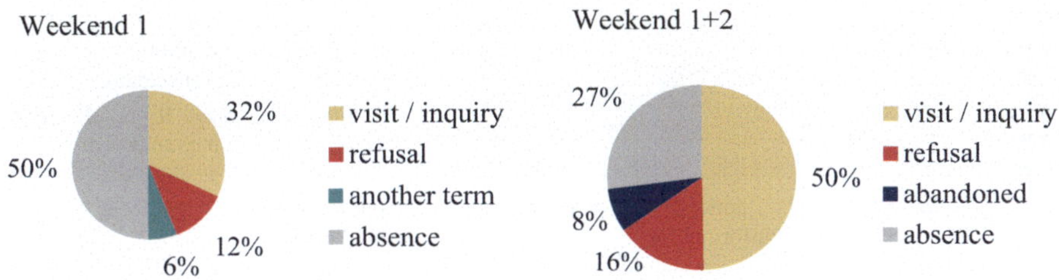


**Fig. 63.2** Effectiveness of approaching the households for the interviews after subsequent series of onsite visits

their heating system, perceived thermal comfort or moisture issues, use of additional heating sources and willingness to change the primary heating system.

Home visits were carried out during two consecutive weekends in January 2018 between 10 a.m. and 3 p.m. The interviews were conducted by 16 trained interviewers (academics, Ph.D. and post-graduate students from Architecture and Environmental Engineering Faculties, WUST) grouped into two-person teams with expertise in building services, such as heating or ventilation, as well as building design. Within each team one person lead the conversation while the other was responsible for taking paper notes of the answers and observations. The notes were subsequently digitalised into a template format. Figure 63.2 shows the effectiveness of the process after the first and second series of visits.

In total, 210 semi-structured interviews were conducted, 32 apartments were deemed uninhabited, in 66 the residents refused to engage and in 210 the inhabitants were away during repeated visits. It was worth repeating the attempted visits over four days of two weekends. Participation rate of 50% in domestic sector is an achievement. The interviews linked with walk-through allowed contextual knowledge of the inhabitants' attitudes and heating-related constraints.

**Data from facility managing (FM) companies**: All 11 FMs relevant to the sample of buildings were approached to provide technical data concerning buildings, operative energy consumption and heating systems within dwellings. Two of these were municipal companies responsible for all social housing units within the quarter, while remaining nine were private enterprises. Research team visited each FM to introduce the details of the project and agreed on data collection scope and procedure, data processing, and so on. Subsequently, a spreadsheet survey customised for each FM with a list of buildings of interest was emailed. The response level to the spreadsheet data collection was low. The research team members needed to arrange subsequent visits and themselves write down data from paper files as only some were in digital form. Sharing anonymised utility data without consent of all households involved was rejected by most FMs. The most complete sets of data except for utility bills were obtained from municipal companies. The efficiency of data collection is presented in Fig. 63.3.

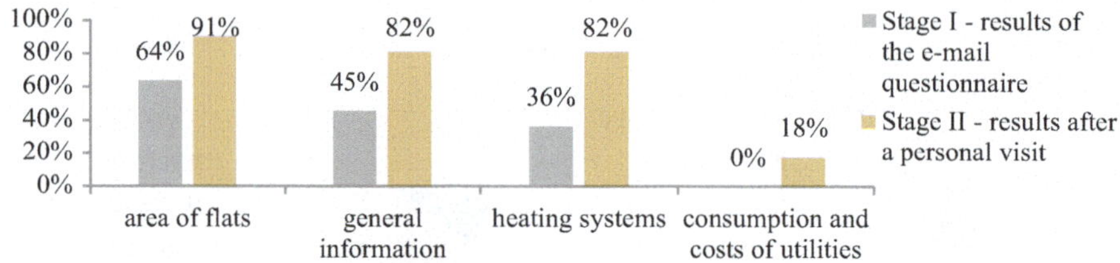


**Fig. 63.3** Type of information sought from FMs and percent of dwellings with responses

## *63.2.3 Key Findings*

Both methods proved highly relevant and complementary in building understanding of the challenge of residential solid fuel combustion. Correlating data from both sources allowed description of heating for 91% of the studied population (Fig. 63.4). Overlap allowed checking reliability of each data source. Notes from home visits and interviews were prioritised over FM derived data.

**Space and water heating**: High total response rate obtained allowed to determine the share of varied heat sources in the tested sample. It was found that 178 dwellings were heated using solid fuel from among 422 (353 if district heating excluded) (Fig. 63.5).

The distribution of responses in the FM derived data and home visits varied slightly. Analysis revealed that more owners engaged in interviews than social housing residents, who tended to refuse or were unavailable more often. However, for social housing the FMs were in possession of primary heating data unavailable for privately owned dwellings (available only if the building was connected to district heating). Therefore, in terms of the key research question the datasets were complementary. It was established that the rate of dwellings using solid fuel for heating was significantly higher in social housing units than in the privately owned flats, that is 40% versus 7%.

Home visits and interviews revealed stark contrast in thermal comfort and living standard between solid fuel and gas/district-heated apartments. The former group was clearly disadvantaged. The key disadvantages not picked up by the FM databases included:

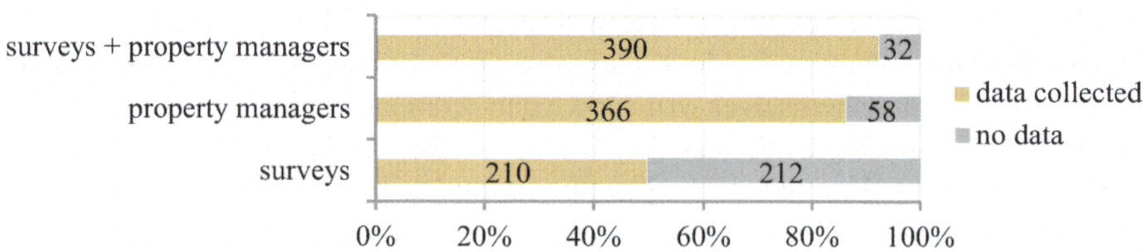


**Fig. 63.4** Data collection rates for different methods for sample of 422 apartments

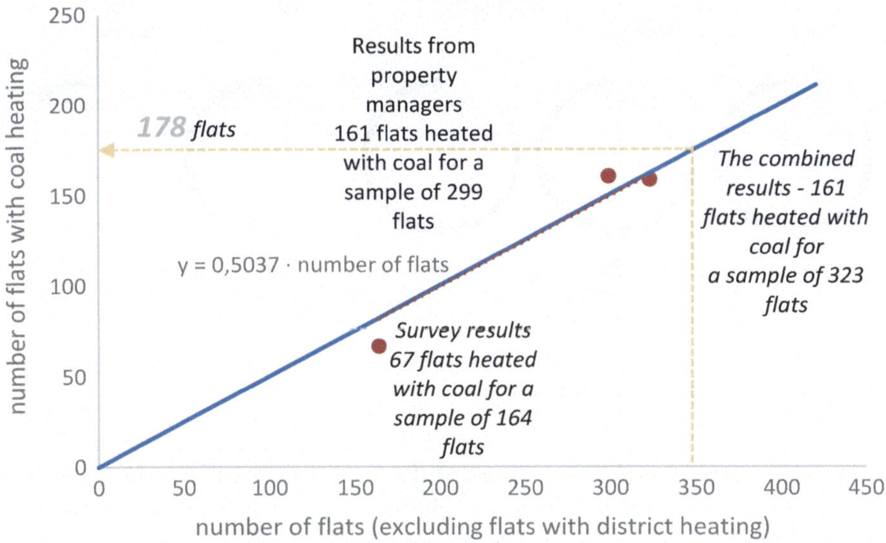


**Fig. 63.5** Correlation of the results obtained from facility managers and surveys

- 56% of solid-fuel-heated apartments needed additional heating, mainly electric, leading to higher utility bills than assumed, when only cost of coal is considered.
- None of the apartments with district heating needed secondary heating source.
- Most of the solid-fuel-heated apartments used inefficient electric water heaters, leading to high utility bills or limited use of hot water.
- 45% of solid-fuel-heated apartments experienced moisture and mould issues, whereas for district or gas heating that was 10 and 9%, respectively.
- 95% of solid-fuel-heated apartments used tiled stoves, located only in some rooms, leaving kitchens, bathrooms and small bedrooms without heating source.
- In buildings where solid fuel was the prevailing source of heating, major heat losses between dwellings were likely due to variations in thermal profiles between adjacent dwellings.

**Inhabitant's feedback**: Change of the heating system was regarded desirable by 64% of the interviewees. The share of households unsatisfied with the status quo varied substantially between different heating sources in use. Vast majority of the inhabitants using solid fuel for heating were looking forward towards change (84%), whereas for apartments with district heating less than 11% would welcome change (Fig. 63.6). It should be noted that in the only building in the sample connected to district heating simultaneously with thermal retrofit, 100% of the interviewed residents were satisfied and none wanted change. The willingness to change was expressed by the inhabitants of tenements with district heating where the building envelope has not been insulated and high energy demand led to high operational costs.

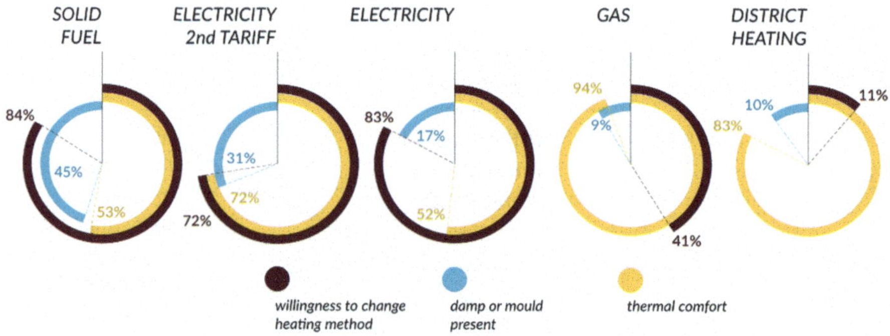


**Fig. 63.6** Share of the residents that want to change the heating system along with their heat comfort perceptions and problems with moisture or mould

Analysis of the digital written copy of the semi-structured interviews allowed contextualised assessment of inhabitants' attitudes regarding change of their heating system, practises related to its use and the perceived thermal comfort. For example, description of daily heating practices and perceived thermal comfort in households with adults working full-time suggested that diurnal thermal profile in apartments using solid-fuel-heated tiled stoves was undesirable: hot through the night and cold in the afternoon. Also solid fuel transport and storage was raised as an issue, in particular onto higher levels and for ageing inhabitants. On the other hand, some solid fuelled households praised resilience and autonomy of their heating, that is independence from power cuts (gas and electric heating would be vulnerable) or centralised heating season (district heating).

Difficulties in obtaining thermal comfort and the moisture problems coincided with the expectation of change for different heat sources (Fig. 63.6). Visits suggested that the interior temperatures in solid-fuel-heated apartments varied, both between dwellings and rooms within a single apartment. The thermal comfort results presented are a robust indication of a general perception of the inhabitants. Further research would be needed to quantitatively assess the severity of thermal comfort problem.

Another theme emerging from the interviews was frustration of many households willing but unable to move away from solid fuel heating. The first choice for most would be district heating, or else gas. The key issue raised with district heating is the need for unanimous decision of all households in a building to apply for a connection to make it financially feasible. The inhabitants are aware of the risk of rise in heating costs. Public understanding of the heating cost does not account for additional expense for subsidiary electric space and water heating. The inhabitants are also aware of decades of lack of investments in the refurbishment work and outstanding needs in this respect. The combined cost of these investments, in particular in grade listed buildings is out of reach of housing communities made up of ca. 10–15 mixed private and social housing units. The combination of technical, organisational and financial barriers results in helplessness and lack of hope for change expressed by many interviewees.

## 63.3 Discussion

**Heating options**: Switching from domestic solid-fuel-based heating in historic tenements is a complex challenge, both for authorities pushing for change and for the inhabitants. In Poland, the typical substitutes to solid fuel heating in urban tenements would be gas, district heating or electric. The interviews indicated prevailing preference for district heating among the residents wanting change from solid fuel. Another option present in the interviews was gas. No one wanted electric heating reckoned as expensive to use. However, unless change is a synchronised investment agreed by all households in a building district, heating is unlikely due to high connection cost. Electric heating is the easiest to install technically and may be introduced irrespective of the neighbour's decisions. Gas heating requires access to ventilation ducting unavailable to many tenement dwellings. Giving up solid fuel heating for gas or electricity has been subsidised by municipality since 2015. Over 2000 households benefitted from the KAWKA programme until 2018. There is also a utility bills subsidy in place to make up for the increase in heating cost. Undoubtedly, there are low emissions as a result of KAWKA, however, there is a drawback. Bigger mix in heating systems among dwellings within a single tenement means it is harder to get enough applicants for district heating to make it financially feasible. The best case scenario as the interview results suggest would be to thermally insulate the historic buildings and convince all households to go for district heating.

**Narrative of change**: Current official narrative advocating change of solid fuel heating systems focuses almost exclusively on air quality [14]. Both at the central and local level, the central message is smog [15]. Using poor quality solid fuel for heating in inefficient unclassed stoves is labelled 'poisoning' in leaflets, even though it is legal to buy such polluting fuel and use inefficient stoves. Health risks associated with poor air quality are most highlighted by the NGOs and urban activists [16, 17]. Interestingly, these themes were completely absent in solid-fuel-heated households statements. Having said that, 84% of the interviewed solid-fuel-heated households welcome change in their heating system. However, the number of those who would need to be convinced might be higher than 16% of the population, as solid fuel was overrepresented in the population that did not take part in the interviews. The negative official narrative about solid fuel heating might be one of the factors underpinning refusals to engage in sharing opinion in that respect.

Surprisingly, the themes repeatedly raised in the interviews such as thermal comfort and mould issues, low living standard or unaccounted for cost of electric water heating associated with solid fuel use in inefficient stoves are totally absent from the public argument advocating change.

## 63.4 Conclusions

The results of the conducted study indicate that mixed data sources and collection methods were needed to reliably indicate the share of different heating systems in occupied apartments. The study revealed that none of the analysed data sources on its own described all the dwellings, and each introduced some error. The error could be diminished by comparing data from different sources.

The approach allowed to indicate quantitatively the occurrence of different heating systems in the studied sample and to gain understanding of the complex factors underpinning inhabitants' attitudes towards heating systems in use and the potential for change. New themes emerged from the study that should broaden the prevailing narrative advocating change of heating systems. Current account focuses on lowering emissions and air pollution. It also assumes that the key barrier to voluntary change of heating systems is the fear of increased operational cost. It is recommended to include in the discourse key areas of concern for solid-fuel-heated households as identified here, for example, lack of thermal comfort, fuel transport issues in particular for the ageing inhabitants or cost of additional heating. Our findings revealed that solid fuel burning households proved to be least satisfied with their heating and most open for change (84%) when compared with inhabitants using different heating systems. This suggests the key barrier to change is not down to negative social attitudes. The building with highest satisfaction rates among the 29 analysed tenements was connected to district heating. Importantly, the change of heating happened simultaneously with in-depth retrofit improving thermal performance of the building envelope. Further research is needed into quantitative performance aspects of that building. Review of policy and research approaches taken internationally, out of scope of this paper due to limited space, might better contextualise the contribution to the field of the reported research approach and findings.

**Acknowledgements** The authors thank the FMs and inhabitants involved in this study for the generous amount of time they provided. The authors gratefully acknowledge the funding provided by Wroclawska Rewitalizacja Sp. z o.o. (Project ID at WUST: 42PT/0001/18).

## References

1. World Health Organisation (WHO): Residential heating with wood and coal: health impacts and policy options in Europe and North America. World Health Organization Regional Office for Europe, Copenhagen (2015). http://www.euro.who.int/__data/assets/pdf_file/0009/271836/ResidentialHeatingWoodCoalHealthImpacts.pdf?ua=1
2. Neira, M., Prüss-Ustün, A., Mudu, P.: Reduce air pollution to beat NCDs: from recognition to action. Lancet **392**, 1178–1179 (2018)
3. World Health Organisation (WHO): Air quality in Europe—2016 report (2016). https://www.eea.europa.eu/publications/air-quality-in-europe-2018. Accessed 20 May 2019
4. European Environment Agency (EEA): Air quality in Europe—2018 report (2018). https://www.eea.europa.eu/publications/air-quality-in-europe-2018. Accessed 20 May 2019

5. EU: Directive 2008/50/EC of the European Parliament and of the Council of 21 May 2008 on ambient air quality and cleaner air for Europe (OJ L 152, 11.6.2008, pp. 1–44) (2008). http://eur-lex.europa.eu/LexUriServ/LexUri-erv.do?uri=OJ:L:2008:152:0001:0044:EN:PDF. Accessed 20 May 2019
6. European Commission: A Europe that protects: clean air for all. COM (2018) 330. Brussels (2018). http://ec.europa.eu/environment/air/pdf/clean_air_for_all.pdf
7. WHO Air Quality Guidelines for particulate matter, ozone, nitrogen dioxide and sulphur dioxide. https://apps.who.int/iris/bitstream/handle/10665/69477/WHO_SDE_PHE_OEH_06.02_eng.pdf?sequence=1. Accessed 20 May 2019
8. HEAL Spalanie węgla w piecach domowych
9. Sejmik Województwa Dolnośląskiego: Uchwała Nr XLI/1405/17, Dz. U.Woj. Doln., Poz.5153, Wrocław (2017). http://bip.umwd.dolnyslask.pl/dokument,iddok,42224,idmp,596,r,r
10. W_R Homepage. http://w-r.com.pl/projekty/modelowa-rewitalizacja-miast/. Accessed 20 May 2019
11. W_R Homepage. http://w-r.com.pl/projekty/cieplozimno/. Accessed 20 May 2019
12. Baborska-Narożny, M.: Building performance evaluation—understanding the benefits and risks for the stakeholders involved: lessons for Poland based on the UK experience. Architectus **1**, 47–61 (2017)
13. Danielewicz, J., Fidorów, N., Jadwiszczak, P., Szulgowska-Zgrzywa, M.: The exploitation costs for various heating systems according to the energetic certification law in Poland. Energy Sources Part B Econ. Planning Policy **9**(3), 301–306 (2014)
14. KOBIZE Homepage. http://www.kobize.pl/uploads/materialy/materialy_do_pobrania/krajowa_inwentaryzacja_emisji/Bilans_emisji_-_raport_podstawowy_2014.pdf. Accessed 20 May 2019
15. ARAW Homepage. https://www.wroclaw.pl/pracownia-miast-wroclaw-bez-smogu. Accessed 20 May 2019
16. Health and Environment Alliance Homepage. https://www.env-health.org/. Accessed 20 May 2019
17. Smog Alarm Homepage. https://krakowskialarmsmogowy.pl/en/rozwiazania/szczegoly/id/94. Accessed 20 May 2019

# Chapter 64
# Behavioural Change Effects on Energy Use in Public Housing: A Case Study

**Andrea Sangalli, Lorenzo Pagliano, Francesco Causone, Giuseppe Salvia, Eugenio Morello and Silvia Erba**

**Abstract** Post-occupancy evaluations of building energy performances after retrofit often show values of energy use higher than calculated during the design phase. This performance gap is due to different factors, among which the user behaviour plays a crucial role. In the context of public housing, in particular, the techno-centric approach which is usually adopted during all the work phases tends to omit the involvement and information of the inhabitants, leading to ways of living not consistent with optimal uses of the building services and components. To tackle this limit, EnerPOP, a project funded by Politecnico di Milano, is developing a methodology for reconciling technical and social aspects in the refurbishment of public housing assets. A case study is adopted to experiment and optimize the methodology, engaging tenants of a public residential building in Milan, deeply renovated in 2014 and inhabited by 500+ people originating from over 30 different countries. The paper reports interim findings, after one year of activities, both in terms of knowledge of the building and its inhabitants and early results in terms of energy saving and social cohesion.

## Nomenclature

HDD Heating degree days

A. Sangalli (✉) · L. Pagliano · F. Causone · S. Erba
end-use Efficiency Research Group, Department of Energy, Politecnico di Milano, via Lambruschini 4, 20156 Milan, Italy
e-mail: andrea.sangalli@polimi.it

G. Salvia · E. Morello
Department of Architecture and Urban Studies, Politecnico di Milano, Piazza Leonardo da Vinci 32, 20133 Milan, Italy

 
J. Littlewood et al. (eds.), *Sustainability in Energy and Buildings*, Smart Innovation, Systems and Technologies 163, https://doi.org/10.1007/978-981-32-9868-2_64

## 64.1 Introduction

Starting from 2016, the Municipality of Milan invested 103.5 million Euros, distributed over 4 years, for the refurbishment of public residential buildings (project "Zero case vuote"—"Zero unhinabited houses") [1]. Energy efficiency measures are a relevant part of this plan; however, the planning, implementation and management of energy efficiency interventions are quite often carried out with a techno-centric approach. This may lead to unintended consequences, such as lower than expected energy savings actually achieved by projects. This "rebound effect" may be better understood if occupants' routines and ways of living are considered [2], since these may not coincide with the expected ones at the design stage.[1] The rebound effect may be framed in very different ways depending on various factors, including the social context, the degree of technological innovation of the solutions adopted in the refurbishment and the configuration of properties (public or private, rental or home ownership) [3, 4].

EnerPOP, a project funded by pre-tax donations for Politecnico di Milano, aims to analyse the technical and social aspects in a public residential building retrofit case study in Milan, in order to identify the causes of the gap between the expected (calculated) energy savings and the actual (measured) ones, and to propose options and perform actions to reduce it. In two previous studies by the same authors [5, 6], the starting situation has been analysed from a technical and social perspective, respectively. This work is intended as a follow-up, with the aim (i) to upgrade the above-mentioned analysis, for a better understanding of the occupants' behaviours demanding energy, and (ii) to show the first results in terms of energy savings and social interactions, after one year from the beginning of the project.

## 64.2 Description of the Case Study

The case study is a large public residential building, owned by the Municipality of Milan, located in a context of marked urban and social periphery. It is a five stories

[1] In this paper, the term "behaviours" is used to refer to occupants' ways of living in the house, including those using energy for heating and achieving thermal comfort in general. Nevertheless, the authors are aware of the ongoing theoretical debate in energy use and comfort studies between approaches informed by psychological approaches related to behavioural change on the one hand; and on the other hand, the sociology informed one addressing social practices and their configurations (cfr. Shove, E. 2010. Beyond the ABC: Climate Change Policy and Theories of Social Change. Environment and Planning A: Economy and Space, 42(6), 1273–1285. https://doi.org/10.1068/a42282). Although relevant, the elements of the debate fall outside the scope of this paper, which aims mainly at sharing preliminary insights on energy monitoring analysis. In this publication the term "behaviours" is preferred for consistency with the contribution in the specific field of the conference and reduce risks of comprehensibility. Forthcoming publications from the EnerPOP project intend to unpack how framing the investigation through Social Practice Theory may provide relevant insights, especially on the ways tenants arrange their routines and the consequences on energy demand.

building, with 154 apartments, about half of which are small two-room apartments of about 40 $m^2$, and the rest four-room apartments with a surface ranging from 65 to 85 $m^2$, for a total of about 11,000 $m^2$ of gross floor area [5].

Since 2013, the building has undergone a major energy refurbishment. The original building was characterized by poor energy performance of the envelope and by obsolete mechanical systems, with resulting thermal discomfort (low air temperatures and radiant temperatures in winter) and low air quality (diffuse presence of mould). The renovation work focused first on improving the quality of the envelope and on the renewal of the heating system, with the centralization of hot water production as well as heating, followed by the connection to the district heating network. In each apartment, a programmable thermostat and thermostatic valves have been installed on the radiators, for control purposes only (billing is not based on individually metered consumption). A centralized mechanical ventilation system was also installed, which operates in extraction only, with extraction vents placed in the bathrooms and fresh air entering the apartments through intake vents placed on the façade. The system works with constant flow at low speed. In 2014, at the end of the retrofit works, the building was certified "class B" according to the Italian energy performance classification, with a calculated total primary energy use for heating of 34 kWh/($m^2$ yr).

Towards the end of 2014, the apartments were progressively assigned to the tenants, which were not the same ones previously occupying the building. At present, the community that resides in the building consists of about 500 people, mostly elderly Italians and families of first-generation immigrants. About 30% of the inhabitants are under the age of 15—a particularly young community compared to the average of the public housing in Milan. About 60% of the tenants belong to families whose head originates from a foreign country rather than Italy, including more than 30 different nationalities. In all families, at least the head of the family is able to speak Italian.

## 64.3 Methodology

In order to gain knowledge about the post-retrofit situation, data on energy use for heating and electrical energy use have been gathered for all the apartments since 2014; whereas data on indoor thermal comfort and indoor air quality have been gathered for a selected subset of apartments since 2016. The thermal energy meters, installed in each apartment on the heating manifold, measure the thermal energy required to reach and maintain the value of air temperature set on the programmable thermostat. The measured energy does not comprise the distribution efficiency of the part of the system placed inside the apartment (between the collector and the radiators) and the emission efficiency of the radiators. The measured value corresponds approximately to the "energy need for space heating" according to the terminology of related standard ISO 52016-1: 2017. Some data are missing over the years in some apartments (due to end of battery charge and sudden sensor failures). Unfortunately, these issues were not solved promptly due to a lack in the maintenance chain between

**Table 64.1** Average values of the energy use for heating (absolute and normalized to HDD) for the 95 apartments

| | Average 2015–18 | 2018–19 (EnerPOP) | Variation (%) |
|---|---|---|---|
| kWh/(m$^2$ yr) | 81 | 67 | −17 |
| kWh/(m$^2$ yr HDD) | 0.040 | 0.035 | −11 |

the building owner and the maintainer. Since 2015, 95 over 154 apartments were continuously monitored without failures. Only these apartments have been considered in the following analysis, in order not to alter the characteristics of the dataset (e.g. the influence of the position of the apartments and households involved). In the considered apartments, residents have not changed substantially over the years; therefore, the results should be comparable.

Table 64.1 summarizes the energy use for heating of all the apartments where the energy meters worked continuously since 2015, showing the average for the heating seasons 2015–16, 2016–17 and 2017–18, compared to the last one, 2017–18. The first available heating season (2014–15) has not been considered because the apartments were progressively assigned since October 2014 and the related lower energy uses depended on non-occupancy. Data are presented as (i) normalized per surface of the available apartments, and (ii) further normalized with respect to the heating degree days (HDD) of the respective seasons.[2] The comparison with the last heating season (2018–19), after the first year of the EnerPOP activities, shows a reduction of 11% in the energy use for heating; the percentage is even greater considering, in the comparison, the last season before EnerPOP (2017–18), instead of the average 2015–18; in this case, the normalized variation is −16%.

Regarding environmental monitoring, since December 2016, temperature, humidity and $CO_2$ sensors have been installed in 17 apartments. They are placed in a barycentric position of the apartment, far from heat sources or thermal dispersions, at a height of 150 cm from the floor, on the side of the programmable thermostat; it can, therefore, be assumed that the monitored air temperature corresponds to that used by the controls of the heating system. In the monitored two-level apartments, an additional sensor was installed on the upper floor. The environmental data are recorded hourly. At the moment, the visualization of the data is not available for the tenants.

Figure 64.1 shows the normalized energy use and the statistical distributions of the temperature values, for the apartments in which both parameters are measured. The values are sorted by increasing energy use, for the heating season 2018–19, together with the corresponding temperatures, whose statistical distributions are represented by a box-and-whisker graph. In this graph, the quartiles are indicated by blue bars (first quartile, q1, second quartile, or median, q2, and third quartile, q3) and the

[2]The HDD have been calculated on an hourly basis, according to ISO 15927-6: 2007. “Heating season” here means the period from October 15 to April 15, according to the Italian laws no. 10 of 9/1/1991 and DPR 412 of 26/8/1993 and subsequent amendments, as a period of operation of heating systems in buildings in the municipalities belonging to the climate zone “E”, like Milan.

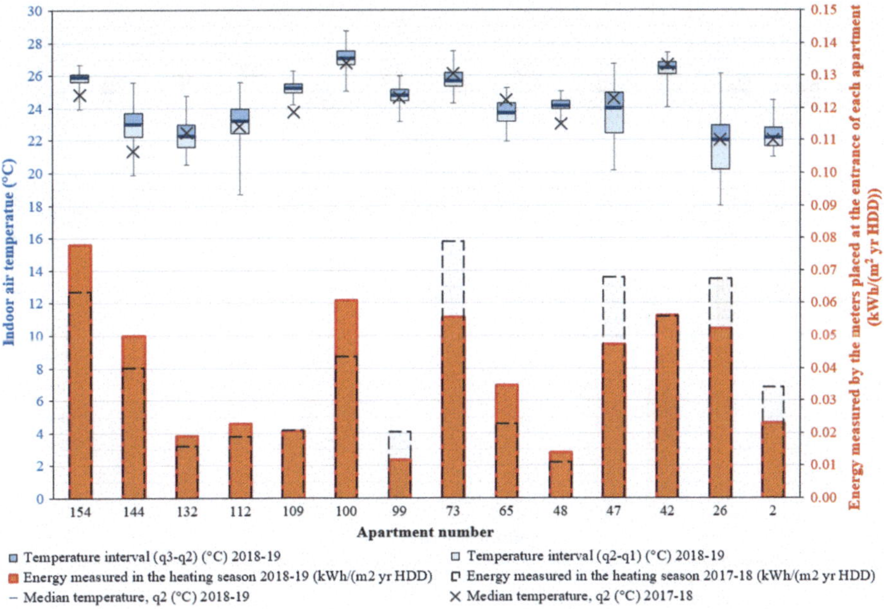


**Fig. 64.1** Heating season 2017–18 and 2018–19: energy measured by the meters placed at the entrance of each apartment and temperatures in the corresponding apartments (the number of which is invented by the authors and do not correspond to actual ones to preserve the tenants' privacy)

maximum and minimum values, included respectively in the interval [q3 + 1.5 * (q3 − q1)] and [q3 − 1.5 * (q3 − q1)], are represented as whiskers above and below the quartiles. The values are compared with the ones of the previous heating season 2017–18, represented with dotted lines, both for the energy use and for the median temperatures.

The variation in energy use for heating over the last two seasons appears to be non-homogeneous: 5 out of 14 apartments show a significant reduction, 5 are quite similar, while 4 increased their values. A certain correlation between the variation of temperatures and energy use is shown in some apartments (48, 112, 47, 144, 73, 42, 100, 154), while the other ones present counter-intuitive trends.

An analysis has been carried out also on the energy use for electricity, comparing, for some apartments, the consumption of two months of winter with those of two months of summer of the same year. Households with higher energy use for electricity in summer probably use active cooling with individual devices, since no active cooling was provided by the municipality. This possibility was suggested also during interviews with tenants. In order to have empirical confirmations, an additional comparison was done with indoor temperatures recorded in June–July, as shown in Fig. 64.2. Except for apartment 65, no particular correlations seem to be drawn at the moment between the two parameters. The reasons may be related to the typical limited and discontinued use of cooling: if only one room is cooled, the temperature recorded by the monitoring system may be not representative, and if cooling is acti-

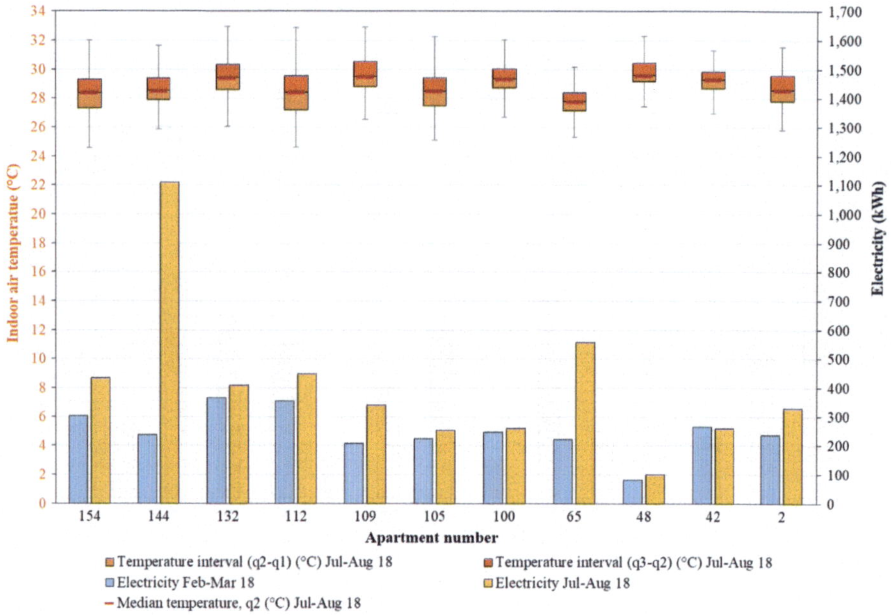


**Fig. 64.2** Energy use for electricity in 2 months (in winter and summer, respectively) in some apartments, compared to the indoor temperatures

vated only in some hours, those few occurrences are outliers which have not been represented in the statistical evaluation of the temperatures over the whole hours of June–July. Additional work is therefore needed with this respect, either installing additional sensors in some apartments, to have a more comprehensive representation of the indoor temperatures over the whole apartment, or trying to draw occupation hours, deriving them from the analysis of indoor $CO_2$ levels over time or from electrical load curves.

A long-term comfort analysis was also performed based on the air temperature values measured in the apartments during the heating season 2018–19, in comparison to those already performed for 2016–17 and 2017–18 [5]. Figure 64.3 presents the comparison between the analysis for the heating season 2017–18 and 2018–19. The graphs show that, although conditions in most of the apartments fall most of the time into category II as required by the standard,[3] there are also several situations characterized by overheating with respect to category II. These occurrences appear to decrease in 2018–19, in some cases, with an increase of undercooling two apartments.

Another long-term analysis was carried out; this time related to indoor air quality, evaluated by measuring the $CO_2$ concentrations in some apartments and comparing

[3]EN 15251 Annex G: "The different parameters for the indoor environment of the building meet the criteria of a specified category when: the parameter in the rooms representing 95% of the occupied space is not more than, for example, 3% (or 5%) of occupied hours a day, a week, a month and a year outside the limits of the specified category [...]".

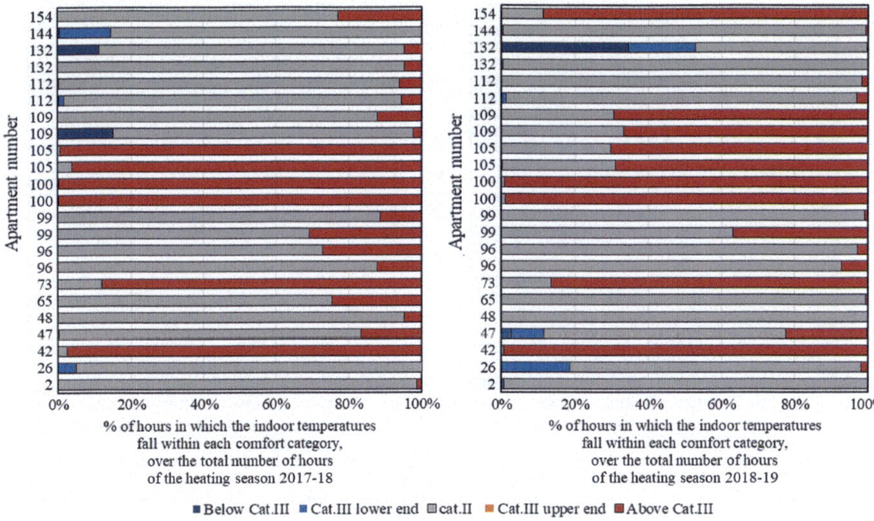


**Fig. 64.3** Percentage distribution of hours in the various comfort categories during the heating period 2017–18 and 2018–19 for the apartments where the indoor environmental sensors were installed. Situations where the temperature is comprised between the upper limit of the cat.N − 1 and the upper limit of cat.N are labelled by "upper end" and situations where the temperature is comprised between the lower limit of the cat.N and the lower limit of the cat.N − 1 are labelled by "lower end". For example, "Cat.II lower end" means that the temperature is between the lower limit of cat.II and the lower limit of cat.I

them with the category limits set by the standard EN 15251. The hours when the $CO_2$ concentrations fall in the various categories were counted, comparing them to the total number of hours in the considered period, thus obtaining the percentages shown in Fig. 64.4.

In parallel to physical measurements, information has been collected about the tenants' habits, by conducting semi-structured interviews to a sample of tenants, as well as conversations at different levels of formalization (e.g. scheduled meetings, unscheduled visits) and diverse participants' observations. The choice of the sample of tenants was initially aiming to an equitable representation for the placement of the apartment (floor, exposure), size of the family (single, couple, family with children) and nationality. The selected people were asked to participate in an interview lasting about 60 min at their home, on a voluntary basis (without reimbursement), and according to their time preferences (date and hour). The analysis of the interviews shows a wide variability of social behaviours and descriptors of thermal comfort, as expected. The difficulty in understanding how the building systems work and how to maintain the ideal temperature at home is common, and it influences the interaction with programmable thermostat, thermostatic valves, windows and accessory devices to achieve the ideal temperature, both in summer and in winter.

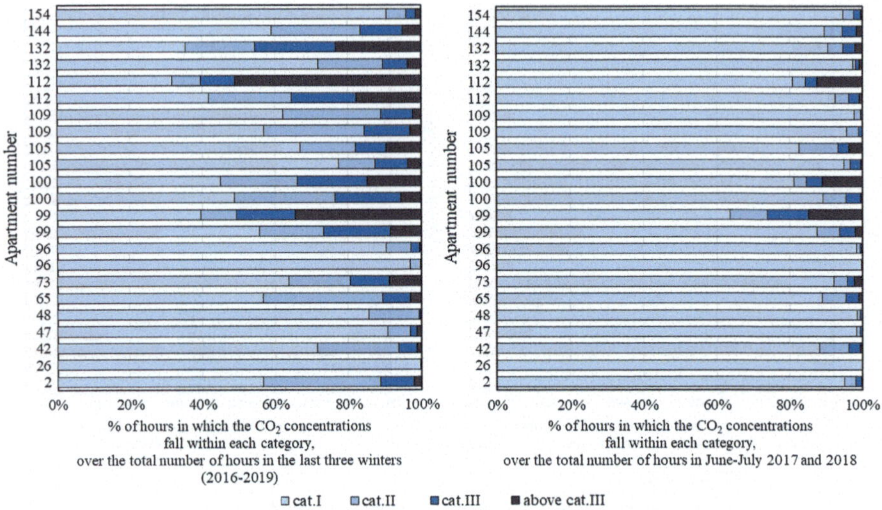


**Fig. 64.4** Percentage distribution of hours in the various categories of $CO_2$ concentrations, for the apartments where the indoor environmental sensors were installed

## 64.4 Discussion of the Results

As shown in Table 64.1, in the last winter 2018–19, the energy use for heating presented an overall reduction of 11% with respect to the average of the previous winters. Since the data are normalized to the HDD, they take into account the different severity of the outdoor conditions over the years. Therefore, other explanations can be sought for this variation, assuming that random variability alone is not enough. Other variations did not take place, such as variation in the social composition, in the billing or in the characteristics of the envelope and of the heating system. Owing to this, the activities of involvement carried out in the first year of the EnerPOP project, may be assumed responsible of a consistent part of the energy savings, because they enabled a rearrangement of tenants' behaviours, which may have generated also a positive, environmentally sustainable impact. The main activities explicitly addressing thermal comfort in winter and energy savings consisted of a campaign informing about the prescription by law concerning the temperature to be set in the buildings, that is 20 °C, and the maximum number of hours in which the heating system can be working, that is 14 h, information for a better use of the programmable thermostat; not keeping windows open to lower the temperature, but doing it with a proper regulation of the programmable thermostat and of the thermostatic valves on the radiators, other than specific interaction by the researcher with some of the tenants , to help them to regulate the programmable thermostat. The campaign consisted of multiple channels including:

1. tutoring, that is researchers supporting tenants onsite;
2. the projection of a video on optimal behaviours (e.g. setting lower temperature, keeping windows closed) co-designed and partly co-created with the tenants, played continuously in a loop, 24 hours a day for several days, in the common spaces at the entrance of the building;
3. a video including instructions on how to set the programmable thermostat, step by step;
4. a leaflet distributed to all the tenants' mailboxes;
5. a notice placed on the communal areas with the link to the online page where the video on the programmable thermostat could be visualized.

Assuming that these actions were effective in a more energy-efficient reconfiguration of behaviours for indoor thermal comfort of the tenants, it is possible to inquire more in-depth about the reasons behind this effectiveness.

First, the multi-channel campaign may have met the more or less evident demand of information and support with regard to the ways of configuring the available devices, mainly the programmable thermostat, which was reportedly assessed as obscure to interpret. In fact, some tenants welcomed the researchers in their houses to solve problems with their radiators, which occasionally were due to the programmable thermostats.

Second, the campaign may have provided information on the legal terms for the heating setting, which were not widely known (e.g. 20 °C max at home, for 14 h maximum). In a couple of interviews held after the launch of the campaign highlighted how children may represent an advantageous means for this, either as they stopped to watch the video together with their parents or because they reported the contents back to the parents afterwards.

Third, the campaign may have inadvertently created the impression of local monitoring which may have discouraged some tenants to abuse of the heating, thus facing risks of receiving a fine. Such case may not be ideal to enable a longer-term transition towards more energy-efficient thermal behaviours; nevertheless, this would suggest that holding initiatives on the topic generates an impact.

Despite the limited number of tenants reached directly by the researchers, possibly due to mistrust, the impact seems to have been larger than expected. This may be due to communications between the tenants by word-of-mouth.

## 64.5 Conclusions

After the first year of field research, it clearly emerges how the actual energy use of a building and the connected energy costs (often paid by the municipality, in public housing contexts) may present significant deviations in comparison to the values estimated during the design phase, even in case of careful design and energy modelling, according to standards and protocols and careful execution of the retrofit work. The actual behaviour carried out in the apartments may differ considerably with respect to the modelled ones, most notably about the time span windows are kept open and shutters rolled up, with related heat dispersion and wasted energy.

The presented analyses show a reduction of 11% in the energy use for heating, compared to the average of the previous three heating seasons, normalized to HDD. In the absence of other explanations, it can be assumed that this result is related to the EnerPOP activities, which started a process of rearrangements of tenants' behaviours towards more environmentally sustainable ways of living.

The current and future activities of the project aim at a more detailed definition of the behaviours conducted in the homes by the tenants, increasing the representativeness of the descriptions provided, and at an identification of patterns of habits, needs, perceptions of comfort, cultural factors which might provide explanatory variables of the variations in energy use. Finally, it is foreseen to develop further actions to support tenants to identify and modify inefficient behaviours, with a process focused on participation and empowerment.

**Acknowledgments** The research has been partially funded by the Polisocial Award 2017 of Politecnico di Milano under the EnerPOP project and by the European Union's Seventh Framework Program for research, technological development and demonstration under grant agreement no. ENER/FP7EN/314632/EU-GUGLE. We are thankful to the partners of the projects and in particular Arch. Manzoni and Arch. Bardeschi of Milan Municipality for their useful collaboration.

## References

1. Retrieved from https://milano.corriere.it/notizie/cronaca/18_gennaio_24/milano-case-popolari-alloggi-sfitti-il-recupero-stanziati-1035-milioni-02fc3aa2-00d0-11e8-b515-cd75c32c6722.shtml
2. Gram-Hanssen, K.: Retrofitting owner-occupied housing: remember the people. Build. Res. Inf. **42**(4), 393–397 (2014). https://doi.org/10.1080/09613218.2014.911572
3. Atkinson, J., Littlewood, J., Karani, G., Geens, A.: Relieving fuel poverty in Wales with external wall insulation. Proc. Inst. Civil Eng. Eng. Sustain. **170**(2), 93–101 (2016)
4. Sousa, Monteiro C., Causone, F., Cunha, S., Pina, A., Erba, S.: Addressing the challenges of public housing retrofits. Energy Proced. **134**, 442–451 (2017)
5. Sangalli, A., Pagliano, L., Causone, F., Salvia, G., Morello, E.: Energy efficiency and occupants' behavior: analysis of a public housing case study. In: Proceeding of AICARR 51st International Conference, Venice, 20–22 Feb 2019
6. Salvia, G., Rotondo, F., Morello, E., Sangalli, A., Pagliano, L., Causone, F.: Sustainability designed with(out) people? Understanding for what energy is (over-)used by tenants in an energy efficient public housing in Milan. In: Proceedings of LENS International Conference, Milan 3 April 2019

# Chapter 65
# Holistic Dwelling Energy Assessment Protocol for Mine Water District Heat Network

**J. R. Littlewood, B. Philip, N. Evans, R. Radford, A. Whyman and P. Jones**

**Abstract** UK buildings and energy infrastructures are heavily dependent on natural gas, and a large proportion is used for domestic space heating; but 50% is imported. Improving energy security and reducing carbon emissions are major government drivers for reducing gas dependency. So, there needs to be a wholesale shift in the energy provision to householders without impacting on thermal comfort levels, convenience or cost of supply. Electrical powered heat pumps are seen as a potential alternative system for heating new dwellings, but will they work in dwellings built prior to 1919? This paper investigates the energy demand of pre-1919 dwellings in Wales, UK as part of a feasibility study to extract water from disused coal mines to supply a district heat network. A holistic surveying protocol providing a more accurate/realistic assessment of total household heat demand is considered. The protocol's techniques include condition surveys, air permeability and thermography tests, and heat loss calculations are discussed. The results were used to predict future (beyond 2019) heat demand after potential retrofit improvements, thereby informing the size of heat pumps required. The findings show estimated heat demand to be in close correlation to household energy bills, and that the use of heat pumps in pre-1919 dwellings is viable, provided sufficient improvement to thermal performance is possible.

## 65.1 Introduction

The thermal performance of the majority of the existing housing stock in Wales bears testimony to the quality of construction from pre-1919, and falls far below

J. R. Littlewood (✉) · N. Evans · R. Radford · A. Whyman
Cardiff Metropolitan University, The Sustainable & Resilient Built Environment Group, Cardiff CF5 2YB, UK
e-mail: Jlittlewood@cardiffmet.ac.uk

B. Philip
Swansea University, Specific, Baglan, Port Talbot SA12 7AX, UK

P. Jones
Swansea University, Active Building Centre, Bay Campus SA1 8EN, UK

 
J. Littlewood et al. (eds.), *Sustainability in Energy and Buildings*, Smart Innovation, Systems and Technologies 163, https://doi.org/10.1007/978-981-32-9868-2_65

the minimum currently set by UK building standards in 2019 [1, 2]. Since 2010, significant sums of money have been invested to improve energy efficiency and combat fuel poverty in some of the most deprived areas of Wales, by retrofitting thousands of dwellings through the Arbed 1, 2 and 3 schemes with exterior wall insulation, roof insulation, new boilers and controls, and windows [3, 4]. Increasing energy efficiency of older properties remains a significant challenge, which cannot be achieved through insulation and air tightness interventions alone, particularly when alterations to historically important architectural features of the building are not permitted [5]. So, alternative solutions are needed, which is explored in this paper; namely, the feasibility of using water from the extensive network of former colliery works as the basis for a district heating network [6]. The paper discusses the context to the need to decarbonise the energy supply for dwellings in Wales; the methodology used in a holistic protocol piloted on eight dwellings in a case study village of 7000 properties as part of the Caerau Heat Network project; and the findings and recommendations from the feasibility study [7].

## 65.2 Context to Energy Demand in the UK in 2019

Heat accounts for nearly half of the energy consumed in the UK and about a third of carbon emissions [8]. Peak gas demand in typical winters exceeds electricity use by around five times, so it will not be possible to deliver UK carbon reduction targets while gas remains the primary source of energy for heating. Substitution of gas heating with direct electrical heating is currently not an option, as there is insufficient capacity in the electrical generation and distribution grid to meet the demand [9]. In spite of the efforts to improve dwellings in the Arbed 1–3 programmes (with circa £110 million grant funding), estimates by the Welsh Government in 2018 suggest that 23% of households in Wales spend more than 10% of income on energy bills and are thereby considered to be in fuel poverty [4, 10]. Many socioeconomic factors contribute to these percentages; however, the problem is exacerbated by the age and consequent poor energy performance of much of the housing stock in Wales and indeed the UK as a whole. Caerau, in the Bridgend County Borough Council (BCBC) region of south-west Wales, was once a proud and prosperous coal mining village, but in 2015 was rated as the fifth most deprived community in Wales [11]. The majority of properties in the village were built in the early 1900s to meet the growing demand for labour in the local colliery, which operated from 1889 to 1979, and at its peak employed 2400 workers.

In 2017, BCBC sought a solution for heating the current community of approximately 7000 inhabitants, while simultaneously addressing the energy trilemma of decarbonisation, increasing the security of supply and reducing the cost of energy [6]. There were no sources of waste heat to be exploited in the area; however, the practicality of using water from the extensive network of former colliery workings as the basis for a district heating network was deemed worthy of further investigation. Measurements from an exploratory bore hole found the temperature of the mine water to be over 20 °C at 220 m (m) depth [12]. Large volume of pumping tests have yet to be completed; therefore the volume of water available and expected temperature drop for a network supplying between 150 and 850 households is currently unknown. However, initial indications are that mine water could provide a better temperature source, that is the 11 °C, than would be expected from a standard ground source heat pump and bore hole [13]. The feasibility of using either a centralised or decentralised heat pump network to provide heat to properties in Caerau was explored. In the centralised model, one or more large heat pumps would be used to heat water at the point where it was extracted from the mine workings, and the hot water would then be circulated around the village in insulated pipes buried underground. Heat exchangers in houses connected to the network would transfer the thermal energy from the heated mine water to a separate hot water circuit to supply the household heating system. The cooled mine water would then be flown back to the mine via the return loop.

The decentralised model would operate similarly; however, in this instance the water would circulate around the village in uninsulated pipes at the temperature that it came out of the mine and would be heated by individual household heat pumps.

Domestic units with thermal outputs of 3 or 6 kW were considered. Both the centralised and decentralised models would require installation of an insulated hot water tank, if not already present in the property. Initial findings suggest that the decentralised model would be the preferred option for the heat network; therefore, space for both a heat pump and storage tank would be required in properties connected to the network. This paper discusses the methodology adopted to provide a clear appraisal of the current heat demand of existing housing stock in Caerau, using a suite of investigative techniques, to enable correct sizing of heat pumps and the proposed heat network. The potential for retrofitting novel, renewable energy and demand management technologies that could complement the proposed district heat network is also considered.

## 65.3 Methodology

A range of complementary techniques were employed as part of a holistic survey protocol to assess heat loss and the level of fabric insulation present, comprising internal and external measured surveys; schedule of condition surveys; air permeability testing; and thermography tests. The information gathered was collated and used to calculate heat loss and heat transfer coefficient based on the methodology from the Standard Assessment Procedure (SAP) 2012, version 9.92 [14]. A visual inspection of the services within the properties was also made, with particular emphasis on the heating infrastructure and whether gas, electricity, or a combination of the two were used for internal temperature control. Where available, data from installed gas and electricity smart meters were collected to compare predicted heat loss figures with actual energy consumption. Many properties in the case study village were built in the early 1900s and the majority have been altered or extended since construction.

Owing to the non-destructive nature of the holistic survey protocol assessment, some assumptions had to be made as to the construction materials used in certain areas of the houses, as definitive identification was not possible.

## 65.4 Results

### *65.4.1 Condition Survey*

The condition survey involved an internal and external measured survey of each property and the production of dimensioned sketches of the floor plans and the elevations. A summary of the type of information captured in the schedule of conditions for one of the properties is as follows: Traditional Welsh three-bedroom mid-terraced house of solid wall construction and with a single-storey extension containing the bathroom and kitchen to the rear. Extension is of solid wall construction. Some thermal upgrades have been undertaken, with modern UPVC double-glazed windows, and insulation applied to the attic spaces. Insulated plasterboard has been applied to the bathroom. Central heating done throughout with a reconditioned condensing boiler and older-style single-panel pressed steel radiators. The condition of the property is generally just adequate. The property has notable damp and ventilation issues, leading to significant black mould growth in some areas of the property. Table 65.1 provides a summary of the property types surveyed, size and current levels of fabric insulation, that is loft, external wall insulation (EWI) and internal wall insulation (IWI). The area of exposed external wall is a key contributor to the overall thermal performance of a building; from this perspective, the properties presented fall into three categories based on the number of external walls: detached—four walls (property a), semi-detached and end terrace—three walls (properties b to d) and

**Table 65.1** Condition survey summary data

| Property title | Property type | Number of bedrooms | Floor area ($m^2$) | Current insulation |
|---|---|---|---|---|
| a | Detached | 5 | 159 | 100 mm loft |
| b | Semi-detached | 4 | 91 | EWI, IWI and 200 mm loft |
| c | End terrace | 4 | 116 | 200 mm loft |
| d | End terrace | 3 | 87 | 100 mm loft |
| e | Mid-terrace | 3 | 85 | EWI to rear |
| f | Mid-terrace | 3 | 78 | 80 mm loft |
| g | Mid-terrace | 3 | 92 | Loft (partial) |
| h | Mid-terrace | 2 | 71 | EWI and 80 mm loft (partial) |

Table 65.2 Results of air permeability, average heat transfer and average heat loss parameter

| Property title | Air permeability $m^3/(h\ m^2)$ @50 Pa | Average heat transfer coefficient (W/K) | Average heat loss parameter ($W/m^2K$) |
|---|---|---|---|
| a | 4.1 | 653 | 4.1 |
| b | 5.7 | 213 | 2.4 |
| c | 8.0 | 392 | 3.4 |
| d | 7.3 | 387 | 4.5 |
| e | 8.5 | 232 | 2.7 |
| f | 6.5 | 239 | 3.1 |
| g | 7.6 | 294 | 3.2 |
| h | 8.1 | 162 | 2.3 |

mid-terrace—two walls (properties e to h). None of the properties surveyed met the recommended minimum level of loft insulation, which is 270 mm of mineral wool quilt (or equivalent) [15].

But one of the properties used a natural gas combination boiler as the primary source for space heating. The detached house had recently been fitted with an air source heat pump. In all cases the size of the central heating radiators would need to be increased in order to deliver heat effectively from a lower temperature—a heat pump-based system.

## 65.4.2 Air Permeability

The air permeability tests were performed using a single fan blower door, in accordance with the Technical Standard L1 from the Air Tightness Testing and Measurement Association [16]. Tests were undertaken within the permitted conditions, that is less than 6 m/s for the wind speed; all external doors and windows closed; all internal doors open; flues and air vents temporarily sealed and gas boilers switched off (if present). The calculated air permeability values are presented in Table 65.2. All properties surveyed were below the current maximum allowable air permeability value for dwellings in UK building regulations of 10 $m^3/h\ m^2$ @50 Pa [1].

## 65.4.3 Heat Loss

The heat loss and heat transfer coefficients were calculated according to the standard assessment procedure [14], inputting data from the measured survey and air permeability tests results. Average annual values are presented in Table 65.2. On comparing the results of properties b, e and h with others in the same categories, the higher thermal resistance afforded by the retrofitted EWI is reflected in the lower average heat transfer coefficient and average heat loss parameters. Properties g and e had conservatories within the normally heated zone, which increase the thermal losses from these properties.

### 65.4.4 Thermography

Thermography was used to highlight the major sources of heat loss and air leakage and also to identify any potential problems associated with ingress of damp and/or thermal bridging, and the SuRBe team at Cardiff Met University have much experience with thermography tests on dwellings since 2010 [17]. False colour thermal maps, or thermograms, were produced and thermally tuned before analysing the results. In order to conduct a meaningful thermographic survey, certain environmental conditions need to be met. Best results are achieved when there is at least a 10 °C temperature difference between the inside and outside of the building; there should be no precipitation during the survey, surfaces should be dry, a maximum wind speed of 10 m/s, and no direct solar radiation on the surfaces for at least 1 hour before the survey. There is also a requirement to be cautious when interpreting thermograms where objects could be reflecting the night sky [18]. Both external and internal qualitative thermographic surveys were conducted to assess the thermal performance of the dwellings, particularly continuity of insulation; thermal bridges; sources of air-leakage at critical construction junctions; and moisture and damp within an element [19, 20]. Owing to space limitations of this paper, only one thermogram and associated digital image from a mid-terrace property is presented in Fig. 65.1. The property presented had the EWI applied to the rear of the house but not to the dressed stone frontage. For internal surveys, any colours indicating extremities of cold entering the

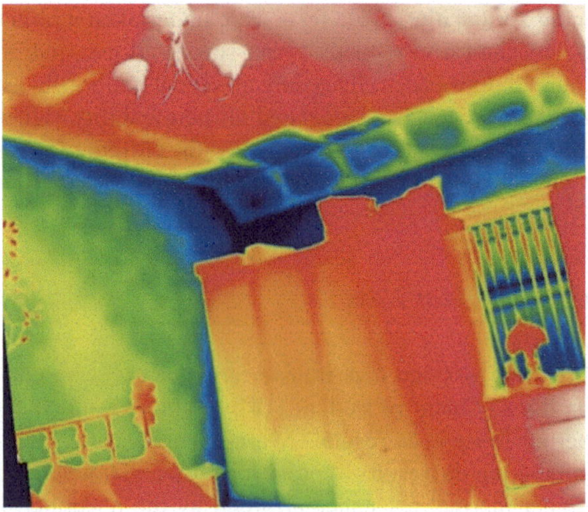

**Fig. 65.1** Thermographic image of the interior of one of the test dwellings

building are potential issues (dark blues and black). The thermogram in Fig. 65.1 shows higher heat loss considerably through the uninsulated, dressed stone exterior wall to the front of the property above the window. The coldest area is in the upper left hand corner above the wardrobe, where humidity levels may be higher due to lack of air circulation.

Heat loss along the raked ceiling at the wall junction may be the result of poor or missing insulation. Thermal bridging at the roof joists is also visible. Heat loss via conduction to the exterior wall and loft space is evident from the cool blue colour at the junction between the party wall (left), the ceiling and the front exterior wall (thermogram temperature scale is at 18–30 °C).

Energy consumption: Using the heat transfer coefficient figures from Table 65.2 and assuming a maximum internal to external temperature difference of 24 °C (assuming an inside temperature of 21 °C and a worst case external temperature of minus 3 °C), the peak heat demand for the properties can be estimated. These are presented in Table 65.3. These values can be used to size the heat pumps required for a distributed heat pump network and the level of additional insulation that would be required to reduce the heat loads below a given threshold for connection to the network. The beneficial effect of additional insulation is highlighted by comparing the current peak loads for properties b and d, which are 5.1 and 9.1 kW, respectively. Property b had the EWI applied to two of the three exterior walls; the dressed stone front wall was not insulated, potentially due to planning restrictions. Internal insulation was also installed in the lounge. Property d had no EWI and minimal loft insulation. A calculation of the theoretical peak heating loads that could be achieved with a high degree of thermal intervention was simulated by adjusting SAP calculations to include triple glazing throughout the properties: 250 mm of EWI on all external walls and 350 mm of loft insulation. The results are presented in Table 65.3. The larger detached, semi-detached and end terrace properties, which have more uninsulated external wall area, show the greatest improvement, as would be expected. These levels of thermal intervention are unlikely to be achieved to Welsh valley terraced housing where dressed stone frontages are an architectural

**Table 65.3** Estimated peak heat load

| Property title | Average heat transfer coefficient (W/K) | Current peak heating load (kW for $\Delta T$ = 24 °C) | Peak heating load after interventions (kW for $\Delta T$ = 24 °C) |
|---|---|---|---|
| a | 653 | 15.6 | 5.4 |
| b | 213 | 5.1 | 3.4 |
| c | 392 | 9.4 | 3.8 |
| d | 387 | 9.1 | 3.8 |
| e | 232 | 5.5 | 3.3 |
| f | 239 | 5.7 | 2.6 |
| g | 294 | 7.0 | 3.3 |
| h | 162 | 3.9 | 2.4 |

feature. Additionally, space constraints where access around properties and in attic spaces would be reduced to unacceptable levels and rerouting of existing services may not be possible. Furthermore, with regard to economics, the costs rise linearly with increasing insulation thickness, while the net benefit diminishes. Similarly the thermal performance improvement of replacing good quality double glazing with triple glazing is small relative to the cost.

## 65.5 Discussion

By far, the greatest opportunity for reducing the heat demand for space heating is to retrofit additional wall and loft insulation. The advantages of EWI as opposed to interior wall insulation include reduced risk of thermal bridging, improved air tightness, thermal mass remains exposed to the internal space to aid control of summer overheating risks, less disruption to occupants, no loss of internal floor area, and internal fixtures and fittings do not have to be relocated [20, 21]. Achieving a complete covering of EWI at critical junctions to prevent all thermal bridging can be challenging, including window and door reveals, wall to roof junctions, any projections such as porches and conservatories and where adjoining buildings meet (as illustrated in Fig. 65.2) [22, 23]. Thermal bridging can lead to increased heat loss and thus a reduction in the overall thermal performance, along with internal surface condensation due to localised lower surface temperatures [24]. To address surface condensation resulting from thermal bridging, occupants can either increase the internal air temperature to raise the internal surface temperature above the dew point temperature of the air, or increase the rate of ventilation to reduce the dew point temperature of the air to below the dew point temperature of the internal surface [25]. However, with either, or a combination, of these approaches, energy use will be increased in order to maintain comfort levels [26], which will undermine the

**Fig. 65.2** Corresponding digital image of the interior of one of the test dwellings

overall effectiveness and purpose of the insulation. Several of the properties surveyed showed signs of damp. These are presented as condensation on the internal surfaces of some external walls and black mould growth in the corners of certain rooms and behind furniture, where air movement was limited. Increasing ventilation improves indoor air quality and the wellbeing of occupants by providing more oxygen for breathing, removing exhaled carbon dioxide, diluting pollutants/odours and reducing moisture build up, and associated damp and mould issues. The most obvious and common approach to increasing ventilation is to open a window, however, this carries with it an energy and cost penalty, as cold fresh air enters the building at the expense of warm air escaping. Retrofitting EWI would be expected to reduce uncontrolled air infiltration, which should be compensated for by additional controlled ventilation. The air permeability test results for properties b and d (5.70 and 7.27 $m^3/h\ m^2$@50 Pa, respectively, Table 65.2) demonstrate this. This was not the case for the two mid-terraced houses which had EWI applied. In both cases poorly finished penetrations for services lead to higher air permeability figures than would have been expected. Based on a heat pump with 6 kW thermal output and rated electrical power consumption of 1.6 kW, the additional load on the local electricity network as a result of transferring the space heating load from gas to electricity would equate to 0.24 MW for 150 connected homes, or 1.36 MW for 850 homes. The impact on the electricity network and the carbon equivalent of the heating solution could be reduced through incorporation of local renewable energy generation and storage. The orientation of the streets in Caerau and the overall topography of the village, in a broad valley which runs approximately south west to north east, lends itself to solar energy generation. The roof area of a typical terraced residence in Caerau is sufficient to accommodate approximately 3 kW (peak) of crystalline silicon photo voltaic (PV) panels. A system of this size would provide a predicted total annual generation of 2770 kWh, based on historical climatic data for the region. Storing this energy for later use could lower grid stress by removing spikes in demand and offset utility costs, particularly if time of use tariffs were introduced. Other options could include electrical storage in batteries; solar thermal generation, an effective means of supplying pre-warmed fresh air to a building; or 'heat batteries' which use excess electricity from PV generation to heat a phase-change material.

## 65.6 Conclusion

A holistic method for assessing household heating energy demand has been presented. Retrofit of thermal efficiency interventions to reduce the overall heat demand of pre-1919 solid wall properties to a level where a mine water-based heat network would be capable of providing the primary heating mechanism were investigated. Complementary renewable solar energy generation and storage technologies that could further reduce heat pump load, offset additional stress on the local electrical distribution grid and alleviate household electricity bills were considered. In general, mid-terraced houses have the lowest peak heat demand, followed by end-terraced/semi-detached and then detached properties; as would be expected with increasing exposed external wall area.

Tight control over the quality of detailing of any EWI needs to be exercised to avoid thermal bridging and water ingress, as observed in the condition and thermographic surveys. In particular, consideration should be given to extending the roof line to ensure a good and permanent seal at the interface between wall and insulation. A moisture-permeable EWI may be more appropriate for solid walled buildings in order to avoid problems arising from interstitial condensation and subsequent dampness which could inadvertently degrade the property over time. Planning restrictions may preclude the installation of EWI and the use of solar generation panels on dressed stone frontages. Clear signs of dampness from a combination of water ingress and insufficient ventilation were apparent in several properties. Somewhat perversely, this problem can be exacerbated by the addition of EWI, which has the effect of reducing uncontrolled air infiltration and increasing temperature variations on internal wall through poor detailing. Appropriate measures to increase ventilation rates should be taken in tandem to ensure that, in striving for increased energy efficiency, the health and wellbeing of residents is not compromised. A combination of solar thermal air heaters and controlled ventilation could help to reduce the risk of condensation on cold surfaces. Orientation/planning restrictions may preclude the use of wall-mounted panels on certain houses. There are a number of technical and economic challenges to enable the use of a mine water district heat network; however, with careful consideration and selection of complementary technologies and close control over quality during installation the desired outcome could be achieved.

**Acknowledgements** The authors would like to thank Bridgend Borough Council, the Engineering and Physical Sciences Research Council, Innovate UK, and the WG for the financial support of this work.

## References

1. WG: Wales Part L: Conservation of fuel and power (2019). https://www.labc.co.uk/professionals/building-regulations-guidance-documents/approved-documents-and-technical-guidance-wales/wales-approved-document-l-conservation-fuel-and-power. Accessed 22 May 2019
2. Atkinson, J., Littlewood, J., Geens, A., Karani, G.: Relieving fuel poverty in Wales with external wall insulation. Eng. Sustain. **170**(2), pp. 93–101 (2017)
3. Welsh Government: Arbed—strategic energy performance investment programme (2013). http://gov.wales/topics/environmentcountryside/energy/efficiency/Arbed/?lang=en. Accessed 22 May 2019
4. Jahic, D., Littlewood, J.R., Karani, G., Atkinson, J., Bolton, D.: Evaluating occupant wellbeing in retrofitted dwellings with the short form 36 questionnaire, in *Vol 131. Smart Innovation, Systems and Technologies series. SEB18 International Conference Australia*, Springer, UK (2018)
5. Atkinson, J., Littlewood, J.R., Geens, A.J., Karani, G.: Did ARBED I save energy in Wales' deprived dwellings. Energy Proced. **63** (2015)

6. Anon: Caerau set to be at heart of £6.5 m green energy revolution (2018). http://welshbusinessnews.co.uk/regions/maesteg-caerau/caerau-set-to-be-at-heart-of-6-5m-green-energy-revolution/. Accessed 1 May 2019
7. BBC: The Welsh county that wants to change the way its residents heat their homes (2019). https://www.walesonline.co.uk/news/local-news/welsh-county-wants-change-way-15865549. Accessed 1 May 2019
8. Carbon Connect: Report: Future Heat Series Part 2—Policy for Heat (2018). https://www.policyconnect.org.uk/cc/research/report-future-heat-series-part-2-policy-heat. Accessed 1 May 2019
9. Anon: UK Energy in Brief 2018 (2018). https://assets.publishing.service.gov.uk/government/uploads/system/uploads/attach-ent_data/file/728374/UK_Energy_in_Brief_2018.pdf. Accessed 1 May 2019
10. WG: Fuel Poverty (2018). https://gov.wales/topics/environmentcountryside/energy/fuelpoverty/?lang=en. Accessed 1 May 2019
11. Anon: Revision to Welsh Index of Multiple Deprivation (WIMD) 2014 Note on key changes (2015). https://gov.wales/sites/default/files/statistics-and-research/2019-05/revision-to-welsh-index-of-multiple-deprivation-wimd-2014-note-on-key-changes.pdf. Accessed 1 May 2019
12. Bridgend Council: Smart System and Heat Programme (2016). https://democratic.bridgend.gov.uk/documents/s8887/160510%20Smart%20System%20and%20Heat%20Programme.pdf. Accessed 1 May 2019
13. Busby, J., Lewis, M., Reeves, H., Lawley, R.: Initial geological considerations before installing ground source heat pumps (2019). https://www.bgs.ac.uk/research/energy/docs/final_paper.pdf. Accessed 1 May 2019
14. BRE: The Government's Standard Assessment Procedure for Energy Rating of Dwellings (2012). https://www.bre.co.uk/filelibrary/SAP/2012/SAP-2012_9-92.pdf. Accessed 1 May 2019
15. Roof and loft (2019). https://www.energysavingtrust.org.uk/home-insulation/roof-and-loft. Accessed 1 May 2019
16. ATTMA: Technical Standard L1, September 2016 edition (2016). https://www.attma.org/attma-release-new-2016-version-of-attma-tsl1/. Accessed 1 May 2019
17. Littlewood, J.R.: Chapter Four—Assessing and monitoring the thermal performance of dwellings, Chapter Four. In: Emmitt, S. (ed.) Architectural Technology: Research & Practice. Wiley Blackwell, Oxford, UK
18. Pearson, C.: Thermal Imaging of Building Fabric. BSRIA, Berkshire (2011)
19. Thomsen, K.E., Rose, J.: Analysis of Execution Quality Related to Thermal Bridges. Danish Building Research Institute, Denmark (2009)
20. Atkinson, J.: Evaluating exterior wall insulation. Unpublished Ph.D. Thesis. Cardiff Metropolitan University, Cardiff, UK (2015)
21. Immendoerfer, A., Houh, C., Andrews, A., Mathias, A.: Fit for the Future: The Green Homes Retrofit Manual—Technical Supplement. Housing Corporation, London (2008)
22. Construction Products Association: An Introduction to Low Carbon Domestic Refurbishment. Construction Products Association, London (2010)
23. Heritage, E.: Energy Efficiency in Historic Buildings—Insulating Solid Walls. English Heritage, London (2010)
24. King, C.W.: Sustainable Refurbishment of Non-Traditional Housing and Pre-1920s Solid Wall Housing. IHS BRE Press, Watford (2010)
25. Ward, T.: Information Paper 1/06: Assessing the Effects of Thermal Bridging at Junctions and Around Openings. BRE, Watford (2006)
26. Burberry, P.: Environment and Services, 8th edn. Pearson Education Limited, Essex (1997)

# Chapter 66
# Privacy in Domestic Building Performance Evaluation—Preliminary Framework for Analysis

**Magdalena Baborska-Narożny**

**Abstract** A framework for the analysis of privacy and domestic building performance evaluation (BPE) shared dimensions is proposed. These conceptualized dimensions, based on non-exhaustive interdisciplinary literature review and BPE process experience, cover focus, aspiration and control as seen through privacy and domestic BPE lens. For each dimension key, directly relating concepts present in privacy discourse and BPE practice are identified. Such an approach allows an overlap in privacy and BPE focus of attention coinciding with contrary aspirations and tension in terms of assigning control over information to be captured. Addressing all these challenges is a crucial part of tacit knowledge of research or consultancy team responsible for the process; however, it is not explicitly reported in BPE literature. The proposed framework is intended to support further development of the domestic BPE ethical procedure, shown here as the key element in responding to privacy concerns. The ethical procedure is obligatory for academic-led BPE while for consultancy it is only recommended. This gap needs to be addressed before some form of domestic evaluations is taken up by mainstream construction. Further research needs are identified into the impact of privacy-related concerns on domestic BPE process.

## 66.1 Introduction

Domestic building performance evaluation (BPE) belongs to real world construction research covering positivist and interpretivist epistemologies [1]. It aims to gain insight and understanding about how and why people interact with and perceive their homes. However, BPE also aims to capture quantitatively, through monitoring and measurement, what is the scale and what are the causes of the performance gap, that is, the difference between design stage energy predictions and actual energy consumption within an inhabited home [2]. The overall objective of the evaluation process is to provide actionable, evidence-based feedback to the building industry,

M. Baborska-Narożny (✉)
Wroclaw University of Science and Technology, ul. B. Prusa 53/55, 50-317 Wroclaw, Poland
e-mail: magdalena.baborska-narozny@pwr.edu.pl



J. Littlewood et al. (eds.), *Sustainability in Energy and Buildings*, Smart Innovation, Systems and Technologies 163, https://doi.org/10.1007/978-981-32-9868-2_66

policy makers and the public, as a step towards lowering the negative impact of buildings on the environment. An actual evaluation of occupied homes as they are lived in is the only way to truly understand and tackle domestic energy consumption [3]. Such objective is in public interest given the climate change challenge and the role assigned to housing in its mitigation. The residential sector is the biggest one, in terms of total energy consumption out of all building sectors in the UK and across the EU countries [4], suggesting it should be prioritized within BPE studies.

Domestic BPE as a basic concept has a long history [5]. However, in its current form, that is based exclusively on voluntary participation and focused around energy consumption issues, it has only emerged in early 1990s in the UK [6], and from the very start was exposed with participants' recruitment challenge. Extensive Stamford Brook case-study project managed to recruit only four participating households out of 100 approached [7]. Gaining representative samples for benchmarking faces a barrier of the difficulty of being allowed access to peoples' homes, 'which are private by their nature' [2]. Almost a decade later, this profound challenge remains unaddressed and is typically not covered by the domestic BPE discourse [8, 9]. In the UK, academic-led domestic BPE includes ethics procedure [e.g. 10] that addresses some aspects of privacy challenge discussed in this paper; however, low participation levels are still an issue, thus access remains a real barrier. Global public interest in broad domestic BPE implementation competes with the recognized right of an individual to refrain from voluntarily signing into the process. Such right derives from privacy concept present in vast academic literature on the subject spanning many disciplines as well as many documents, for example, in Article 12 of the 'Universal Declaration of Human Rights' [11].

This paper addresses a gap identified in domestic BPE discourse by focusing on exploring its multifaceted and unavoidable connection with privacy. First, privacy is defined for the purpose of domestic BPE based on a non-exhaustive interdisciplinary literature review. This is followed with a proposed framework for analysis of privacy and domestic building performance evaluation (BPE) shared dimensions. The components of BPE ethical procedure are then discussed against the proposed framework with key gaps identified. Finally, the limitations of the paper and the recommendations for further research are developed.

## 66.2 Privacy in the Domestic BPE Process

### *66.2.1 Privacy Non-exhaustive Interdisciplinary Literature Review*

There are multiple and in-conclusive attempts to define privacy within different disciplines, such as psychology, philosophy, philosophical anthropology, law or sociology, and recently also information technology. The conditional nature of the concept of privacy is described by some as chaotic [12], while others note that privacy 'is a concept in disarray' [13]. Limited space in this paper justifies an introduction of the highlights provided by the Oxford English Dictionary. Here six different past usages of the word 'privacy' are distinguished and three of these are of particular interest in the context of domestic BPE. Privacy means [14]:

- *'the state or condition of being alone, undisturbed, or free from public attention, as a matter of choice or right; seclusion; freedom from interference or intrusion'*,
- *'the secrecy, concealment, discretion; protection from public knowledge or availability'*,
- *'a private place, a place of concealment or retreat, a private apartment'*.

These concise definitions identify key challenges that domestic BPE needs to tackle in order to fulfil its goals, as highlighted in the introduction. The phrase '… as a matter of choice' suggests that privacy 'depends upon conditions' [13]—a crucial point illustrating that the concept of privacy is both complex and contingent on a number of social and physical conditions.

**Privacy process model in the context of BPE**. Based on earlier privacy theories, in particular on Burgoon's four privacy dimensions [15], Dienlin [16] proposes privacy process model (PPM), including an objective description of privacy context, as a measure of an individual detachment through four independent dimensions:

- informational privacy—the amount of information collection taking place in a given situation,
- social privacy—the number of people present and the how acquainted they are,
- psychological privacy—the extent to which people present in a situation engage in intimate and personal, or trivial and impersonal conversations,
- physical privacy—the physical distance or seclusion.

Understanding of a domestic BPE situation, through an in-depth case study in particular, suggests that this is objectively low privacy context in each dimension: a lot of information is collected, recorded or logged; the one-to-one encounters between the researcher who is a stranger to the recruited participant are typical; the participant is encouraged to unilaterally share private or intimate information (e.g. home occupancy patterns or number of showers per week); and the researcher enters and observes the space of home otherwise protected from observation. However, the PPM observes it is not the context itself but how it is perceived that is decisive for privacy behaviour, that is, for readiness to self-disclose. Self-disclosing covers facts, thoughts, feelings and experiences [17]. All of these may be of interest in the BPE process as a part of inhabitant feedback from occupancy stage. Dienlin understands that the context perception is impossible to adjust but assumes the context and behaviour can be controlled. Thus, if there is a difference between an individual's perceived and desired privacy status, people aim to change either privacy context or privacy behaviour. The ability to control context or behaviour is crucial for satisfactory privacy status according to PPM.

For domestic BPE this psychological approach to privacy sheds light on two issues. First, it is to be expected that objectively low privacy context of the domestic BPE will also be perceived as such by the approached potential participants, typically inclining them to refrain from agreeing to self-disclose. Second, the role of ethical procedure proves fundamental in terms of responding to an individual's need for controllability of privacy context and behaviour. Through the ethical procedure the participant is given the right to withdraw from the study at any point without any consequences, to withdraw part of the data or all the data from the study, not to agree for publication of images, and so on [18]. However, the effort of the BPE team is to plan and conduct the research in a way that minimizes the risk of privacy context modification; that is, agreed home visits will take place and all the planned data collected. Same applies to modifying behaviour, that is, refraining from self-disclosure by the recruited participants; for example, refusing to answer questions or not allowing access to previously agreed spaces. This would result in gaps in data difficult to manage at the analysis stage. Surprisingly, ethical procedure is still only a recommended but not obligatory part of the domestic BPE process.

### *66.2.2 Defining Privacy Through Domestic BPE Lens*

The above privacy model is helpful in understanding BPE recruitment stage difficulties, the researcher–participant interaction or the necessity of an ethical procedure. However, it does not cover all aspects of privacy—BPE entanglement. It does not cover questions such as why the need for privacy should be respected or why BPE should be allowed to challenge it. Also, it does not cover the collective aspect of domestic BPE, that is involving and potentially affecting not only individuals but also formal or informal institutions, and gaining knowledge from multiple stakeholders about preselected sets of participants. The voluntary nature of BPE participation and a range of stakeholders involved [19] requires acknowledging that privacy behaviour understood as self-disclosure may be refused not only by human persons but also by institutions. To address these gaps a BPE-focused privacy definition is proposed, followed by a preliminary framework for analysis of the relationship between privacy and domestic BPE.

Through the lens of domestic BPE, privacy is the expectation of an individual or an institution to have the right to conceal or avoid judgement by being in control of information regarding home, the way it is being inhabited, the environmental impact of one's lifestyle, and also the physical performance of a dwelling's fabric and systems.

## 66.3 Privacy and Domestic BPE Dimensions Framework

A framework for the analysis of privacy and BPE shared dimensions is proposed responding to questions underpinning each of them, namely 'What is the object of concern/study?', 'Why?' and 'What is involved?'. The responding dimensions are conceptualized as: 'focus', 'aspiration' and 'being in control' as seen through privacy and domestic BPE lens. For each dimension key, directly relating concepts present in the discourse of respective discipline are identified (Table 66.1).

**Focus: person—inhabitant, place—home, institution—target group**. As highlighted earlier, privacy is focused on the state or the right of an individual—a person. It also prioritizes place, namely home, as an area requiring protection. An individual needs to briefly withdraw from all social roles and tensions to relax and regain strength, and home is seen as a place where such withdrawal should be possible. Privacy needs its 'place' and 'In none is the zone of privacy more clearly defined then when bounded by the unambiguous physical dimensions of an individual's home.' [13]. The 'sanctity of home', an expression used in the twenty-first century US court verdict to stress homes privileged status [20], is a concept derived from ancient Greece and Rome. The God's protection over the home established its status as a place particularly sheltered from the enemies and entering someone's home with bad intentions was perceived as sacrilege [13]. The amalgam of home and family as the sphere requiring special protection with restricted access is a well-established theme in Anglo-American liberal discourse [18], known in law discourse as 'castle doctrine' [21] rooted in sixteenth century [22], which led to the well-known English expression 'A man's home is his castle.'

The person and place privacy concepts have direct counterparts in domestic BPE focus on the inhabitant and dwelling. BPE explores individual practices, motivations, preferences or skills. However, it also monitors fabric and systems performance or interface usability evaluating the physical context of inhabitant actions [23].

Last but not the least, the group dimension represented by the term institution exists in privacy legal debate, and for BPE discourse a 'target group' term is proposed. Institutions, formal and informal, cover all scales and types of social interaction, ranging from family or household, through a co-housing community, residents'

**Table 66.1** Privacy and BPE shared questions and dimensions versus domain-specific concepts

| Shared privacy/BPE | | Domain-specific concepts | |
|---|---|---|---|
| Dimensions | Related questions | Privacy | Domestic BPE |
| Focus on… | What is considered? | Person | Inhabitant |
| | | Place | Dwelling |
| | | Institution | Target group |
| Aspiration | Why? | Dignity | Validity and reliability of findings |
| | | Non-judgement | Benchmarking and feedback |
| | | Autonomy | Learning opportunity |
| Being in control of… | What is involved? | Sharing information | Data collection and dissemination |
| | | The right to conceal | The right to withdraw data |
| | | Audience selection | Dissemination |

association within a specific development, or organizations such as developers or building managing companies. Each institution may have their own agenda and plan for outreach activities and communication channels with the outside world, for example, neighbours, public or private landlords, or clients. Institutions are too diverse to be covered by a common privacy rights analysis. However, the theoretical debate on privacy interpersonal rights is on-going and the US legal practice related to allowing institutions such as family, associations or corporations to claim privacy concerns when refraining from sharing information are not being automatically rejected but rather considered on a case-to-case basis [24, 25]. This is a crucial reflection point for BPE research: an organization may be granted legal permission not to reveal (even when otherwise obliged to share certain information—whereas in BPE it would always be voluntary sharing) in case of a risk of compromising the privacy rights of its members. In domestic BPE the target group is understood as all either directly approached for information in the evaluation process or potentially affected by BPE findings. The mosaic effect resulting from multiple data sources may actually lead to exposing inconsistencies in communication, underperformance of buildings that might be perceived as a risk to property market price directly affecting the participants–owners or undermine a developer's marketing strategy.

**Aspiration: dignity—validity and reliability of findings, freedom from judgement—benchmarking, autonomy—learning opportunity**. To understand why privacy is worthy of protection, it is important to understand the aspirations underpinning the concept of privacy. Also, to accept that domestic BPE is worthy to relinquish, some privacy requires recognizing its aspirations.

The discourse in philosophy and philosophical anthropology introduces the concepts of human dignity, autonomy and freedom from someone else's judgement as those laying foundation for legal protection of privacy [26, 27]. The argument raised is that human lifelong potential for change and 'work-in-progress' condition would be undermined and limited if exposed to instantaneous judgement or critique. Privacy enables safety zone for all aspects of human trial and error and uninterrupted individual reflection. Domestic BPE in contrast aspires to capture and diagnose the status quo of building delivery process and home occupancy in order to identify cause–effect relationship for the observed patterns of buildings-related resources consumption. It aims to bring viable and reliable evidence-base reflecting the complexities of occupancy stage energy consumption. Privacy aims to secure a hiding space while BPE needs to reveal the fine-grain findings in hope of addressing problems observed in bigger scale, for example, city, country or globally. Also best-practice guidance in relation to the private sphere of home use and life-style choices must be sensitive in relation to the autonomy of inhabitant's cultural, economic or physical context.

The notion of home as a sphere free of judgement of others introduces the normative dimensions at play in BPE projects, due to the use of benchmarks, rankings and contextualization for energy and water use levels or practices. Benchmarking in relation to privacy is a sensitive topic. It can come as a nasty surprise to some people that their water or energy use is 'above average' as this carries an inherent judgement, if for some reason they have been given access to this personalized data. This may be covered by ethical processes which require anonymity, but becomes a dilemma in action research when the inhabitants are encouraged to use individual data feedback to help improve the performance of their home through behaviour change. Individual data may also be presented against some benchmarks to improve performance in relation to 'nudge' theory which states that social norms can provide a strong incentive to achieve more [28]. Anonymity would also be challenging in self-assessment effort undertaken by some low-carbon community groups [29].

**Being in control: sharing information—data collection and dissemination, the right to conceal—the right to withdraw data, audience selection—dissemination**. Being in control of what is shared with whom forms a key privacy dimension for conducting domestic BPE. The flow and content of information between parties is critical in this regard. The vulnerable position of the business stakeholders of new-built or refurbished developments (developers, architects, contractors, suppliers) in the BPE process is a particular concern. Before the research starts they are the main sources of the development-related information that reaches the public and their clients—the inhabitants. As a result of the BPE, they lose monopoly on knowledge related to prior occupancy stages of building life-cycle and, consequently, on communication in this respect.

The expected level of inhabitants' self-disclosure level is decided at the research design stage. Methods selection defines the fieldwork that will be undertaken and the degree of control that participants will have in relation to visiting times and duration and any disruption to the home life. Most importantly, specific BPE methods give the inhabitants different level of control over information sharing—in some they presume to have full control of what they reveal (e.g. in a survey questionnaire) and in others they have none (e.g. temperature monitoring).

Key ethical assumption is that the participants 'have a sufficient awareness of what they are disclosing' [30]. However, a skilled researcher may discover more than a participant would consciously wish to reveal to him or her. The in-depth BPE collecting and cross-comparing data gathered through physical monitoring, ethnographic observation and inhabitants' feedback has a high potential to question claims made in the interviews or surveys [3]. Even if data is disseminated in an anonymized form, it may be disturbing for a participant to recognize own statements in a publication and learn that the researcher has identified lack of consistency between the claims made and the actual practices. The right to withdraw data confirmed in BPE ethical procedure is the key instrument for an inhabitant to express frustration with the study.

The available monitoring techniques allow data acquisition without actively engaging the inhabitants, thereby not giving them the opportunity to conceal and filter the information they want to share. In its urge to understand the link between behaviour and energy consumption, some BPE research deploys techniques that allow real-time tracking of the inhabitants' location around a dwelling and identifying appliances they are using [31]. As the researchers claim: 'Disaggregated occupancy and energy data has the potential to estimate personalized energy use by answering the questions "who, where and when?"' [32]. Such fine grain surveillance of daily domestic practices has been found acceptable for voluntary participants incentivized to live for a limited period of time in demonstration homes equipped with all the sensors to carry out the intensive monitoring. Interestingly, according to the research team 'such a system is a specialized research tool, and is not intended for mass deployment due to the system's high capital cost and the need for occupants to wear location devices' [32]. Highest level of privacy intrusion of this Panopticon approach has not been identified as a barrier for wider application of this method in domestic BPE, which reveals a gap in understanding privacy issues in the domestic context. Fears of loss of control over information have already been raised when the details of appliances used become available to utility companies on a mass scale as a result of installation of smart metering technology [33, 34]. Utilities and grid operators need the fine grain data to optimize energy supply and demand and challenge the problem of peak load and intermittent supply from renewables. Opting out of 'smart meters' despite their introduction as a useful support tool for household domestic energy management [35] indicates that the existing desire to stay in control over sharing home-related information.

Information sharing vector in the BPE is not continuously directed from residents or business stakeholders towards the researchers. At feedback and dissemination stage, it points from the researchers back to all participants and then also to the wider public. The occupants will only have some degree of control over this stage, once they have agreed to images, quotes and any other data agreed being used, since their permission may not be needed for publication, provided their quotes and images have been anonymized. It is a moot point whether an invasion of privacy still exists when people have been quoted without their knowledge but have given general consent that quotes may be used. Interestingly, there is also a dynamic balance between inhabitants, stakeholders and researchers in terms of their joint and individual knowledge and understanding about how the homes are performing. In some cases the researchers know more than the stakeholders and inhabitants about what is going on in the home, particularly in relation to energy use patterns or physical aspects of the home. The researcher is then in control of what to share with the inhabitants and the target group and whether to share this during the fieldwork stage itself or at the end of the project. Not all needs to be shared, although there will be a duty of care to highlight to all relevant parties any issues which jeopardize health and safety.

Audience selection for self-disclosure covers social and psychological privacy dimensions in the PPM theory. The data collection situation involving interviews or surveys may be objectively assessed using these dimensions. However, the interviews are conducted to give insights shared with wider audiences. The direct interview privacy context needs to be understood through the lens of the future public dissemination. Anonymizing data mitigates the fear of being personally exposed in an unwanted way. However, a desire to present oneself at a chosen angle applies not only to individuals but also to all the business stakeholders. At the signing agreement stage they aim to safeguard the reporting levels and agreeing of the content of feedback going to participants as well as craft the dissemination that reaches the wider public. Their particular vulnerability at the feedback stage derives from the fact that all development-level findings are shared in an un-anonymized form; it is feasible to anonymize data for individual participants and the development as a whole when disseminating to the public, but not to the participating inhabitants.

Developers may feel rightly anxious about litigation from inhabitants if the research leads to revealing any design faults or installation and commissioning issues. Fear of 'bad news' may lead to reluctance to involve in the research [36]. Research-driven domestic BPE must be careful with the pitfall of findings-related legal actions, and with the exemption of earlier mentioned duty of care, rather aim towards preserving good relations with all the business stakeholders involved. The overall aim of the research-driven BPE is not to punish in any way the participants involved but provide them and the wider industry with useful lessons for the future. The potentially threatening BPE capacity needs to be efficiently mitigated while retaining the researchers' freedom to disseminate all findings they find relevant in an anonymized form. Reaching a mutually satisfying agreement between the researchers and the main organizational stakeholder involved and regulating information sharing policy become the crucial threshold in the preparatory stage that opens an access to the site and allows participants' recruitment.

## 66.4 Conclusions

The proposed privacy and domestic BPE dimensions framework is intended to provide discussion points for domestic BPE discourse. The concepts linked with each dimension indicate an overlap in terms of privacy and domestic BPE focus, on contrary to aspirations and risks that need addressing when control over information is examined through the lens of privacy or BPE. The privacy-related challenge for domestic BPE is still unaddressed systematically and is largely tackled through the tacit knowledge of the research team involved. The domestic BPE ethical procedure is the key means to secure balanced control over information between the participants and researchers as well as much needed explanation why is the self-disclosure behaviour sought. However, so far it is a standard practice for academic research only while for consultancy-led evaluations it is only recommended. This gap needs to be addressed before some forms of domestic evaluations are taken up by mainstream construction as recommended by the UK design professional bodies. The framework points towards research gaps related to systematic review of reporting recruitment procedures and its success rates among disseminated BPE projects, participants' dropout rates or reception among the domestic BPE participants of the feedback and dissemination resulting from the study. Also, the perception of current ethical procedure among domestic BPE participants needs to be evaluated to guide its future development, in particular in counties such as Poland that do not have BPE procedures in place.

**Acknowledgements** The author gratefully acknowledges the funding provided by Polish Ministry of Science (WUST Statutory Funding: 0401/0063/18).

## References


1. Schweber, L.: Putting theory to work: the use of theory in construction research. Constr. Manag. Econ. **33**(10), 840–860 (2015)
2. Stevenson, F., Leaman, A.: Evaluating housing performance in relation to human behaviour: new challenges. Build. Res. Inf. **38**(5), 437–441 (2010)
3. Gram-Hanssen, K.: Residential heat comfort practices: understanding users. Build. Res. Inf. **38**(2), 175–186 (2010)
4. UK Government: National Energy Efficiency Action Plan (2017), https://ec.europa.eu/energy/sites/ener/files/documents/uk_neeap_2017.pdf. Accessed 28 March 2019
5. Hafetz, J.L.: 'A Man's Home is His Castle?': reflections on the home, the family, and privacy during the late nineteenth and early twentieth centuries. Willam & Mary J Women Law **8**(2), 175–242 (2002)
6. Lowe, R., Bell, M.: Building regulation and sustainable housing. Part 2: technical issues. Struct. Surv. **18**(2), 77–88 (2000)
7. Wingfield, J., Bell, M., Miles-Shenton, D., South, T., Lowe, B.: Evaluating the impact of an enhanced energy performance standard on load-bearing masonry domestic construction Understanding the gap between designed and real performance: lessons from Stamford Brook. Department for Communities and Local Government, London (2011)
8. Bordass, B., Cohen, R., Standeven, M., Leaman, A.: Assessing building performance in use 3. Build. Res. Inf. **29**, 114–128 (2001)
9. Lowe, R., Chiu, L.F., Oreszczyn, T.: Socio-technical case study method in building performance evaluation. Build. Res. Inf. 46, 469–484 (2017); Hargreaves, T., Wilson, C., Hauxwell-Baldwin, R.: Learning to live in a smart home. Build. Res. Inf. **46**(1), 127–139 (2018)
10. Nooraei, M., Littlewood, J.R., Evans, N.I.: Feedback from occupants in 'as designed' low-carbon apartments, a case study in Swansea, UK. Energy Proced. **42**, 446–455 (2013)
11. United Nations: The Universal Declaration of Human Rights (2015), http://www.un.org/en/udhrbook/pdf/udhr_booklet_en_web.pdf. Accessed 28 March 2019
12. Inness, J.C.: Privacy, Intimacy and Isolation. Oxford University Press, New York (1992)
13. Solvoe, J.D.: Understanding Privacy. Harvard University Press, Cambridge (2009)
14. Oxford English Dictionary, www.oed.com. Accessed 28 March 2019
15. Burgoon, J.K.: Privacy and communication. In: Burgoon, M. (ed.) Communication Yearbook 6 (206–249). Routledge, Beverly Hills, CA (1982)
16. Dienlin, T.: The psychology of privacy: analyzing processes of media use and interpersonal communication. Doctoral Thesis, University of Hohenheim (2017). http://opus.uni-hohenheim.de/volltexte/2017/1315/. Accessed 28 March 2019
17. Taddicken, M.: The 'privacy paradox' in the social web: The impact of privacy concerns, individual characteristics, and the perceived social relevance on different forms of self-disclosure. J. Comput.-Med. Commun. **19**(2), 248–273 (2014)
18. Sharp, T.: Ethical issues in domestic building performance evaluation studies. Build. Res. Inf. **47**(3), 318–329 (2019)
19. Baborska-Narożny, M.: Building performance evaluation—understanding the benefits and risks for the stakeholders involved. Architectus **1**(49), 47–61 (2017)
20. Legal Inf. Inst., https://www.law.cornell.edu/supct/html/99-8508.ZO.html. Accessed 28 March 2019
21. Legal Inf. Inst., https://www.law.cornell.edu/wex/castle_doctrine, Accessed 28 March 2019
22. Titelman, G.Y.: Random House Dictionary of Popular Proverbs and Sayings, p. 229 (1996)
23. Baborska-Narożny, M., Stevenson, F., Grudzinska, M.: Overheating in retrofitted flats: occupant practices, learning and interventions. Build. Res. Inf. **45**(1–2), 40–59 (2017)
24. Pollman, E.: Corporate Right to Privacy. Univ. Minn. Law Rev. **99**, 27–88 (2014)
25. Orts, E.W., Sepinwall, A.: Privacy and organizational persons. Univ. Minn. Law Rev. **99**(6), 2275–2322 (2015)
26. Floridi, L.: On human dignity as a foundation for the right to privacy. Philos. Technol. **29**, 307 (2016). https://doi.org/10.1007/s13347-016-0220-8

27. Mokrosinska, D.: Privacy and Autonomy: On Some Misconceptions Concerning the Political Dimensions of Privacy. Law Philos. **37**(2), 117–143 (2018)
28. Hobson, K., Hamilton J., Mayne, R.: Monitoring and evaluation in UK low-carbon community groups: benefits, barriers and the politics of the local. Local Environ. Int. J. Justice Sustain. https://doi.org/10.1080/13549839.2014.928814 (2014)
29. Homan, R.: The ethics of Social Science Research. Longman, Harlow, UK (1991)
30. Platt, R., Cook, W., Pendleton, A.: Green streets, strong communities. Technical Report (2011). http://observgo.uquebec.ca/observgo/fichiers/59058_green-streets-strong-communities_July2011_7703.pdf. Accessed 28 March 2019
31. Spataru, C., Gillott, M.: The use of intelligent systems for monitoring energy use and occupancy in existing homes. Intell. Build. Int. **3**(1), 24–31(8) (2011)
32. Gillot, M. Spataru, C.: Mapping occupancy and energy use in homes: an advanced real-time location and energy tracking system. In: Loveday, D.L., Vadodaria, K. (eds.), CALEBRE: Consumer Appealing Low Energy technologies for Building Retrofitting. Loughborough Univ., pp. 25–29 (2013). http://www.lboro.ac.uk. Accessed 28 March 2019
33. Horne, Ch., Darras, B., Bean, E., Srivastava, A., Frickel, S.: Privacy, technology, and norms: the case of smart meters. Soc. Sci. Res. **51**, 64–76 (2015)
34. Reinhardt, A., Englert, F., Christin, D.: Averting privacy risk of smart metering by local data preprocessing. Pervasive Mob. Comput. **16**, 171–183 (2015)
35. King, N.J., Jessen, P.W.: For privacy's sake: consumer 'opt outs' for smart meters. Comput. Law Secur. Rev. **30**, 530–539 (2014)
36. Benndorf, V., Kübler, D., Normann, H.T.: Privacy concerns, voluntary disclosure of information, and unraveling: an experiment. Eur. Econ. Rev. **75**, 43–59 (2015)

# Author Index





J. Littlewood et al. (eds.), *Sustainability in Energy and Buildings*, Smart Innovation, Systems and Technologies 163, https://doi.org/10.1007/978-981-32-9868-2

**Z**

The manufacturer's authorised representative in the EU is Springer Nature Customer Service Centre GmbH, Europaplatz 3, 69115 Heidelberg, Germany. If you have any concerns regarding our products, please contact ProductSafety@springernature.com

Printed and bound by CPI Group (UK) Ltd, Croydon, CR0 4YY
25/09/2026
02222886-0015